18TH LOWELL WAKEFIELD FISHERIES SYMPOSIUM

HERRING
EXPECTATIONS FOR A NEW MILLENNIUM

FRITZ FUNK, JAMES BLACKBURN, DOUGLAS HAY, A.J. PAUL,
ROBERT STEPHENSON, REIDAR TORESEN, AND DAVID WITHERELL,
EDITORS

PROCEEDINGS OF THE SYMPOSIUM
HERRING 2000: EXPECTATIONS FOR A NEW MILLENNIUM

ANCHORAGE, ALASKA, USA
FEBRUARY 23-26, 2000

UNIVERSITY OF ALASKA SEA GRANT COLLEGE PROGRAM
AK-SG-01-04

PRICE $40.00

Elmer E. Rasmuson Library Cataloging-in-Publication Data

International Herring Symposium (2000 : Anchorage, Alaska)
    Herring : Expectations for a new millennium : herring2000 / edited by Fritz Funk, [et al.] – Fairbanks, Alaska: University of Alaska Sea Grant College Program, 2001.

    800 p. :   ill. ;   cm. – (University of Alaska Sea Grant College Program; AK-SG-01-04)

    "Proceedings of the Symposium Herring 2000: Expectations for a New Millennium, Anchorage, Alaska, February 23-26, 2000."

    Includes bibliographical references and index.

    1. Herring—Congresses. 2. Herring—Growth—Congresses. 3. Herring—Reproduction—Congresses. 4. Herring fisheries—Congresses. I. Title. II. Funk, Fritz. III. Series: Lowell Wakefield Fisheries symposium series (18th : 2000 : Anchorage, Alaska). IV. Series: Alaska Sea Grant College Program report ; AK-SG-01-04.

SH351.H5H47 2000

ISBN 1-56612-070-5

Citation for this volume is: 2001. F. Funk, J. Blackburn, D. Hay, A.J. Paul, R. Stephenson, R. Toresen, and D. Witherell (eds.), Herring: Expectations for a new millennium. University of Alaska Sea Grant, AK-SG-01-04, Fairbanks. 800 pp.

Credits

This book is published by the University of Alaska Sea Grant College Program, which is cooperatively supported by the U.S. Department of Commerce, NOAA National Sea Grant Office, grant no. NA86RG-0050, project A/151-01; and by the University of Alaska Fairbanks with state funds. The University of Alaska is an affirmative action/equal opportunity institution.
    Sea Grant is a unique partnership with public and private sectors combining research, education, and technology transfer for public service. This national network of universities meets changing environmental and economic needs of people in our coastal, ocean, and Great Lakes regions.

University of Alaska Sea Grant
P.O. Box 755040
205 O'Neill Bldg.
Fairbanks, Alaska 99775-5040
Toll free (888) 789-0090
(907) 474-6707 Fax (907) 474-6285
http://www.uaf.edu/seagrant/

Contents

About the Symposium ix

The Lowell Wakefield Symposium Series ix

Proceedings Acknowledgments x

Nomenclature Note x

Feeding, Competition, and Predation

The Role of Herring Investigations
in Shaping Fisheries Science
Robert L. Stephenson 1

Temperature Effects on Zooplankton
Assemblages and Juvenile Herring
Feeding in Prince William Sound, Alaska
Robert J. Foy and Brenda L. Norcross 21

Is the Production of Coho Salmon in the Strait of
Georgia Linked to the Production of Pacific Herring?
R.J. Beamish, G.A. McFarlane, and J. Schweigert 37

Common Factors Have Opposite Impacts on
Pacific Herring in Adjacent Ecosystems
G.A. McFarlane, R.J. Beamish, and J. Schweigert 51

Trophic Position of Pacific Herring in Prince William
Sound, Alaska, Based on Their Stable Isotope Abundance
Thomas C. Kline Jr. 69

Pacific Herring in the Western
Bering Sea Predatory Fish Diet
Alexei M. Orlov and Sergei I. Moiseev 81

Herring Abundance, Food Supply, and Distribution
in the Barents Sea and Their Availability for Cod
*Emma L. Orlova, Elena I. Seliverstova,
Andrey V. Dolgov, and Valentina N. Nesterova* 91

Distribution and Feeding Habits of Juvenile
Herring (*Clupea pallasii*) in Northern Japan
Masayoshi Sasaki, Ryotaro Ishida, and Takayanagi Shiro 101

Predation by Smelt (*Hypomesus japonicus*) on
Herring Larvae (*Clupea pallasii*) in Karagin Bay
Petr M. Vasilets 117

Estimating Whole Body Energy Content
for Juvenile Pacific Herring from Condition
Factor, Dry Weight, and Carbon/Nitrogen Ratio
A.J. Paul, J.M. Paul, and T.C. Kline 121

Growth, Reproduction, and Production

Biological Characteristics of Atlantic Herring as
Described by a Long-Term Sampling Program
Michael J. Power and T. Derrick Iles 135

Is the Decrease in Growth Rate of Atlantic
Herring in the Baltic Sea Density-Dependent?
A Geostatistical Application
M. Cardinale and F. Arrhenius 153

Growth Pattern of Baltic Herring
in Relation to Spawning Time
Jan Eklund, Marjut Rajasilta, and Päivi Laine 155

Anatomy of a Strong Year Class: Analysis
of the 1977 Year Class of Pacific Herring
in British Columbia and Alaska
D.E. Hay, M.J. Thompson, and P.B. McCarter 171

Spawning Bed Selection by Pacific Herring
(*Clupea pallasii*) at Atsuta, Hokkaido, Japan
*Hiroshi Hoshikawa, Ken-ichiro Tajima,
Tadashi Kawai, and Tomohiro Ohtsuki* 199

Relationship between Date of Hatching and Growth
of Herring Larvae and Juveniles in Ishikari Bay
*Ryotaro Ishida, Masayoshi Sasaki,
Shiro Takayanagi, and Hideo Yoshida* 227

Decline of the Sakhalin-Hokkaido Herring
Spawning Grounds near the Sakhalin Coast
Elsa R. Ivshina 245

Norwegian Spring-Spawning Herring

Factors Influencing Location and Time of
Spawning in Norwegian Spring-Spawning Herring:
An Evaluation of Different Hypotheses
Aril Slotte 255

Norwegian Spring-Spawning Herring (*Clupea harengus*)
and Climate throughout the Twentieth Century
Reidar Toresen and Ole Johan Østvedt 279

Oocyte Degeneration in Female Recruits of Norwegian
Spring-Spawning Herring (*Clupea harengus*)
G.P. Mazhirina and E.I. Seliverstova 285

The Relevance of a Former Important Spawning
Area in the Present Life History and
Management of Norwegian Spring-Spawning Herring
Ingolf Røttingen and Aril Slotte 297

Hydrography and Environmental Impacts

Spawning Stock Fluctuations and Recruitment Variability
Related to Temperature for Selected Herring (*Clupea harengus*) Stocks in the North Atlantic
Reidar Toresen 315

Effect of Herring Egg Distribution and
Environmental Factors on Year-Class
Strength and Adult Distribution: Preliminary
Results from Prince William Sound, Alaska
Evelyn D. Brown and Brenda L. Norcross 335

Herring Occurrence in the Sound (ICES SD23)
in Relation to Hydrographical Features
*J. Rasmus Nielsen, Bo Lundgren, Torben F. Jensen,
and Karl-Johan Stæhr* 347

The Norwegian Spring-Spawning Herring:
Environmental Impact on Recruitment
*R. Sætre, R. Toresen, T. Anker-Nilssen,
and P. Fossum* 357

Disease and Population Assessment of Pacific
Herring in Prince William Sound, Alaska
*Terrance J. Quinn II, Gary D. Marty,
John Wilcock, and Mark Willette* 363

Stock Assessment

Taking Stock: An Inventory and Review
of World Herring Stocks in 2000
  D.E. Hay and others 381

Survival of Pacific Herring Eggs on
Giant Kelp in San Francisco Bay
  Sara Peterson 455

Seasonal Variation in Herring Target Strength
  E. Ona, X. Zha, I. Svellingen, and J.E. Fosseidengen 461

Estimates of Egg Loss in Pacific Herring Spawning
Beds and Its Impact on Stock Assessments
  Jake Schweigert and Carl Haegele 489

Herring Tilt Angles, Measured through Target Tracking
  E. Ona 509

Latitudinal Difference in Recruitment
Dynamics of Clupeid Fishes: Variable to
the North, Stable to the South
  *Yoshiro Watanabe, Norio Shirahuji, and
  Masayuki Chimura* 521

Estimation of First-Year Survival of Pacific Herring
from a Review of Recent Stage-Specific Studies
  Brenda L. Norcross and Evelyn D. Brown 535

Stock Identification, Stock Structure, and Biodiversity

Herring Stock Structure, Stock Discreteness,
and Biodiversity
  *R.L. Stephenson, K.J. Clark, M.J. Power,
  F.J. Fife, and G.D. Melvin* 559

Biological Characteristics and Stock Enhancement
of Lake Furen Herring Distributed in Northern Japan
  Tokimasa Kobayashi 573

Morphometric Variation among Spawning Groups
of the Gulf of Maine–Georges Bank Herring Complex
  Michael P. Armstrong and Steven X. Cadrin 575

Herring of the White Sea
    G.G. Novikov, A.K. Karpov, A.P. Andreeva,
    and A.V. Semenova 591

A Tagging Experiment on Spring-Spawning
Baltic Herring (Clupea harengus membras)
in Southwest Finland in 1990-1998
    J. Kääriä, M. Naarminen, J. Eklund,
    N. Jönsson, G. Aneer, and M. Rajasilta 599

Microsatellite Polymorphism and Population
Genetic Structure of Atlantic Herring in the
Baltic and Adjacent Seas
    Carl André, Fredrik Arrhenius,
    Mats Envall, and Per Sundberg 611

Microsatellite Population Structure in
Herring at Three Spatial Scales
    Arran A. McPherson, Christopher T. Taggart,
    Paul W. Shaw, Patrick T. O'Reilly, and Doug Cook 615

Intermingling of Herring Stocks in the Barents Sea Area
    K.E. Jørstad and others 629

Gizhiga-Kamchatka Herring
Stock Level and Catch Potential
    Andrey A. Smirnov 635

Fishery Management

Management of North Sea Herring
and Prospects for the New Millennium
    J.H. Nichols 645

A New Approach to Managing a Herring Fishery:
Effort vs. Quota Controls
    Denis Tremblay 667

Industry Acoustic Surveys as the Basis for
In-Season Decisions in a Comanagement Regime
    Gary D. Melvin, Robert L. Stephenson,
    Mike J. Power, F.J. Fife, and Kirsten J. Clark 675

Present State of the Okhotsk Herring Population
after Large-Scale Fishery Resumption
    V.I. Radchenko and I.V. Melnikov 689

Baltic Herring Fisheries Management in Estonia:
A Biological, Technical, and Socioeconomic Approach
    Tiit Raid and Ahto Järvik 703

Changing Markets for Alaska Roe Herring
    Terry Johnson and Gunnar Knapp 721

Social and Economic Impacts
Linking Biological and Industrial
Aspects of the Finnish Commercial
Herring Fishery in the Northern Baltic Sea
    *Robert Stephenson, Heikki Peltonen, Sakari Kuikka,
    Jukka Pönni, Mika Rahikainen, Eero Aro,
    and Jari Setälä* 741

Participants 761

Index 767

About the Symposium

Herring 2000: An International Symposium on Expectations for a New Millennium is the eighteenth Lowell Wakefield symposium. The program idea was suggested by Fritz Funk of the Alaska Department of Fish and Game as a follow-up to the 1980 Alaska Herring Symposium and the 1990 International Herring Symposium. The meeting was held February 23-26, 2000, in Anchorage, Alaska. Contributors gave 45 oral and 14 poster presentations.

The symposium was organized and coordinated by Brenda Baxter, University of Alaska Sea Grant College Program, with the assistance of the organizing committee. Organizing committee members are Fritz Funk, Alaska Department of Fish and Game; Doug Hay, Department of Fisheries and Oceans, Canada; A.J. Paul, University of Alaska Fairbanks, Institute of Marine Science; Connie Ryan, California Department of Fish and Game; Rob Stephenson, Department of Fisheries and Oceans, Canada; Reidar Toresen, Marine Research Institute, Norway; Dave Witherell, North Pacific Fishery Management Council; and Mei-Sun Yang, U.S. National Marine Fisheries Service, Alaska Fisheries Science Center.

Symposium sponsors are Alaska Department of Fish and Game; North Pacific Fishery Management Council; U.S. National Marine Fisheries Service, Alaska Fisheries Science Center; and Alaska Sea Grant College Program, University of Alaska Fairbanks.

The Lowell Wakefield Symposium Series

The University of Alaska Sea Grant College Program has been sponsoring and coordinating the Lowell Wakefield Fisheries Symposium series since 1982. These meetings are a forum for information exchange in biology, management, economics, and processing of various fish species and complexes as well as an opportunity for scientists from high latitude countries to meet informally and discuss their work.

Lowell Wakefield was the founder of the Alaska king crab industry. He recognized two major ingredients necessary for the king crab fishery to survive—ensuring that a quality product be made available to the consumer, and that a viable fishery can be maintained only through sound management practices based on the best scientific data available. Lowell Wakefield and Wakefield Seafoods played important roles in the development and implementation of quality control legislation, in the preparation of fishing regulations for Alaska waters, and in drafting international agreements for the high seas. Toward the end of his life, Lowell Wakefield joined the faculty of the University of Alaska as an adjunct professor of fisheries where he influenced the early directions of the university's Sea Grant Program. This symposium series is named in honor of Lowell Wakefield and his many contributions to Alaska's fisheries. Three Wakefield symposia are planned for 2002-2004.

Proceedings Acknowledgments

This publication presents 49 symposium papers. Each full-length paper was reviewed by two peer reviewers, extended abstracts had one review each, and papers were revised according to recommendations by associate editors who generously donated their time and expertise: Fritz Funk, Jim Blackburn, Doug Hay, A.J. Paul, Rob Stephenson, Reidar Toresen, and Dave Witherell.

University of Alaska Sea Grant thanks the following individuals for reviewing papers: Michael P. Armstrong, Fredrik Arrhenius, Bill Atkinson, Greg Bargmann, Nicholas J. Bax, Jim Blackburn, Kevin Brennan, David Brewer, Charles Burkey Jr., Steven X. Cadrin, Max Cardinale, Mark G. Carls, James Carscadden, Kirsten Clark, Geir Dahle, Are Dommasnes, Jan Eklund, W.J. Fletcher, Juha Flinkman, Robert J. Foy, Jeffrey T. Fujioka, Fritz Funk, John M. Gates, Carl Haegele, Pete Hagen, Cornelius (Nils) Hammer, Sture Hansson, Doug Hay, Susan Hills, Jens Christian Holst, Ahto Järvik, Arne Johannessen, B. Alan Johnson, Knut E. Jørstad, Juha Kääria, Thomas C. Kline Jr., Jon Bent Kristoffersen, Sakari Kuikka, Dan Lane, Jason Link, Tomasz B. Linkowski, Alec MacCall, Steve Mackinson, Ole A. Mathisen, Bruce McCarter, Michael F. McGowan, Gary D. Melvin, Christian Möllmann, Henrik Mosegaard, Franz Mueter, Gunnar Naevdal, John Nichols, Brenda L. Norcross, Chris W. Oliver, Egil Ona, Gudmundur J. Oskarsson, Ted Otis, William Overholtz, A.J. Paul, Stacey Paul, Bob Piorkowski, Mike Power, Terry Quinn, Tiit Raid, Ingolf Røttingen, Katherine Rowell, Sue Sanders, Jake Schweigert, Linda R. Shaw, Michael Sinclair, Aril Slotte, Henrik Sparholt, Rob Stephenson, David K. Stevenson, Kevin D.E. Stokesbury, Gary Thomas, Denis Tremblay, Cemal Turan, Lauri Urho, Erik Williams, Bruce Wright, Mei-Sun Yang, Phil Yund, Christopher Zimmermann, and Jie Zheng.

Copy editing is by Kitty Mecklenburg of Pt. Stephens Research Associates, Auke Bay, Alaska; and Sue Keller, University of Alaska Sea Grant. Layout and format by Kathy Kurtenbach, and cover design by Tatiana Piatanova, both of University of Alaska Sea Grant.

Nomenclature Note

The specific epithet for Pacific herring is correctly spelled with a double i: *Clupea pallasii* Valenciennes, 1847. The name has appeared in literature with or without the second i, depending on prevailing opinion. The *International Code of Zoological Nomenclature*, fourth edition (1999), which took effect January 1, 2000, clarifies that the spelling of a species name given in the original description of the species is to be retained. Therefore, the spelling *pallasii* is retained herein.

The Role of Herring Investigations in Shaping Fisheries Science

Robert L. Stephenson
Department of Fisheries and Oceans, Biological Station, St. Andrews, New Brunswick, Canada

Abstract

Herring (Atlantic, Baltic, and Pacific) have contributed to major fisheries and have been important to coastal communities for centuries. As a result, herring has been the subject of considerable scientific study, and it is not surprising that herring fisheries and herring research have been at the core of several major developments in fisheries science and fisheries management. Early theories regarding migration, the development of the population/stock concept, and tracking and quantification of year classes were based largely on herring investigations. More recently herring research has been key to development of hydroacoustics methods, hypotheses on population regulation and abundance, the link between fisheries dynamics and hydrography, and a number of innovative approaches to fisheries regulation and management. These developments have shaped fisheries science and had an impact on science generally. In spite of severe depletion or collapse of several major stocks at some point in the last few decades, herring remain the subject of major fisheries and of research. Future research developments might be expected, as in the past, to reflect the major initiatives and general themes of fisheries science and management modified by specific issues concerning herring fisheries.

Introduction

Not only has the herring from earliest times been the commercial fish of northern waters (after stealing the crown from the cod, *Gadus morhua*, perhaps as early as the tenth century), but it was in northern Europe that taxonomy, fishery science and biology first took on a modern aspect . . . contributions to

---

R.L. Stephenson was visiting research professor at the Finnish Game and Fisheries Research Institute, Helsinki, Finland, at the time this paper was written.

ichthyology tended to come from countries . . . where herring was exploited and from time to time held the fate of empires. (Whitehead 1985, pp. 3, 4)

As a recruit to fisheries research in the early 1980s, I became intrigued by the evolution in fisheries science and management. I knew in working on herring assessment and research that I was working in support of a prominent fishery, and that this species seemed to have its fair share of literature. I was following in the footsteps of a number of revered scientists who had worked on various herring fisheries. I remember, for example the excitement of first reading A.G. Huntsman's papers on Bay of Fundy herring (Huntsman 1917, 1918), realizing that he had been in the same location and looking at the same fishery more than 70 years earlier. I then remember becoming somewhat dismayed to see how many of the obvious questions (1) had been thought of, (2) had been worked on by my predecessors, and (3) apparently could not be resolved by some of the most prominent fisheries scientists of the past. I became aware of the fact that some of what has been done and published has, sadly, been forgotten and ignored, and that much research in science is spent rediscovering and revisiting some basic biological questions. It was obvious, however, that there had certainly been some great breakthroughs in the past century—and that many of these had been the direct result of work on herring.

In this paper I attempt to evaluate how large an impact studies of herring and herring fisheries have had. I make the claim that work on herring, more than that on any other fish, has shaped the development of fisheries biology and fisheries management. Further, I contend that this trend should continue in the future. I suggest that some reflection on historical herring research reminds us of some basic characteristics of fisheries research—including the persistence of several major themes and recurrent interest in ideas, some of which were generated more than a century ago.

Over a little more than a century there have been huge developments in fisheries science and fisheries management. This period has been marked by industrialization of fisheries, with rapid development in fishing technology and expansion of fisheries. It has also been a period of profound development in concepts (including recognition of the concept of overfishing), the emergence of fisheries management (and in the last few decades the "modern fisheries management experiment" [Stephenson and Lane 1995]), and improved analytical capabilities.

Cushing (1988) in describing the historical development of fisheries and fisheries science claims that fisheries science has two roots. First, there was development of marine biology in the nineteenth century, resulting from a Victorian interest in the sea, and characterized by an increase in descriptive natural history, and the great expeditions of the 1800s. Second, the first period of industrialization, in which there was an increasing interest in exploitation (and eventually in overfishing), led to the development of quantitative fisheries science as shown, for example in the figure summarizing the development of assessment methods which appears in Caddy (1999, p. 7).

The period beginning about 1875 appears to have been especially significant, for it was about that time that there was a change from descriptive to quantitative scientific investigation, the development of major research programs on fisheries, and the beginnings of fisheries science as the entity we know today. The first paragraph in Tim Smith's book *Scaling Fisheries* (1994) sets the scene in this period:

> In 1864 the Norwegian government asked George Ossian Sars, son of the pioneering Norwegian marine biologist Michael Sars, to determine why the cod catches from the Lofoten Islands in northern Norway fluctuated so greatly.... A few years later, after several visits to the coastal fisheries, Sars asked for and, to his surprise, was loaned a ship to extend his studies offshore. Within twenty years Norway had established a scientific agency to study the fluctuations in its fisheries, and had outfitted it with a ship, laboratories, and a fish hatchery. By the turn of the century many other countries had joined Norway in establishing agencies for the scientific study of their fisheries, many of which joined in an international research organisation in 1901. What was it about fisheries that justified the creation, and continues to justify the funding, of national and international research programs?

> The problem that Sars began to address more than 100 years ago [why fish catches vary] was important then, and remains important today.

Herring Have Supported Important Fisheries for Centuries

Mitchell (1864) stated, "Until salt was used, it was probable that the commerce in herrings was inconsiderable," but cited a number of early written accounts of the use of herring dating to the eleventh century and some earlier. The importance of herring to the economics and culture of coastal communities of Europe for many centuries is well documented, and is reflected, for example in the persistence of the Baltic herring markets of Turku and Helsinki, Finland, which have taken place each autumn for more than 200 years.

Whitehead's introductory paper from the proceedings of the 1983 International Symposium on the Biological Characteristics of Herring and Their Implication for Management, summarized the prominent position of *Clupea harengus* or "king herring" (Whitehead 1985):

> Lacepede ... lavished upon it suitably flowery phrases (Lacepede 1803, p. 429): [translated] "Herring is one, the use of whose natural production has decided the destiny of empires. The bean of coffee, the leaves of tea, the spices of the tropics, the worms that make silk, are of smaller influence on the nations richness than the herring of the Atlantic Ocean. A luxury or whim comes first to mind; but the real place is claimed by herring."

> Linneaeus (1758, p. 318) said it all in two words *copiosissimus* piscis [= the most prolific of fish]. *Copiosissimus* the fish, which by the twelfth century had become a staple in the European diet.... *Copiosissimus* too the literature, which reflects the number of studies, the sheer man-hours, the salaries, the equipment, the running costs of institutes and ships—the capital investment whose mainspring (if not ultimate goal) is the greater exploitation of *Clupea* as a resource.

The historical importance of herring in Europe is affirmed by Alan Davidson in the new *Oxford Companion to Food* (Davidson 1999), "Herring ... of all fish probably the one which has most influence on the economic and political history of Europe."

Themes in Herring Research

There is a rich and diverse literature concerning herring. As in other fields of science, there has been a continuous expansion in literature, particularly during the twentieth century. It seems convenient to divide herring research into the periods before and after 1875.

Prior to 1875 Herring Studies Were among the Earliest Marine Scientific Investigations

It was inevitable that the economic importance of herring led to a great deal of attention and interest. Prior to about 1875, the issue was almost exclusively explanation of the variability in catches which had so much impact on social and economic affairs of some nations. Since the beginning, herring fisheries had to cope with large variations in catches, caused by large variations in abundance, and this fueled concern, speculation, documentation, and eventually scientific study. Among the longest of any fisheries data series available are some Swedish (Bohuslan) and Norwegian herring fisheries for which records go back some 500 years (Csirke 1988).

During the eighteenth century, accounts of herring feature in early writings on marine fisheries such as astronomer Edmond Halley (of Halley's Comet fame; 1728), *Atlas Maritimus et Commercialis;* Johann Anderson (1746), *Accounts of Iceland, Greenland and the Davis Strait;* James Anderson (1785), *Account of the Present State of the Hebrides and Western Coasts of Scotland*. It was the subject of at least one entire book, *Natural History of Herring*, by British naval physician Solas Dodd (1752).

Johann Anderson wrote in the preface to his book (1746) that he thought the most important part of the book was his herring migration theory. This is now referred to as the "polar migration theory," and indeed this was one of the major fisheries science ideas generated from observations on herring.

Observations and accounts of this type continued through most of the nineteenth century. Explanation of the variability in catches remained the issue. There seems to have been little further development, but much observation, that provided the context for later developments.

After 1875 Herring Studies Contributed to the Development of Fisheries, Fisheries Science, and Fisheries Management

The progression of major themes in herring research is evident from an examination of the literature since 1875 (Table 1). For the period since

Table 1. **Decadal chronology of major themes in published herring research.**

Ancient times	Herring fisheries documented as early as 240 A.D.
1860s	The existence of winter and summer herring is noted. These are not considered different enough to be called different species.
	Food, feeding and behavior of Atlantic herring from Europe and eastern North America.
	Migration patterns observed.
	Examination of spawning beds and substrates. Attempts to incubate eggs.
1880s and 1890s	Heinke's theory of races in fish stocks.
1910s	Hjort's publications on stock fluctuations and dynamics and the existence of races of herring in northern Europe and eastern Canada.
1920s	Stock and population structure and dynamics—using races, vertebral counts and spawning season to distinguish races.
	Attempts to distinguish races using vertebral counts and spawning season.
	Use of scales for aging.
	Studies of food and feeding behavior of adults and larvae.
	Length and age at maturity.
	Assessment methodology—attempts to introduce a mathematical or statistical approach.
	Migration of shoals.
	Hypotheses about the cyclical nature of population fluctuations.
	Examination of the influence of quantity of spawning herring on the size of the stock of following year.
1930s	Tags used to study migration. Comparisons of the recovery rates of external tags and internal tags in the body cavity. Use of electromagnetic detectors to collect tags in fish processed for fish meal.
	Stock and population structure and dynamics—attempts to distinguish races (including spring vs. fall spawners) using morphological features, vertebral counts and growth increments from scales.
	Maturity in relation to age and length.
	Factors affecting year-class strength, including hydrography.
	Lunar influence on fishing.

Table 1. (Continued.) Decadal chronology of major themes in published herring research.

1930s (Contd.)	Spawning waves observed with the oldest fish spawning first and leaving earlier than the younger fish.
1940s	Examination of geographic variations in vertebral counts: northern vs. southern.
	Use of vertebral variations to distinguish populations.
	Studies of age and length at maturity and the relationship between age and length and fecundity.
	Stomach content analysis in relation to the distribution of plankton populations.
	Tagging in the Pacific.
	Concern over overfishing in Alaska.
1950s	Double helix structure of DNA molecule published in 1953.
	Methods published for the isolation of DNA from herring roe and sperm.
	Sonic sounders used to locate pelagic fish and to study the behavior and size of herring schools.
	Lab studies on herring behavior, feeding, and shoaling and on fertilization and development of herring eggs.
	Behavior studied in response to hydrographic conditions and light.
	Continued interest in distinguishing populations based on vertebral counts and morphological variations.
	Studies of oogenesis and spermatogenesis.
	Comparisons of the effectiveness of different gear types.
1960s	Forecasting catches based on larval recruitment and age composition.
	Maturity scale for gonadal development recommended by Herring Committee to ICES in 1965.
	Continued interest in the use of tags, morphometrics, and meristics to determine homing, migration patterns for spawning, and feeding and population mixing on the offshore feeding areas.
	Fish pumps.
	Characterization and identification of spawning grounds and substrates.
	Use of acoustics to locate herring and to study the size of herring schools.
	Internal and external tags used in studies of migration and homing.
1970s	Larval surveys and larval abundance indices used for estimating stock size.
	Larval retention areas.

Table 1. (Continued.) Decadal chronology of major themes in published herring research.

	Concern over the mismanagement of herring stocks.
	Concern with toxicological issues—pollution, oil, PCBs, and heavy metals.
	Tagging using external tags, internal tags, and parasites as biological tags (*Anisakis* spp.).
	Distinguishing stocks and populations using otoliths, morphometrics, and meristics.
	Effect of hydrographic, oceanographic, and environmental factors on eggs and embryonic development.
1980s	Tagging using external tags, internal tags and parasites as biological tags.
	Multispecies and environmental models as well as single species models.
	Study of effects of hydrographic, oceanographic, and environmental factors on egg and larval development and on spawning.
	Continued refining of acoustic techniques for locating and measuring herring schools.
	Comanagement.
	Locating spawning grounds and categorizing spawning behavior, duration and intensity.
	Concern with toxicological issues particularly in the Baltic—pollution, oil, heavy metals, and radiation fallout from Chernobyl.
	Continued attempts to distinguish populations with tools such as genetics, otolith structure, tagging, biological tags, morphometrics, and meristics.
	The effect of climate and the environment—El Niño, Mount St. Helens, and the increase in atmospheric CO_2.
1990-1999	The arrival of the "precautionary approach."
	Interactions between herring fisheries and the environment.
	Environmental approach and multispecies models.
	Acoustics—multi-beam and side-scan sonar.
	Genetics—microsatellite DNA.
	Fisheries management science and comanagement of resources with industry.
	Concern with toxicological issues, particularly in the Baltic and western North America—pollution, heavy metals, and oil (*Exxon Valdez*).
	Continued attempts to distinguish populations with some new tools (genetics, elemental analysis of otoliths) and some old (otoliths, vertebral counts).
	Consideration of role of herring as forage.

1920, this is based on topics of papers on Pacific, Baltic, and Atlantic herring (with herring in the title), which were listed in *Biological Abstracts* (1920s to 1970s) or in *Aquatic Sciences and Fisheries Abstracts* (since 1975). For the earlier period it is based largely on historical summaries and existing syntheses of the development of fisheries science, including an examination of the herring research preceding and during the early years of the International Council for the Exploration of the Sea (Stephenson and Clark 2002).

The literature indicates that herring research seems to have been influenced by three general factors:

1. It has reflected and contributed to general developments in fisheries research and management, for example:
   - Migration patterns as the explanation of fluctuation in catch
   - Existence of separate populations/stocks
   - Aging, year classes, stock fluctuations
   - Concept of overfishing
   - Stock assessment methodologies
   - Acoustic methods
   - 1950s interest in genetics following Watson and Crick
   - Fisheries management
   - Feeding, trophic relationships, multispecies aspects

2. It has reflected issues or changes in the use and demand for herring, including:
   - Practical aspects of management of herring fisheries
   - Development of major fish meal markets
   - Use of herring in animal/livestock feed
   - Herring as feed in aquaculture

3. It has responded to unique or local issues and perturbations, such as:
   - *Exxon Valdez* oil spill in Alaska
   - Release of radioactivity at Chernobyl
   - Eruption of Mount St. Helens
   - El Niño events

Herring Studies Have Contributed to Several Major Developments in Fisheries Science and Management

Modern taxonomy, fishery science, and biology were born in Europe. This, together with the long-standing economic importance of the herring, has given *Clupea harengus* (and its close relative *C. pallasi*) a prestige and literature that quite outshadows those of any other. (Whitehead 1985)

There have been a number of major developments in fisheries science and management based largely on herring work. These include:

A. Polar Migration Theory

The earliest major development in fisheries science based primarily on herring appears to be the herring migration theory put forward by Johann Anderson (1746). According to Anderson's theory (Wegner 1993 and pers. comm.):

> The north Polar Sea is the real home of the herring. Beneath the permanent ice cover the herring is safe from its "enemies," such as different fishes, whales (who cannot come up for air under ice), gulls, and fishermen of all nations. Below the ice the herring is undisturbed and grows rapidly reaching a maximum size due to plentiful food. However, because of the increasing population and diminishing food supply the main shoal of herring leave this safe area and migrate south early (January) each year.
>
> As soon as the shoal reaches the open waters its "enemies" pounce on it and drive the fish permanently further south. Fishermen catch large amounts of herring in certain well-known areas according to the seasons. . . .
>
> The remaining herring which have not been eaten by predators or caught by fishermen at various places along the way, join up with the juvenile herring from different spawning grounds toward the end of the year and return again to the "real north" to reproduce the stock beneath the ice.

Wegner points out that the polar migration theory was able to answer the most important questions concerning Atlantic herring at the time. Herring were different lengths and weights in different places because of the distance away from the plentiful food of the north, and the energy needed to escape predators. The absence of herring on a well-known fishing ground was due to predation. Changing migration patterns were a reflection of changing predator behavior. In addition to the main spawning beneath the polar sea there was spawning at different times and locations during the migration.

Toward the end of the nineteenth century, there was considerable speculation about the large variations in abundance of herring and views that challenged the polar migration theory. William Marshall, for example in the book *Die Deutschen Meere und ihre Bewohner* (1895) summarizes the theories related to why herring come to spawning grounds some times and not others (translated by C. Hammer):

> Since man is very much involved in this matter, he has long been looking for an explanation for this awkward phenomenon. In older times there was talk about punishment from God for the fishermen and merchants who have become rich and "ubermutig." From the Baltic the herring are reported to have vanished since cannons producing considerable noise were used in the sea battle. For this reason one claims to have noticed unusual migrations of the herring during the Swedish Russian war in 1789, and then again it is reported that the water is poisoned by the remains of culling of whales and seals and

the cooking of the oil. The trawl nets are accused for destroying the seabed and at the same time killing eggs and herring larvae. Then the steamships have been made responsible for creating too much water movement, as if turbulence in the water body has not been there before mankind has thought about steamships. The Dutch fishermen talk about a king of herring who leads the schools, apparently they assume several kinds of non-herring fish for this purpose [the German name of John Dory (*Zeus faber*) is Heringskonig, i.e. king of herring, who is coming into the North Sea in summer and forages on herring; remarks by Hammer]. If this fish is caught and not set free, his "folk" disintegrate into small bunches and get lost, not having the appropriate knowledge.

The polar migration theory was broadened to include other species, and developed to include the influence of oceanographic conditions (in addition to predation) on migration routes and remained popular until undermined by the work on herring by Heincke at end of the nineteenth century (Sinclair 1997). It was still a popular working hypothesis at the beginning of the twentieth century (Sinclair 1997) and the influence is seen, for example in Huntsman's conclusions regarding migratory "processions" of herring around the Bay of Fundy (Huntsman 1918).

About the end of the twentieth century, there was a great increase in the documentation of fisheries, in scientific study, and literature. The field of fisheries science began to develop at a great rate. Herring was an important species, and remained the subject of several further developments.

B. Population Thinking

Sinclair and Solemdal (1988), and Sinclair (1988, 1997) chronicle the important development of "population thinking" which took place in fisheries biology between 1878 and 1930. During this period, research on the interannual variability in abundance and on intraspecific variability in body form led to a major change in thinking—from consideration of species as types (later known as the essentialist species concept) to consideration of the species made up of groups of populations (component populations).

This shift had a great deal to do with herring, through the work of Heincke. Sinclair (1997) discusses how Heincke (1878, 1882, 1898) undermined the polar migration theory:

> He [Heincke] initiated his work in the early 1870s within the context of the herring overfishing debate in the Baltic. There were two schools of thought. Based on the work of the Swedish natural historian Nilsson, it was believed that there were many local forms of herring in the Baltic, each form having a limited distributional area with short seasonal migrations. Under this interpretation there was considerable scope for overfishing, even with the simple harvesting technology of the time. The second school of thought, championed by the Danish scientist Henrik Krøyer, favored the traditional interpretation based on Anderson's "migration" theory, with local overfishing being impossible. The public debate was heated, but unresolved based on the methodologies used at the time to identify the putative local forms.
>
> Heincke approached the issue by carrying out a detailed life history study of two herring spawning components off Kiel—the spring and autumn spawners.

He sampled spawning fish, the early larval stages, the postlarvae and the juveniles of the two components. . . . He concluded, using primitive multivariate statistics, that the two herring spawning groups were separate races of the same species, and that the racial differences of herring become expressed during the transition from the larval stage to the definitive (juvenile) stage.He hypothesised that the migrations of herring races were based on both reproductive and feeding instincts. The time and location of spawning was selected in relation to seasonal plankton dynamics and the transport of larvae.

The development of population thinking involved the shift from the species to the population as the appropriate unit of study. The work of Hjort (synthesized in the paper of 1914) generalized the findings of Heincke on herring to other commercial species, and clearly identified the significance of "population thinking" to fisheries management. This development, which began with Heincke's 1878 paper, represents the second major advance based largely on herring, and one which was of great significance:

> The early debate concerning the existence of several self-sustaining populations within the distributional limits of marine species centered on Atlantic herring in the northeast Atlantic. Fr. Heincke convincingly resolved the controversy by extensive sampling of herring and sprat for the analysis of a large number of meristic and morphometric characters, and the development of new statistical methodology (including rudimentary multivariate statistics). (Sinclair and Solemdal 1988)

Reconstructing the lineage of historical ideas is difficult. It is clear, for example, that as early as the 1860s J.M. Mitchell (*The Herring: Its Natural History and National Importance,*1864) disagreed with the prevailing migration theory, and anticipated the concept of populations that appeared about 1914. Mitchell proposed that herring live in the seas adjacent to the coasts or bays where they spawn, and after spawning they return to sea in the neighborhood where they feed until spawning again. The fry continue near the spawning ground until they are of sufficient size. He proposed 11 pieces of evidence to support this:

1. Every year, at a certain period of the year, a particular size of herring always resorts to the same place (i.e., herring caught off Stadtland, Norway, is much larger than those on the west of Scotland). Because of this size differentiation, they can't all be from one large group.

2. Quality. The fish caught off Shetland were not as fat as those from Thurso. There are marked differences in appearance and quality between fish of different areas.

3. Time of appearance. For example, herring are caught in Loch Fyne before any are caught near Cape Wrath.

4. There was no well authenticated instance of herring being seen approaching south from a high north latitude.

5. No shoals of herring were known from Greenland seas or found in the stomachs of herring predators from those areas.

6. Whales that feed principally on herring frequent the shores of Britain and Norway year-round.
7. Bloch (no reference) had established that fish of similar size to herring could not make, from spring to autumn, the long voyage attributed to herring.
8. Herring may be found in some localities year-round.
9. No shoals of herring have ever been seen returning to the north from the south.
10. Why would the smallest herring from a large school go to the Baltic and larger ones to the North Sea?
11. Other fish have similar habits in spawning: salmon, sprat, shad, and pilchards.

This is one of several instances of ideas in the literature that seem to pre-date the sources normally cited.

C. Age Determination, Year-Class Strength, and Prediction

Studies of age structure in herring populations in the early 1900s represent the third major advancement in fisheries science based on herring. In 1907 J. Hjort proposed applying the approaches of vital statistics (human demography) to fish using rings in the scales of herring at a meeting of the International Council for the Exploration of the Sea (ICES) (Hjort 1908 cited by Smith 1988). Although the idea was not embraced totally by ICES, Hjort and colleagues continued their studies. They noted that the relative number at successive ages varied in both herring and cod, and in 1929 Hodgson (1957) was able to use this method to forecast the impact of the strong 1924 year class on the herring fishery.

Much of the discussion and synthesis of the work of Heincke and of Hjort took place in committees of the newly formed ICES (see review by Stephenson and Clark 2002). The combined impact of the last two advancements was a very large change in thinking—a replacement rather than a modification of ideas—and has been termed by Sinclair (1997) a true paradigm shift. The work originating largely from observation of herring certainly set the course for fisheries science and management which persists to this day. The concept of self sustaining stocks or populations forms the basis of assessment and management, and age-based assessment methods are prominent. Interestingly, the prominence of herring in the further evolution and application of the age-based assessment technique was replaced by groundfish species which seemed to offer better case studies. It is noteworthy that herring assessments of this type often suffer from poor abundance indices and lack of recruitment indices, and actually do not perform very well (Stephenson 1997).

D. Hydroacoustic Estimation of Fish Abundance

Undoubtedly, the development of hydroacoustic estimation of fish distribution and abundance is a major scientific development based on herring research. An interesting and important early study was that carried out with an echo sounder by Skipper Ronald Balls when fishing for herring in the North Sea during the 1930s (Thomasson 1981). Balls, a drifter skipper, undertook pioneering work using a Marconi 424 echo sounder during fishing operations on his own vessel *Violet and Rose*. He kept detailed records of his observations which he published, among other places, in the *ICES Journal du Conseil* (Balls 1948):

> It was through a hint from Dr. Hodgson of the Fisheries Laboratory, Lowestoft, that I first became interested in the idea of using the echo sounder for spotting herring shoals, and in 1933 I fitted an instrument to my drifter Violet and Rose YH 757.
>
> The seven years' experience with the Echometer in herring fishing, on which this account is based, was gained, not in a research vessel, but under ordinary commercial drifter conditions. My idea was to find out whether or not the machine could increase catching efficiency; therefore my working notes had to be absolutely unbiased, and whereas a properly organized experiment in this field would have had the advantage of more copious and detailed notes, this is perhaps balanced in my amateur attempt by keenness of observation and the length of time covered.

Balls reported that the Echometer worked with "steadily increasing efficiency" for locating herring on 400 nights on the summer fishing grounds but not well in the East Anglian autumn fishery, suggesting "that some change in local conditions on the two fishing grounds is the cause."

This account is of interest not only because it helped establish the validity of hydroacoustics, but also for the role of the fishing industry in the development. Balls addresses this issue in the paper, first talking about his crew, then about the role of the fishing industry in research:

> Some of these men had sailed with me for ten years, and would handle a plankton indicator with the same professional discernment as they mended or shot a net. But they still maintained a chilly silence at most "scientific stuff" I mentioned. And it was with a source of secret amusement and some wonder to see one of them calmly watching the sounder as I steered.
>
> For the scientist, there is a source of much material on herring behaviour in the fishermen's knowledge if it could be properly tapped.

This field has developed greatly, especially with technical developments of the past 35 years (including echo integration, improved scientific sounders, development of linearity principle, and studies of target strength), and acoustic surveys are used as the basis for several major herring assessments (e.g., Ona 2001). In a recent development industry (purse seine) vessels are making quantitative surveys using their own equipment (Stephenson et al. 1999, Melvin et al. 2001).

E. Hypotheses on Population Regulation and Abundance

> Notwithstanding a rich literature on population regulation and its central role in mathematical population genetics and evolutionary theory, an understanding of the control of abundance has not materialized. (Sinclair 1988, p. 2)

Debate has continued throughout much of the past century both in fisheries and general ecology on a number of issues surrounding the question of population regulation. Sinclair (1988) identifies four components to this issue:

1. What determines the differences between species in population richness?
2. Why are the component populations of a particular species distributed in the observed geographical patterns?
3. What processes control the absolute abundance of the populations (or what controls their means)?
4. What processes control the temporal fluctuations in abundance of the individual populations (or what controls their variances)?

In fisheries, the factors contributing to recruitment variability have been (and still are) a popular topic of research. Herring studies have contributed to the notions of a critical period (Hjort 1914) and to the further development and elaboration of this in Cushing's "match/mismatch" hypothesis (Cushing 1969, 1995). But the major recent advancement in this field is linked explicitly to herring. In the "herring hypothesis" of Iles and Sinclair (1982) observations on herring distribution and abundance are linked to physical oceanographic conditions, as follows: "The number of herring stocks and the geographic location of their respective spawning sites are determined by the number, location and extent of geographically stable larval retention areas."

According to this hypothesis (commonly called the "larval retention hypothesis") the very existence of a population depends on the ability of larvae to remain aggregated. Iles and Sinclair attributed the persistence of herring populations to the presence of locations where the physical oceanography was such that larvae from a spawning group, with the use of specific behavior (such as appropriately timed vertical movement) could maintain an aggregated distribution. Further, Iles and Sinclair argued that population abundance is a function of the physical system underlying larval retention—pointing out that small stocks are associated with small hydrographic features and large stocks with large features. The herring hypothesis addressed both the geographical or spatial basis of herring populations and the control of mean abundance.

Sinclair (1988) extended and generalized the herring hypothesis in his "member/vagrant" hypothesis in an attempt to account for the four questions mentioned above in a wide range of sexually reproducing populations. The herring hypothesis has generated considerable discussion and research, especially in the area of linkages between fish and hydrographic conditions.

Research in herring continues to contribute to the issue of the structure of fish populations. There has been continued discussion about stock complexity and the exact nature of herring populations, the metapopulation concept, and the relevance of the scale of the population to intraspecific biodiversity and to management (McQuinn 1997, Stephenson 1999, Stephenson et al. 2001, Smedbol and Stephenson 2001).

F. Linking Fisheries Dynamics with Hydrography

Hay et al. (2001) point out that the impact of ocean environment or long term climate change on herring abundance and distribution is a common theme among the world's herring fisheries. Although the mechanisms are not fully understood, the link between fisheries dynamics and hydrography has been clearly established in herring. Zebdi and Collie (1995) demonstrated large-scale coincidence in recruitment patterns of herring throughout the North Pacific in relation to climate forcing. Diagnoses of the major herring stock collapses have pointed not only to the obvious impact of high fishing mortality, but also to periods of low recruitment which are ultimately linked to environmental conditions (Stephenson 1997), and papers at this Herring 2000 symposium (Toresen and Ostvedt 2001, Ivshina 2001), documented the relationship between environmental change and large historical fluctuations in the abundance of the major Norwegian spring-spawning and Sakhalin-Hokkaido stock complexes.

Work on herring has led to major recent advancement in modeling the interaction of larval fish and hydrography (summarized by Heath 1992, Stephenson and Clark 2002), including the first published three-dimensional hydrodynamic model of the advection of fish larvae (Bartsch et al. 1989).

G. Advances in Fisheries Management

The historical importance of herring fisheries led to some of the earliest restrictions in fishing practices. In the northwest Atlantic herring fisheries, for example, there have been closed seasons and restrictions on use of torches to attract herring for well over a century (Scattergood and Tibbo 1959). While there was almost no management restriction of international fisheries involved in the major herring collapses of the late 1960s and early 1970s (Jakobsson 1985), since that time herring fisheries have featured in the development of several aspects of modern fisheries management including early quotas (in both the Pacific [Parsons 1993] and Atlantic [Iles 1993]), minimum landing size, seasonal and area closures, limited entry, and other effort restrictions. More recently, herring fisheries have also been the subject of innovative management schemes involving individual vessel quotas (Stephenson et al. 1993), comanagement and within season management regimes (Stocker 1993, Iles 1993, Stephenson et al. 1999).

The Potential Role of Herring Studies in Future

In spite of severe depletion or collapse of several major herring stocks at some point in the last few decades (see for example Jakobsson 1985, Stephenson 1997) most stocks have recovered and herring remain the subject of major fisheries and of research. Research developments might be expected, as in the past, to reflect the major initiatives and general developments in fisheries science and management modified by specific issues concerning the harvesting and use of herring and local events. Prediction of the future role of herring studies requires anticipation of the future direction of fisheries science and management generally.

Stephenson and Lane (1995) presented a critique of the current state of fisheries science and management. Fishery failures have continued in spite of recent management. There has been the realization that fisheries are complex, multidisciplinary systems requiring a more comprehensive approach to management than has been used to date. At the same time, there have been changing views on management structures and strategies, and evolving international standards. They proposed a more integrated approach involving fisheries science, fisheries management, and techniques from the field of management science in a new "fisheries management science." According to this view future management should focus on integrated fisheries, rather than just fish populations. It requires a more holistic view, including a combination of biological, social, and economic consideration of the fishery system.

The recent paper by Caddy (1999) also expresses the growing concern at the "uncertain effectiveness of most fisheries assessment and management approaches" and calls for an examination and change in management. The paper points out that recent international agreements provide the basis for construction of appropriate management frameworks. It discusses, among other things, the need for broader, multidisciplinary approaches, consideration of the broader ecosystem, the complications of overlapping jurisdictions, and the consideration of non-exploitive benefits.

I would suggest that herring studies may help to advance scientific thinking at least in the following areas:

The ecology of herring:
- Population structure, especially the nature of complex populations
- Intraspecific biodiversity and sustainability
- Biological reference points for conservation
- Interaction between herring and hydrography

The role of herring in the broader ecosystem:
- Pelagic fishery system structure and function
- Herring as a food item for other species
- Responses of fish populations to toxins and anthropogenic effects

Management of herring fisheries:
- Innovative methods of pelagic fish survey and assessment
- Evaluation of herring fishery systems
- Methods of integrating biological with social and economic aspects
- Increased role of industry in science and management
- Management in accordance with increasing standards implied by recent concepts such as the "precautionary approach" and "convention on biological diversity"

I suggest that herring studies will continue to help define and resolve major concepts in fisheries science. At the moment, this seems to be happening in the area of the structure and regulation of populations. I further suggest that herring studies may help equate fisheries concepts with those of terrestrial ecology, for example in work on complex population structure and metapopulations.

Acknowledgments

I thank the organizers of the Herring 2000 symposium for the invitation to present this overview as the keynote address in that symposium. I am grateful to Kirsten Clark who helped find and review historical documents, and to Cornelius Hammer and David Robichaud who assisted with translations. My interest in historical developments in fisheries science has been stimulated by close association with Drs. Derrick Iles and Mike Sinclair who together, and separately, have done much to synthesize and make sense of what has gone on in herring research and indeed in fisheries science generally. Much of the reading and thinking for this paper was undertaken while I was visiting the Finnish Game and Fisheries Research Institute, and I am grateful to both FGFRI and DFO for that opportunity.

References

Anderson, James. 1785. An account of the present state of the Hebrides and western coasts of Scotland. Being the substance of a report to the Lords of Treasury. Printed by M. Graisberry for Messrs. White, Byrne, McKenzie, Marchbank, Jones and Moore, 1786. Dublin.

Anderson, Johann. 1746. Nachrichten von Island, Grönland und der Strasse Davis, zum wahren Nutzen der Wissenschaften und der Handlung. [Accounts of Iceland, Greenland and the Davis Strait.] Georg Christian Grund, Hamburg.

Balls, R. 1948. Herring fishing with the Echometer. J. Cons. 15(2):193-206.

Bartsch, J., K. Brander, M. Heath, P. Munk, K. Richardson, and E. Svendsen. 1989. Modelling the advection of herring larvae in the North Sea. Nature 340(6235):632-636.

Caddy, J.F. 1999. Fisheries management in the twenty-first century: Will new paradigms apply? Rev. Fish Biol. Fish. 9:1-43.

Csirke, J. 1988. Small shoaling pelagic fish stocks. In: J.A. Gulland (ed.), Fish population dynamics, 2nd edn. Wiley, New York, pp. 271-302.

Cushing, D.H. 1969. The regularity of the spawning season of some fishes. J. Cons. Cons. Int. Explor. Mer 33:81-97.

Cushing, D.H. 1988. The provident sea. Cambridge University Press, New York.

Cushing, D.H. 1995. Population production and regulation in the sea: A fisheries perspective. Cambridge University Press, New York. 354 pp.

Davidson, A. 1999. The Oxford companion to food. Oxford University Press. 892 pp.

Dodd, J.S. 1752. An essay towards a natural history of the herring. London.

Halley, E. 1728. Atlas Maritimus et Commercialis. London.

Hay, D.E., et al. 2001. Taking stock: An inventory and review of world herring stocks in 2000. In: F. Funk, J. Blackburn, D. Hay, A.J. Paul, R. Stephenson, R. Toresen, and D. Witherell (eds.), Herring: Expectations for a new millennium. University of Alaska Sea Grant, AK-SG-01-04, Fairbanks. (This volume.)

Heath, M.R. 1992. Field investigations of the early life stages of marine fish. Adv. Mar. Biol. 28:1-174.

Heincke, Fr. 1878. Die Varietaten des Harengs I. Jahres. Komm. Unt. d. Deutsch. Meere in Kiel 4-6:37-132.

Heincke, Fr. 1882. Die Varietaten des Harengs II. Jahres. Komm. Unt. d. Deutsch. Meere in Kiel 7-11:1-186.

Heincke, Fr. 1898. Naturgeschichte des Harengs I. Die Localformen und die Wanderungen des Harengs in den Europaischen Meeren. Abhanglungen der Deutschen Seefischereivereins, Ver 2, s cxxxvi u. 128 S.

Hjort, J. 1914. Fluctuations of the great fisheries of northern Europe. Rapp. Cons. Perm. Int. Explor. Mer 20.

Hodgson, W.C. 1957. The herring and its fishery. Routledge and Kegan, London.

Huntsman, A.G. 1917. Herring investigations: Spawning, behaviour and growth of the young, summer of 1917. Manuscript Reports of the Biological Stations, Fisheries Research Board of Canada 352. 25 pp.

Huntsman, A.G. 1918. Growth of the young herring (so-called sardines) of the Bay of Fundy: A preliminary report. Canadian Fisheries Expedition 1914-1915. Kings Printer, Ottawa, pp. 165-171.

Iles, T.D. 1993. The management of the Canadian Atlantic herring fisheries. In: L.S. Parsons and W.H. Lear (eds.), Perspectives on Canadian marine fisheries management. Can. Bull. Fish. Aquat. Sci. 226:123-150.

Iles, T.D., and M. Sinclair. 1982. Atlantic herring: Stock discreteness and abundance. Science 215:627-633.

Ivshina, E.R. 2001. Decline of the Sakhalin-Hokkaido herring spawning grounds near the Sakhalin coast. In: F. Funk, J. Blackburn, D. Hay, A.J. Paul, R. Stephenson, R. Toresen, and D. Witherell (eds.), Herring: Expectations for a new millennium. University of Alaska Sea Grant, AK-SG-01-04, Fairbanks. (This volume.)

Jakobsson, J. 1985. Monitoring and management of northeast Atlantic herring stocks. Can. J. Fish. Aquat. Sci. 42(Suppl. 1):207-221.

Marshall, W. 1895. Die Deutschen Meere und ihre Bewohner. Leipzig.

McQuinn, I.H. 1997. Metapopulations in the Atlantic herring. Rev. Fish Biol. Fish. 7:297-329.

Melvin, G.D., R.L. Stephenson, M.J. Power, F.J. Fife, and K.J. Clark. 2001. Industry acoustic surveys as the basis for in-season decisions in a comanagement regime. In: F. Funk, J. Blackburn, D. Hay, A.J. Paul, R. Stephenson, R. Toresen, and D. Witherell (eds.), Herring: Expectations for a new millennium. University of Alaska Sea Grant, AK-SG-01-04, Fairbanks. (This volume.)

Mitchell, J.M. 1864. The herring: Its natural history and national importance. Edmonston and Douglas, Edinburgh. 372 pp.

Ona, E. 2001. Herring tilt angles measured through target tracking. In: F. Funk, J. Blackburn, D. Hay, A.J. Paul, R. Stephenson, R. Toresen, and D. Witherell (eds.), Herring: Expectations for a new millennium. University of Alaska Sea Grant, AK-SG-01-04, Fairbanks. (This volume.)

Parsons, L.S. 1993. Management of marine fisheries in Canada. Can. Bull. Fish. Aquat. Sci. 225. 763 pp.

Scattergood, L.W., and S.N. Tibbo. 1959. The herring fishery of the northwest Atlantic. Bull. Fish. Res. Board Can. 121. 42 pp.

Sinclair, M. 1988. Marine populations. An essay on population regulation and speciation. University of Washington Press, Seattle. 252 pp.

Sinclair, M. 1997. Prologue. Recruitment in fish populations: The paradigm shift generated by ICES Committee A. In: R.C. Chambers and E.A Trippel (eds.), Early life history and recruitment in fish populations. Chapman and Hall, London, pp 1-27.

Sinclair, M., and P. Solemdal. 1988. The development of "population thinking" in fisheries biology between 1878 and 1930. Aquat. Living Resour. 1:189-213.

Smedbol, R.K., and R.L. Stephenson. 2001. The importance of managing within-species diversity in modern fisheries, with examples from the northwestern Atlantic. J. Fish Biol. 59(Suppl. A). (In press.)

Smith, T.D. 1988. Stock assessment methods: The first fifty years. In: J.A. Gulland (ed.), Fish population dynamics, 2nd edn. Wiley, New York, pp. 1-34.

Smith, T.D. 1994, Scaling fisheries: The science of measuring the effects of fishing, 1855-1955. Cambridge University Press, New York. 392 pp.

Stephenson, R.L. 1997. Successes and failures in the management of Atlantic herring: Do we know why some have collapsed and others survived? In: D.A. Hancock, D.C. Smith, A. Grant and J.P. Beumer (eds.), Developing and sustaining world fisheries resources: The state of science and management: 2nd World Fisheries Congress proceedings. CSIRO Australia, pp. 49-54.

Stephenson, R.L. 1999. Stock complexity in fisheries management: A perspective of emerging issues related to population sub-units. Fish. Res. 43:247-249.

Stephenson, R.L., and K. Clark. 2002. The role of ICES herring investigations in shaping fisheries science and management. ICES Mar. Sci. Symp. 215. (In press.)

Stephenson, R.L., and D.E. Lane. 1995. Fisheries management science: A plea for conceptual change. Can. J. Fish. Aquat. Sci 52:2051-2056.

Stephenson, R.L., D.E. Lane, D.G. Aldous, and R. Nowak. 1993. Management of the 4WX Atlantic herring (*Clupea harengus*) fishery: An analysis of recent events. Can. J. Fish. Aquat. Sci. 50:2742-2757.

Stephenson, R.L., K. Rodman, D.G. Aldous, and D.E. Lane. 1999. An in-season approach to management under uncertainty: The case of the SW Nova Scotia herring fishery. ICES J. Mar. Sci. 56:1005-1013.

Stephenson, R.L., K.J. Clark, M.J. Power, F.J. Fife, and G.D. Melvin. 2001. Herring stock structure, stock discreteness, and biodiversity. In: F. Funk, J. Blackburn, D. Hay, A.J. Paul, R. Stephenson, R. Toresen, and D. Witherell (eds.), Herring: Expectations for a new millennium. University of Alaska Sea Grant, AK-SG-01-04, Fairbanks. (This volume.)

Stocker, M. 1993. Recent management of the British Columbia herring fishery. In: L.S. Parsons and W.H. Lear (eds.), Perspectives on Canadian marine fisheries management. Can. Bull. Fish. Aquat. Sci. 226:267-293.

Thomasson, E. (ed.). 1981. Study of the sea: The development of marine research under the auspices of the International Council for the Exploration of the Sea. Fishing News Books, Farnham, Surrey, England. 256 pp.

Toresen, R., and O.J. Ostvedt. 2000. Variation in abundance of Norwegian spring-spawning herring (*Clupea harengus*, Clupeidae) throughout the 20th century and the influence on climatic fluctuations. Fish and Fisheries 2000 1(3):231-256, Blackwell Science.

Wegner, G. 1993. Ein Mamburger Burgermeister und eine Heringstheorie (1746). Deutsche Hydrographische Zeitschrift Erganzungschaft, Reihe E, No. 25. 14 pp.

Whitehead, P.J.P. 1985. King herring: His place amongst the clupeoids. Can. J. Fish. Aquat. Sci. 42(Suppl. 1):3-20.

Zebdi, A., and J.S. Collie. 1995. Effect of climate on herring (*Clupea pallasi*) population dynamics in the northeast Pacific Ocean. In: Climate change and northern fish populations. Can. Spec. Publ. Fish. Aquat. Sci. 1221:277-290.

Temperature Effects on Zooplankton Assemblages and Juvenile Herring Feeding in Prince William Sound, Alaska

Robert J. Foy
University of Alaska Fairbanks, Fishery Industrial Technology Center, Kodiak, Alaska

Brenda L. Norcross
University of Alaska Fairbanks, Institute of Marine Science, Fairbanks, Alaska

Abstract

The Pacific herring (*Clupea pallasii*) population in Prince William Sound, Alaska, declined in 1993, prompting questions regarding feeding ecology of juveniles and their nursery areas. To study the feeding ecology of juvenile herring we investigated prey availability and herring diets within four nursery areas from May 1996 to March 1998. Quality of nurseries was evaluated with respect to zooplankton community structure and variables affecting that community. Zooplankton distribution was significantly influenced by seasonal and annual environmental factors among bays in Prince William Sound. The diets of juvenile herring were examined and related to prey availability. Prey densities, incidence of feeding, and prey taxa richness increased during the spring months and were minimal during the winter. Prey selection was highest during the winter months when prey availability was lowest. Lower zooplankton densities in the fall of 1997 than in 1996 were correlated to warmer water temperatures. Consequently, feeding by juvenile herring decreased in the fall and winter of 1997-1998. Increased temperatures may be responsible for higher growth rates and higher winter survival of juvenile herring observed that year in Prince William Sound.

Introduction

The importance of successful feeding in early stages of fish development is well documented (Hjort 1914; Lasker 1975, 1978; Houde 1987; Anderson 1988). Food limitation and/or quality can increase density-dependent mortality, thus affecting year-class strength (Walters and Juanes 1993). Fish growth rates affected by prey availability and temperature regulate recruitment by influencing mortality rates (Ware 1975, Shepherd and Cushing 1980, Anderson 1988). Atlantic herring (*Clupea harengus*) are prey-limited in the winter months and must reach a satisfactory level of condition prior to fasting during the winter (Blaxter and Holliday 1963). The juvenile development stage is especially critical for Pacific herring < 250 mm (*Clupea pallasii* Valenciennes 1847), which incur a substantial amount of mortality during their first winter (Paul et al. 1998; Foy and Paul 1999; V.E. Patrick, manuscript in review; K.D.E. Stokesbury, University of Massachusetts Dartmouth, manuscript in review).

We investigated juvenile herring feeding in Prince William Sound (PWS), Alaska, to find possible mechanisms for decreased adult biomass from 1993 to 1999. The cause of the population decline has been speculated to be the result of high incidence of disease in adults (Marty et al. 1998). Disease was likely a mechanism of mortality within the herring population that was directly linked to fish condition and nutritional status (Pearson et al. 1999).

In 1996 and 1997, feeding by juvenile herring and availability of prey varied seasonally among bays in PWS and were at their peak in May (Foy and Norcross 1999a). Prince William Sound zooplankton abundance and biomass were highest from May to July in herring nursery areas (Foy and Norcross 1999b, and manuscript in review). Shifts in prey species composition and decreased prey abundance prior to winter suggested that herring feeding would be minimal (Foy and Norcross, manuscript in review).

Environmental variables influence fish condition directly by affecting growth rates and indirectly by altering the community structure of the prey. Juvenile herring growth rates in PWS were significantly correlated to average water temperatures in 1996 and 1997 (Stokesbury et al. 1999). Zooplankton species composition and abundance was significantly correlated to temperature and salinity in the same time period (Foy and Norcross, manuscript in review). The objective of this study was to examine the response of juvenile herring feeding behavior to zooplankton availability and environmental conditions in PWS.

Materials and Methods

Prince William Sound is a large, fjord-type estuarine system consisting of numerous shallow bays, fjords, and tidewater glaciers located on the southern coast of Alaska in the North Pacific Ocean (Niebauer et al. 1994; Fig. 1). We sampled 15 times between March 1996 and March 1998 in two deep

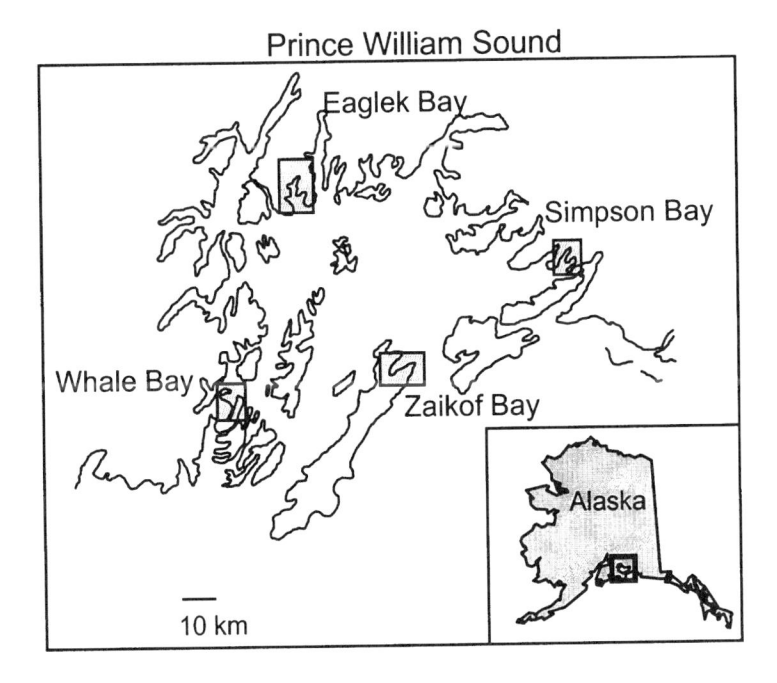

Figure 1. Location of study sites in Prince William Sound, Alaska. Simpson, Eaglek, Whale and Zaikof bays were sampled from 1996 until 1998.

fjords with depths > 250 m (Eaglek and Whale bays) and two estuaries with depths < 100 m (Simpson and Zaikof bays; Table 1). Data from the four bays were pooled for this study due to consistent interannual trends among bays.

Temperature loggers (Onset Computer Corporation) were deployed at a central location in each of the four bays at a depth of 5 m. Temperatures were sampled every 30 minutes from October 1996 until March 1998. Prior to October 1996, at the same sites temperature loggers were deployed, conductivity and temperature at depth (CTD) measurements were taken at 1 m intervals with a SeaBird SEACAT SBE 19 instrument. Tows occurred once per month and were from the surface to 1 m from the maximum depth. Only CTD data from a depth of 5 m were used in this study. An analysis of variance and a post hoc Tukey test were used to test for differences in the average temperature among months (Zar 1996).

Zooplankton were collected by vertical tows ($n = 535$) with a 0.5-m-diameter 300-mm-mesh ring net. Three to five stations located equidistant from the head to the mouth of each bay were sampled to determine small-scale heterogeneity in zooplankton community structure. A net was

Table 1. Numbers of zooplankton tows and juvenile herring stomachs sampled from March 1996 to March 1998 in Prince William Sound.

Month	Number of zooplankton tows	Number of stomachs
Mar 1996	0	335
May 1996	31	391
Jun 1996	34	371
Jul 1996	12	135
Aug 1996	33	257
Oct 1996	35	270
Nov 1996	10	144
Dec 1996	7	80
Feb 1997	12	80
Mar 1997	70	153
May 1997	65	294
Jul 1997	63	267
Aug 1997	62	298
Oct 1997	47	105
Mar 1998	49	102
Total	530	3,282

lowered to 30 m and then retrieved at a constant speed (0.5 m per second) to avoid bias associated with net avoidance. Volume filtered was determined by multiplying the area of the ring net mouth by the depth of the tow. Wire angle was noted in order to maintain consistency on depth sampled between tows and to calculate volume sampled. Samples were immediately preserved in a 10% buffered formaldehyde solution. All taxa from subsamples split with a Folsom splitter were identified to species and life stage when possible.

Juvenile herring schools ($n = 193$) were targeted acoustically and collected with a commercial purse-seine vessel with a 250-m × 34-m or 250-m × 20-m, 150-mm stretch-mesh anchovy seine net or a trawl vessel with a 40-m × 28-m, 150-mm-mesh midwater wing trawl net. Herring < 250 mm were considered to be nonspawning juveniles less than 3 years old (Stokesbury et al. 2000). Between 15 and 20 fish were randomly chosen from each catch and preserved in a 10% buffered formaldehyde solution for at least 24 hours prior to being transferred to 50% isopropanol for diet analysis. Post-processing analyses included determining taxa richness (number of taxa), prey energy density, and prey selectivity (see Foy and Norcross 1999a, submitted b).

Juvenile herring food preferences were estimated by Chesson's selectivity (a) index (Chesson 1978, 1983):

$$\alpha_i = \frac{r_i / n_i}{\sum_{j=1}^{m} r_j / n_j}$$

where r_i is the proportion of taxa i in the herring diet, n_i is the proportion of taxa i in the environment, m is the number of prey taxa, and j represents each taxa. Diet proportions, r_i, were calculated for the stomach contents of each fish sampled. Zooplankton proportions in the environment, n_i, were calculated from samples collected from multiple sites within the four bays during the same sampling periods that the fish were caught. The selectivity index ranges from 0 (avoidance) to 1 (selection) and a value of $1/m$ represents neutral selection. An analysis of variance and a post hoc Tukey test were used to test for differences in the average selectivity of all taxa among months (Zar 1996). For selection information of particular prey species see Foy and Norcross (submitted a).

Results

Temperatures in the nearshore surface waters ranged from 4.3° to 13.3°C in 1996, 3.5° to 14.8°C in 1997, and was 5.0°C in March 1998 (Fig. 2). Temperatures at 5 m were significantly different among months ($F = 155.4$, d.f. = 14, $P < 0.01$). Temperatures were coolest in March in 1997 and 1998 and warmest in August in 1996 and 1997. The fall of 1997 and the spring of 1998 were significantly warmer than those of the previous year.

The zooplankton density in the upper 30 m was seasonally and interannually variable. Zooplankton density was highest in June 1996 at 3,166 zooplankters per m³ (Fig. 3). Zooplankton abundance decreased in the winter of 1996-1997 to less than 90 zooplankters per m³. Zooplankton densities were significantly higher in May 1997 than May 1996 ($F = 27.2$, d.f. = 94, $P < 0.01$). Zooplankton densities in July, August, and October 1997 were all significantly lower than in the same months in 1996 ($P < 0.01$). No sampling occurred in June 1997 to compare to 1996. The zooplankton species richness was highest in May of both 1996 (41 taxa) and 1997 (34 taxa) and lowest in October 1996 (28 taxa) and 1997 (23 taxa; Fig. 4). There was an overall decreasing trend in the number of zooplankton taxa between May of 1996 and October 1997.

The number and diversity of prey taxa in the herring diets per fish varied among months and between 1996 and 1997. The number of prey per herring stomach in 1996 increased from 91 in March to a peak of

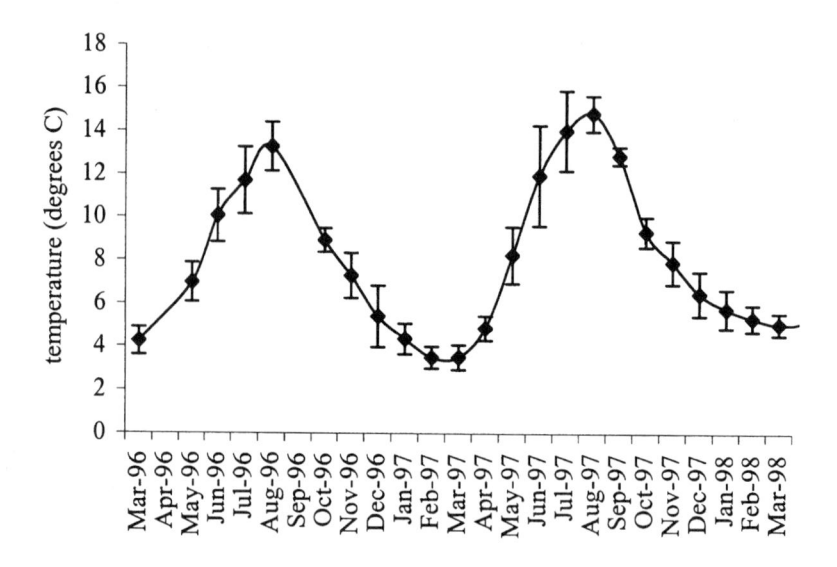

Figure 2. Average temperature at depth of 5 m from fixed temperature loggers and CTD casts in four Prince William Sound bays.

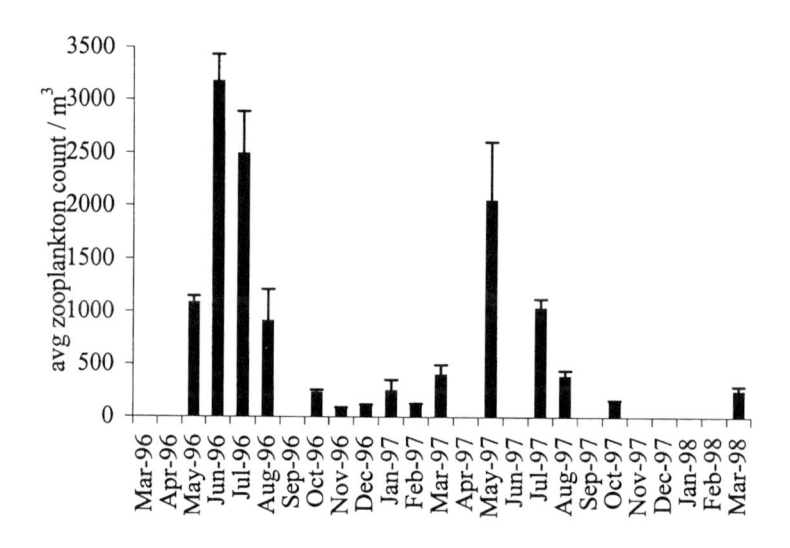

Figure 3. Average (standard error) zooplankton density (count per m³) for each month from March 1996 to March 1998 in four Prince William Sound bays. Blank spaces represent months that were not sampled.

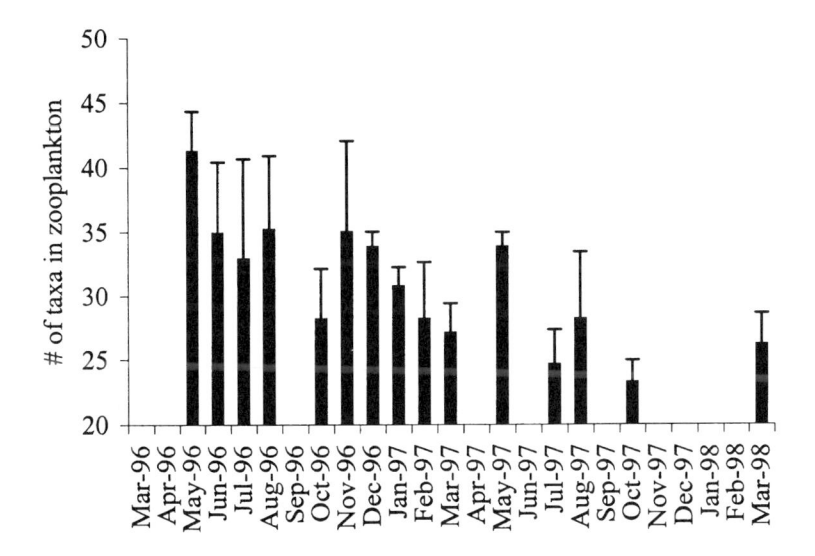

Figure 4. Number of zooplankton taxa collected each month from March 1996 to March 1998 in four Prince William Sound bays. Blank spaces represent months that were not sampled.

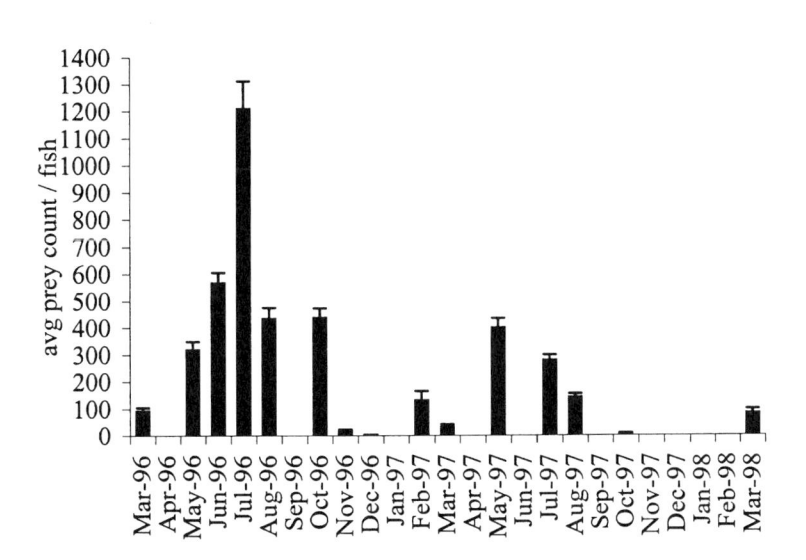

Figure 5. Average (standard error) prey density (count per fish) in juvenile herring diets from March 1996 to March 1998 in four Prince William Sound bays. Blank spaces represent months that were not sampled.

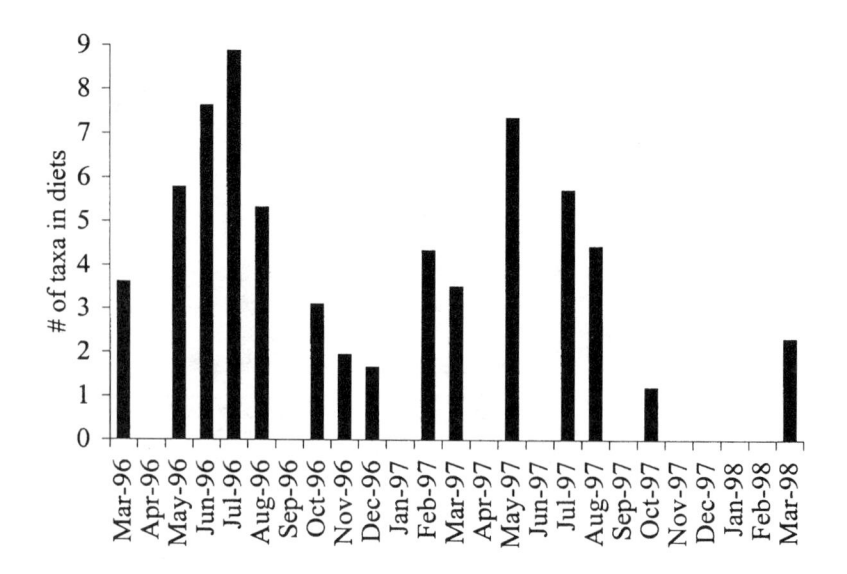

Figure 6. Number of prey taxa in juvenile herring diets from stomachs collected from March 1996 to March 1998 in four Prince William Sound bays. Blank spaces represent months that were not sampled.

1,209 in July and then declined to 3 in December (Fig. 5). Species composition consisted mostly of small calanoid copepods and Cirripedia from March to June whereas Cladocera and Larvacea became important between June and October (Table 2). In 1997 the average number of prey was larger in May (398 prey per fish) than July (278 prey per fish). The dominant species in the diets varied among small copepods, large copepods, Cirripedia, Euphausiacea, and Larvacea in 1997.

Juvenile herring had significantly more prey per fish in May 1997 than in 1996 corresponding to the increased availability of prey in May 1997 ($F = 12.3$, d.f. = 354, $P < 0.01$). The number of prey per fish for every other month in 1997 was significantly lower than in 1996 ($P < 0.01$). The number of taxa in the diets of the herring was highest in July 1996 (9 taxa) and May 1997 (7 taxa; Fig. 6). The lowest number of taxa in the diets occurred in December 1996 (2 taxa) and October 1997 (1 taxon). All months in 1997 except May had fewer taxa in the diets than in 1996. The number of empty stomachs was highest in winter (November to March) and lowest from June to August (Fig. 7). The percentage of empty stomachs in October 1997 was three times greater than in October 1996.

Juvenile herring were more selective during the winter months when prey was scarce than in the summer months when prey was abundant

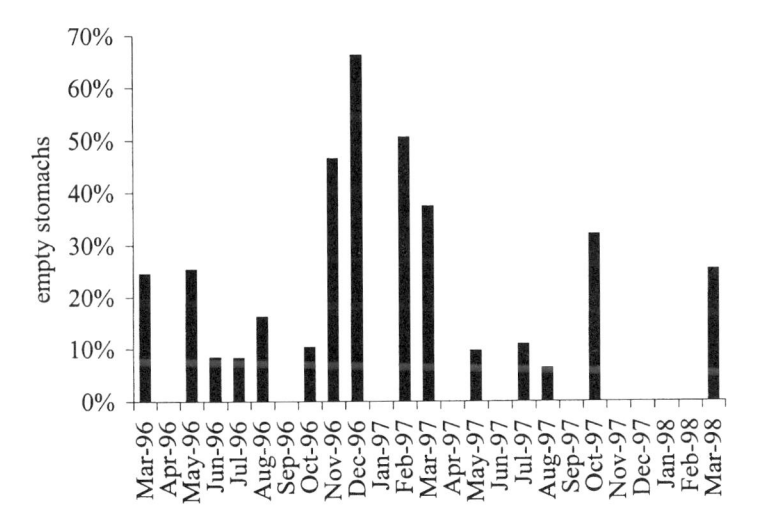

Figure 7. Percentage of juvenile herring stomachs that were empty from March 1996 to March 1998 in four Prince William Sound bays. Empty spaces represent months that were not sampled.

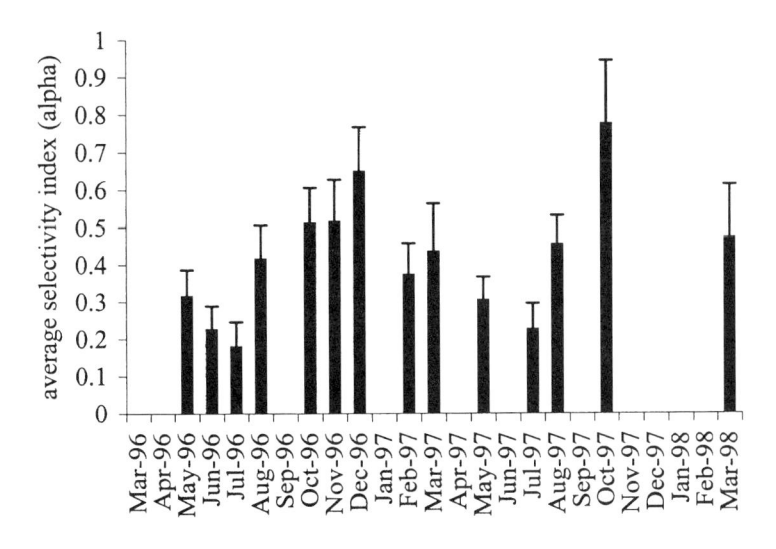

Figure 8. Average (standard error) selectivity index (alpha) of prey in juvenile herring diets from March 1996 to March 1998 in four Prince William Sound bays. Blank spaces represent months that were not sampled.

Table 2. Composition by taxon of stomach contents and number of prey per fish for each month sampled from March 1996 to March 1998. Only values greater than 0.5% of the total sum of taxa abundance from each month are reported.

Tax	1996 Mar	May	Jun	Jul	Aug	Oct	Nov	Dec	1997 Feb	Mar	May	Jul	Aug	Oct	1998 Mar
Arthropoda, Copepoda, Calanoida															
<2.5 mm	38	129	125	740	138	98	2	1	28	9	88	64	72	2	1
>2.5 mm	4	53	45	9	71		3	1	87	13	103	5		1	
Cyclopoida					2										
Harpacticoida	1							1	2						
Cirripedia cyprid and nauplius	29	83	52								184	11	1	1	61
Malacostraca															
Euphausiacea	12	12	4	7		3	1	1	1	3		41	3	3	1
Mysidacea							1	1						1	
Amphipoda					29			1					8		
Branchiopoda															
Cladocera		11	277	119	22						2	26	9		9
Decapoda megalops and zoea	1	4	3									29	23		
Ostracoda										4					
Mollusca															
Bivalvia larvae			23	55	3							12			
Gastropoda juveniles	0		7	12			3	0				15			
Annelida															
Polychaeta juveniles		4													
Urochordata															
Larvacea				143	144	326	8		9	2		66	7		2
Bryozoa															
Cyphonautes larvae							2						13		
Chaetognatha													1		
Miscellaneous															
Fish egg (1.0 mm)	3									1					5
Fish larvae									1						
Invertebrate egg (<0.2 mm)	1	11	24	109	15	3						3		1	1

($F = 4.79$, d.f. $= 13$, $P < 0.01$; Fig. 8). Selectivity index values ranged from 0.17 to 0.65 in 1996 (monthly neutral value range = 0.06-0.14) and from 0.22 to 0.77 in 1997 (monthly neutral value range = 0.04-0.22). Juvenile herring were significantly more selective in October 1997 than in October 1996 ($P < 0.01$).

Discussion

This study analyzed the response of herring feeding behaviors to the availability of zooplankton affected by environmental fluctuations over a short temporal scale. Changes in the interaction between predator and prey communities have been studied in relation to environmental conditions (Ware 1991, McGowan et al. 1998). Seasonal patterns of feeding responses were expected due to increased productivity in Prince William Sound in April (Goering et al. 1973). Feeding by herring during the winter is not well documented (Hay et al. 1988, Rudstam 1988) but is minimal in Prince William Sound (Foy and Paul 1999). Feeding during the summer and fall is then likely critical for energy storage prior to overwintering (Quast 1986, Paul et al. 1998, Paul and Paul 1998a, Foy and Paul 1999).

Zooplankton availability varies seasonally and annually among and within bays in PWS (Foy and Norcross, submitted a). Zooplankton species composition and abundance are dependent on multiple biological and environmental factors. The seasonal variability in zooplankton density encountered in this project was typical given the strong seasonal production cycles in PWS. The decline in zooplankton density and number of taxa between 1996 and the fall of 1997 was, however, not expected. There was also a shift in species composition in the zooplankton community occurring particularly in the fall between 1996 and 1997 (Foy and Norcross, manuscript in review). We encountered large densities of *Paracalanus parvus* in all the bays sampled in the fall of 1997. This species is not usually encountered north of the Queen Charlotte Islands in British Columbia (Giesbrecht 1892 as cited in Cameron 1957, Ermakova 1994), suggesting that higher temperatures observed in 1997 were responsible for its presence in PWS.

The decline in zooplankton abundance observed in the summer of 1997 coincides with increased temperatures that occurred at the same time (Foy and Norcross, submitted a). Temperatures were 2° warmer in the fall and winter of 1997 and may have been instrumental in indirectly reducing zooplankton density and composition due to predation and species succession. Temperature has been found to affect the availability of zooplankton prey for Bering Sea larval herring (Maksimenkov 1982). Although higher temperatures may enhance zooplankton production, we speculate that the higher temperatures may have increased the demand by herring on its prey population due to increased growth. These events combined with factors affecting stratification in the water column such as increased temperatures and freshwater runoff, may have limited nutrient input to the euphotic zone in the fall, inhibiting productivity that could support secondary production. Our zooplankton sampling did not continue into the winter of 1997-1998 to study the effects of a warmer than normal winter in PWS.

The reduced incidence of feeding and fewer prey taxa ingested observed in the juvenile herring in 1997 may be a response to the lower zooplankton availability. Food composition changed from being dominated by larvaceans and small calanoid copepods in the fall of 1996 to only small calanoids in 1997 (Table 2). If the densities of prey had been higher in 1997, having only small copepods (with higher energy density) to eat may have been positive for the condition of herring prior to overwintering. Studies estimating the assimilation rates given in situ prey compositions found that the smallest juvenile herring are liable to fall below basal metabolic demands in a year with temperatures similar to 1996 (Foy and Norcross 1999a). Warmer temperatures in the fall of 1997 increased growth rates of juvenile herring (Stokesbury et al. 1999). Consequently, herring predation pressure on the zooplankton community increased and led to lower prey concentrations in the fall. Lower feeding occurrence in the fall caused the herring to have a lower fall weight at length than in previous years (Stokesbury et al. 1999). Despite this, herring were in better energetic condition in the fall of 1997 than in 1996 (Paul and Paul 1998b) and consequently, a larger number of smaller fish survived through the winter of 1997-1998 than in 1996-1997. The average length and weight of the fish that survived the 1997-1998 winter were smaller and lower than in previous years (K.D.E. Stokesbury, Center for Marine Science and Technology, University of Massachusetts Dartmouth, pers. comm.), providing evidence that the smallest fish did not die from starvation during the winter as has been speculated in previous years (Paul et al. 1998, Foy and Paul 1999).

In conclusion, we have speculated on how lower prey availability affects the feeding dynamics of juvenile herring. Evidence suggests that environmental conditions may have contributed by affecting the prey resource base for herring in 1997. Consequently, herring growth rates increased and the energy density of herring was high by fall 1997. However, we hypothesize that a combination of lower feeding in the fall and warm winter temperatures meant that the average condition of surviving fish in the spring was lower than in previous years. Although no direct correlations have been made, we can speculate that lowered spring condition factor in herring may have enabled diseases to proliferate in the herring population (Marty et al. 1998, Pearson et al. 1999), ultimately leading to another herring crash in 1999 (Alaska Department of Fish and Game, Commercial Fisheries Division, Cordova, pers. comm.).

Acknowledgments

This project was funded by the Exxon Valdez Oil Spill Trustee Council through the Sound Ecosystem Assessment project. The findings presented by the authors are their own and not necessarily the Trustee Council position. We thank P. Lovely and C. Stark for sample sorting.

References

Anderson, J.T. 1988. A review of size dependent survival during pre-recruit stages of fishes in relation to recruitment. J. Northwest Atl. Fish. Sci. 8:55-66.

Blaxter, J.H.S., and F.G. Holliday. 1963. The behaviour and physiology of herring and other clupeids. Adv. Mar. Biol. 1:261-393.

Cameron, B.E. 1957. Some factors influencing the distribution of pelagic copepods in the Queen Charlotte Islands area. J. Fish. Res. Board Can. 14:165-202.

Chesson, J. 1978. Measuring preference in selective predation. Ecology 59:211-215.

Chesson, J. 1983. The estimation and analysis of preference and its relationship to foraging models. Ecology 64:1297-1304.

Ermakova, O.O. 1994. Distribution and dynamics of the copepod *Paracalanus parvus* (Copepoda, Calanoida) in Amurskiy Bay (Sea of Japan). Russ. J. Mar. Biol. 20:189-195.

Foy, R.J., and B.L. Norcross. 1999a. Spatial and temporal differences in the diet of juvenile Pacific herring (*Clupea pallasi*) in Prince William Sound, Alaska. Can. J. Zool. 77:697-706.

Foy, R.J., and B.L. Norcross. 1999b. Feeding behavior of herring (*Clupea pallasi*) associated with zooplankton availability in Prince William Sound, Alaska. In: Ecosystem approaches for fisheries management. University of Alaska Sea Grant, AK-SG-99-01, Fairbanks, pp. 129-135.

Foy, R.J., and A.J. Paul. 1999. Winter feeding and changes in somatic energy content for age 0 Pacific herring in Prince William Sound, Alaska. Trans. Am. Fish. Soc. 128:1193-1200.

Giesbrecht, W. 1892. Systematik und Faunistik der pelagischen Copepoden des Golfes von Neapel und der angrenzenden Meeresabschnitts. Fauna und Flora des Golfes von Neapel 19. Berlin.

Goering, J.J., W.E. Shiels, and C.J. Patton. 1973. Primary production. In: D.W. Hood, W.E. Shiels, and E.J. Kelley (eds.), Environmental studies of Port Valdez. Institute of Marine Science, Fairbanks, pp. 253-279.

Hay, D.E., J.R. Brett, E. Bilinski, D.T. Smith, E.M. Donaldson, G.A. Hunter, and A.V. Solmie. 1988. Experimental impoundments of prespawning Pacific herring (*Clupea harengus pallasi*): Effects of feeding and density on maturation, growth, and proximate analysis. Can. J. Fish. Aquat. Sci. 45:388-398.

Houde, E.D. 1987. Fish early life dynamics and recruitment variability. Am. Fish. Soc. Symp. 2:17-29.

Hjort, J. 1914. Chapter 1: The herring stock in Norwegian coastal waters. In: Fluctuations in the great fisheries of northern Europe viewed in the light of biological research. Conseil Permanent International pour l'Exploration de la Mer, Copenhagen, Denmark, pp. 14-228.

Lasker, R. 1975. Field criteria for survival of anchovy larvae: The relation between inshore chlorophyll maximum layers and successful first-feeding. Fish. Bull., U.S. 73:453-462.

Lasker, R. 1978. The relation between oceanographic conditions and larval anchovy food in the California current: Identification of factors contributing to recruitment failure. Rapp. P.-V. Reun. Cons. Int. Explor. Mer 173:212-230.

Maksimenkov, V.V. 1982. Correlation of abundance of food (for herring larvae) zooplankton and water temperature in the Korfo-Karaginsk region of Bering Sea. Sov. J. Mar. Biol. 8:132-136.

Marty, G.D., E.F. Freiberg, T.R. Meyers, J. Wilcock, T.B. Farver, and D.E. Hinton. 1998. Viral hemorrhagic septicemia virus, *Ichthyophonus hoferi*, and other causes of morbidity in Pacific herring, *Clupea pallasi*, spawning in Prince William Sound, Alaska, USA. Dis. Aquat. Org. 32:15-40.

McGowan, J.A., D.R. Cayan, and L.M. Dorman. 1998. Climate-ocean variability and ecosystem response in the Northeast Pacific. Science 281:210-217.

Niebauer, H.J., T.C. Royer, and T.J. Weingartner. 1994. Circulation of Prince William Sound. J. Geophys. Res. 99:14,113-14,126.

Paul, A.J., and J.M. Paul. 1998a. Comparisons of whole body energy content of fasting age zero Alaskan Pacific herring (*Clupea pallasi*) and cohorts overwintering in nature. J. Exp. Mar. Biol. Ecol. 226:75-86.

Paul, A.J., and J.M. Paul. 1998b. Spring and summer whole-body energy content of Alaskan juvenile Pacific herring. Alaska Fish. Res. Bull. 5(2):131-136.

Paul, A.J., J.M. Paul, and E.D. Brown. 1998. Fall and spring somatic energy content for Alaskan Pacific herring (*Clupea pallasi*) relative to age, size and sex. J. Exp. Mar. Biol. Ecol. 223:133-142.

Pearson, W.H., R.A. Elston, R.W. Bienert, A.S. Drum, and L.D. Antrim. 1999. Why did the Prince William Sound, Alaska, Pacific herring (*Clupea pallasi*) fisheries collapse in 1993 and 1994? Review of hypotheses. Can. J. Fish. Aquat. Sci. 56:711-737.

Quast, J.C. 1986. Annual production of eviscerated body weight, fat, and gonads by Pacific herring, *Clupea harengus pallasi*, near Auke Bay, southeastern Alaska. Fish. Bull., U.S. 84:705-721.

Rudstam, L.G. 1988. Exploring the dynamics of herring consumption in the Baltic: Applications of an energetic model of fish growth. Kiel. Meeresforsch. Sonderh. 6:312-322.

Shepherd, J.G., and D.H. Cushing. 1980. A mechanism for density dependent survival of larval fish as the basis of a stock-recruitment relationship. J. Cons. Cons. Int. Explor. Mer 39:160-167.

Stokesbury, K.D.E., R.J. Foy, and B.L. Norcross. 1999. Spatial and temporal variability in juvenile Pacific herring, *Clupea pallasi*, growth in Prince William Sound, Alaska. Environ. Biol. Fishes 56:409-418.

Stokesbury, K.D.E., J. Kirsch, E.D. Brown, G.L. Thomas, and B.L. Norcross. 2000. Spatial distributions of Pacific herring, *Clupea pallasi*, and walleye pollock, *Theragra chalcogramma*, in Prince William Sound, Alaska. Fish. Bull., U.S. 98:400-409.

Walters, C.J., and F. Juanes. 1993. Recruitment limitation as a consequence of natural selection for use of restricted feeding habitats and predation risk taking by juvenile fishes. Can. J. Fish. Aquat. Sci. 50:2058-2070.

Ware, D.M. 1975. Relation between egg size, growth and natural mortality of larval fish. J. Fish. Res. Board Can. 32:2503-2512.

Ware, D.M. 1991. Climate, predators and prey: Behaviour of a linked oscillating system. In: Long term variability of pelagic fish populations and their environment. Pergamon Press, Oxford, pp. 279-291.

Zar, J.H. 1996. Biostatistical analysis. Prentice Hall, New Jersey. 662 pp.

Is the Production of Coho Salmon in the Strait of Georgia Linked to the Production of Pacific Herring?

R.J. Beamish, G.A. McFarlane, and J. Schweigert
Pacific Biological Station, Nanaimo, British Columbia, Canada

Abstract

There was no relationship between the trends of adult coho salmon abundance and adult herring biomass from 1952 until the present. In particular, coho abundance trends increased during the period of extremely low herring biomass in the late 1960s and early 1970s. In contrast, coho abundance was low during the period of high herring biomass in the 1990s.

A change in the ocean environment in the 1990s increased coho marine mortality, possibly due to reduced growth rates of coho during their first marine summer. Juvenile herring and juvenile coho consume some similar prey. Thus, it is possible that when the ocean environment changed in the 1990s, the marine carrying capacity for coho was reduced, and herring affected coho survival by competing for similar prey.

Introduction

In recent years there has been a dramatic reduction in the abundance of coho salmon (*Oncorhynchus kisutch*) at the southern limit of its distribution (Beamish et al. 2000). In British Columbia, the abundance of some stocks is so low that restrictions were placed on all salmon fishing (DFO 1998) contributing to the lowest total commercial catches in history (Fig. 1). Maturing coho traditionally feed on Pacific herring (*Clupea pallasii*), a behavior that does not go unnoticed by recreational fishermen. This predator and prey association is believed by some to be related to the low abundance of coho in the late 1990s.

A recent hypothesis explaining the natural regulation of coho abundance identifies an early marine predation period and summer growth rates as the mechanisms that regulate the final abundance (Beamish and Mahnken 1999). Traditionally, marine survival of coho was seen to be a predation-related mechanism (Pearcy 1992). A diversity of predators spread

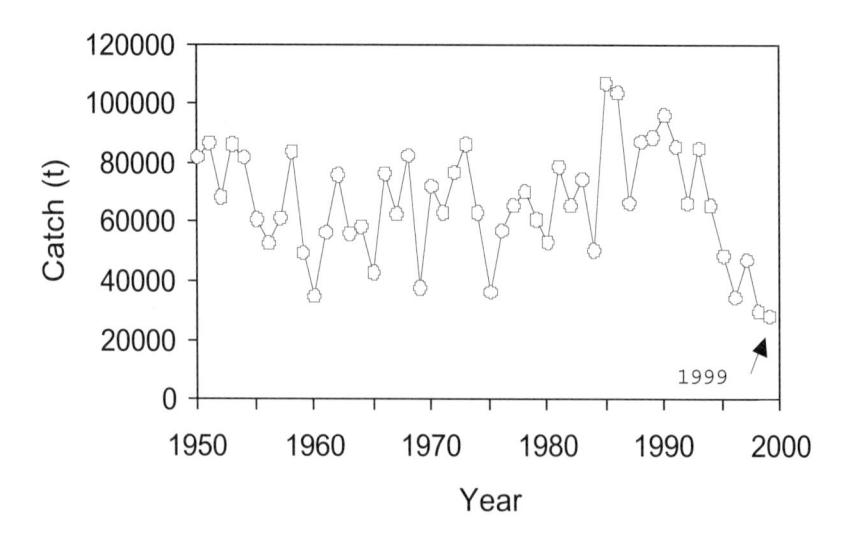

Figure 1. Total Canadian commercial catch of salmon (all species) from 1950 to 1999.

out over the range of coho salmon killed between 85% and 98% of the coho smolts shortly after they entered the ocean. According to a recent hypothesis (Beamish and Mahnken 1999), relatively large numbers of coho escape this early predation period. A second period of major mortality occurs in the fall and is related to the ability of an individual to grow to a critical size. Coho not reaching this critical size are unlikely to survive. This hypothesis links climate and carrying capacity and emphasizes the importance of the availability of preferred food for juvenile coho survival. If the critical size, critical period hypothesis is valid, interspecific competition for food could contribute to the late summer and winter mortalities.

In this study we examine the trends of herring biomass in relation to the total returns of coho to determine if trends in one species are related to trends in the other. There have been some extreme fluctuations in abundances that facilitate the comparison. We use information on the diets of juvenile coho salmon and herring to examine the possibility that Pacific herring in recent years may actually contribute to the decreased marine survival of coho through competition. Because competition may be more critical during some climate regimes than others (Beamish et al. 1999b), we also examine recent changes in abundance to recent climate.

Methods

Herring biomass was determined by constructing year-class strengths using age estimates of spawning fish and estimates of the density of eggs (Schweigert and Fort 1999). Adult coho abundance was reconstructed using a combination of estimates of catch, exploitation rates, marine survivals, and hatchery and wild percentages. From 1952 to 1974, year of entry into the sea (1950 to 1972 brood years), the total commercial catch in the Strait of Georgia was obtained from the Department of Fisheries and Oceans (DFO) catch database which was described by Kuhn (1988). Sport catch from 1953 to 1974 (1952 to 1973 year to sea) was from Argue et al. (1983). We used the 1953 sport catch to represent the 1952 catch. We used exploitation rates of 50% and 75% which approximate the range reported by Argue et al. (1983). Because only a portion of coho remain in the Strait of Georgia, we added the returns from coho that move out of the Strait of Georgia in their first marine year and are fished off the west coast of Vancouver Island (Beamish et al. 1999a) to estimate the total number of adults produced. We used an average percentage of 41.5% from Beamish et al. (1999a) that migrated offshore from 1973 to 1987 (year-to-sea) to estimate the average percentage that moved offshore from 1954 to 1974. The total returns, therefore, were the total catch multiplied by the inverse of the estimated average exploitation rates.

Estimates of total returns from 1975 to 1999 were reconstructed by estimating the total number of hatchery and wild smolts entering the Strait of Georgia and using estimates of their marine survival. Smolt releases from Canadian hatcheries into the Strait of Georgia were obtained from the DFO catch database (Kuhn 1988). Fry releases were not included because marine survival is believed to be very low. The percentage of hatchery and wild coho from 1975 to 1994 was from Kadowaki et al. (1994) and from R.M. Sweeting (manuscript in review) for the years 1996, 1997, and 1998. Marine survival rates for coho were from Beamish et al. (2000). Total returns were the total hatchery and wild smolt production multiplied by the marine survival.

Abundance trends of coho and herring were compared using least squares regression analysis (Zar 1984).

Diet contents of ocean-age juvenile coho and juvenile Pacific herring age 1+ and 2+ were analysed by Haegele (1997) from samples collected in the Strait of Georgia in June-July and August-September, 1990-1993. The results of this study were compared with the diet of juvenile coho collected from the Strait of Georgia in June-July and September of 1997 and 1998 in our study (King and Beamish 1999). The two studies identified prey categories differently. To compare the two studies, we combined the prosobranchs, pteropods, cladocerans, ostracods, barnacles, and larvaceans from Haegele (1997) into the "other" category and combined crab and shrimp from Haegele (1997) into the "decapod" category. Haegele (1997) reported diet in numbers of prey items in stomachs. The diets from our 1997 and 1998 studies were reported as the percentage volume of each prey category. To compare the diets from the two studies we measured 100 individual widths and lengths of *Pseudocalanus minutus* and *Neocalanus plumchrus*, and 30 widths and lengths of *Euphausia pacifica* to estimate the relative volumes among the three species.

Environmental Indices

Vertical temperature profiles were taken 8-20 times each month in the deep water at the Nanoose Bay Naval Underwater Weapons Test Range approximately midway between the north and south boundaries of the Strait of Georgia. The temperature profiles were collected using several different continuous temperature and depth instruments (for a history of the Nanaimo site, see Beamish et al. 1995).

Fraser River flow data were obtained from the Water Resources Branch, Environment Canada, and are shown as average annual and average April flow in m^3 per second (Beamish et al. 1995, 1999a). The annual period is expressed as an anomaly from the long-term mean for both data series. The annual period is from April 1 to March 31.

Results

Adult herring biomass estimates are shown beginning in 1951 (Fig. 2). The abundance prior to the mid-1960s was large, as were the removals from fishing (Fig. 2). Abundance declined in the mid-1960s and the fishery was closed from 1968 to 1971. The abundance of herring increased in the late 1970s and has remained high relative to the levels in 1950s although there were reductions in the mid-1980s. A small fishery started in 1972 and has remained at low levels relative to the 1950s and early 1960s.

The total returns of coho using a 50% exploitation rate resulted in approximately 50% higher total returns from 1952 to 1976 than the 75% exploitation rate, but the trend is identical to the 75% exploitation rate (in Fig. 3 we show only the reconstructed total returns using the 75% exploitation rate). Total returns of coho calculated using a 50% exploitation rate fluctuated from 1952 to 1992 around a mean of 1.6 million coho (S.D. ± 580,000). Large abundances were indicated for 1969 and 1970, the years immediately following the lowest herring biomass in the time series. The abundance of coho increased in the 1970s and 1980s, prior to the decline in the mid-1990s (Fig. 3A). If we used a 75% exploitation rate prior to 1975, coho abundance increased substantially immediately after the collapse of the herring stocks. Beginning in 1993 there was a dramatic and unprecedented decline in the total coho returns. The decline was precipitous without any of the variation in the abundance trend common in the earlier part of the time series.

Herring: Expectations for a New Millennuim 41

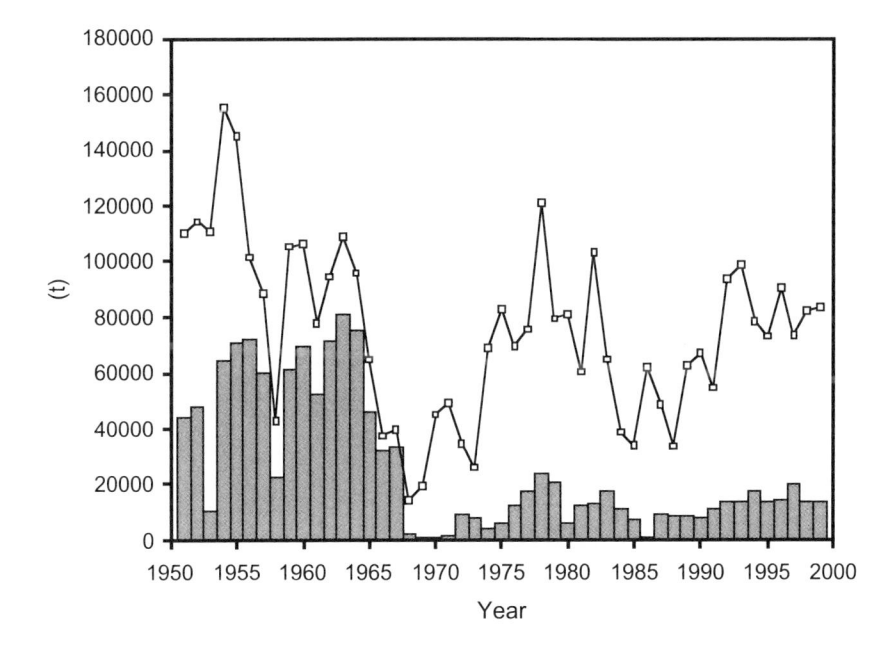

Figure 2. Pacific herring biomass (solid line) and catch (shaded bars) from 1950 to 1999.

There was no change in the releases of hatchery coho but the total number of hatchery and wild coho declined slightly to an average of 16.2 million after 1993 compared to the average up to 1993 of 19.3 million (Fig. 4). The total number of smolts entering the Strait of Georgia increased in the 1980s. Wild smolt abundance declined in the 1990s from an average of 13.6 million from 1975 to 1992 to 7.1 million from 1993 to 1999. Currently the percentage of wild smolts (offspring of naturally spawning parents) is the lowest in the data series.

There was no relationship between the trends in total coho returns and adult herring biomass (Fig. 3). The dramatic decline of herring in the mid-1960s was associated with large and consistent adult coho abundances. The high abundances of herring in the 1990s were associated with dramatic declines in numbers of adult coho. For both the 50% and 75% exploitation rates there was no significant relationship ($P > 0.40$, $r^2 < 0.01$). The use of the range in exploitation rates shows that in general coho abundance fluctuated around the changing trends of herring and not in association with the trends. We lagged the relationship of coho with herring by 1, 2, and 3 years to test the possibility that large herring abundances affected coho adult production through improved escapements and egg to smolt survival resulting in larger coho abundances 1-3 years later. For the 75% exploitation rate the r^2 was <0.001, 0.04, and 0.03 for all three

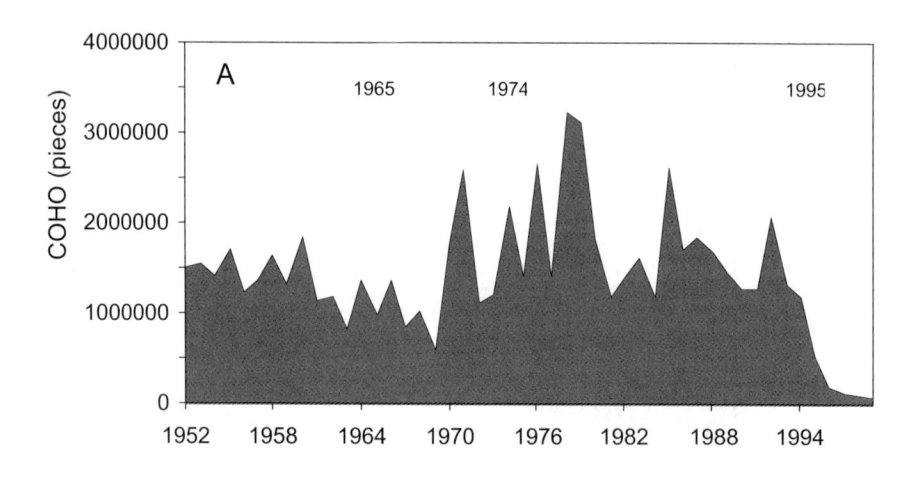

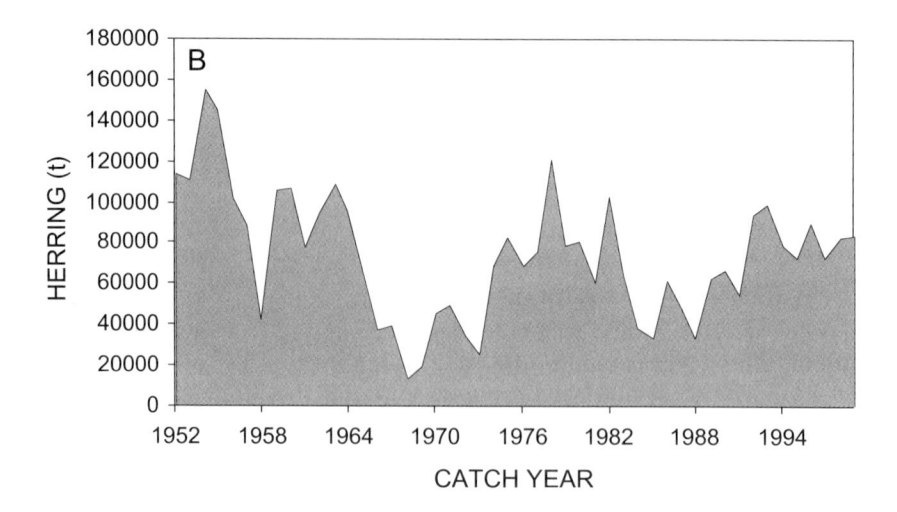

Figure 3. Abundance of adult coho (A) and herring (B) in the Strait of Georgia from 1952 to 1999. Coho abundance (A) was estimated using a 75% exploitation rate from 1952 to 1976.

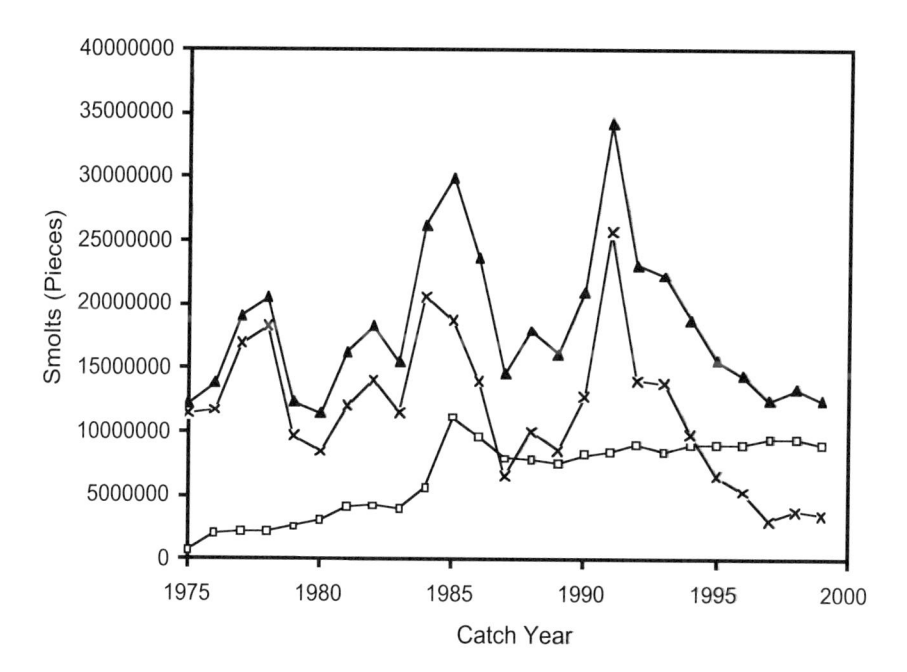

Figure 4. Hatchery (open square), wild (×), and total (triangle) coho salmon smolt abundances in the Strait of Georgia from 1975 to 1999. Wild coho estimates were based on hatchery numbers correlated from percentage of hatchery fish in the catch and known hatchery releases.

lags with P values of 0.99, 0.20 and 0.23, respectively. For the 50% exploitation rate, r^2 was 0.001, 0.025, and 0.011 with P values of 0.78, 0.29, and 0.48 for the three lags. There was an indication of lower coho abundances in the early to mid-1960s, but this was during a period of high herring abundance (Fig. 2) and not at the time of reduced herring abundances in the late 1960s and early 1970s.

The diets of ocean-age 0 coho and age 1+ and 2+ herring (Table 1) show the preference of herring for copepods and small eggs of invertebrates and the preference of coho for small fish as prey. However decapods, amphipods, and euphausiids are prey common to both diets. Based on measurements of length and width of copepods and euphausiids we estimated that 1 euphausiid equals approximately 20 *P. minutus* and 10 *N. plumchrus*. Haegele (1997) reported that invertebrate eggs and copepods constitute approximately 75% of the herring diet (by numbers). Our estimates of the relative volume between copepods and euphausiids indicate that in the summer samples the volume of euphausiids in the stomachs of herring would be approximately 2 times the volume of eggs and copepods combined. In the fall samples, euphausiids account for about half the vol-

Table 1. Diet of age 1+ and 2+ herring and ocean age-0 juvenile coho, 1990-1993 (Haegele 1997) and diet of ocean age juvenile coho, 1997 and 1998 (King and Beamish 1999).

	Herring 1990-93 Jun-Jul N	Coho 1990-93 Jun-Jul N	Coho 1997 Jun-Jul Volume (%)	Coho 1998 Jun-Jul Volume (%)	Herring 1990-93 Aug-Sep N	Coho 1990-93 Aug-Sept N	Coho 1997 Aug-Sep Volume (%)	Coho 1998 Aug-Sep Volume (%)
Invertebrate eggs	57,803	75	—	—	12,697	25	—	—
Copepods	79,213	428	0	0.34	92,969	248	0.05	0.25
Decapods	3,339	12,535	52.65	45.76	543	9,282	1.48	3.68
Amphipods	12,027	1,546	0.62	8.86	1,817	10,389	10.79	1.16
Euphausiids	26,519	85	1.03	6.39	5,629	2,520	51.72	60.20
Insects	22	16	0.1	0.06	232	91	0.5	10.72
Teleosts	119	858	45.56	38.45	4	10	35.28	14.05
Other	8,594	83	0	0.14	18,222	1,326	0.06	0.42
Unknown	—	—	—	—	—	—	0.05	—
Chaetognaths	—	—	—	—	—	—	—	0.13
Total prey (N)	187,636	15,626	—	—	132,113	23,891	—	—
Fish examined (N)	1,991	222	272	573	1,375	148	652	733.00
Fish with food (N)	1,235	163	257	545	818	93	548	564.00
Prey per fish (N)	152	96	—	—	162	257	—	—

A dash indicates no data.

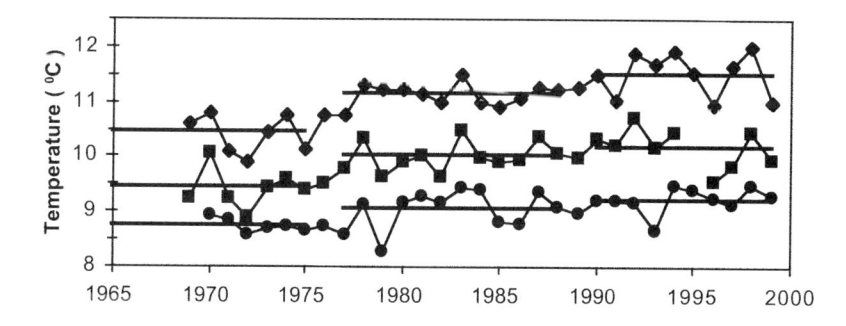

Figure 5. Central Strait of Georgia temperatures (°C) from 1969 to 1999. Temperatures are shown for surface, 10 m, and bottom (395 m). Data from 1969 to 1989 from Fissel et al. (1991), and from 1990 to present courtesy of W. Funk (Nanoose Bay Naval Base).

ume of copepods and invertebrate eggs. As euphausiids, decapods, and amphipods are common items in the diets of herring and coho, a large portion of the diet of the two fishes is similar. The diet composition of coho in our study was quite similar to the diets observed by Haegele (1997, Table 1). In both studies, decapods, amphipods, euphausiids and fish (teleost) accounted for most of the diet, confirming that the data from the earlier study can be combined with our study to show that the two species overlap in their feeding habitats.

Climate Indicators

In this report we show three indicators of climate and climate change in the Strait of Georgia. There are other indicators of large-scale climate change events in the subarctic Pacific (Mantua et al. 1997, Beamish et al. 1999b, McFarlane et al. 2000a) which indicate trends in the climate and ocean environment changes in 1977 and 1989.

Temperature on the surface, at 10 m, and near the bottom has increased since the 1970s (Fig. 5). Surface temperatures averaged almost 1°C higher in the 1990s than in the early 1970s. The increases in temperatures are shown as step increases because in previous reports we showed that significant step-ups in surface temperature occurred (Beamish and Neville 1999). Increases in temperature occurred at other depths although the increases were smaller than at the surface.

Total discharge from the Fraser River increased from 1950 to 1976, decreased from 1977 until 1995, and in recent years has undergone extreme fluctuations (Fig. 6A). In the 1990s the spring freshet began earlier. April flows increased in recent years (Fig. 6B). We used the date at which the average March flow doubled as an indicator of the onset of the spring flows. In the 1980s, the average March flow was 1,012 m^3 per second. In April, during the 1980s this value doubled by April 19 (on average). From

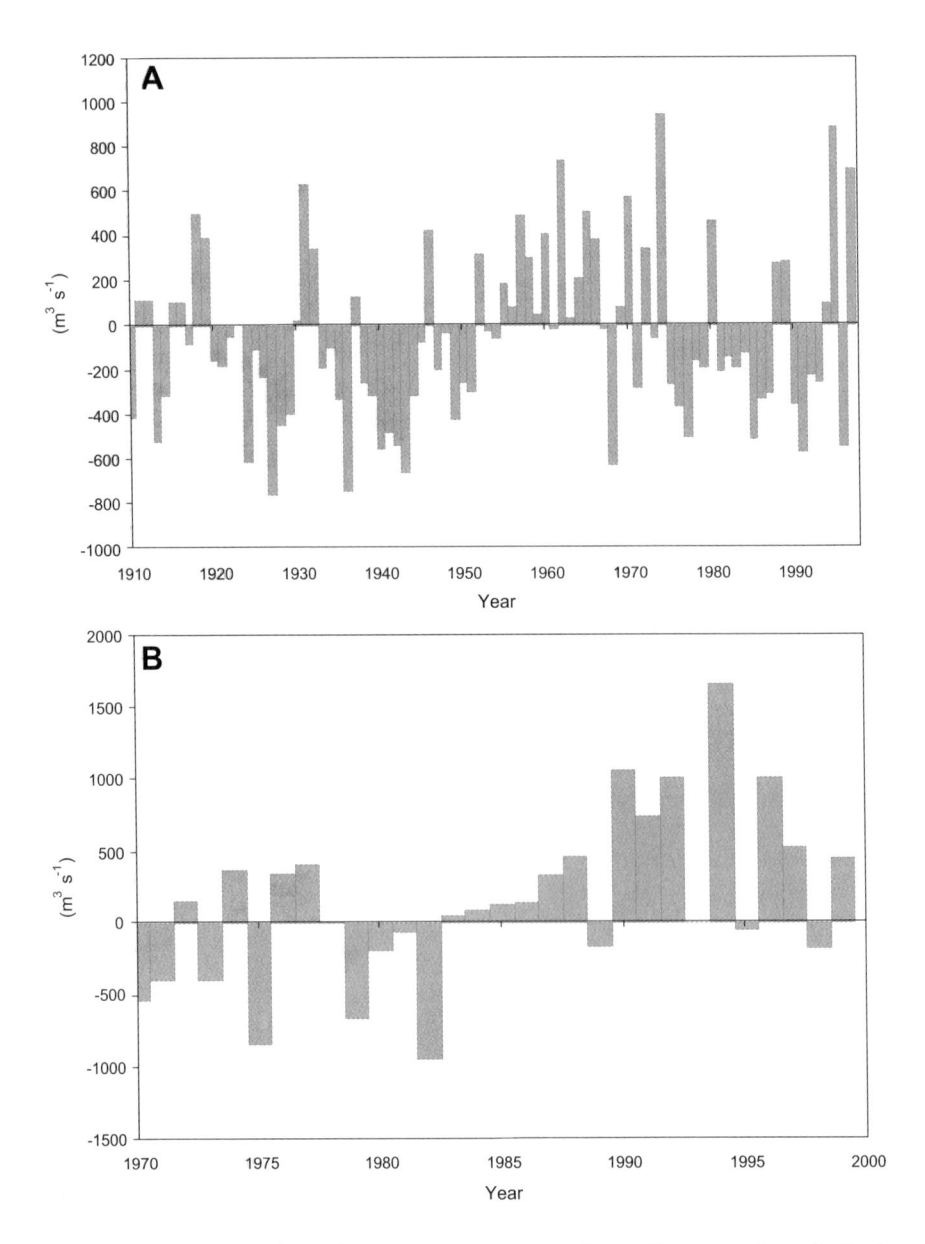

Figure 6. Flow rates from the Fraser River at Hope (Water Resources Branch, Environment Canada) shown as anomalies from a long-term mean. (A) Average annual (April-March) flow (m^3 per second) from 1910 to 1999. (B) Average April flow (m^3 per second) from 1970 to 1999.

1990 to 1997 the March average flow was 1,044 m^3 per second. The doubling of this value was reached an average 10 days earlier (April 9). In 1998 the March value doubled on April 24, and in 1999 on April 19.

Discussion

Coho in their second marine year feed extensively on herring (Healey 1976). Because of this preference, there is concern that catching too many herring will affect coho survival. We show that in the Strait of Georgia there is no relationship between the 50-year trends in herring and adult coho abundance. From 1967 through to 1971 when the herring population collapsed there was no impact on coho abundance. These results are consistent with the findings of Healey (1976), who concluded that the collapse of the herring stocks in the 1960s did not affect the production of any of the salmon species. A reverse situation occurred in the 1990s when herring abundance was high and coho abundance declined dramatically. Despite the large herring abundance, a herring fishery, and a relatively large and constant supply of coho smolts, the abundance of coho declined to historical low levels.

The testing of the abundance trends showed no correlation, but it is important to note that our time series for coho is a reconstruction based on reported catches, estimated hatchery and wild percentages, and two average exploitation rates that represented the possible ranges at a period when catch data are missing or approximate. There are other reconstructions of total coho returns for the Strait of Georgia stocks. The total returns estimated by Argue et al. (1983) are lower than ours from 1952 to 1976. They identified a decrease in coho returns in the early 1960s, and a return to higher abundances in the late 1960s through to the mid-1970s. Despite being lower, the Argue et al. (1983) estimates corroborate our observations that coho abundance increased during the low biomass period for herring in the late 1960s and early 1970s. Another time series of total coho returns is from 1984 to 1999, by Holtby et al. (2000). Their estimated total returns for the late 1990s are about 3 times larger than ours. However, both time series show that total coho returns were extremely low, relative to total returns in the 1980s and early 1990s. Thus both reconstructions show a dramatic decline in coho production at a time when herring biomass was at very high levels.

The decline in coho abundance was associated with a decline in marine survival from approximately 15% or higher in the 1970s and 1980s to less than 2% in the late 1990s (Beamish et al. 2000). Over the period of decline, smolt releases from hatcheries remained constant but there was a decline in wild smolt production. The decline in coho survival was synchronous throughout the southern distribution of coho, indicating that the decline resulted from a common factor (Beamish et al. 2000). A logical factor would be climate. There was a shift in the climate and ocean state in 1989 (Beamish et al. 1999b, Watanabe and Nitta 1999, McFarlane et al. 2000) and we identify a change at about this time in our study. The decline in coho in the 1990s appears to be associated with a less favorable marine habitat or a reduced capacity of the marine ecosystem resulting in reduced growth of coho in the summer of their first marine year (Beamish and Mahnken 1999). If this critical size, critical period hypothesis is valid, species competing for prey of coho could reduce their growth rate. The changes after 1989 were responsible for a new behavior in which virtually all coho left the Strait of Georgia (Beamish et al. 1999a). The response was related to an increase in sea level height and changes in surface winter salinities (Beamish et al. 1999a). This change is also associated with an earlier movement of copepods into the surface layers (Bornhold et al. 1998) and improved survival of larval hake and herring (Beamish and McFarlane 1999).

Invertebrates such as euphausiids, decapods, and amphipods common to herring diets were also an important part of the coho diet. Juvenile herring were an important part of the diet of juvenile coho. If the amount of growth over the first marine summer is related to brood year strength, and herring are an important prey, why is there no evidence of a relationship between herring abundance and coho abundance? Herring abundance was high during the 1990s and we know that relatively large numbers of coho juveniles survive until the fall of the first ocean year (Beamish et al. 2000). Perhaps coho direct their feeding during the first marine summer toward invertebrates such as euphausiids and feed opportunistically on juvenile herring. Juvenile herring would be both prey and competitor and the importance of this competition would be related to the favorability of the oceanic conditions.

The Strait of Georgia was warmer in the 1980s and 1990s, and the surface waters were fresher in the winters in the 1990s (Beamish et al. 1999a). There also were substantially more competitors such as herring and hake (McFarlane et al. 2001) and chum salmon (*O. keta*) (Beamish and Folkes 1998). The warmer ocean required that coho consume more food to maintain metabolic requirements (Jobling 1994). We speculate that it is the availability of preferred invertebrate prey that is important for coho growth rates during the first marine summer. Factors that reduce the availability of such prey contributed to the increased marine mortality of coho. While this is speculation, our point is that it is clear that looking at the dynamics of only a single species overlooks the ecosystem impacts that influence the abundance trends of the species. All species compete for habitat and natural processes provide the selection that limits the size of their populations. Human intervention into this natural process should be assessed more as an aggregate impact than a single species impact.

Acknowledgments

We appreciate the assistance of Chrys-Ellen Neville, Korey Poier, and Dr. Rusty Sweeting.

References

Argue, A.W., R. Hilborn, R.M. Peterman, M.J. Staley, and C.J. Walters. 1983. Strait of Georgia chinook and coho fishery. Can. Bull. Fish. Aquat. Sci. 211. 91 pp.

Beamish, R.J., and M. Folkes. 1998. Recent changes in the marine distribution of juvenile chum salmon off Canada. North Pacific Anadromous Fish Commission Bull. 1:443-453.

Beamish, R.J., and C. Mahnken. 1999. Taking the next step in Fisheries Management. In: Ecosystem approaches for fisheries management. University of Alaska Sea Grant, AK-SG-99-01, Fairbanks, pp. 1-21.

Beamish, R.J., and G.A. McFarlane. 1999. Applying ecosystem management to fisheries in the Strait of Georgia. In: Ecosystem approaches for fisheries management. University of Alaska Sea Grant, AK-SG-99-01, Fairbanks, pp. 637-644.

Beamish, R.J., and C.M. Neville. 1999. Large-scale climate-related changes in the carrying capacity for salmon in the Strait of Georgia and northern North Pacific ecosystems. In: K. Sherman and Q. Tang (eds.), Large marine ecosystems of the Pacific Rim: Assessment, sustainability, and management, Blackwell Science, pp. 27-41.

Beamish, R.J., G.A. McFarlane, and R.E. Thompson. 1999a. Recent declines in the recreational catch of coho salmon (*Oncorhynchus kisutch*) in the Strait of Georgia are related to climate. Can. J. Fish. Aquat. Sci. 56:506-515.

Beamish, R.J., B.E. Riddell, C-E.M. Neville, B.L. Thomson, and Z. Zhang. 1995. Declines in chinook salmon catches in the Strait of Georgia in relation to shifts in the marine environment. Fish. Oceanogr. 4:243-256.

Beamish, R.J., D. Noakes, G.A. McFarlane, L. Klyashtorin, V.V. Ivanov, and V. Kurrashov. 1999b. The regime concept and natural trends in the production of Pacific salmon. Can. J. Fish. Aquat. Sci. 56:516-526.

Beamish, R.J., D. Noakes, G.A. McFarlane, W. Pinnix, R. Sweeting, and J. King. 2000. Trends in coho marine survival in relation to the regime concept. Fish. Oceanogr. 9:114-119.

Bornhold, E., D. Mackas, and P. Harrison. 1998. Interdecadal variations in developmental timing of the copepod *Neocalanus plumchrus* (Marukawa) in the Strait of Georgia. Eos 79(1). (Abstract.)

Department of Fisheries and Oceans (DFO). 1982. Strait of Georgia coho salmon planning process and recommendations. South Coast Initiative Final Report, March 1992. Fisheries and Oceans, Canada. 54 pp.

Department of Fisheries and Oceans (DFO). 1998. Coho salmon, final report. Coho Response Team, Fisheries and Oceans, Pacific Region, Canada. 508 pp.

Fissel, D.B., J.R. Birch, and R.A.J. Chave. 1991. Measurements of temperature, salinity and sound velocity at the Nanoose Bay Naval Underwater Weapons Test Range, 1967-1984 (updated March 1991 to include 1985-1989, Appendix 4). Arctic Sciences Ltd., Sidney, B.C. 368 pp.

Haegele, C.W. 1997. The occurrence, abundance and food of juvenile herring and salmon in the Strait of Georgia, British Columbia in 1990 to 1994. Can. Manuscr. Rep. Fish. Aquat. Sci. 2390. 124 pp.

Healey, M.C. 1976. Herring in the diets of Pacific salmon in Georgia Strait. Fish. Res. Board Can. Manuscr. Rep. 1382. 38 pp.

Holtby, B.L., K. Simpson, R.W. Tanasichuk, and J.R. Irvine. 2000. Forecast for southern British Columbia coho salmon in 2000. PSARC (Pacific Scientific Advice Review Committee) Working Paper 500-03. 26 pp.

Jobling, M. 1994. Fish bioenergetics. Chapman and Hall, London. 309 pp.

Kadowaki, R., J. Irvine, B. Holtby, N. Schubert, K. Simpson, R. Bailey, and C. Cross. 1994. Assessment of Strait of Georgia coho salmon stocks (including the Fraser River). PSARC (Pacific Scientific Advice Review Committee) Working Paper S94-9.

King, J.R., and R.J. Beamish. 1999. Diet comparisons indicate a competitive interaction between ocean age-0 chum and coho salmon. North Pacific Anadromous Fish Commission Bull. 47:147-169

Kuhn, B.R. 1988 The MRP-reporter program. A data extraction and reporting tool for the Mark Recovery Program database. Can. Tech. Rep. Fish. Aquat. Sci. 1625. 145 pp.

Mantua, N.J., S.R. Hare, Y. Zhang, J.M. Wallace, and R.C. Francis. 1997. A Pacific interdecadal climate oscillation with impacts on salmon production. Bull. Am. Meteorol. Soc. 78:1069-1079.

McFarlane, G.A., J.R. King, and R.J. Beamish. 2000. Have there been recent changes in climate? Ask the fish. Prog. Oceanogr. 47:147-169.

McFarlane, G.A., R.J. Beamish, and J. Schweigert. 2001. Common factors have opposite impacts on Pacific herring in adjacent ecosystems. In: F. Funk, J. Blackburn, D. Hay, A.J. Paul, R.Stephenson, R. Toresen, and D. Witherell (eds.), Herring: Expectations for a new millenium. University of Alaska Sea Grant, AK-SG-01-04, Fairbanks. (This volume.)

Pearcy, W.G. 1992. Ocean ecology of North Pacific salmonids. University of Washington Press, Seattle. 179 pp.

Schweigert, J., and C. Fort. 1999. Stock assessment for British Columbia herring on 1999 and forecasts of the potential catch in 2000. Canadian Stock Assessment Secretariat, Research Document 99. 178 pp.

Watanabe, M., and T. Nitta. 1999. Decadal changes in the atmospheric circulation and associated surface climate variations in the Northern Hemisphere winter. J. Climate 12(2):494-510.

Zar, J.H. 1984. Biostatistical analysis, 2nd edn. Prentice-Hall, New Jersey. 718 pp.

Common Factors Have Opposite Impacts on Pacific Herring in Adjacent Ecosystems

G.A. McFarlane, R.J. Beamish, and J. Schweigert
Pacific Biological Station, Nanaimo, British Columbia, Canada

Abstract

The biomass of Pacific herring off the west coast of Vancouver Island was low in the 1990s as a result of Pacific hake predation. The biomass of Pacific herring in the Strait of Georgia was high in the 1990s despite large biomasses of Pacific hake. The Strait of Georgia is connected to the west coast of Vancouver Island by the Juan de Fuca Strait and herring move freely between the two areas. The opposite trends in the biomasses of herring in these two adjacent ecosystems were a consequence of different restructuring of the ecosystems in response to a common climate-related change.

Introduction

There is evidence that climate and climate change profoundly impact fish populations (Kawasaki and Omori 1988, Beamish and Bouillon 1993, Mantua et al. 1997, Beamish et al. 1999b, Clark et al. 1999, McFarlane et al. 2000). For example, Pacific halibut (*Hippoglossus stenolepis*) abundance follows trends related to climate and ocean conditions that persist on a decadal time scale (McCaughran 1999). There are other examples of synchronous changes in the dynamics of species that are linked to large-scale changes in climate ocean conditions (McFarlane et al. 2000). Evidence is accumulating to show that fish dynamics and environmental conditions can be stable on decadal scales (regimes), and shift abruptly from one steady state to another (Isaacs 1975, Beamish et al. 1999b). Regime shifts in 1925, 1947, 1977 (Francis and Hare 1994, Minobe 1997, Beamish et al. 1999b), and 1989 (Beamish et al. 1999b, McFarlane et al. 2000) have been identified. An important theoretical consideration in the regime concept is that the responses of fish populations can be specific to particular ecosystems (Beamish et al. 1999b) rather than specific to a surrogate such as temperature.

Regimes and regime shifts could be an important factor regulating the abundance of Pacific herring (*Clupea pallasii*). This contrasts with the view that the abundance of herring stocks is controlled primarily by fishing. In this study we compare the response of herring stocks in the Strait of Georgia and off the southwest coast of Vancouver Island (La Perouse Bank area) to climate-related changes that occurred in their ecosystems, particularly after 1989. Proving this relationship requires a better understanding of ecosystem dynamics as well as an awareness that past interpretations probably were an oversimplification of the population dynamics of herring. We can, however, show that an explanation used to explain low herring abundance in one area can be applied to an adjacent ecosystem to explain high herring abundance.

Materials and Methods

Study Area

The Strait of Georgia is located between Vancouver Island and the British Columbia mainland (Fig. 1). It is a semi-enclosed sea connected to the Pacific Ocean in the north by Johnstone Strait and in the south by Juan de Fuca Strait. For a detailed description see Thomson (1981) and Beamish and McFarlane (1999).

The La Perouse Bank region (Fig. 1) is located off the west coast of Vancouver Island, at the northern terminus of the coastal upwelling domain (McFarlane et al. 1997). This production system is characterized by a relatively narrow continental shelf, intense wind-induced upwelling in summer, and high phytoplankton and euphausiid biomass.

Key Species

In the Strait of Georgia echo-integration surveys were conducted on spawning concentrations of Pacific hake (*Merluccius productus*) during February-March 1981, 1988, 1993, and 1996. Echo-integration was conducted on predetermined parallel tracklines spaced 6 km apart. Calibrated acoustic systems were used on all surveys. Fishing was used to identify targets and echograms were examined to eliminate bottom interference, plankton concentrations, and nontarget fish. Standard target strength length relationships were used to convert acoustic backscatter to fish density. Geographic information system (GIS) procedures were used to convert fish density measurements to the biomass estimates reported in Kieser et al. (1999). Abundance in 1977 was estimated using biomass at age data (Saunders and McFarlane 1999) from 1981 to 1983 and removing the contribution of the strong 1977 and 1978 year classes.

Echo-integration surveys have been conducted along the U.S. and Canadian west coasts on a triennial basis since 1977 (Wilson and Guttormsen 1997). The procedures were similar to the Strait of Georgia procedures. The surveys are conducted during summer when hake distribution is at the northern limit. In addition, echo-integration surveys have been con-

Herring: Expectations for a New Millennium 53

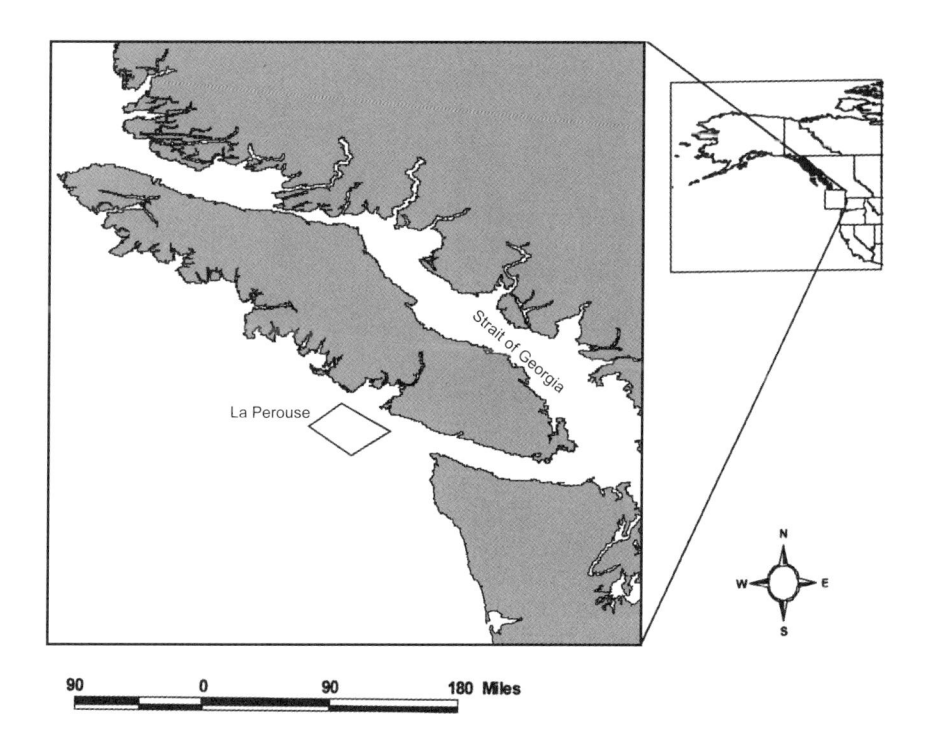

Figure 1. The location of the Strait of Georgia and the La Perouse area off the west coast of Vancouver Island.

ducted annually since 1990 off the west coast of Canada during August. Hake biomass in the La Perouse area was also estimated using midwater trawl surveys for 1970, 1974, and 1975 (Ware and McFarlane 1995).

Pacific herring stock biomass in the Strait of Georgia and off the west coast of Vancouver Island was obtained from Schweigert and Fort (1999) using two assessment models: an age-structured model and an escapement model.

Biological Data

Biological data were collected from hake and herring captured during commercial and research trawling operations in the Strait of Georgia and off the west coast of Vancouver Island. Fork length, sex, and maturity were recorded for all hake sampled and paired otoliths were collected for age determination from fish from selected trawl sets.

Stomach contents for all hake sampled were identified to lowest taxonomic group possible, and volume (cc) estimated for each prey item. All herring were counted and where possible, measured for length. In the

Strait of Georgia, stomach contents were analyzed for 3 years in the 1970s, 3 years in the 1980s, and 5 years in the 1990s. Off the west coast of Vancouver Island stomach contents were analysed in 1983 and 1985 to 1997.

Environmental Data

In this report we use water temperature and the Aleutian low pressure system as climate indicators. We chose the Aleutian low pattern because it is the dominant winter atmospheric feature, and is related to wind circulation patterns, ocean circulation patterns, and temperatures. For a detailed description of climate change in the subarctic Pacific see Beamish et al. (1999b), Mantua et al. (1997), and McFarlane et al. (2000).

In the Strait of Georgia vertical temperature profiles were taken 8-20 times each month in the deep water at the Nanoose Bay Naval Underwater Test Range approximately midway between the north and south ends of the Strait of Georgia. The temperature profiles were collected using several different continuous temperature and depth instruments (for a history of the Nanoose site, see Beamish et al. 1995). Sea surface temperature data from the west coast of Vancouver Island were accessed through the Institute of Ocean Sciences website (http:/www.ios.bc.ca/ios/osap/data). Temperature data at depth were obtained from R.E. Thomson (Institute of Ocean Sciences, Sidney, B.C.).

We obtained the Aleutian Low Pressure Index (ALPI) from the Pacific Biological Station website (http://www.pac.dfo-mpo.gc.ca/sci/sa-mfpd/english/clm-indxl.htm) which has been updated from Beamish et al. (1997).

Data Analysis

We used intervention analysis, which accounts for unusual events in a time series that modify a Box-Jenkins ARIMA (Auto Regressive Integrated Moving Average) model (Box and Jenkins 1976, Noakes 1986), to test for significant regime-shift signals (1977 and 1989) in the sea surface temperature index (SST) in both areas and the ALPI. Interventions in the form of abrupt steps are introduced to the model and their significance tested as a change in mean in the time series at that point.

We used ANCOVA with stepwise regression analysis to test for the best description of the change in female hake size at age 4. We used 1993, 1994 and 1995 as years to describe a possible change in mean size. ANCOVA tested for a change in mean size (intercept) in 1993, 1994, or 1995 and for differences in the rate of change (slope) of size before and after those shift years.

Results

Strait of Georgia

From 1977 to 1980 the biomass of Pacific hake increased from approximately 60,000 t to 80,000 t (Fig. 2a). The estimates in 1988 and 1993 indicate a higher abundance than in 1977 with 1993 as the largest biomass. The 1981 and 1988 estimates were acoustic estimates. Swept volume estimates were similar (Shaw et al. 1990). There was a small fishery throughout the study period (average 3,200 t annually) with maximum removals of 11,859 t in 1995 (Saunders and McFarlane 1999).

There were large decreases in the size of age 4 hake during this period (Fig. 3a). From 1979 to 1981, the mean size of 4-year-old fish declined 7.0% in length and 16.3% in weight. The next shift in size trends occurred in the mid-1990s. ANCOVA identified a change in mean size in 1994 as the best descriptor for the change in hake size at age 4 ($T = 6.84$, $r^2 = 0.72$, $P < 0.0001$).

The new mean was 13.0% lower in length and 51.2% lower in weight. The sizes from 1995 to 1998 remained about constant (Fig. 3a). The change in size was apparent in the mean length of all age groups in the commercial fishery averaged for the mid-1980s and mid-1990s (Fig. 4a).

Annual catches of herring averaged 13,300 t from the mid-1970s to the mid-1980s, and 14,400 t in the 1990s (Fig. 5a). The abundance of herring increased abruptly from 1974 to 1978 to levels approaching those that supported the high catches of the mid-1950s to 1960s (Schweigert and Fort 1999). Despite an almost constant removal since the mid-1970s (Fig. 5a), abundance declined in the mid-1980s to low levels of 40,000 to 50,000 t and then increased to approach historical high levels (mean = 90,000 t) in the early 1990s (Fig. 6a). This high abundance was maintained throughout the 1990s by above-average survival of year classes in 1989, 1990, 1991, 1993, 1994, and 1995 (Schweigert and Fort 1999).

Herring was a common prey item in the diet of Pacific hake in the Strait of Georgia in the 1970s and 1980s, but disappeared completely from the diet in the 1990s (Table 1). D. Hay (Pacific Biological Station, Nanaimo, B.C., pers. comm.) found a similar percentage of herring (13%) in hake diets in a 1983 study. In another study (Shaw et al. 1985), it was shown that hake outside the main concentrations and closer to shore, had the highest percentage of herring in their diets. In March and April 1981, hake were estimated to consume 4,000-8,000 t of herring (McFarlane and Beamish 1985) at a time when the biomass of herring was estimated to be 60,000 t, and the fishery took 12,000 t (Schweigert and Fort 1999).

West Coast of Vancouver Island

Hake abundance in the La Perouse area has fluctuated around 210,000 t since 1977 (Fig. 2b). However, the total abundance of hake off the west coast of Canada increased dramatically in the 1990s associated with an increased migration north of the LaPerouse area beginning in the early 1990s (Fig. 2b). There was no trend in size at age during the 1980s or 1990s (Fig. 3b) and the mean length in the commercial fishery has not changed (t-test, $P > 0.05$) (Fig. 4b).

Levels of catch of Pacific herring averaged 16,000 t in the late 1970s and about 5600 t since (Fig. 5b). Herring abundance in the early 1970s

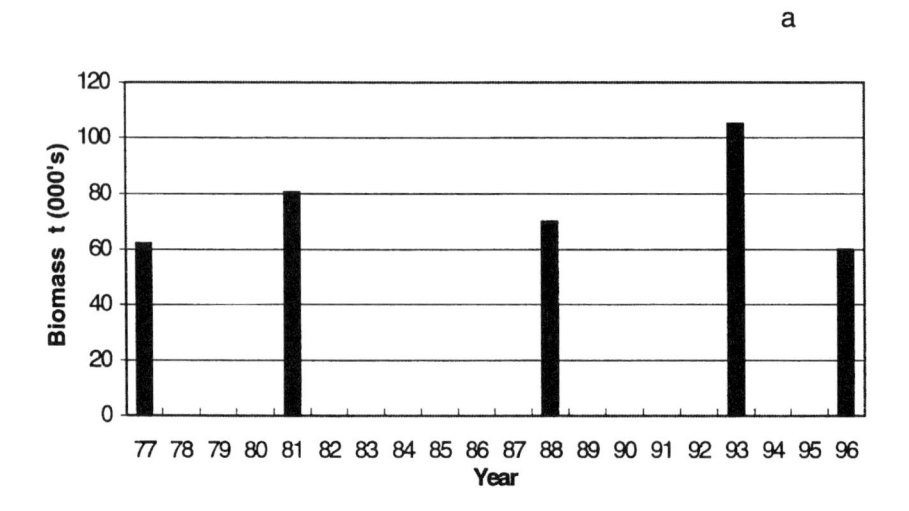

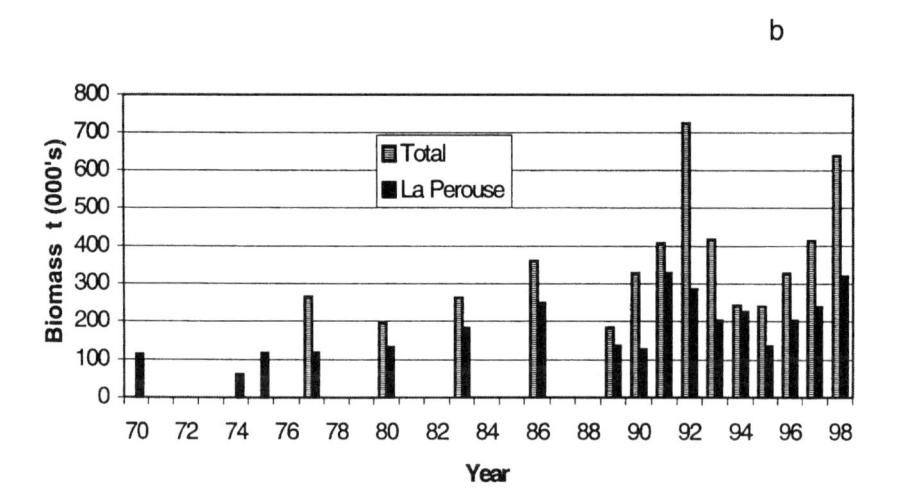

Figure 2. Biomass estimates (t) of Pacific hake in (a) the Strait of Georgia and (b) the La Perouse area.

Herring: Expectations for a New Millennium 57

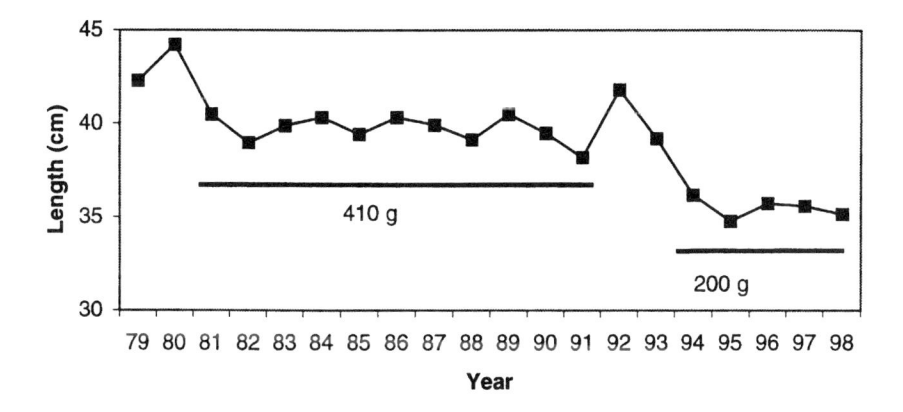

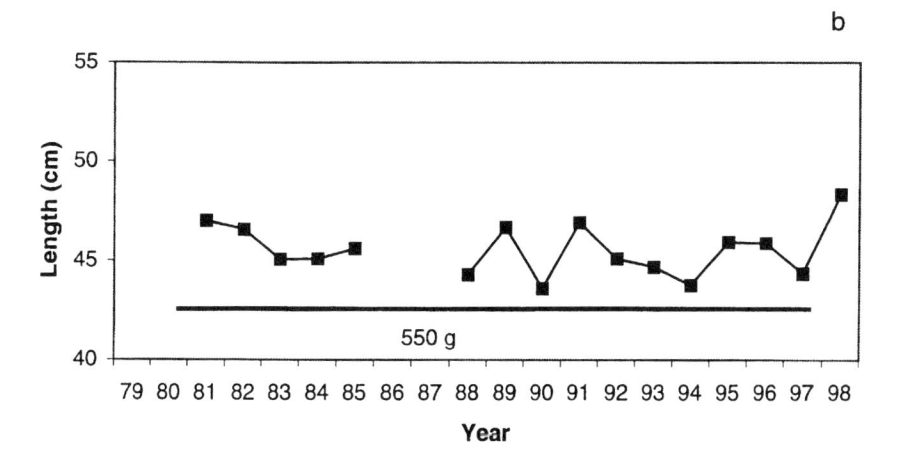

Figure 3. Trends in mean length and weight at age 4 for female Pacific hake from (a) the Strait of Georgia (1979-1998) and (b) the La Perouse area (1981-1998).

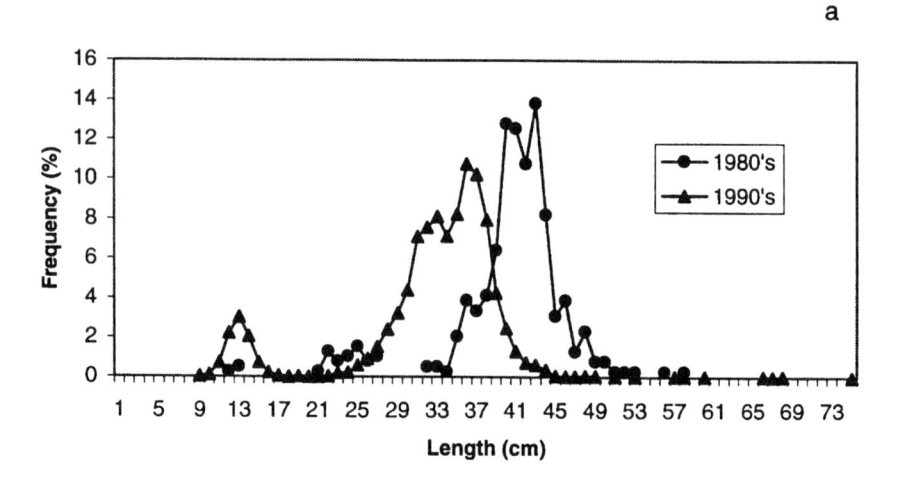

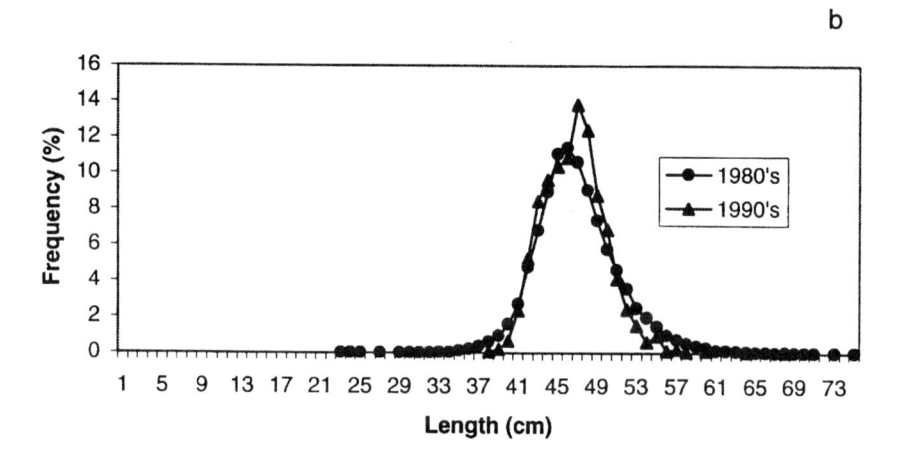

Figure 4. Mean size of Pacific hake in the commercial fishery during the 1980s and 1990s from (a) the Strait of Georgia and (b) the La Perouse area.

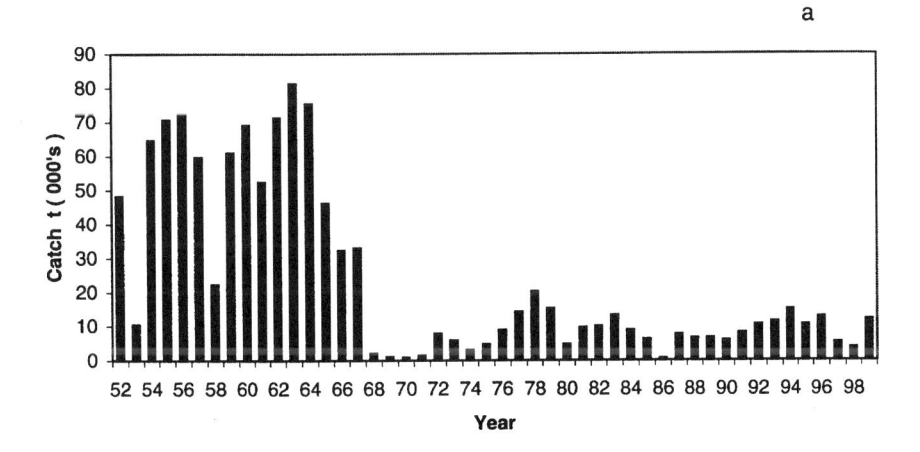

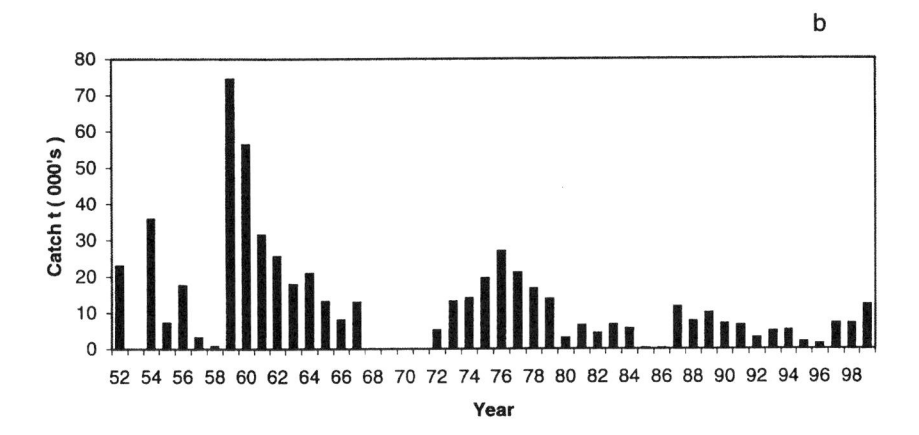

Figure 5. Annual catch (t) of Pacific herring (1952-1999) from (a) the Strait of Georgia and (b) the west coast of Vancouver Island.

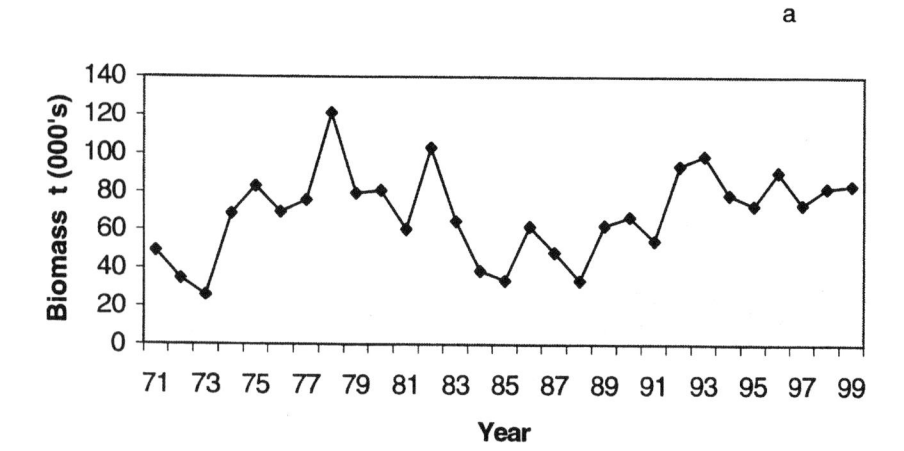

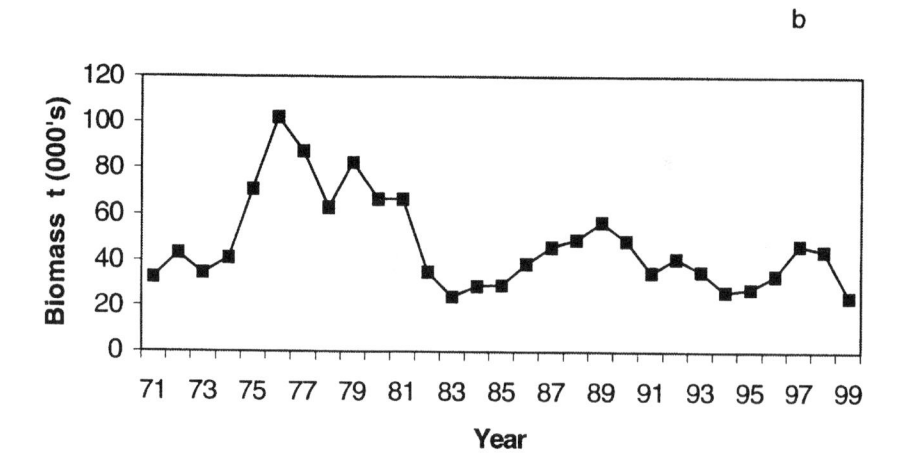

Figure 6. Biomass estimates (t) of Pacific herring in (a) the Strait of Georgia and (b) the west coast of Vancouver Island.

Table 1. Percentage of herring (volume) in the diet of hake in the Strait of Georgia.

Year	Month	Number sampled	% herring in diet
1974	Feb	601	10.0
1975	Jan-May	3,293	23.3
1976	Jan-May	2,552	9.9
1981	Feb-May	2,072	5.7
1983	Feb and Apr	2,597	30.3
1985	Mar	607	2.1
1995	Feb-Apr	572	0.0
1996	Feb-Mar; Oct-Nov	570	0.0
1997	Feb; Sep-Oct	447	0.0
1998	Feb-Mar; Sep	307	0.0

ranged from 42,963 t to 102,160 t (1972-1976) and declined to approximately 30,000 t in the mid-1980s and has fluctuated between 24,000 t and 57,000 t since. Since 1990 the mean annual biomass decreased to 35,000 t (Fig. 6b). The percentage of herring in the stomachs of hake ranged from 9.2% to 58.7% from 1983 to 1997 (Table 2).

Environment

Throughout the study period the surface, 10 m, and bottom temperatures in the Strait of Georgia increased (Fig. 7a). We showed the increases as step increases. Intervention analysis indicated these stepwise increases were significant (Beamish et al. 1999b). The average surface temperature changed from 10.45°C prior to 1977 to 11.3°C in the 1980s and 11.7°C in the 1990s (Fig. 7a). The sea surface temperature (Fig. 7b) in the La Perouse area showed a significant increase in mean trends after 1977 and another increase (not significant) after 1989 (Beamish et al. 2000).

The Aleutian low pressure index (ALPI) showed a significant upward shift ($P < 0.05$) after 1977 followed by a decline after 1989 (Fig. 8). Because the ALPI is a standardized anomaly and is a measure of the area of the winter low pressure (December to March), an upward shift in the index is an indication that the area of low pressure has increased. After 1989 the area of low pressure is about average for the time series. The Aleutian low is the dominant climatic feature in the North Pacific in winter. Its intensity and location determines atmospheric and ocean circulation patterns and temperatures (McFarlane et al. 2000). These basinwide changes have consequences for coastal areas like the west coast of Vancouver Island and the Strait of Georgia.

Table 2. Percentage of herring (volume) in the diet of hake off the west coast of Vancouver Island (La Perouse).

Year	Month	Number sampled	% herring in diet
1983	Jul-Aug	1,377	57.0
1985	Jul-Aug	820	40.0
1986	Jul-Aug	2,386	9.2
1987	Jul-Aug	1,824	28.0
1988	Jul-Aug	3,219	58.7
1989	Jul-Aug	1,148	11.8
1990	Jul-Aug	998	30.5
1991	Jul-Aug	1,105	25.8
1992	Jul-Aug	1,663	36.2
1993	Jul-Aug	953	58.5
1994	Jul-Aug	907	18.6
1995	Jul-Aug	916	39.2
1996	Jul-Aug	836	26.8
1997	Jul-Aug	462	17.3

Discussion

Pacific hake is the dominant resident species in the Strait of Georgia (Beamish and McFarlane 1999). Off the west coast of Vancouver Island, in the La Perouse Bank area, Pacific hake are part of a large migratory stock which, during the 1960s, 1970s, and 1980s spawned off Baja California in the winter and migrated north to summer feeding grounds (Francis 1983). Prior to 1990, approximately 25-30% of the mature biomass moved into Canadian waters. Since the early 1990s approximately 40% of the stock has been present in the Canadian zone (Wilson and Guttormsen 1997, Wilson et al. 2000).

The fishery for Pacific herring dominated catches in the Strait of Georgia and off the west coast of Vancouver Island from the early 1950s until the mid-1960s (Schweigert and Fort 1999). The fishery collapsed in the mid-1960s and was closed from 1967 to 1971. The fishery reopened in 1972 and has been less than 15% (Strait of Georgia) and 30% (west coast of Vancouver Island) of the previous fishing.

Herring are now at high levels of abundance in the Strait of Georgia and low levels of abundance off the west coast of Vancouver Island. Predation on herring by hake off the west coast of Vancouver Island increased in direct relation to the increased northward migration of Pacific hake (Ware and McFarlane 1995). The warmer surface water is associated with the increased hake abundance. Off the west coast the high percentage of

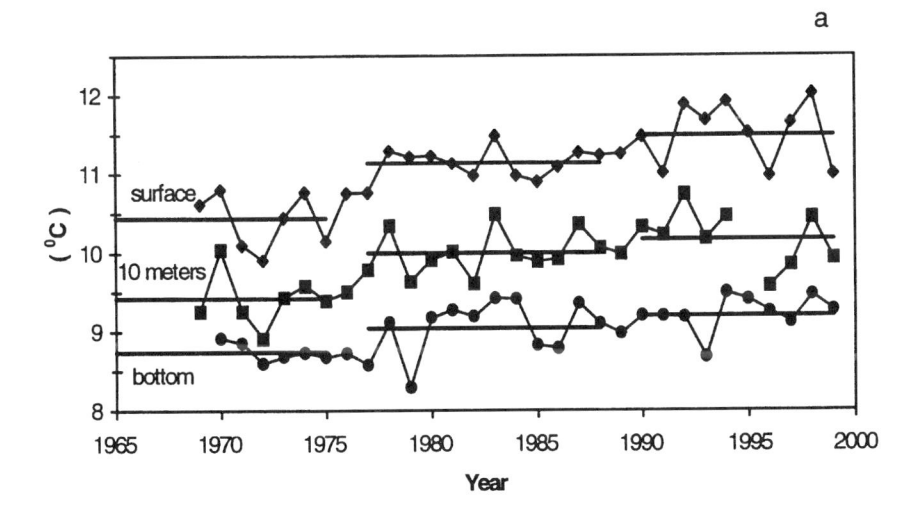

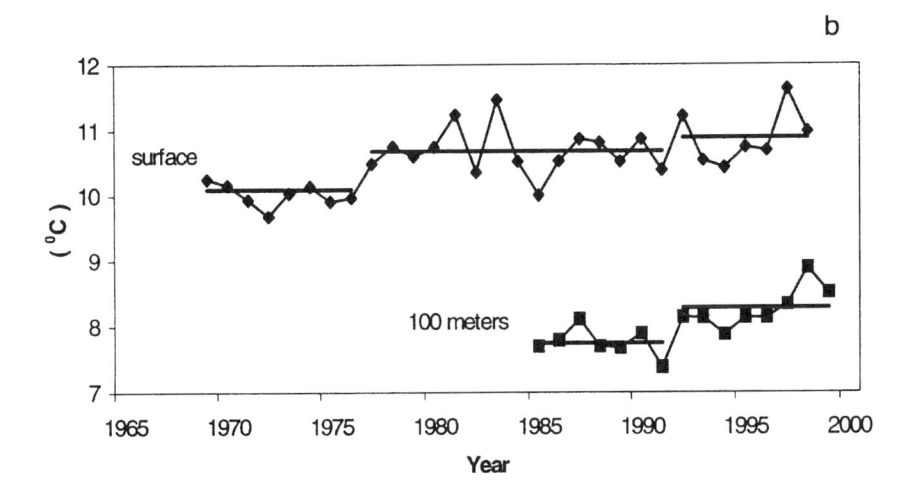

Figure 7. The average annual seawater temperature at (a) surface, 10 m, and bottom in the Strait of Georgia; and (b) surface and 100 m off the west coast of Vancouver Island. Solid horizontal lines indicate average temperature for regime.

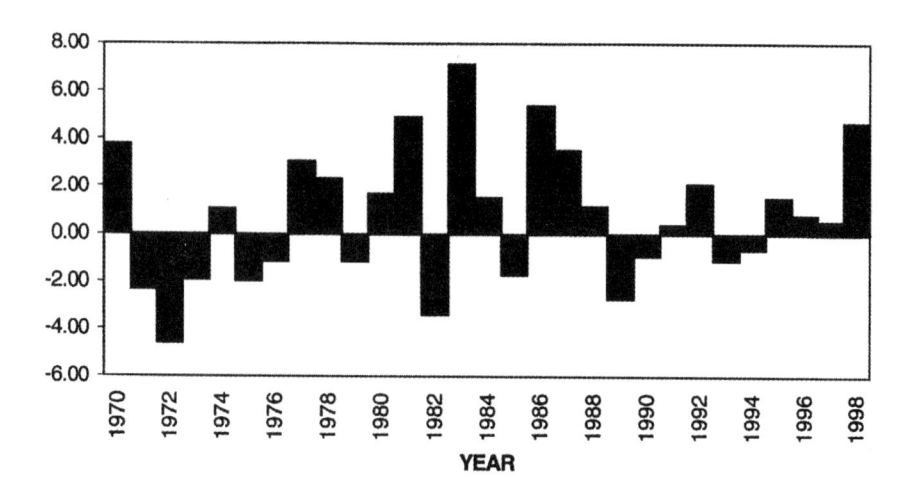

Figure 8. The Aleutian low pressure index (ALPI), 1970-1998. It is calculated as the mean area (km^2) with sea level pressure ≤ 100.5 kPa and is expressed as an anomaly from the 1950-1997 mean.

herring in the diet of hake (approximately 37% annually) is clear evidence of the preference of hake for herring as a prey. This preference for herring and the large biomass of hake has been shown to be the principal cause of the low abundance of herring in this area (Ware and McFarlane 1995). Hake in the Strait of Georgia reduced their predation on herring despite having a high biomass. After 1989, there was a shift to higher mean sea surface temperatures in both areas that was part of a large-scale shift in climate/ocean conditions as seen in the change in the pattern of the Aleutian low.

Herring in the Strait of Georgia have a high biomass even though the surface waters are also warmer. The increased temperature in the Strait of Georgia was not associated with a reduction in herring abundance as reported for the west coast of Vancouver Island. Instead it was associated with improved hake survival, reduced individual hake growth, and the elimination of predation on herring. Our observations of the elimination of herring in the diet of hake in the 1990s is consistent with observations of Tanasichuk et al. (1991), that hake less than 40 cm prey almost exclusively on euphausiids.

The factor common to both ecosystems was climate and climate change. The well-known climate shift of 1977 (Francis and Hare 1994, Minobe 1997, Beamish et al. 1999b) was associated with an increase in temperature in both ecosystems and an intensification of the Aleutian low pressure system in the subarctic Pacific. Our proposal of a 1989 regime shift (Beamish et al. 1999b, McFarlane et al. 2000) has now been identified in other studies (Hare and Mantua 2000) and appears to be receiving cautious acceptance. In the Strait of Georgia, this change was associated with an increase in ocean temperatures. Other changes such as increases in sea level heights (Beamish et al. 1999a), an earlier spring freshet (Beamish and McFarlane 1999) in the Fraser River, and changes in dominant wind direction (Beamish et al. 1999a) have also been noted. Off the west coast of Vancouver Island, sea surface temperature increased and the Aleutian low pressure system was reduced in intensity. In other publications (Beamish et al. 1999a, McFarlane et al. 2000) we suggest that a common factor may trigger these hemispheric shifts in climate/ocean ecosystems.

It is the nature of the "reorganization" of the ecosystem after a regime shift that determines the impacts on a species. A measure of the change, such as temperature, is only one factor affecting the dynamics of populations. The climate changes in 1977 and 1989 elicited opposite responses from herring in two adjacent ecosystems. This was despite similar temperature trends in these two systems. This suggests that physical factors need to be related to ecosystem dynamics and not just to the observed effect on a single species. It also means that once we see indications that climate/ocean conditions are changing we need to identify the new "ecosystem organization" and adapt our management strategies to this new reality. While the relationships we propose in this report may be viewed as speculative, they identify the need to reevaluate past explanations. How speculative is this explanation relative to other interpretations? A conclusion that probably is commonly accepted is that herring management should consider herring as a component of ecosystem processes that are affected by the natural and fishing effects on herring.

References

Beamish, R.J., and D.R. Bouillon. 1993. Pacific salmon production trends in relation to climate. Can. J. Fish. Aquat. Sci. 50:1002-1016.

Beamish, R.J., and G.A. McFarlane. 1999. Applying ecosystem management to fisheries in the Strait of Georgia. In: Ecosystem approaches for fisheries management. University of Alaska Sea Grant, AK-SG-99-01, Fairbanks, pp. 637-664.

Beamish, R.J., G.A. McFarlane, and R.J. King. 2000. Fisheries climatology: Understanding decadal scale processes that naturally regulate British Columbia fish populations. In: T. Parsons and R. Harrison (eds.), Fisheries oceanography: A science for the new millennium. Blackwell Science Ltd., Osney Mead, Oxford, pp. 94-145.

Beamish, R.J., G.A. McFarlane, and R.E. Thomson. 1999a. Recent declines in the recreational catch of coho salmon on the Strait of Georgia are related to climate. Can. J. Fish. Aquat. Sci. 56:506-515.

Beamish, R.J., C.E. Neville, and A.J. Cass. 1997. Production of Fraser River sockeye salmon (*Oncorhynchus nerka*) in relation to decadal-scale changes in the climate and the ocean. Can. J. Fish. Aquat. Sci. 54:543-554.

Beamish, R.J., B.E. Riddell, C.M. Neville, B.L. Thomson, and Z. Zhang. 1995. Declines in chinook salmon catches in the Strait of Georgia in relation to shifts in the marine environment. Fish. Oceanogr. 4:243-256.

Beamish, R.J., D. Noakes, G.A. McFarlane, L. Klyashtorin, V.V. Ivonov, and V. Kurashov. 1999b. The regime concept and natural trends in the production of Pacific salmon. Can. J. Fish. Aquat. Sci. 56:516-526.

Box, G.E.P., and G.M. Jenkins. 1976. Time-series analysis: Forecasting and control. Holden-Day, San Francisco.

Clark, W.G., S.R. Hare, A.M. Parma, P.J. Sullivan, and R.J. Trumble. 1999. Decadal changes in growth and recruitment of Pacific halibut (*Hippoglossus stenolepis*). Can. J. Fish. Aquat. Sci. 56:242-252.

Francis, R.C. 1983. Population and trophic dynamics of Pacific hake (*Merluccius productus*). Can. J. Fish. Aquat. Sci. 40:1925-1943.

Francis, R.C., and S.R. Hare. 1994. Decadal-scale regime shifts in the large marine ecosystems of the North-east Pacific: A case for historical science. Fish. Oceanogr. 3:379-391.

Hare, S., and N. Mantua. 2000. Empirical evidence for North Pacific regime shifts in 1977 and 1989. Prog. Oceanogr. 47:103-145.

Issaacs, J.D. 1995. Some ideas and frustrations about fishery science. Calif. Coop. Ocean Fish. Investig. Rep. 18:34-43.

Kawasaki, T., and M. Omori. 1988. Fluctuations in the three major sardine stocks in the Pacific and the global temperature. In: T. Wyatt and G. Larranenta (eds.), Long term changes in marine fish populations: A symposium in Vigo, Spain, 18-21 Nov. 1986. Imprenta REAL, Bayona, Spain, pp. 37-73.

Kieser, R., M.W. Saunders, and K. Cooke. 1999. Review of hydro acoustic methodology and Pacific hake biomass estimates for the Strait of Georgia, 1981 to 1998. Canadian Stock Assessment Secretariat, Reseach Document. 99/15. 53 pp.

Mantua, N.J., S.R. Hare, Y. Zhang, J.M. Wallace, and R.C. Francis. 1997. A Pacific interdecadal climate oscillation with impacts on salmon production. Bull. Am. Meteorol. Soc. 78:1069-1079.

McCaughran, D.A. 1999. Seventy-five years of halibut management success. In: D.A. Hancock, D.C. Smith, A. Grant, and J.P. Beumer (eds.), Developing and sustaining world fisheries resources. Proceedings of the Second World Fisheries Conference, Brisbane, Australia, 1996. CSIRO, Australia.

McFarlane, G.A., and R.J. Beamish. 1985. Biology and fishery of Pacific whiting, *Merluccius productus*, in the Strait of Georgia. NMFS Mar. Fish. Rev. 47(2):23-34.

McFarlane, G.A., J.R. King, and R.J. Beamish. 2000. Have there been recent changes in climate? Ask the fish. Prog. Oceanogr. 47:147-169.

McFarlane, G.A., D.M. Ware, R.E. Thomson, D.L. Mackas, and C.L.K. Robinson. 1997. Physical, biological and fisheries oceanography of a large ecosystem (west coast of Vancouver Island) and implications for management. Oceanol. Acya 20:191-200.

Minobe, S. 1997. A 50-70 year climate oscillation over the North Pacific and North America. Geophysical Research Letters 24(6):682-686.

Noakes, D.J. 1986. Quantifying changes in British Columbia Dungeness crab (*Cancer magister*) landings using intervention analysis. Can. J. Fish. Aquat. Sci. 43:634-639.

Saunders, M.W., and G.A. McFarlane. 1999. Pacific hake. Strait of Georgia stock assessment for 1999 and recommended yield options for 2000. Canadian Stock Assessment Secretariat, Reseach Document. 99/08. 10 pp.

Schweigert, J., and C. Fort. 1999. Stock assessment for British Columbia herring in 1999 and forecasts of the potential catch in 2000. Canadian Stock Assessment Secretariat, Reseach Document. 99/01. 178 pp.

Shaw, W., G.A. McFarlane, and R. Kieser. 1990. Distribution and abundance of the Pacific Hake (*Merluccius productus*) spawning stocks in the Strait of Georgia, British Columbia, based on trawl and acoustic surveys in 1981 and 1988. Int. North Pac. Fish. Comm. Bull. 50:121-134.

Shaw, W., G.A. McFarlane, I. Barber, K. Rutherford, A. Cass, M. Saunders, R. Scarsbrook, and M. Smith. 1985. Biological survey of Pacific hake, walleye pollock and spiny dogfish in selected inshore areas and open waters of the Strait of Georgia, R/V *G.B. Reed*, February 7-28 and April 18-24, 1983. Can. Data Rep. Fish. Aquat. Sci. 500. 240 pp.

Tanasichuk, R.W., D.M. Ware, W. Shaw, and G.A. McFarlane. 1991. Variations in diet, daily ration, and feeding periodicity of Pacific hake (*Merluccius productus*) and spiny dogfish (*Squalus acanthias*) off the lower west coast of Vancouver Island. Can. J. Fish. Aquat. Sci. 48:2118-2128.

Thomson, R.E. 1981. Oceanography of the British Columbia coast. Can. Spec. Publ. Fish. Aquat. Sci. 56. 291 pp.

Ware, D.M., and G.A. McFarlane. 1995. Climate induced changes in hake abundance and pelagic community interactions in the Vancouver Island Upwelling System. In: R.J. Beamish (ed.), Climate change and northern fish populations. Can. Spec. Publ. Fish. Aquat. Sci. 121:590-521.

Wilson, C.D., and M.A. Guttormsen. 1997. Echo integration–trawl survey results for Pacific whiting (*Merluccius productus*) along the Pacific coast of the U.S. and Canada during summer, 1995. NOAA Tech. Memo. NMFS-AFSC-74.

Wilson, C.D., M.A. Guttormsen, K. Cooke, M.W. Saunders, and R. Kieser. 2000. Echo integration–trawl survey of Pacific hake, *Merluccius productus*, off the Pacific coast of the United States and Canada during July-August, 1998. NOAA Tech. Memo. NMFS-AFSC-118. 103 pp.

Trophic Position of Pacific Herring in Prince William Sound, Alaska, Based on Their Stable Isotope Abundance

Thomas C. Kline Jr.
Prince William Sound Science Center, Cordova, Alaska

Abstract

Acting as a conveyor of energy between the plankton and piscivorous organisms, Pacific herring (*Clupea pallasii*) play an important ecological role in Prince William Sound and other coastal waters of the northeast Pacific Ocean. Because of a regional isotopic gradient and the fidelity of consumer isotopic composition with that of their diet, natural carbon and nitrogen stable isotope abundance was used as a tool to determine Prince William Sound herring trophic position. Nitrogen stable isotope abundance was used in this study to estimate and provide evidence for consistency in herring trophic level. Some seasonal variability during the 1994 to 1998 study period was noted. Assuming that energy available for the highest trophic levels is determined primarily by the number of trophic steps, trophic level consistency suggests that energy per unit of biomass is propagated in proportion to its variability at the food web base. Previously, herring carbon isotopic composition suggested the latter to vary on annual time scales. Forage levels for herring consumers may thus fluctuate in accordance with herring food chain length and variability in planktonic productivity patterns. Measurable trophic level shifts would be significant given the narrow-ranging well-defined herring trophic level found in this study.

Introduction

The failure of several Prince William Sound (PWS), Alaska, vertebrate species to recover from population crashes following the 1989 T/V *Exxon Valdez* oil spill has raised concerns that shifts in food web structure may have occurred. Of particular concern is recruitment of *Clupea pallasii* (Pacific herring), presently at a historical low in abundance in PWS, a fjord-like inland sea that receives oceanic water from the Gulf of Alaska (GOA) via the Alaska coastal current (Niebauer et al. 1994). Furthermore, Pacific herring play a key role in the subarctic Pacific pelagic ecosystems by being in an intermediary trophic position between plankton and consumers of herring such as other fishes, birds, and mammals including humans. Recently, Kline (1999b) has shown through the use of natural stable carbon isotope abundance ($^{13}C/^{12}C$) that herring in PWS may be dependent upon carbon generated in the Gulf of Alaska and that the degree of this dependency fluctuates from year to year.

Stable isotope ratios effectively provide empirical evidence of trophic relationships in marine food webs because of their predictable relationship when comparing their abundance in consumers and diet. Nitrogen stable isotope ratios provide excellent definition of relative trophic level (Fry 1988). The heavy isotope of nitrogen, ^{15}N, is enriched by about 0.34% (or 3.4 per mil in conventional delta units, see Materials and Methods) with each trophic level (Minagawa and Wada 1984) and has been shown to accurately indicate the relative trophic status of species within an ecosystem (Cabana and Rasmussen 1994). Trophic level as used here refers to the average number of feeding steps between predators and their prey. For example, if the diet of predator A was 100% prey B, there would be 1.0 trophic levels between them. However, if A also consumed C, and C was also part of the diet of B, the trophic level difference between A and B would most likely ≠ 1.0. The effective trophic level difference between A and B would then depend upon the relative contribution of B and C in the composition of A's diet as well as the relative trophic level differences between B and C.

Carbon stable isotope ratios ($^{13}C/^{12}C$) can effectively be used to trace multiple sources of carbon into food webs once it is established that an ecosystem has carbon sources with distinctive isotopic signatures (Fry and Sherr 1984). This latter point has been established since GOA and PWS carbon sources for PWS had distinctive carbon isotope signatures of about –23 and –19 (in $\delta^{13}C'$ units), respectively (see Kline 1999b). Accordingly, stable carbon isotopes measured in herring and other biota suggested that in some years the GOA may supply the majority of carbon for the PWS pelagic ecosystem (Kline 1999b).

An Ecopath (Christensen and Pauly 1992) model and $^{15}N/^{14}N$ natural abundance data predicted corroborating trophic levels for pelagic biota including herring from PWS (Kline and Pauly 1998) but did not provide details relating patterns of variability of relative trophic level between herring and secondary pelagic producers and between herring and potential teleost consumers. Kline's (1999a) $^{15}N/^{14}N$ inventory for herring from 1994 and 1995 suggested only slight interannual and intersite variability while Kline and Pauly (1998) determined trophic levels of 3.3 and 3.33 (where herbivores are trophic level = 2) based on Ecopath and isotopic analyses of $N = 459$ herring, respectively. Since these publications, a more extensive isotopic database has been generated for herring ($N > 2{,}000$) and other pelagic biota of PWS for the period 1994-1998 from which patterns of GOA carbon were found from $^{13}C/^{12}C$ data (Kline 2001). The purpose of this study was to use the extensive $^{15}N/^{14}N$ data to assess the trophic position of PWS herring during this period seeking potential ontogenetic, spatial, and temporal patterns.

Assessing the trophic position of PWS herring was accomplished through three tasks: (1) by assessing effects of $^{15}N/^{14}N$ variability of PWS herbivorous copepods used as the trophic baseline (Kline 1999b); (2) by assessing herring ontogenetic $^{15}N/^{14}N$ shifts and hence trophic level shifts based on fish size; and (3) by assessing PWS herring temporal and spatial $^{15}N/^{14}N$ variability from a four-bay time series.

Materials and Methods

Data Generation

The rationale for, and detailed description of methods used to generate the isotopic database for this study form a major portion of Kline (1999b) and so are only given briefly here. Copepods were picked in the field from plankton samples that were made with either ring nets or a MOCNESS system and in the case of the large-sized species, *Neocalanus cristatus*, analyzed as individuals. Other *Neocalanus* spp. were analyzed by pooling two individuals together. Fishes were collected with seines, traps, and trawls. Herring data for those up to a length of about 150 mm were largely a part of a collaborative effort with A.J. Paul who measured whole-body energetic content of the same individual fish (Kline 1999b). Herring and other fishes were analyzed in replicate and data averaged to provide one isotopic datum per fish for each element (N and C). Only stable isotopes of N are considered here. N stable isotope ratios, $^{15}N/^{14}N$ are expressed in delta (δ) units as the per mil deviation from the $^{15}N/^{14}N$ content of air N_2.

Data Selection

The mean $\delta^{15}N$ values of terminal feeding stages of *Neocalanus cristatus* and *Neocalanus* spp. were pooled from PWS and GOA spring (1995) or May (1996-1997) oceanographic sampling stations (Kline 1999b). *Neocalanus* spp. included *N. cristatus, N. plumchrus,* and *N. flemingeri* (1996-1997) from the GOA or PWS. A portion of these data were extracted from published data (Kline 1999a,b) while others are reported here for the first time. Obtained similarly were the pooled $\delta^{15}N$ values of diapause stage *Neocalanus cristatus* and *Neocalanus* spp. from the appropriate deep habitat in PWS (Kline 1999b).

All herring sampled from April 1994 to March 1998 for which length data (standard length) were available were pooled and analyzed for trend using a scatterplot and regressed for best fit using R^2 and P values. Commencing in the fall of 1995, juvenile herring approximately < 15 cm in length were sampled at four sites (see Norcross and Brown 2001, this volume) within PWS at approximately 1.5 month intervals between March 1996 and October 1997 (Kline 2000). This four-bay time series enabled assessing finer scale temporal variability in several parameters (whole-body and diet energetics, growth, diet composition) for age 0 and 1 PWS herring than had been previously examined (Norcross et al. 1996). Approximately $N = 25$ from each site-sampling which were isotopically analyzed are reported here with S.D. estimation. Age 0 and 1 herring were pooled for this analysis since it was not possible to split the data by age and also retain both the temporal and spatial coverage. Thus ontogenetic effects were examined using a separate data selection that also benefited by including data from additional PWS sites (Kline 1999b).

Comparing the trophic level of herring to other organisms from PWS was facilitated by pooling $\delta^{15}N$ data of herbivorous copepods and large teleosts to reflect herbivores (potential herring forage) and piscivores (potential herring predators), i.e., expected to have, respectively, lower and higher trophic levels. Diapausing *Neocalanus cristatus* were selected as a proxy for the PWS herbivore trophic level based upon their previous use and the observation that the fraction originating from the GOA matched the fraction of GOA carbon in PWS herring (Kline and Pauly 1998, Kline 1999b). Potential piscivores consisted of larger-sized (> 20 cm) taxa. Juvenile stages (< 20 cm) and taxa known to have similar trophic level as herring (Kline and Pauly 1998) were therefore excluded. The N for copepods and teleosts selected were approximately 700, sufficient for generation of histograms for comparison with herring $N \cong 2,000$.

Trophic level was determined by comparing $\delta^{15}N$ values to a reference value (Vander Zanden et al. 1997). The $\sigma^{15}N$ of higher trophic levels were calculated by adding the trophic enrichment factor, 3.4 (Minagawa and Wada 1984, Kline 1997), to the reference value. The herbivorous copepod *Neocalanus cristatus*, i.e., trophic level = 2, was used as the reference (Kline and Pauly 1998, Kline 1999b).

The following formula was used to calculate trophic level:

$$TL_i = (\delta^{15}N_i - \delta^{15}N_H / 3.4) + 2$$

where TL_i is the trophic level of organism i, $\delta^{15}N_i$ is the mean $\delta^{15}N$ value of organism i, and $\delta^{15}N_H$ is the mean herbivore $\delta^{15}N$ value.

Statistical analysis of the data was facilitated by using the Statview 4.5 (Abacus Concepts, Inc., Berkeley, CA) computer program while the Deltagraph 3.1 (Delta Point, Inc., Monterey, CA) computer program was used to generate the figures.

Table 1. Mean $\delta^{15}N$ and S.D. of terminal feeding and diapause stages of the large (length $\cong$ 9 mm) herbivorous copepod *Neocalanus cristatus* (except as noted) from PWS and the GOA; data came from indicated sources.

Sample	$\delta^{15}N$	S.D.
Feeding		
Prince William Sound		
1995[a]	8.6	1.1
1996[b]	8.3	0.8
1997[c]	8.9	1.2
1996[c,d]	8.6	0.7
1997[c,d]	8.9	0.9
Gulf of Alaska		
1995[a]	7.3	0.7
1996[b]	8.5	1.1
1997[c]	7.8	1.3
1996[b,c,d]	8.8	1.0
1997[c,d]	7.6	1.0
Diapause		
Prince William Sound		
1994[a]	8.7	1.6
1995[a]	7.8	2.8
1996[b]	8.6	1.5
1996[b,c,d]	8.5	1.1
1994-1996[c]	8.4	2.0

[a]Kline 1999b.
[b]Kline 1999a.
[c]This study.
[d]All *Neocalanus* spp., otherwise only *N. cristatus*.

Results

There were only slight variations in the mean $\delta^{15}N$ values for the species *Neocalanus cristatus* and for the three *Neocalanus* species. (Table 1). The lowest $\delta^{15}N$ value of 7.3 was found for 1995 feeding stage GOA *N. cristatus* while the highest $\delta^{15}N$ value of 8.9 was found for both feeding stage PWS *N. cristatus* and PWS *Neocalanus* spp. mean values in 1997. Diapausing copepods had a smaller range in values reflecting the integration of GOA and PWS copepod sources also suggested through their $\delta^{13}C$ values (Kline 1999b). Note, however, that the mean diapausing value for *N. cristatus* in

1995 and 1996, 7.8 and 8.6, respectively, more closely matched those from the GOA, 7.3 and 8.5, respectively, rather than those from PWS, 8.6 and 8.3, respectively, corroborating the significant contribution of GOA-generated production for PWS deduced from $\delta^{13}C$ (Kline 1999b). The mean $\delta^{15}N$ for all 1996 diapausing species, 8.5, was ^{15}N-depleted compared to both PWS and GOA but more closely matched PWS at 8.6 than the GOA at 8.8. The collective mean for *N. cristatus* sampled diapausing from the entire study, 8.4 ± 2.0 (S.D.), is discussed in greater detail below in relation to herring and potential piscivorous teleosts.

There was only a slight shift in $\delta^{15}N$ and hence trophic level (TL) as a function of size for PWS herring (Fig. 1). Only herring < 100 mm were TL < 3.0 while very few herring at any size were TL > 3.5. The slight $\delta^{15}N$ shift with size was reflected in the regressions which only explained 16% of the variability. There appears to be a trophic level break for herring < 70 mm as these were mostly TL < 3.0 whereas the majority of the larger herring were between TL = 3 and TL = 3.5.

Spatial and temporal variability of PWS herring $\delta^{15}N$ and hence trophic level was only slight (Fig. 2). The low trophic levels during the summer months can, in part, be explained by age-0 fish < 70 mm. Although some of the variability among sites might be explained by unequal representation of the two year classes (Norcross et al. 1996), age-1 herring at sites with lowest mean TL had lower TL than those from sites with higher mean TL.

There was slightly more than one trophic level difference between herring and herbivores while there was slightly less than one trophic level difference between herring and their potential predators (Fig. 3). Diapausing *Neocalanus cristatus* copepods ($N = 669$) which were used as the trophic level = 2.0 (herbivore) baseline had a mean $\delta^{15}N = 8.39$ with a standard deviation of 2.01. They were slightly skewed to higher values (skewness = 0.79) and were leptokurtic (kurtosis = 3.46). Large teleosts (> 20 cm; $N = 727$) had a mean $\delta^{15}N = 14.66$ (interpreted trophic level = 3.85) with a standard deviation of 1.22 (0.36 trophic levels) were similarly leptokurtic (kurtosis = 3.37) and positively skewed (skewness = 0.73). Herring ($N = 2,084$), which had a mean $\delta^{15}N = 12.47$ (interpreted trophic level = 3.20) with a standard deviation = 0.60 (0.18 trophic levels), were more strongly leptokurtic (kurtosis = 4.72) and skewed to lower $\delta^{15}N$ (skewness = –1.02). The negative skewness was due to the lower trophic level of smaller herring while the high kurtosis and narrow standard deviation suggests a well-defined $\delta^{15}N$-based trophic level for herring. The difference of 1.20 trophic levels between herring and herbivores is consistent with herring also consuming carnivorous zooplankters. The difference of 0.65 trophic levels between herring and large teleosts is consistent with prey-switching by larger fishes. For example, PWS walleye pollock (*Theragra chalcogramma*), a major constituent taxon, consume zooplankters (including *Neocalanus* spp.) as well as herring and other prey. However, note the positive skewness and sample distribution indicating that teleosts can feed as high as trophic level five.

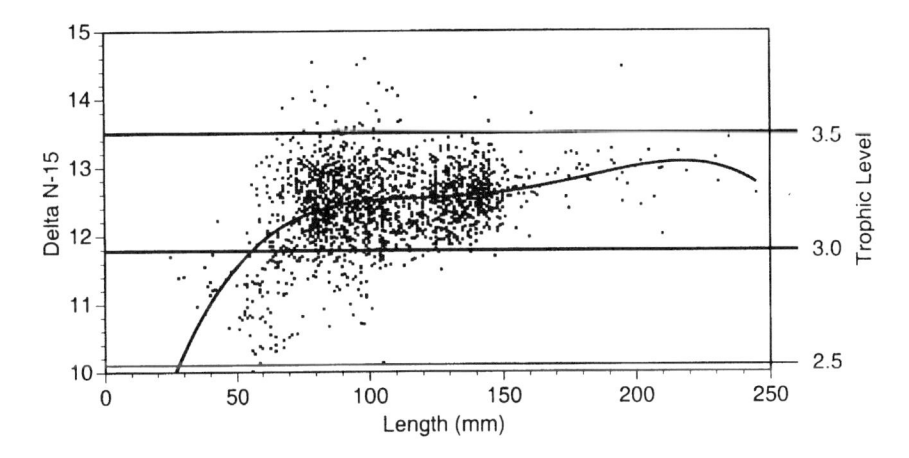

Figure 1. Scatterplot of $\delta^{15}N$ versus standard length for Prince William Sound, Alaska herring sampled between April 1994 and March 1998 with interpreted trophic levels. The fourth order polynomial shown provided a better fit to the data than a linear regression, explaining 16% of the variability compared to 10%, respectively (coefficient and regression P < 0.0001).

Discussion

Herring production levels is of concern for recruitment of herring, presently at a historical low in abundance in PWS. In particular, forage levels for herring consumers may fluctuate in accordance with herring food chain length and variability in planktonic productivity patterns if these factors are important to herring productivity. While carbon stable isotope ratios provided strong evidence for large interannual-scale shifts in source of production supporting PWS, measurable trophic level shifts would be significant given the narrow-ranging well-defined herring trophic level found in this study. The narrow $\delta^{15}N$ range of herring compared to herbivores conferred a well-defined trophic position during the period of this study.

Potential errors in estimating herring trophic level from $\delta^{15}N$ were likely to be less than about ⅓ trophic levels based on the relatively small temporal and spatially dependent departures of mean herbivore $\delta^{15}N$ from a value of 8.4. The sources of $\delta^{15}N$ variation arise in the phytoplankton, varying among phytoplankton species because of size and species-dependent differences in isotopic fractionation and isotopic effects arising from their selective draw-down of ^{14}N in the dissolved inorganic nitrogen pool (see review by Owens 1987). Kline 1999b estimated that 40% of $\delta^{15}N$ variability of PWS zooplankton could be ascribed to nitrogen dynamics while spatial gradients accounted for 1 per mil differences, equivalent to the trophic fractionation of about ⅓ trophic levels. However, the spatial gra-

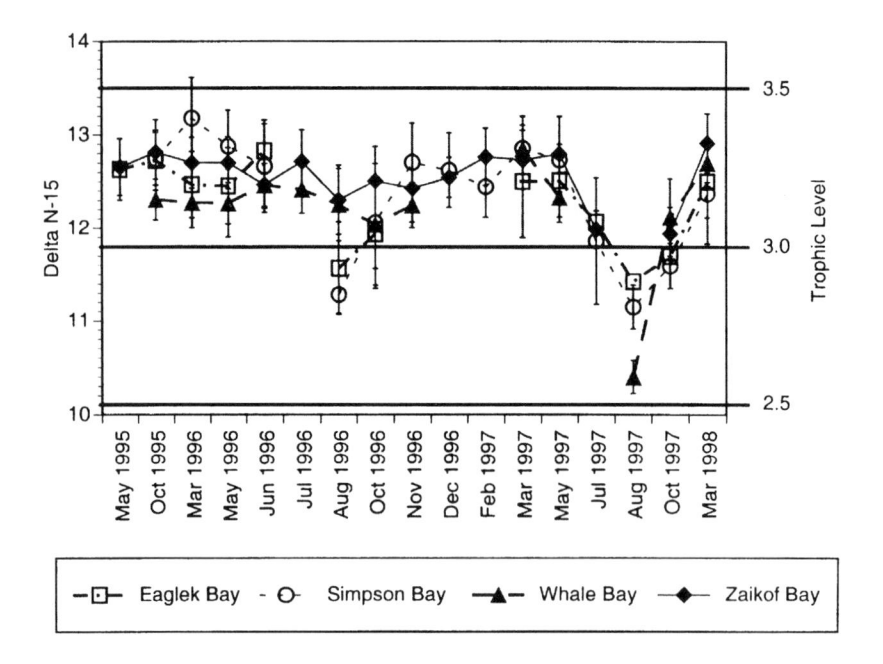

Figure 2. Temporal and spatial variability in $\delta^{15}N$ and interpreted trophic level for Prince William Sound herring. Error bars depict standard deviations. Lower trophic levels occurred during summer when new recruits were small. Trophic level variability was otherwise slight. Norcross and Brown (2001, this volume) show the locations and provide a description of the four-bay sampling program.

dient was weaker in later years of the study so $\delta^{15}N$-based errors should be less than 1/3 trophic level overall.

There were a few copepods with anomalous $\delta^{15}N$ values for a herbivore. One interpretation for copepods with very high $\delta^{15}N$ and hence a high trophic level is feeding on micro-heterotrophs dependent upon fish detritus, either spawned-out salmon carcasses or anthropogenic fish wastes such as from the numerous PWS processing plants. Since these were small in number they had only negligible effect for this study since the central value of a large sample size was used for the calibration point ($\delta^{15}N = 8.4$). Furthermore, carnivorous copepods and omnivorous euphausiids and amphipods had appropriate $\delta^{15}N$ values, consistently higher than 8.4 (Kline 1999b).

The relatively slight shifts of herbivore $\delta^{15}N$ cannot explain the pattern of herring seasonal shifts $\delta^{15}N$ up to near 3 per mil. Therefore, a seasonal shift to lower trophic level feeding interpretation is more valid. The pre-

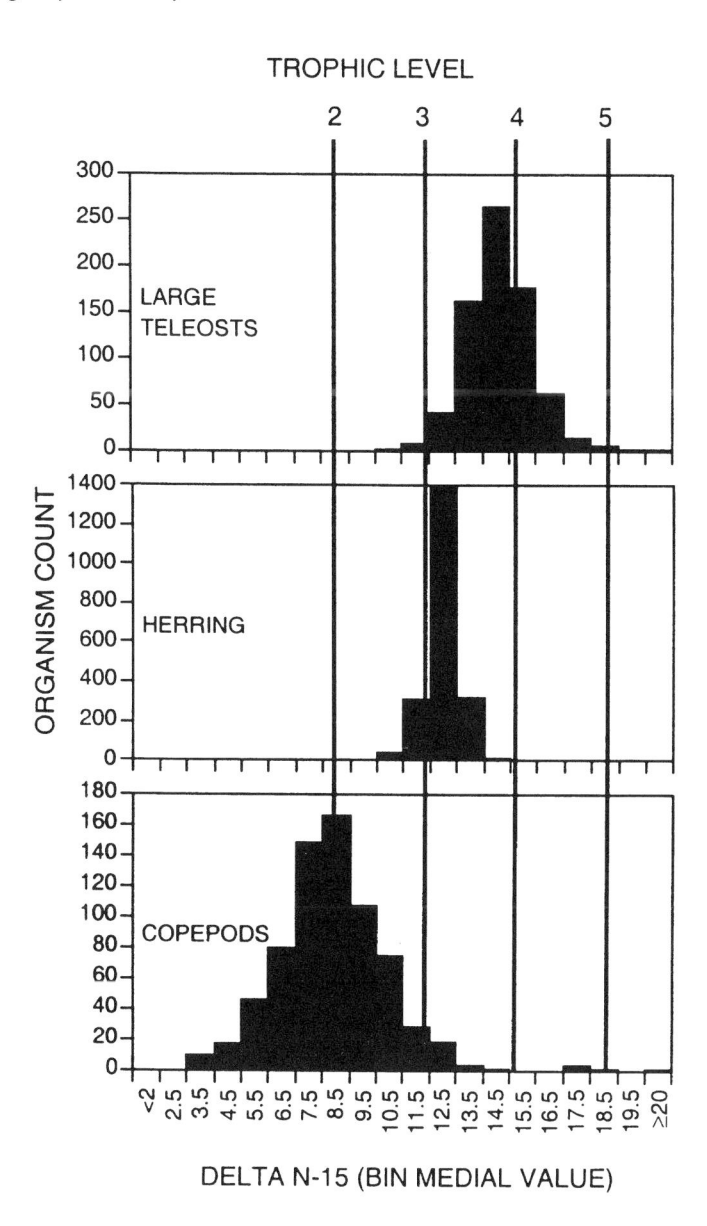

Figure 3. Comparison of $\delta^{15}N$ and interpreted trophic level variability through histograms of a large database.

ponderance of smaller newly recruited herring in summer and their suggested ontogenetic trophic level shift at ~ 7 cm length found in summer can only explain a part of this shift since both age classes apparently shifted to lower trophic levels during summer of 1997.

An Ecopath model for the same time period as these data conferred trophic levels of 3.10 and 3.03 for adult (fish > 18 cm) and juvenile herring, respectively (Okey and Pauly 1998: table 74), slightly lower than the over all trophic level estimate of 3.2 for all stages estimated here. These differences could be explained in part by seasonal shifts since Alaska field data (used for Ecopath input) tend to have a "summer-bias" which is when trophic levels are lower. The less than 0.1 trophic level difference between adult and juvenile herring found by Ecopath is closely reflected by the $\delta^{15}N$-based trophic levels. However, the range in herring trophic level independent of fish length as well as variability in space-time was greater than the small differences found for the two herring Ecopath functional groups. Given that the number of functional groups that can be incorporated into an Ecopath model is limited to 50 (Okey and Pauly 1998), it would be prudent to pool herring into one. Given a standard deviation of 0.18 trophic levels for the $\delta^{15}N$-based trophic level of PWS, the Okey and Pauly (1998), Ecopath-based trophic level was in good agreement, closely reflecting the previous Ecopath-isotope comparison of Kline and Pauly (1998). This assessment thus increases our confidence that we have made a good estimation of the trophic status of PWS herring.

Nitrogen stable isotope abundance thus provided a good estimate for herring trophic level. There was very little ontogenetic trophic change in herring larger than about 70 mm. Herring trophic level was generally constant over the 4-year time period of this study although some seasonal variability was noted. If trophic level shifted on longer time scales, e.g., decadal, these shifts should be considered significant given the narrow-ranging, well-defined herring trophic level found in this study.

Acknowledgments

This project was funded by the *Exxon Valdez* Oil Spill Trustee Council. Nevertheless, the findings presented by the author are his own and not necessarily the Trustee Council's position. SEA colleagues from various institutions collected fish specimens. J.M. Paul and P. Shoemaker did the initial herring sample preparation. J. Williams, E. Stockmar, J. Murphy, B. Barnett, and T. Cooney assisted with *N. cristatus* sampling. John Williams and Kim Antonucci did the laboratory processing of samples for stable isotope analysis. Bruce Barnett and Norma Haubenstock at the University of Alaska Fairbanks Stable Isotope Facility performed the stable isotope analysis. This paper benefited from the suggestions made by J. Blackburn and two anonymous reviewers.

References

Cabana, G., and J.B. Rasmussen. 1994. Modeling food chain structure and contaminant bioaccumulation using stable nitrogen isotopes. Nature 372:255-257.

Christensen, V., and D. Pauly. 1992. ECOPATH II—A software for balancing steady-state models and calculating network characteristics. Ecol. Modelling 61:169-185.

Fry, B. 1988. Food web structure on Georges Bank from stable C, N, and S isotopic compositions. Limnol. Oceanogr. 33:1182-1190.

Fry, B., and E.B. Sherr. 1984. $\delta^{13}C$ measurements as indicators of carbon flow in marine and freshwater ecosystems. Contrib. Mar. Sci. 27:13-47.

Kline Jr., T.C. 1997. Confirming forage fish food web dependencies in the Prince William Sound ecosystem using natural stable isotope tracers. In: Forage fishes in marine ecosystems. University of Alaska Sea Grant, AK-SG-97-01, Fairbanks, pp. 257-269.

Kline Jr., T.C. 1999a. Monitoring changes in oceanographic forcing using the carbon and nitrogen isotopic composition of Prince William Sound pelagic biota. In: Ecosystem approaches for fisheries management. University of Alaska Sea Grant, AK-SG-99-01, Fairbanks, pp. 87-95.

Kline Jr., T.C. 1999b. Temporal and spatial variability of $^{13}C/^{12}C$ and $^{15}N/^{14}N$ in pelagic biota of Prince William Sound, Alaska. Can. J. Fish. Aquat. Sci. 56(Suppl. 1): 94-117.

Kline Jr., T.C. 2001. Evidence of biophysical coupling from shifts in abundance of natural stable carbon and nitrogen isotopes in Prince William Sound, Alaska. In: G.H. Kruse, N. Bez, A. Booth, M.W. Dorn, S. Hills, R.N. Lipcius, D. Pelletier, C. Roy, S.J. Smith, and D. Witherell (eds.), Spatial processes and management of marine populations. University of Alaska Sea Grant, AK-SG-01-02, Fairbanks, pp. 363-376.

Kline Jr., T.C., and D. Pauly. 1998. Cross-validation of trophic level estimates from a mass-balance model of Prince William Sound using $^{15}N/^{14}N$ data. In: F. Funk, T.J. Quinn II, J. Heifetz, J.N. Ianelli, J.E. Powers, J.F. Schweigert, P.J. Sullivan, and C.-I. Zhang (eds.), Fishery stock assessment models. University of Alaska Sea Grant, AK-SG-98-01, Fairbanks, pp. 693-702.

Minagawa, M., and E. Wada. 1984. Stepwise enrichment of ^{15}N along food chains: Further evidence and the relation between $\delta^{15}N$ and animal age. Geochim. Cosmochim. Acta 48:1135-1140.

Niebauer, H.J., T.C. Royer, and T.J. Weingartner. 1994. Circulation of Prince William Sound, Alaska. J. Geophys. Res. 99:113-126.

Norcross, B.L., and E.D. Brown. 2001. Estimation of first-year survival of Pacific herring from a review of recent stage-specific studies. In: F. Funk, J. Blackburn, D. Hay, A.J. Paul, R. Stephenson, R. Toresen, and D. Witherell (eds.), Herring: Expectations for a new millennium. University of Alaska Sea Grant, AK-SG-01-04, Fairbanks. (This volume.)

Norcross, B.L., E.D. Brown, K.D.E. Stokesbury, and M. Frandsen. 1996. Juvenile herring growth and habitat. In: R.T. Cooney, Sound ecosystem assessment (SEA): An integrated science plan for the restoration of injured species in Prince William Sound, Alaska, *Exxon Valdez* Oil Spill Restoration Project Annual Report (Restoration Project 95320). Alaska Department of Fish and Game, Anchorage.

Okey, T.A., and D. Pauly. 1998. A trophic mass-balance model of Alaska's Prince William Sound ecosystem, for the post-spill period 1994-1996. University of British Columbia Fisheries Centre Research Reports 6:4. Vancouver, Canada. 144 pp.

Owens, N.J.P. 1987. Natural variations in ^{15}N in the marine environment. Adv. Mar. Biol. 24:389-451.

Vander Zanden, M.J., G. Cabana, and J.B. Rasmussen. 1997. Comparing trophic position of freshwater fish calculated using stable nitrogen isotope ratios ($\delta^{15}N$) and literature dietary data. Can. J. Fish. Aquat. Sci. 54:1142-1158.

ured to Cape Navarin in the north and to southeastern Kamchatka in the
Pacific Herring in the Western Bering Sea Predatory Fish Diet

Alexei M. Orlov and Sergei I. Moiseev
Russian Federal Research Institute of Fisheries and Oceanography (VNIRO), Moscow, Russia

Extended Abstract

Pacific herring, *Clupea pallasii,* inhabiting the western Bering Sea belong to the Korf-Karagin population distributed in Korf and Karagin bays and adjacent waters. In the past that population was one of most abundant in the North Pacific. During periods of high abundance, herring were distributed to Cape Navarin in the north and to southeastern Kamchatka in the south. The maximal annual harvest (268,000 metric tons) was recorded in 1961. As a result of climatic change and overexploitation the abundance of herring in the western Bering Sea sharply declined (Kachina 1986a). Now the herring stock is in a depressed condition and there is no herring fishery.

Investigations were conducted aboard the Japanese trawler *Kayomaru No. 28* during May-July 1997 on the western Bering Sea continental slope from the eastern part of Olyutorsk Bay (168ºE) to the border of the U.S. Exclusive Economic Zone (177º30′W) east of Cape Navarin. A total of 170 bottom trawl hauls (47 during the bottom trawl survey and 123 during bottom trawl commercial fishing operations) were made using bottom trawls with horizontal and vertical mouth openings of about 25-30 m and 5-7 m, respectively. Mesh size was 100 mm. During the investigations, data on spatial distribution, size composition, and stomach contents of the most abundant species were obtained. Stomach sampling and processing were conducted using standard methods (Yang 1993, 1996). Below we characterize some features of spatial distribution and size composition of Pacific herring in the western Bering Sea during summer 1997, and importance of herring in the diets of some fishes.

During summer 1997 Pacific herring were caught only in the central part of Olyutorsk Bay (Figs. 1 and 2), which is one of the usual feeding areas for herring in the western Bering Sea (Kachina and Prokhorov 1966). Maximum catches consisted of more than 1,500 specimens per hour of trawling. Herring size (Fig. 3) ranged in fork length from 24 to 33 cm, with

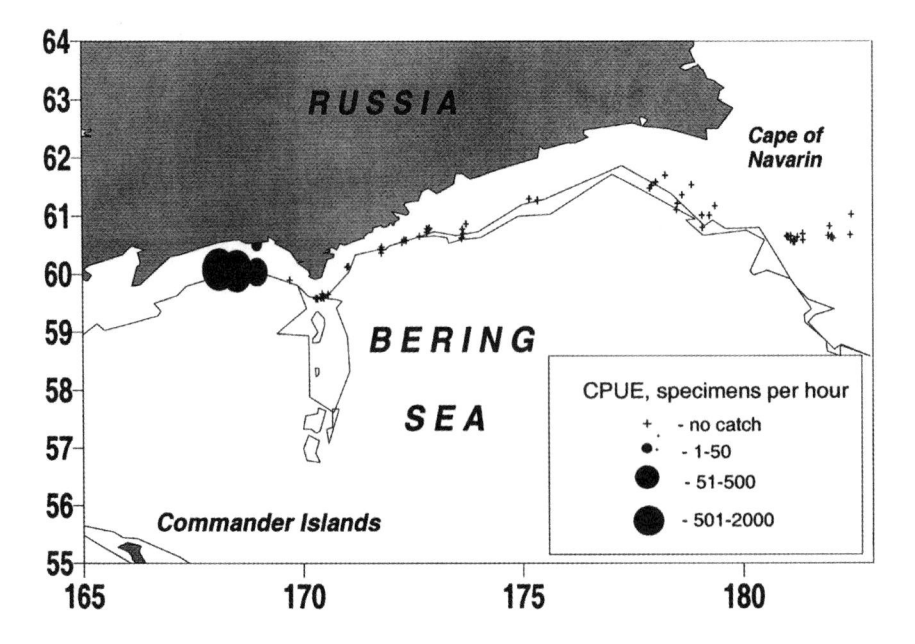

Figure 1. Distribution and relative abundance of Pacific herring categorized by CPUE (kg per hour trawling) in May-July 1997 in the western Bering Sea (thin lines are isobaths).

a mean and S.D. of 28.79 ± 0.12, which is very similar to 1971 to 1975, when herring fork length was 24 to 33 cm and mean length was 28.4 cm (Kachina 1986b). In 1938-1943 herring in the western Bering Sea were larger, from 25 to 33 cm with a mean length of about 32.0 cm (Panin 1950). Herring body weight in 1997 ranged from 180 to 460 g with a mean and S.D. of 309.0 ±10.86 g, which was considerably larger than in 1958-1968, when mean body weight did not exceed 286 g (Kachina 1970). During the period of our research, herring fed intensively and most of the fish (83.9% of the females and 54.2% of the males) were in postspawning condition.

Herring were found in fish stomachs from 21 bottom trawl hauls (mostly in the central part of Olyutorsk Bay), while herring were caught in only 8 hauls (Fig. 1). In addition, feeding on herring by fishes was noted in two catches located outside of Olyutorsk Bay (Fig. 2). This probably occurred because near-bottom herring schools generally cannot be caught by bottom trawls. However, predatory fishes (halibuts, shortraker rockfish, skates, large pollock, etc.) may forage some distance from the bottom (Orlov 1997, 1998). During our investigations herring were found in the stomachs of 10 fish species. According to percent by weight herring were very important (Table 1) in the diets of Alaska skate, *Bathyraja parmifera*

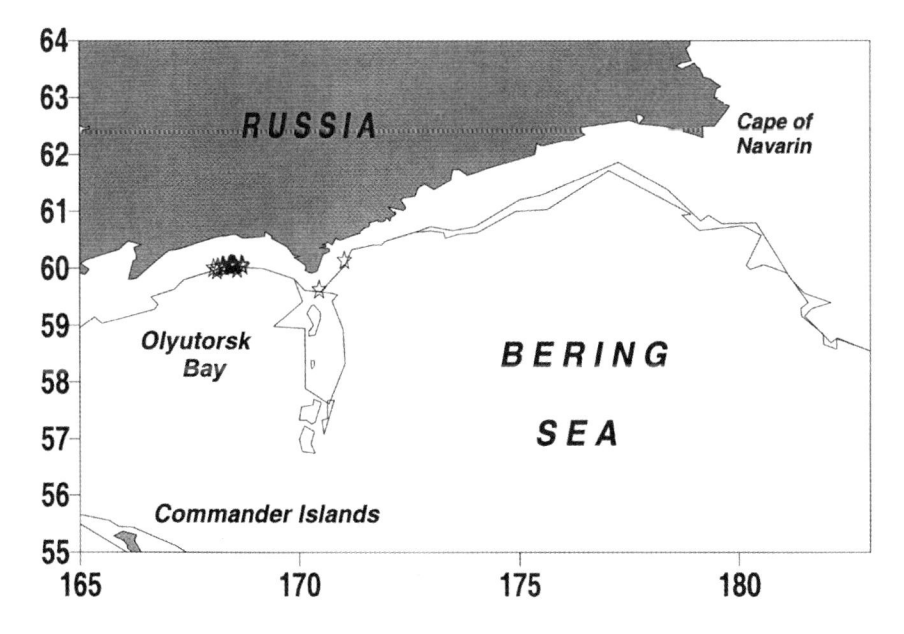

Figure 2. Map showing bottom trawl stations (stars) at which herring were found in predator stomachs (thin lines are isobaths).

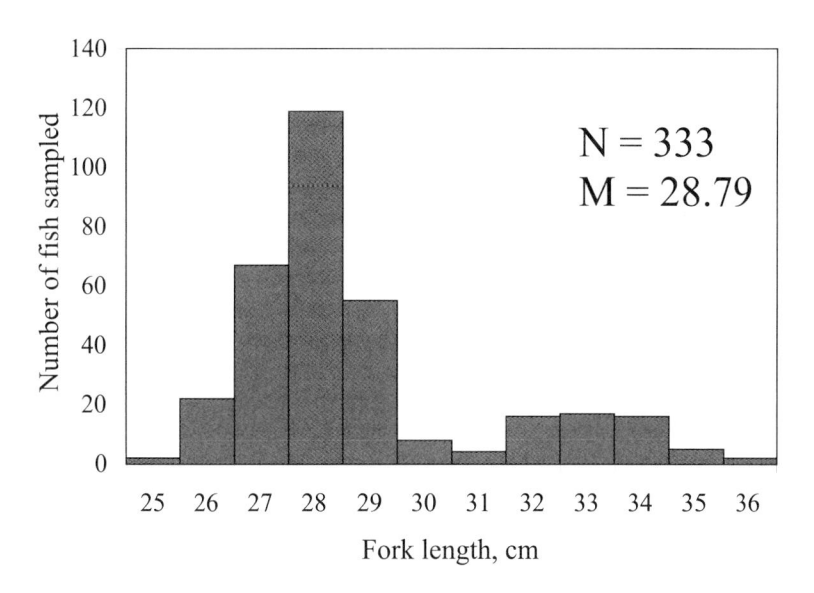

Figure 3. Length frequencies of Pacific herring in the western Bering Sea in May-July 1997 (n = number of fish measured; M = mean length, cm).

Table 1. Importance of Pacific herring and occurrence of fishes in stomachs of some predators that consumed Pacific herring in the western Bering Sea in 1997.

Prey category	1	2	3	4	5	6	7	8	9	10
Number of stomachs analyzed	151	113	68	139	93	2,113	301	262	446	590
Number of stomachs contained food	127	86	58	123	84	1,673	107	206	181	411
Predator's mean length (SL), cm	139.87± 2.74	80.00 ± 1.74	69.82± 0.56	75.38± 2.98	63.82± 1.09	51.72± 0.15	61.30± 0.64	73.92 ± 0.76	54.83 ± 0.37	69.30 ± 0.043
Predator's mean weight, kg	26.035 ± 2.294	4.607 ± 0.285	1.936± 0.041	4.596± 0.481	1.557± 0.064	0.973± 0.008	4.323± 0.125	5.250 ± 0.271	1.828 ± 0.271	3.573 ± 0.086
Total prey weight, g	46,057	9,089	2,048	8,050	2,515	na	2,880	31,715	10,794	80,316
Herring weight, g	100	3,055	235	475	30	na	648	3,665	1,116	7,123
Herring weight, % of stomach contents	0.22	33.61	11.47	5.90	1.19	na	22.50	11.56	10.34	8.87
Frequency of occurrence, %										
Pacific lamprey, *Entosphenus tridentatus*							0.9			
Pacific herring, *Clupea pallasii*	0.8	23.4	8.6	4.1	1.2	0.2	7.5	15.5	8.8	12.4
Pacific blacksmelt, *Bathylagus pacificus*					0.8					
Northern smoothtongue, *Leuroglossus schmidti*			1.7	0.8		7.0	3.7			0.2
Chum salmon, *Oncorhynchus keta*	1.6									
Salmon unidentified, *Oncorhynchus* spp.	1.6									
Northern lampfish, *Stenobrachius leucopsarus*	0.8	1.2					1.9		0.6	0.5
Garnet lampfish, *S. nannochir*		1.2	1.7	2.4	1.2	0.2	12.2			2.2

Table 1. (Continued)

Prey category	1	2	3	4	5	6	7	8	9	10
Lanternfish unidentified, Myctophidae			1.7	0.8		4.2	20.6			4.9
Giant grenadier, *Albatrossia pectoralis*	7.9							0.5		0.5
Popeye grenadier, *Coryphaenoides cinereus*	7.9							0.5		0.2
Walleye pollock, *Theragra chalcogramma*	3.9	10.5		8.1	2.4		0.9	18.5	13.8	19.0
Threespine stickleback, *Gasterosteus aculeatus*		1.2								
Shortraker rockfish, *Sebastes borealis*		1.2								
Atka mackerel, *Pleurogrammus monopterygius*			1.7							
Porehead sculpin, *Icelus canaliculatus*		1.2								
Armorhead sculpin, *Gymnocanthus detrisus*				0.8						
Spinyhead sculpin, *Dasycottus setiger*										0.2
Darkfin sculpin, *Malacocottus zonurus*	0.8	1.2						1.5	1.1	
Blackfin poacher, *Bathyagonus nigripinnis*								1.0		
Japanese dog poacher, *Percis japonica*				0.8						

na = not available

(1 = Pacific sleeper shark, *Somniosus pacificus*; 2 = Alaska skate, *Bathyraja parmifera*; 3 = whitebrow skate, *B. minispinosa*; 4 = Aleutian skate, *B. aleutica*; 5 = Okhotsk skate, *B. violacea*; 6 = walleye pollock, *Theragra chalcogramma*; 7 = shortraker rockfish, *Sebastes borealis*; 8 = Pacific halibut, *Hippoglossus stenolepis*; 9 = Kamchatka flounder, *Atheresthes evermanni*; 10 = Greenland turbot, *Reinhardtius hippoglossoides*.)

Table 1. (Continued.) Importance of Pacific herring and occurrence of fishes in stomachs of some predators that consumed Pacific herring in the western Bering Sea in 1997.

Prey category	1	2	3	4	5	6	7	8	9	10
Forktail snailfish, *Careproctus furcellus*	2.4					0.1				0.2
Dimdisc snailfish, *Elassodiscus tremebundus*		1.6							1.0	
Snailfish unidentified, Liparidae					0.8					0.2
Longsnout prickleback, *Lumpenella longirostris*		2.3			4.9				1.1	
Ebony eelpout, *Lycodes concolor*	0.8						0.9	0.5	1.1	
Shortfin eelpout, *L. brevipes*		1.2						0.5	1.1	0.7
Black eelpout, *L. diapterus*								0.5		0.7
Kamchatka flounder, *Atheresthes evermanni*	2.4									
Flathead sole, *Hippoglossoides elassodon*								0.5		0.2
Unidentified fish	8.7	2.3	5.2	9.8	3.6	0.1	15.9	4.9	13.8	5.1
Fishery offal	10.2	19.8	20.7	12.2	9.5	8.7	3.7	29.6	40.9	43.3

na = not available

(1 = Pacific sleeper shark, *Somniosus pacificus*; 2 = Alaska skate, *Bathyraja parmifera*; 3 = whitebrow skate, *B. minispinosa*; 4 = Aleutian skate, *B. aleutica*; 5 = Okhotsk skate, *B. violacea*; 6 = walleye pollock, *Theragra chalcogramma*; 7 = shorttraker rockfish, *Sebastes borealis*; 8 = Pacific halibut, *Hippoglossus stenolepis*; 9 = Kamchatka flounder, *Atheresthes evermanni*; 10 = Greenland turbot, *Reinhardtius hippoglossoides*.)

(33.61% by weight); whitebrow skate, *B. minispinosa* (11.47%); Pacific halibut, *Hippoglossus stenolepis* (11.56%); Greenland turbot, *Reinchardtius hippoglossoides* (8.87%); Kamchataka flounder, *Atheresthes evermanni* (10.34%); and shortraker rockfish, *Sebastes borealis* (22.50%). According to frequency of occurrence, which was estimated as the number of stomachs that contained that food item divided by the total number of stomachs examined, herring occupied the second place after walleye pollock, *Theragra chalcogramma,* in most predator diets. During the period of herring high abundance (1950s and 1960s) the frequency of occurrence in stomachs of Pacific halibut in the western Bering Sea according to Novikov (1964, 1974) was 3.1-4.0% (15.5% in our study), and its weight percentage according to Gordeeva (1954) was 2-3 % (11.56% in our study). Other fishes that consumed herring in the past according to Gordeeva (1954) and Novikov (1974) were shortraker rockfish (frequency of occurrence 10.7 and 7.5% in the 1950s to 1960s and 1997, respectively); sablefish, *Anoplopoma fimbria* (frequency of occurrence 0.6 and 0.0%, respectively); and Pacific cod, *Gadus macrocephalus* (0.8 and 0.0% by weight, respectively). Published information on diets of Pacific sleeper shark, skates, and walleye pollock in the western Bering Sea in the 1950s and 1960s is lacking.

Pacific herring also play an important role in the diets of many commercially important fishes of the eastern Bering Sea (Lang et al. 1991, Livingston et al. 1993), where its main consumers are large walleye pollock and Pacific cod. Herring composed (by weight) 0.6-1.8% of the diet of walleye pollock, and 1.0-6.4% of the Pacific cod diet (Lang et al. 1991, Livingston et al. 1993, Lang and Livingston 1996). Other predatory fishes consumed herring too. Herring (by weight) made up 3.2% of the diet for sablefish; 3.1-7.9% for arrowtooth flounder, *Atheresthes stomias;* 0.6-0.7% for Greenland turbot; and 0.6-4.8% for Pacific halibut (Brodeur and Livingston 1988, Yang and Livingston 1986, Lang et al. 1991, Livingston et al. 1993, Lang and Livingston 1996, Mito et al. 1999). Herring are also of considerable importance in the diet of Pacific halibut in British Columbia waters (Best and St.-Pierre 1986). Some predators consumed herring in the Aleutian Islands area, where its weight in stomach contents made up 0.52% for Pacific cod, 0.49% for arrowtooth flounder, and 1.72% for Pacific halibut (Yang 1996). In the Gulf of Alaska the percentage by weight of herring in predator stomachs was 0.20% for walleye pollock, 0.37% for Pacific cod, 9.15% for arrowtooth flounder, and 2.21% for sablefish (Yang 1993).

Results of our study showed that consumption of Pacific herring in the western Bering Sea by predators in the present period considerably increased in comparison with that of the 1950s and 1960s, when herring were more abundant. On the other hand, most western Bering Sea predators fed on herring more intensively than in areas of the northeastern Pacific although the present level of Korf-Karagin herring abundance is very similar to that of herring belonging to eastern Bering Sea and Gulf of Alaska populations (Naumenko 2000). It seems contradictory. However, we suggest that the reason for these oppositions were not related to the level of herring abundance in the western Bering Sea and were associated with the fact that our study coincided with herring feeding, when they formed dense schools. The biological condition of herring and decline of predatory fish abundance within Olyutorsk Bay from May to July may confirm the above suggestion.

References

Best, E.A., and G. St.-Pierre. 1986. Pacific halibut as predator and prey. IPHC Tech. Rep. 21:1-27.

Brodeur, R.D., and P.A. Livingston. 1988. Food habits and diet overlap of various eastern Bering Sea fishes. NOAA Tech. Memo. NMFS F/NWC-127. 76 pp.

Gordeeva, K.T. 1954. Food habits of halibuts in the Bering Sea. Izv. TINRO 39:111-134. (In Russian.)

Kachina, T.F. 1970. Regularity of abundance dynamics of Korf-Karaginsky herring. Ph.D. thesis. TINRO, Vladivostok. 21 pp. (In Russian.)

Kachina, T.F. 1986a. Condition of the Korf-Karaginsky herring stock. Rybn. Khoz. 2:24-27. (In Russian.)

Kachina, T.F. 1986b. Pacific herring. Biological resources of the Pacific Ocean. Nauka, Moscow, pp. 146-156. (In Russian.)

Kachina T.F., and V.G. Prokhorov. 1966. Korf-Karaginsky herring. Rybn. Khoz. 11:14-16. (In Russian.)

Lang, G.M., and P.A. Livingston. 1996. Food habits of key groundfish species in the eastern Bering Sea slope region. NOAA Tech. Memo. NMFS-AFSC-67. 111 pp.

Lang, G.M., P.A. Livingston, R. Pacunski, J. Parkhurst, and M.-S. Yang. 1991. Groundfish food habits and predation on commercially important prey species in the eastern Bering Sea from 1984 to 1986. NOAA Tech. Memo. NMFS-F/NWC-207. 240 pp.

Livingston, P.A. 1993. Importance of predation by groundfish, marine mammals and birds on walleye pollock *Theragra chalcogramma* and Pacific herring *Clupea pallasi* in the eastern Bering Sea. Mar. Ecol. Prog. Ser. 102:205-215.

Livingston, P.A., A. Ward, G.M. Lang, and M.-S. Yang. 1993. Groundfish food habits and predation on commercially important prey species in the eastern Bering Sea from 1987 to 1989. NOAA Tech. Memo. NMFS-AFSC-11. 192 pp.

Mito, K., A. Nishimura, and T. Yanagimoto. 1999. Ecology of groundfishes in the eastern Bering Sea, with emphasis on food habits. Dynamics of the Bering Sea. University of Alaska Sea Grant, AK-SG-99-03, Fairbanks, pp. 537-579.

Naumenko, N.I. 2000. Biology and fisheries of the Far East marine herrings. Doctoral thesis. VNIRO, Moscow. 45 pp. (In Russian.)

Novikov, N.P. 1964. Basic biological features of Pacific halibut (*Hippoglossus hippoglossus stenolepis* Schmidt) in the Bering Sea. Izv. TINRO 51:167-207. (In Russian.)

Novikov, N.P. 1974. Commercially important fishes of the continental slope of the North Pacific Ocean. Pishchevaya Promyshlennost', Moscow. 308 pp. (In Russian.)

Orlov, A.M. 1997. Ecological characteristics of the feeding of some Pacific predatory fish of southeastern Kamchatka and he northern Kuril Islands. Russ. J. Aquat. Ecol. 6(1-2):59-74.

Orlov, A.M. 1998. The diets and feeding habits of some deep-water benthic skates (Rajidae) in the Pacific waters off the northern Kuril Islands and southeastern Kamchatka. Alaska Fish. Res. Bull. 5(1):1-17.

Panin, K.I. 1950. Materials on the herring biology off the northeastern Kamchatka coast. Izv. TINRO 32:3-36. (In Russian.)

Yang, M.-S. 1993. Food habits of the commercially important groundfishes in the Gulf of Alaska in 1990. NOAA Tech. Memo. NMFS-AFSC-22. 150 pp.

Yang, M.-S. 1996. Diets of the important groundfishes in the Aleutian Islands in summer 1991. NOAA Tech. Memo. NMFS-AFSC-60. 105 pp.

Yang, M.-S., and P.A. Livingston. 1986. Food habits and diet overlap of two congeneric species, *Atheresthes stomias* and *Atheresthes evermanni*, in the eastern Bering Sea. Fish. Bull., U.S. 82(3):615-623.

Herring Abundance, Food Supply, and Distribution in the Barents Sea and Their Availability for Cod

Emma L. Orlova, Elena I. Seliverstova, Andrey V. Dolgov, and Valentina N. Nesterova
Knipovich Polar Research Institute of Marine Fisheries and Oceanography (PINRO), Murmansk, Russia

Abstract

The role of herring in the food base of Arcto-Norwegian cod in the 1980s and 1990s was examined by a review of historical data. The main factor influencing the availability of herring to cod is its distribution in the Barents Sea. When herring abundance is relatively low, its distribution is spatially restricted. The prey of herring also varies spatially: herring mainly feed on copepods in the west and on euphausiids in the central and eastern areas. Capelin is the main competitor for the food of herring. Fluctuations in the abundance of capelin affect the availability of food to herring, which affects their seasonal feeding rhythms, fat accumulation, and vertical distribution. Capelin abundance also affects the consumption of herring by cod.

Introduction

After a stock decline in the 1970s, Norwegian spring-spawning herring recovered and regained their previous place in the Barents Sea ecosystem in the 1980s and 1990s. During the period of decline the role of herring as prey for cod changed. In the 1970s, following the sudden decrease in abundance, herring were not available to cod. At that time cod fed mainly on capelin, which had increased in abundance (Ponomarenko and Yaragina 1985). In the period from 1984 to 1997, the issue of the role of herring as food for cod arose because of the relatively low abundance of herring and because of sharp fluctuations in capelin numbers (ACFM 1998). Also, during this period, temperatures increased and this also affected the distribution of fishes. Distribution also related to the availability of planktonic food organisms.

The most favorable periods for cod are those when wintering herring occupy bottom depths and are not very mobile. The duration of this period can vary from 2 to 4 months (between November and February). The duration depends on wintering conditions and the level of accumulated fat, which does not usually exceed 10-20% or an index of fatness of 2-3 in young herring (Rudakova 1966). Feeding intensity of herring is indicated by the level of fat reserves from January to March. At this time herring migrate to the middle and deeper depths where there are concentrations of overwintering small crustaceans. From February on they migrate vertically following the plankton (Tikhonov 1939, Boldovsky 1941). This behavior makes it possible for herring to be consumed by cod in the late winter and spring and also during the autumn feeding from August to October. The availability of herring is also influenced by their horizontal migrations within the Barents Sea, with seasonal movements to the west at the end of winter and to the east in summer. These variations can cause the migrations of cod and herring to be either coincident or independent.

The present paper considers the interaction of plankton condition, year-to-year changes in feeding intensity and fatness of herring, and the characteristics of the vertical and local distribution of herring which influenced their availability for cod in the 1980s-1990s.

Data Sources and Methods

Herring data were taken from catches of research vessels and bycatch of fishing vessels. Zooplankton data were obtained from annual assessment surveys in autumn-winter and spring. Data on plankton condition were obtained for 1985-1993, and data on the distribution, feeding, and fatness of herring, as well as data on cod feeding, was collected from 1985 to 1996. A field analysis of feeding (by the frequency of occurrence of particular food organisms) shown as a percentage of the total number of all fish with food, was made using the determination of index of stomach fullness and degree of fatness. Data on the consumption of food by cod was examined by quantitative analyses of weights of gut contents (about 90,000 specimens) and field observation based on incidence of prey species in guts (>55,000 specimens).

Year-to-Year and Seasonal Variations of Herring Distribution in the Barents Sea

In the 1950s-1960s herring distribution was related to their total abundance and to oceanographic factors. Marti (1956) and Shutova-Korzh (1960, 1962) concluded that the duration of the presence of herring in the Barents Sea depends on abundance and year-class strength. During periods with poor year classes, herring remain only for 2-3 years, whereas during periods of abundant year classes, herring may remain up to 7 years. Further,

Table 1. Absolute abundance of herring (billions) in the Barents Sea from 1984 to 1997 (ACFM 1998).

1984	1985	1986	1987	1988	1989	1990	1991	1992	1993	1994	1995	1996	1997
21.4	10.9	3.0	0.0	0.0	0.0	4.4	29.5	52.3	130.0	85.5	17.3	4.2	6.4

during warm years, herring are broadly distributed to the northwestern, eastern, and southeastern areas and to the entrance into the White Sea, but not in the western and central areas. In cold years, herring distribution was limited by the central and coastal branches of the North Cape Current. Table 1 shows the absolute abundance of herring (billions) in the Barents Sea from 1984 to 1997.

When the area of distribution of herring suddenly decreased in concert with the decrease in abundance, temperature may have played a smaller role in their distribution. This was more apparent in the western and central areas in 1986, 1987, 1990, and 1992 and especially in the eastern areas where herring were infrequent even in the very warm years. Herring were almost absent in the northeastern areas, where there was an increase in abundance and habitat of capelin, the main food competitor.

Herring distribution is strongly linked to its food base in the main habitats. In the western areas the spring phytoplankton bloom begins earlier. This area also supports a high density and the bloom has a long duration. Herring migrate here after wintering, and the concentrations of herring mainly depend on the biomass of copepods. Alternately, in the central and eastern areas, where the summer and autumn feeding occurs, herring depend on the density of euphausiid concentrations (Manteifel 1941).

The residence time of herring varies among different areas. In cold years in the western areas of the Barents Sea, they are limited to 1-2 seasons (often January-February and March-April or March-April only), when herring constitute 56-83% of the total fish biomass in the area. In warm years large herring concentrations are observed for a longer period (two to three seasons) in these areas. A large proportion of the herring remain to overwinter in this area. In cold years in the central sea, herring occur mainly in winter-spring and winter seasons, but in warm years this area became more important for feeding in the summer and autumn.

The seasonal distribution of herring was more clearly defined in the eastern areas where overwintering of herring also occurred in warm years. In this area there is a good correspondence between the depth distribution of herring, the level of plankton development, and the biological condition (fatness) of herring. The onset of migration and its duration in different areas vary and depend directly on conditions of herring feeding and level of their fatness.

Feeding of Herring: Food Composition, Dynamics of Feeding, and Fatness

In 1985, a year of moderate temperatures and in spite of the high fat index (>2) in 2-year-old herring, their feeding activity in the central area wintering grounds increased unusually early, in the middle of January, and at great depths. This was probably connected with the increased demand for food of these young herring. The low biomass of plankton in 1985, exacerbated by the increased early consumption by herring, led to a sharp decrease of herring fatness (~0) by mid-February. This promoted more feeding activity and a gradual movement of herring to shallower depths, as well as feeding on euphausiids (in addition to *Calanus*). Herring fed intensively on euphausiids until mid-March. In the majority of western and coastal areas, where herring were at different depths, their feeding and fatness in the spring-winter period was variable. Herring became fat in September-October because of intensive summer-autumn feeding. In spite of the average fatness level, some fish continued to consume food. From October to December, herring migrated to the central areas for wintering after feeding (but did not move to the greatest depths) and had a high index of fatness (1.57-2.44).

In 1986, a year with features peculiar to a cold year, the broad distribution and a low feeding activity promoted a slower expenditure of fat reserves in age-3 herring. These fish began feeding actively in the wintering grounds of the west and central areas when their fatness index ranged from 1.7 to 2.2. This occurred from mid- to late February at depths of 200-250 m. In spite of a relatively low level of plankton development in some areas (e.g., euphausiids in the western sea), the food supply of herring was high because of the sudden decrease of capelin abundance. Herring fed predominantly on euphausiids, and that promoted a high level of fatness between February and March-May. Later, these herring followed the euphausiids and *Calanus* up in the water column, staying at depths of 150-180 m from March to the middle of May, and at the 100-m depth at the end of May. In contrast, in the coastal waters of Murmansk, the main food of herring in the end of May was the deeper small crustaceans. Herring made daily vertical migrations to feed on these at depths of 210-265 m. Because of this intensive feeding, which lasted until October and which was mainly on a high biomass of euphausiids, the herring fat index was between 2.2 and 3.0 by September. Feeding also continued in October in the upper layers (20-30 m). Herring fatness at the end of the season fluctuated greatly (from 1 to 2.4) depending on the food supply in different areas. It was lowest in the extreme western areas, where the maturing herring were concentrated and spending their energy on gonad maturation. During this warm period herring feeding also was influenced by such factors as increased metabolism (an effect of high temperature) and the expansion of herring distribution as well as the behavior and abundance of plankton organisms, and the plankton consumption by other fish species which also had increased in abundance.

In 1991, there was a sudden increase of capelin but their impact on plankton abundance was not as great. Herring feeding was good due to high biomass levels of *Calanus* and euphausiids. At this time, however, there were only limited data available on herring feeding so only the main seasonal stages of feeding are described. On the overwintering grounds, the feeding was early, beginning in February, and at shallow depths (100-150 m). The later summer-autumn feeding period in western areas ended in early October when herring had a high fatness index (2-3). In the central areas, at a large range of depths (140-240 m), small concentrations of herring with low fatness were found.

In 1992-1993, when the abundance of capelin and herring was high, the biomass of euphausiids was particularly low. The biomass of *Calanus*, however, was relatively high, and that determined their important role in the food of herring in that year. Herring distribution between years was similar to that of other years, although it varied seasonally. In the anomalous warm year of 1992, as with other warm years, the beginning of active feeding of herring in the western areas was in mid-February (at 140 m depth) when the fatness index of herring was high. Their food was diverse, with a slight preponderance of euphausiids and *Calanus*. The subsequent spring-summer feeding of herring was very poor, however, because of the rapid sinking of *Calanus* to the great depths. As a result, herring fatness in August-September was low and did not exceed 1.1-1.2. In October in some areas, herring did not eat. At the end of the feeding season, which lasted till November, herring sank to depths of 150-250 m, and herring fatness fluctuated between 1 and 1.7. In the extreme western areas fatness was higher.

In 1993, the main changes in herring feeding, fatness dynamics, and distribution were associated with *Calanus*, the biomass of which increased two-fold compared to 1992, probably because of the reduction in capelin abundance. Herring fatness did not exceed 1 in January-February because it was low prior to the overwintering period. In the majority of areas the feeding began early: late January in the central sea, and early February in the western areas. In central areas, a large number of fish with low fatness were in the upper 50-m layer from the end of the previous year. In most herring habitats they fed consistently on *Calanus* and a little on euphausiids at different depths (60-220 m) during the whole winter-spring period. Exceptional herring habitats were found in some western and, especially, coastal areas, where feeding on euphausiids was intensive (in the latter case it took place at 50-m depths).

Regular (though moderate) feeding on *Calanus* promoted high fatness of herring until the end of March. However, variable summer feeding (on *Calanus*, mainly) caused reduction and fluctuation of fatness in western and coastal areas. September was a turning point in the year for herring, since some fish then ceased feeding, which led to highly variable fatness in the population. The most stable situation was observed in the extreme western part of the sea where herring with high fatness (2-3) continued feeding. In the other areas, local and moderate feeding on *Calanus* also continued at a fatness level of 1.1-2.15. At the end of September-October in most of the area, herring ceased eating almost completely. Their fatness fluctuated greatly in the coastal areas and not very much in the western and central areas (from 1.32 to 2.34 in total). In many cases, herring continued to stay in the upper layers. The fish migrated to greater depths only at the end of December where they overwintered at a fatness level of 1.55-2.43.

In subsequent years (1994-1996), capelin abundance was lower and herring feeding changed. In spite of the limited data for those years, it seems that the situation changed back to that seen in previous years, where normal herring seasonal feeding and fat accumulation were promoted by the increased abundance of plankton, especially euphausiids. Under such conditions the herring fat reserves were spent more slowly, and the herring stayed longer on the overwintering grounds.

The Availability of Herring and Their Consumption by Cod

Since the 1930s, it has been known that the importance of herring as food for cod was related to the abundance of herring (Zatsepin and Petrova 1939). In some years (1936, 1937) cod fed on herring during the whole year, but maximally in winter-spring and autumn-winter. Herring accounted for 50-70% of the food of cod, and capelin a much smaller amount. In those years, the occurrence of herring in the summer feeding of cod was also very high (to 30-40%).

In the 1950s-1960s herring were important in the diet of cod (Shutova-Korzh 1966). Cod ate herring more often in the western and central areas, where herring comprised between 28-58% and 37-66%, respectively, of stomach contents. There was no apparent relationship between the intensity of cod feeding on herring. In spite of large year-to-year fluctuations in capelin between the 1930s and 1960s, the incidence of capelin as food for cod was rather stable at 20-23%. After a sudden reduction of herring abundance, in the 1970s, this value began to rise to 40%, on average (Ponomarenko and Yaragina 1985). In the 1980s-1990s, capelin were more abundant than herring (ACFM 1998) and were also more dominant in the food of cod. Nevertheless, in some seasons and years there were conditions when herring were more available. This was especially true in the western areas, where the occurrence of herring in cod stomachs was 46-98%. The heightened consumption of herring occurred at times of both high and low herring abundance. This tendency was seen in the central and coastal areas as well. The characteristics of cod feeding on herring were influenced by several factors, including year-to-year fluctuations of the feeding rhythm, fat accumulation, and feeding behavior of herring, the direction and dates of their feeding and wintering migrations, as well as the availability of capelin for cod.

The clearest link between cod and herring is seen in years of abundant year classes of herring, such as 1983 and 1991-1992. The unusual data for 1984 (Orlova et al. 1989) proved that cod consumed herring yearlings as early as May, but mainly later in the year (September-October) over a broad range within the central areas. In 1985-1986, in spite of an early increase of herring feeding activity and their migration to middle depths, good conditions for cod to feed on herring still existed. In 1986, herring were intensively consumed by cod as a result of the overlap of concentrations of cod and herring (in the western and central areas). The consumption of herring was intensive on the slopes of the Murmansk Bank and Central Plateau in February during the period that herring feed on euphausiids. A very important factor in that period was the presence of some cod, feeding on herring, at medium depths. These cod were found in catches taken by a pelagic trawl (Ajiad 1990). The percentage of herring in cod food in those catches was 26.6% by weight versus 4.8% in cod caught by a bottom trawl. In the areas of main cod concentrations (Finnmarken and Demidov banks and Murmansk Tongue) cod fed mainly on capelin during the period of their limited local spatial overlap. Capelin and herring were found in equal numbers in cod stomachs (1 specimen per stomach, on average) despite the larger size of the herring (up to 19 cm). In March, the area of cod feeding on herring extended to the west (Murmansk Tongue and Demidovskaya Bank) where the incidence of herring in cod stomachs was 13-14% (or 1 herring for every 2 feeding cod). This coincided with herring feeding on *Calanus* and euphausiids as they did in February. Also in 1986, a low level of cod feeding on herring was found in the coastal areas as well (in February-April and June-August). In autumn the level of cod feeding on herring was low because of their migration to the west. The total weight of herring consumed by cod fluctuated from 2% (1984) to 5.3% and 5% (1985 and 1986, respectively).

In 1992, the main consumption of herring by cod was at the end of the year when large concentrations of migrating cod overlapped those of wintering herring (in the western and central areas). The percentage of young herring in the diet of young cod ranged between 9% and 17% in the northern part of the Novaya Zemlya shallows and the Kanin Bank.

The diet of cod in 1993 was influenced by the prolonged duration of the main concentrations of capelin on the overwintering grounds. This led to an earlier start of cod feeding on herring. It began in February and was most intensive (up to 62% and 23%, respectively, on the Finnmarken and Malangen banks. In other areas, where cod fed on capelin in March-April, only single specimens of herring were found in cod. In May, the feeding of cod was diverse. Herring occurred in cod diets in the area from Nordkyn Bank to the northwestern slope of Murmansk Bank, and in large quantities (to 18-40%) in some areas. The weights of herring in cod diets in

1992-1993 were high, 9.6% in each. In more recent years (1994-1996) capelin abundance decreased, and the boundary of the feeding area of cod extended to the east. As a result, cod fed regularly on herring despite the drop in herring abundance. The feeding took place during different seasons: in 1994, in winter-spring, mainly; and 1995-1997, in spring and autumn-winter seasons, and sometimes in summer. However, the weight of herring in cod stomach contents sharply decreased over the period: from 6.8% and 7.3 % (in 1994 and 1995, respectively) to 1.4% and 0.1% (in 1996 and 1997, respectively).

The total annual consumption of herring by cod (in absolute values) was not large, especially when compared to the consumption of capelin. Even in the years of increased cod feeding on herring it did not exceed 190,000-235,000 t, and the maximum was 383,000 t in 1992. Thus, it is evident that the level of herring consumption in the 1980s-1990s was much lower than that in the 1930s-1950s compared to capelin consumption, and in some years was even smaller than that of other major food items such as shrimp, polar cod, and euphausiids.

Summary and Conclusions

The role of Barents Sea herring in the food of cod was assessed from the results of long-term investigations on the relationship of the horizontal and vertical distribution of herring, with herring feeding conditions in the Barents Sea, and on the general availability of food for cod. In the 1980s-1990s, when herring abundance was low, the main areas of their distribution and feeding were the western and central sea. In the western areas, the distribution of herring was mainly associated with *Calanus* biomass, whereas in the central and eastern areas the association was with euphausiid abundance.

Large fluctuations in abundance of capelin, the main food competitor of herring, impacted the availability of herring for cod. In some years, capelin abundance was enhanced by fluctuations in plankton abundance, and this varied between cold and warm periods. Herring in their 2nd and 3rd year (ages 1+ and 2+) were relatively abundant in the Barents Sea during moderately cold (1985-1986) and warm (1991-1993) periods. In those years, food conditions changed the traditional seasonal pattern of herring feeding, fattening, and vertical distribution.

The diversity of seasonal rhythms of feeding and fat accumulation in herring is associated with the variability of their food base, and this results in large differences in their behavior, including vertical migration. These migrations affect the availability of herring for cod, although the main factor affecting the consumption of herring is its relative abundance compared to capelin. This was especially seen in those years when a sudden change in the ratio of herring and capelin occurred in the Barents Sea. At the same time, independent of the level of consumption of capelin, the availability of herring to cod is more associated with the longer duration of herring in the bottom and middle depths during the winter-spring and autumn-winter periods. At the present level of abundance, herring do not occupy a dominant position in the diet of cod. However, given the generally low abundance of fish prey for cod, herring do add some stability to the cod diet.

References

ACFM. 1998. Report of the Northern Pelagic and Blue Whiting Fisheries Working Group. ACFM 18. 276 pp.

Ajiad, A.M. 1990. Variability in stomach contents of cod, collected by demersal and pelagic trawl in the southern part of the Barents Sea. ICES C.M. 1990/G:3, pp. 3-7.

Boldovsky, G.V. 1941. Food and feeding of herring in the Barents Sea. Tr. PINRO 1:219-286. (In Russian.)

Manteifel, B.P. 1941. Plankton and herring in the Barents Sea. Tr. PINRO 7:125-218. (In Russian.)

Marti, Yu. Yu. 1956. The main stages of the life cycle of the Atlanto-Scandian herring. Tr. PINRO 9:5-61. (In Russian.)

Orlova, E.L., E.G. Berestovsky, S.G. Antonov, and others. 1989. The main food interrelations between fishes of the polar seas. Nauka, Leningrad, pp. 182-198. (In Russian.)

Ponomarenko, I. Ya., and N.A. Yaragina. 1985. Seasonal and multi-year dynamics of capelin occurrence in the food of cod in the Barents Sea. In: Feeding and availability of food for fish at different stages of development as a factor of formation of their abundance, growth and concentrations. Moscow, pp. 3-19. (In Russian.)

Rudakova, V.A. 1966. Conditions and main laws of feeding of Atlanto-Scandian herring (*Clupea harengus harengus* L.) in the Norwegian Sea (1951-1962). Tr. PINRO 17:5-54. (In Russian.)

Shutova-Korzh, I.V. 1960. Distribution, growth and maturation of herring of different year-classes in the Barents Sea. In: The Soviet fishery investigations in the seas of the European north. VNIRO-PINRO, pp. 361-370. (In Russian.)

Shutova-Korzh, I.V. 1962. The main results of studying of the Murman herring between 1947 and 1960. Tr. PINRO 14:81-93. (In Russian.)

Shutova-Korzh, I.V. 1966. Changes of areas of many-vertebra and little-vertebra herring in the Barents Sea. Tr. PINRO 17:209-222. (In Russian.)

Tikhonov, V.N. 1939. Winter distribution of herring (*Clupea harengus harengus*) in the southern part of the Barents Sea. Tr. PINRO 4:5-39. (In Russian.)

Zatsepin, V.I., and N.S. Petrova. 1939. Feeding of commercial stocks of cod in the Barents Sea. Tr. PINRO 5. 169 pp. (In Russian.)

Distribution and Feeding Habits of Juvenile Herring (*Clupea pallasii*) in Northern Japan

Masayoshi Sasaki, Ryotaro Ishida, and Takayanagi Shiro
Hokkaido Central Fisheries Experimental Station, Yoichi, Hokkaido, Japan

Abstract

Distribution and feeding habits of wild and artificially produced juvenile herring are described for the Ishikari subpopulation of Pacific herring on the west coast of Hokkaido in Japan. Herring were collected using a beach seine, larva net, small beam trawl, and fishing efforts at the coast of Atsuta, and with set nets in the Ishikari River, in the summers of 1996-1999. Large numbers of wild juveniles, ranging from about 30 to 50 mm in total length, were collected in the surf zone near the river mouth of the Ishikari River (salinity range 0.03-33.65 psu) in late June-early July. From early to middle July larger wild juveniles (50-70 mm) were found in the Ishikari River. These results suggest that in the Atsuta estuary, herring juveniles gather in the surf zone at about 30 mm in length and move to the river when they grow to about 50 mm in total length. Small coastal and brackish copepods (*Paracalanus parvus, Oithona similis, O. atlantica, Clausocalanus pergens, Eurytemora* spp.), harpacticoids, and cladocerans composed the majority of food organisms in the guts of juveniles (including some artificially produced herring) up to 60 mm. However, the composition differed between sampling locations and dates and juveniles greater than 80 mm (all artificially produced herring) fed on larger organisms such as mysids, amphipods, and fishes.

Introduction

For stock enhancement of the Ishikari subpopulation of Pacific herring on the west coast of Hokkaido, artificially produced young herring (averaging 50-74 mm) have been released since 1996. About 100,000 young herring were released on 23 June 1996, 400,000 in 1997, and about 1 million in 1998 and 1999. The releasing was done in the area just outside the port of Kotan (Fig. 1), and the first release dates were 24 June in 1996, 16 June in 1997, 26 May in 1998, and 28 May in 1999. Natural spawning beds have also been found in the study area (Hoshikawa et al. 2001).

To assess the effectiveness of these stock enhancement efforts, it is necessary to examine the suitable body size, timing, and placement of release with reference to the ecology of wild juvenile herring. Few ecological reports on wild juvenile herring are available for the west coast of Hokkaido, so we investigated the distribution and feeding habits of sampled juvenile herring using a beach seine, larva net, and small beam trawl at the coast of Atsuta village, and with set nets in the Ishikari River in summer, from 1996 to 1999.

Material and Methods

Study Area

Our study area is located on the coast of Atsuta in Ishikari Bay. Several small rivers discharge into the area, the largest being the Ishikari River (yearly mean quantity of flow is 200 m^3 per second and maximum is 3,000 m^3 per second) (Fig.1).

Sampling of Herring

In the summer, from 1996 to 1999, larvae and juveniles were collected using a beach seine, larva net, and small beam trawl (Fig.1). The beach seine measured 6.5 × 1.5 m, with 30-m ropes and a mesh size from 5.0 mm to 1.0 mm (mesh size in the cod-end was 1.0 mm). With this sampler we swept from the water surface to a depth of 1.5 m and up to about 50 m from the shoreline. We investigated salinity in sea area in every study transect (Fig. 1) and salinity varied between 0.03 and 34.24 psu.

The larva net (mouth diameter, 130 cm; length, 450 cm; aperture, 0.33 mm) was pulled by oblique tows for about 5-10 minutes from a depth of 5 m to the surface at 1.5 knots in the years 1996 and 1997. A small beam trawl net (with the same dimensions as the beach seine and mesh size in the cod-end of 1.0 mm) was towed for about 5-10 minutes at 1.5 knots on the sea bottom along the shoreline. Fishing using small hooks (about 3 mm) and line by hand was also conducted at the ports of Kotan and Atsuta.

From 1997 onward, a fisherman caught samples of herring larvae and juveniles with small set nets at the mouth of the Ishikari River and about 4 km upstream in the river. In these study transects salinity varied normally between 0 and 33.3 psu. In the surface layer (0-5 m) the salinity was 0.1 psu but in deeper layers up to 33.3 psu (M. Yamaguchi, Hokkaido Central Fisheries Experimental Station, unpubl. data).

Fish Treatment

All samples were immediately preserved in 5% formalin solution at sea. In the laboratory, the total length of herring was measured to the nearest 0.1 mm. If the number of larvae was more than 100, the 100 individuals to be

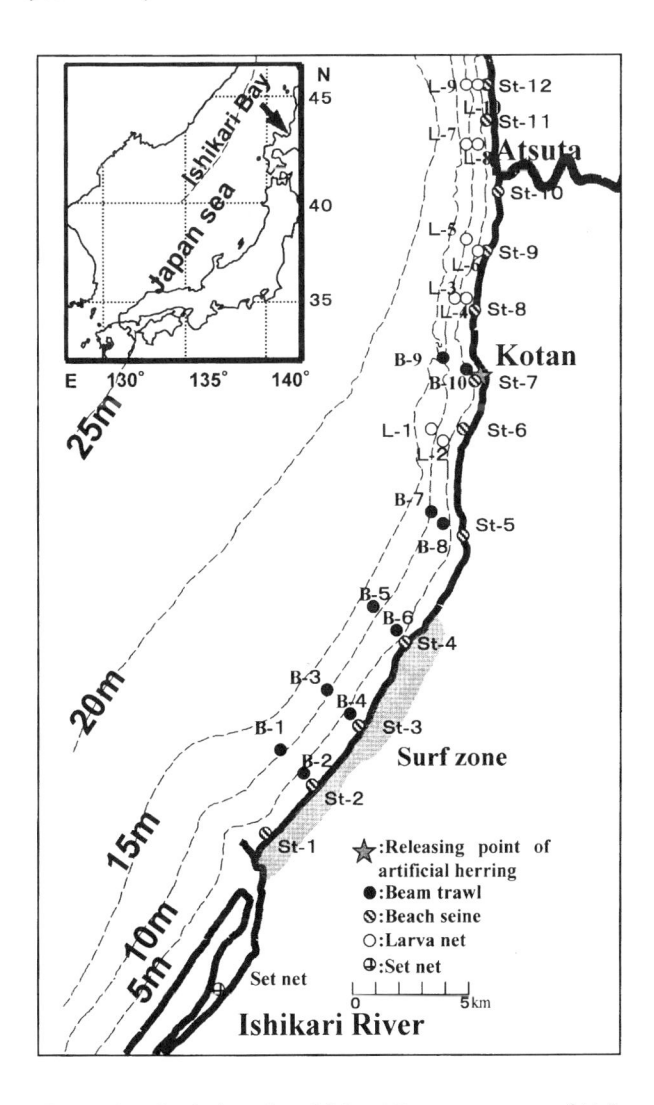

Figure 1. Study location, Ishikari Bay, west coast of Hokkaido Island, Japan.

measured were selected randomly; if the number was less than 100, all individuals were measured. The lengths presented were not corrected for shrinkage in formaldehyde.

Distinction between Artificially Produced Herring and Wild Herring

We distinguished artificial and wild herring by investigating ALC (alizarin complexone) marking on the otoliths through a fluorescence microscope. In some fish ($n = 2,320$) otoliths were removed to observe ALC. The method is presented in detail by Ishida et al. (2001).

Stomach Contents

The food of herring was investigated by examining the stomach contents of 98 samples, including artificially produced herring. The relative volume of organisms of each food species was estimated by inspection through a microscope, taking into account the number of individuals. The contents were examined under 20× to 400× magnification. Prey organisms were identified to major taxonomic groups. Copepods were identified, when possible, to species. Lengths of intact copepods and other prey items were recorded. Fish having empty digestive tracts were omitted from all calculations.

Results

Collection of Herring

No juvenile herring were caught with the small beam trawl net and larva net. Herring were caught mainly with a beach seine in the surf zone near the mouth (Figs. 1 and 2) of the Ishikari River from late June to early July. From late June to mid-July, herring were also collected with set nets in the Ishikari River. From mid-June to late July herring were also caught by fishing at the ports of Kotan and Atsuta (Table 1, Fig. 1).

Size Composition of Wild Herring Caught

Herring caught in the surf zone from late June to early July ranged from about 30 to 50 mm in total length and herring caught with set nets in the Ishikari River ranged about 50 to 70 mm in total length. Herring fished in late July were mainly 70-90 mm in length (Table 1, Fig. 2).

Composition of Diet

Species composition and number of food items per juvenile in the stomach contents from the 11 herring samples are presented in Table 2. A total of 40 food items were found, including diatoms, protozoans, cladocerans, copepods, amphipods, brachiopods, decapods, gastropods, appendicularians, cirripeds, mysids, fish eggs, and fish. The food composition and number of food items per juvenile varied with season and locality.

Herring: Expectations for a New Millennium 105

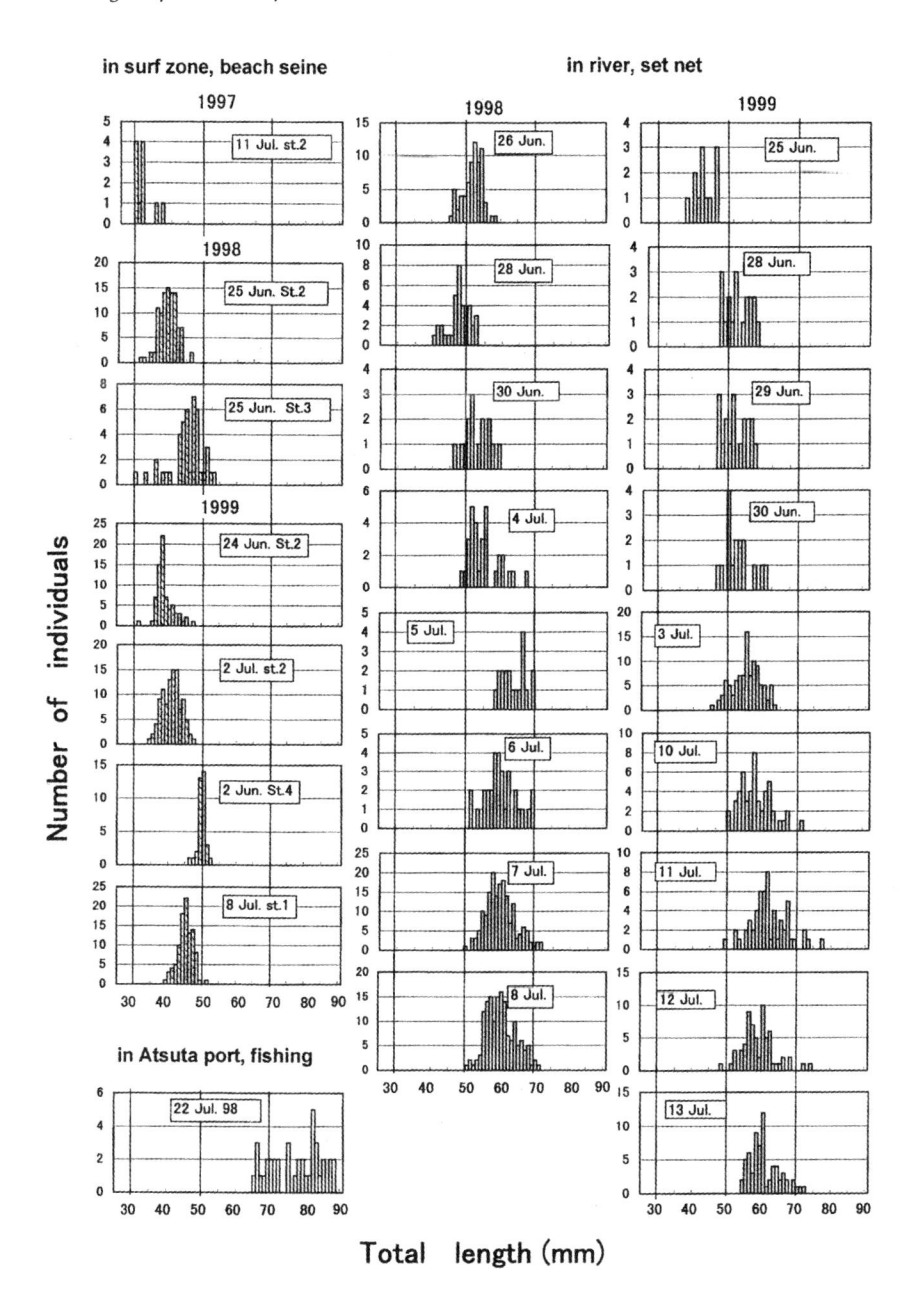

Figure 2. Length distributions of wild herring juveniles collected on the coast at Atsuta village and in the Ishikari River.

Table 1. Number of herring collected with each gear.

		st-1	st-2	st-3	st-4	st-5	st-6	st-7	st-8	st-9	st-10	st-11	st-12	Kotan	Atsuta	River
							Beach seine							Fishing		Set net
1996	25 Jun	–	–	–	–	–	–	–	–	–	–	–	–	–	–	–
	2 Jul	–	–	–	–	–	0(0)	0(0)	0(0)	51(0)	–	–	–	4(43)	24(0)	–
	9 Jul	–	–	–	–	0(0)	0(0)	31(4)	0(0)	51(0)	0(0)	–	–	1(26)	–	–
	12 Jul	–	–	–	–	–	–	–	–	–	–	–	–	–	9(7)	–
1997	27 May	–	0(0)	0(0)	0(0)	0(0)	–	–	–	–	–	–	–	–	–	–
	28 May	–	–	–	–	–	0(0)	0(0)	0(0)	0(0)	0(0)	0(0)	–	–	–	–
	4 Jun	–	–	–	–	–	–	–	–	0(0)	0(0)	–	–	–	–	–
	5 Jun	–	–	–	–	–	–	–	–	–	–	–	–	–	–	–
	17 Jun	–	–	–	–	–	0(0)	0(0)	0(0)	0(0)	0(0)	0(0)	0(0)	–	–	–
	18 Jun	–	0(0)	0(0)	0(1)	0(395)	0(0)	0(0)	0(0)	0(0)	0(0)	0(0)	–	–	–	0(11)
	23 Jun	–	–	–	0(0)	0(0)	–	–	–	–	–	–	–	–	–	–
	1 Jul	–	–	–	–	–	0(0)	0(0)	0(0)	0(0)	0(0)	0(0)	0(0)	–	–	0(10)
	2 Jul	–	–	–	–	0(0)	0(0)	0(0)	0(0)	0(0)	0(0)	0(0)	–	–	–	0(10)
	3 Jul	–	–	–	–	–	–	–	–	–	–	–	–	–	–	0(4)
	5 Jul	–	–	–	–	–	–	–	–	–	–	–	–	–	–	–
	6 Jul	–	–	–	–	–	–	–	–	–	–	–	–	–	–	–
	10 Jul	–	–	–	0(0)	–	0(0)	0(0)	0(0)	0(0)	0(0)	–	0(0)	–	0(24)	0(7)
	11 Jul	–	11(0)	–	–	–	–	–	–	–	–	–	–	–	–	0(4)
	13 Jul	–	–	–	–	–	–	–	–	–	0(479)	–	–	–	–	–
	23 Jul	–	–	–	–	0(0)	0(0)	0(0)	–	0(3)	0(0)	0(0)	0(0)	–	1(13)	–
	25 Jul	–	0(0)	0(0)	0(0)	0(3)	–	0(1)	–	0(0)	0(0)	–	–	–	1(3)	–
1998	27 May	–	0(0)	0(0)	0(0)	0(0)	–	0(0)	–	0(0)	0(0)	–	–	–	–	–
	3 Jun	–	1(0)	0(0)	0(0)	0(0)	–	0(0)	–	–	–	–	–	–	–	–
	10 Jun	–	1999(0)	43(1)	7(3)	0(0)	–	–	–	–	–	–	–	–	–	–
	25 Jun	–	–	–	–	–	–	–	–	–	–	–	–	5(24)	–	68(5)
	26 Jun	–	–	–	–	–	–	–	–	–	–	–	–	–	–	35(0)
	28 Jun	–	–	–	–	–	–	–	–	–	–	–	–	–	–	–
	30 Jun	–	–	–	–	–	–	–	–	–	–	–	–	–	–	15(1)
	2 Jul	–	2(1)	0(0)	3(1)	0(0)	–	0(0)	–	0(0)	0(0)	–	–	2(2)	–	–
	3 Jul	–	–	–	–	–	–	–	–	–	–	–	–	–	–	–

Dash indicates no sampling. Left figures are wild herring and figures in parentheses are artificially produced herring.

Table 1. (Continued.) Number of herring collected with each gear.

		st-1	st-2	st-3	st-4	st-5	Beach seine st-6	st-7	st-8	st-9	st-10	st-11	st-12	Fishing Kotan	Atsuta	Set net River
	4 Jul	–	–	–	–	–	–	–	–	–	–	–	–	–	–	31(7)
	5 Jul	–	–	–	–	–	–	–	–	–	–	–	–	–	–	19(10)
	6 Jul	–	–	–	–	–	–	–	–	–	–	–	–	–	–	31(4)
	7 Jul	–	–	–	–	–	–	–	–	–	–	–	–	–	–	173(9)
	8 Jul	–	–	–	–	–	–	–	–	–	–	–	–	–	–	152(8)
	16 Jul	–	2(0)	0(0)	0(0)	–	–	0(0)	–	–	0(0)	–	–	–	–	–
	21 Jul	–	–	–	–	–	–	–	–	–	–	–	–	–	39(2)	–
	22 Jul	0(0)	0(0)	–	0(0)	–	–	0(0)	–	–	0(0)	–	–	–	–	–
	30 Jul	–	–	–	–	–	–	–	–	–	–	–	–	–	8(0)	–
	5 Aug	–	–	–	–	–	–	–	–	–	–	–	–	–	2(0)	–
1999	26 May	–	0(0)	0(0)	0(0)	0(0)	–	0(0)	–	–	0(0)	–	–	–	–	12(1)
	8 Jun	–	0(8)	0(5)	0(3)	0(0)	–	0(0)	–	–	0(0)	–	–	–	–	4(13)
	24 Jun	0(0)	123(0)	0(2)	0(66)	0(0)	–	0(0)	–	–	0(0)	–	–	–	–	11(1)
	25 Jun	–	–	–	–	–	–	–	–	–	–	–	–	–	–	18(1)
	27 Jun	–	–	–	–	–	–	–	–	–	–	–	–	–	–	16(4)
	28 Jun	–	–	–	–	–	–	–	–	–	–	–	–	–	–	12(29)
	29 Jun	–	–	–	–	–	–	–	–	–	–	–	–	–	–	–
	30 Jun	–	–	–	–	–	–	–	–	–	–	–	–	–	–	–
	1 Jul	–	–	–	–	–	–	–	–	–	–	–	–	–	–	–
	2 Jul	–	1,665(0)	0(0)	35(0)	–	–	0(0)	–	–	0(0)	–	–	–	–	1,748(84)
	3 Jul	–	–	–	–	–	–	–	–	–	–	–	–	–	–	4(10)
	4 Jul	–	–	–	–	–	–	–	–	–	–	–	–	–	–	8(10)
	7 Jul	–	–	–	–	–	–	–	–	–	–	–	–	–	–	–
	8 Jul	2,777(0)	4(0)	0(0)	3(1)	–	–	0(0)	–	–	0(0)	–	–	–	–	68(7)
	9 Jul	–	–	–	–	–	–	–	–	–	–	–	–	–	–	51(9)
	10 Jul	–	–	–	–	–	–	–	–	–	–	–	–	–	–	57(5)
	11 Jul	–	–	–	–	–	–	–	–	–	–	–	–	–	–	65(3)
	12 Jul	–	–	–	–	–	–	–	–	–	–	–	–	–	–	67(5)
	13 Jul	–	–	–	–	–	–	–	–	–	–	–	–	–	–	–
	15 Jul	1(0)	1(0)	0(0)	0(0)	–	–	–	–	–	0(0)	–	–	–	–	–
	6 Aug	0(0)	0(0)	–	–	–	–	–	–	–	0(0)	–	–	–	–	–

Dash indicates no sampling. Left figures are wild herring and figures in parentheses are artificially produced herring.

Table 2. Species composition of food organisms from the guts of juvenile herring.

Sampling area or st.no	st-9	st-7	In river	In river	In river	In river	st-2	In river	st-2	st-3	st-4
Sampling date	2-Jul-96	2-Jul-96	3-Jul-97	5-Jul-97	6-Jul-97	11-Jul-97	11-Jul-97	13-Jul-97	25-Jun-98	25-Jun-98	25-Jun-98
No. examined	10	13	10[b]	10[b]	4[b]	7[b]	10	4[b]	10	10	9
Range of total length (mm)	46-56	44-72	75.1-91.9	80.8-98.6	85.1-98.3	81.8-104.4	29.4-37.1	94.8-104.1	33.7-41.8	27.8-61.3	35.7-61.7
Percent empty guts	0.0	0.0	0.0	50.0	100.0	60.0	0.0	0.0	0.0	0.0	0.0
Mean number of food organisms per herring	118.2	38.2	40.5	32.5	0.0	7.3	439.4	5.3	514.4	773.9	389.1
Mesocalanus tenuicornis cf copepodid										1.3 40.0	2.0 22.2
Paracalanus parvus[a]	11.5 100.0	1.0 1.5					3.3 36.4		31.9 100.0	2.0 10.0	44.1 77.8
Calanoida[a] unidentified	21.1 100.0	5.0 53.8	1.0 20.0	11.0 20.0			18.2 100.0		38.3 100.0	92.7 90.0	42.4 100.0
Pseudocalanus sp.		1.0 7.7					142.5 100.0			54.4 100.0	
Clausocalanus pergens							29.9 90.1		4.9 100.0	21.1 80.0	7.1 77.8
Eurytemora pacifica	64.4 100.0	1.0 15.4					2.0 18.2				
Eurytemora affinis	2.8 100.0	1.0 7.7					14.2 100.0		1.3 30.0		
Eurytemora herdmanii							8.5 100.0			1.0 10.0	
Eurytemora spp.	5.3 80.0			19.5 20.0					1.0 10.0	2.0 30.0	1.0 11.1
Eurytemora larvae									5.0 20.0		
Centropages abdominalis	1.0 10.0						2.0 9.1				
Acartia omorii										2.0 10.0	2.0 22.2
Acartia steueri							16.3 100.0				5.0 11.1
Tortanus discaudatus			5.0 10.0				1.0 9.1				
Tortanus derjugini			5.0 10.0							1.0 20.0	117.1 77.8
Tortanus sp.							186.3 100.0		283.3 100.0	287.6 100.0	105.7 77.8
Oithona similis							2.4 72.7		56.2 100.0	41.8 100.0	3.6 77.8
Oithona atlantica	1.0 10.0	1.0 7.7		2.0 10.0					5.8 100.0	11.8 80.0	
Cyclopoida unidentified & larvae									10.3 60.0		
Microsetella sp.		16.5 100.0									16.9 88.9
Harpacticoida	9.1 90.0	1.0 7.7					12.0 100.0		24.0 100.0	74.6 100.0	13.3 88.9
Gammaridae	1.0 20.0								33.2 100.0	119.6 100.0	10.8 88.9
Cladocera unidentified		1.5 15.3							2.1 100.0	8.0 100.0	1.0 11.1
Podon spp.										1.0 10.0	
Evadne spp.		1.0 7.7								1.0 10.0	
Crustacea nauplius	1.0 10.0	2.2 38.4							3.5 20.0		
Pelecypoda larvae		1.5 15.4									
Gastropoda larvae		1.0 7.7									
Cirripedia larvae (Cypriform)					6.3 42.9			1.0 25.0			
Mysidacea adult larvae			2.0 20.0		1.0 14.3				4.7 70.0	7.0 10.0	
Amphipoda			4.0 20.0							3.0 30.0	1.0 22.2
Decapoda larvae									2.6 100.0	38.6 90.0	14.4 88.9
Oikopleura spp.		2.0 20.0							5.0 10.0		
Polychaeta									1.2 50.0		
Eggs			23.5 20.0				1.0 27.3			2.6 50.0	1.7 33.3
Fish								4.3 100.0			
Diatoms		1.5 15.3									

[a] *Paracalanus* sp. and *Clausocalanus* sp., considered primarily *Paracalanus parvus* by size and form.
[b] All artificially produced herring.
Numerical value on the left and right side of each position represents the number of food items per juvenile and the percentage of juvenile feeding, respectively.

On 2 July 1996, all fish fed on *Eurytemora pacifica* at station 9 and harpacticoids at station 7 near Kotan.

The number of food items per juvenile collected at the river in 1997 was low (0-40.5 individuals) and the percentage of empty guts was high. On 3 July 1997 herring fed on larger copepods (*Tortanus discaudatus* and *T. derjugini*) and larger organisms like mysids, fish, and amphipods. On 5 July 1997 *Eurytemora* spp. were the most abundant food items and on 6 July the stomachs of all herring were empty. On 11 July 1997 artificial herring fed on larger organisms, such as mysids, fish, and amphipods. On the other hand, also on 11 July 1997, smaller wild herring caught at station 2 (surf zone) fed on *Oithona similis* and *Paracalanus parvus* (including unidentified Calanoida). Over 100 individuals of these items were ingested per juvenile, and *Clausocalanus pergens* was also ingested with a mean value of 30 individuals per juvenile.

In late June 1998, there were about 400-790 individuals food items per juvenile in each sample. Abundant food items were *Paracalanus parvus* (including unidentified Calanoida), *Oithona similis*, *O. atlantica*, cladocerans, and *Oikopleura* spp., and especially *Oithona similis* (over 100 individuals in each sample) and *Paracalanus parvus* (over 50 individuals).

Size of Food Organisms

Size of food eaten by herring is presented in Fig. 3. The mean size of *Oithona similis* eaten was 0.4 mm in length. *Paracalanus parvus* and *Clausocalanus pergens* were 0.5 mm in length. Harpacticoids, *Acartia* spp., *Evadne* sp., *Podon* sp., and *Oithona atlantica* were similar sizes (0.6-0.7 mm in length). The mean lengths of *Eurytemora pacifica* and *E. affinis* were about 0.8 mm and 0.9 mm, respectively. *Tortanus discaudatus* and *T. derjugini* were about 1.6 mm and 1.8 mm in length, respectively. Amphipods averaged 2.0 mm, mysids 7.0 mm, and fish 9.2 mm.

Food Organisms of Herring Size Classes

The relation of food items by size in 12 size classes and their relation to fish size is presented in Figs. 4 and 5. Of herring examined for stomach contents, most of the juveniles up to 60 mm were wild (only one artificial herring in the 55-60 mm group), whereas most beyond 60 mm were produced artificially (only one wild herring in the 60-70 mm group). The major food items of the fish in the 25-30 mm, 30-35 mm, and 35-40 mm size classes were *Oithona similis* and *Paracalanus parvus*. *Acartia* spp. and *Clausocalanus pergens* were also common. In the 40-45 mm size class, *Oithona similis* and *Paracalanus parvus* were still the largest component, but harpacticoids were also a large component. In the 45-50 mm group, *Eurytemora pacifica, Oithona atlantica, Oithona similis,* harpacticoids, and cladocerans were the largest component. *Oikopleura* sp., *Microsetella* sp., and harpacticoids were about 10%. In 50-55 mm herring *Eurytemora pacifica, Oithona atlantica,* and cladocerans were the largest components and the

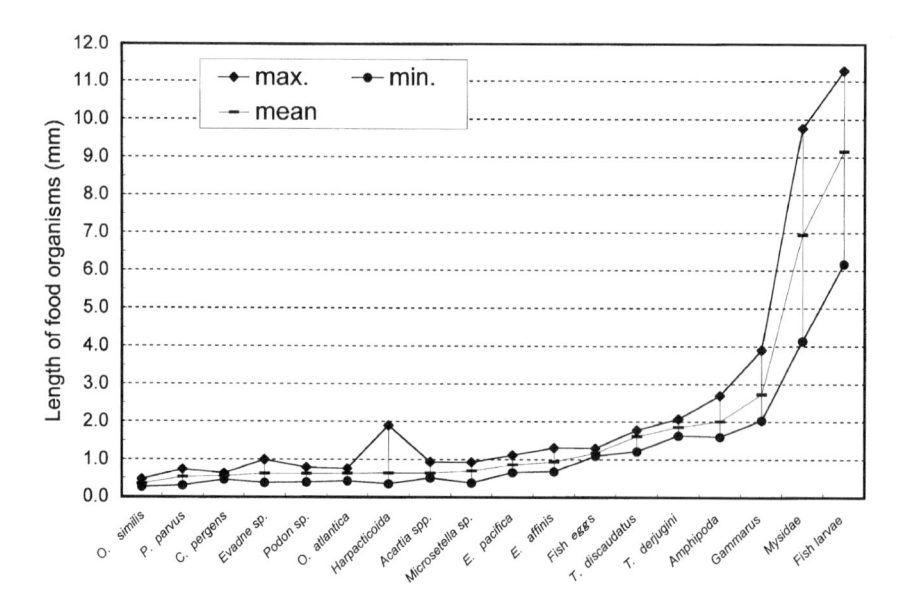

Figure 3. Size of the main food organisms taken by herring juveniles.

percentage of *Oithona similis* and *Paracalanus parvus* decreased. In the 60-70 mm size class harpacticoids dominated, whereas *Oithona similis*, *Tortanus discaudatus*, and harpacticoids were the largest components for 70-80 mm herring. In size classes larger than 80 mm *Tortanus derjugini*, fish, mysids, and amphipods were dominant.

Relation between the Size of Food Organisms and the Size of Herring

The size range in body length of the food organisms eaten by herring is presented in Fig. 5. It is apparent that for herring up to 42 mm in length, the maximum length of food organisms was about 1.0 mm. Thereafter, the maximum length of the food organisms in fish of 40-60 mm increases stepwise; 47-48 mm herring prey have a maximum length of about 1.5 mm, and 53-54 mm herring prey have a maximum length of about 2.0 mm. For herring greater than 70 mm in length the maximum length of the food organism increased greatly.

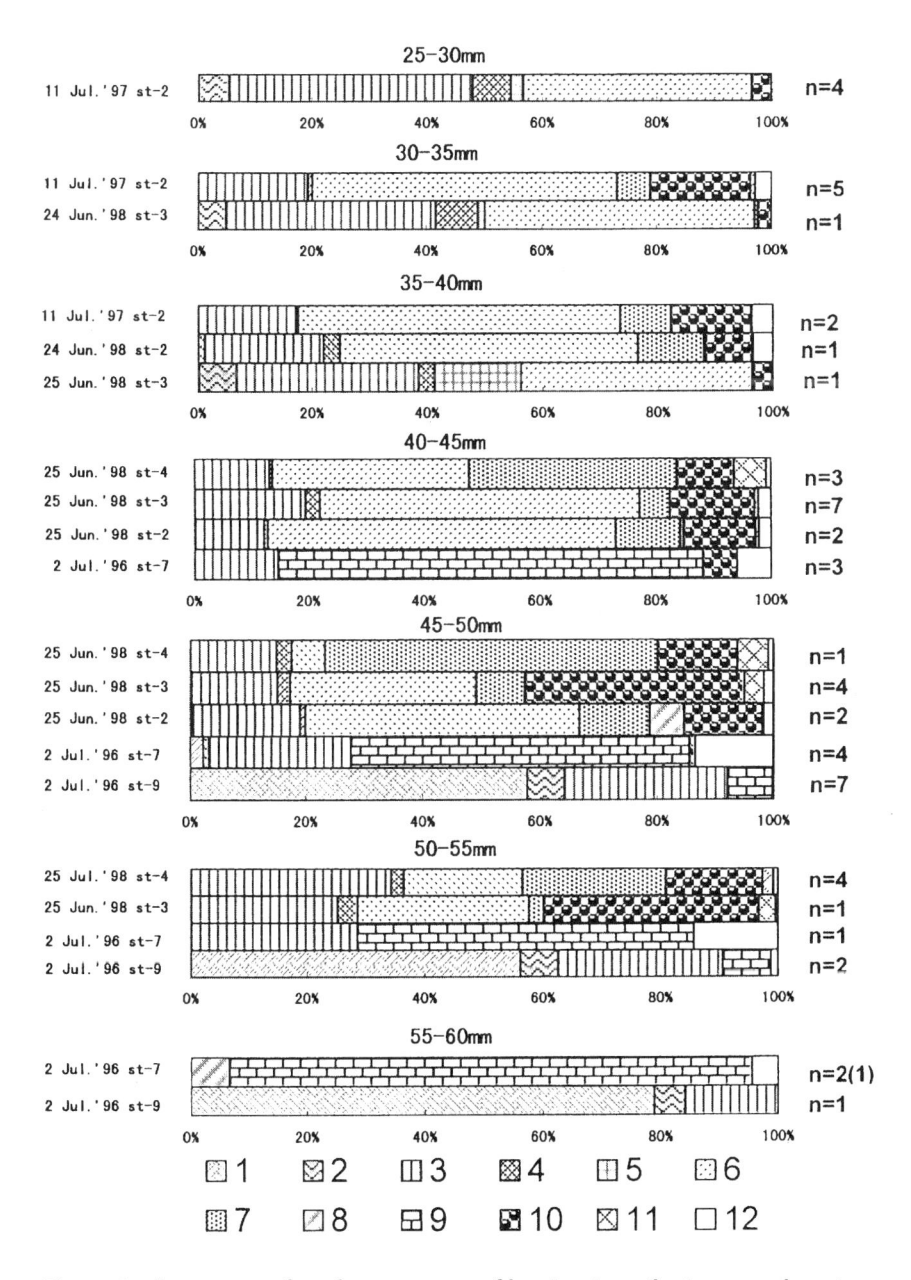

Figure 4. Percentage of total gut contents of herring juveniles in seven class sizes by 5-mm intervals. n = total number measured; in parentheses is number of artificially produced herring. (1) *Eurytemora pacifica*; (2) *Eurytemora spp.*; (3) *Paracalanus parvus*; (4) *Clausocalanus pergens*; (5) *Acartia spp.*; (6) *Oithona similis*; (7) *Oithona atlantica*; (8) *Microsetella sp.*; (9) *Harpacticoida*; (10) *Cladocerans*; (11) *Oikopleura sp.*; (12) others.

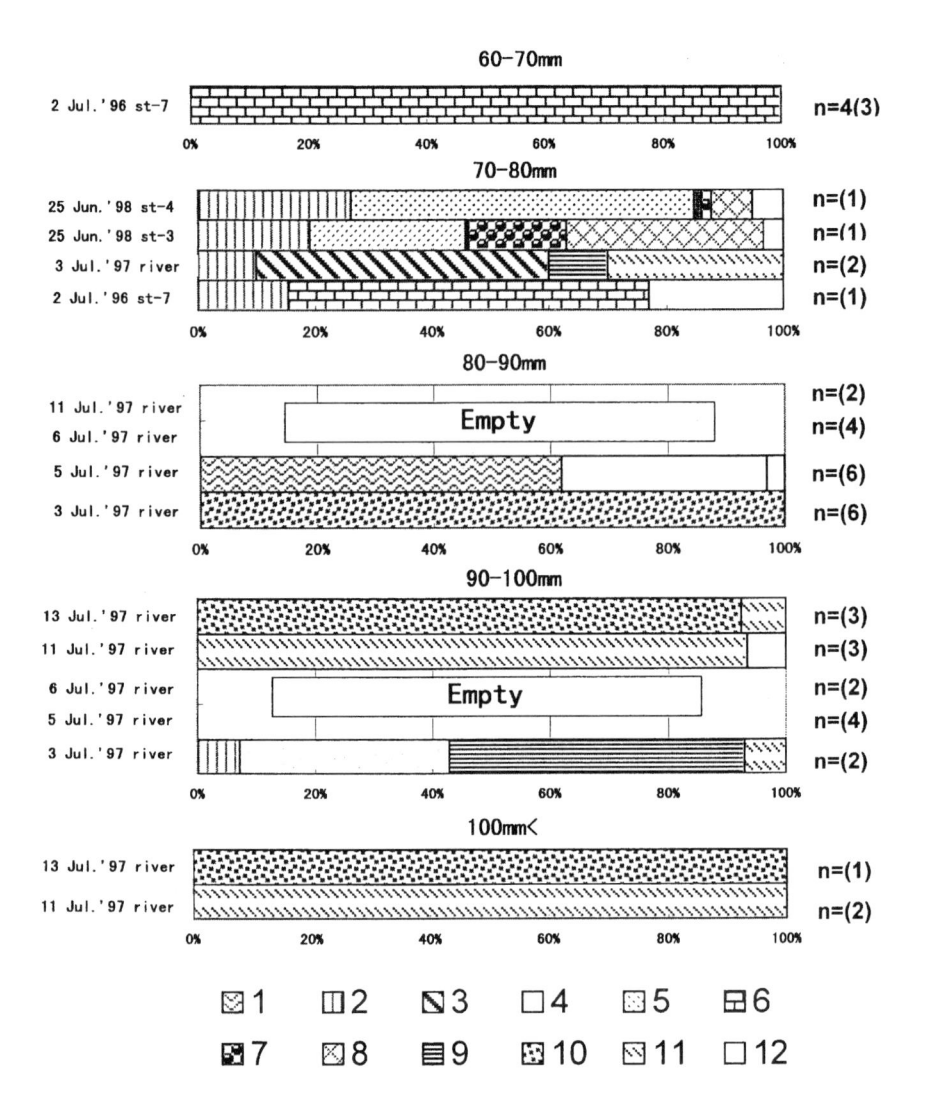

Figure 5. Percentage of total gut contents of herring juveniles in five classes by 10-mm intervals. n = total number measured; numbers in parentheses are artificially produced herring. (1) Eurytemora spp.; (2) Paracalanus parvus; (3) Tortanus discaudatus; (4) Tortanus derjugini; (5) Oithona similis; (6) Harpacticoida; (7) Cladocerans; (8) Oikopleura sp.; (9) Amphipoda; (10) Fish; (11) Mysidae; (12) others.

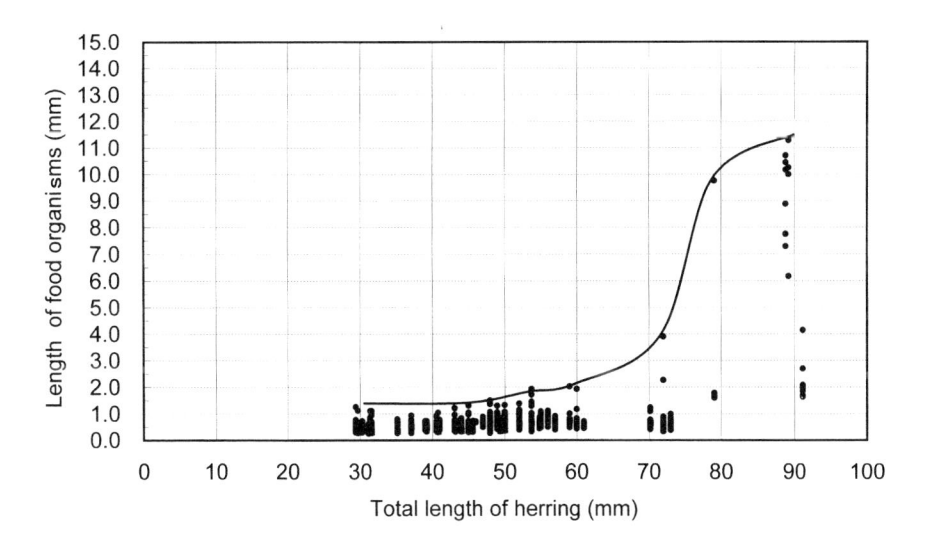

Figure 6. Size range in body length of food organisms taken from the guts of herring juveniles.

Discussion

Distribution of Larvae and Juvenile Herring

In our investigations, juveniles 30-50 mm in length were mainly caught in the surf zone of the sandy beach near the Ishikari River from late June to early July (Table 1, Fig. 2). Thereafter juveniles were not caught in this area, and in early to mid-July larger juveniles (50-70 mm) were found in the Ishikari River (Table 1, Fig. 2). Ishida (Hokkaido Central Fisheries Experimental Station, Hokkaido, pers. comm., Dec. 1999) confirmed that the hatching date of juveniles caught in the surf zone and juveniles caught in the river were almost the same. Thus we suppose that in the Atsuta estuary juvenile herring gather in the surf zone at about 30 mm in length, stay for 1 or 2 weeks there, and then move to the river when they grow to about 50 mm in total length.

This appearance in the estuary of juveniles 30 mm in length was reported from the northwest coast of Sakhalin (Yamamoto 1949), the east coast of the Korean Peninsula (Uchida et al. 1958), and British Columbia in Canada (Hourston 1958, Hay and McCarter 1997). We therefore suggest that Pacific herring migrate to estuaries when they grow to about 30 mm in length. A high amount of juvenile herring is also found in shallow waters in the Baltic Sea (Urho and Hildén 1990; Winkler 1996; J. Määriä, University of Turku, Finland, manuscript in review) and in the Atlantic Ocean (Beyst et al. 1999).

Feeding Habits

As reported by Wailes (1936), Pokrovskaya (1954), and Sherman and Honey (1971), the food composition and number of food items per juvenile herring differed between sampling location and date, as in our study (Table 2). As the juveniles grew the size of the food organisms they consumed increased and size and type of food varied greatly (Figs. 3-6). In juveniles 30-45 mm in length the smallest food organisms such as *Paracalanus parvus* and *Oithona similis* were abundant in different areas in successive years (Fig. 4), so these species appear to be important food organisms. The principal prey of 45-60 mm juvenile herring varied, but the food organisms were almost the same size (Figs. 3 and 4) and were very common in brackish or coastal water. The principal prey items for 70-mm juveniles were harpacticoids, *Tortanus discaudatus,* and *T. derjugini*; these are larger than copepods and were fed on mainly by juveniles up to 60 mm (Figs. 3 and 5). For juveniles greater than 80 mm in length the number of food items decreased and the prey shifted from small crustaceans such as copepods, harpacticoids, and cyclopoids to larger organisms like mysids, fish, and amphipods (Figs. 3 and 5). This suggests that herring juveniles select prey of a particular size, as mentioned by Pokrovskaya (1954) and Sherman and Honey (1971), and that the feeding habits change drastically between 70 mm and 80 mm in size.

Acknowledgments

This research was funded by the project "Stock enhancement of neritic herring in Japan Sea of Hokkaido" of Hokkaido prefectural government. The authors thank the staff of Ishikari fishery extension office for their help during field studies at Atsuta and Mr. T. Sonoda for assistance with sampling in the Ishikari River. We also thank K. Hirano for valuable suggestions for the identification of food organisms. We also thank Dr. A. J. Paul and an anonymous referee who provided constructive comments on the manuscript. We are also grateful to Dr. Juha Kääriä for valuable comments.

References

Beyst, B., J. Mees, and A. Cattrijsse. 1999. Early postlarval fish in the hyperbenthos of the Dutch Delta (southwest Netherlands). J. Mar. Biol. Assoc. U.K. 79:709-724.

Hay, D.E., and P.B. McCarter. 1997. Larval distribution, abundance, and stock structure of British Columbia herring. J. Fish Biol. 51:155-175.

Hoshikawa, H., K. Tajima, T. Kawai, and T. Ohtsuki. 2001. Spawning bed selection by Pacific herring (*Clupea pallasii*) at Atsuta, Hokkaido, Japan. In: F. Funk, J. Blackburn, D. Hay, A.J. Paul, R. Stephenson, R. Toresen, and D. Witherell (eds.), Herring: Expectations for a new millennium. University of Alaska Sea Grant, AK-SG-01-04, Fairbanks. (This volume.)

Hourston, A.S. 1958. Population studies on juvenile herring in Barkley Sound, British Columbia. J. Fish. Res. Board Can. 15(5):909-960.

Ishida, R., M. Sasaki, S. Takayanagi, and H. Yoshida. 2001. Relationship between date of hatching and growth of herring larvae and juveniles in Ishikari Bay. In: F. Funk, J. Blackburn, D. Hay, A.J. Paul, R. Stephenson, R. Toresen, and D. Witherell (eds.), Herring: Expectations for a new millennium. University of Alaska Sea Grant, AK-SG-01-04, Fairbanks. (This volume.)

Pokrovskaya, I.S. 1954. Young herring feeding along southwestern Sakhalin coast. Izv. TINRO 43:202-204. (In Russian.)

Sherman, K., and K.A. Honey. 1971. Seasonal variations in the food of larval herring in coastal waters of central Maine. Rapp. P.V. Reun. Cons. Int. Explor. Mer 160:121-124.

Uchida, K., S. Imai, S. Mito, Y. Huzita, M. Ueno, Y. Shouzima, T. Tida, S. Tahuku, and K. Dotu. 1958. Studies on the eggs, larvae and juveniles of Japanese fishes. Ser. Second Lab. Fish. Biol. Fish. Dep. Fac. Agric. Kyushu Univ., Fukuoka. 89 pp. (In Japanese.)

Urho, L., and M. Hildén. 1990. Distribution patterns of Baltic herring larvae, *Clupea harengus* L., in the coastal waters off Helsinki, Finland. J. Plankton Res. 12:41-54.

Wailes, G.H. 1936. Food of *Clupea pallasii* in southern British Columbia waters. J. Biol. Board Can. 1(6):477-486.

Winkler, H. 1996. Peculiarities of fish communities in estuaries of the Baltic Sea: The Darss-Zingst-Bodden chain as an example. Limnologica 26:199-206.

Yamamoto, K. 1949. Ecological study of fishes in northern Japan. Sci. Rep. Hokkaido Fish. Hatch. 4(1):16-26. (In Japanese.)

Predation by Smelt (*Hypomesus japonicus*) on Herring Larvae (*Clupea pallasii*) in Karagin Bay

Petr M. Vasilets
Kamchatka Research Institute of Fisheries and Oceanography (KamchatNIRO), Petropavlovsk-Kamchatsky, Russia

Abstract

Food composition of Japanese surf smelt (*Hypomesus japonicus*) was examined from collections taken in Karagin Bay on 10-11 June 1996. The major prey of smelt were Pacific herring larvae (*Clupea pallasii*) of 13-18 mm in length, which on average made up 97% of stomach content mass. The maximum number of larvae in a stomach was 700 (averaging 276 herring larvae over 22 surf smelt). This was the first documentation of predation by Japanese surf smelt on Pacific herring larvae in Karagin Bay.

Introduction

Karagin Bay is a spawning area for a large population of Pacific herring (*Clupea pallasii*) and the location of early embryogenesis. (Naumenko 1996). It is known that one of the factors affecting survival of herring is predation (Karpenko and Maksimenkov 1991). The Japanese surf smelt (*Hypomesus japonicus*) is one of the highly abundant species inhabiting Karagin Bay (S.G. Korostelev, Kamchatka Research Institute of Fisheries and Oceanography, 18 Naberezhnaya, Petropavlovsk-Kamchatsky, 683600, Russia, pers. comm., Mar. 1999). Prior investigations showed that the fish component of the surf smelt diets was often significant (Karpenko and Vasilets 1996). Therefore, smelt inhabiting Karagin Bay might be an important fish predator due to its high abundance in certain years. Because there are no reports of predation by Japanese surf smelt on Pacific herring larvae in Karagin Bay in the literature, we investigated this phenomenon.

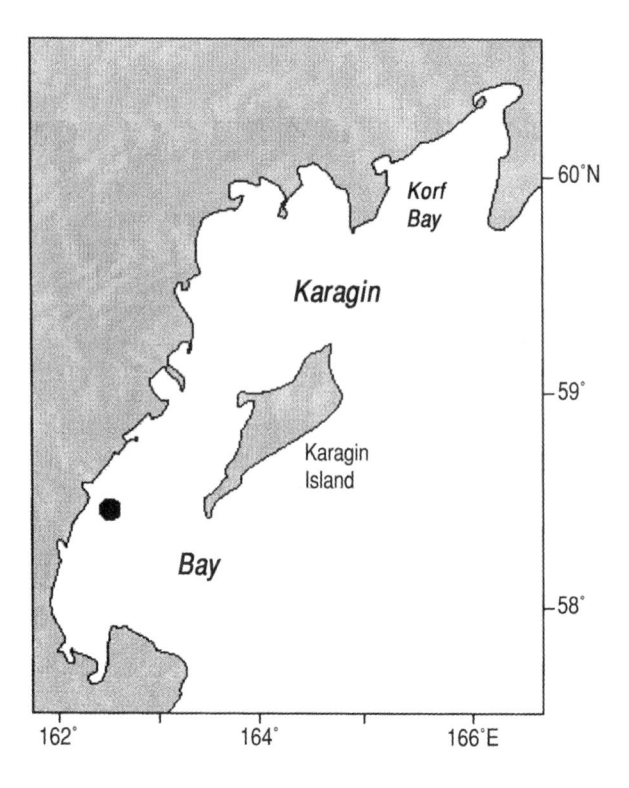

Figure 1. Location of sampling.

Materials and Methods

Twenty-two smelt were caught during a research fishery for sockeye salmon by purse seine in Karagin Bay on June 10-11, 1996 at 58° 42′N and 162° 29′ E (Fig. 1). Biological analysis of the fish was performed just after capture. Fish stomachs were preserved in 10% formalin. We weighed and identified the stomach contents for each individual. Prey specimens were identified to species. We measured the number, length, and weight of each prey item. Average frequency *(F)* and average fraction weight *(W)* were estimated for each prey item. We also calculated the stomach content index *(SCI)* as an indicator of stomach fullness (Takahashi et al. 1999) using the following equation:

$$SCI = SW/BW \times 100,$$

where *SW* is the wet weight (g) of the stomach contents and *BW* is the wet weight (g) of the body of the fish.

Table 1. Food composition of Japanese surf smelt.

Food item	W	F
Copepods	0.06	9.1
Gammarids	1.30	4.5
Shrimps	0.02	9.1
Crab larvae	9.80	36.4
Herring larvae	87.82	95.5
Sole larvae	0.66	22.7
Pisces, unidentified	0.34	18.2
Stomachs examined (no.)	22	
Fish feeding (%)	95.5	
Stomach content index (%)	42.1	
Average standard length (mm)	193	

W = weight of food items (%), F = frequency of food items in stomachs (%).

Results

The 22 surf smelt caught on 10-11 June ranged in size from 163 to 204 mm and from 46 to 82 g (averaging 193 mm and 66 g). All but one of the 22 smelt had food in their stomachs. Maximum SCI was 15% (average 4.21%). Herring larvae made up 87.8% of stomach contents by mass. Other prey items were other fish larvae, and larvae of crabs, shrimps, gammarids, and copepods (Table 1). All of the smelt we examined fed on herring larvae, with the exception of the one fish with an empty stomach. The maximum number of larvae observed in a stomach was 700 (averaging 276).

Although this predation rate is high, there has not yet been reliable evidence that a decrease of herring abundance is associated with removal of herring larvae by Japanese surf smelt. More data should be collected on the abundance of Japanese surf smelt itself and the duration and rate of consumption of herring larvae.

References

Karpenko, V.I., and V.V. Maksimenkov. 1991. The impact of Pacific salmon on herring survival in the western Bering Sea. Proceedings of the International Herring Symposium. University of Alaska Sea Grant, AK-SG-91-01, pp. 445-449.

Karpenko, V.I., and P.M. Vasilets. 1996. Biology of smelt (Osmeridae) in the Korf-Karagin coastal area of the southwestern Bering Sea. In: O.E. Mathisen and K.O. Coyle (eds.), Ecology of the Bering Sea: A review of Russian literature. University of Alaska Sea Grant, AK-SG-96-01, Fairbanks, pp. 217-235.

Naumenko, N.I. 1996. Stock dynamics of western Bering Sea herring. In: O.E. Mathisen and K.O. Coyle (eds.), Ecology of the Bering Sea: A review of Russian literature. University of Alaska Sea Grant, AK-SG-96-01, Fairbanks, pp. 169-175.

Takahashi, K., T. Hirose, and K. Kawaguchi. 1999. The importance of intertidal sand-burrowing peracarid crustaceans as prey for fish in the surf-zone of a sandy beach in Otsuchi Bay, northeastern Japan. Fish. Sci. 65(6):856-864.

Estimating Whole Body Energy Content for Juvenile Pacific Herring from Condition Factor, Dry Weight, and Carbon/Nitrogen Ratio

A.J. Paul and J.M. Paul
University of Alaska Fairbanks, Institute of Marine Science, Seward Marine Center, Seward, Alaska

T.C. Kline
Prince William Sound Science Center, Cordova, Alaska

Abstract

During 1997 the seasonal changes in whole body energy content (WBEC) of Pacific herring (*Clupea pallasii* Valenciennes 1847) from Simpson Bay, Prince William Sound, Alaska, were examined. For fish ≤150 mm standard length (SL) the relationship between SL and whole body weight (wt) was strongly related ($r^2 \geq 0.94$). However, seasonally WBEC (kJ/g wet wt or kJ/g dry wt) exhibited a wide range of values and a weak relationship to wet weight condition factor ($r^2 \leq 0.56$). The WBEC showed a strong linear relationship between energy density and whole body percent dry weight ($r^2 \geq 0.94$). The results show that seasonal estimates of WBEC cannot be accurately derived from wet weight values or the condition factor, but changes in energy density can be estimated from the whole body percent dry weight data with potential errors of ≤0.5 kJ/g. The relationship between whole body carbon/nitrogen ratio and WBEC wet weight was linear ($r^2 = 0.79$). The percent carbon showed some relationship ($r^2 \approx 0.7$) to WBEC and percent dry weight but the nitrogen content of tissues was not as predictable ($r^2 \approx 0.4$).

Introduction

In Prince William Sound, and throughout much of Alaska, the Pacific herring, *Clupea pallasii,* has supported important fisheries for roe, bait, and food, and is a major prey for several other species. Recently an ecosystem study, Sound Ecosystem Assessment (SEA), examined the somatic energy content of herring (Paul et al. 1997, 1998; Paul and Paul 1998a, 1998b, 1999). In Prince William Sound herring spawn in April and May and metamorphose about July (Paul and Paul 1998b) when the water column is stratified and primary productivity is generally nutrient-limited. It is during the summer season that the young-of-the-year herring play a major role as forage species supporting a variety of seabirds, marine mammals, and other fishes. In the winter, the juveniles feed, but rely primarily on stored energy to survive (Foy and Paul 1999).

Currently there are several models that map the energy flow of forage species to the upper trophic level organisms. Unfortunately, the measurement of the energetic content of prey species is time consuming and expensive. In most bony fish the WBEC (whole body energy content) can be estimated from the body's percent dry weight (Hartman and Brandt 1995) once the formulas relating these two parameters have been derived. In this study we derived these formulas for juvenile Pacific herring collected during different months. We also examined the relationship between wet weight condition factor vs. the carbon/nitrogen ratio, WBEC, and percent dry weight for Pacific herring ≤150 mm SL.

Methods

During February, March, May, July, August, and October of 1997 juvenile herring were captured for WBEC from Simpson Bay, Prince William Sound, Alaska (Fig. 1). Day of capture and sample sizes by age appear on the figures. Juvenile herring were captured with 50-m-diameter by 4-m-deep purse seines with 3-mm stretch mesh. At each collection site the nets were set three times to capture specimens. After capture all fish were immediately frozen in seawater and kept frozen until they were processed.

In the laboratory the fish were partially thawed and all fish were measured for SL to the nearest 1 mm, then weighed to the nearest 0.1 g. Condition factor was determined with the equation: CF = (whole body wet weight in grams $\times$ $100/SL^3$). Individual herring were freeze dried whole until no moisture was apparent. Next they were dried further in a convection oven at 60°C until they reached a constant weight. Individual tissue wet and dry weight values were used to calculate the moisture content of every fish. Dried tissues were ground in a mill and measurements of caloric content were made using bomb calorimetry. All calorimetric samples were weighed to the 0.0001 g with a single sample burned per fish. The WBEC of each individual was determined (kJ/g wet wt and kJ/g dry wt).

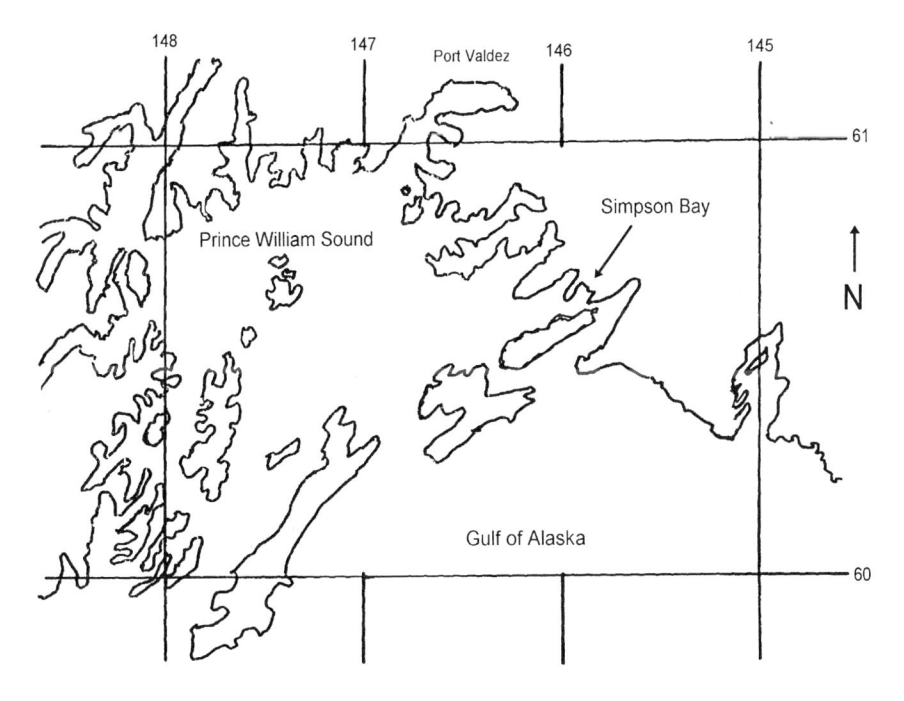

Figure 1. Map of the study area in Prince William Sound, Alaska, showing Simpson Bay where herring were captured for analysis of somatic energy content.

The ash weight of 72 herring captured in Prince William Sound during 22-26 October 1997 was measured. Dried ground whole body samples from each fish were weighed to 0.00001 g in a weighed porcelain crucible with a loose top and a subsample combusted gradually over 4 hours to 600°C.

The tissue from the herring used for WBEC analysis was also analyzed for carbon (C) and nitrogen (N) content. For this report all the fish from various sampling dates were combined for C or N data presentation. Powdered samples of juvenile herring used in the WBEC analysis ($N = 588$) were sent to the University of Alaska Fairbanks Stable Isotope Facility where they were analyzed for C and N stable isotope ratios in duplicate using a continuous flow isotope ratio mass spectrometer interfaced with an automated sample combustion and purification unit (Europa Scientific 20/20 equipped with a Roboprep unit). Quality assurance and quality control protocols included analyses of laboratory standards before and after every five samples as well as the sample duplication. Poorly replicated samples (difference in delta values 0.6 ⁰/oo) were re-run. A single isotopic analysis generated the following data: $^{13}C/^{12}C$ and $^{15}N/^{14}N$ ratios expressed in standard delta units, $\delta^{13}C$ and $\delta^{15}N$, respectively; and percent C and percent N. The percent C and percent N data were used to calculate C/N atomic ratios. The data presented consist of mean $\delta^{13}C$, $\delta^{15}N$, and C/N of replications. The fish used for C/N measurements were aged by reading the scales and only juvenile fish were included.

Results

The dry weight/wet weight ratio was not related to length (Figs. 2 and 3). The relationship between wet condition factor and WBEC (Figs. 2 and 3) was generally not a credible one. This poor relationship reflects the wide range in values for moisture content and WBEC of individuals from a single collection. When WBEC and condition factor were related the correlation coefficient r^2 was ≤0.2 during March, May, August, and October, while in February and July values for r^2 ranged from 0.42 to 0.56. Thus, WBEC cannot be accurately estimated from wet weight condition factor. In Pacific herring growth in length and mass are influenced by different factors, so there is a notable variability in mass-at-age (Figs. 2 and 3; Tanasichuk 1997).

The relationship between ash weight and length was curvilinear ($r^2 = 0.86$), but there was variability in the ash weights for any given length (Fig. 4). This diversity of ash weights for a given length reflects individual differences in the robustness of the skeleton.

There was a noticeable fluctuation in the whole body percent dry weight with the lowest values (17.6 and 18.3%) in August and May and highest values (22.3 and 22.6%) in July and October (Table 1). During all months there was a strong linear relationship between WBEC and percent dry weight with $r^2 \geq 0.94$ (Table 1). The relationship of WBEC and the percent dry weight in the body is shown for fish from one net set in Fig. 5. Note that fish with 20-21% dry weight values have WBEC of 3.8-4.3. In estimating the energy transfer to predators, or the potential of young-of-the-year herring to survive the winter fast, errors of 0.5 kJ/g wet wt would markedly influence the results.

The relationship between whole body C/N ratio and WBEC wet weight was linear, $r^2 = 0.79$ (Fig. 6). A reasonable relationship between C/N and WBEC has been seen in other pelagic fish such as walleye pollock (Harris et al. 1986). However, at any WBEC there was a wide range of C/N ratios. The percent C showed a linear relationship ($r^2 \approx 0.7$) to WBEC (Fig. 7) and percent dry weight (Fig. 8), but the percent N content of tissues was not so predicable ($r^2 \approx 0.4$; Figs. 7 and 8).

Discussion

The ability to estimate seasonal WBEC from the percent dry weight appears to be common in several species of bony fish (Hartman and Brandt 1995). For Clupeiformes in general the relationship between these two parameters is WBEC kJ/g wet wt = 0.3286 (% dry wt) – 2.522; $r^2 = 0.82$ (Hartman and Brandt 1995). Using this formula and assuming dry weight

Herring: Expectations for a New Millennium 125

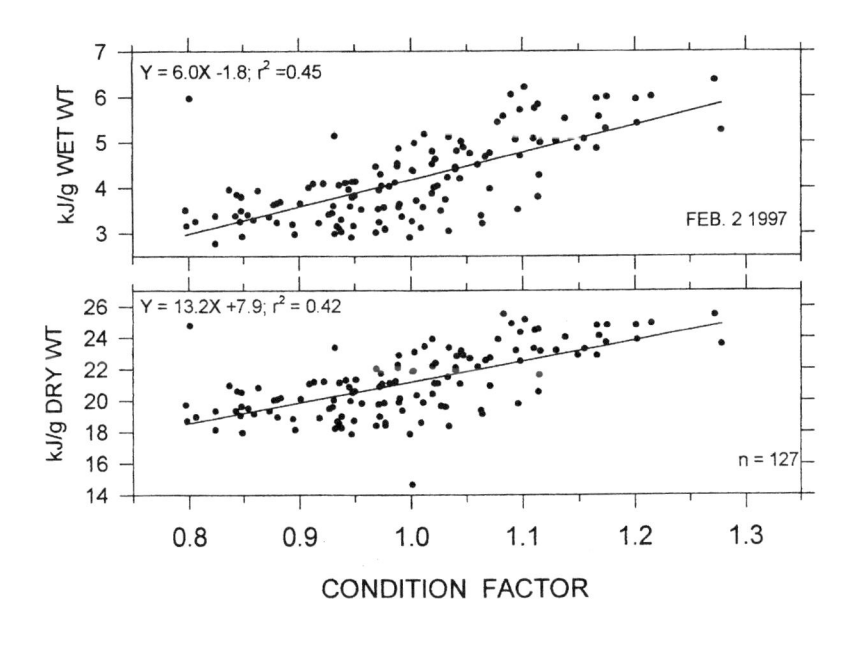

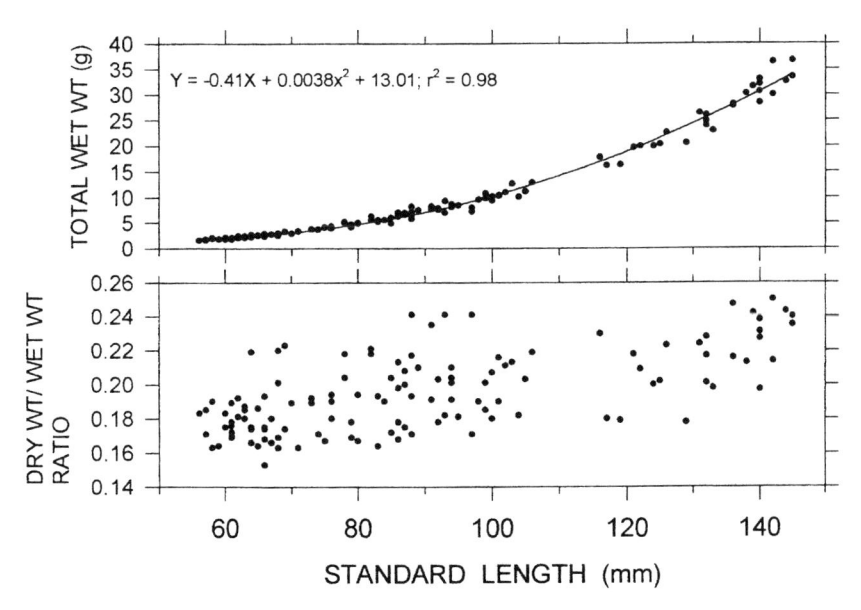

Figure 2. Whole body energy content kJ/g wet weight (top) or kJ/g dry weight (second from top) vs. wet weight condition factor; standard length vs. whole body wet weight (third from top), and the percentage of whole body weight that is dry matter (bottom) for individual Simpson Bay herring captured in February.

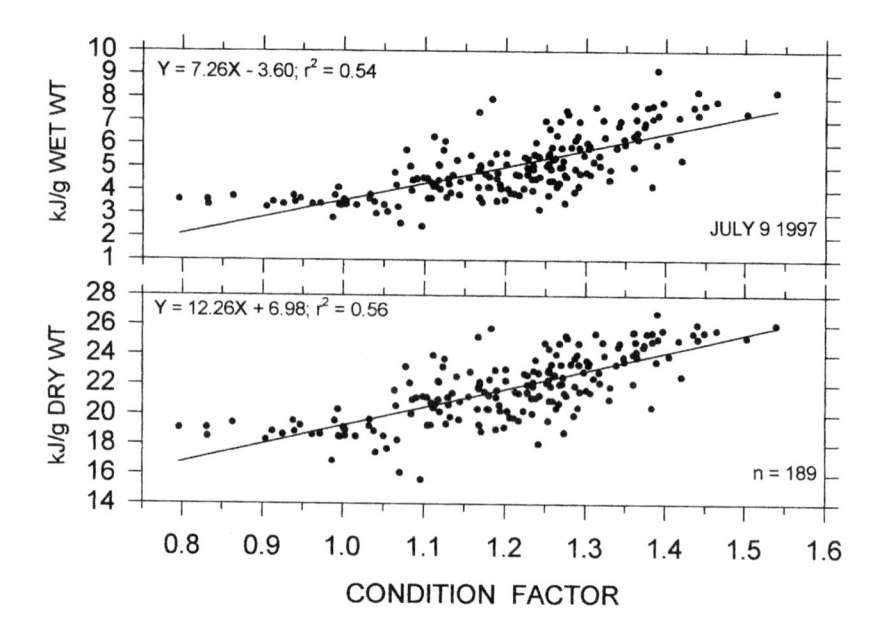

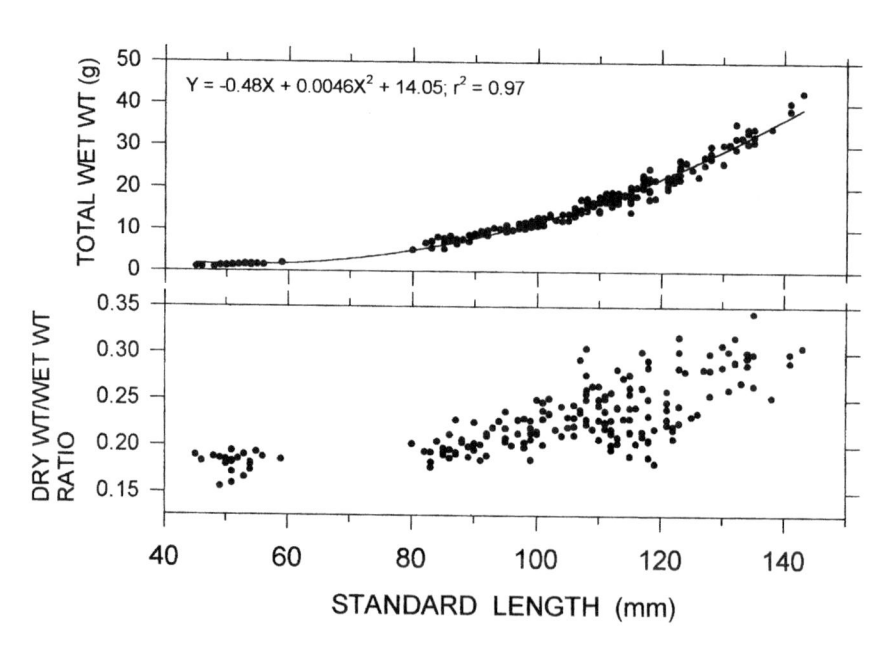

Figure 3. Whole body energy content kJ/g wet weight (top) or kJ/g dry weight (second from top) vs. wet weight condition factor; standard length vs. whole body wet weight (third from top), and the percentage of whole body weight that is dry matter (bottom) for individual Simpson Bay herring captured in July.

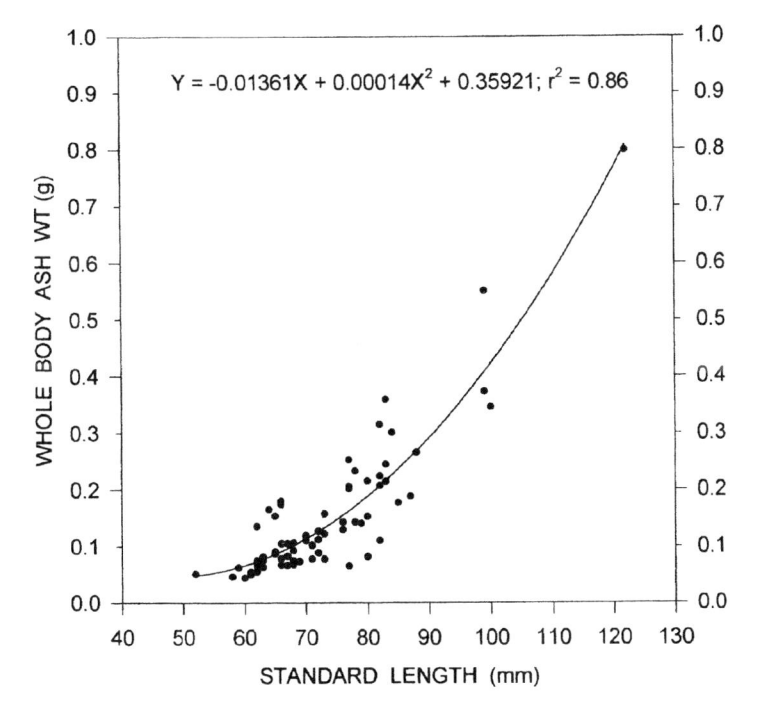

Figure 4. Whole body ash weight (g) relative to standard length (mm) for Pacific herring captured in Prince William Sound, Alaska, during October.

Equation on figure: $Y = -0.01361X + 0.00014X^2 + 0.35921; r^2 = 0.86$

Table 1. Equations for estimating whole body energy content (kJ/g wet weight) for juvenile Pacific herring with standard lengths ≤150 mm collected in Simpson Bay, Prince William Sound.

Date	Body % dry wt mean (S.D.)	a	b	r^2	Mean whole body energy content (kJ/g wet wt)
02/10/97	19.5 (2.3)	−3.4	0.38	0.96	4.01
03/05/97	20.3 (1.7)	−2.8	0.34	0.94	4.10
05/05/97	18.3 (2.0)	−2.4	0.32	0.96	3.45
07/09/97	22.3 (3.8)	−3.2	0.36	0.96	4.82
08/12/97	17.6 (1.5)	−1.8	0.29	0.97	3.30
10/22/97	22.6 (1.8)	−2.7	0.33	0.97	4.75
All fish	20.4 (4.2)	−2.8	0.34	0.95	4.24

Whole body energy content (Y) is estimated from percent dry weight of the whole body. All models are of the linear form Y (kJ/g wet wt) = $a + b$ (percent dry weight) where a and b are empirically derived constants.

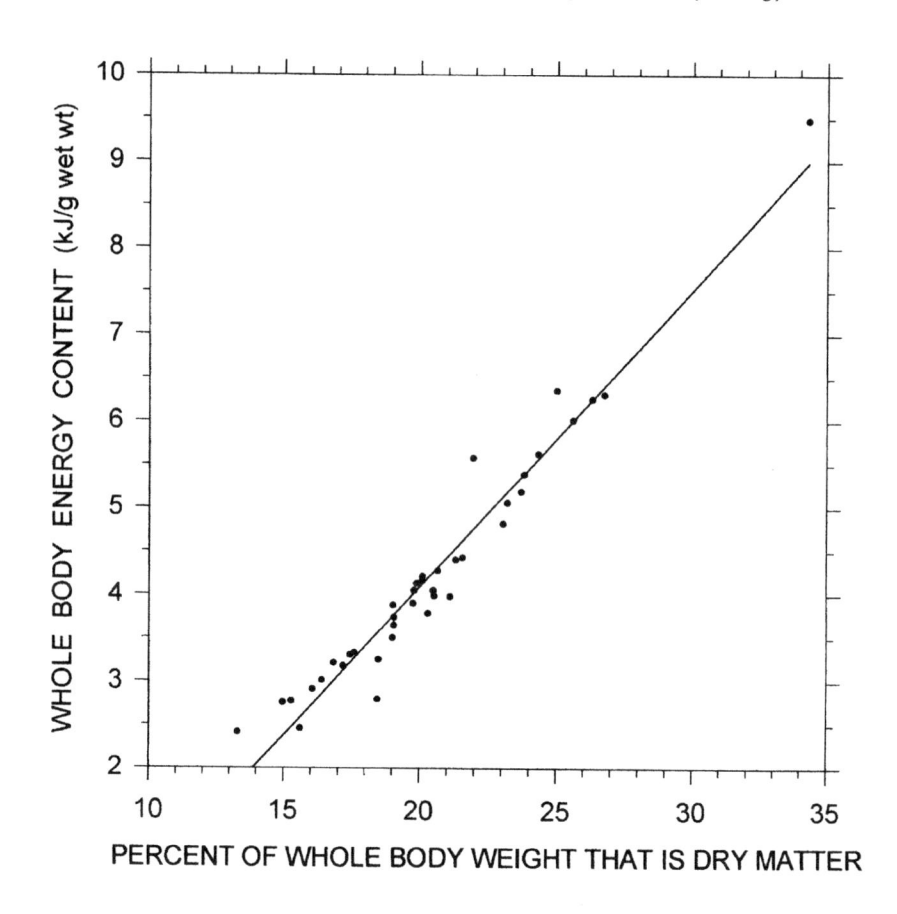

Figure 5. The relationship between the percent dry weight of the whole body and whole body energy content (kJ/g wet weight) for Pacific herring captured in Simpson Bay, Prince William Sound, Alaska.

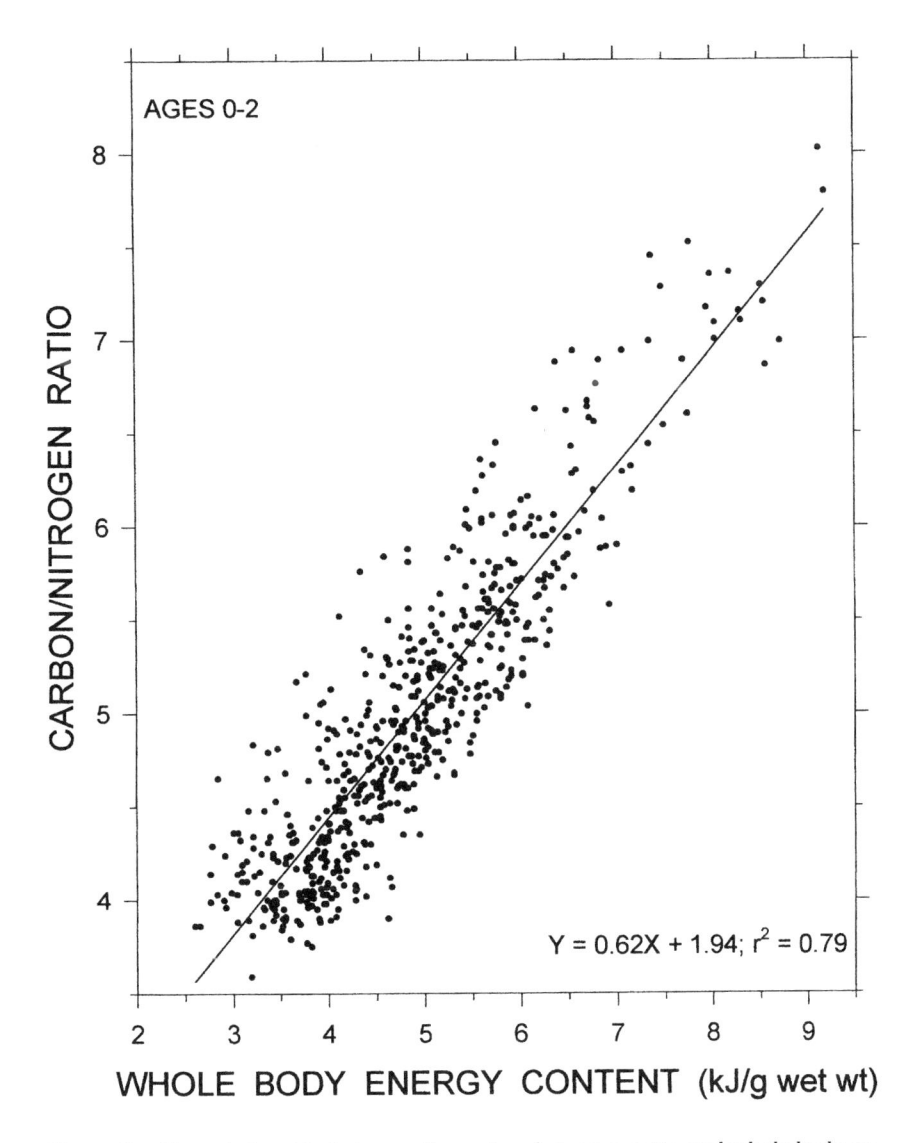

Figure 6. The relationship between the carbon/nitrogen ratio and whole body energy content (kJ/g wet weight) for juvenile Pacific herring captured in Prince William Sound, Alaska.

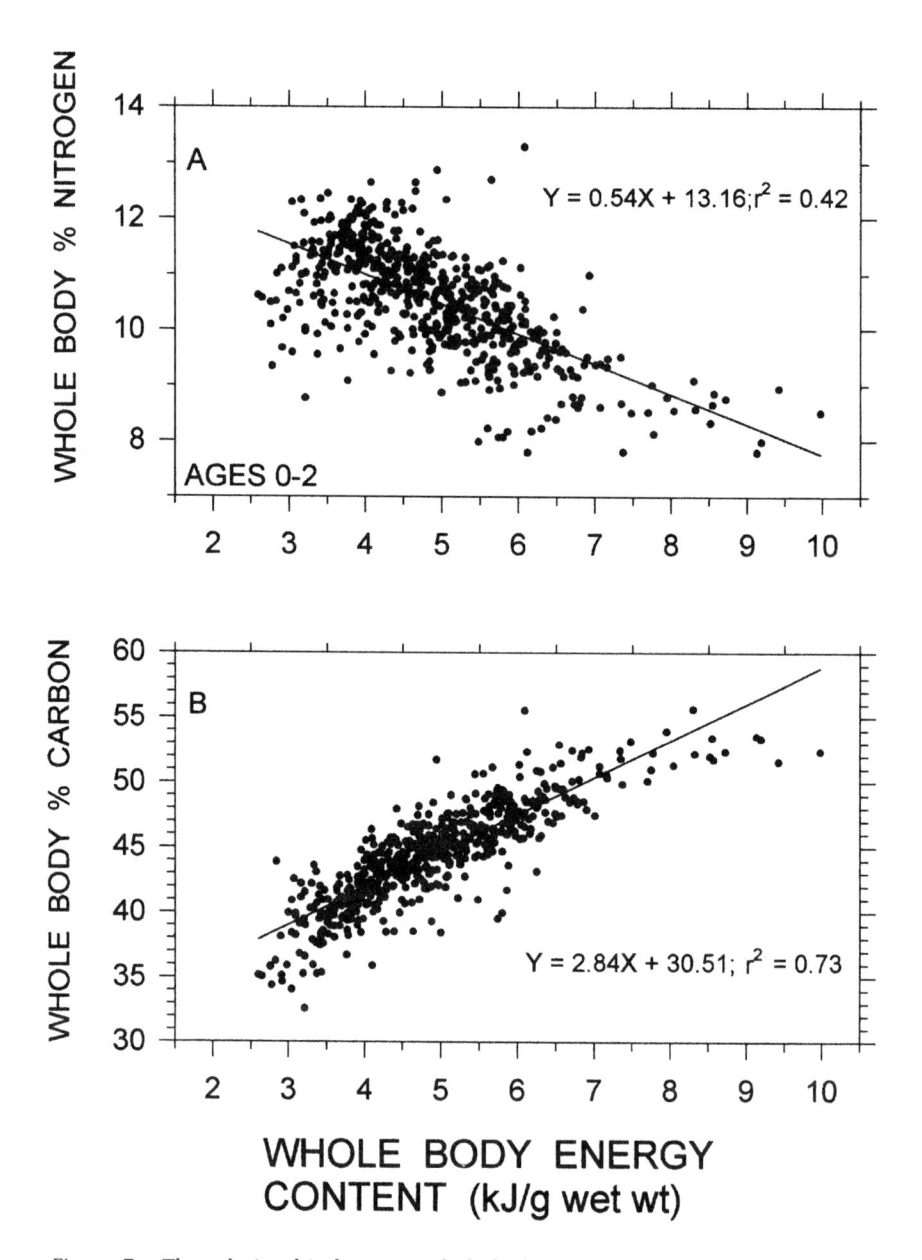

Figure 7. The relationship between whole body energy content (kJ/g wet weight) and the nitrogen (A) or carbon (B) content (percent of tissue) for juvenile Pacific herring captured in Prince William Sound, Alaska.

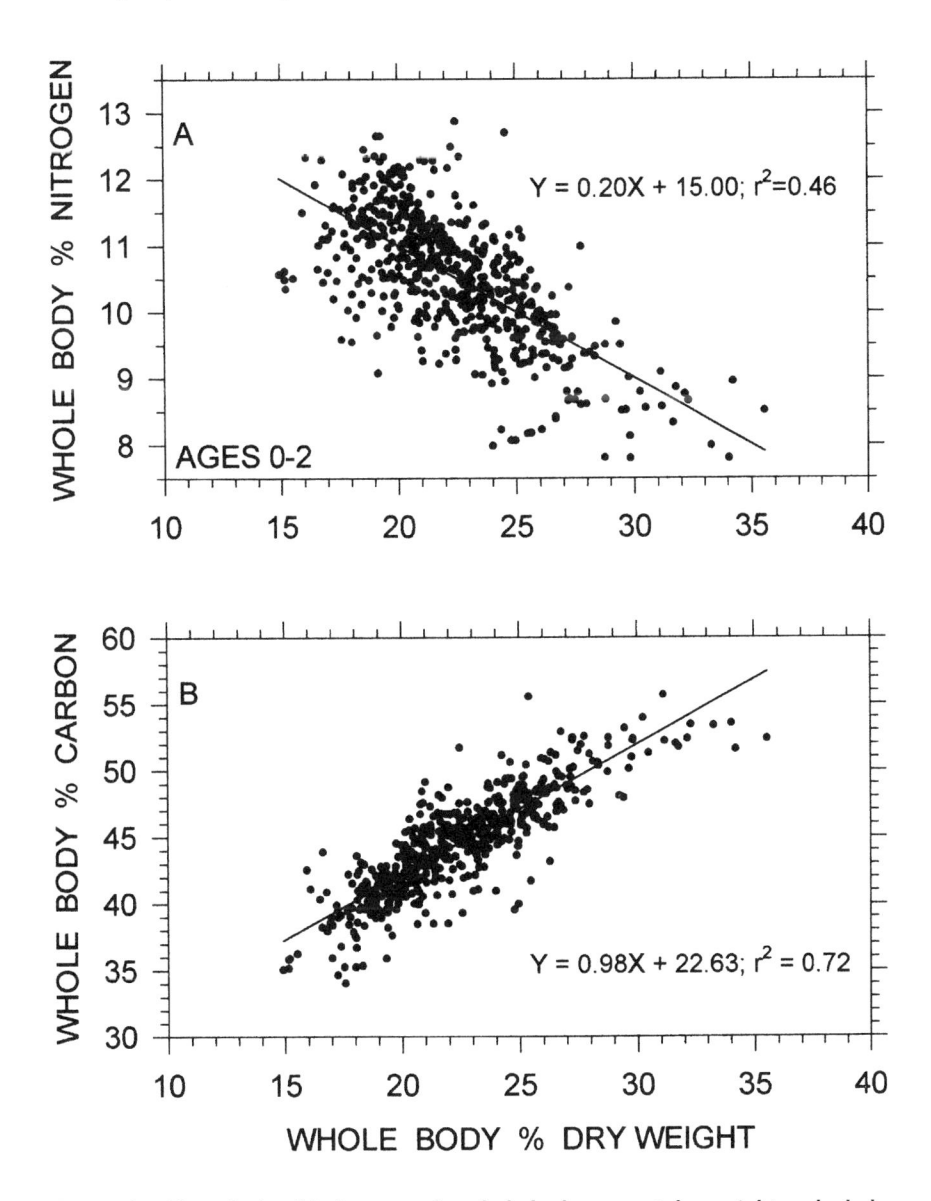

Figure 8. The relationship between the whole body percent dry weight and whole body percent nitrogen (A) or percent carbon (B) for juvenile Pacific herring captured in Prince William Sound, Alaska.

was 20.4% of live weight, a generalized clupeid would have a WBEC of 4.18 kJ/g wet wt. This would compare to 4.24 kJ/g wet wt for the average of all Pacific herring we processed (Table 1, last row).

The wide variation in WBEC for a given sampling period (Figs. 2 and 3) also occurs in Atlantic herring (Blaxter and Holliday 1963) and Canadian Pacific herring (Tanasichuk 1997). This range in WBEC may result from several interacting factors such as birth date, regional prey abundance, and intraspecific or interspecific competition for food (Blaxter and Holliday 1963). Trophic models must account for this large variability in WBEC. In one study of foraging murres *(Uria aalge)* and sooty shearwaters *(Puffinus griseus)* in Canada, the WBEC value assigned to all age 0-2 herring prey was 7.03 kJ/g wet wt (Logerwell and Hargreaves 1997). This value is considerably higher than that found in Prince William Sound herring (Figs. 2 and 3). The murres lived in the British Columbia study area all year and the shearwaters about 180 days. Over such long periods the WBEC of herring should change considerably (Figs. 2 and 3; Table 1). Obviously an average WBEC value injects large errors in estimates of prey needs. For example, in the February samples (Fig. 2) fishes with condition factors of 1.0-1.2 had WBEC of 3.0 to 6.0 kJ/g wet wt. If the seabirds were targeting the weaker herring they would be consuming those with the lowest energy density. Conversely, they may have the ability to seek energy-rich prey.

Fisheries biologists usually record the age, length, and wet weight of fishes. Likewise, the identification of fish prey in predator stomachs typically gives taxa, length, and sometimes weight. Even though there is typically a strong correlation between herring length and wet body weight, WBEC cannot be accurately estimated using wet weight values or condition indices. In any group of herring there is a wide variation in WBEC, and if the somatic energy content values are needed, it is best to measure WBEC directly. However, if it is impractical to measure WBEC directly, using a conversion estimate based on percentage of body weight that is dry weight could be substituted. Determining the percent dry weight is time consuming, but it involves relatively low cost, easy-to-use technology.

Acknowledgments

This project was funded by the Exxon Valdez Oil Spill Trustee Council through the Sound Ecosystem Assessment project. Several SEA investigators and staff aided with juvenile fish collection. J. Williams, M. McEwen, C. Adams, P. Shoemaker, and R. Schmidt helped with laboratory procedures.

References

Blaxter, J.H., and F.G. Holliday. 1963. The behavior and physiology of herring and other clupeids. Adv. Mar. Biol. 1:261-393.

Foy, R.J., and A.J. Paul. 1999. Winter feeding and changes in somatic energy content for age 0 Pacific herring (*Clupea pallasi*) in Prince William Sound, Alaska. Trans. Am. Fish. Soc. 128:1193-1200.

Harris, R.K., T. Nishiyama, and A.J. Paul. 1986. Carbon, nitrogen and caloric content of eggs, larvae, and juveniles of the walleye pollock, *Theragra chalcogramma*. J. Fish Biol. 29:87–98.

Hartman, K., and S. Brandt. 1995. Estimating energy density of fish. Trans. Am. Fish. Soc. 124:347-355.

Logerwell, E.A., and N.B. Hargreaves. 1997. Seabird impacts on forage fish: Population and behavioral interactions. In: Forage fishes in marine ecosystems. University of Alaska Sea Grant, AK-SG-97-01, Fairbanks, pp. 191-195.

Paul, A.J., and J.M. Paul. 1998a. Comparisons of whole body energy content of captive fasting age zero Alaska Pacific herring (*Clupea pallasi* Valenciennes) and cohorts over-wintering in nature. J. Exp. Mar. Biol. Ecol. 226:75-86.

Paul, A.J., and J.M. Paul. 1998b. Spring and summer whole-body energy content of Alaskan juvenile Pacific herring. Alaska Fish. Res. Bull. 5:131-136.

Paul, A.J., and J.M. Paul. 1999. Interannual and regional variation in body length, weight and energy content of age-0 Pacific herring from Prince William Sound, Alaska. J. Fish Biol. 54:996-1001.

Paul, A.J., J.M. Paul, and E.D. Brown. 1997. Ovarian energy content of Pacific herring from Prince William Sound, Alaska. Alaska Fish. Res. Bull. 3:103-111.

Paul, A.J., J.M. Paul, and E.D. Brown. 1998. Fall and spring somatic energy content for Alaskan Pacific herring (*Clupea pallasi* Valenciennes 1847) relative to age, size and sex. J. Exp. Mar. Biol. Ecol. 223:133-142.

Tanasichuk, R.W. 1997. Influence of biomass and ocean climate on the growth of Pacific herring (*Clupea pallasi*) from the southwest coast of Vancouver Island. Can. J. Fish. Aquat. Sci. 54:2782-2788.

Biological Characteristics of Atlantic Herring as Described by a Long-Term Sampling Program

Michael J. Power and T. Derrick Iles
Department of Fisheries and Oceans, Biological Station, St. Andrews, New Brunswick, Canada

Abstract

A fishery sampling program was initiated in 1969 at the St. Andrews Biological Station in support of the assessment and management of the herring stocks of the Gulf of Maine. This program has provided the basis for a number of management initiatives (including the first total allowable catch limits and individual transferable quotas) and scientific developments (including the larval retention hypothesis and conservation of population subunits). The sampling program was based on standard protocols for the measuring and recording of key biological characteristics. All life history stages were sampled but with a strong focus on the prespawning and spawning stages. There has been a continuity of personnel and methodology since 1969 that has maintained a consistency of interpretation and there has been complete coverage of the fishery.

The resultant information has been organized into a well-edited and easily accessible database with a total of some 400,000 individuals with basic biological characteristics and over 3.8 million fish measured for length frequency. This represents some of the most comprehensive information for any fish stock in the literature, and perhaps for any animal, covering as it does five or six complete generations of this species. This publication describes some key biological attributes of this herring stock and their variation over three decades.

Introduction

Practical fisheries research is based largely on monitoring, and the fisheries themselves have become the primary research tool. The proper documentation of catches and the biological analysis of representative samples are crucial for the assessment of populations leading to management advice. This research also generates data for various ecological and evolutionary analyses and for other biological insights.

East Atlantic herring has been a major economic and political entity for most of the last millennium, and has been a major focus of scientific research since fisheries science began about a century ago (Stephenson 2001). West Atlantic herring, while never of the same socioeconomic impact as East Atlantic herring, has been important to coastal communities and has been the subject of increasing scientific studies and management developments, especially since the 1960s.

Atlantic herring proved to be highly vulnerable to the excess fishing pressure that developed after the second world war and a number of herring fisheries and stocks collapsed and even disappeared, especially in the 1950s and 1960s. Concern for herring within the International Commission of Northwest Atlantic Fisheries (ICNAF) led in 1972 to the first agreement within any of the world's multinational fisheries for total allowable catch (TAC) limits. This initiative played an important role in the adoption of fisheries management principles generally, and in the extension of coastal state management in 1977 (Iles 1993).

The herring fisheries of the Bay of Fundy and Scotian Shelf regions of eastern Canada (Fig. 1) boast a number of management firsts. They were among the first (since 1970) to have commercial marine fisheries regulated by limited entry, the first in 1972 to come under TAC limits, and have had individual vessel quotas for purse seiners since 1976 (Iles 1993). In addition, research on these stocks has led to developments in scientific thought, most notably the herring (or larval retention) hypothesis of Iles and Sinclair (1982), and more recently, work on the importance and conservation of population subunits may change our views of stock structure (Stephenson et al. 2001a).

There has been a herring research program at the St. Andrews Biological Station for most of its history of nearly 100 years. The program of monitoring and research was expanded in 1969 as the basis for the evaluation of these fisheries (stock assessment) and for targeted research, in support of assessment and linked to the advance of management initiatives. Major elements of the program are extensive sampling of the commercial fisheries, as well as various research surveys including larval herring and acoustic surveys.

The herring fisheries of this area are currently the largest in the western Atlantic with annual landings in the order of 100 kt (Stephenson et al. 2000). The fishery is mainly by purse seine (80 kt), coastal traps (30 kt), and some by gillnet gear types (<10 kt). Both juveniles and adults of the Scotia-Fundy herring stocks are fished in this area in feeding, prespawning, spawning, and overwintering aggregations. Sample collection over the last decade has covered a wide area including Georges Bank (5Z), the Gulf of Maine (5Y), and the Cape Breton area (4Vn) (Fig. 1). The focus of this analysis is on the samples in areas 4X and 4W that are done primarily in support of the stock assessment for the southwest Nova Scotia/Bay of Fundy stock.

Herring: Expectations for a New Millennium 137

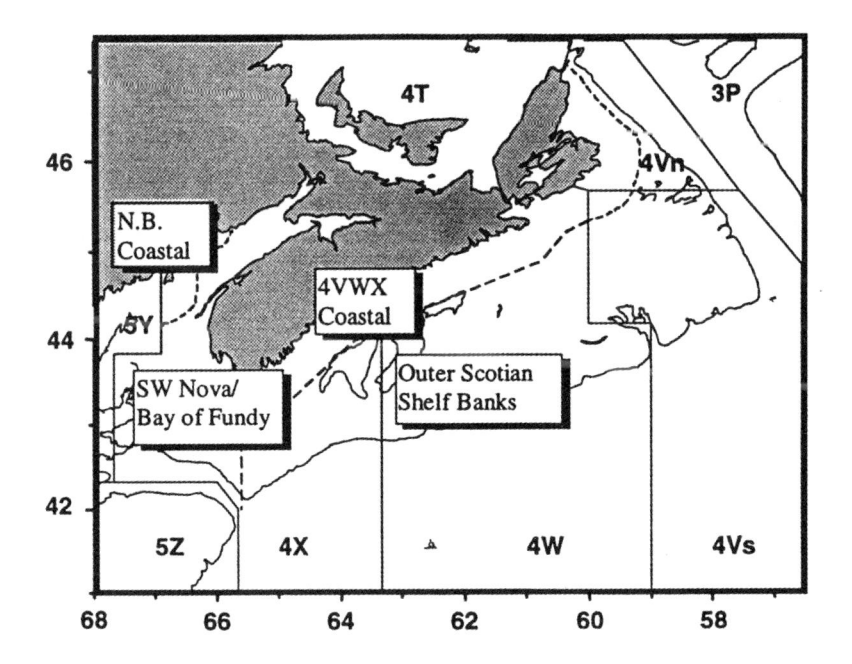

Figure 1. *Fishing unit areas and locations of management units for herring fisheries in the Scotia-Fundy area.*

This paper describes the nature and extent of the historical sampling program for herring fisheries of the Scotia-Fundy region of Canada and presents some key biological attributes of these herring.

Materials and methods

Since 1969 the research program has maintained a continuous strategy and methodology of sampling the commercial fisheries. The biological sampling program used at St. Andrews is based on the international protocol set out by the Herring Committee of the International Council for Exploration of the Sea (ICES) in 1962 and described by Parrish and Saville (1965). This allows direct comparison with similar information from other herring stocks in particularly the east Atlantic. Some aspects have been modified as knowledge was acquired, and techniques and equipment improved, but there has been a continuity of staff and maintenance of standards and basic methods over the whole period to ensure consistency (Hunt 1987).

The main body of biological information involves the measurement of length, weight, gonad stage with gonad weights, and the determination of age by the interpretation of annual ring formation on the otolith for individual herring. Other information, including stomach fullness and contents, fat content, and fin ray and vertebral counts have been gathered on an ad hoc basis over the series.

Biological sampling for these stocks has followed a two-stage design (Hunt 1987). Random length frequency samples of 150-200 fish, taken from as many landings as possible, have been measured for total length (to the nearest 0.5 cm). In addition, subsamples of 2 fish per length group have been selected from a subset of the length frequency samples in each area, month, and gear sector and measured for length, weight, age, sex, maturity stage, and gonad weight. These data are stored in an easily accessible Pelagic Samples database on the Oracle database server at the St. Andrews Biological Station. In recent advancements the data are entered directly to the database, allowing immediate checks for consistency and errors.

In this time period there have been some 24,000 samples involving measurement of 4 million herring and collection of 425,000 fish for more detailed analysis. The fish taken for detailed analysis and lab processing extend back to 1957 (Fig. 2). There was a large increase in 1971 with the implementation of the expanded program and a peak in 1977 when the Gulf of St. Lawrence area was also part of the sampling responsibility. There have been reductions in recent years, as the program has been rationalized and the sampling efforts more focused on areas of concern to the stock assessments in the Scotia-Fundy area. There are presently about 5,000-10,000 fish examined for detailed biological attributes each year.

In the recent decade (1990-1999) about 9,000 samples were collected, 1.4 million measured, and 80,000 sampled for detailed biological attributes (Fig. 3). The spatial distribution of these samples over the past decade reflects the distribution of the fishery and coverage by broad-scale research surveys. The distribution of samples by month through the year (Fig. 4) reflects the seasonal nature of the herring fishery that is more active during the summer months. All parts of the herring growth cycle are sampled.

Recognizing that vessels and processing plants also measure fish on a routine basis for their own marketing purposes and responding to a willingness and interest of industry to become involved and contribute to the scientific program, an industry sampling program was implemented in 1996. Crew members on vessels and plant workers have been trained to collect length frequencies and to select and freeze subsamples for later analysis at the St. Andrews lab. The fishing industry now collects the majority of length frequency samples and frozen subsamples from these fisheries (Table 1).

Herring: Expectations for a New Millennium 139

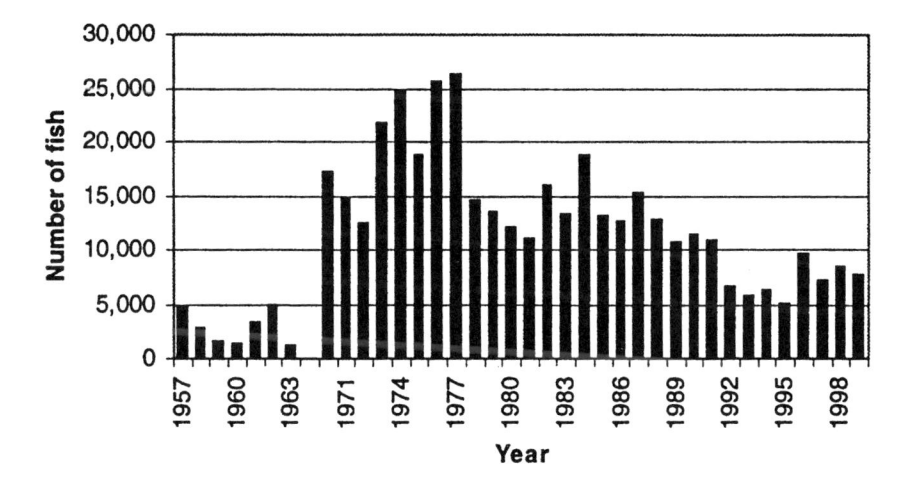

Figure 2. Numbers of Atlantic herring processed (by year) for detailed biological characteristics study in the Scotia-Fundy area.

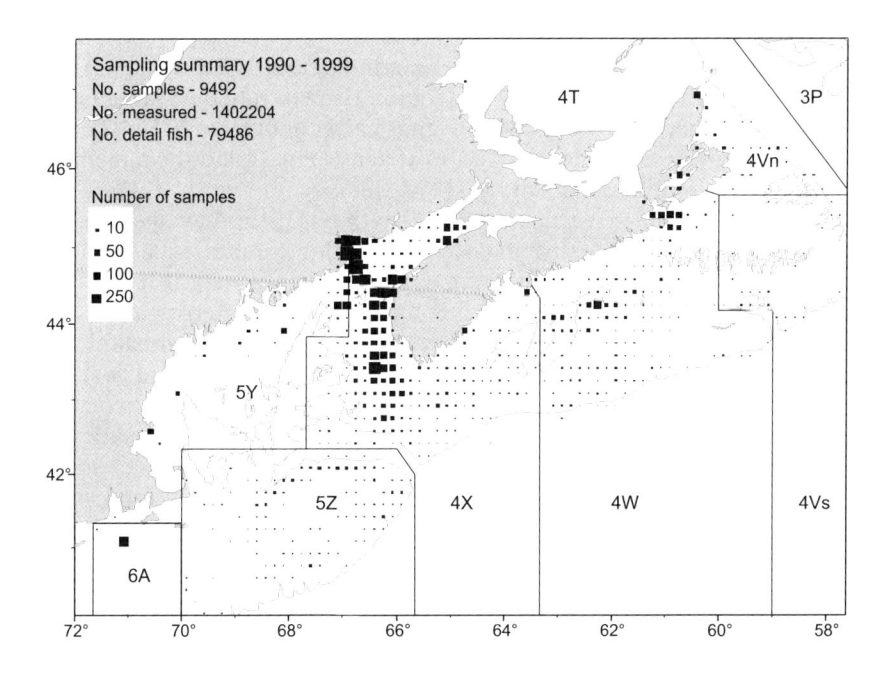

Figure 3. Scotia-Fundy biological sampling summary for herring from 1990 to 1999; numbers of samples by 10-mile-square grid area.

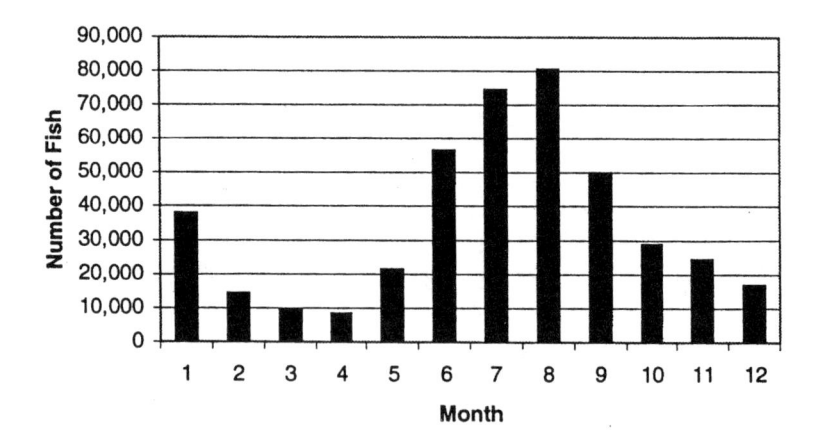

Figure 4. Numbers of Atlantic herring processed (by month) for detailed biological characteristics study in the Scotia-Fundy area.

Results

Herring from 4WX record a maximum length of about 375 mm (Fig. 5) and weight of about 400 g by age 10-11 (Fig. 6). Growth rate decreases exponentially and there is little growth in length after about age 6 (Fig. 7). The maximum growth rate in length occurs at age 1 (Fig. 7) and the maximum growth rate in weight occurs at age 3 (Fig. 8).

Maturity ogives were calculated by length (Fig. 9) and by age (Fig. 10) for the entire data set using males and females combined. Maturation takes place between 200 and 300 mm (Fig. 9) or ages 2-6 (Fig. 10). The 50% maturity occurs at a length of 252 mm and at an age of 3.42 years.

The mean length at age for 4WX herring (Fig. 11) has been stable for most of this time period although there is an indication of a small decreasing trend in recent years in the older ages. The younger ages are more variable over time and this may be reflective of their higher growth rate for the first few years of life as previously shown by the growth increment.

The mean weight at age shows a decrease in recent years for most older ages (4-10) in comparison to length at age over time (Fig. 12). For example, an 8-year-old that was about 350 g in 1980 was only about 300 g in 1999. The younger ages (<4), however, have remained relatively stable. Perhaps this difference is related to the different juvenile distributions from the adult population in the Bay of Fundy. A similar phenomenon (reduction in adult but stable juvenile growth rate) has been noted over the past decade in Baltic herring (Stephenson et al. 2001b).

Table 1. Number of herring samples collected by DFO personnel sampling of commercial fishing (Commercial), independent observer program (Observer), DFO research surveys (Research), and members of the fishing industry (Industry).

Year	Commercial	Observer	Research	Industry	Total
1970	333				333
1971	221				221
1972	237				237
1973	628				628
1974	850				850
1975	650				650
1976	668				668
1977	775				775
1978	800				800
1979	796				796
1980	747				747
1981	184				184
1982	998		2		1,000
1983	957				957
1984	794		105		899
1985	699		19		718
1986	543		12		555
1987	525		13		538
1988	617		36		653
1989	569		41		610
1990	385	185	41		611
1991	310	268	38		616
1992	287	205	44		536
1993	182	421	1		604
1994	223	228	14		465
1995	138	244	108		490
1996	125	49	71	868	1,113
1997	78	0	114	1,443	1,635
1998	225	0	98	1,376	1,699
1999	49	89	198	1,388	1,724
Total	16,901	1,796	1,042	5,666	25,405

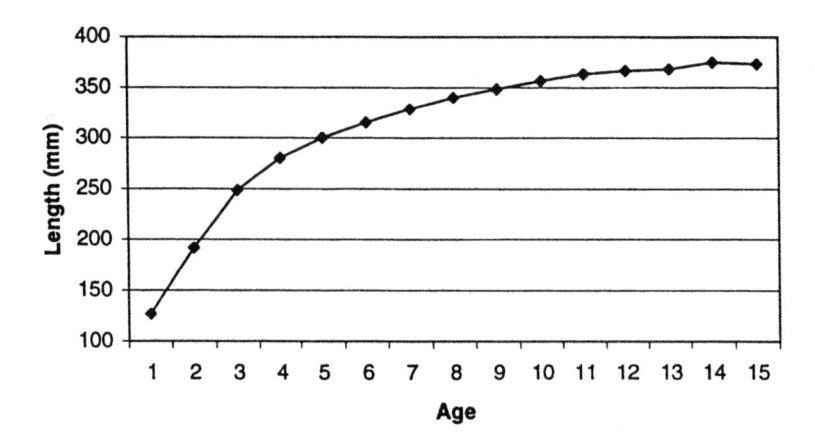

Figure 5. 4WX herring mean length at age (September to December data selected for years 1970-1999).

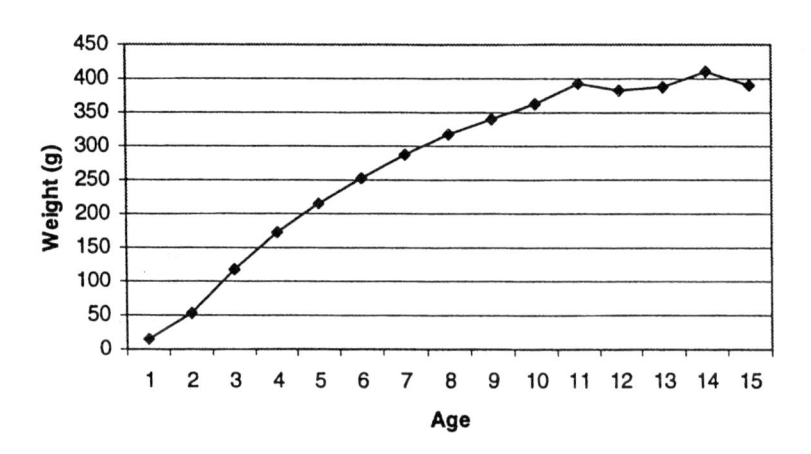

Figure 6. 4WX herring mean weight at age (September to December data selected for years 1970-1999).

Figure 7. 4WX herring mean length increment at age (September to December data selected for years 1970-1999).

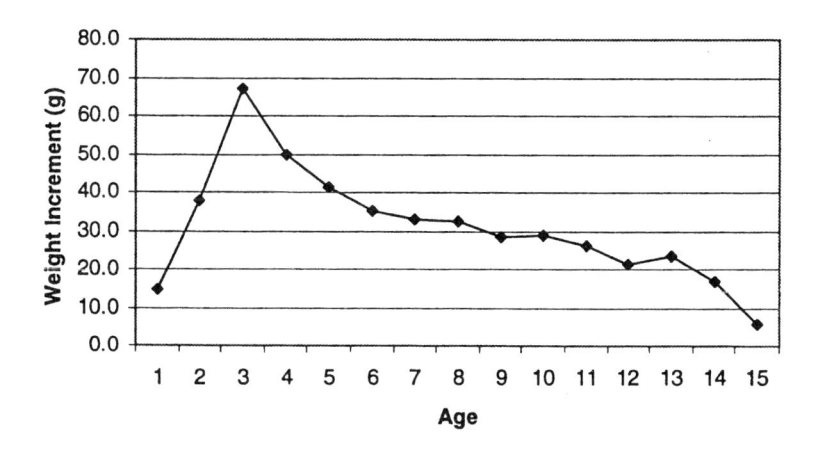

Figure 8. 4WX herring mean weight increment by age (September to December data selected for years 1970-1999).

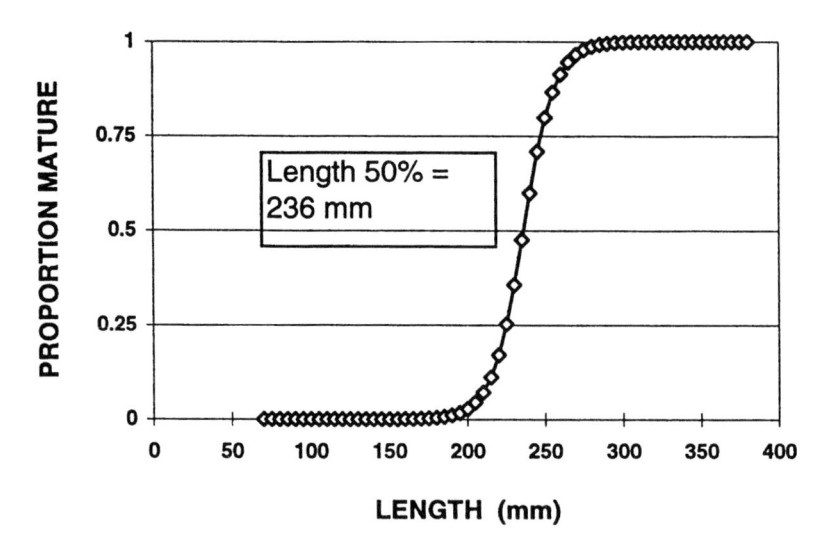

Figure 9. 4WX herring maturation ogive by length (1970-1999 data for males and females combined).

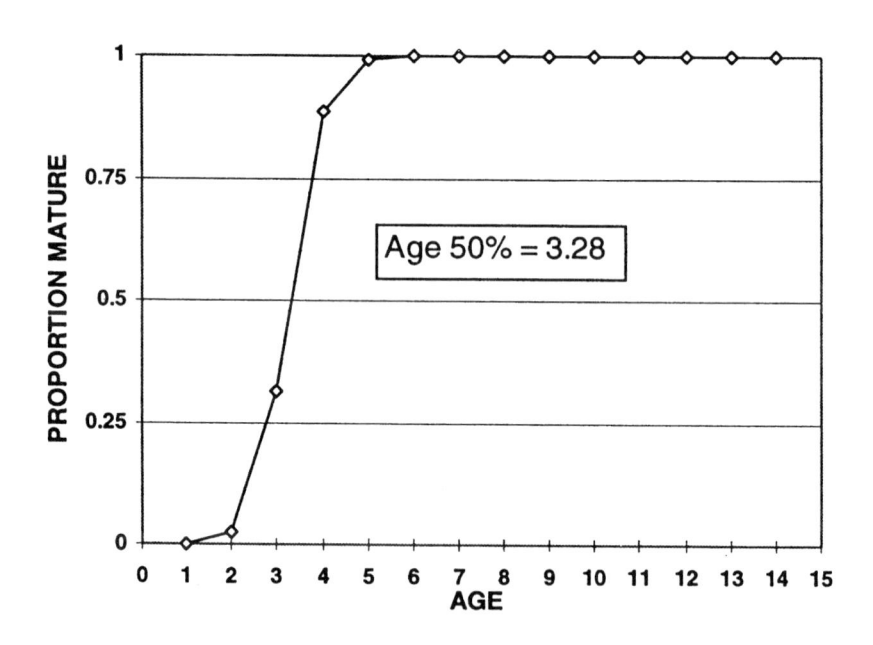

Figure 10. 4WX herring maturation ogive by age (1970-1999 data for males and females combined).

Herring: Expectations for a New Millennium 145

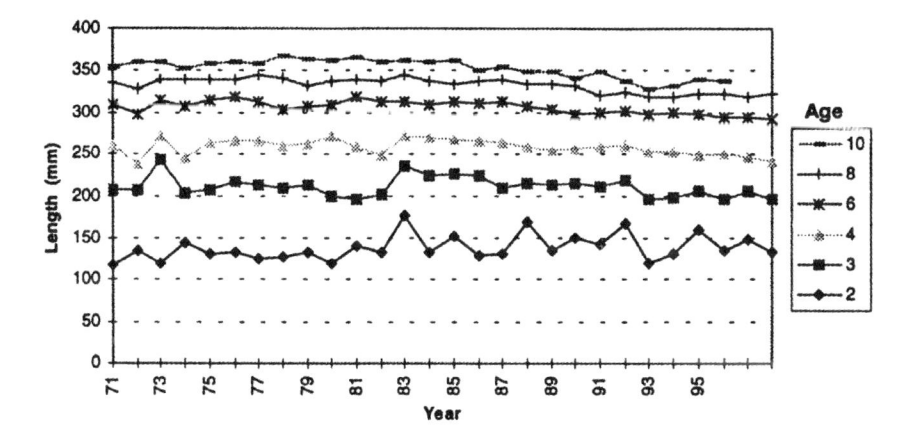

Figure 11. 4WX herring mean length at age from 1970 to 1999.

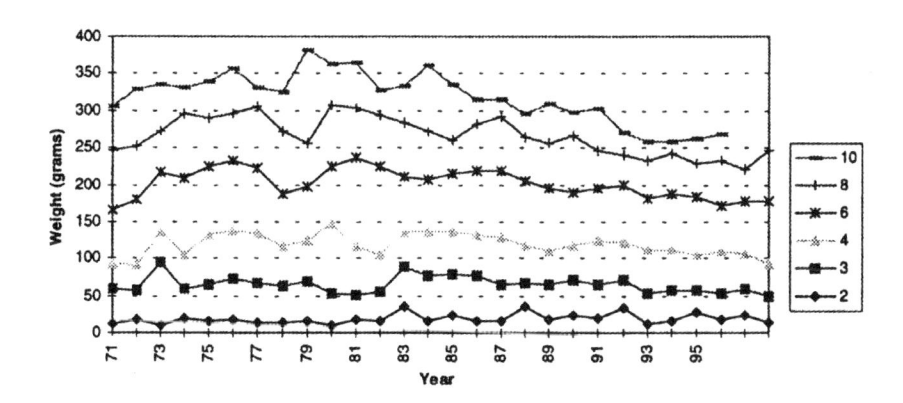

Figure 12. 4WX herring mean weight at age from 1970 to 1999.

The length-weight relationship is a standard calculation used in the analysis of the catch data and is used to adjust the length frequency samples of the catch up to the total weight of the catch (Doubleday and Rivard 1983). The data are fitted to the exponential relationship $W = a \times L^b$ and are usually calculated by month each year for use in the assessment. Typical fitted relationships for 4WX herring for recent decades show a high correlation (Fig. 13) and small differences with the weight at length generally smaller for the 1970s, higher in the 1980s, and in between for recent years.

The sex ratio for herring is thought to be 50:50. For 4WX herring a higher proportion of males was recorded (Fig. 14) for smaller herring <140 mm TL, and a higher proportion of females for herring >300 mm TL. For smaller herring, which have not yet reached sexual maturity, this might be explained by a misclassification of sex as it is sometimes difficult to distinguish the sexes based on the visual criteria being used. For larger fish, where classification is reliable, the reason is unknown but the higher fraction of females may be of significance in the life history for this stock.

Sex ratio by year was also calculated (Fig. 15) in order to ascertain whether there have been trends or selection by the fishery, resulting perhaps from the roe fishery in recent years where there is a preference for female herring. There are many variations over the time period but no persistent trend is seen for recent years.

The mean condition factor by month (Fig. 16) reflects the seasonal feeding, growth, and maturation cycle that peaks in July and August, with the maximum condition corresponding to the peak in spawning. The mean condition factor by year was calculated for the summer months, when it is typically near the maximum and the winter months when it is near the low (Fig. 17). There is considerable year-to-year variation in condition factor. Winter condition fluctuates without much trend, but summer condition was relatively high during the 1980s.

The proportion of gonad weight to body weight (Fig. 18) is an important character for the roe herring market, which attempts to get the best yield from the fishery by timing the catches to the peak of the prespawning period. The best yield of about 20% occurs for males and females 27-31 cm in length.

Discussion

Perhaps the largest development in this program in the last few years has been an increase in industry involvement in all aspects of the assessment program. Melvin et al. (2001) show the increased involvement of industry in acoustic surveys, and there has also been the major development of industry participation in sampling summarized in this paper (Table 1). This has been positive in several ways. The samples, now more than ever, reflect the full temporal and spatial aspects of the fishery; the industry

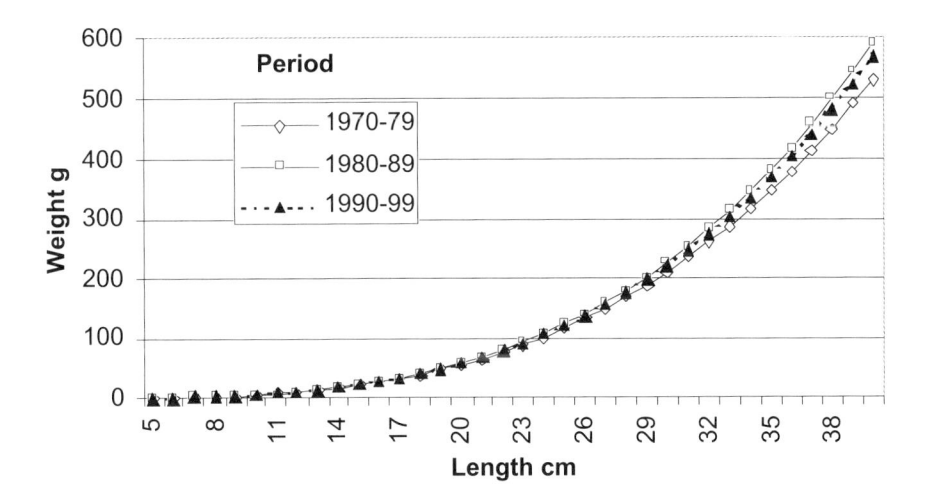

Figure 13. 4WX herring length weight for all months combined for the periods 1970-1979, 1980-1989, and 1990-1999.

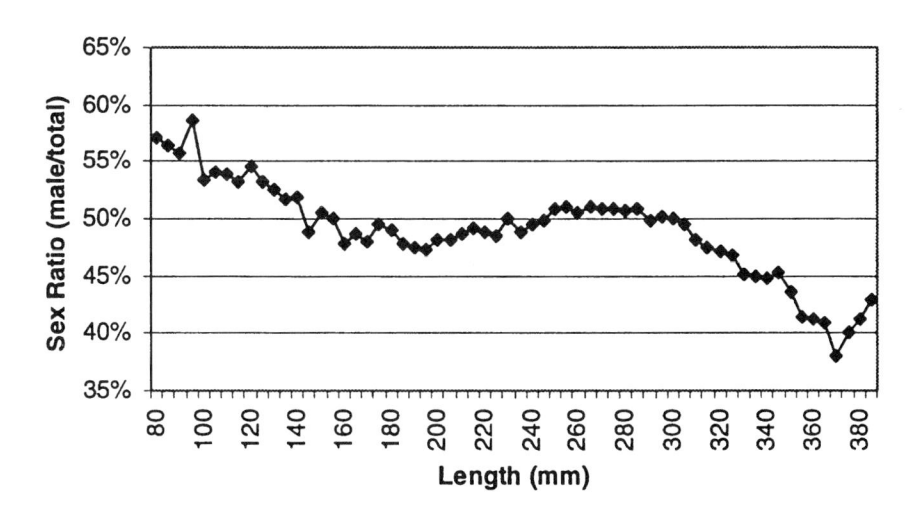

Figure 14. 4WX herring sex ratio (proportion of males) by length (mm) from 1970 to 1999.

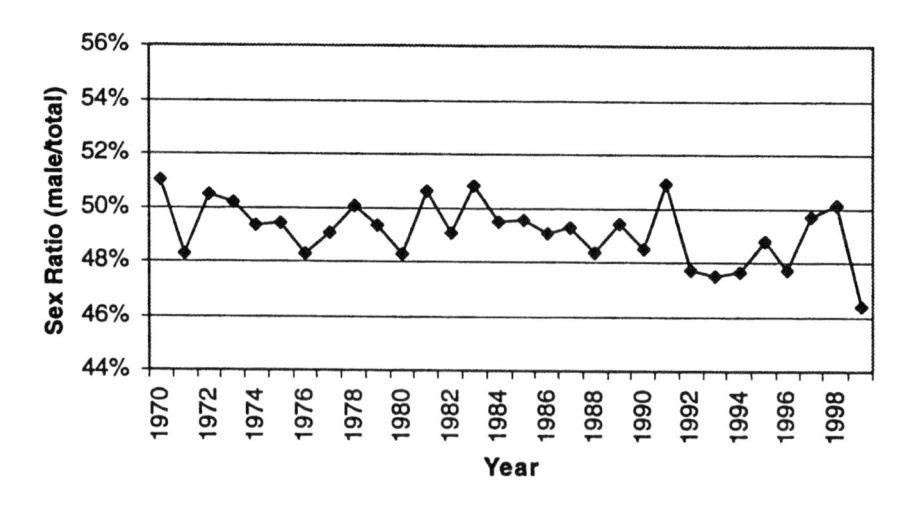

Figure 15. 4WX herring sex ratio (proportion of males) from 1970 to 1999.

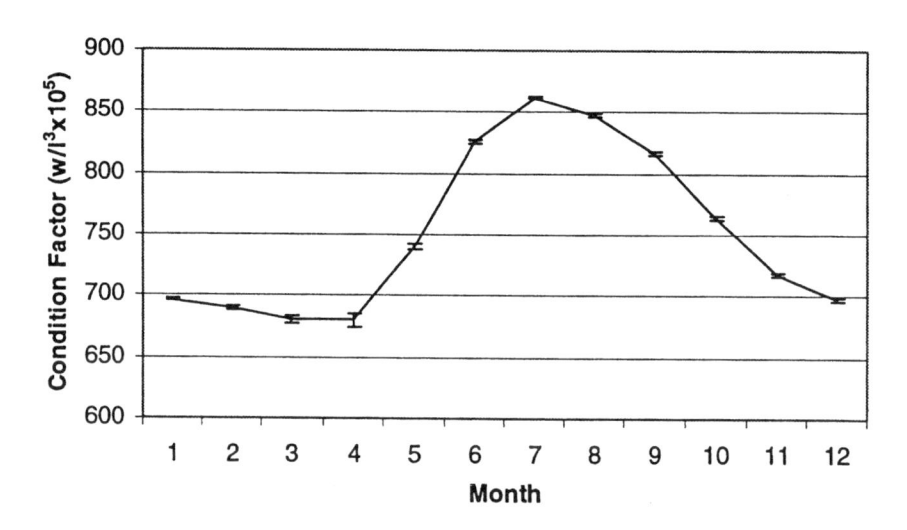

Figure 16. 4WX herring average condition factor by month from 1970 to 1999 (selected mature fish from 25 to 30 cm).

Herring: Expectations for a New Millennium 149

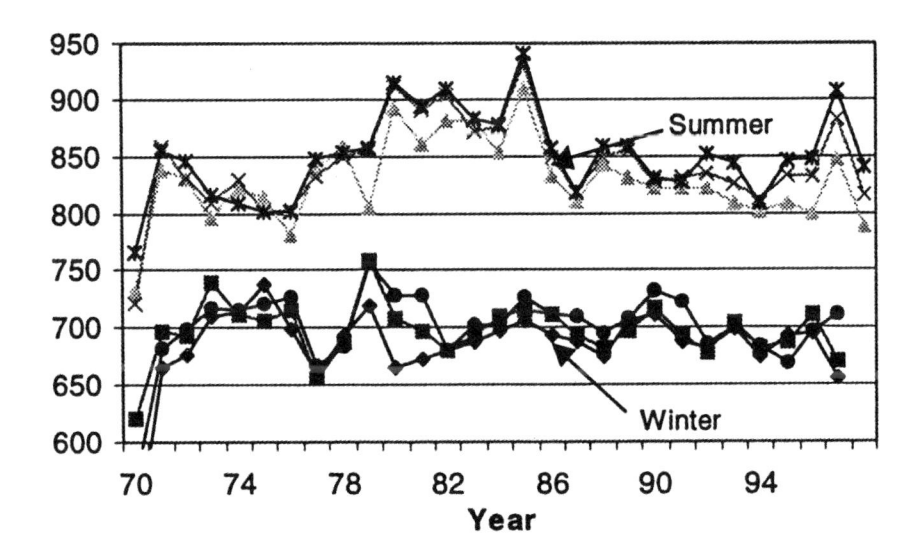

Figure 17. *4WX herring condition factor for ages 4-6 for summer (August-September) and winter (January-February) months for the period 1970-1999.*

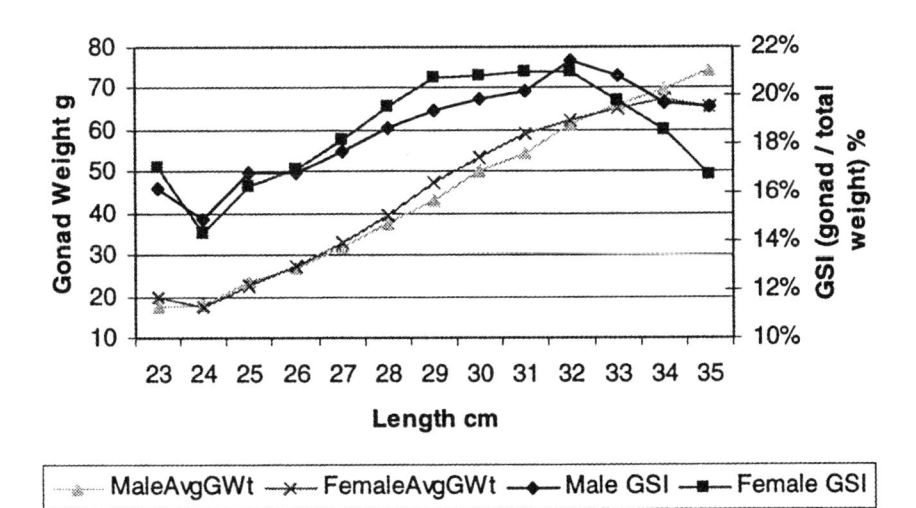

Figure 18. *4WX herring average mature gonad weight (g) and GSI (%) for ripe gonads (stage 5) from males and females separately from 1970 to 1999.*

feels considerable involvement and ownership in the assessment process by providing critical information; and scientific staff are spending time dealing with data and analyzing information rather than just collecting it. These recent sampling and survey initiatives have been an integral part of the development of a more responsive, collaborative management structure for Scotia-Fundy herring that involves in-season management (Stephenson et al. 1999).

The monitoring data sets for 4WX herring are among the most complete describing any herring population and perhaps any animal, covering as it does five or six generations for this species. These data have provided the basis for assessment and management of a complex of herring stocks in the Scotia-Fundy area. Historical performance of the primary southwest Nova Scotia/Bay of Fundy spawning component, based on these data (Stephenson et al. 2000), shows that the stock has had some substantial fluctuations (characteristic of herring stocks), but by close monitoring, to this point at least, has not collapsed like many other herring stocks.

The knowledge of the fishery resulting from the information in the extensive sampling program has allowed a number of other substantial developments and new concepts. The Iles and Sinclair herring (larval retention) hypothesis (Iles and Sinclair 1982) and critical observations on vertical movement as a mechanism in larval retention (Stephenson and Power 1988) have come from this program. Changing views of stock structure and the importance of maintaining discrete subunits (i.e., spawning grounds) have also been developed here (Stephenson 1999, Stephenson et al. 2001a). The involvement of industry in sampling and surveying and a great involvement in various forms of co-management (in the 1970s [Iles 1993] and today) have been a key feature of this stock.

Several fisheries labs have "assessment and associated research" as their mandate and data sets gathered under such circumstances are invaluable for scientific study. Much more can be gleaned from this and other extensive data sets. There is a need to engage in some "data-mining" and in making data available to others (recent advancements on the Web make this easily possible) and in comparison between areas (Hay 2001).

Acknowledgments

The authors wish to thank the numerous samplers who have participated in the collection of herring samples over the past decades. We also thank our industry colleagues for their continued support in this regard. The authors also would like to express their gratitude to Dr. Rob Stephenson for his encouragement and support with the preparation of this paper.

References

Doubleday, W.G., and D. Rivard (eds.). 1983. Sampling commercial catches of marine fish and invertebrates. Can. Spec. Publ. Fish. Aquat. Sci. 66. 290 pp.

Iles, T.D. 1993. The management of the Canadian Atlantic herring fisheries. In: L.S. Parsons and W.H. Lear (eds.), Perspectives on Canadian marine fisheries management. Can. Bull. Fish. Aquat. Sci. 226:123-150.

Iles, T.D., and M. Sinclair. 1982. Atlantic herring: Stock discreteness and abundance. Science 215:627-633.

Hay, D. 2001. Taking Stock: An Inventory and review of the world herring stocks in 2000. In: F. Funk, J. Blackburn, D. Hay, A.J. Paul, R. Stephenson, R. Toresen, and D. Witherell (eds.), Herring: Expectations for a new millennium. University of Alaska Sea Grant, AK-SG-01-04, Fairbanks. (This volume.)

Hunt, J.J. 1987. Herring sampling program in the Scotia-Fundy Region, 1975-85. Can. Manuscr. Rep. Fish. Aquat. Sci. 1923. 21 pp.

Melvin, G.D., R.L. Stephenson, M.J. Power, F.J. Fife, and K.J. Clark. 2001. Industry acoustic surveys as the basis for in-season decisions in a comanagement regime. In: F. Funk, J. Blackburn, D. Hay, A.J. Paul, R. Stephenson, R. Toresen, and D. Witherell (eds.), Herring: Expectations for a new millennium. University of Alaska Sea Grant, AK-SG-01-04, Fairbanks. (This volume.)

Parrish, B.B., and A. Saville. 1965. The biology of the north-east Atlantic herring populations. Oceanogr. Mar. Biol. Ann. Rev. 3:323-373.

Stephenson, R.L. 1999. Stock complexity in fisheries management: A perspective of emerging issues related to population subunits. Fish. Res. 43:247-249.

Stephenson, R.L. 2001. The role of herring investigations in shaping fisheries science. In: F. Funk, J. Blackburn, D. Hay, A.J. Paul, R. Stephenson, R. Toresen, and D. Witherell (eds.), Herring: Expectations for a new millennium. University of Alaska Sea Grant, AK-SG-01-04, Fairbanks. (This volume.)

Stephenson, R.L., and M.J. Power. 1988. Semidiel vertical movements in Atlantic herring *Clupea harengus* larvae: A mechanism for larval retention? Mar. Ecol. Prog. Ser. 50:3-11.

Stephenson, R.L., K. Rodman, D.G. Aldous, and D.E. Lane. 1999. An in-season approach to management under uncertainty: The case of the SW Nova Scotia herring fishery. ICES J. Mar. Sci. 56:1005-1013.

Stephenson, R.L., K.J. Clark, M.J. Power, F.J. Fife, and G.D. Melvin. 2001a. Herring stock structure, stock discreteness, and biodiversity. In: F. Funk, J. Blackburn, D. Hay, A.J. Paul, R. Stephenson, R. Toresen, and D. Witherell (eds.), Herring: Expectations for a new millennium. University of Alaska Sea Grant, AK-SG-01-04, Fairbanks. (This volume.)

Stephenson, R.L., H. Peltonen, S. Kuikka, J. Ponni, M. Rahikainen, and E. Aro. 2001b. Linking biological and industrial aspects of the herring fishery in the northern Baltic Sea. In: F. Funk, J. Blackburn, D. Hay, A.J. Paul, R. Stephenson, R. Toresen, and D. Witherell (eds.), Herring: Expectations for a new millennium. University of Alaska Sea Grant, AK-SG-01-04, Fairbanks. (This volume.)

Stephenson, R.L., M.J. Power, K.J. Clark, G.D. Melvin, F.J. Fife, T. Scheidl, C.L. Waters, and S. Arsenault. 2000. 2000 evaluation of 4VWX herring. DFO Canadian Stock Assessment Secretariat Res. Doc. 2000/65. 114 pp.

Is the Decrease in Growth Rate of Atlantic Herring in the Baltic Sea Density-Dependent? A Geostatistical Application

M. Cardinale and F. Arrhenius
National Board of Fisheries, Institute of Marine Research, Lysekil, Sweden

Extended Abstract

Different hypotheses have been formulated to explain the decrease in weight-at-age and condition of Atlantic herring (*Clupea harengus*) in the Baltic Sea. However, different authors have recently stressed that a decrease in zooplankton biomass in relation to stock density could play a crucial role in determining the observed phenomena (for a review see Cardinale and Arrhenius 2000). We used geostatistical and kriging techniques to test the hypothesis of negatively density-dependent growth of herring in the Baltic proper. In accordance with the density-dependent hypothesis, areas of high pelagic fish abundance should correspond to areas of a larger decrease of herring growth rates. We mapped density of zooplanktivorous fish and the proportion of herring in the pelagics mix (herring and sprat) in the Baltic proper during the last decade and zooplanktivorous fish (sprat and herring) abundance was tested for correlation with zooplankton abundance. The result of this study showed that there was a strong increase in abundance of sprat, the other pelagic fish species in the Baltic, from 1990 and onward essentially in the areas north-northwest of the Baltic proper (Subdivisions 27-29S). Those areas corresponded to the areas of larger decrease of both weight-at-age and herring condition (fish specific mass) (see Cardinale and Arrhenius 2000 for details). Zooplankton biomass was significantly negatively correlated with pelagic fish numbers in the same areas. This suggested that growth of herring in the Baltic proper is resource-limited. Top-down processes at work in the Baltic Sea were discussed. We argued that, when Baltic cod biomass, due to high fishing mortality and severe abiotic conditions, declined in the end of the 1980s, top-down processes went to work in the Baltic. This determined an increase in zooplanktivorous fish biomass and, as a consequence, also the predation pressure on zooplankton increased. Thus, the zooplankton biomass decreased, mirroring the decline of pelagic fish growth rate. This hypothesis represents, so far, the most plausible one.

Acknowledgment

A report based on this study has been submitted to the journal *Fisheries Research* (Amsterdam).

Reference

Cardinale, M., and F. Arrhenius. 2000. Decreasing weight-at-age of Atlantic herring (*Clupea harengus*) from the Baltic Sea between 1986 and 1996: A statistical analysis. ICES J. Mar. Sci. 57:1-12.

Growth Pattern of Baltic Herring in Relation to Spawning Time

Jan Eklund, Marjut Rajasilta, and Päivi Laine
University of Turku, Archipelago Research Institute, Vaasa, Finland

Abstract

Baltic herring spawn from early spring to late autumn. The main spawning season varies from March-April in the southern Baltic to June-July in the north. Growth of individual herring commences in spring, after spawning for early spawners and before spawning for late spawners. The growth pattern of individual fish is thus linked to the time of spawning. Consequently, growth characteristics of the scale and otolith have been used for separating spring- and autumn-spawning fish.

In this study, otolith increments in spawning fish were measured. Fish were collected from the Archipelago Sea in 1987-1991 and from five locations in the Baltic Sea (northern and southern Bothnian Bay, Bothnian Sea, Archipelago Sea, and Rügen) in 1996 and 1998. Parameters examined were first- and second-year increment and last increment before spawning.

Between Baltic subareas, first-year growth was largest in the south and decreasing toward the north. The second-year increment showed the opposite trend, being smallest in the south and increasing toward the north. In the Archipelago Sea, first- and second-year increments were both significantly smaller than in other areas. Over the spawning season, first-year otolith increment decreased from April to August while the second-year increment increased. The last increment prior to spawning showed no trend.

The differences in otolith growth between areas reflected the differences in growth season and spawning time. The first-year growth indicated that in the Archipelago Sea, early spawning fish are the progeny of early spawning parents. Low growth in the first year is compensated by second-year growth, indicating a tradeoff between somatic and gonadal growth. Poor growth in the Archipelago Sea may be a sign of local environmental factors limiting herring growth.

Introduction

In the Baltic Sea, herring (*Clupea harengus* L.) live in an environment much different from that in the oceans. The Baltic Sea is brackish, with salinity ranging from 25 ppt in the south to less than 2 ppt in the innermost parts of the Gulf of Bothnia and Gulf of Finland. Moreover, environmental conditions are highly variable compared to those in the ocean: salinity, temperature, and other environmental parameters vary over short time spans (days or weeks) depending on weather conditions, and salinity also over a time scale of decades, depending on the inflow of saline water from the North Sea (e.g., Matthäus 1985). Still, herring live and spawn in the entire Baltic basin.

The main spawning time of Baltic herring varies with latitude, extending from April to September (Ojaveer 1981, Blaxter 1985). The production cycle in the sea starts in March in the southern Baltic and further north after the ice break in April, May, or June (review by Lenz 1985). Like in other temperate fish, the reproduction and growth cycle of Baltic herring is linked to the production cycle (Iles 1974, Cushing 1975). In the seasonal growth pattern of herring, rapid somatic growth does not coincide with rapid development of gonads or with spawning (Iles 1974).

The growth pattern of herring can be interpreted from scales or otoliths, and used for separating fish of different geographical origin (deBarros and Holst 1995) or spawning at different times (Dahl 1907, Hellevaara 1912, Runnström 1936). In Baltic herring, growth patterns of the otoliths have been widely used to discern between spring and autumn spawning fish (Kompowski 1971). The distinct annual growth cycle and the coupling of somatic growth and spawning time, together with the relatively clear annual marks on the otoliths and scales of Baltic herring provide an opportunity for studying the growth strategy and allocation of resources of the fish.

In the southern Baltic, because of early spawning and a long growth season, herring are expected to grow more in their first year than fish in the northern Baltic. Similarly, in any area, fish hatching from an early spawn are expected to grow more in their first year than those hatching later in the season. Because of the energy demands of gonad production (Iles 1974), growth differences in the first year are also expected to influence the second year's growth. In this study, otolith growth in the first and second year of life and in the last year prior to sampling is used to investigate differences in growth between herring from different parts of the Baltic and between herring spawning at different times in the Finnish Archipelago Sea.

The growth pattern of herring has consequences also for stock assessment and management. If subgroups of the population, e.g., early and late spawners or spawners in different areas, have differing growth patterns, then the relative contribution of these subgroups to the exploited stock will directly influence the assessment. The decrease in growth of Baltic herring experienced since the early 1980s (Nordic Council of Ministers 1994) underlines the importance of a correct understanding of the species' growth patterns.

Material and Methods

In the Archipelago Sea, commercial catches from trap nets set on the spawning grounds were sampled throughout the spawning season from April to August in 1989-1991 (Table 1). Random samples of about 100 fish were drawn from the catch and the fish were measured, weighed, and sexed. Gonad stage was determined (stage 4-5 = ripening, 6 = ripe, 7 = running) and the sagittal otoliths were removed. For storage and reading, otoliths were fixed onto clear slotted plastic chips with Euparal (ASCO Laboratories, 52 Levenshulme Rd., Gorton, Manchester M18 7NN) or varnish. The sampling program and sample treatment are described in more detail by Rajasilta et al. (1999).

In May-July 1996, samples were taken at four locations on the Finnish coast (Airisto in the Archipelago Sea, Sundom in the Bothnian Sea, Kalajoki and Simo in the Bothnian Bay) and in April 1998, one sample on the German coast (Thiessow on Rügen). These samples were taken on or by the spawning ground from commercial trap net or, in one case, from trawl catches. Samples were treated as above, but otoliths were stored in glycerol on slotted plastic trays. Sampling stations are shown in Fig. 1 and sample data in Table 1.

Otoliths were aged and measured in reflected light under a binocular microscope. Age was read as the number of winter rings (narrow, translucent rings). Growth in the sampling year had not yet commenced. For each fish, the distance from the otolith nucleus to each annulus and to the otolith margin were measured. The 1989-1991 otoliths were aged and measured on an ocular grid scale at 16× magnification along the clearest radius. Unclear or poorly mounted otoliths were discarded. The 1996 and 1998 samples were read at about 40× magnification and measured digitally (in pixels) with a Calcomp 2500 digital meter fitted to the binocular microscope. Measurements were taken on the post- or pararostrum radius, and unclear otoliths were discarded.

All otolith measurements were transformed into millimeters. For the analysis, the following parameters were calculated: increment during first growth season (INC1), increment during second season (INC2), and increment during the year prior to sampling (LSTINC).

Prespawning fish and non-numerous age groups were excluded from the data. Gonad stages 6-7 and age groups 3-7 were retained in the 1987-1991 data, and stages 4-7 and ages 3-6 in the 1996 and 1998 data. In all, the two data sets consisted of 2,741 fish from the Archipelago Sea in 1989-1991 and 277 fish from the Finnish and German coasts in 1996 and 1998.

The relationship between otolith radius and fish length was described by linear regression: radius = $a + b$ × length. Mean increment values by age group and month or area were calculated and means were compared by Tukey's multiple comparison. Differences in otolith increments between geographical area and spawning month were analyzed by analysis of variance (ANOVA). Statistical analyses were performed with the SAS/STAT software (SAS Institute Inc., Cary, NC 27512-8000).

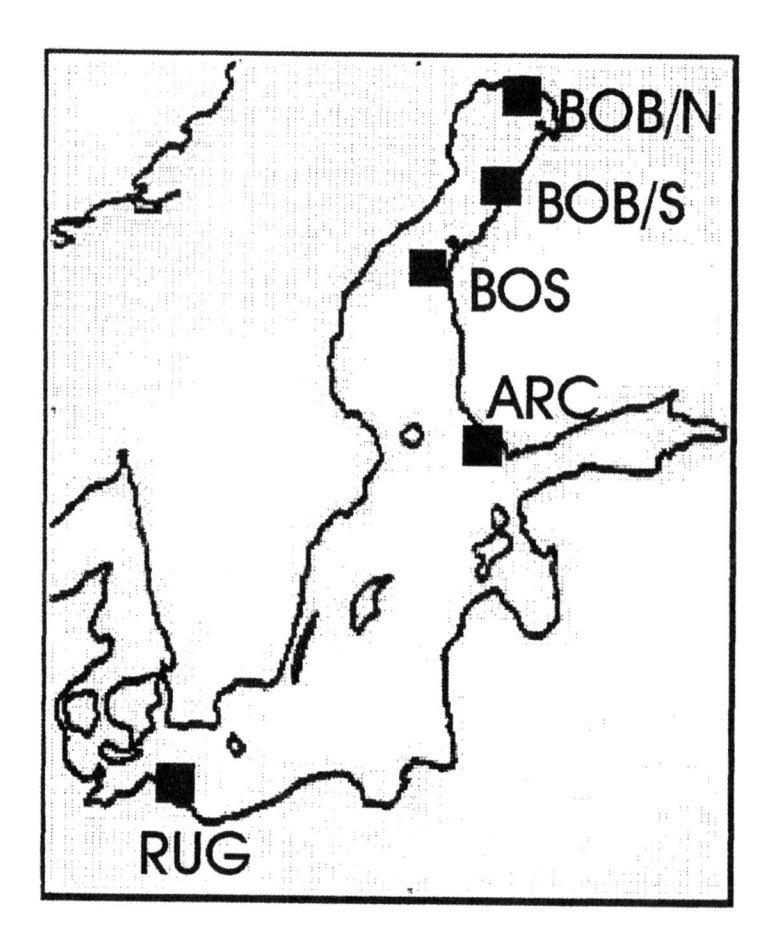

Figure 1. Baltic Sea sampling locations. RUG = Rügen, Thiessower Endhaken; ARC = Archipelago Sea, several locations; BOS = Bothnian Sea, Sundom; BOB/S = Bothnian Bay, Kalajoki; and BOB/N = Bothnian Bay, Simo.

Table 1. Sampling data.

Location	Abbreviation	Date	Gear	N	n	Length (cm) (min-max)	Age mean (min-max)
Rügen, 1998	RUG	02 Apr	Trap net	100	94	26.0 (20.9-29.7)	4.5 (3-6)
Archipelago Sea, 1996	ARC	23 Jun	Trap net	85	43	16.6 (14.3-19.3)	4.6 (3-6)
Bothnian Sea, 1996	BOS	24 May	Trap net	73	60	17.8 (13.9-21.5)	4.4 (3-6)
Bothnian Bay/S, 1996	BOB/S	17 Jun	Trap net	100	27	18.7 (16.6-22.5)	5.0 (3-6)
		01 Jul	Trawl				
Bothnian Bay/N, 1996	BOB/N	30 Jun	Trap net	159	19	18.0 (15.7-19.4)	5.3 (4-6)
Archipelago Sea,		Apr	Trap net	184	167	20.0 (14.7-34.2)	6.0 (3-7)
1989-1991		May	Trap net	1,203	825	18.4 (12.1-29.0)	3.0 (1-13)
		Jun	Trap net	1,795	1,390	17.9 (12.6-32.5)	4.0 (2-11)
		Jul	Trap net	443	286	16.8 (13.4-24.2)	2.0 (1-11)
		Aug	Trap net	87	74	18.2 (15.0-21.8)	4.0 (2-8)

Length and age means given for spawning fish in 1996 and 1998, length means and age mode for entire sample in 1989-1991. N = number of fish in sample, n = number of age 3-7 spawning fish. Locations are shown on Fig. 1.

Table 2. Linear regression of otolith radius distance on fish length.

Area	d.f.	r^2	a	b	$P(b=0)$
Archipelago Sea, 1989-1991	2,741	0.52	0.28	0.050	
Finnish and German coast, 1996 and 1998	276	0.77	0.64	0.047	
Rügen, 1998	54	0.27	0.49	0.058	<0.001
Archipelago Sea, 1996	59	0.67	0.47	0.057	<0.001
Bothnian Sea, 1996	36	0.48	0.08	0.077	<0.001
Bothnian Bay/S, 1996	30	0.43	0.31	0.065	<0.001
Bothnian Bay/N, 1996	93	0.57	0.30	0.060	<0.001

d.f. = degrees of freedom, a = intercept, b = angular coefficient, r^2 = correlation coefficient.

Results

Otolith Radius Measure and Fish Length

The relation between otolith radius measure and fish length was approximately linear, although with high variability. Correlation coefficients were small, varying from 0.27 in the Rügen sample to 0.67 in the Archipelago Sea 1996 sample. The angular coefficients differed significantly from zero and varied between areas from 0.057 to 0.077 (Table 2).

Geographic Variation in Otolith Growth

Otolith growth showed a distinct geographical pattern of variation over the locations sampled. In the analysis of variance, area contributed significantly to the variation in first, second, and last year's increment, while variation between age groups was significant for the last increment only (Table 2).

The otolith increment formed during the first growth season (INC1) was largest in the southern Baltic (Fig. 2). At Rügen, the first increment was about 2 mm, in the Bothnian Sea 1.5 mm, and in the Bothnian Bay about 1 mm. In all age groups tested, the Rügen value was significantly larger than the others and the Bothnian Sea and Bothnian Bay values differed significantly from each other. In the Archipelago Sea sample, the first increment was significantly smaller than in other areas, about 0.7 mm (Table 3). The second increment (INC2) showed the opposite pattern. Here, Rügen and Bothnian Sea values were 0.7-0.8 mm with no significant difference, while the Bothnian Bay had significantly higher values above 1 mm (Table 3). Again, the Archipelago Sea values of about 0.3 mm were significantly lower than in other areas. The Archipelago Sea increment values were of the same magnitude in both the 1989-1991 and the 1996 data (Tables 3 and 4), although the 1996 sample consisted of very small fish (Table 1).

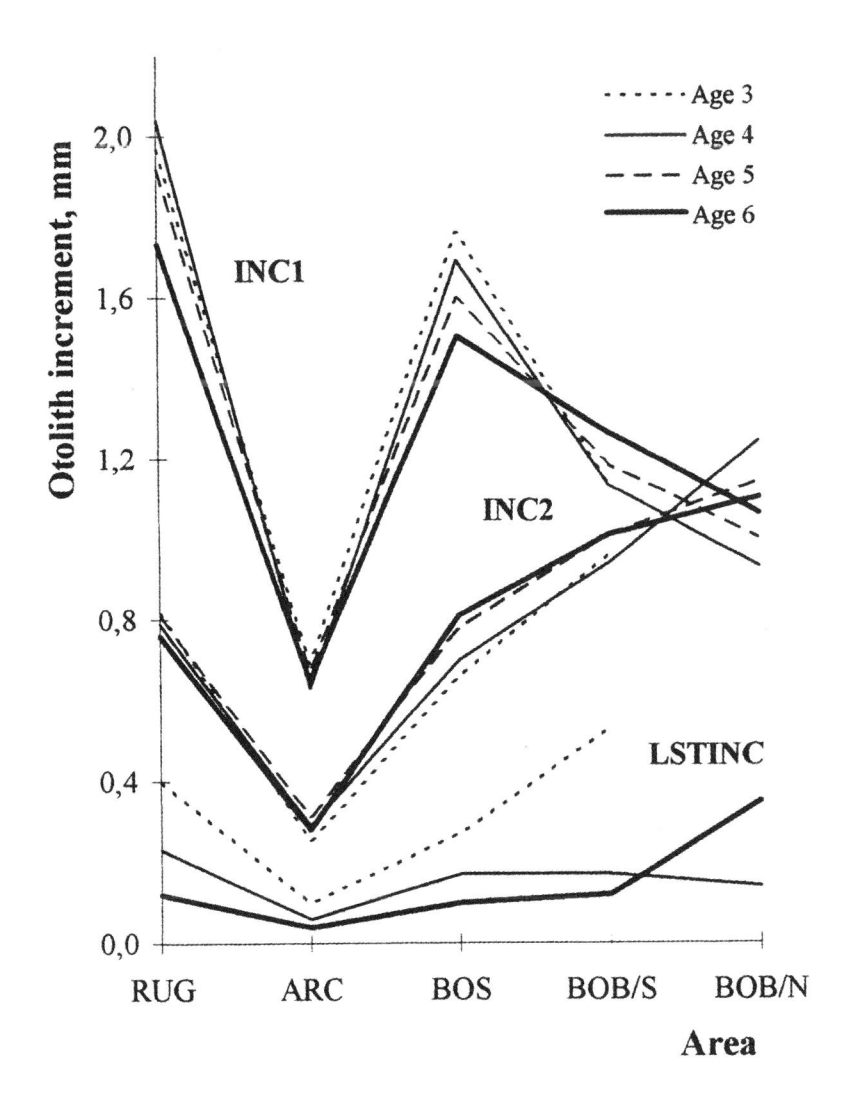

Figure 2. Geographical variation of otolith increment. RUG = Rügen, southern Baltic, 1998; ARC = Archipelago Sea, 1996; BOS = Bothnian Sea, 1996; BOB/S = Bothnian Bay, Kalajoki, 1996; and BOB/N = Bothnian Bay, Simo, 1996. INC1 = first-year increment, INC2 = second-year increment, LSTINC = increment in year prior to sampling.

Table 3. Otolith increment by area.

	ANOVA		
INC1			
Source	**d.f.**	**F**	**P > F**
Area	4	360.7	<0.001
Age	3	2.2	0.09
Area × age	11	2.0	0.03
INC2			
Source	**d.f.**	**F**	**P > F**
Area	4	152.9	<0.001
Age	3	0.73	0.8
Area × age	11	0.78	0.7
LSTINC			
Source	**d.f.**	**F**	**P > F**
Area	4	7.7	<0.001
Age	3	5.9	<0.001
Area × age	11	1.8	<0.001

Age		RUG	ARC	BOS	BOB/S	BOB/N
3	INC1	**1.97**	0.69	**1.76**	*1.13*	
	INC2	**0.82**	0.25	**0.66**	**0.96**	
	LSTINC	**0.40**	*0.10*	0.27	**0.53**	
4	INC1	**2.04**	*0.63*	**1.69**	**1.13**	**0.93**
	INC2	*0.79*	0.29	**0.70**	**0.94**	1.24
	LSTINC	**0.23**	0.06	**0.17**	**0.17**	*0.14*
5	INC1	**1.92**	0.67	**1.60**	**1.18**	*1.00*
	INC2	*0.81*	0.31	**0.78**	**1.01**	1.14
	LSTINC	**0.16**	0.04	**0.12**	**0.12**	0.13
6	NC1	**1.73**	*0.65*	**1.50**	*1.26*	**1.06**
	INC2	*0.76*	0.28	**0.81**	1.01	1.10
	LSTINC	0.12	0.04	0.10	0.12	0.35

Sample means and Tukey's multiple comparison of means. Means significantly ($P < 0.05$) **larger** or *smaller* than at least one other mean in the same row are denoted by bold type and italics, respectively. A mean that is both larger than and smaller than another mean in the same row is in bold italics. RUG = Rügen, 1998; ARC = Archipelago Sea, 1996; BOS = Bothnian Sea, 1996; BOB = southern and northern Bothnian Bay, 1996. INC1 = first-year increment, INC2 = second-year increment, LSTINC = increment in year prior to sampling. Increments in millimeters.

Otolith increment in the last growth season prior to sampling (LSTINC) was, in all age groups, significantly smaller in the Archipelago Sea than in other areas. For most age groups, Rügen had significantly larger values than the other areas. Bothnian Sea and Bothnian Bay values were intermediate, except for the large value for age group 3 in the Bothnian Bay, and did not differ significantly from the Rügen values (Table 3).

Temporal Variation in Otolith Growth

There was a strong temporal trend in first- and second-year increment (Fig. 3) and no trend in the last increment prior to sampling. Month and age both contributed significantly to the observed variation of all three variables (Table 4.)

Over the spawning season in the Archipelago Sea, there was a marked decline of the first-year increment from April to August (Fig. 3). In all age groups, April and May values were significantly larger than for other months. Also, in age group 3, the July value was significantly larger than for August (Table 4).

The second increment, instead, showed an increasing trend over time (Fig. 3). For age groups 4-6, the months June-August had significantly larger values than April and May, and for ages 3 and 7, the June value was significantly larger than for the preceding months (Table 4). The last increment prior to spawning increased slightly over time (Fig. 3), but there were no significant differences between the monthly means (Table 4).

Discussion

The use of otoliths to study fish growth is susceptible to errors in aging and measurement of the otolith and to biological variation in the relationship between otolith and fish size. That age estimates vary between readers is well known (Beamish and McFarlane 1995, Campana et al. 1995, Eklund et al. 2000). To minimize errors, fish of seemingly uncertain age were discarded from the data. Also, including only age groups 3-7, the presumably larger aging error for older age groups is avoided. The Archipelago Sea otoliths from 1989-1991 were read at low magnification and measured with an ocular grid scale. This method of measurement is prone to errors caused by the position of the eye and the otolith, compared to the more accurate measurements of the Baltic data subset. The two data subsets (1989-1991 and 1996/1998) were aged and measured by the same reader using consistent methods within the subsets and errors are thought to be unbiased with respect to sampling time and location.

Generally, otolith size seems to closely reflect fish size (Campana and Neilson 1985). However, the ratio between otolith and fish size is influenced by somatic growth rate (Campana 1990), which results in relatively larger otoliths in slow-growing fish (Mosegaard et al. 1988). In our study, samples from the northern Baltic consisted of smaller and somewhat older fish than southern Baltic samples. The uncoupling of otolith and body

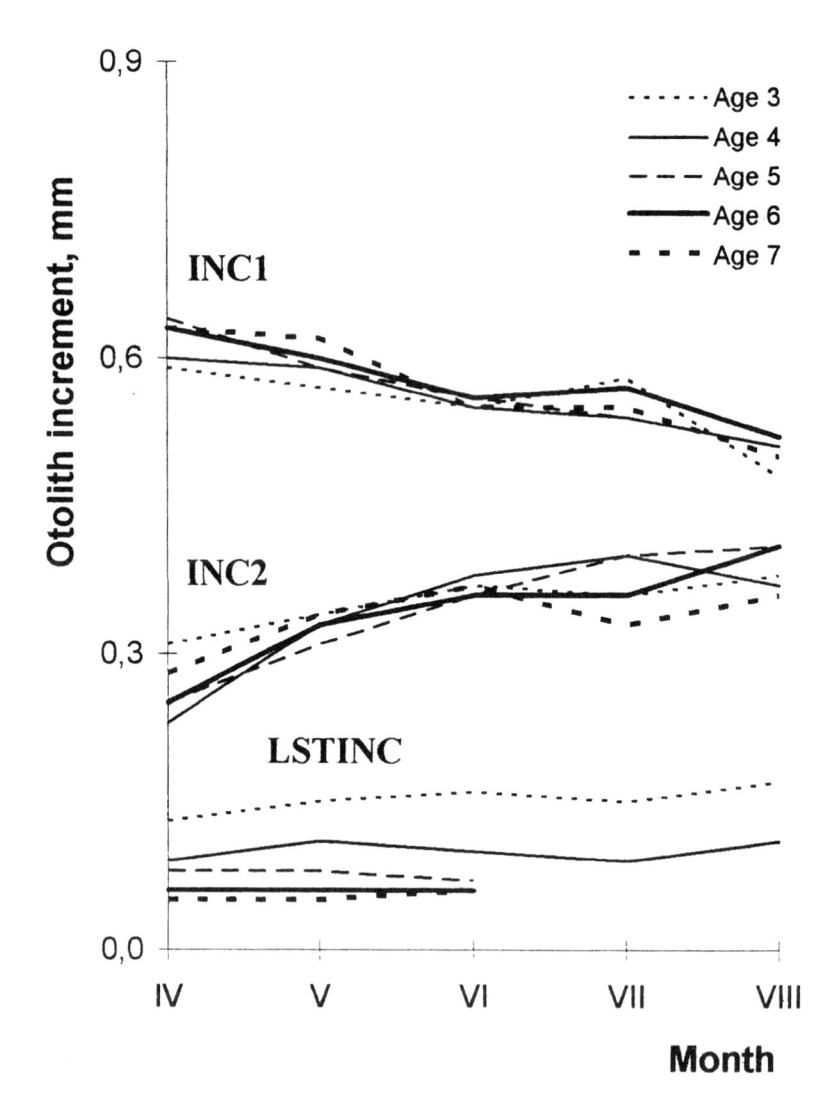

Figure 3. Temporal variation of otolith increment, Archipelago Sea, 1989-1991. INC1 = first-year increment, INC2 = second-year increment, LSTINC = increment in year prior to sampling.

Table 4. Otolith increment by spawning month, Archipelago Sea, 1989-1991.

ANOVA			
INC1			
Source	**d.f.**	**F**	**P > F**
Month	4	45.9	<0.001
Age	6	76.9	<0.001
Month × age	23	1.6	0.03
INC2			
Source	**d.f.**	**F**	**P > F**
Month	4	53.1	<0.001
Age	6	26.4	<0.001
Month × age	23	1.9	0.01
LSTINC			
Source	**d.f.**	**F**	**P > F**
Month	4	55.7	<0.001
Age	6	801	<0.001
Month × age	23	49.1	<0.001

Age		April	May	June	July	August
				Monthly mean		
3	INC1	**0.59**	**0.57**	0.55	**0.58**	*0.48*
	INC2	0.31	*0.34*	**0.37**	0.36	0.38
	LSTINC	0.13	0.15	0.16	0.15	0.17
4	INC1	**0.60**	**0.59**	*0.55*	*0.54*	*0.51*
	INC2	*0.23*	**0.33**	0.38	**0.40**	**0.37**
	LSTINC	0.09	0.11	0.10	0.09	0.11
5	INC1	**0.64**	*0.59*	0.56	*0.54*	*0.51*
	INC2	*0.25*	*0.31*	0.36	**0.40**	**0.41**
	LSTINC	0.08	0.08	0.07		0.08
6	INC1	**0.63**	**0.60**	0.56	0.57	*0.52*
	INC2	*0.25*	**0.33**	0.36	0.36	**0.40**
	LSTINC	0.06	0.06	0.06		0.06
	INC1	**0.63**	**0.62**	*0.55*	*0.55*	*0.50*
	INC2	*0.28*	*0.34*	**0.37**	0.33	0.36
	LSTINC	0.05	0.05	0.06		

Sample means and Tukey's multiple comparison of means. Means significantly (*P* < 0.05) **larger** or *smaller* than at least one other mean in the same row are denoted by bold type and italics, respectively. A mean that is both larger than and smaller than another mean in the same row is in bold italics. INC1 = first-year increment, INC2 = second-year increment, LSTINC = increment in year prior to sampling. Increments in millimeters.

size would be expected to bias results, giving northern herring a larger otolith increment value. This bias may to some extent mask the decline in first-year increment from southern to northern sampling locations and the differences between the two Bothnian Bay locations in Fig. 2.

The large first increment in the southern Baltic compared to other areas reflects the early spawning, long growth season and high productivity of the sea in this area (Ojaveer 1981, Lenz 1985). In the Bothnian Sea and Bothnian Bay, increments were expectedly smaller. The very small increment of Archipelago Sea herring cannot be explained by climatological or oceanographic factors. The Archipelago Sea (60°N) is situated about halfway between the Rügen (54°N) and northern Bothnian Bay (66°N) locations. In the Archipelago Sea, a decline in herring growth has been observed since the late 1980s, leading to a reduction in spawning fish size of about 30% (Nordic Council of Ministers 1994, Rajasilta et al. 1999). The decline in herring growth has been attributed to a decline in salinity, acting on prey abundance (Vuorinen et al. 1998) or on herring osmoregulation (Rajasilta et al. 1999).

In the Archipelago Sea, the relatively larger first increment of early spawners is an indication that these fish have experienced a long first growth season and themselves were hatched from an early spawn. The difference in spawning time does not in itself indicate any degree of genetic or reproductive separation between early or late spawners. In the Archipelago Sea, herring spawning is continous over several months (Rajasilta et al. 1992). As noted previously by Hellevaara (1912) from the growth pattern of Archipelago Sea herring scales, individual fish may shift their time of spawning, i.e., from early to late spawning or vice versa. The Baltic spring and autumn spawning herring have traditionally been regarded as separate populations (e.g., Ojaveer 1981). The difference in first-year growth between early and late spawners found here supports the hypothesis of Aneer (1985) that spawning time is determined by food availability (and, consequently, growth) prior to spawning. Assuming there is no systematic difference in feeding success or physiological efficiency between fish hatching at different times, the larvae of early spawners will recruit as early spawners themselves.

Second-year growth was inversely related to growth in the first year. Over the spawning season in the Archipelago Sea data, there was an increasing trend in second-year growth, seemingly compensating the decreasing trend in first-year growth. Also between areas, herring from the northern locations in the Gulf of Bothnia had comparatively large otolith increments in the second year of life. As suggested by Aneer (1985), the variable spawning time of Baltic herring can be explained by food availability and, hence, by growth history. In the Archipelago Sea, the youngest spawning fish are 2 years old and 3-year-old fish are probably fully recruited to the spawning population (Hellevaara 1912, Rajasilta 1992). Good growth in the first year will enable the fish to invest energy in gonad development and to recruit spawn earlier than fish with less growth.

Herring exhibit a regular pattern of body growth, with smoothly declining annual increments, where the investment into gonad growth does not show up as retarded somatic growth (Iles 1974, 1984). Recruit spawners are 2 or 3 years old in all parts of the Baltic (Hellevaara 1912, Ojaveer 1981), even though they differ much in size. As suggested by Pauly (1994), spawning is not directly related to size or age, but instead regulated by oxygen availability and gill surface. According to this theory, somatic growth tends toward a certain, individual size at which gonad development commences. Large first-year growth is then expected to be followed by modest second-year growth, followed, in the case of Baltic herring, by gonad development and spawning in the third year of life. Conversely, small first-year growth is expected to be compensated by large growth in the second year, followed by gonad development and spawning in the third or fourth year of life. As pointed out by Wootton (1985), fish actively partition resources between gonad and soma, which together with individual differences in foraging success will bring about variation of growth and reproduction parameters.

Compared to other areas, Achipelago Sea herring had small increments, and in contrast to other areas, Archipelago Sea herring had poor growth both in the first and in the second year. This study gives no clue to the reason for this aberrant growth pattern. Possibly, growth in the two first years of life is so small that the partition of resources between soma and gonads by the mechanism outlined above commences only in the third year. The difference in growth and growth pattern between the Archipelago Sea and other areas indicates the existence of some local source of disturbance, calling for intensified environmental monitoring. Poor growth of individual fish is a sign of low production of the fish stock, which may call for management action. Obviously, this study yields no management advice, but points to the importance of monitoring individual herring growth for management purposes.

Growth completed in the year prior to spawning did not vary significantly between spawning months. Between areas, only the small increment of Archipelago Sea fish differed from other locations. The single high value for age group 6 in the northern Bothnian Bay may be a chance result of aging inaccuracy. According to the Pauly "oxygen hypothesis" referred to above, the release of gametes in spawning creates a surplus oxygen capacity available for somatic growth, and a new cycle of gonad development again starts when the surplus has been used up. By this hypothesis, no systematic difference in growth during the year prior to spawning is expected.

Acknowledgments

We thank Dr. Nils Jönsson, Dr. Thomas Lorenz, Mr. Ralf Granlund, Mr. Alpo Huhmarniemi, and Mr. Atso Romakkaniemi for their assistance with the samples. Sötvattenslaboratoriet Drottningholm kindly assisted us with laboratory facilities and measuring equipment. The financial assistance to Jan Eklund from Suomen Luonnonvarain Tutkimussäätiö is gratefully acknowledged. We also thank the two referees, whose helpful criticism much improved this paper. This work was in part financed by the European Union as part of the DG XIV study no. 96/98.

References

Aneer, G. 1985. Some speculations about the Baltic herring (*Clupea harengus membras* L.) in connection wih the eutrophication of the Baltic Sea. Can. J. Fish. Aquat. Sci. 42(Suppl.1):83-90.

Beamish, R.J., and G.A. McFarlane. 1995. A discussion of the importance of aging errors and an application to walleye pollock: The world's largest fishery. In: D.H. Secor, J.M. Dean, and S.E. Campana (eds.), Recent developments in otolith research. University of South Carolina Press, Columbia, pp. 545-566.

Blaxter, J.H.S. 1985. The herring: A successful species? Can. J. Fish. Aquat. Sci. 42 (Suppl.1):21-30.

Campana, S.E. 1990. How reliable are growth back-calculations based on otoliths? Can. J. Fish. Aquat. Sci. 47:2219-2227.

Campana, S.E., and J.D. Neilson. 1985. Microstructure of fish otoliths. Can. J. Fish. Aquat. Sci. 42:1014-1032.

Campana, S.E., M.C. Annand, and J.I. McMillan. 1995. Graphical and statistical methods for determining the consistency of age determinations. Trans. Am. Fish. Soc. 124:131-138.

Cushing, D.H. 1975. Marine ecology and fisheries. Cambridge University Press, Cambridge. 278 pp.

Dahl, K. 1907. The scales of the herring as a means of determining age, growth and migration. Rep. Norw. Fish. Mar. Invest. 2(6).

deBarros, P., and J.C. Holst. 1995. Identification of geographic origin of Norwegian spring-spawning herring (*Clupea harengus* L.) based on measurements of scale annuli. ICES J. Mar. Sci. 52:863-872.

Eklund, J., R. Parmanne, and G. Aneer. 2000. Between-reader variation in herring otolith age and effects on estimated population parameters. Fish. Res. (Amst.) 46:147-154.

Hellevaara, E. 1912. Tutkimuksia Lounais-Suomen silakasta (Studies of SW Finland herring). Suomen Kalatalous 1:19-59. (In Finnish.)

Iles, T.D. 1974. The tactics and strategy of growth in fishes. In: F.R. Harden Jones (ed.), Sea fisheries research. Elek Science, London, pp. 331-345.

Iles, T.D. 1984. Allocation of resources to gonad and soma in Atlantic herring (*Clupea harengus* L.) In: G.W. Potts and R. Wootton (eds.), Fish reproduction. Academic Press, London, pp. 331-347.

Kompowski, A. 1971. Typy otolitow sledzi poludniowego Baltyku (Types of otoliths in herring from the southern Baltic). Prace Morskiego Instytutu Rybackiego 16 (Ser. A):109-141. (In Polish, with English abstract.)

Lenz, J. 1985. Plankton. In: G. Rheinheimer (ed.), Meereskunde der Ostsee. Springer Verlag, Berlin, pp. 136-150. (In German.)

Matthäus, W. 1985. Temperatur, Salzgehalt und Dichte (Temperature, salinity and density). In: G. Rheinheimer (ed.), Meereskunde der Ostsee. Springer Verlag, Berlin, pp. 75-81. (In German.)

Mosegaard, H., H. Svedäng, and K. Taberman. 1988. Uncoupling of somatic and otolith growth rates in Arctic char (*Salvelinus alpinus*) as an effect of differences in temperature response. Can. J. Fish. Aquat. Sci. 45:1514-1524.

Nordic Council of Ministers. 1994. Growth changes of the herring in the Baltic. Temacord, Copenhagen, p. 532.

Ojaveer, E. 1981. Marine pelagic fishes. In: A. Voipio (ed.), The Baltic Sea. Elsevier Scientific Publishing Co., Amsterdam, pp. 276-292.

Parmanne, R. 1990. Growth, morphological variation and migrations of herring (*Clupea harengus* L.) in the northern Baltic Sea. Finnish Fish. Res. 10:1-48.

Pauly, D. 1994. On the sex of fish and the gender of scientists. Chapman and Hall, London. 248 pp.

Rajasilta, M. 1992. Timing of spawning in the Baltic herring (*Clupea harengus membras*) in the Archipelago Sea, SW Finland: Regulatory mechanisms and consequences for offspring production. Dissertation, University of Turku, Turku, Finland. Acta Univ. Turkuensis Ser. A81. 19 pp.

Rajasilta, M., J. Eklund, P. Laine, N. Jönsson, and T. Lorenz. 1999. Intensive monitoring of spawning populations of the Baltic herring (*Clupea harengus membras* L.). Final Report of study project No. 96 068, 1997 1998. University of Turku, Finland, 74 pp.

Rajasilta, M., J. Eklund, M. Hänninen, M. Kurkilahti, J. Kääriä, P. Rannikko, and M. Soikkeli. 1992. Spawning of herring (*Clupea harengus membras* L.) in the Archipelago Sea. ICES J. Mar. Sci. 50:233-246.

Runnström, S. 1936. A study of the life history and migrations of the Norwegian spring-herring based on analysis of the winter rings and summer zones of the scale. Rep. Norw. Fish. Mar. Invest. 5(2):5-103.

Vuorinen, I., J. Hänninen, M. Viitasalo, U. Helminen, and M. Kuosa. 1998. Proportion of copepod biomass declines with decreasing salinity in the Baltic Sea. ICES J. Mar. Sci. 55:767-774.

Wootton, R.J. 1985. Energetics of reproduction. In: P. Tytler and P. Calow (eds.), Fish energetics: New perspectives. Croom Helm, London, pp. 231-254.

Anatomy of a Strong Year Class: Analysis of the 1977 Year Class of Pacific Herring in British Columbia and Alaska

D.E. Hay, M.J. Thompson, and P.B. McCarter
Department of Fisheries and Oceans, Pacific Biological Station, Nanaimo, British Columbia, Canada

Abstract

In 1977 an exceptionally strong year class of herring (*Clupea pallasii*) developed in many parts of British Columbia (B.C.) and Alaska. Within B.C. the strong 1977 year class occurred among geographically separated stocks, with different spawning times. In B.C. the number of eggs deposited in 1977 was approximately normal. The 1977 year class recruited to the B.C. herring fishery in 1980 as age 3 and dominated catches for more than 5 years. From analysis of sample data, we determined that the age-specific size of juveniles from the 1977 year class was normal but in later years size at age decreased relative to that of year classes from other years. We confirmed this through analyses of early growth from archival collections of herring scales. Scale growth of the 1977 year class during ages 1 and 2 was normal, indicating robust somatic growth of 1977-year-class juveniles in 1978 and 1979. Postrecruit growth, however, was below normal. The changes in age-specific difference in growth correspond approximately to different habitats used by pre- and postrecruit herring. In general, prerecruits occupy shallow, nearshore waters for the first 2 years of life. Robust juvenile growth within a strong year class is evidence of an abundant food supply for juveniles in 1978 and 1979. In contrast, the relative growth rates of 1977-year-class adults in offshore waters were much slower in postrecruits relative to that of preceding years, perhaps reflecting density-dependent limitations to growth. We speculate that a climate-based enhancement of juvenile food supply, particularly copepods, could have led to increased juvenile growth and decreased predation on young herring stages. This is consistent with the observation that the 1977 year class was strong in several other species.

Introduction

The year class of herring spawned in 1977 was strong throughout much of northern British Columbia (B.C.), parts of southeastern and central Alaska, and the Bering Sea (Hollowed et al. 1987, Hollowed and Wooster 1995). A strong 1977 year class also occurred in several other species, including blackcod and lingcod (Hollowed et al. 1987). The year 1977 was also recognized as a time of significant change in oceanographic conditions in the Gulf of Alaska (Trenberth and Hurrell 1995, Beamish et al. 1999). In this paper we review the geographic distribution of this exceptional 1977 herring year class among different areas in B.C. and Alaska. For B.C. stocks we examine historical data on the distribution and timing of herring spawning, and growth or size-at-age data that are collected annually in support of catch monitoring. We examine these results in the context of other key observations about herring biology, as well as observations about 1977 year classes in other species, to comment on factors affecting the development of this year class in herring.

In B.C., herring mature sexually at age 3 (Hay 1985) so that catch samples from the herring roe fishery consist mainly of age-3 fish and older. Some samples, however, contain some prerecruit fish at age 2, or about 24 months of age. In 1979, when the 1977 year class was age 2, most 2-year-old herring were about the same size, or even larger than age-2 herring from other years. This indicates that individuals within this strong 1977 year class grew normally, or more rapidly than those of other years. Such an observation, however, could have resulted from a sampling bias that led to preferential selection of the fastest-growing 2-year-old fish. This could have occurred in 1979 and 1980 (when the 1977 year class was age 2 and age 3, respectively). If so, this could give the misleading impression that this year class was growing faster. To examine this further, we analyzed patterns of growth from scale collections to retrospectively compare growth during the first 2 years of life with that of other years.

In 1977, a strong 1977 year class developed in widely separated herring populations with different spawning times. Therefore in 1977 conditions favoring high survival occurred over a broad range of space (thousands of kilometers) and time (at least months and perhaps up to 2 years). Evidence from spawn survey data in B.C. indicates that the number of eggs deposited in 1977 was normal. Recent information on the distribution of larval and juvenile (< 2 years old) Pacific herring (Haegele 1997, Hay and McCarter 1997) indicate that rearing habitats are mainly in shallow, nearshore waters. Further, we know that copepods are usually the dominant food organisms of juvenile Pacific herring (Wailes 1936, Haegele 1997). These simple observations, combined with the observations that the year class was synchronous over a broad spatial and temporal range, indicate that conditions suitable for good survival existed for a long time, over a broad geographical area, with positive effects felt in inshore, shallow waters. In this paper we briefly but systematically consider alternate explanations for the development of this strong year class. We tentatively conclude that a simple and direct explanation for the 1977 year class was an exceptional production of zooplankton, probably copepods, that must have been available to early life history stages of herring and other species between 1977 and 1979. This increase in food supply would have reduced predation rates on eggs, larvae, and juveniles of herring between 1977 and 1979. If such an event occurred, and if it were prolonged and geographically widespread it would explain all our key observations (listed in the discussion) about increased survival and normal juvenile growth of herring, as well as synchronously strong, 1977 year classes in other species.

Methods
Fish Sampling Data and Analyses

Age-frequency data from Alaska were taken from published data, mainly from several technical report series from the Alaska Department of Fish and Game. Given a choice of data from different gear sources (i.e., purse seines or gillnets) we chose purse-seine data. Data from southeastern Alaska were based on cumulative estimates of published analyses of raw data for individual samples (Blankenbeckler and Larson 1987). All sources of Alaskan data are cited. Size and age data for B.C. herring were available from routine samples taken throughout the B.C. coast, mainly during spring months prior to spawning (March and April). These data include length, weight, sex, gonad weight, and age (determined from scales). Usually 400-500 samples of about 100 fish each are taken annually from all parts of the coast. The data are stored digitally (Microsoft Access™) on computers at the Pacific Biological Station, Nanaimo, B.C. Extracted data were analyzed using Minitab™ statistical software. Data from other areas of the eastern North Pacific were taken from published reports that present data on frequency of age-3 fish in 1980 (the 1977 year class). Estimates of the total numbers at age were taken from the 1999 annual assessment report (Schweigert et al. 1998).

Age-frequency analysis was limited to samples collected by purse-seine nets between January 1 and April 31. Restricting the samples to purse seines eliminated all gillnet samples that select for larger fish (Hay et al. 1986). Also, gillnet samples were taken closest to the spawning areas. Size-at-age analysis was limited to samples collected by purse-seine nets or midwater trawls between November 1 and April 31. Some might argue, however, that the age percentages of each age class do not constitute evidence of a strong year class because total numbers of individuals may change over time. We examined this by a comparison of estimates of numbers at age, for four different areas: Togiak, Alaska, and three northerly regions of British Columbia. Data were provided by F. Funk (Alaska Department of Fish and Game) and J. Schweigert (Canada Department of Fisheries and Oceans). Some of these estimates are published in stock assessment reports (e.g., Schweigert et al. 1998).

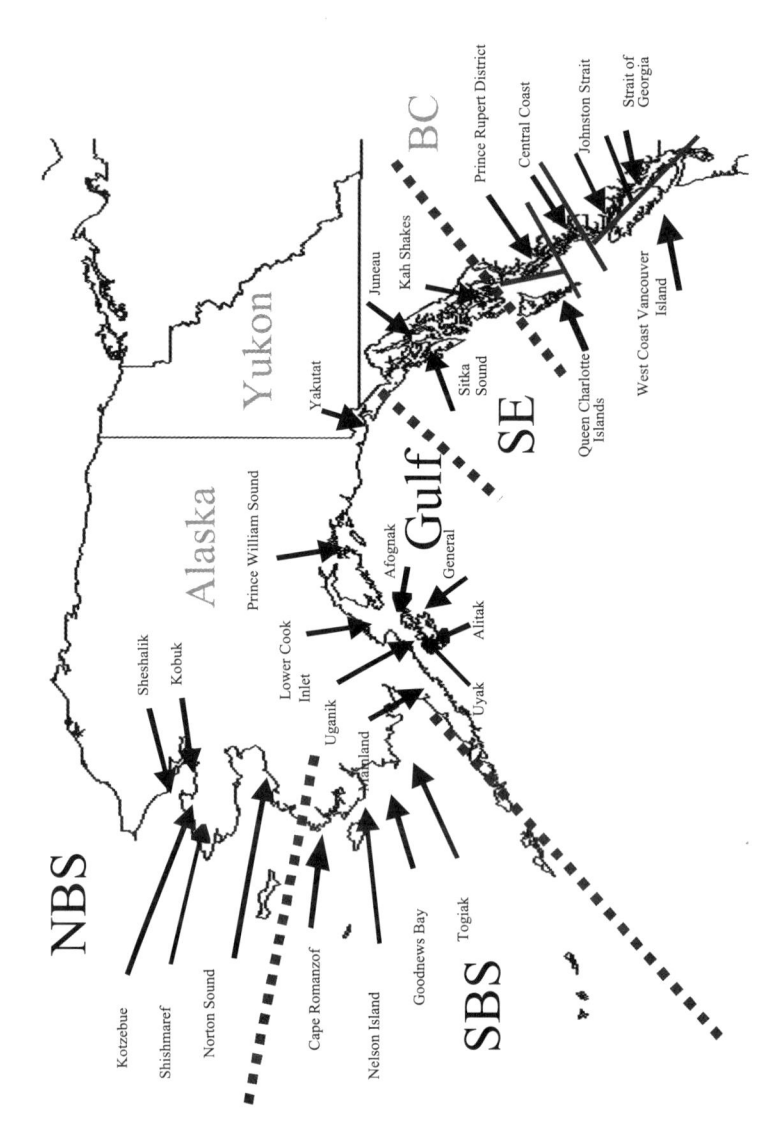

Figure 1. The Alaskan and British Columbia coast showing approximate locations of names used in the text and major geographical divisions. NBS and SBS refer to northern and southern Bering Sea, Gulf refers to the Gulf of Alaska, and SE refers to southeastern Alaskan waters.

Herring: Expectations for a New Millennium

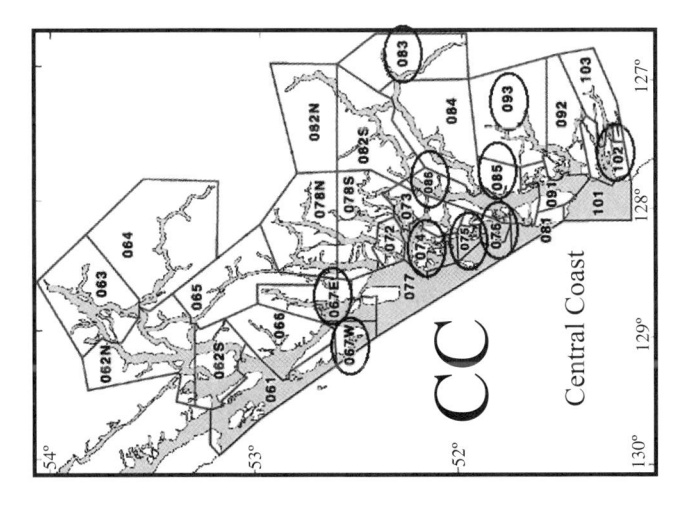

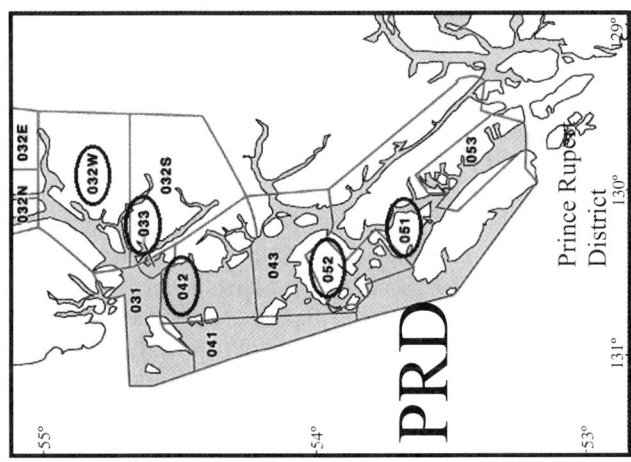

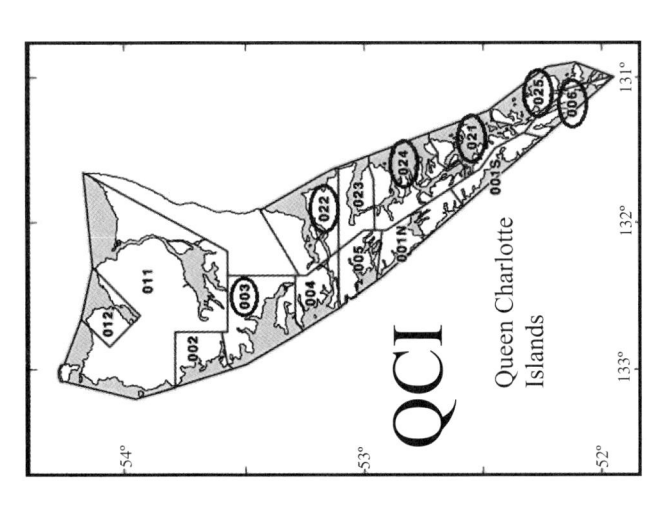

Figure 1. (Continued.) Three inserts show more detail of the British Columbia coast for each of three regions: the Queen Charlotte Islands, Prince Rupert District, and Central Coast. The numbers with circles represent different herring sections noted in the text.

Geographical Definitions

Alaskan waters are differentiated into the southeastern Alaska (SE), Gulf of Alaska (Gulf), southern Bering Sea (SBS), and northern Bering Sea (NBS) areas, as shown in Fig. 1. The B.C. coast is divided into six different regions: Queen Charlotte Islands (QCI), Prince Rupert District (PRD), Central Coast (CC), Johnstone Strait (JS), Strait of Georgia (SOG), and West Coast of Vancouver Island (WCVI). Each region can be divided further into statistical areas (Fisheries and Oceans Canada) and herring subareas called sections (Haist and Rosenfeld 1988). In total, there are 108 different herring spawning sections in B.C. of approximately equal size. Figure 1 shows sections only for the QCI, PRD, and CC northerly regions.

Spawn Indices

Herring spawn is quantified annually as the product of spawn length (in meters) by area-specific spawn coefficients based on median spawn width and egg layers (Hay and Kronlund 1987, Hay and McCarter 1999). Spawning does not occur at each location every year, but for those years in which spawning does occur, the spawn index for a location is an estimate of the total spawning area (in square meters), adjusted by the density of spawn. The index for larger geographical groupings (sections, statistical areas, and regions) is simply the sum of the index of the component areas.

Scale Analyses

Annual sampling of herring catches involves routine collection of scales for age determination (Hamer 1989). Scales were collected mainly during the period immediately preceding the roe fishery. These scales were mounted on glass slides and are archived and organized so that they can be individually related to data stored in the catch sampling Access™ database. We took various scale measurements from digitized images using an Olympus™ (BH2) microscope with a Pixera™ camera and SigmaScan™ image analysis software. A calibrated scale from an ocular micrometer was used to convert measurements made on the screen to 0.01-mm units with a measurement precision of about 0.05 mm. We examined 1,677 scales of age-3, age-4, and age-5 herring collected between 1972 and 1997 in the QCI, PRD, and CC regions. This included 80 scales from the 1977 year class, 40 collected as age-3 fish in 1980 and 40 as age-5 fish in 1982. We measured the distance between the focus (the point where scale growth starts) to the end of the first and second annuli and the edge of the scale (i.e., the third annulus). Measurements always were made in the anterior-posterior axis (Fig. 2) corresponding to the axis used to measure fish length (from nose to tail). Fish length data represent standard lengths, from the tip of the snout to the end of the hypural plate in the tail. We did not make back-calculated estimates of length at age because of possible problems with this approach (Francis 1990). Instead we used the distance between the focus and scale annuli as an index of fish length. The assumption that

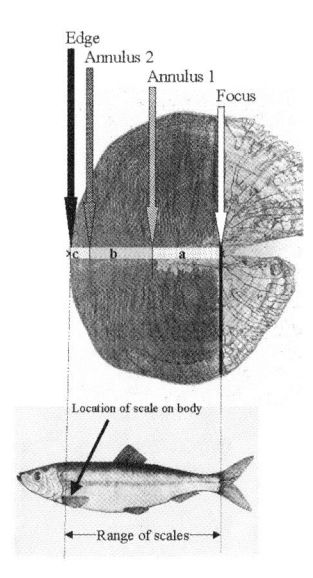

Figure 2. A scale from an age-3 fish (approximately 35 months of age from hatching) collected from the 1977 year class in March 1980. The large arrows identify the scale focus, first annulus, second annulus, and scale edge (corresponding to the end of growth following the third summer of life). The line between the focus and edge is distinguished as zones a, b, and c, corresponding to growth in the first, second, and third year of life, respectively. The relationship of the scale to the corresponding area of growth of the fish is indicated by the inserted figure of a herring.

sub-sections of the scale correspond to lengths of early ages is based on a positive linear relationship between the total scale length (focus to edge) and total fish length, and we provide evidence of this relationship.

Results
Geographical Distribution of the 1977 Year Class in B.C. and Alaska

The strong 1977 year class occurred in most areas of the northern and southern Bering Sea, parts of the Gulf of Alaska, southeastern Alaska, and the QCI, PRD, and CC regions of B.C. The broad geographical distribution of the strong year class is shown in Table 1 as the proportion of the 1977 year class age-4 fish in 1981 from samples taken throughout Alaska. Table 2 shows the proportion of the 1977 year class taken as age-3 fish from B.C. waters. In most areas the 1977 year class dominated, often comprising 50% or more of the population, and representing more than 90% in

Table 1. List of herring age-frequency data for various locations by purse seine in Alaska, highlighting the 1977 year class as age-4 herring in 1981. The sample from Yakutat represents age-5 herring in 1982.

Location	Number of fish	1	2	3	4	5	6	7	8+
Chukchi Sea[a]									
Shishmaref Area (Table 18)	166				**28.9**	4.8	9.0	20.5	36.7
Kotzebue Area (Table 19)	840		0.6	7.5	**68.0**	5.7	1.8	3.2	13.2
Kobuk Area (Table 20)	843		0.7	7.8	**69.3**	8.7	1.2	1.5	10.8
Shehalik Area (Table 21)	618		0.3	1.6	**64.8**	9.2	1.6	3.3	19.0
Lower Cook Inlet[b]	31			6.5	**74.2**	16.1	3.2		
Eastern Bering Sea[c]									
Norton Sound (Table 39)	3,158			6.5	**74.5**	9.5	1.3	6.2	2.0
Cape Romanzof District (Table 31)	588			7.8	**55.3**	13.2	1.5	10.4	11.8
Nelson Island (Table 29)	716			2.1	**37.3**	14.2	1.5	24.3	20.5
Goodnews Bay (Table 25)	1,249			1.0	**74.5**	11.0	1.4	7.8	4.1
Togiak[c] (Table 5)	1,917				**61.9**	8.1	1.1	16.1	11.6
Kulukak District (Table 1)	182			1.6	**87.9**	9.9			0.5
Nunavachek District (Table 2)	397			2.5	**57.6**	7.8	1.3	17.8	12.8
Hagemeister District (Table 4)	845			0.7	**53.7**	8.9	1.6	19.4	15.6
Prince William Sound[d]				2.1	**19.2**	**71.1	6.7	0.9	
Kodiak Area[e]									
Afognak	598		1.7	11.9	**34.4**	8.5	12.4	27.3	3.8
Uganik	572		7.0	23.4	**51.4**	10.5	3.3	3.8	0.5
Uyak	663		19.8	20.5	**43.0**	7.5	3.2	4.2	1.8
General	403		0.6	13.8	**28.5**	22.2	10.3	24.3	0.2
Mainland	663		0.3	12.7	**57.5**	21.3	3.0	2.7	2.6
Southeastern[f]									
Kah Shakes	557			4.7	**65.0**	10.4	11.3	3.8	4.8
Juneau	231			6.9	**71.4**	18.7	1.3	1.7	
Sitka	297			1.3	**15.2**	**68.0	14.2	1.3	
Yakutat	101			8.0	**22.0**	49.0	21.0	1.0	

All frequencies greater than age 7 are pooled as age 8+. Age frequencies for the Togiak area are shown as a pooled summary for the composite area and for component areas. Two areas with strong 1976 year classes are noted with double asterisks.
[a]Whitmore et al. 1983.
[b]Schroeder 1989 (Table 50).
[c]Fried et al. 1982.
[d]Funk and Sandone 1990 (percentages calculated from data presented in Table B, page 33).
[e]Burkey and Ried 1988 (Table 33).
[f]Blankenbeckler and Larson 1987.

Table 2. List of herring age-frequency data for various locations (or sections) within major regions in northern British Columbia, highlighting the 1977 year class as age-3 herring in 1980.

Regions and sections	Number	Age 2	3	4	5	6	7	8+
Queen Charlotte Islands								
3	267	0.37	**76.03**	13.11	4.49	3.37	1.87	0.75
6	1,188	1.35	**77.78**	4.88	6.23	3.45	2.10	4.21
21	449	2.45	**90.87**	3.56	0.67	1.56	0.67	0.22
22	86	—	**84.88**	10.47	1.16	3.49	—	—
23	356	10.11	**80.90**	1.12	0.84	2.81	0.84	3.38
24	224	0.45	**87.05**	5.80	4.02	1.79	0.89	—
25	1,830	0.33	**84.48**	4.21	5.63	2.57	1.91	0.88
All	4,400	1.61	**82.66**	4.82	4.66	2.75	1.66	1.84
North Coast								
32	88	1.14	**84.09**	5.68	4.55	1.14	1.14	2.27
33	707	3.11	**73.27**	9.90	5.09	3.82	2.83	1.98
42	277	—	**60.65**	7.22	13.72	6.50	6.86	5.05
51	185	—	**16.22**	23.78	15.14	20.00	16.22	8.64
52	2,200	2.50	**85.14**	4.68	2.95	2.18	1.59	0.96
All	3,457	2.26	**77.03**	7.00	4.95	3.79	3.04	1.94
Central Coast								
67	1,167	2.06	**83.98**	4.20	3.08	2.23	3.17	1.29
74	1,383	5.64	**70.43**	8.39	9.11	3.69	1.74	1.01
75	182	5.49	**69.23**	6.04	12.64	4.95	0.55	1.10
76	81	—	**51.85**	12.35	19.75	12.35	2.47	1.23
83	180	21.67	**54.44**	5.00	5.56	3.89	5.00	4.44
85	596	15.44	**61.41**	6.71	9.23	4.19	2.35	0.68
86	78	5.13	**58.97**	10.26	15.38	7.69	1.28	1.28
93	162	8.64	**70.99**	8.64	7.41	1.85	1.85	0.62
102	243	—	**13.99**	67.49	5.35	11.52	1.65	—
All	4,072	7.25	**71.49**	6.63	7.81	3.46	2.23	1.13

All data are from sampling records maintained by Fisheries and Oceans Canada. All frequencies greater than age 7 are pooled as age 8+. All samples were collected by purse seine between January and April of each year.

parts of the QCI region of B.C. The 1977 year class was not strong in Prince William Sound (central Gulf of Alaska) or Sitka (Southeast Alaska), although there was a strong year class in those locations in 1976. Within B.C. the year class was particularly strong in the QCI, PRD, and CC regions. It also was relatively strong in parts of the West Coast of Vancouver Island but not exceptional. The strong 1977 year class may not have developed in some herring populations in coastal fjords, but these populations usually are small and samples are not available for all years. There was a broad geographical distribution of populations with a strong 1977 year class, identified as those with high proportions of age-4 fish in 1981 in Alaska or age-3 fish in 1980 in B.C. (Fig. 3).

Estimates of the numbers at age for the 1977 year class, relative to those of the same age for a 10-year period from 1978 to 1987, are shown in Table 3. For these four stock assessment areas, the numbers of individuals in the 1977 year class exceed those of the other year classes from 1978 to 1987. This provides corroborative evidence that the 1977 year class was very strong and that the consistently high percentage of the 1977 year class in age frequencies is a consequence of this strong year class and not an artifact of sampling. The relative strength of the 1977 year class can be compared over a greater period by comparing the numbers of fish at age 4 from these same populations (Fig. 4). Based on estimated numbers at age 4, the 1977 year class exceeds all others from 1975 to 1996 in all areas except the CC region of B.C.

Variation in the Timing and Amount of Spawning in 1977

The herring spawn index (Hay and Kronlund 1987, Hay and McCarter 1999) is estimated annually for all areas. The cumulative index of spawn, shown for each of the three regions (Fig. 5A), was not unusually high or low in any region in 1977. This indicates that the 1977 year class was not generated from an unusual abundance of eggs. Further, there was nothing remarkable about the timing of 1977 herring spawnings in any region (Fig. 5 B-D). This indicates that the 1977 year class appeared to have a normal starting time relative to other years, and with an initial egg number that was not unusually high.

Age-Specific Growth of the 1977 Year Class

The length at age for seven age classes from age 2 to age 8 is shown by year from 1951 to 1997 for the QCI, PRD, and CC regions (Fig. 6). The 1977 year class stands out because the size at age of older age classes was smaller than that of many other years. The important aspects of these comparisons in Fig. 6A-C are as follows: (1) the 1977-year-class size at age was smaller than similar ages in adjacent years, and there appeared to be some depression of size at age, appearing almost as a "trough" in Fig. 7, as the year class passed through the populations; (2) there was a decrease in the size at age (in ages 3 to 8) in all areas since the early 1980s; and (3) the

Herring: Expectations for a New Millennium 181

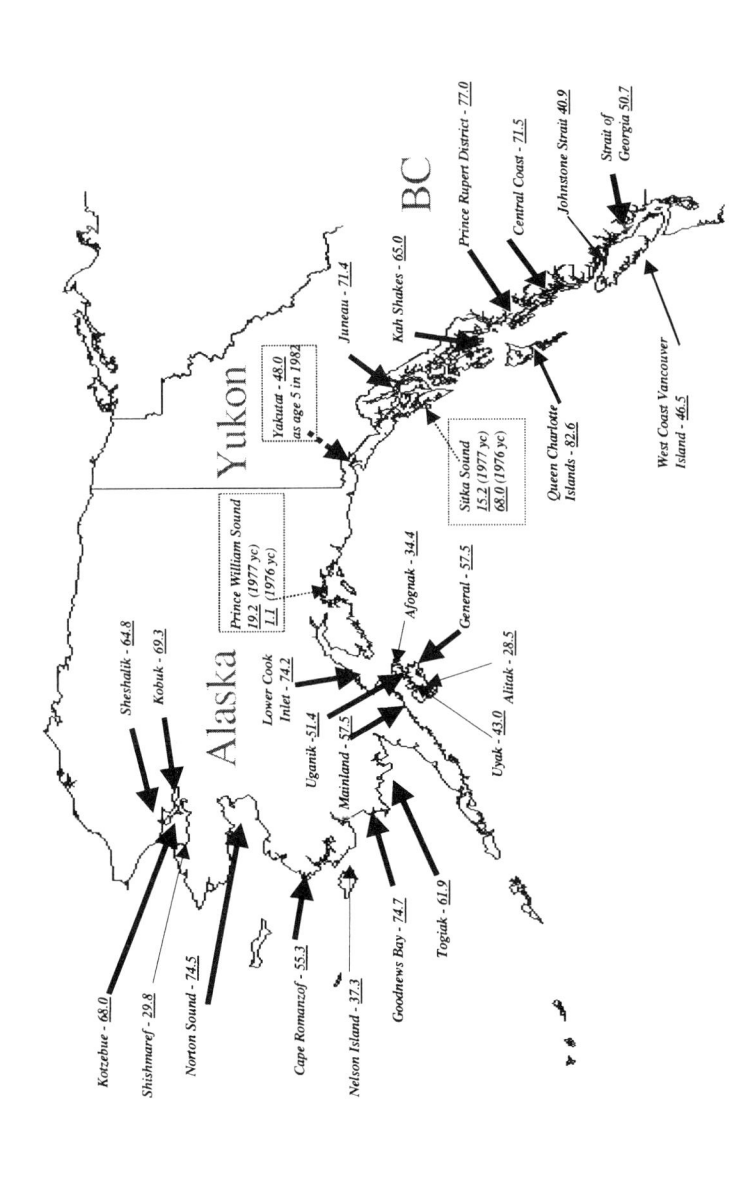

Figure 3. The distribution and relative strengths of the 1977 year class by location in Alaska and British Columbia. The arrows distinguish between locations where the 1977 year class was greater (thick arrow) or less (thin arrow) than 50% of the population. Prince William Sound and Sitka (southeastern Alaska) had strong 1976 year classes. The estimate of the 1977 year class in Yakutat was determined from ages 5 fish in 1982.

Table 3. Estimated pre-fishery abundance of herring, in millions of fish, shown for ages 4-8 for Togiak, Alaska and ages 3-8 for the QCI, PRD, and CC regions of British Columbia.

Year	3	4	5	6	7	8
Togiak, Alaska						
1978	—	415	309	84	1	13
1979	—	44	325	234	60	0
1980	—	93	34	243	174	44
1981	—	**1,329**	73	25	170	119
1982	—	1,122	**1,045**	56	19	126
1983	—	410	882	**799**	42	14
1984	—	73	324	682	**604**	31
1985	—	211	58	254	527	**461**
1986	—	72	167	45	195	398
1987	—	236	57	131	35	149
Queen Charlotte Islands						
1978	179	84	38	69	38	24
1979	101	97	42	16	25	14
1980	**1,997**	62	47	15	5	8
1981	145	**1,240**	38	25	8	2
1982	59	90	**745**	21	13	4
1983	64	36	55	**446**	13	7
1984	433	39	22	32	**253**	7
1985	191	261	23	12	18	**142**
1986	44	116	152	13	7	9
1987	91	27	69	87	7	4
Prince Rupert District						
1978	95	104	26	37	52	31
1979	100	52	50	10	13	18
1980	**1,728**	56	26	22	4	5
1981	313	**1,009**	31	13	10	2
1982	390	183	**577**	17	7	5
1983	403	228	106	**332**	10	4
1984	1180	240	136	63	**197**	6
1985	222	695	140	77	35	**104**
1986	208	130	396	76	39	16
1987	783	120	72	206	37	19
Central Coast						
1978	65	36	47	48	15	8
1979	45	39	17	12	6	1
1980	**455**	34	29	13	9	4
1981	87	**344**	26	21	9	6
1982	94	66	**252**	17	13	5
1983	41	68	46	**158**	10	7
1984	35	30	47	29	**95**	6
1985	134	24	19	27	15	**46**
1986	56	93	16	11	15	8
1987	106	40	62	10	7	9

The estimates are shown for a 10-year period from 1978 to 1987, and the numbers at age of the 1977 year class are shown in bold. The data for B.C. herring were provided by J. Schweigert and are similar to published estimates (Schweigert et al. 1998). Alaskan data were provided by F. Funk, Alaska Department of Fish and Game.

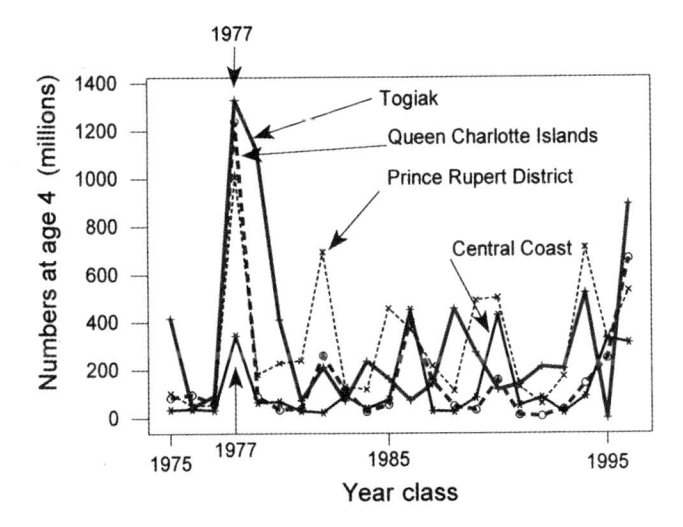

Figure 4. The numbers of fish at age 4 between 1975 and 1996 for Togiak, Alaska and three regions in British Columbia. The figure shows that the 1977 year class was strong in all of the populations and that it was the strongest year class observed in three of the populations between 1975 and 1996. The numbers are estimated from age-structured analyses used for stock assessment and were provided by Fritz Funk, Alaska Department of Fish and Game and J. Schweigert, Canada Department of Fisheries and Oceans. The thick gray line represents Togiak, Alaska.

size at age of juvenile (age 2) herring shows no particular positive or negative trend in time. These observations support the view that decreases in size at age occurred mainly in older age groups and not in juveniles. This observation is confirmed in Fig. 7 which shows the mean length of herring at each age, with the 1977 year class highlighted, relative to the length at age for other years. The size at age 2 of the 1977 year class was normal, or slightly above normal, but sharply lower at older ages in later years.

Temporal changes in age-specific growth were described previously (Tanasichuk 1997, Tanasichuk and Schweigert 1998), but only for fish of age 3 and older. The stability of size at age 2, relative to the changes in older groups, however, has not been described. Figure 8 shows the temporal sequence of size at age for ages 2 to 5 for the Prince Rupert District. The mean length at age (Fig. 8A) and weight at age (Fig. 8B) of fish are shown according to birth year, so that each year represents a single year class. The interannual differences in size at age are most pronounced in older ages. Both length and weight at age 2 are relatively similar over all years, and it appears that weight at age 2 increased from the mid-1970s to the 1990s during the time that age-specific length and weight of age-3 to age-8 herring declined sharply.

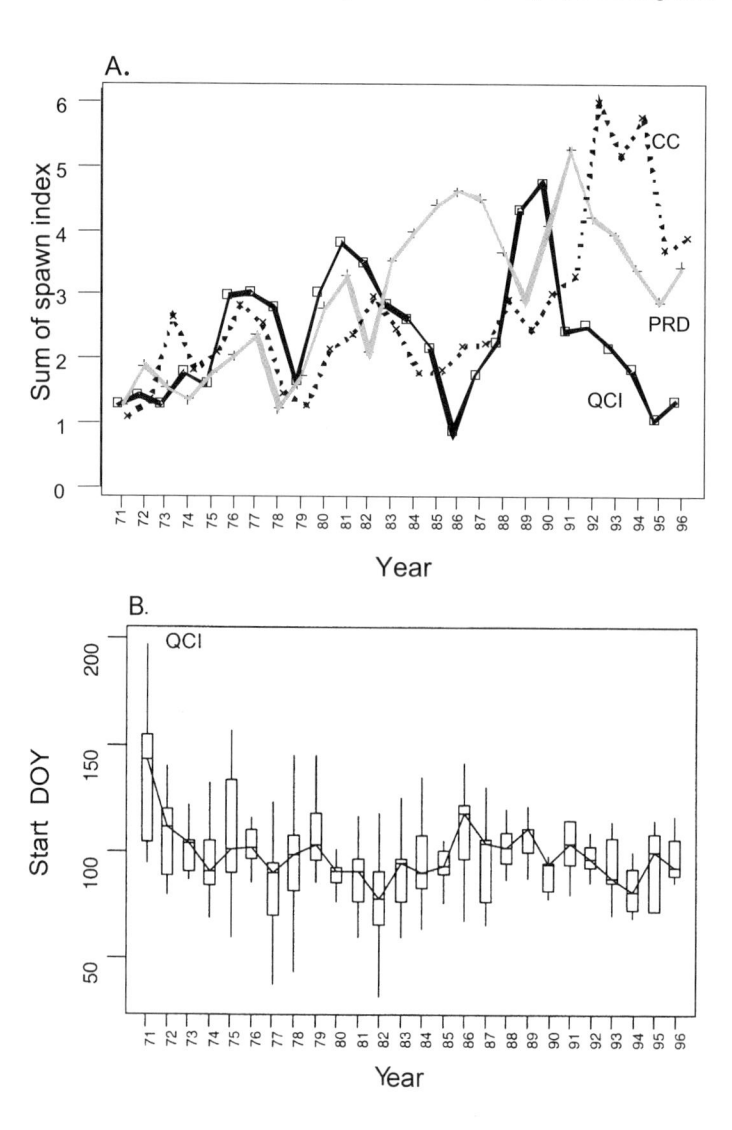

Figure 5. The spawn index and timing of spawning in northern British Columbia in 1977—shown by region, 1971 through 1996 with 1997 indicated. (A) Cumulative spawn index for each region in 1977, with the year 1977 highlighted as a vertical line. (B-D) Mean, 5% CI, and range of spawning days by DOY (day of the year) for each region from 1971 to 1996. The figure shows that the dates and amount of spawning were not exceptional in 1977.

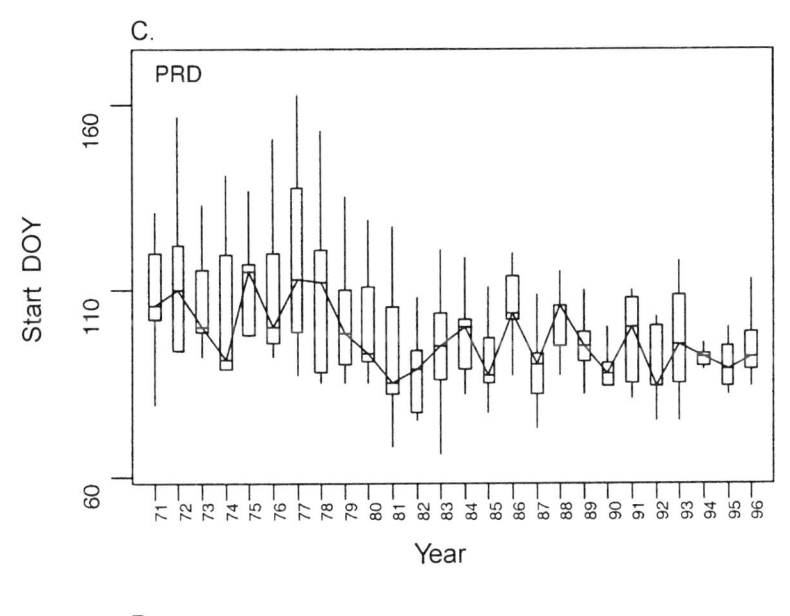

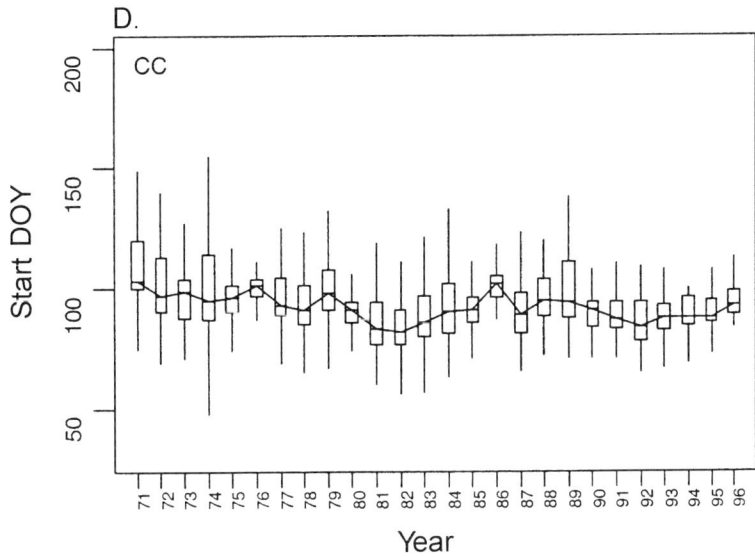

Figure 5. (Continued.)

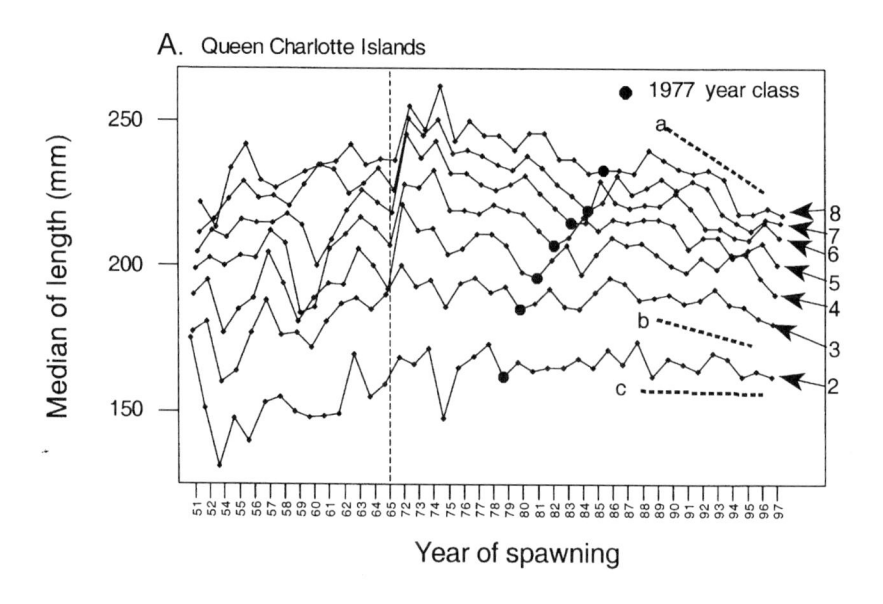

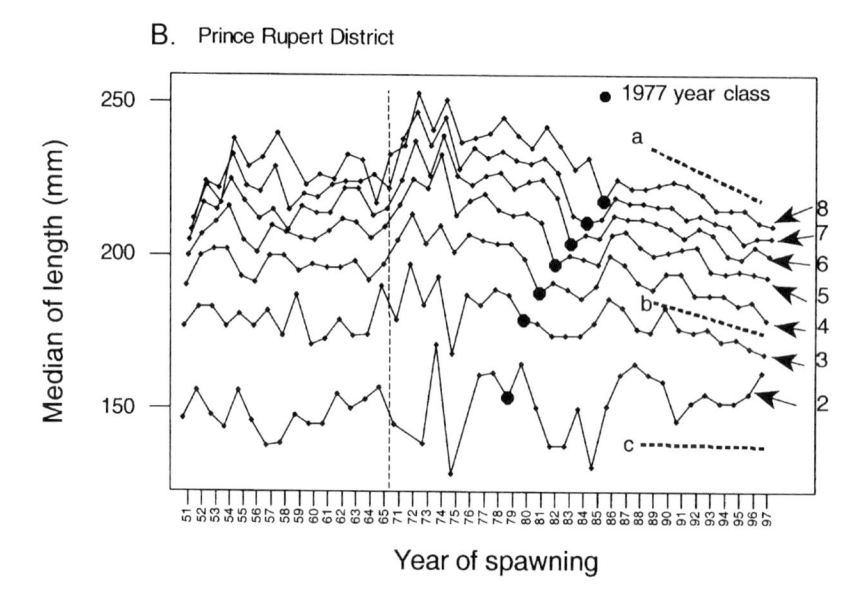

Figure 6. The size at age of herring for three regions, 1951-1997. The length of the 1977 year class, as it progressed through the populations, is indicated by large dots. (The vertical dashed lines indicate periods when there is a gap in the records.) The diagonal dashed lines show the decline in size at age that began in the early 1980s. Note that the decline seems steeper in older fish (diagonal lines a and b) relative to little change for age 2 fish (indicated by line c).

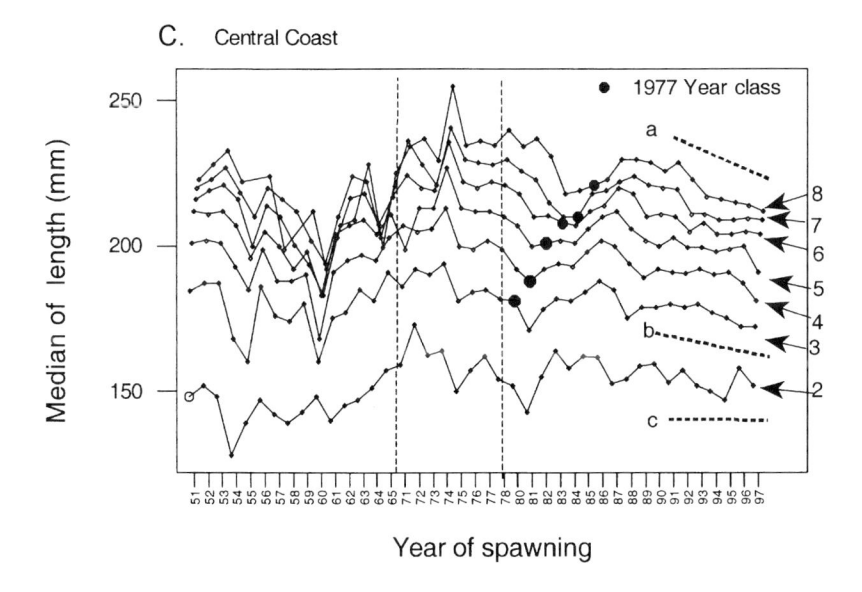

Figure 6. (Continued.)

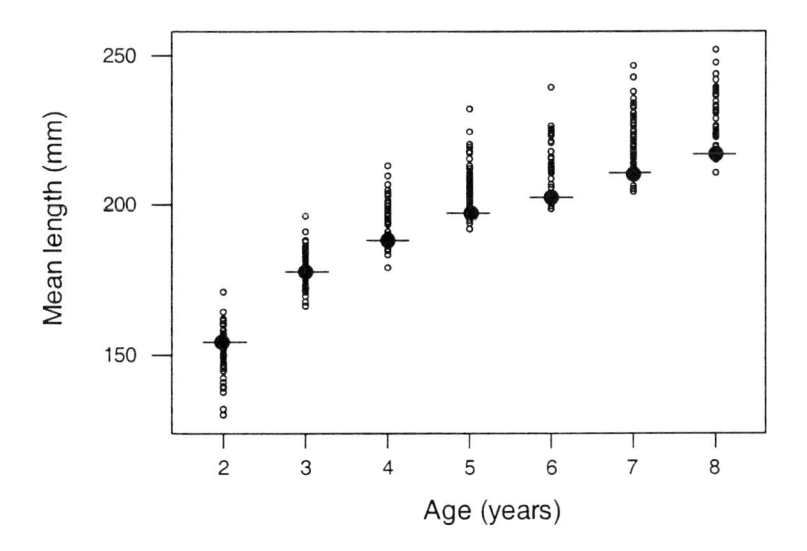

Figure 7. The mean length at age for the 1977 year class (large symbol with short horizontal line) relative to the size at age for all other year classes from 1951 to 1996, shown for the Prince Rupert District only. Note that the mean length at age for the 1977 year class is slightly larger than average at age 2, but becomes smaller than average in subsequent years.

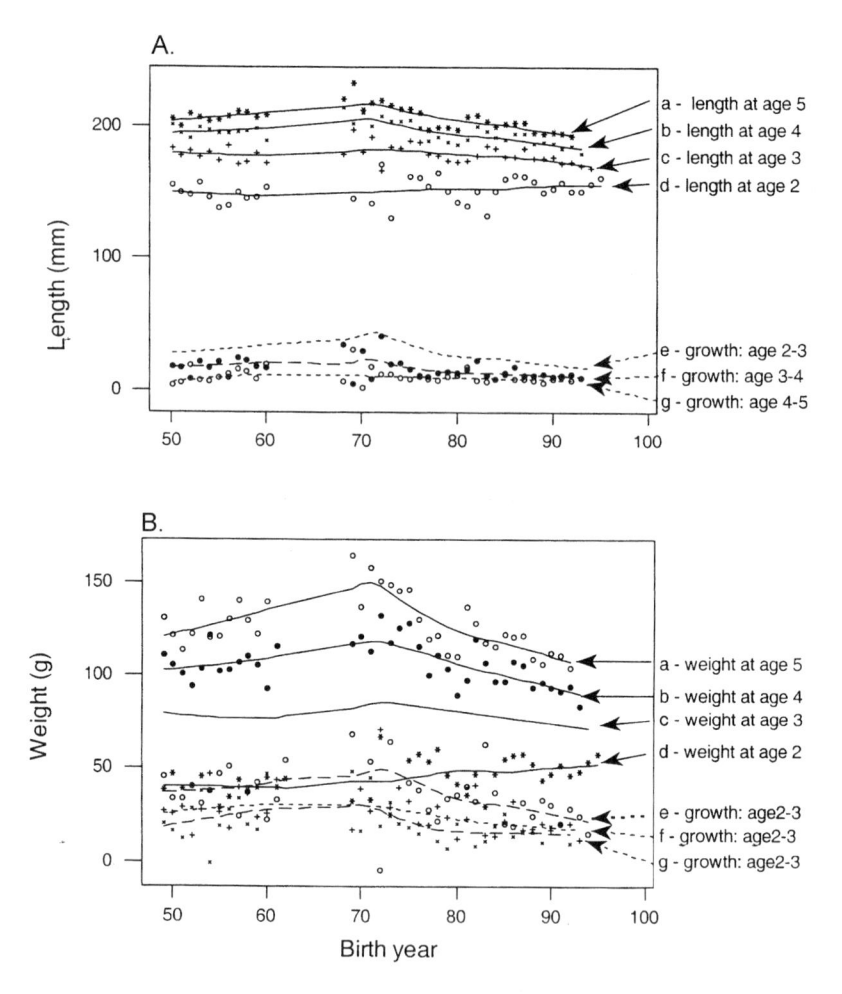

Figure 8. Comparison of length (millimeters) at age for herring from the Prince Rupert District between 1951 and 1996. (A) The four lines at the top show smoothed trends determined from a locally weighted smoothing procedure (LOWESS) in the statistical analysis software (Minitab™) that uses approximately half of the available data points to determine the position of a weighted point. The lines show the smoothed trends of individual years for age 5 (line a), age 4 (line b), age 3 (line c), and age 2 (line d). The lengths are shown according to the year of birth, so each cohort is represented in a vertical line. The differences between the lines: a minus b, b minus c, and c minus d represent the growth increment for fish in their fourth, third, and second years, respectively. The smoothed (LOWESS) differences are shown as lines e, f, and g. Note that line d (length at age 2) is not decreasing, but all growth increments are decreasing. (B) Comparisons of weight (grams) at age for herring from the Prince Rupert District between 1951 and 1996. The explanation for the lines is the same as that for Fig. 8A. Note, however, that the weight at age of age 2 herring is increasing since 1980.

Prince Rupert District

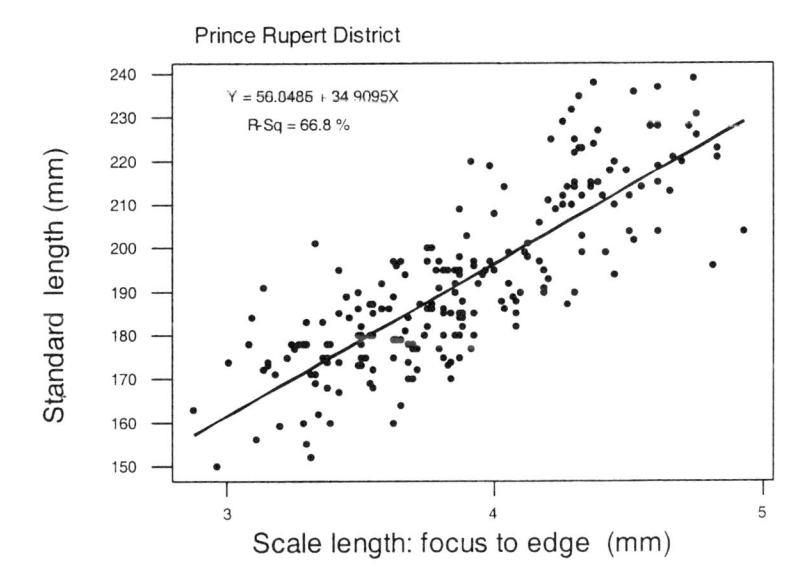

Figure 9. Relationship between scale length (focus to edge in millimeters, Fig. 3) and total length (millimeters) for herring from northern British Columbia, shown as a linear regression. The regression is significant. The figure indicates that scale measurements are a reasonable proportional approximation of total fish length.

Early Scale Growth

The relationship between scale length and fish length is linear and significant (Fig. 9). Mean scale length corresponding to growth in the first year of life, from the focus to annulus 1, ranges between 1.5 mm and 2 mm (Fig. 10A). This distance, when compared among years, indicates that the 1977 year class had normal scale growth during the first and second years of life. Fig. 10B shows the same analyses but restricted to age-5 fish. Again these fish had normal growth during the first and second years of life. This indicates that the normal, or slightly superior, growth of age-2 fish, as determined by sample analyses (i.e., as seen in Figs. 6 and 7), is not necessarily explained by selective sampling of faster-growing individuals in 1977. Rather, based on scales, it appears that juvenile growth during the first 2 years of life in 1977 was normal in B.C. In contrast to the normal juvenile scale sizes of the 1977 year class, scale sizes for 1977 year-class age-3 and age-5 fish decreased (Fig. 10C and D). This indicated that the same individuals that had normal scale growth as juveniles, had relatively depressed scale growth between age 2 and age 3 (i.e., mainly during the summer, fall and winter of 1979). Scale growth during the 3 years between 1979 and

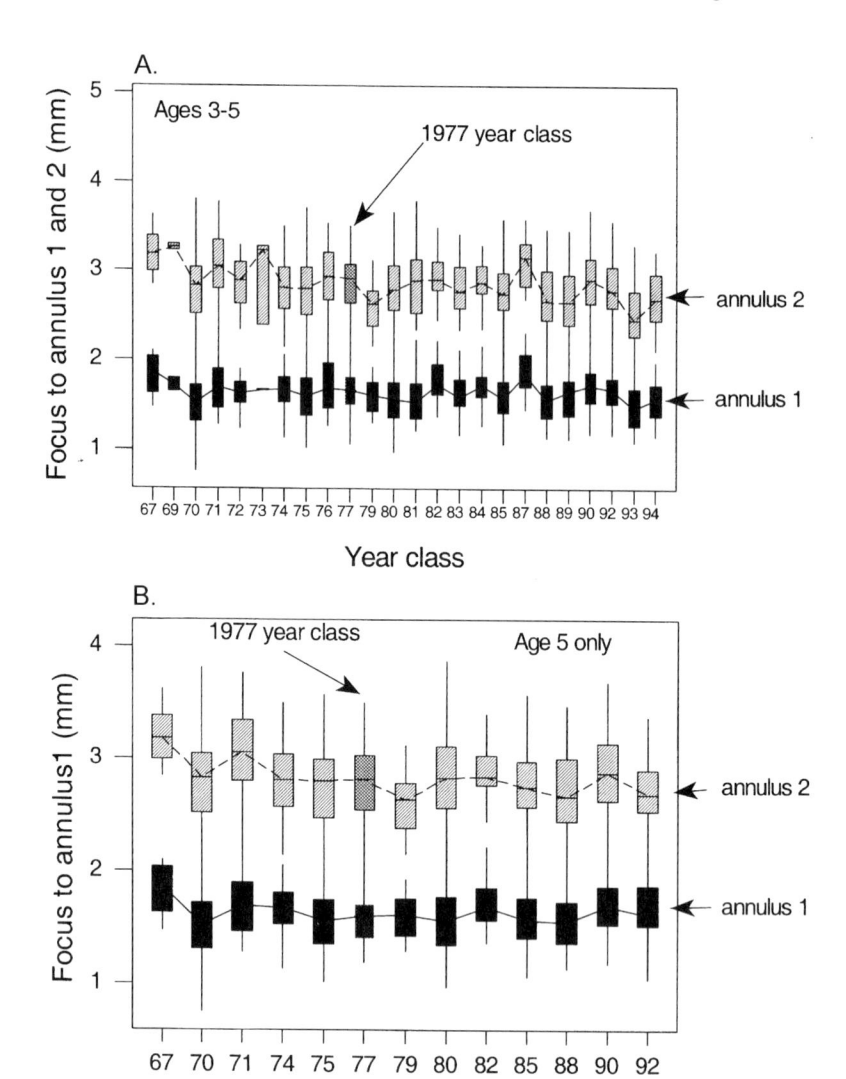

Figure 10. Boxplots, showing mean, 95% CI (box) and range (vertical line) of total scale measurements, from the focus to annulus 1 and 2, shown for different year classes from 1967 to 1994. All scales were from herring collected from northern British Columbia. All herring were either 3, 4, or 5 years of age. (A) Scale measurements (millimeters) from the focus to annulus 1 are indicated as dark boxes and from the focus to annulus 2 as shaded boxes. (B) Scale measurements as per Fig. 10A, except all fish were age 5. In each case, there is no obvious trend for scale lengths, corresponding to the end of juvenile growth at age 2, to change with time.

Herring: Expectations for a New Millennium 191

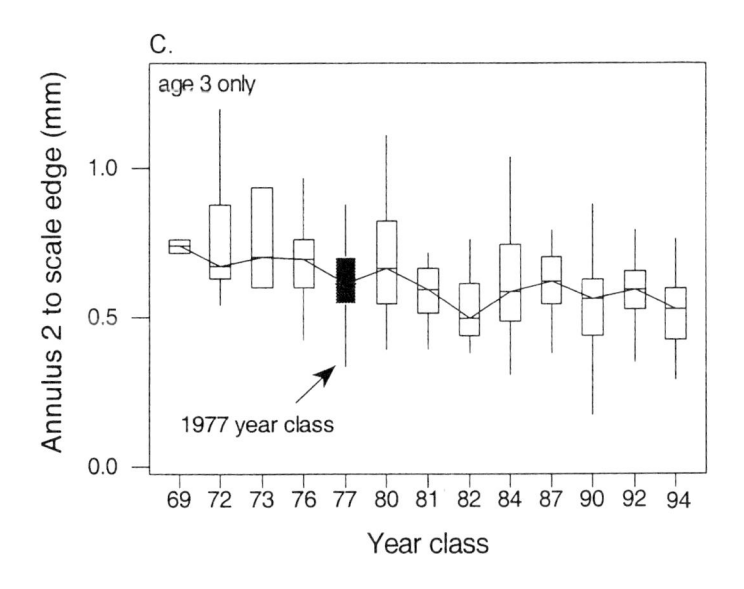

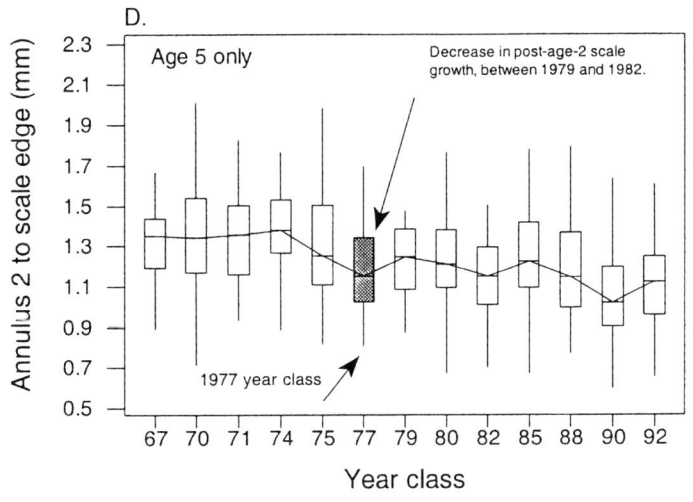

Figure 10. (Continued.) (C) Scale measurement at age 3 (focus to edge) first annulus. The total scale length is shorter in 1980 (the 1977 year class) and subsequent years, as indicated by the shaded boxes. (D) Scale measurement corresponding to growth increment within the third year (second annulus to scale edge). After 1980, the scale growth in the third year is distinctly less (indicated by shaded boxes). Note that this figure corresponds closely to the results presented in Fig. 9.

1982 also was lower (Fig. 10D). These results provide independent corroboration of the size-at-age trends shown in Fig. 8, which show a trend for a decrease in postjuvenile growth in the 1977 year class, and also in many of the subsequent year classes.

Discussion
Biological Significance of the Results
The strong 1977 year class of herring occurred over a broad geographic area of the northeast Pacific (Tables 1 and 2). This is based on the high proportion of age-3 fish in 1980 or age-4 fish in 1981 in Alaska and B.C., and is confirmed with comparisons of numbers at age from age-structure analysis (Table 3). This 1977 year class was produced by different stocks that had different spawning times (Figs. 3 and 4). It was particularly strong among all northern B.C. locations (Table 2). In general, spawning times were normal in 1977, when the year class began as eggs (Fig. 5). Spawning times also were normal in 1980, when most individuals within the 1977 year class spawned for the first time, although mean spawning times among different local populations varied by up to 2 months (Fig. 5). Analyses of size at age from the catch sampling database and from the analysis of scales indicate that the size at age 2 of the 1977 year class (in 1979) was approximately normal relative to other years (Fig. 7). Indeed, the size at age 2 was relatively constant with time and has not shown the same decline in size at age observed in older age classes (Fig. 6). In general, for ages of 3 or greater, size at age of most year classes has declined since 1980. When the 1977 year class reached age 3, its size at age was approximately normal, but declined sharply thereafter (Figs. 6 and 8).

In the following discussion we consider our results and observations in light of other information relevant to the recruitment of herring. We present these as five key observations, of which the first two observations are based on the results in this report. We then comment on probable factors that did, or did not, affect the development of the strong 1977 year class.

Key Observations and Implications about the Strong 1977 Herring Year Class
Key observation 1
The 1977 year class occurred over a wide temporal and geographic range. It was strong in different areas within B.C. and parts of southeastern and central Alaska and the Bering Sea (Hollowed and Wooster 1995). Also the 1977 year class occurred in different populations with different spawning times: spawning time for herring over this range extends for nearly 3 months. Therefore, the factor(s) that promoted the strong year class were widely distributed in space and time. This observation has implications for the critical period hypothesis (reviewed by May 1974) or the match-

mismatch hypothesis (Cushing 1972, 1975) as explanations for the production of strong year classes. The large spatial and temporal range of the 1977 year class of herring and other species indicates that in 1977 there was a wide window of opportunity for the production of a strong year class. This is contrary to the common interpretation of the match-mismatch and critical period hypotheses which both imply that temporal opportunities for strong year classes are narrow.

Key observation 2

In B.C., the size at age of age-1 and age-2 juveniles of the 1977 year class was relatively large when compared to other years. As they aged, however, the size at age of the 1977 year class declined in relation to other years. The size at age 3 (or 35-36 months) of the 1977 year class was normal in most areas in 1980. Age 3 is the first year that the age group is present in the fishery samples, except in some years there may be some age-2 (24-month) fish. Growth rates were normal, and perhaps enhanced, for the first 2 or 3 years of life of the 1977 year class.

Key observation 3

A strong 1977 year class also occurred in several other species, including blackcod and lingcod (Hollowed and Wooster 1995). Climate-related changes, but not necessarily increases in abundance, also occurred in other marine species, including salmonids (Beamish et al. 1999) and pollock *Theragra chalcogramma* (Ohtani and Azumaya 1995). Further, there are a number of periods when there has been synchrony of strong year classes among different species in the North Pacific (Hollowed et al. 1987), which is strong evidence of environmental influence on the production of year-class strength.

Key observation 4

The habitats occupied by age-1 and age-2 herring are mainly inshore (Haegele 1997), whereas most of the older age groups (age 3 and older) tend to occupy shelf waters. From our present understanding of herring life history, there is little opportunity for interaction between the adult stock (age 3+ and older), and the recruiting year class. The first opportunity for direct interaction is during the third winter of life, at age 2+, when (B.C.) herring start to mature sexually and join the adult stock. However, winter feeding is minimal and growth is slight.

Key observation 5

In B.C. egg deposition in 1977 was not exceptional, and we are not aware of any reports indicating abnormal spawning in Alaska in 1977. Therefore it follows that in 1977 survival from eggs to the juvenile and recruit stage, between 1977 and 1980, was relatively higher (or mortality was lower) than most other years. For some populations one could invoke changes in immigration or emigration rates, as larvae or juveniles, as possible explanations for the 1977 year-class formation, and although this might apply to single populations, it would not explain the broad, coastwide changes we saw in 1977. Therefore we exclude these as potential explanations.

If lower mortality of early life history stages is part of the explanation for the formation of the 1977 year class, how did this occur? Presumably it was from a decrease in the mortality rates associated with one or more of the following factors: starvation, disease, or predation. We discuss each of these below.

1. Starvation. We have no direct evidence about starvation of the 1977 year class but we know that juvenile growth of the 1977 year class was about normal compared to other years or year classes. Such growth is unlikely if starvation were common.

2. Disease. Epizootics are known to occur in herring (Meyers et al. 1986) and there is evidence that disease can lead to catastrophic mortality. If disease were associated with the formation of the 1977 year class, however, we would instead be looking for an effect that would result in widespread *survival*. Therefore, while we have no evidence that such a change in the incidence of disease occurred we suggest that it is unlikely that this would be an explanation for the development of the 1977 year class.

3. Predation. A decrease in predation between the egg/larval stages and prerecruit stage could occur if there were (a) fewer predators, or (b) if the predators "switched" or decreased predation on herring for a different prey species. Was there a decrease in the numbers of available predators? Although we have no data on the interactions of predators and prey of herring in 1977 we do observe that some common herring predators (lingcod and blackcod) also had strong 1977 year classes. Nor, to our knowledge, was there any decrease in the numbers of piscivorous salmon (coho and chinook) predators that year. Therefore it seems improbable that there was a decrease in the potential community of herring predators between 1977 and 1980. Rather, it may have increased. If so, the remaining explanation for the development of the strong 1977 year class was a general decrease in predation because the main herring predators had alternate prey. We arrive at this explanation by a process of elimination of alternate explanations, and acknowledge that the evidence for elimination of some alternatives is thin. Nevertheless, we suggest that an explanation of predator switching during early life history stages provides an adequate explanation for all of the key observations in this report, and some previous reports (Hollowed et al. 1987). If the main predators of young herring had alternate food sources then this could account for a decrease in predation on herring, and perhaps juvenile fish in general. Further, if the source of alternate prey were sufficiently abundant, and if the prey were available both to herring and their predators, then this would explain key observations (made above) that there was normal growth of the strong year class.

Outstanding Questions about Biological Mechanisms

What could explain the widespread abundance of an alternate food source, and what is such a food source? If the cause of the strong 1977 year class is the same in all geographic areas, from northern B.C. to the Bering Sea, and if this was from decreased predation associated with availability of an alternate food source, then clearly the factors which promoted this alternate food source were widespread. There have been some suggestions (Hollowed and Wooster 1992, Polovina et al. 1995) that there can be such occurrences, although the biological mechanisms for changes in productivity or food abundance on shelf and inshore waters, resulting from mid-gyre changes, are not well understood. What are the possible alternate prey species? Again, we do not know but we can make the following observations. The main food for herring at ages 1 and 2 is copepods (Haegele 1997, Foy and Norcross 1999, Hay and McCarter 2000). If the abundant 1977 year class ate mainly copepods at ages 1 and 2, as do herring of the same age but different year classes (Haegele 1997), then copepods must have been abundant in nearshore northern waters in both 1977 and 1978. There is a precedent for assuming that an abundant source of an alternate zooplankton prey species can reduce predation on herring. Ware and McFarlane (1995) showed that increased euphausiid production resulted in decreased hake predation on herring off the west coast of Vancouver Island. Their work involved adult herring, but we suggest that the same mechanisms could operate at the juvenile stages. Would the factors promoting a strong year class of herring also support strong year classes of other species, such as blackcod and lingcod? Again, the answer is a tentative yes. Both of these species have early life stages (first several years of life) in nearshore waters. Hollowed et al. (1987) noted a number of occurrences of strong year-class synchrony among many species in the North Pacific. It seems highly probable that these changes are related to annual environmental factors, but the biological mechanisms that operate between open ocean changes and survival and growth of fish in nearshore areas are not clear. In the case of herring, and other animals that inhabit mainly shallow nearshore waters, the pathways may be complex and our understanding of the detail of potential mechanisms is lacking. Perhaps potential solutions may be found in comprehensive analyses of juvenile fish habitat requirements combined with comparisons among different populations of herring, as we tried to do in this study, as well as interspecific comparisons. In particular, if we can better determine the distributions and habitat requirements at various stages of the life history, then detailed comparisons of population dynamics may point to common ecological requirements affecting recruitment.

References

Beamish, R.J., D.J. Noakes, G.A. Mcfarlane, L. Klyashtorin, V.V. Ivanov, and R. Kurashov. 1999. The regime concept and natural trends in the production of Pacific salmon. Can. J. Fish. Aquat. Sci. 56:516-526.

Blankenbeckler, D., and R. Larson. 1987. Pacific herring (*Clupea harengus pallasi*) harvest statistics, hydroacoustical surveys, age, weight, and length analyses, and spawning ground surveys for southeastern Alaska, 1980-1983. Alaska Dept. Fish Game Tech. Fish. Rep. 202. 121 pp.

Burkey, C., and J. Ried. 1988. Statistics of the commercial fishery for Pacific herring from the Kodiak area, Alaska. Alaska Dept. Fish Game Tech. Fish. Rep. 88-11. 113 pp.

Cushing, D.H. 1972. The production cycle and the numbers of marine fish. Symp. Zool. Soc. London 29:213-232.

Cushing, D.H. 1975. Marine ecology and fisheries. Cambridge University Press, London.

Foy, R.J., and B.L. Norcross. 1999. Spatial and temporal variability in the diet of juvenile Pacific herring (*Clupea pallasi*) in Prince William Sound, Alaska. Can. J. Zool. 77:697-706.

Francis, R.I.C.C. 1990. Back-calculation of fish length: A critical review. J. Fish Biol. 36:883-902.

Fried, S.M., C. Whitmore, and D. Bergstrom. 1982. Age, sex, and size composition of Pacific herring, *Clupea harengus pallasi*, from eastern Bering Sea coastal spawning sites. Alaska Dept. Fish Game Tech. Rep. 78. 40 pp.

Funk, F.C., and G.J. Sandone. 1990. Catch-age analysis of Prince William Sound, Alaska herring, 1973-1988. Alaska Dept. Fish Game Fish. Res. Bull. 90-01. 36 pp.

Haegele, C.W. 1997. The occurrence, abundance and food of juvenile herring and salmon in the Strait of Georgia, British Columbia in 1990 to 1994. Can. Manuscr. Rep. Fish. Aquat. Sci. 2390. 124 pp.

Haist, V., and L. Rosenfeld. 1988. Definitions and codings of localities, sections, and assessment regions for British Columbia herring data. Can. Manuscr. Rep. Fish. Aquat. Sci. 1944. 123 pp.

Hamer, L. 1989. Procedures for collecting and processing British Columbia herring samples. Can. Manuscr. Rep. Fish. Aquat. Sci. 2030. 27 pp.

Hay, D.E. 1985. Reproductive biology of Pacific herring (*Clupea harengus pallasi*). Can. J. Fish. Aquat. Sci. 42:111-126.

Hay, D.E., and A.R. Kronlund. 1987. Factors affecting the distribution, abundance, and measurement of Pacific herring (*Clupea harengus pallasi*) spawn. Can. J. Fish. Aquat. Sci. 44:1181-1194.

Hay, D.E., and P.B. McCarter. 1997. Larval distribution abundance stock structure of British Columbia herring. J. Fish Biol. 51:155-175.

Hay, D.E., and P.B. McCarter. 1998. Distribution and timing of herring spawning in British Columbia. Canadian Stock Assess. Secretariat Res. Doc. 99:14. 44 pp.

Hay, D.E., and P.B. McCarter. 2000. Spatial, temporal and life-stage variation in herring diets in British Columbia. In: B.A. Mergy, B.A. Taft, and W.T. Peterson. PICES-GLOBEC International Program on Climate Change and Carrying Capacity. Report of the 1999 MONITOR and REX workshops, and the 2000 MODEL workshop on lower trophic level modelling. North Pacific Marine Science Organization (PICES) Scientific Report 15:95-97.

Hay, D.E., K.D. Cooke, and C.V. Gissing. 1986. Experimental studies of Pacific herring gillnets. Fish. Res. 4:191-211.

Hollowed, A.B., and W.S. Wooster. 1992. Variability of winter ocean conditions and strong year-classes of Northeast Pacific groundfish. ICES Mar. Sci. Symp. 195:433-444.

Hollowed, A.B., and W.S. Wooster. 1995. Decadal scale variations in the eastern subarctic Pacific: II. Response of Northeast Pacific fish stocks. In: R.J. Beamish (ed.), Climate change and northern fish populations. Can. Spec. Publ. Fish. Aquat. Sci. 121:375-386.

Hollowed, A.B., K.M. Bailey, and W.S. Wooster. 1987. Patterns in recruitment of marine fishes in the northeast Pacific Ocean. Biol. Oceanogr. 5:99-131.

May, R.C. 1974. Larval mortality in marine fishes and the critical period concept. In: J.H.S. Blaxter (ed.), The early life history of fish. Springer-Verlag, New York, pp. 3-19.

Meyers, T.R., A.K. Hauck, W.D. Blankenbeckler, and T. Minicucci. 1986. First report of viral erythrocytic necrosis in Alaska, USA, associated with epizootic mortality in Pacific herring, *Clupea harengus pallasi* (Valenciennes). J. Fish Dis. 9:479-491.

Ohtani, K., and T. Azumaya. 1995. Influence of interannual changes in ocean conditions on the abundance of walleye pollock (*Theragra chalcogramma*) in the eastern Bering Sea. In: R.J. Beamish (ed.), Climate change and northern fish populations. Can. Spec. Publ. Fish. Aquat. Sci. 121:87-95.

Polovina, J.J., G.T. Mitchum, and G.T. Evans. 1995. Decadal and basin-scale variation in mixed layer depth and the impact on biological production in the Central and North Pacific 1960-1988. Deep-Sea Res. 42:1701-1716.

Schroeder, T.R. 1989. A summary of historical data for the lower Cook Inlet, Alaska, Pacific herring roe fishery, December 1988. Fish. Res. Bull. 89-04. 116 pp.

Schweigert, J.F., C. Fort, and L. Hamer. 1998. Stock assessment for British Columbia herring in 1997 and forecasts of the potential catch in 1998. Can. Tech. Rep. Fish. Aquat. Sci. 2217. 64 pp.

Tanasichuk, R.W. 1997. Influence of climate and ocean biomass on the growth of Pacific herring (*Clupea pallasi*) from the southwest coast of Vancouver Island. Can. J. Fish. Aquat. Sci. 54:2782-2788.

Tanasichuk, R.W., and J.F. Schweigert. 1998. Variation in size at age of Pacific herring *(Clupea pallasi)* and its implications for quota recommendations and fisheries management. PSARC (Pacific Scientific Advice Review Committiee, Canada) Working Paper H98-3. 24 pp.

Trenberth, K.W., and J.W. Hurrell. 1995. Decadal coupled atmosphere-ocean variations in the north Pacific Ocean. In: R.J. Beamish (ed.), Climate change and northern fish populations. Can. Spec. Publ. Fish. Aquat. Sci. 121:15-24.

Wailes, G.H. 1936. Food of *Clupea pallasii* in southern British Columbia waters. J. Biol. Board Can. 1:477-486.

Ware, D.M., and G.A. McFarlane. 1995. Climate-induced changes in Pacific hake *(Merluccius productus)* abundance in the Vancouver Island upwelling system. In: R.J. Beamish (ed.), Climate change and northern fish populations. Can. Spec. Publ. Fish. Aquat. Sci. 121:509-521.

Whitmore, D.C., S.M. Fried, and D.J. Bergstrom. 1983. Age, sex, and size composition of Pacific herring, *Clupea harengus pallasi*, from southeastern Chukchi Sea coastal sites, Alaska, 1977-1982. Alaska Dept. Fish Game Tech. Rep. 86. 31 pp.

Spawning Bed Selection by Pacific Herring (*Clupea pallasii*) at Atsuta, Hokkaido, Japan

Hiroshi Hoshikawa
Hokkaido Central Fisheries Experimental Station, Hokkaido, Japan

Ken-ichiro Tajima
Hokkaido Institute of Mariculture, Hokkaido, Japan

Tadashi Kawai
Hokkaido Nuclear Energy Environmental Research Center, Hokkaido, Japan

Tomohiro Ohtsuki
Wakkanai Fisheries Experimental Station, Hokkaido, Japan

Abstract

Spawning bed selection by a local population of Pacific herring (*Clupea pallasii*) was studied at Atsuta, Ishikari Bay on the west coast of Hokkaido, in northern Japan, using scuba diving. We surveyed the vegetation, macrobenthos, and character of the bottom in 1996 and 1997. Occurrence of herring eggs was studied from 1997 to 1999. The spawning bed characteristics such as water temperature and topography of spawning beds were measured during spawning season in 1998 and 1999.

Rich vegetation covered most of the shore and shallow sublittoral, but spawning beds were limited to the southern part of study area. In one of these restricted areas herring spawned in exactly the same beds in 1998 and 1999, and eggs were found repeatedly during the same spawning period.

Most of the eggs were attached to seagrass, *Phyllospadix iwatensis*, which was a dominant species along with *Laminaria* spp. and *Cystoseira hakodatensis*. Egg densities varied both spatially and temporally from several hundred to a maximum of 1.6 million per square meter. Egg density was also high along the narrow grooves running perpendicular to the flat, platform-like rocky shore.

The main spawning period in 1998 was in late March, while in 1999, two main spawning periods occurred in early March and early April. The spatial distribution of eggs laid in the second spawning didn't overlap with that of the first spawning in 1999. The spawning bed selection at Atsuta is discussed in relation to the vegetation, topography of the rocky shore, and wave conditions during the spawning season.

Introduction

Herring generally spawn on vegetation in the Pacific, the Atlantic, the Baltic Sea, the Okhotsk Sea, and the Sea of Japan (Tamura and Okubo 1953a; Tamura et al. 1954; Galkina 1959, 1960; Aneer et al. 1983; Haegele and Schweigert 1985; Kääriä et al. 1988; Rajasilta et al. 1993). According to Stacey and Hourston (1982) and Obata et al. (1997), herring select the substratum for spawning with their pelvic and pectoral fins. Thus the character of the bottom is one of the important factors for spawning bed selection (Kääriä 1999).

In Pacific and Atlantic herring, spawning beds were not exactly predictable from year to year (Tyurnin 1973, Hourston and Rosenthal 1976, Messieh 1987, Hay and McCarter 1997). However, herring spawned at almost the same beds in the Baltic Sea each year (Oulasvirta et al. 1985, Rajasilta et al. 1993, Kääriä 1999).

On the west coast of Hokkaido, northern Japan, the catch of Hokkaido-Sakhalin herring decreased gradually from 9.73×10^5 tons in 1897 to several hundred tons owing to the disappearance of predominant year classes in relation to the rise in sea temperature due to climatic regime shifts (Tanaka 1991). At present, a local population (called the Ishikari Bay herring) is caught and probably spawns at several spawning beds along the coast of Atsuta and Rumoi (Fig. 1).

Clarification of the mechanisms of spawning bed selection is very important for the preservation and the enhancement of spawning grounds and the creation of new ones (see Aneer 1989, Kääriä 1999). In this study we examined the distribution of eggs along the coast of Atsuta in relation to the vegetation and estimated the spawning periods.

Materials and Methods

Field investigations were conducted at Atsuta, Ishikari Bay, located on the west coast of Hokkaido, northern Japan from 1996 to 1999 (Fig. 1 and Table 1).

Vegetation, Macrobenthos, and Topographic Surveys

Distribution and biomass of vegetation and macrobenthos were determined from collections from 209 sites along 19 transect lines set (no. 1 to 19 from north to south and sites set at 10 m intervals from 0 to 100 m along the transect lengths) in May 28 and 29, 1996 and from 128 sites along 8 transect lines (no. 4, 5, 9, 10, 15, 16, 18, and 19 and sites set at 10 m

Herring: Expectations for a New Millennium 201

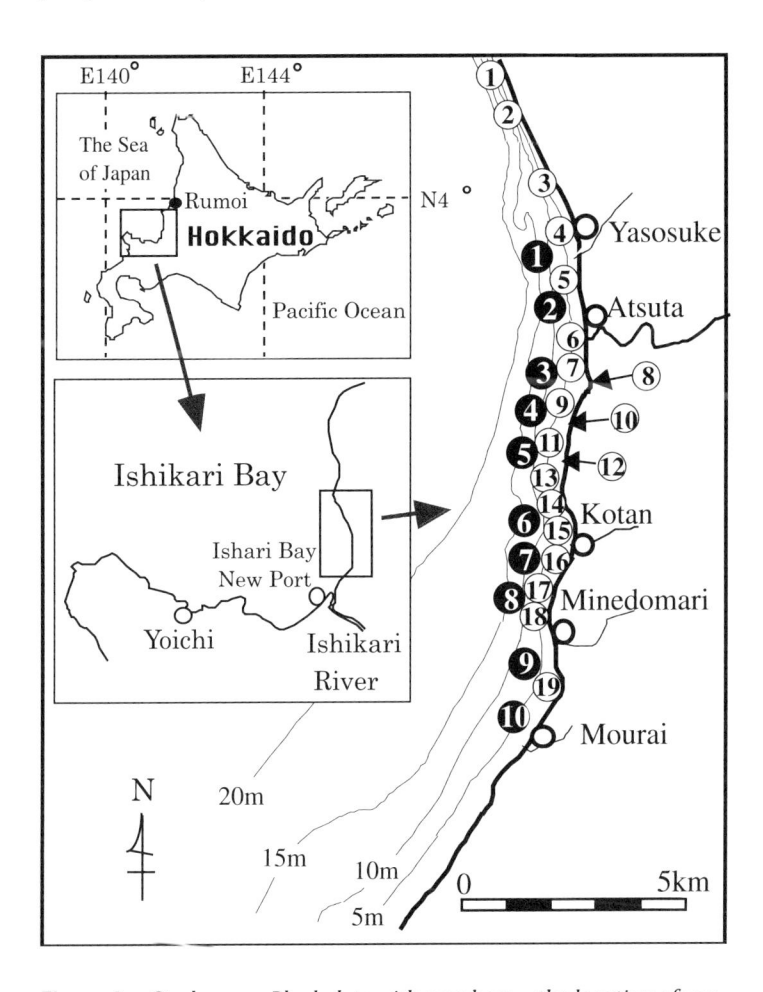

Figure 1. Study area. Black dots with numbers = the location of vegetation, macrobenthos, and herring egg surveys in 1996 and 1997 by scuba. Circles with numbers = the location of herring egg surveys by scuba in 1998.

Table 1. Research activities at Atsuta in 1996, 1997, 1998, and 1999 (see also Fig. 1.).

Study objective	1996	n	1997	n	1998	n	1999	n
Vegetation and macrobenthos along the coast of Atsuta	May 28,29	209	Jun 12	128	—		—	
Occurrence of herring eggs along the coast of Atsuta								
Scuba and skin diving	—		Apr 14	44	Mar 5	20	Apr 9	2
					Mar 26	20	Apr 16	3
					Apr 9	10		
Washed up algae and seagrass	—		—		Mar 26	2	Mar 26	11
					Apr 9	1	Apr 27	2
Study at the Minedomari spawning beds								
Egg development and hatching	—		—		Apr 10	39	Mar 5	2
					Apr 20	6	Mar 15	2
					Apr 28	7	Mar 25	34
							Apr 9	3
							Apr 16	41
							Apr 22	11
							Apr 27	11
							May 7	29
Egg density	—		—		Apr 10	30	Mar 25	34
							Apr 16	28
							May 7	29
Temperature in spawning bed continuously	—		—		Feb-Mar Apr-May	[a]	Feb-May	

n = Number of samples or surveyed sites. Dash = not studied.
[a] Temperature measured at Kotan.

intervals from 0 to 100 m and 20 m intervals from 120 to 200 m) on June 12, 1997. These transects were selected in order to cover the whole study area. The first basis of selection was the mouth area of small rivers and, second, the amount of herring catch (by professional fishermen) in the area studied. These studies were conducted in late spring (after the spawning season) when the biomass of algae shows the maximum value in all seasons in this area. Scuba divers sampled and recorded vegetation in 0.25 m² and benthos in 1 m² plots and the sediment type and depth at each site.

The number of individuals and total wet weight of each species of large algae, seagrass, and macrobenthos were recorded. Total wet weight was recorded for each species of small algae.

Search for Spawning Beds

The occurrence of herring eggs was examined by scuba divers (Fig. 1, Table 2) at 4 transect lines (altogether 44 sites in No. 4, 9, 14, and 19 set at 10 m intervals from 0 to 100 m perpendicular to the shore along each study transect) on April 14, 1997.

On March 5 and 26, 1998 10 transect lines (Fig. 1, Table 2) and on April 9, 1998, 5 transect lines were studied. Scuba divers collected vegetation and searched eggs in each transect line in 1997 and in 1998 both in shallower (from 0 to 3 m depth zone) and deeper (from 3 to 12 m zone) areas. The occurrence of eggs and the substratum type was determined in the laboratory (except one sample in April 9, 1998, which was determined on a boat).

Algae and seagrass that had washed ashore were also examined for the occurrence of eggs at Kotan and Mourai in March 26, at the shore of Minedomari in April 9, 1998 and along the whole shore from Yasosuke to Minedomari (about 10 km) on March 26, 1999. Scuba diving surveys for the occurrence of eggs were conducted where eggs on algae were observed at Minedomari in April 9, 1998, and at the north part of Kotan on April 9, 1999. Underwater observations by skin diving were also conducted at Mourai in April 16, and at the north part of Minedomari in April 27, 1999.

Estimate of Spawning Period and Hatching Period

Herring eggs were collected at Minedomari, to identify the spawning period and hatching period based on the developmental stage of eggs in each year. The sampling schedule is presented in Table 1. Samples were taken on three days in April 1998, and on eight days between March 5 and May 7 in 1999. All samples of eggs with seagrass and algae were preserved with 5% formalin.

Stage of development of eggs was used to determine the time of spawning. At least 100 eggs were measured from each sample. The developmental stages of eggs (Table 3) were classified into fourteen stages based on McMynn and Hoar (1953), Kuwatani et al. (1978), and Kimmel et al. 1995.

We used two methods to estimate spawning times of herring. One was to back-calculate the date of fertilization based on the relationship between the incubation period and the average water temperature (Alderdice and Velsen 1971). Another was to estimate spawning time using number of days from the fertilization to each developmental stage (Yamaguchi 1926, Fujita and Kokubo 1927, McMynn and Hoar 1953, Kuwatani et al. 1978). The live eggs collected at Minedomari were brought back to the Hokkaido Central Fisheries Experimental Station in Yoichi, which is located about 80 km southwest of Atsuta. The eggs were kept in a tank under running seawater directly from the nearby sea at ambient temperature, to determine the exact hatching period. The water temperature was measured every day.

Table 2. Field observations on herring eggs using scuba diving, and the wet weight of vegetation surveyed for herring eggs.

Year	Date	Depth	Number of study sites	Vegetation (kg)	Results
1997	Apr 14[a]	0.3-3 m	4 transect lines × 11 sites = 44 sites	49.7	No eggs
1998	Mar 5[b]	2-12 m	10 transect lines × 2 sites = 20 sites	74.1	No eggs
	Mar 26[b]	1-7 m	10 transect lines × 2 sites = 20 sites	7.6	No eggs
	Apr 9[c]	1-5 m	5 transect lines × 2 sites = 10 sites	No data	No eggs

[a]See text and Fig. 1.
[b]Area no. 1-10 shown in Fig. 1.
[c]Area no. 6-10 shown in Fig. 1.

Distribution of Eggs in the Spawning Beds at Minedomari

The distribution of eggs in the spawning beds found at Minedomari was surveyed with line transects on April 10, 1998, and March 25, April 16, and May 7, 1999. A rock platform, which extended from the shore to the edge from 60 to 150 m in distance, was observed at Minedomari. The edge of the platform was traced using aerial photography. The depth of the edge was from 0.7 to 1 m. There were several grooves on the platform running from northwest to southeast. Eight transect lines (line A to H) set on the platform from the end of the breakwater at 25 to 50 m intervals along the coast of Minedomari were used to survey the distribution of eggs on the platform (Figs. 2 and 3). Quantitative egg samples with vegetation were taken using a small quadrat (25 cm × 25 cm) at 20 m (in 1998) and 10 m (in 1999) intervals along the lines by scuba diving, skin diving, and walking. Observations on the occurrence of eggs near the sampling points were also made along each line.

Each specimen of algae and seagrass in the samples was weighed and the number of leaves counted in the case of *Phyllospadix iwatensis*. The eggs on the algae and seagrass in each small quadrat were counted and used to estimate total numbers of eggs per one square meter.

Vertical Distribution of Eggs on Seagrass and Algae

The vertical distribution of eggs on *Phyllospadix iwatensis* and *Cystoseira hakodatensis* (the dominant algal and seagrass species with *Laminaria*

Table 3. Developmental stage of herrings eggs.

No.	Developmental stage
1	2 cells
	4 cells
	8 cells
2	Morula
3	Blastula
4	Gastrula
5	Embryonic shield expanded
6	Formation of Kupfer's vesicle
7	Kupfer's vesicle clear
8	Lenses develop in eyes
9	Differentiation of tail
10	Body rounds 1.2 times around yolk
11	Body rounds 1.5 times around yolk
12	Eye color dark, body rounds 2 times around yolk
13	Eye color black, pigmentation on body
14	Starts hatching

spp. (*L. religiosa, L. ochotensis,* and *L. cichorioides*) at Atsuta) was examined to determine how herring use them as a spawning substratum. The length of the algae and seagrass were measured from the bottom to the top, and cut into 5 or 10 cm strips. The algae and seagrass were weighed and the eggs were counted in each of the height layers. In the case of *P. iwatensis*, the number of leaves was also counted in each layer.

Water Temperature

Water temperature at Minedomari was measured continuously using a temperature data recorder (Optic Stow Away Temp logger) set on the bottom from February to March in 1998, but after March 1998 temperature was recorded only at Kotan (Figs. 1 and 4) at 1 m depth. In 1999 temperature was measured continuously from February to May at Minedomari.

Results

Vegetation, Macrobenthos, and Topographic Features along the Atsuta Coast

A total of 42 species of algae and seagrass were found and among them many annual species of small algae were abundant (Table 4). The biomass of vegetation was dominated by *Laminaria* spp., (57.6% in 1996 and 74.5%

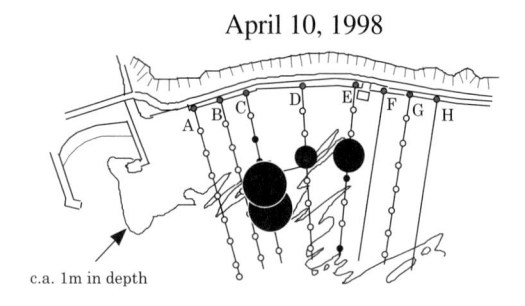

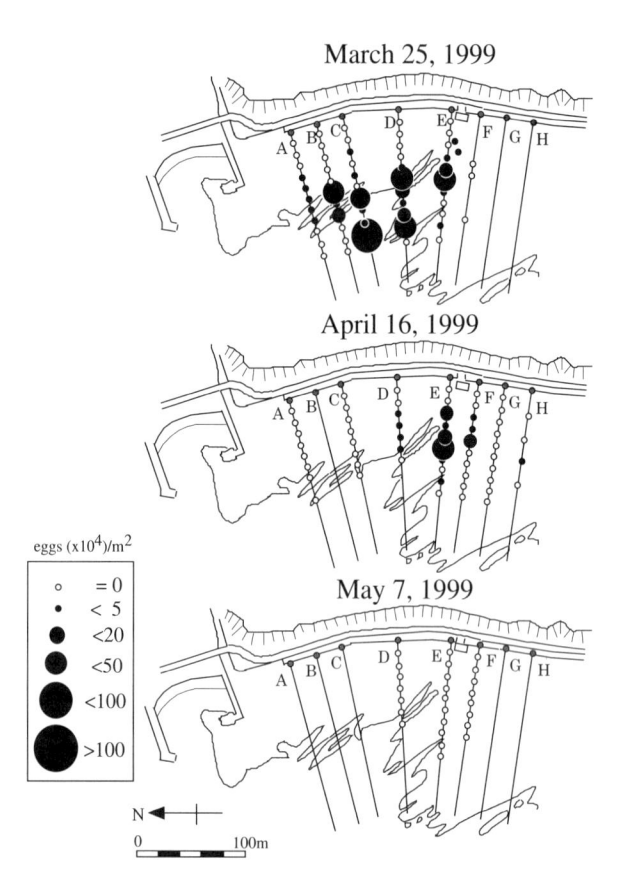

Figure 2. Distribution of herring eggs at Minedomari, Atsuta, in 1998 and 1999. O = eggs not found.

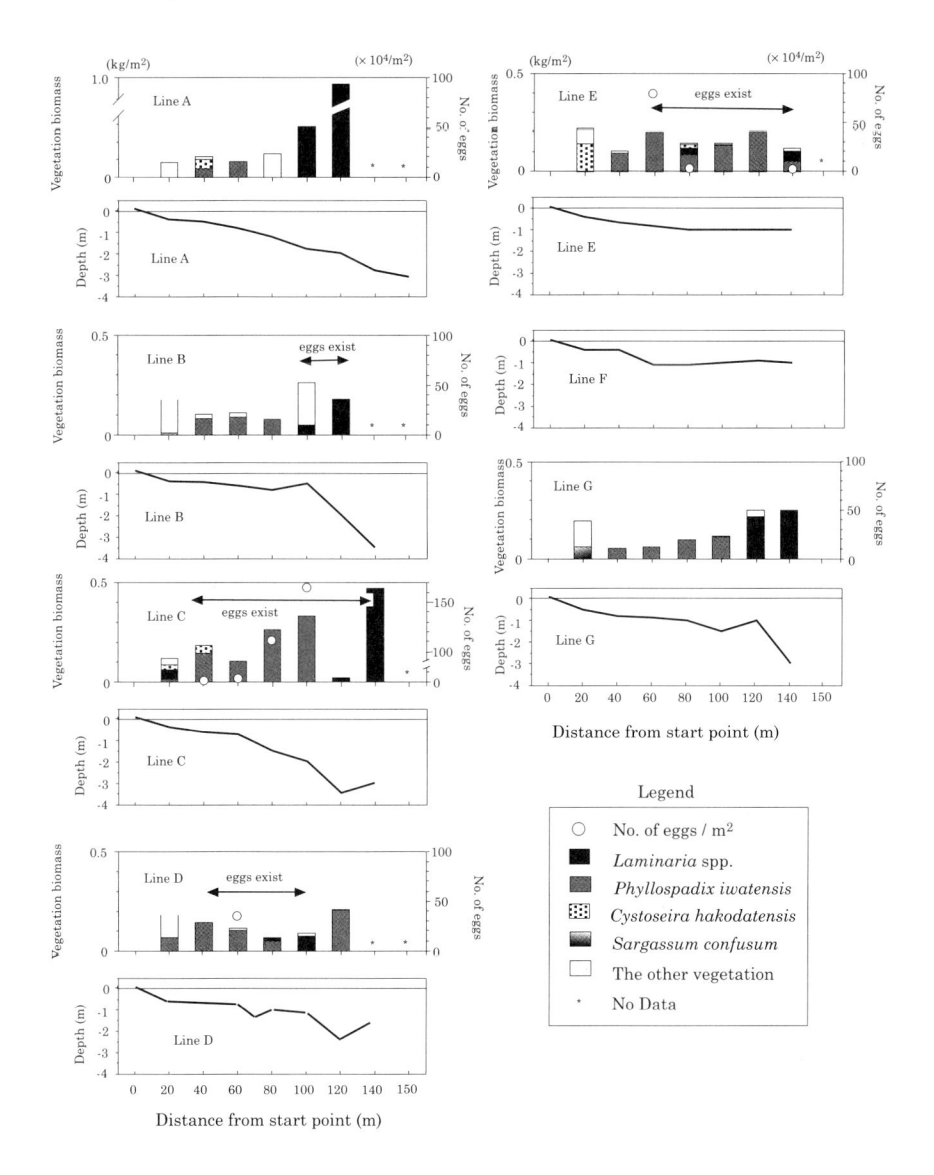

Figure 3. Vegetation and profile of each line at Minedomari, and herring egg density in April 10, 1998.

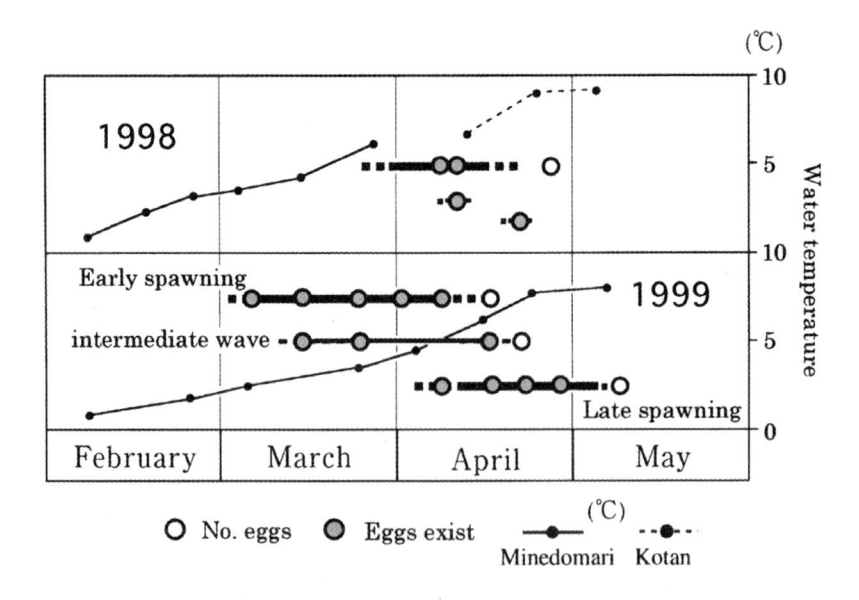

Figure 4. Spawning and hatching periods of herring during 1998 and 1999.

in 1997; Fig. 5) followed by *Cystoseira hakodatensis* and *Phyllospadix iwatensis*. These three accounted for 87.8% (1996) and 91.8% (1997) of the biomass. *Laminaria* spp. was distributed in the exposed zone (e.g., the edge of platform), *C. hakodatensis* was found in the sheltered zone (like the base of platform), and *P. iwatensis* was found in areas of intermediate exposure. The macrobenthos collected included sea urchins, abalone, starfishes, crabs, snails, limpets and sea cucumbers (Table 4). The biomass of vegetation was low (under 2.5 kg per m^2) in the northern part and high (max over 5 kg per m^2) in the southern part along the coast of Atsuta in 1996 and 1997 (Figs. 6 and 7). The reverse tendency was observed in the macrobenthos. The amount of macrobenthos was lower in the southern transects.

The topographic features of each line also changed. In the northern part boulders covered the bottom, the slope was steeper, and the depth was 3-5 m at 100 m and respectively 5-7 m at 200 m from shoreline. In the southern part of the study area a rock platform was well developed and the slope was gentler with a depth less than 2 m at 100 m and 3.4-6.4 m at 200 m from shore (Figs. 1, 2, and 3). Most macrobenthos inhabited depths greater than 2 m on the northern coast of Atsuta. However, there was a small amount of macrobenthos at the same depth along the southern coast.

Table 4. List of vegetation and benthos collected in the line transect survey at Atsuta in 1996 and 1997 (Figs. 5 and 6).

Annual small algae	Annual large algae	Echinodermata
Ulva pertusa	Laminaria spp.[a]	Strongylocentrotus intermedius
Monostroma nitidum	Undaria pinnatifida	S. nudus
Grateloupia turuturu	Costaria costata	Hemicentrotus pulcherrimus
G. filicina	Desmarestia lingulata	Glyptocidaris crenularis
G. okamurae	D. viridis	Asterias amurensis
Heterosiphonia japonica		Asterina pectinifera
Ceramium kondoi	**Perennial small algae**	Henricia sp.
Scytosiphon lomentaria	Dictyopteris divaricata	Stichopus japonicus
Chrysymenia wrightii	Rhodoglossum japonicum	
Porphyra pseudolinearis	Dictyota dichotoma	**Mollusca**
Congregatocarpus	Laurencia saitoi	Haliotis discus hannai
pacificus	Gymnogongrus	Acmaea (N.) pallida
Chondrus ocellatus	flabelliformis	Neptunea arthritica
C. elatus	Neorhodomela aculeata	Necella freycineti
C. pinnualatus	Neodilsea yendoana	Omphalius rusticus
Symphyocladia latiuscula	Ceratostoma sp.	
Polysiphonia morrowii		**Crustacea**
Hyalosiphonia caespitosa	**Perennial large algae**	Pugettia quadridens quadridens
Acrosorium yendoi	Sargassum confusum	Pagurus spp.
Petalonia fascia	S. thunbergii	
Punctaria latifolia	S. miyabei	
Grateloupia okamurae	Cystoseira hakodatensis	
Palmaria palmata		
Lomentaria catenata	**Perennial large seagrass**	
	Phyllospadix iwatensis	

[a] *Laminaria* spp. contains *L. religiosa*, *L. ochotensis*, and *L. cichorioides*.

Distribution of Spawning Beds

No eggs were observed during field investigation in 1997 and 1998 in study sites chosen in advance (Table 2). On April 14, 1997, *Laminaria* spp. was dominant in the samples (79.7% of wet weight) and *P. iwatensis* was the second dominant species (15.6%). On March 5, 1998 *Laminaria* spp. was also dominant (60.9% and the second dominant species was *P. iwatensis* (13.8%). In March 26, a lot of suspended matter was present in the water and divers could not observe the bottom. Only 7.6 kg of seagrasses were collected and *P. iwatensis* was the dominant species (70% in wet weight).

Some herring eggs attached to *P. iwatensis* were found washed up on the shore at Kotan and Mourai in March 26, 1998, but no eggs were found by skin divers in bottom areas nearby. A lot of eggs were found on *P. iwatensis* washed up on the shore of the south side of Minedomari, a distance of about 70 m along the coast on April 9, 1998. Based on the skin diving observation near the shore, we found the high-density spawning

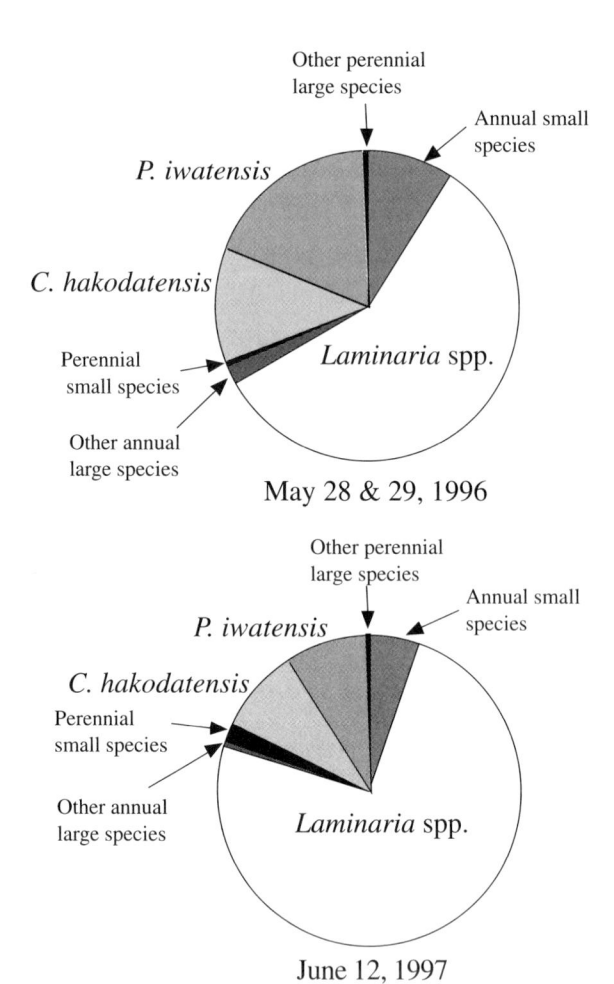

Figure 5. Species composition of vegetation in biomass.

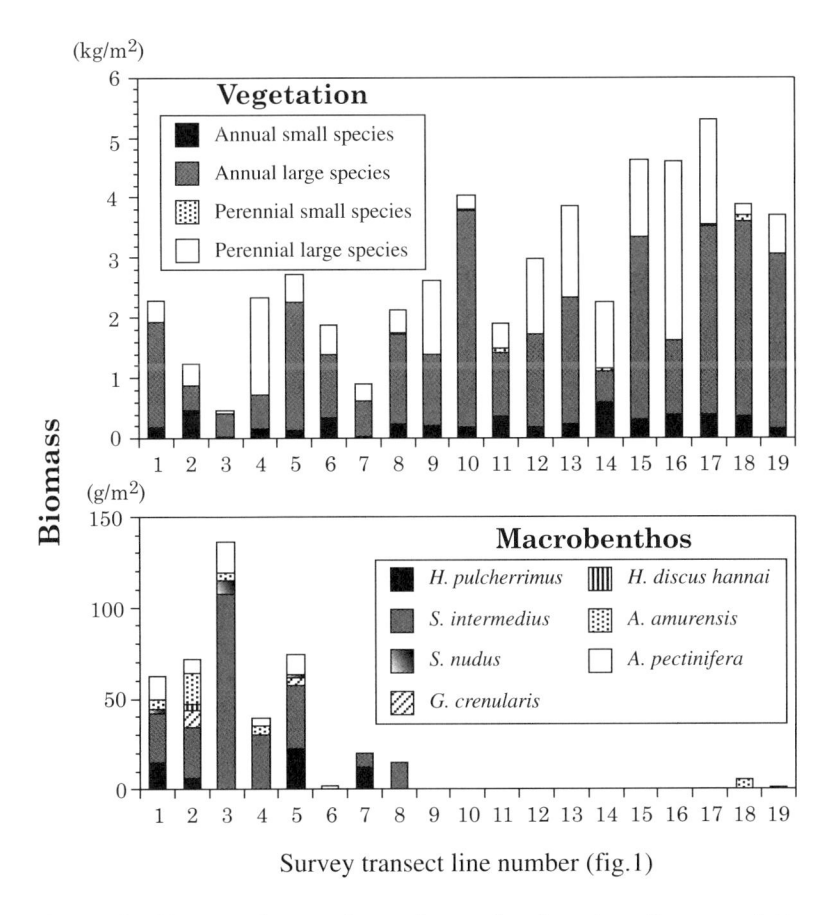

Figure 6. Biomass of vegetation and macrobenthos along the coast of Atsuta in 1996. List of species is presented in Table 4.

beds at Minedomari. Herring spawned in exactly the same restricted area also in 1999. Eggs were found repeatedly during the spawning period in 1999 (Figs. 2, 4, and 8).

A lot of eggs attached to seagrass were also washed ashore north and south of Kotan on March 26, 1999 (Fig. 8). We found two relatively small spawning beds in front of the sites where *P. iwatensis* with herring eggs had been washed up at the northern part of Kotan using scuba diving on April 9, 1999 (Fig. 8). The area was 6,250 m² (maximum depth was 1.7 m) and 450 m² (maximum depth was 0.7 m). *P. iwatensis* was the main substratum for eggs but some eggs were also attached to *C. hakodatensis*. Numbers of eggs per leaf of *P. iwatensis* in these beds were less (4 to 82 eggs per leaf) than that on the main spawning beds found in Minedomari

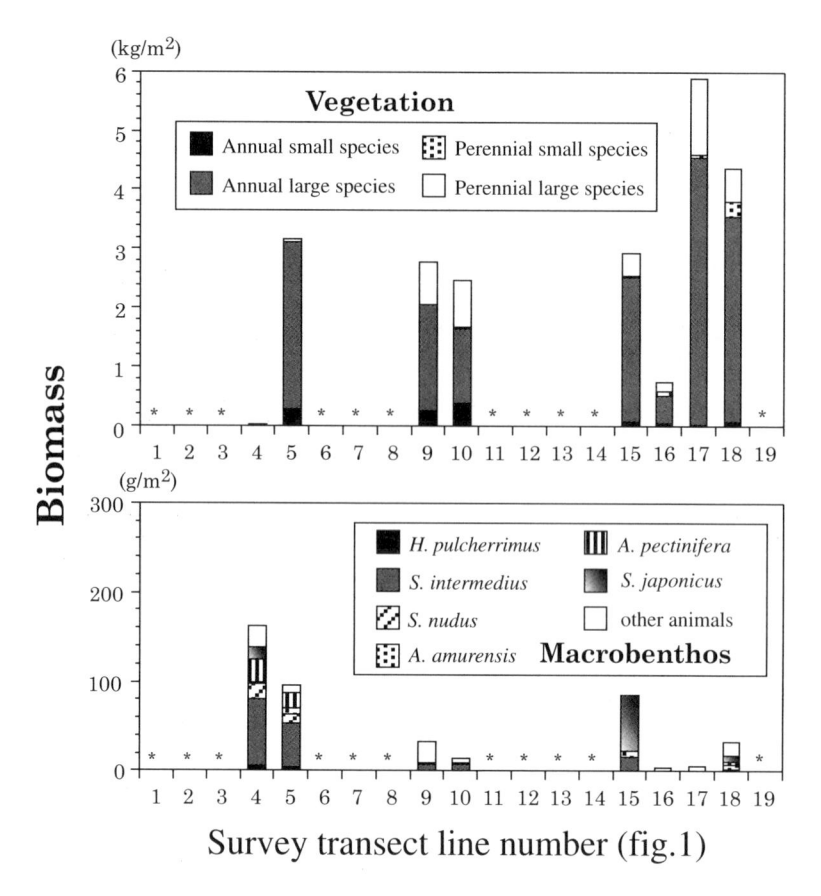

Figure 7. Biomass of vegetation and macrobenthos along the coast of Atsuta in 1997. List of species is presented in Table 4. * = no data.

(0 to 412 in 1998 and 39.3 to 70.7 in 1999). Two small additional spawning beds (the area of 25 m² and 30 m², respectively) were found using skin-diving at Mourai, on April 16, 1999. *P. iwatensis* grew in some small groups at these beds in Mourai and the number of eggs per a leaf were few (6 to 10 eggs per leaf). Few eggs (5 eggs per leaf) were found at the northern part of Minedomari in April 27, 1999. The developmental stages of eggs in those small spawning beds were almost the same as at Minedomari (Fig. 9). Because of the high egg density and the repeated spawning during the same spawning period, we studied the spawning bed at Minedomari in more detail.

Spawning and Hatching Periods in Spawning Beds

Estimated spawning and hatching periods and water temperature in Minedomari in 1998 and 1999 are shown in Fig. 4.

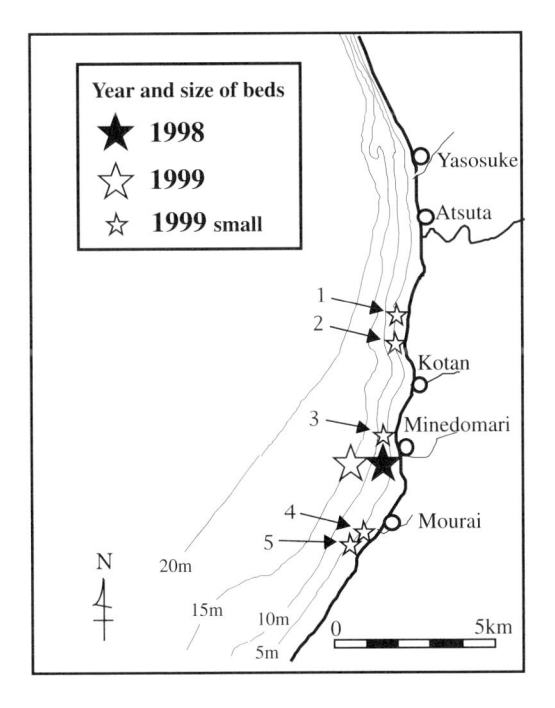

Figure 8. Distribution of the spawning beds. See also Fig. 9.

In 1998, a lot of eggs were collected on April 9 and 10, and a few more eggs were observed April 20. No eggs were observed in April 28 probably because of low transparency in water.

The eggs collected on April 10, 1998, contained two different developmental stages (Fig. 10). One was the embryo having dark eyes (developmental stage 12, Table 3), and another was an embryo with no eyes (stage 7). The ratios of the stages were 80% in stage 12 and 15% in stage 7, and the young embryo was limited at 40 to 60 m from the shoreline (line C in Fig. 2). Based on the report by Kuwatani et al. (1978), the fertilization period of the embryo with dark eyes was estimated at about 16 days, and the no eyes stage 6 days before at temperatures from 5 to 6°C.

Live eggs (developmental stage 12) collected from Minedomari in April 10, 1998, and kept at 6.9-7.5°C in a tank in the laboratory hatched on April 16, 1998. Based on temperatures measured at Minedomari and in the tank, and a predictive equation given by Alderdice and Velsen (1971) the period from fertilization to hatching on April 16, 1998, was calculated as 22 days. Thus the spawning periods of eggs observed in April 10 at Minedomari were estimated to have been late March and early April. The developmen-

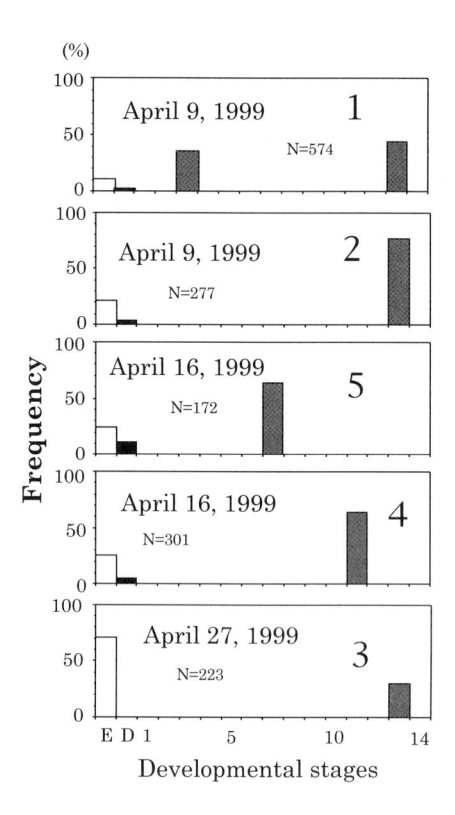

Figure 9. Developmental stages of herring eggs in field surveys at Kotan, Mourai, and the northern part of Minedomari, in 1999. See Fig. 8.

tal stage of eggs collected at April 20, 1998, was blastula (stage 3 in Table 3) and the stage was estimated to be just one day or less from fertilization. Based on the field observation of developmental stages, it was concluded that at least three different spawning waves occurred at Minedomari in 1998. The first spawning wave seemed to be the biggest one according to the amount of eggs found (Fig. 10).

Eggs collected on March 5, 1999, were the late gastrula (stage 4). According to the results of Yamaguchi (1926), the periods from fertilization to these stages were from 2 to 3 days at 6.2°C. Thus it seemed that the actual time period since fertilization at Minedomari was somewhat longer.

On March 16, 1999, the two different stages were observed suggesting an additional spawning probably occurred after the first observation. Spawning periods were estimated to have occurred 8 days and 12 to 14 days before March 16 when the survey was carried out, based on the egg

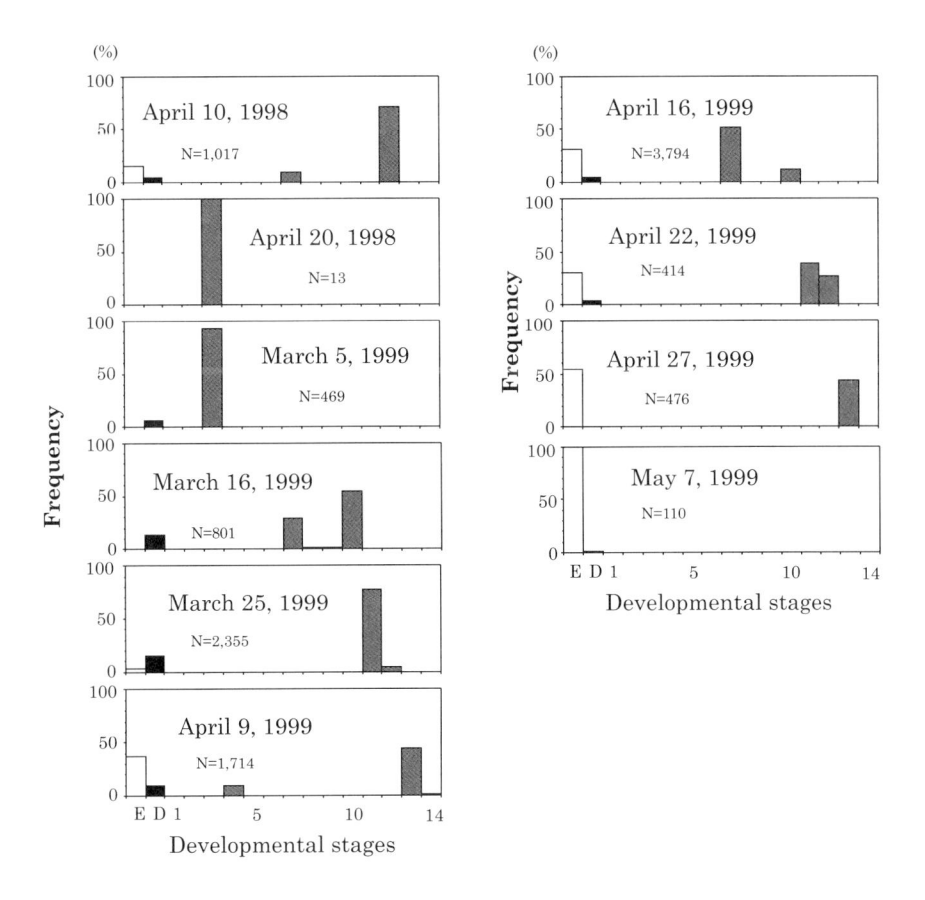

Figure 10. Developmental stages of herring eggs in field surveys at Minedomari, in 1998 and 1999. See developmental stages in Table 3. E: Empty eggs, D: Dead eggs, N: Number of eggs studied.

development rate data of Kuwatani et al. (1978). Most of the embryos observed on March 25 had eyes but no pigmentation or tails. The developmental stage was almost the same in that survey, compared to two different groups observed on March 16. It should be noted that samples were not taken exactly in the same spots in the spawning bed.

On April 9, the larvae hatched with slight stimulation to the *P. iwatensis* on which the eggs were attached. A lot of late gastrula stage eggs also were observed on the same day representing repeated spawning. According to Yamaguchi (1926) and Fujita and Kokubo (1927), the time period required to reach late gastrula stage was 2 to 3 days. The temperature during their experiments was higher than it was at Minedomari, and the actual period of egg development may have been somewhat longer.

Two different developmental stages were observed also on April 16, 1999. A more developed group was spawned in early April, and a later group was spawned just after April 9. On April 22, there were also two different groups corresponding to those groups observed on April 16. On April 27, empty eggs appeared and juveniles just hatching out were also observed, and on May 7 no more living eggs were found. From these field observations, we concluded that Pacific herring at Minedomari in 1999 spawned primarily in early March and early April. Field observations showed that some spawning also occurred during mid-March and mid-April, but only a small numbers of eggs were produced. Hatching periods of early spawning and late spawning groups were estimated as just after April 9 and 27, respectively.

Egg Distribution in Spawning Beds

The distributions of eggs in the spawning bed at Minedomari on April 10, 1998, and on March 25, April 16, and May 7, 1999, are shown in Figs. 2, 3, and 11.

On April 10, 1998, the area where eggs were observed was about 12,500 m^2 (Fig. 2). Eggs were more abundant in the middle of the bed at the edge of the rocky platform or in the grooves on lines C, D, and E (Fig. 2). Eggs were found at depths from 0.5 to 3.0 m and seemed to be more abundant at sites shallower than 2 m (Fig. 3). Egg density ranged from 0 to 1,622,000 per m^2, with the average value of 242,000 per m^2 (Table 5).

Laminaria spp. was dominant at Minedomari, but no eggs were found on *Laminaria* spp. Eggs were attached mainly on *P. iwatensis*, which was also abundant in spawning beds, but they were also on *C. hakodatensis*. A small amount of eggs were also observed on small algae *(Palmaria palmata* and *Mazzaella japonica)*.

On March 25, 1999, the distribution pattern of herring eggs in the bed of Minedomari was almost the same as in 1998, but eggs were found over a wider area that extended to the north. The area of the spawning bed was about 11,900 m^2 (170 m along the coast, maximum 70 m from the shoreline). Eggs were abundant in the middle of the bed on lines B, C, D and E, at the edge of the platform, or in the grooves (Figs. 2 and 11). The density

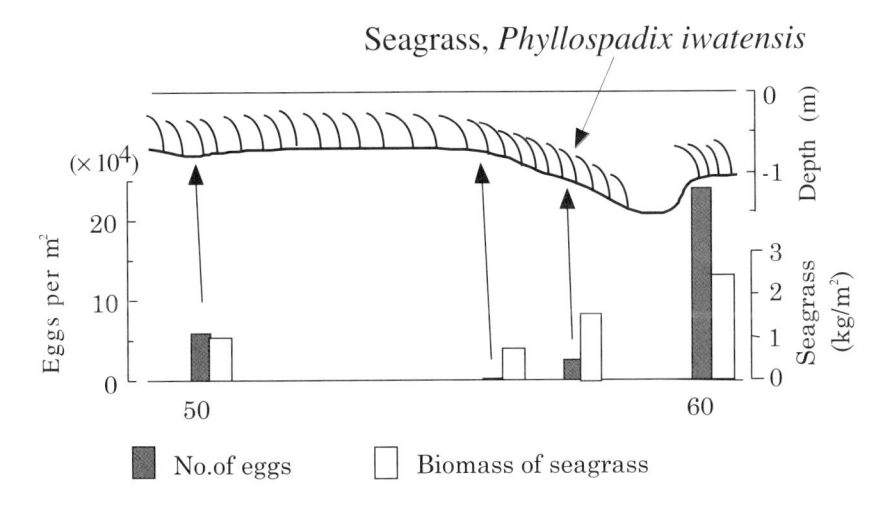

Figure 11. Distribution of herring eggs (on line D in Fig. 2) in relation to the bottom topography observed on March 25, 1999

Table 5. Egg density in the spawning beds at Minedomari in 1998 and 1999; see also Fig. 2.

Date	Eggs ($n \times 10^4$)/m² Max	Min	Ave	S.D.	n	Spawning bed Area (m²)	Total eggs ($n \times 10^4$)
April 10, 1998	162.2	0	24.2	49.3	17	13,250	320,650
March 25, 1999	53.8	0	9.6	14.1	34	10,650	102,240
April 16, 1999	23.8	0	4.5	7.5	14	4,500	20,250

of eggs was lower than in 1998, with a range of 0 to 538,000 per m² and average value of 96,000 per m².

On April 16, many empty egg membranes but no eggs were observed on the lines from A to C because of the hatch-out of early spawning group mentioned above. The newly spawned eggs found on April 9 (Fig. 2) were distributed farther south than these of March 25. Eggs collected on line D were in a later stage of development than those on lines E to H, indicating that the spawning period on line D was intermediate between the early spawning event which produced eggs that were hatched just after April 9 and the late spawning event which produced eggs found on the same day. Eyes of larvae in eggs were observed clearly on April 27 and only empty egg membranes were found on May 7, at the same sites along lines D to F. Therefore, the hatching period of the late spawning group was thought to have been from April 27 to May 7.

Eggs were attached mainly to *P. iwatensis* and *C. hakodatensis*. The vertical distribution pattern of eggs on *P. iwatensis* and *C. hakodatensis* was different (Figs. 12 and 13). Eggs were thicker on the middle or on top sections of *P. iwatensis,* but eggs were thicker on the bottom of *C. hakodatensis*. *C. hakodatensis* has air floats on its leaves and remains erect while *P. iwatensis* is pushed down near the bottom by wave action almost all the time. Thus the vertical distribution pattern of eggs indicates that herring spawn near the bottom.

Discussion

The character of the bottom, especially the type of vegetation, is an important factor in spawning bed selection by herring (Kääriä 1999). Along the coast of Atsuta, we identified 42 species of algae and seagrass and the primary species were *Laminaria* spp., *Cystoseira hakodatensis* and *Phyllospadix iwatensis*. The biomass of vegetation varied between 0.5 and 6.0 kg per m² in the study area representing clearly higher values compared to the southwest coast of Hokkaido Island. Coraline flats are typical in that area because of overgrazing by sea urchins (*Strongylocentrotus nudus*) (Agatsuma 1997) and normally the biomass of algae and seagrass is less than 0.5 kg per m².

Herring eggs were most prevalent on *P. iwatensis* and the second most important substratum was *C. hakodatensis*. Though *Laminaria* spp. was dominant in biomass, practically no eggs were found attached to *Laminaria* spp. Obata et al. (1997) observed that herring in a tank selected a softer fragmented substratum made of hemp palm for spawning. The leaves of *P. iwatensis* are long, flat, and narrow (2 to 5 mm in width) and they are well suited for the attachment of herring eggs. In contrast, the surface of *Laminaria* spp. is covered by mucilage and is very slimy. Tamura and Okubo (1953a) observed that a lot of herring eggs attached to *Laminaria* just after spawning had been dislodged easily by a storm from the surface of the kelp. Galkina (1960) observed herring eggs at Tunguzskaya Bay on the Okhotsk Peninsula, attached to red algae, *Halosaccion* sp. and on pebbles or rocks on the bottom, but not on *Laminaria*. According to Tada (2000) *P.*

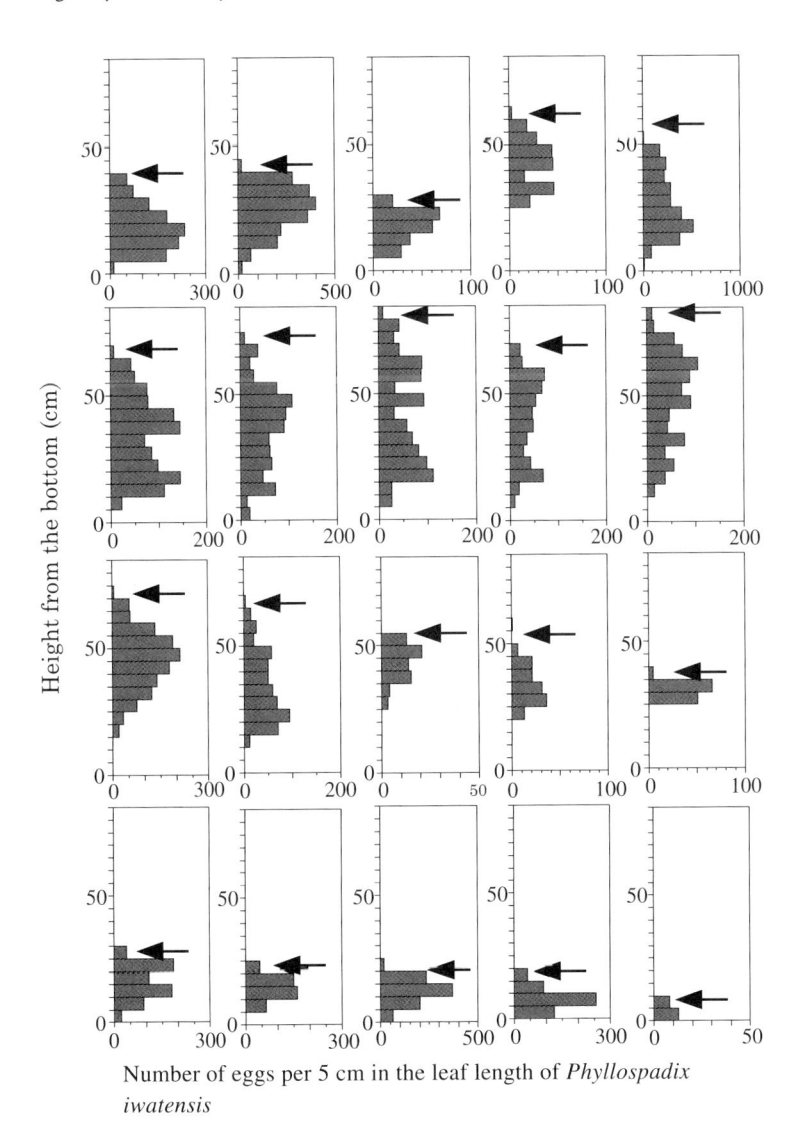

Figure 12. Vertical distribution of herring eggs on the seagrass Phyllospadix iwatensis, at Minedomari, on April 10, 1998. Arrows show the height of the leaf tip.

Wet weight of *Cystoseira hakodatensis* (g /10 cm : —●—)

Number of herring eggs per 10 cm in the height of *Cystoseira hakodatensis*

Figure 13. Vertical distribution of herring eggs on the sargassum Cystoseira hakodatensis, at Minedomari, on April 16, 1999. Arrows show the height of the algal tip.

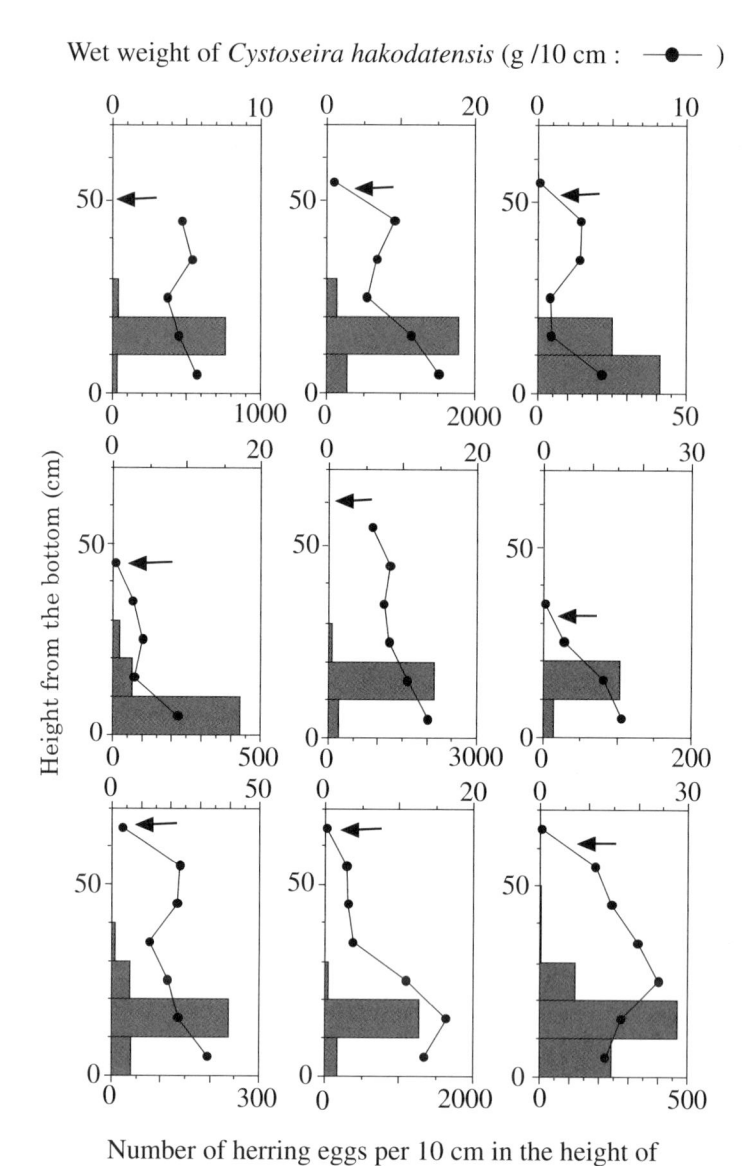

iwatensis, Sargassum confusum, and *Coccophora langsdorfii* (but not *Laminaria* spp.) were important spawning substratum in Rumoi (80 km to north of the area studied in this paper).

Along the coast of Atsuta, the vegetation pattern in the littoral and sublittoral zone is almost the same in each area and in each year. However, dense concentrations of herring eggs occur at restricted areas such as Minedomari. Abundant vegetation is an important factor for the spawning success of herring, but other factors also appear to affect the selection of spawning beds.

The other probable factor for spawning bed selection is bottom topography. The bottom declines steeply in the northern part of the Atsuta coast and there is a wide shallow rocky platform in the southern part (Fig. 1). The bottom at Mourai in the most southern part of Atsuta is sandy. The platform is well developed between Atsuta and Minedomari, but the grooves continued from deeper areas to the platform observed at Minedomari have not been recognized in the other areas studied. The distribution of eggs in the spawning bed at Minedomari suggests that the topographical character of the platform affects spawning bed selection. Egg density was high along the narrow grooves on the platform, but lower on the vegetation offshore or far from the grooves. In southwest Finland, one common feature of the spawning ground is that they are situated near the deep fracture lines running on the sea bed (Rajasilta et al. 1993, Kääriä et al. 1997) and herring are using these routes when migrating into the area. The platform edge in our study area also acts as a breakwater reducing the effects of waves. Therefore, it is possible that herring use the grooves to enter the shallower area for spawning. Based on the field observations at Minedomari, most of the waves break at the edge of the platform or at the mouth of the grooves. Haegele and Schweigert (1985) reported that Pacific herring spawning is restricted to sheltered areas rather than in the surf zone, and speculated that the selection of sheltered shore for spawning is an adaptation to minimize egg loss. The shallow depths of the Minedomari spawning ground probably also reduces predation by fish like the Arubeskee greenling (*Pleurogrammus azanus*), which feeds actively on herring eggs in Hokkaido (Tamura and Okubo 1953b).

The main spawning period in 1998 was in late March, and in 1999 two spawning peaks were found in early March and early April. Herring fisheries using gillnet starts normally from mid-February and continue to the end of April in Atsuta. According to Takayanagi and Tanaka (2000) there was one peak of fishing from early to mid-March in 1998 and two peaks in late February and mid-March in 1999. The number of spawning waves corresponded well to the fisheries peaks in each year but the spawning periods shifted later compared with the peak of the fishery. Gonadal index was maximum in late February and the value was almost the same at the end of March in each year (Takayanagi and Tanaka 2000). These facts indicate that herring stocks were mature enough to spawn during the fishing period but actual spawning was delayed compared to peaks in the herring fishery and spawning beds were limited to the shallower area in Atsuta as mentioned before. Thus it seems that the opportunity to enter such shallow area for spawning is limited by the wave condition and topographical features.

The spatial distribution of eggs laid in the second spawning didn't overlap with that of the first spawning in 1999. Avoiding the overlap of spawning beds among spawning waves reduced the layers of eggs in each blocks. Haegele et al. (1981) indicated that herring distinguish between the differences of substratum types. Hay (1985) suggested the possibility that herring could distinguish the herring eggs on the substratum. Therefore, it is possible that the late spawning wave identified the eggs spawned by the early wave and avoided these by shifting to a more southern area.

Egg densities varied from several hundred to a maximum of 1.6 million per square meter and the egg mortality was under 18% in every sample studied. In the study area the total amount of herring caught in 1998 was 32.4 t and consisted of almost entirely of 2-year-old fishes, while the amount of fish was 56.8 t in 1999, consisting of 3-year-old (early peak) and 2-year-old (late peak) fishes (Takayanagi and Tanaka 2000). Size of later shoals was smaller than earlier ones in 1999 (Takayanagi and Tanaka 2000) and the total amount of fish in the later wave was also lower than in the earlier spawning wave. The difference between early and late stocks would be the main reason for the small amount of eggs in the late spawning wave in 1999. The maximum density of eggs in 1998 was higher than 1999, but the size of herring was smaller than that of the early wave in 1999 because of younger age composition. So it seemed that a higher amount of herring concentrated to spawn at Minedomari in 1998.

Food supply and temperature prior to the spawning period are considered the most important factors in the control of spawning time (Aneer 1985, Hay and Brett 1988, Rajasilta 1992, Rajasilta et al. 1996). The temperature during the spawning period in 1998 was about 1°C in early February, and rose to 5°C in late March or early April and in late April it was 8 to 9°C. Although the start of spawning in 1999 was earlier than in 1998, the reason seemed to be the difference of the fishing peak as mentioned above. The temperature during spawning season in 2-year-old herring in both years was approximately 5°C.

In 1999, we found five additional small spawning beds besides Minedomari. Because the developmental stage of eggs in such small beds was the same as found at Minedomari, spawning time of each bed was also about the same. It is considered that the water temperature affects the period from fertilization to hatching (Alderdice and Velsen 1971, Kuwatani et al. 1978). The reason that eggs disappeared in late April 1998 was an early hatching period related to the higher temperature in 1998. The period from spawning to hatching in each wave in 1999 was about 40 days and 30 days respectively. So, eggs from late spawning would hatch out after a shorter period because of higher temperature.

Herring spawn in shallow areas (under 2 m) along the coast of Atsuta. In many other important spawning beds eggs have also been found in shallow areas (e.g., Hay 1985, Käariä et al. 1988, Oulasvirta and Lehtonen 1988, Aneer 1989, Rajasilta et al. 1993). According to Tamura et al. (1953a 1954 and 1955) the depth of spawning beds for Hokkaido-Sakhalin herrings was usually deeper than 4 m. Rajasilta et al. (1993) discussed the reason why spawning of Baltic herring takes place in shallow water in relation with egg-hatching success. They said that spawning in shallow water reduced the incubation time as temperature rises faster in spring. Also, in our study area vegetation as spawning substratum is limited in shallower areas because of overgrazing by sea urchins after the 1960s (Agatsuma 1997) and sand cover on the bottom in the southern part of Atsuta. Tamura et al. (1953a, 1954, 1955) observed that herring spawned not only on vegetation but also on pebbles and rocks. In our study, herring spawned only on vegetation in Atsuta. However, we have insufficient data to discuss whether preference for the substratum differs between the Hokkaido-Sakhalin herring and Ishikari Bay herring.

The effect of low salinity on the fertilization and hatching success of herring has been studied by Alderdice and Velsen (1971), Kuwatani et al. (1978) and Griffin et al. (1998). These studies show that in Pacific herring, fertilization, survival rates to yolk sac absorption, and hatching rate were higher in optimum salinity (between 16 and 24 psu). Since salinity presumably varies along a gradient from north to south in Atsuta in relation to the location of the Ishikari and small rivers, it is necessary in the future to survey the salinity continuously in spawning beds. Although the topographic characteristics are thought to be important to the spawning bed selection of herring in relation with wave condition, there are no data on how wave height or bottom current speed is possibly related the spawning behavior of herring. Surveying these environmental factors during the spawning season could help us to understand why in one of these restricted areas in Minedomari herring spawned year after year in exactly the same spawning bed.

Acknowledgments

The authors thank the staff of the Ishikari fishery extension office for help during field studies at Atsuta. We also thank Dr. R. Stephenson and one anonymous reviewer for helpful comments on the manuscript. We are also grateful to Dr. Juha Käariä for a critical review of the manuscript and for valuable comments. This research was funded by the project "Stock Enhancement of Neritic Herring in Japan Sea of Hokkaido" of the Hokkaido prefectural government.

References

Agatsuma, Y. 1997. Ecological studies on the population dynamics of the sea urchin *Strongylocentrotus nudus*. Sci. Rep. Hokkaido Fish. Exp. Stn. 51:1-66.

Alderdice, D.F., and F.P.J. Velsen. 1971. Some effects of salinity and temperature on early development of Pacific herring (*Clupea pallasi*). J. Fish. Res. Board Can. 28:1545-1562.

Aneer, G. 1985. Some speculations about the Baltic herring (*Clupea harengus membras*) in connection with the eutrophication of the Baltic Sea. Can. J. Fish. Aquat. Sci. 42:83-90.

Aneer, G. 1989. Herring (*Clupea harengus* L.) spawning and spawning ground characteristics in the Baltic Sea. Fish. Res. 8:169-195.

Aneer, G., G. Florell, U. Kautsky, S. Nellbring, and L. Sjostedt. 1983. In-situ observations of Baltic herring (*Clupea harengus membras*) spawning behaviour in Askolandsort area, northern Baltic proper. Mar. Biol. 74:105-110.

Fujita, T. and S. Kokubo. 1927. Studies on herring. Bull. School of Fish. Hokkaido Univ. 1(1):1-127.

Galkina, L.A. 1959. Reproduction of herring in Gizhiga Bay. Izvest. Pac. Inst. Fish. Oceanogr. (TINRO) 47:86-99. (Transl. by the Fisheries Agency of Jap. Govern. 3:63-87, 1967.)

Galkina, L.A. 1960. Reproduction and development of Okhotsk herring. Izvest. Pac. Inst. Fish. Oceanogr. (TINRO) 46:3-40. (Transl. by the Fisheries Agency of Jap. Govern. 3:1-62, 1967.)

Griffin, F., M. Pillai, C. Vines, J. Kääriä, T. Hibbard-Robbins, R. Yanagimachi, and G.N. Cherr. 1998. Effects of salinity on sperm motility, fertilization, and development in the Pacific herring, *Clupea pallasi*. Biol. Bull. 194:25-35.

Haegele, C.W., and J.F. Schweigert. 1985. Distribution and characteristics of herring spawning grounds and description of spawning behavior. Can. J. Fish. Aquat. Sci. 42:39-55.

Haegele C.W., R.D. Humphrey, and A.S. Hourston. 1981. Distribution of eggs by depth and vegetation type in Pacific herring (*Clupea harengus pallasi*) spawning in southern British Columbia. Can. J. Fish. Aquat. Sci. 38:381-386.

Hay, D.E. 1985. Reproductive biology of Pacific herring (*Clupea harengus pallasi*). Can. J. Fish. Aquat. Sci. 42:111-126.

Hay, D.E., and J.R. Brett. 1988. Maturation and fecundity of Pacific herring (*Clupea harengus pallasi*): An experimental study with comparisons to natural populations. Can. J. Fish. Aquat. Sci. 45:399-406.

Hay, D.E., and P.B. McCarter. 1997. Larval distribution, abundance, and stock structure of British Columbia herring. J. Fish Biol. 51:155-175.

Hourston, A.S., and H. Rosenthal. 1976. Sperm density during active spawning of Pacific herring (*Clupea harengus pallasi*). J. Fish. Res. Board Can. 33:1788-1790.

Kääriä, J. 1999. Reproduction of the Baltic herring (*Clupea harengus membras* L.): Factors affecting the selection of spawning beds in the Archipelago sea, in SW Finland. Ph.D. thesis, Annales Universitatis Turkuensis, Ser. AII Tom. 116:1-88.

Kääriä, J., M. Rajasilta, M. Kurkilahti, and M. Soikkeli. 1997. Spawning bed selection by the Baltic herring (*Clupea harengus membras* L.) in the Archipelago of SW Finland. ICES J. Mar. Sci. 54:917-923.

Kääriä, J., J. Eklund, S. Hallikainen, R. Kääriä, M. Rajasilta, K. Ranta-aho, and M. Soikkeli. 1988. Effects of coastal eutrophication on the spawning grounds of the Baltic herring in the SW Archipelago of Finland. Kieler Meeresforsch. Sonderh. 6:348-356.

Kimmel, C.B., W.W. Ballard, S.R. Kimmel, B. Ullmann, and T.F. Schilling. 1995. Stages of embryonic development of the zebrafish. Dev. Dyn. 203:253–310.

Kuwatani, Y., S. Shibuya, T. Wakui, and T. Nakanishi. 1978. Study on culturing herring egg and fry. Rep. Tech. Dev. Herring Propagation Cult. 1971-1974, pp. 11-71. (In Japanese.)

McMynn, R.G., and W.S. Hoar. 1953. Effects of salinity on the development of the Pacific herring. Can. J. Zool. 31:417-432.

Messieh, S.N. 1987. Some characteristics of Atlantic herring (*Clupea harengus*) spawning in the southern Gulf of St. Lawrence. NAFO Sci. Council Studies. 11:53-61.

Obata, H., K. Yamamoto, and T. Matubara. 1997. Induction of spawning using an artificial spawning bed in captive Pacific herring *Clupea pallasi*. Saibai Giken 25(2):75-80.

Oulasvirta, P., and H. Lehtonen. 1988. Effects of sand extraction on herring spawning and fishing in the Gulf of Finland. Mar. Pollut. Bull. 19:383-386.

Oulasvirta, P., J. Rissanen, and R. Parmanne. 1985. Spawning of Baltic herring (*Clupea harengus* L.) in the western part of the Gulf of Finland. Fin. Fish. Res. 5:41-54.

Rajasilta, M. 1992. Timing of spawning in the Baltic herring (*Clupea harengus membras*) in the Archipelago Sea, SW Finland: Regulatory mechanisms and consequences for offspring production. Ph.D. thesis, Annales Universitatis Turkuensis. 164 pp.

Rajasilta, M., J. Kääriä, P. Laine, and M Soikkeli. 1996. Is the spawning of the herring in the northern Baltic influenced by mild winters? In: Proceedings of the 13th Symposium of the Baltic Marine Biologists. University of Latvia, Riga, pp. 185-191.

Rajasilta, M., J. Eklund, J. Hänninen, M. Kurkilahti, J. Kääriä, P. Rannikko, and M. Soikkeli. 1993. Spawning of herring *(Clupea harengus membras* L.) in the Archipelago Sea. ICES J. Mar. Sci. 50:233-246.

Stacey, N., and A.S. Hourston. 1982. Spawning and feeding behavior of captive Pacific herring. Can. J. Fish. Aquat. Sci. 39:489-498.

Tada, M. 2000. Study on the vegetation species composition in the spawning beds of herring at Rumoi. In: Progress report of stock enhancement of neritic herring in Japan Sea of Hokkaido. Wakkanai Fisheries Experimental Station, Wakkanai, pp. 56-65. (In Japanese.)

Takayanagi, S., and N. Tanaka. 2000. Basic study on the resource management. In: Progress report of stock enhancement of neritic herring in Japan Sea of Hokkaido. Wakkanai Fisheries Experimental Station, Wakkanai, pp. 106-158. (In Japanese.)

Tamura, T., and S. Okubo. 1953a. Some observations on the natural spawning of the spring herring in the western coast of Hokkaido. Sci. Rep. Hokkaido Fish Hatchery 8:21-32.

Tamura, T., and T. Okubo. 1953b. Study on the food of Atkafish (*Pleurogrammus azonus* Jordan & Mets) caught on the breeding ground of herring in Hokkaido. Sci. Rep. Hokkaido Fish Hatchery 7:93-103.

Tamura, T., S. Ohigashi, and T. Hirobe. 1955. Some observations on the natural spawning of the spring herring in the western coast of Hokkaido, III. Fishing condition of spring herring in 1955 and underwater observations on the natural spawning ground of the herring at Tomamai. Sci. Rep. Hokkaido Fish Hatchery 10:115-131.

Tamura, T., S. Okubo, T. Fujita, and T. Hirobe. 1954. Some observations on the natural spawning of the spring herring in the western coast of Hokkaido, II. Observations made by diving into water on the natural spawning ground of the herrings. Sci. Rep. Hokkaido Fish Hatchery 9:95-112.

Tanaka, I. 1991. Long-term changes of coastal sea temperature on the west coast of Hokkaido, Japan, with respect to herring resources. In: T. Kawasaki, S. Tanaka, Y. Toba, and A. Taniguchi (eds.), Long-term variability of pelagic fish populations and their environment. Pergamon Press, Oxford, pp. 395-396.

Tyurnin, B.V. 1973. The spawning range of Okhotsk herring. Proc. Pacific Inst. Fish. Oceanogr. 86:12-21. (Translated by the Transl. Bur., Dept. of the Secretary of State, Canada.)

Yamaguchi, M. 1926. Study on the habit of herring. Fisheries Research Report 17. 280 pp. (In Japanese.)

Relationship between Date of Hatching and Growth of Herring Larvae and Juveniles in Ishikari Bay

Ryotaro Ishida, Masayoshi Sasaki, Shiro Takayanagi, and Hideo Yoshida
Hokkaido Central Fisheries Experimental Station, Hokkaido, Japan

Abstract

As the first step to investigate the causes of fluctuation in the Ishikari Bay herring stock, the growth of larvae and juveniles was examined by otolith analysis. Herring larvae and juveniles of the Ishikari Bay population were collected from the west coast of Hokkaido, Japan, from May through July in 1998 and 1999. Daily increment formation was verified in the laboratory by comparing the number of increments between alizarin complexion (ALC) marking on the 19th and 33rd day after hatching. A linear relationship ($Y = 0.935X - 8.14$) fitted between the actual age and the number of increments indicated that the first increment was formed on day 8 after hatching. This is similar to a previous report on the Pacific herring (Moksness and Wespestad 1989).

Peaks in the estimated day of hatching of wild herring larvae and juveniles have shown good agreement with results of direct research by scuba diving for in situ observations of herring eggs at Atsuta (Hoshikawa et al. 2001). This indicates that larvae and juveniles originated from the eggs spawned around Atsuta.

The relationship between the days after hatching and the standard length showed that growth of the late-hatched group was faster than the early-hatched group. The larvae and juveniles that experienced higher temperatures grew faster.

Introduction

The herring in Ishikari Bay are considered to be oceanic herring which spawn in Ishikari Bay and have a small migration area (Kobayashi et al. 1990, Kobayashi 1993).

In the coastal waters of Ishikari Bay, the annual landing of matured herring between 1985 and 1996 ranged from 1 to 4 t, but in 1997 markedly increased to 24.8 t and thereafter gradually increased (41.1 and 87.6 t in 1998 and 1999, respectively). Takayanagi et al. (1998) suggested that the marked increase of Ishikari Bay herring in landings since 1997 might be due to the occurrence of strong year class.

This herring grows faster than other populations in the Japan Sea and reaches sexual maturity in 2 years after hatching (Ishida et al. 1997). It is likely that the year-class strength is determined within a short period after hatching. However, very little is known of the early life history of the Ishikari Bay herring stock. A recent study on the distribution of the Ishikari Bay herring larvae and juveniles made it possible to study the growth of larvae and juveniles.

For the first step to investigate the causes of the Ishikari Bay herring stock fluctuation, the growth of larvae and juveniles was examined by otolith analysis. Moksness and Wespestad (1989) reported that otolith increments of the Pacific herring were formed on a daily basis from the end of the yolk-sac stage (age 8 days). The daily otolith increments, however, may be influenced by water temperature, genetic differences, and conditions of the otolith observations such as the magnification and investigator (Campana and Moksness 1991). We also investigated the relationship between the number of increments and the actual ages of Ishikari Bay herring larvae in a laboratory experiment.

Materials and Methods

Validation of Otolith Daily Increment

Spawning herring were collected in Atsuta, Hokkaido, Japan on 25 February 1997 (Fig. 1). Herring eggs were artificially fertilized at the Ishikari Fisheries Extension Office. They were transported to the Hokkaido Central Fisheries Experimental Station in Yoichi, Japan and incubated in an indoor tank with a running filtered seawater supply at temperatures between 5.4° and 5.8°C. Larvae hatched on 25 March 1997. They were divided into two 100-liter open-system tanks and fed a continuous supply of rotifers and *Artemia* enriched with omega-3 highly unsaturated fatty acid (HUFA) and beta-carotene.

Water temperature was maintained at 5-12°C to be much the same as the temperature of coastal Yoichi, which approximates the temperature of the spawning ground in Atsuta.

On 13 and 27 April, 19 and 33 days after hatching, respectively, larvae in one of the two tanks were immersed in 200 mg per liter ALC solution for 24 hours. After ALC treatment, 10 larvae were sampled and ALC marking on the

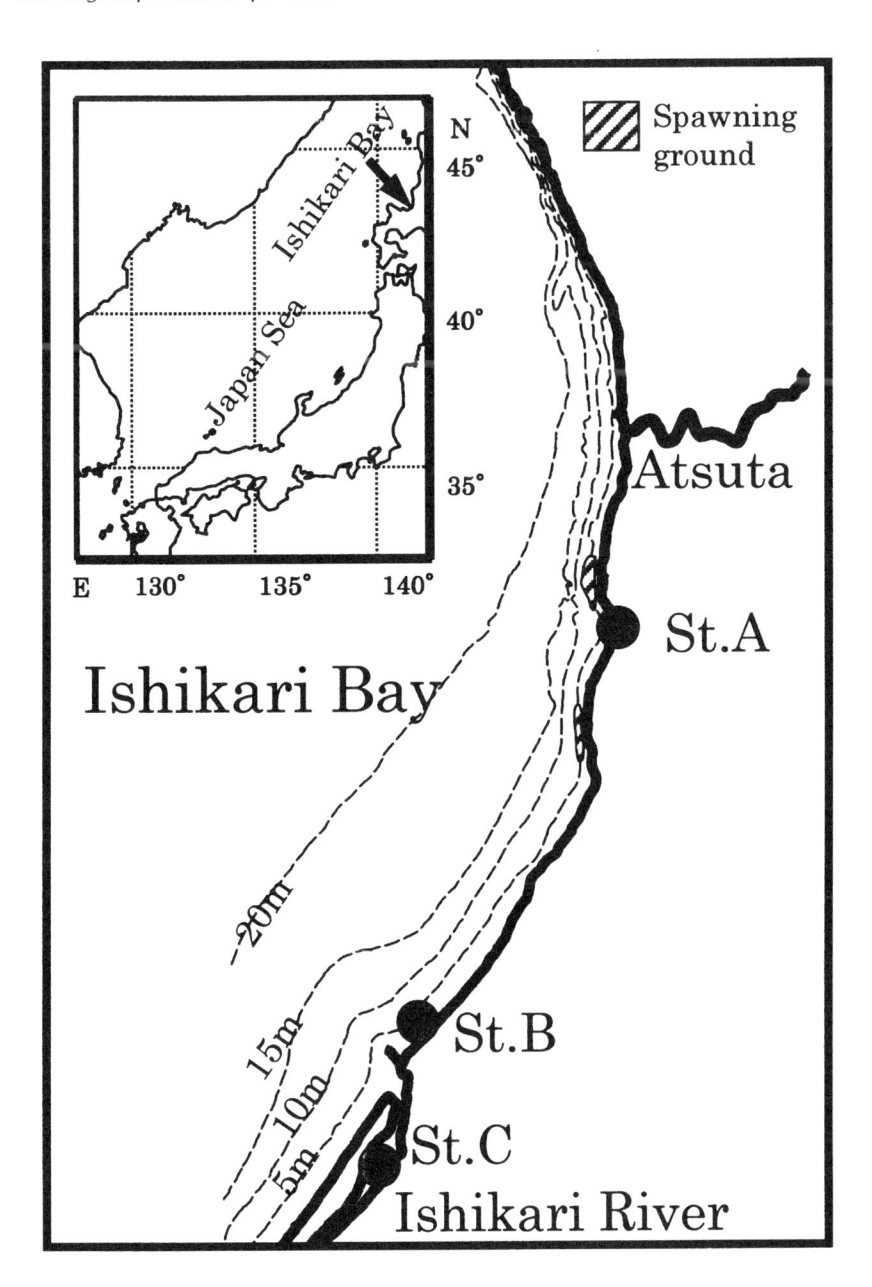

Figure 1. Map of Ishikari Bay with the spawning ground and sampling stations. (Hoshikawa in press)

edge of the sagittae was confirmed. Fifty larvae were sacrificed on days 6, 7, 17, 19, 33, and 36 after hatching and immediately preserved in 70% ethanol. Larvae sampled on days 19 and 33 after hatching were used for the otolith analysis.

Wild Herring Larvae and Juveniles

The Ishikari Bay herring larvae and juveniles were collected from Ishikari Bay on the west coast of Hokkaido during April to July of 1998 and 1999 (Fig.1) and were preserved in 70% ethanol immediately after capture.

Sampling information on the larvae and juveniles used on this study is shown in Table 1. At Station A, samples were collected with a fish-attracting light and hand net at nighttime. Samples at stations B and C were obtained from Sasaki et al. (2001). Although the study of Sasaki et al. (2001) covered a large area and was conducted over a long period, only samples that caught over 31 individuals in a day were used for this study. Methods for collection of samples at stations B and C are reported by Sasaki et al. (2001).

Preparation for Otolith Analysis

More than 6 weeks after capture, the body length (standard length, SL) of each larva were measured to the nearest 0.01 mm and the largest otoliths (the sagittae) from each larva were removed and mounted with the convex side up in clear nail varnish on a glass slide. Otoliths of herring over 22 mm SL were polished with aluminum oxide lapping films (60-0.3 mm grit size) to expose increments on a Ziess compound microscope at ×1,600 magnification. ALC markings in otoliths were observed under an ultraviolet light. Increments were counted using a video system. Aging of field-caught juveniles over 60 mm SL became difficult to determine.

Artificially inseminated herring juveniles (to approximately 62 mm SL) reared using ALC solution on day 0 or 10 after hatching have been released from Atsuta since 1996 for restocking and investigation of the migration of herring larvae and juveniles. Numbers of herring juveniles (×10,000) and release dates were: 16.1 on 14 June 1996; 24.8 during 16-24 June 1997; 74.3 during 16-17 June 1998; and 116.8 during 27 May-14 June 1999. We could distinguish between wild and artificially inseminated fishes by observation of the ALC markings on the otoliths.

Water Temperature of Coastal Ishikari Bay

Water temperature of coastal Ishikari Bay at the spawning ground is given in Hoshikawa et al. (2001).

Results

Validation of Daily Otolith Increments

Comparison of the growth of the control and ALC marked larvae is shown in Fig. 2. There was no significant difference between the control and the

Table 1. Sampling gear, average standard length (SL), and number of wild herring caught at Atsuta, Hokkaido, Japan.

Station	Gear	Sampling date	Average SL (mm)	S.D. (mm)	Number of fish caught
A	Fish-attracting light	26 May 1998	22.5	3.1	28
B	Beach seine	25 Jun 1998	39.6	2.6	1,999
C	Set net	26 Jun 1998	52.3	4.6	68
C	Set net	4 Jul 1998	55.5	4.5	31
C	Set net	7 Jul 1998	60.7	4.7	173
A	Fish-attracting light	11 Jun 1999	29.4	4.2	128
B	Beach seine	24 Jun 1999	39.2	2.6	123
B	Beach seine	2 Jul 1999	41.0	2.7	1,665
B	Beach seine	8 Jul 1999	45.4	2.2	2,777
C	Set net	3 Jul 1999	55.5	3.5	1,244
C	Set net	13 Jul 1999	60.9	4.3	67

ALC marked larvae (t-test, $t = 0.77$, d.f. $= 248$), indicating that there was no effect of the ALC treatment on growth.

Relationship between the number of increments and the actual ages is shown in Fig. 3. A linear relationship was fitted as follows, where Y is estimated age and X is real age:

$$Y = 0.935X - 8.14 \ (n = 50 \text{ and } R^2 = 0.8944).$$

However, the average rate of increment formation was 0.94, and the deposition rate was significantly different from 1 (t-test, $t = 1.83$, d.f. $= 48$). Standard deviations of the estimated age from the real age on days 19 and 33 after hatching were 2.7 and 2.5, respectively, although expected was within 1. The expected number of increments between ALC markings was 13, and the counts ranged from 11 to 14. The mean and standard deviation were 12.8 and 0.6, respectively (Figs. 4 and 5). Thus, the number of increments between the two marks with ALC corresponded with the number of days elapsed (13 days) between the first and second treatment of ALC, indicating the rate of increment formation was 1 day (t-test, $t = 1.50$, d.f. $= 21$). In the Atlantic herring, it was found that the detection of increments during the first 15-20 days after hatching are probably too fine to be resolved (Campana et al. 1987). Lough et al. (1982) also reported that the average number of days after hatching of 3-increment larvae was approximately 22 days. In the Pacific herring, increments were formed on a daily basis from the end of the yolk-sac stage (8 days after hatching)

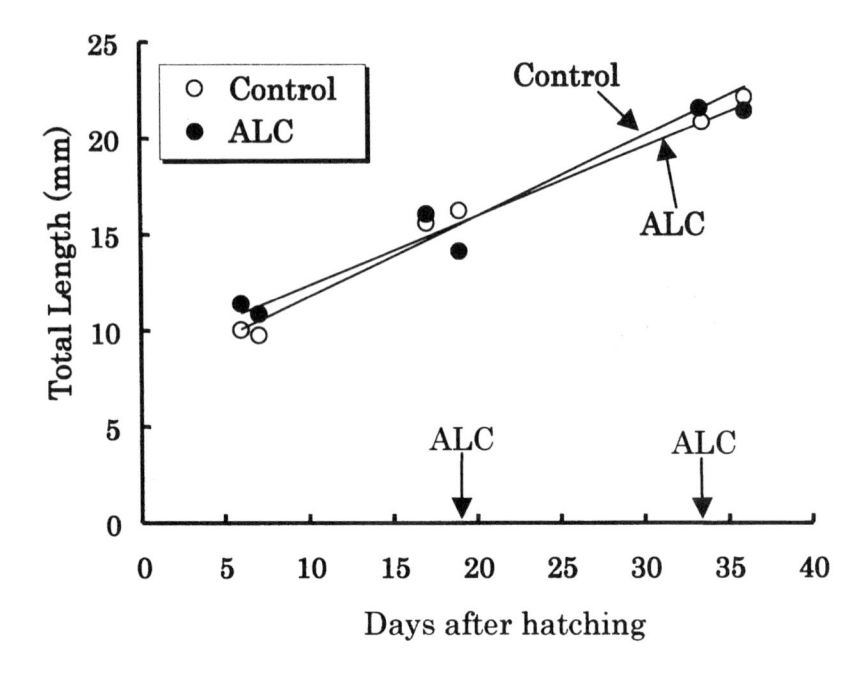

Figure 2. Effect of the ALC treatment on the growth of Ishikari Bay herring. The ALC marking was conducted on days 19 and 33 after hatching.

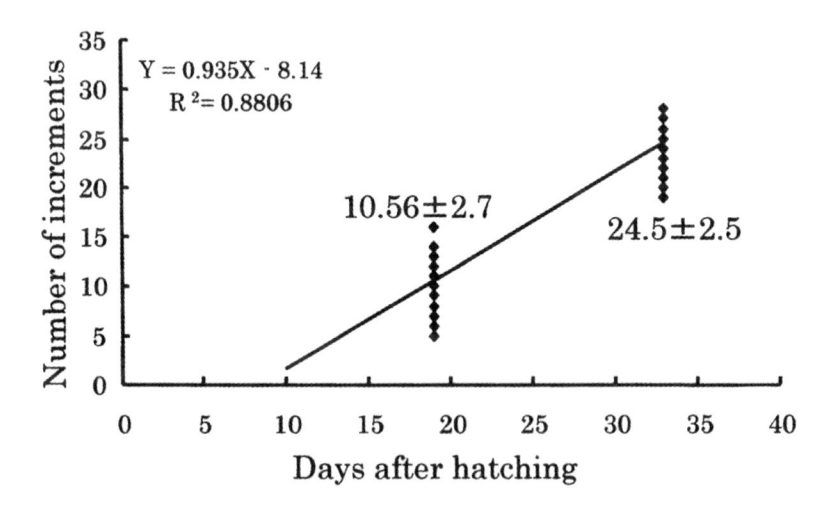

Figure 3. Relationship between the actual age in days and the number of otolith increments.

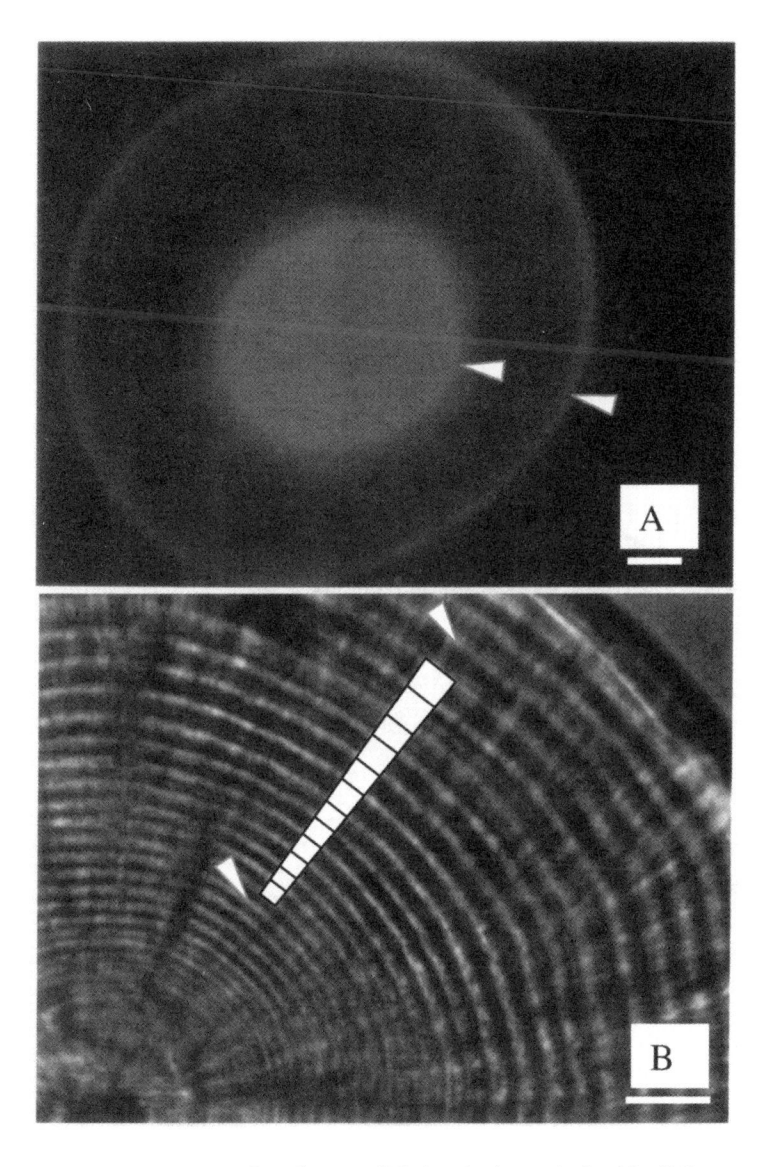

Figure 4. Micrographs of an otolith (sagitta) marked with ALC on days 19 and 33 after hatching. (A) Micrograph under ultraviolet light. (B) Micrograph under normal transmitted light. Scale bar = 10 mm.

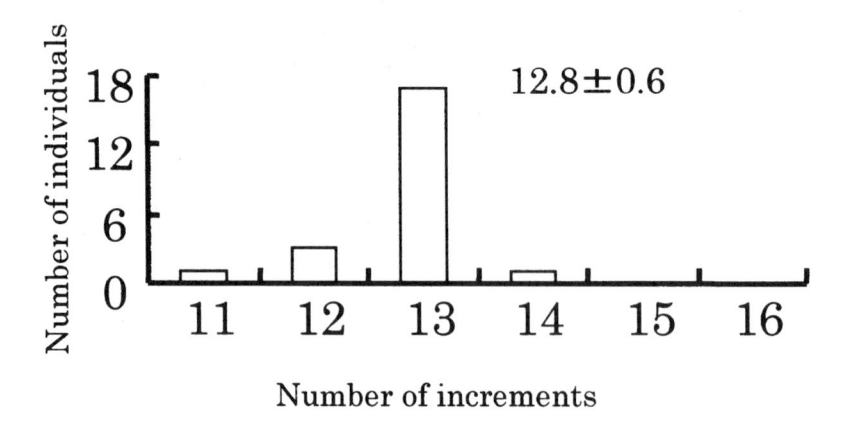

Figure 5. The number of increments between the two marks with ALC corresponded with the number of days elapsed (13 days) between the first and second treatment of ALC, indicating that the rate of increment formation was 1 day.

(Moksness and Wespestad 1989). In the Ishikari Bay herring, mean day of the first increment formation was 8 days after hatching. This was similar to a previous study on the Pacific herring (Moksness and Wespestad 1989).

Estimation of Hatching Date of Wild Herring

It was found that the day of increment formation was 8 days after hatching on average in this study. Consequently, a correction factor of 8 was added to the number of increments counted to establish the estimated age in days after hatching. Estimated date of hatching of larvae and juveniles from otolith analysis is shown in Figs. 6 and 7. In 1998 the estimated hatching day sampled at station A on 26 May started in early April, and reached a peak in the middle of April. At station B on 25 June the hatching period was different from station A on 26 May, and started and reached a peak in late April. At station C on 26 June and 4 July most of them hatched in early April. On 7 July hatching showed a bimodal shape. First and second peaks were similar to that on 26 May and 25 June. In 1999 at station A on 11 June showed a bimodal shape peaked in early April (early-hatched larvae and juveniles) and early May (late-hatched larvae and juveniles). At station B on 24 June, early-hatched juveniles were dominant, but most of the juveniles sampled on 2 July and 8 July were late-hatched. At station C on 3 July, early-hatched juveniles were dominant although late-hatched juveniles were at station B in this period. On 13 July at station C hatching showed a bimodal shape; first and second peaks were similar to that on 11 June.

Growth of Wild Herring Larvae and Juveniles

A lot of ALC marked artificial juveniles released from Atsuta, over 15 km distance from the estuary, were recaptured in the Ishikari River estuary (Sasaki et al. 2001). Wild juveniles over 41 mm SL were distributed in the estuary and juveniles under 41 mm were in the surf area, around station B (Sasaki et al. 2001). These results indicated that the estuary is a suitable area for sampling because juveniles approximately over 41 mm immigrate into the estuary. The hatching day distributions of the estuary (station C) in the present study showed a bimodal shape in both years (Figs. 6 and 7). Growth curves were examined separately before and after on 22 April and 25 April on 1998 and 1999, respectively. The late-hatched individuals obviously grew faster than early hatched ones in both years (Fig. 8)

Relationship between Growth and Water Temperature

A continuous increase in water temperature of coastal Ishikari Bay in 1998 and 1999 was observed from hatching to the juvenile period in both years (Fig. 9), indicating that late-hatched individuals certainly experienced higher temperatures than early-hatched ones. To investigate an effect of temperature on growth of Ishikari Bay herring larvae and juveniles, relationships between the hatching days, used to indicate the temperature range that certain larvae and juveniles experienced, and the average standard lengths were compared (Fig. 10). This result showed that the late-hatched individuals grow faster than the early ones, indicating that water temperature affected the apparent growth of Ishikari Bay herring larvae and juveniles.

Discussion

Moksness and Wespestad (1989) and Wespestad and Moksness (1990) showed that daily growth increments in Pacific herring were true daily increments at a similar growth rate observed in nature and that mesocosm experiments were an appropriate environment for studying herring because they produce a similar growth rate as in nature. In the present study, in the laboratory-reared experiment, larvae 33 days after hatching reached 22 mm. Even though small tanks were used, this result is in good agreement with the growth of wild larvae and juveniles, although Moksness and Wespestad (1989) and Geffan (1982) suggested that larvae reared in small tanks were smaller than in the wild, supposing that otolith increment formation in the present study was similar to nature. Here, the cause of the similar growth to nature in small tanks may be due to the enrichment of the diet.

The number of increments between two marks with ALC corresponded with the number of days elapsed between the first and the second ALC treatments with a small standard deviation. In the Pacific herring this is the first time the chemical marking method confirmed that the deposition rate is one with a low standard deviation. However, the deposition rate estimated from the relationship between the actual number of days after

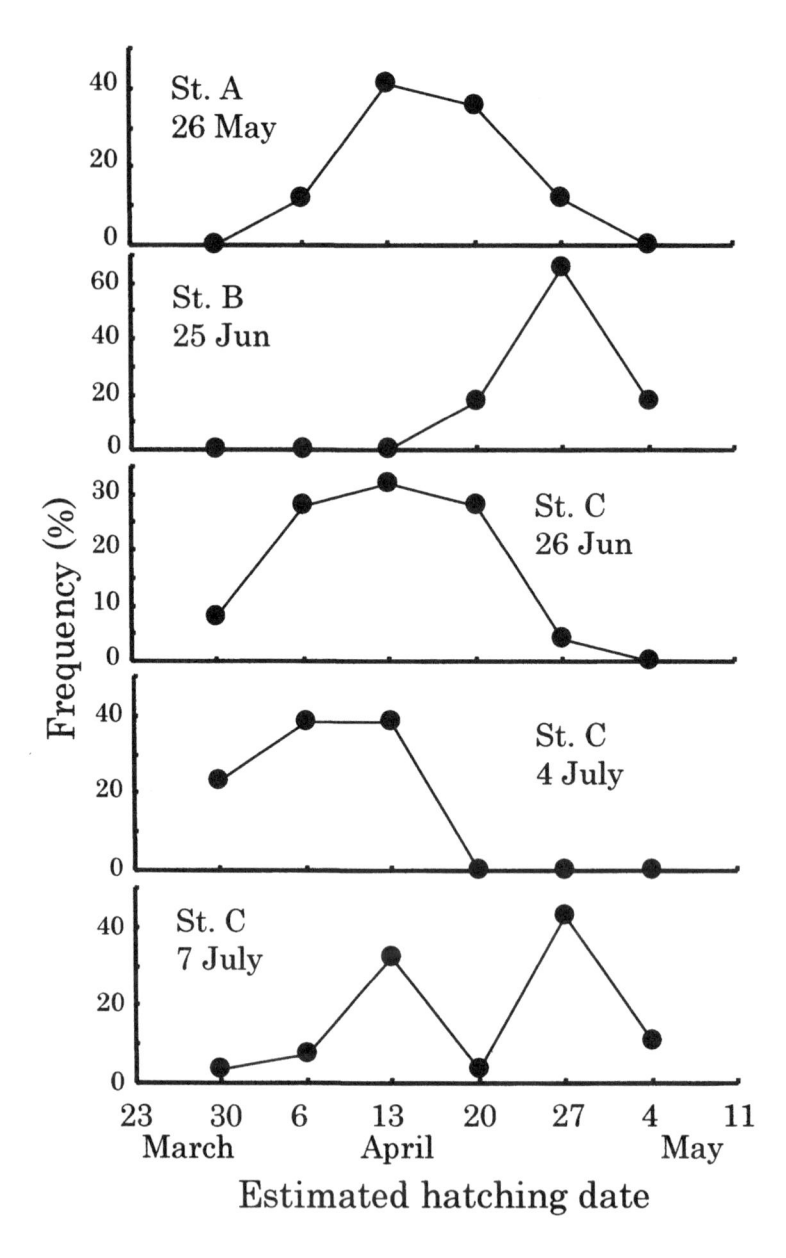

Figure 6. Hatching distribution of the Ishikari Bay herring larvae and juveniles in 1998.

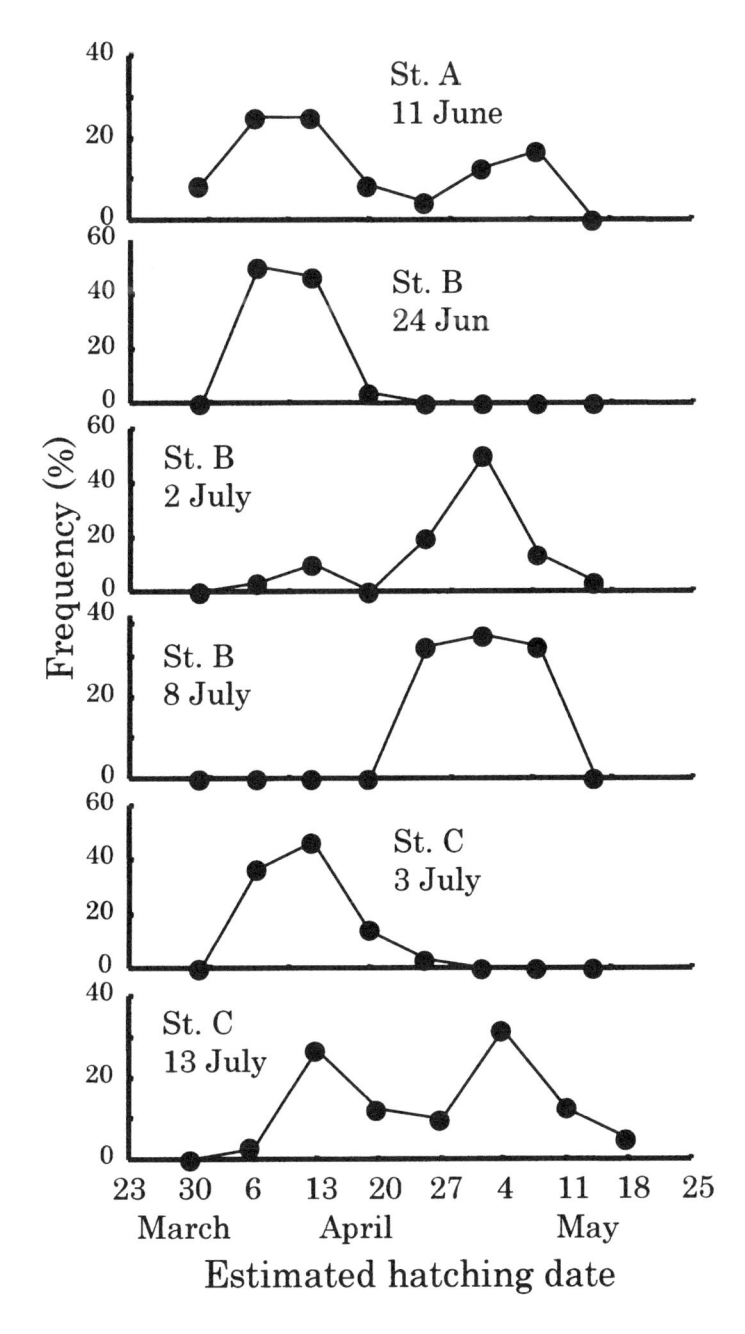

Figure 7. Hatching distribution of the Ishikari Bay herring larvae and juveniles in 1999.

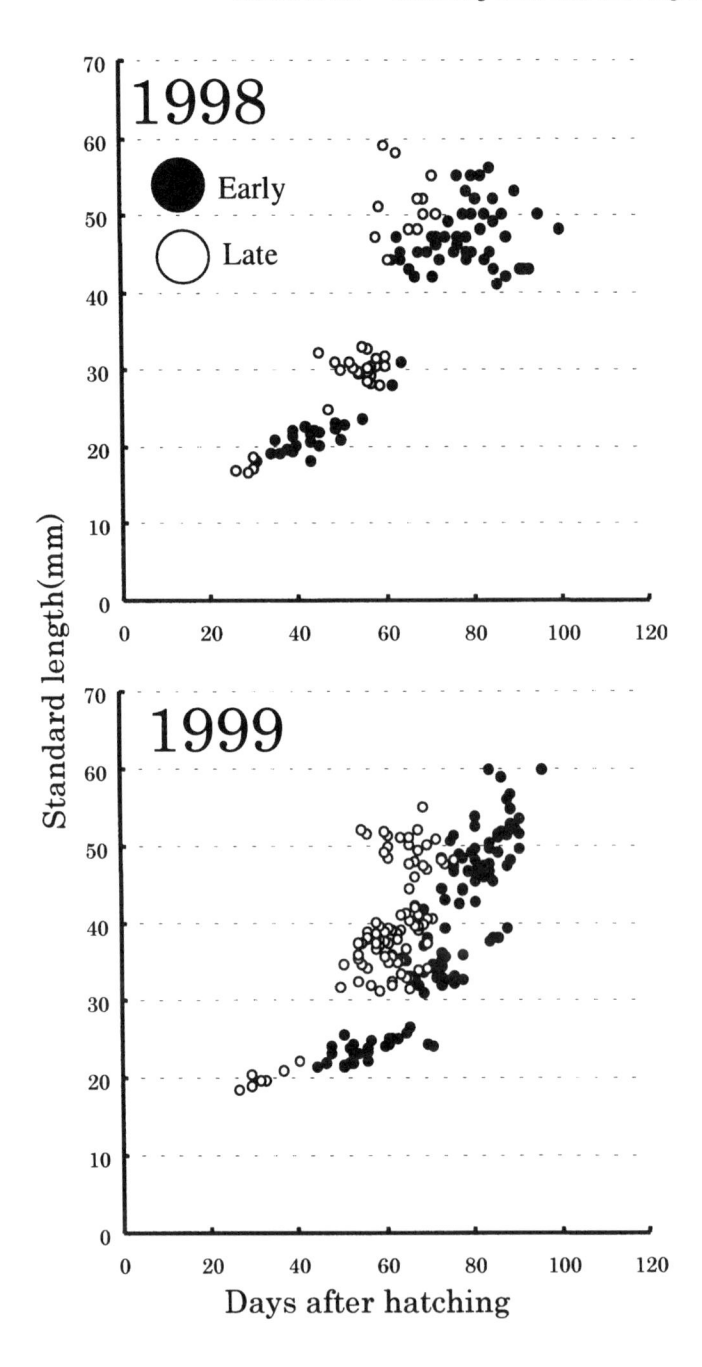

Figure 8. Relationship between the number of days after hatching and the standard length.

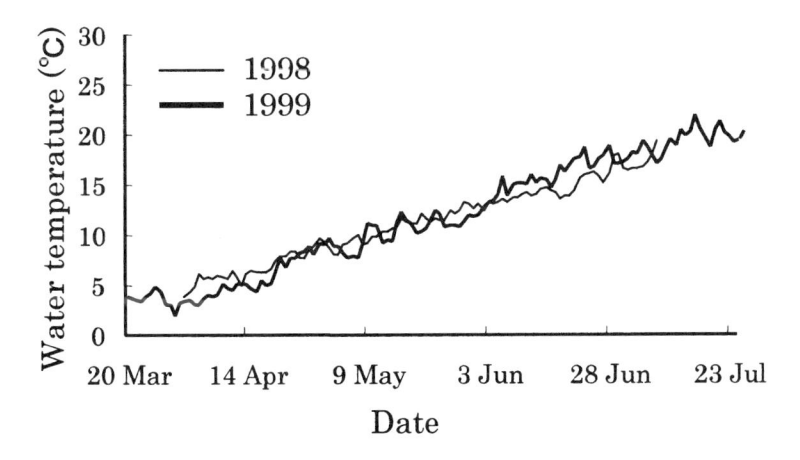

Figure 9. Water temperature of coastal Ishikari Bay (Hoshikawa et al. 2001).

hatching and the number of increments counted was close to 1, but significantly different from 1 increment per day. The standard deviation of the estimated age from the real age was very high, which is in accordance with previous studies on Pacific herring (Moksness and Wespestad 1989), and Atlantic herring (Moksness 1992a, Campana et al. 1987). These results indicate that the rate of increment formation was 1 day from at least 19 days after hatching. A previous study on Atlantic herring showed that the precision of age and hatching date estimation was poor for larvae less than 15-20 days of age (Campana and Moksness 1991). Campana et al. (1987) on Atlantic herring found narrow increments (0.2-0.3 mm), which are too narrow to be observed under a light microscope, in the area close to the hatching check. Moksness (1992a) reported that 7.9% of the estimated hatching days based on the relationship were underestimated by more than 5 days of the true age, and the suggested cause of this was that those larvae initially had a low growth rate and therefore had very narrow increments close to the hatching check. A possible explanation for the large standard deviation of the estimated age from the real age (3 days) in the Ishikari Bay herring is the individual differences in the day the first increment, which is visible under a light microscope, is formed. Consequently an error of 3 days or so for the estimation of the number of days after hatching of Ishikari Bay herring under light microscope is considered unavoidable.

Juveniles (26-41 mm) hatched around Atsuta are distributed in the surf area (station B) and at approximately over 41 mm in the Ishikari River estuary (station C) (Sasaki et al. 2001). Estimated hatching day distribution, especially in 1999, showed that the early-hatched individuals mi-

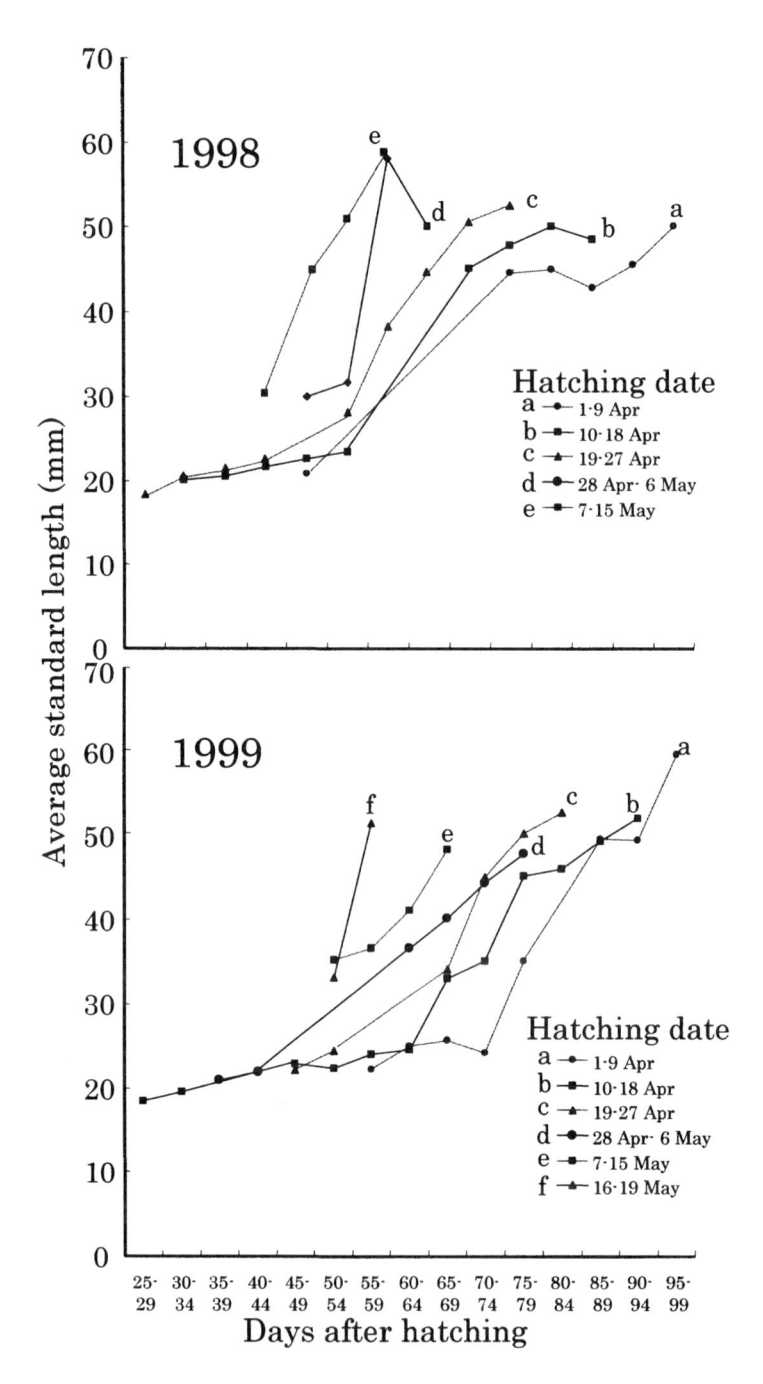

Figure 10. Effect of water temperature.

grated earlier to the surf area (station B) and the estuary than the late-hatched ones in both years. In 1998, there were no catches of early-hatched at station B on 25 June. The cause of this is that either the research was conducted with less frequency in 1998 than 1999 or the juveniles stayed for a shorter period around station B. Migration of adult herring for spawning to the coastal Atsuta peaked twice, in early February and late February, in 1999 (Takayanagi et al. 1999). Results of direct observation on herring egg distributions by scuba diving showed that hatching peaked on 9-10 April and 5-6 May in 1999 (Hoshikawa et al. 2001). These results indicate that sampling in this study took place at suitable points, and larvae and juveniles used in this study originated from the eggs spawned around Atsuta.

The present study shows that the larvae and juveniles that experienced higher temperature grew faster. Jones (1985) also reported within-season difference in growth of Atlantic herring larvae. The estimated date of hatching of Atlantic herring larvae sampled in the Sheepscot River estuary showed that there were at least two age groups, and larvae distributed in the estuary appeared at about 4 weeks old (Townsend and Graham 1981). These results indicate that growth of larvae and juveniles, even if in the same cohort, do not have the same growth patterns and it is necessary to estimate the hatching day to investigate the growth of Ishikari Bay herring larvae and juveniles.

We cannot, however, conclusively specify the reason for the differences in the within-season growth (apparent growth) shown in this study, because we have no data to describe it. Two possible explanations are (1) water temperature and (2) size-selective mortality. McGurk (1984) reported a higher growth rate for batches of Pacific herring larvae developing at higher temperatures. Moksness and Fossum (1992) suggested that higher growth rate in Atlantic herring was caused by higher temperature. On the other hand, size-selective mortality of Atlantic herring larvae in mesocosms was found from the result of continuous sampling and from observations on the distance from the nucleus to the hatch check (Folkvord et al. 1997). Campana and Moksness (1991) on Atlantic herring and Moksness and Wespestad (1989) on Pacific herring concluded that the use of otolith microstructure (e.g., otolith radius and total length) to estimate the back-calculated growth rate gives very accurate and precise estimates. To confirm whether mortality is size-selective or not, it will be necessary to compare apparent growth and estimated individual growth trajectory based on increment width.

In either case, it is obvious that apparent growth of larvae and juveniles is influenced by water temperature. Moksness and Fossum (1991) separated spring- and autumn-spawned herring larvae by measuring the distance from the nucleus to the hatch check and the increment widths in the otolith microstructure. Stenevik et al. (1996) and Moksness and Fossum (1991) on Norwegian spring-spawned herring larvae and Moksness (1992b) on North Sea herring larvae reported that larvae spawned in different grounds were identified from the difference in otolith microstructure.

Results of these studies indicate that otolith microstructure analysis during the larval and juvenile period is a good way to identify components of larvae that have experienced different environmental conditions.

Furthermore, Zhang and Moksness (1993) have shown that otoliths of spawning Atlantic herring can be treated to expose the increment width pattern in the larval period. We also find that the otolith of the adult Ishikari Bay herring can be treated to expose the increments under a scanning electron microscope (Ryotaro Ishida, Hokkaido Fisheries Experimental Station, pers. comm., Sep. 1997). In the next study, we will analyze the otolith microstructure, the back-calculated growth, of the spawning herring and of the juveniles used in this study to distinguish between the progeny of early- and late-hatched herrings as juveniles and adults.

Acknowledgments

We would like to thank Dr. A.J. Paul and two anonymous reviewers for their helpful comments and correction of the English text in previous versions of the manuscript, and Dr. Tokimasa Kobayashi at the Hokkaido National Fisheries Research Institute Fisheries Agency of Japan and Mitsuhiro Sano at the Hokkaido Central Fisheries Experimental Station for giving us an opportunity to attend this symposium. We also thank staff at the Ishikari Fisheries Extension Office for assistance in obtaining spawning herring.

References

Campana, S.E., and E. Moksness. 1991. Accuracy and precision of age and hatch date estimates from otolith microstructure examination. ICES J. Mar. Sci. 48:303-316.

Campana, S.E., J.A. Gagn, and J. Munro. 1987. Otolith microstructure of larval herring (*Clupea harengus*): Image or reality? Can. J. Fish. Aquat. Sci. 44:1922-1929.

Folkvold, A., K. Rukan, A. Johannessen, and E. Moksness. 1997. Early life history of herring larvae in contrasting feeding environments determined by otolith microstructure analysis. J. Fish. Biol. 51(Suppl. A):250-263.

Geffan, A.J. 1982. Otolith ring deposition in relation to growth rate in herring (*Clupea harengus*) and turbot (*Scophthalmus maximus*) larvae. Mar. Biol. (Berl.) 71:317-326.

Hoshikawa, H., K. Tajima, T. Kawa, and T. Ohtsuki. 2001. Spawning bed selection by Pacific Herring (*Clupea pallasii*) at Atsuta, Hokkaido, Japan. In: F. Funk, J. Blackburn, D. Hay, A.J. Paul, R. Stephenson, R. Toresen, and D. Witherell (eds.), Herring: Expectations for a new millennium. University of Alaska Sea Grant, AK-SG-01-04, Fairbanks. (This volume.)

Ishida, R., S. Takayanagi, and M. Sasaki. 1997. Growth and maturity of Ishikari Bay herring. Abstracts for the meeting of the Japanese Society of Fisheries Science, September 27-30, 1997:13. (In Japanese.)

Jones, C. 1985. Within-season difference in growth of larval Atlantic herring, *Clupea harengus harengus*. Fish. Bull., U.S. 83:289-298.

Kobayashi, T. 1993. Biochemical analyses of genetic variability and divergence of populations in Pacific herring. Bull. Natl. Res. Inst. Far Seas Fish. 30:1-77.

Kobayashi, T., M. Iwata, and K. Numachi. 1990. Genetic divergence among local spawning population of Pacific herring in the vicinity of northern Japan. Nippon Suisan Gakkaishi 56:1045-1052.

Lough, R.G., M. Pennington, G.R. Bolz, and A.A. Rosenberg. 1982. Age and growth of larval Atlantic herring *Clupea harengus* L., in the Gulf of Maine-Georges Bank region based on otolith growth increments. Fish. Bull., U.S. 80:187-199.

McGurk, M.D. 1984. Effect of delayed feeding and temperature on the age of irreversible starvation and on the rates of growth and mortality of Pacific herring larvae. Mar. Biol. (Berl.) 84:13-26.

Moksness, E. 1992a. Validation of daily increments in otolith microstructure of Norwegian spring-spawning herring (*Clupea harengus* L.). ICES J. Mar. Sci. 49:231-235.

Moksness, E. 1992b. Difference in otolith microstructure and body growth rate of North Sea herring (*Clupea harengus* L.) larvae in the period 1987-1989. ICES J. Mar. Sci. 49:223-230.

Moksness, E., and P. Fossum. 1991. Distinguishing spring- and autumn-spawned herring larvae (*Clupea harengus* L.) by otolith microstructure. ICES J. Mar. Sci. 48:61-66.

Moksness, E., and P. Fossum. 1992. Daily growth rate and hatching-date distribution of Norwegian spring-spawning herring (*Clupea harengus* L.). ICES J. Mar. Sci.49:217-221.

Moksness, E., and V. Wespestad. 1989. Aging and back calculating growth of Pacific herring, *Clupea pallasi*, larvae by reading daily otolith increments. Fish. Bull., U.S. 87:509-513.

Sasaki, M., R. Ishida, and S. Takayanagi. 2001. Distribution and feeding of jvenile herring (*Clupea pallasii*) in Northern Japan. In: F. Funk, J. Blackburn, D. Hay, A.J. Paul, R. Stephenson, R. Toresen, and D. Witherell (eds.), Herring: Expectations for a new millennium. University of Alaska Sea Grant, AK-SG-01-04, Fairbanks. (This volume.)

Stenevik, E.K., P. Fossum, A. Johannessen, and A. Folkvord. 1996 Identification of Norwegian spring herring (*Clupea harengus* L.) larvae from spring grounds off western Norway applying otolith microstructure analysis. Sarsia 80:285-292.

Takayanagi, S., M. Sasaki, and R. Ishida. 1998. Characteristic of spawning migration of Ishikari Bay herring (*Clupea pallasi*). Abstracts for the meeting of the Japanese Society of Fisheries Science, September 26-29, 1999:42. (In Japanese.)

Townsend, D.W, and J.J. Graham. 1981. Growth and age structure of larval Atlantic herring, *Clupea harengus harengus*, in the Sheepscot River estuary, Maine, as determined by daily growth in otolith. Fish. Bull., U.S. 79:123-130.

Wespestad, V.W., and E. Moksness. 1990. Observations of growth and survival during the early life history of Pacific herring *Clupea pallasi* from Bristol Bay, Alaska, in a marine mesocosm. Fish. Bull., U.S. 88:191-200.

Zhang, Z., and E. Moksness. 1993. A chemical way of thinning otoliths of adult Atlantic herring (*Clupea harengus*) to expose the microstructure in the nucleus region. ICES J. Mar. Sci. 50:213-217.

Decline of the Sakhalin-Hokkaido Herring Spawning Grounds near the Sakhalin Coast

Elsa R. Ivshina
Sakhalin Research Institute of Fisheries and Oceanography (SakhNIRO), Yuzhno-Sakhalinsk, Russia

Abstract

The Sakhalin-Hokkaido herring population was historically the most abundant herring population in the Far East seas. At present, the population is in long and deep depression. Climatic conditions and the fishery contributed to the stock decrease.

This paper presents the results of a study of the Sakhalin-Hokkaido herring spawning grounds at the southwestern Sakhalin coast (Sea of Japan, Tatar Strait) from 1948 to 1998. It shows the considerable reduction of stock biomass, catch, and spawning areas.

During the period of high abundance, Sakhalin-Hokkaido herring spawned throughout a vast area: near the coasts of southwestern, southern, and southeastern Sakhalin, western and northeastern Hokkaido, and the southern Kuril Islands. The reduction of spawning grounds appeared in the 1910-1920s, beginning from the southern part of the spawning area. In the 1940-1950s, herring spawned widely despite the fact that a decrease in the Sakhalin-Hokkaido herring stock was evident. Then, in the 1980s-1990s herring eggs occurred only in places that had previously been considered secondary spawning grounds. The main spawning grounds were empty, and egg density on substratum had decreased. No great variations were observed in the algae community and there was no decrease in the state of potential spawning substratum for herring. Rather, the reduction of spawning area was a consequence of the decrease in herring abundance.

Introduction

The Sakhalin-Hokkaido herring was the most abundant stock among all Far Eastern herring populations. It was unique in its commercial catch, which reached almost 1 million tons in 1897. No other Far Eastern herring population gave such high annual catches. But none of the other populations exhibited such a deep depression as the Sakhalin-Hokkaido stock did. The main reasons for the depression were likely the change of oceanographic conditions (mainly, sea warming), and influence of the fishery (Svetovidov 1952; Probatov 1958; Kondo 1963; Motoda and Hirano 1963; Pushnikova 1996, 1997).

At present, the depression of the Sakhalin-Hokkaido herring is not only continuing, but is even expanding. The spawning and feeding areas have been reduced greatly. The immature Sakhalin-Hokkaido herring is distributed in limited areas only in Aniva and Terpeniya bays, and the northeastern coast of Japan. Mature herring concentrate for feeding mainly in the Bay of De Langle, whereas in the period of high abundance they were widely recorded from Moneron Island to the northern part of Tatar Strait in the Sea of Japan (Druzhinin 1964). The reduction of spawning areas is one of the most evident consequence of the catastrophic decrease in abundance.

During the period of high abundance the Sakhalin-Hokkaido herring spawned in a vast area of the Sea of Japan and the Sea of Okhotsk. Herring migrated to spawn from the northwestern Sakhalin coast to Honshu Island in the Sea of Japan, and in some years, down to the southern part of this island. In the Okhotsk Sea, herring spawned in Aniva Bay, along the northeastern Hokkaido coast, and near Kunashir Island (southern Kurils). Such large spawning areas remained until the end of he 19th century. Then herring abundance decreased and there was considerable reduction of spawning grounds near Honshu Island, and near southern and eastern Hokkaido Island. The displacement of the main spawning stocks northward was likely caused by oceanographic factors. In the middle 1930s, an abrupt reduction in total abundance of Sakhalin-Hokkaido herring took place, and this was immediately reflected by the weakening of the spawning run to the Hokkaido coast.

Despite the decrease in herring spawning near Hokkaido Island in the 1930s, it increased near the Sakhalin coast, in the northern part of the spawning area. Furthermore, the abundance of spawning herring remained relatively stable at the end of the 1930s near the Sakhalin coast. It decreased much more after the mid-1950s compared to the level of the previous years (Piskunov 1952, Probatov 1954, Probatov and Darda 1957, Probatov 1958, Rumyantsev 1958, Kondo 1963, Motoda and Hirano 1963). Based on the analysis of spawning runs, fishery statistics, and biological parameters of Sakhalin-Hokkaido herring, A.N. Probatov supposed in 1959 that in future it would be possible to expect a further decrease of abundance as well as short-term and more local runs along the western coast. (Probatov 1959). Unfortunately, this forecast has been realized completely. The purpose of this work was to determine the scales and rates of reduction of the Sakhalin-Hokkaido herring spawning grounds near the Sakhalin coast.

Materials and Methods

The data from monitoring of the Sakhalin-Hokkaido herring spawning grounds near the southwestern Sakhalin coast during April-May 1948-1998 are presented in this paper. Diving surveys were used to determine the number of eggs and limits of the spawning area. Diving surveys were conducted annually during 1948-1952 and 1978-1998. During 1953-1977 they were conducted once every 1-3 years. In the 1950s, the research covered the places of the most intensive spawn. A total length of spawning grounds was determined by direct observations of the herring spawning run to the coast, and by observation of substrata with herring eggs on the shore. During the 1960s-1990s, a period of low abundance, a survey was conducted at all places of potential and actual spawning grounds. About 100-150 diving surveys were carried out annually, and biostatistics and commercial catch information were collected.

Results

The spawning grounds of the Sakhalin-Hokkaido herring population were historically located over a vast area of the Japan and Okhotsk seas (Fig. 1). The western Sakhalin coast was the northern boundary of the spawning area. Unpublished data indicate that the 1930s appeared to be the most characteristic period of the northern type of herring distribution, when the spawning grounds extended to the northwestern coast (V. Bogaevsky). The western coast of southern Sakhalin became the main spawning ground of this herring stock after the Hokkaido coast (Fig. 2).

In the 1940s, the herring spawn was usually intensive, especially in the area from Cape Lopatin to Cape Slepikovsky. Spawning of most of the herring stock began in the first week of April and continued through May. But, at that time, some researchers noted that spawning did not occur along the northwest coast "... because [of] the evident depression of the Sakhalin-Hokkaido herring stock" (V. Bogaevsky, unpubl., 1946).

In the early 1950s, the region of spawning herring distribution remained similar to that of the 1940s. Egg concentration at some places was also at a high level (9-10 million per m^2). However, the extent of places with intensive herring run (southward from Cape Slepikovsky) was slightly reduced. In Aniva Bay the herring spawn was observed in traditional areas practically throughout the coast (Fridland 1951).

The mid-1950s showed the next considerable decrease in abundance of Sakhalin-Hokkaido herring (Fig. 3) (Probatov 1958). In 1953 the total number of herring spawners was estimated to have been only a quarter of those in 1950-1951. The decrease in abundance resulted in localization and shortening of herring spawn. In that year (1953) and later the majority of spawning grounds were not used. A gradual reduction of spawning areas and lower intensity of spawning run occurred each year after 1953 (Rumyantsev 1963).

In the 1960s a further reduction of spawning areas near the Sakhalin coast was recorded, along with the decrease in abundance of herring stocks. In Aniva Bay herring spawning was observed only in its northern part. The basic spawning areas along the western coast (to the south of Cape

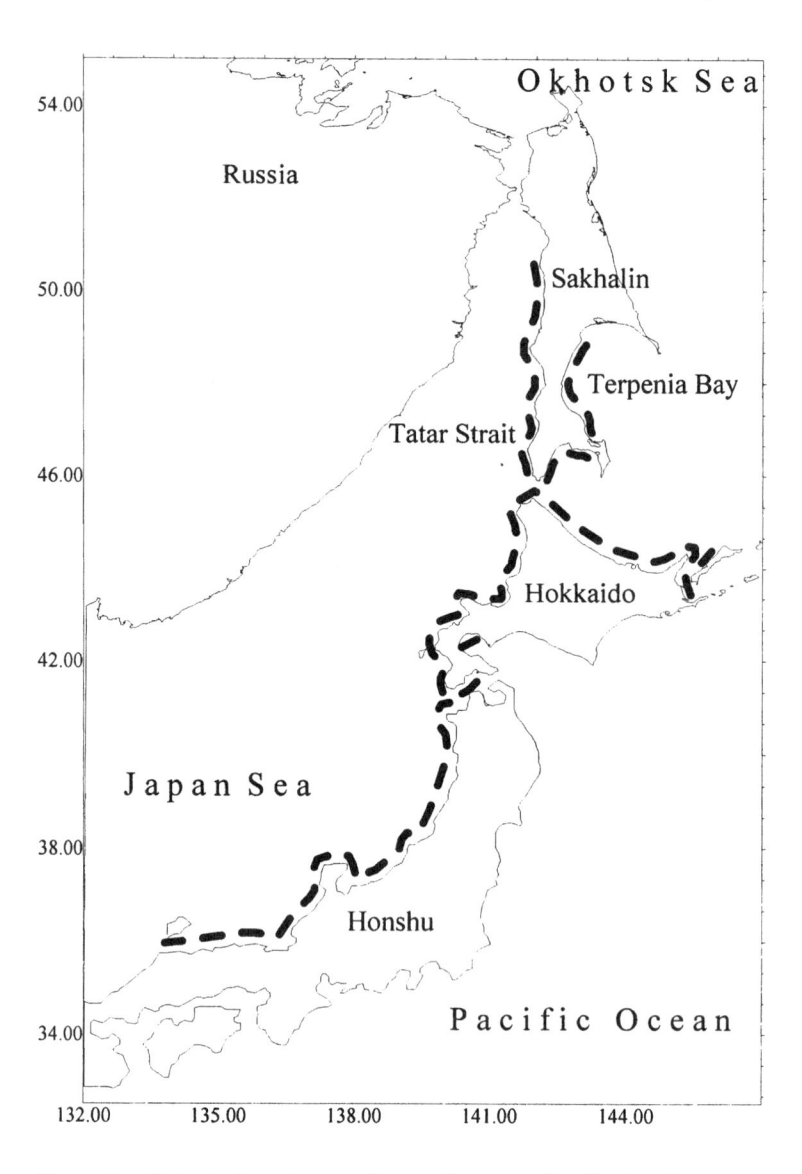

Figure 1. Historical maximum of spawning area distribution (interrupted line) of Sakhalin-Hokkaido herring (after Probatov 1954).

Herring: Expectations for a New Millennium 249

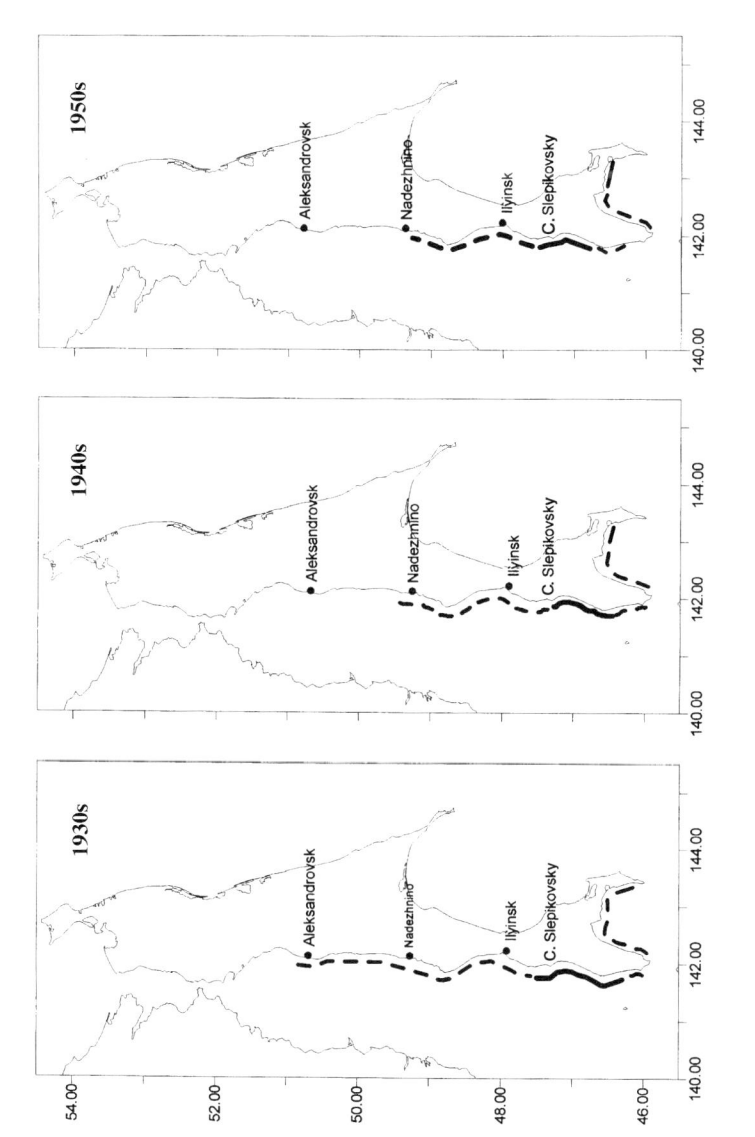

Figure 2. Spawning grounds of Sakhalin-Hokkaido herring in the 1930s to 1950s. The continuous line indicates sites of greatest spawning.

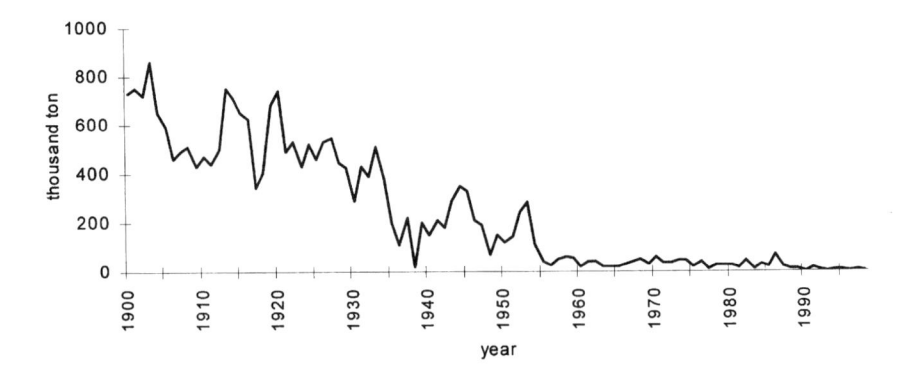

Figure 3. Catch of Sakhalin-Hokkaido herring.

Slepikovsky) were not used. In the late 1960s, the basic spawning grounds were located only northward from Cape Slepikovsky (in the region of Ilyinsk and Nadezhdino), but their area reduced annually: from 525,000 m^2 in 1967, to 45,000 m^2 in 1968, and 42,500 m^2 in 1969 (Fig. 4) (Darda 1969).

The same trend continued during the next 10 years. In 1972, the total spawning area was 18,800 m^2 and in 1974 only 16,900 m^2. In 1975 no spawning of Sakhalin-Hokkaido herring was recorded in Aniva Bay. Despite the appearance of the abundant 1973 year class, causing the increase of the total stock abundance in the second half of the 1970s (see Fig. 3), the spawning areas were not enlarged. By the end of the 1970s few large spawning grounds for Sakhalin-Hokkaido herring were observed (Fig. 4).

During the last 20 years herring eggs occurred most frequently near the western Sakhalin coast at the 2-3 m depth on *Sargassum pallidum, S. miyabei, Cystoseira crassipes,* and *Phyllospadix iwatensis.* In the 1980s-1990s, *Phyllospadix,* which has been characteristic of the best spawning grounds in the past (Fridland 1951), reached an average of 2.25 kg per m^2. In the past, the biomass was only 0.5 kg per m^2 in favorable and unfavorable sites (Fig. 5). The best spawning grounds for herring coincided with high biomass of the preferred substratum in the 1950s and 1970s-1990s. During the period of decline of the Sakhalin-Hokkaido herring stock, algae suitable to be a substratum for spawn was at a rather high abundance. Therefore, substratum was not a limiting factor for herring abundance.

Nevertheless, in the 1980s-1990s, the vast coastal areas with a high biomass of algae and seagrass remained empty during the herring spawning period. The Sakhalin-Hokkaido herring used the limited area with a weakly developed algae belt. The length of the spawning grounds was less than 2 km long on the average (Fig. 4).

Along with the reduction of the extent of spawning grounds, other parameters of spawning areas have changed too: maximum depth from

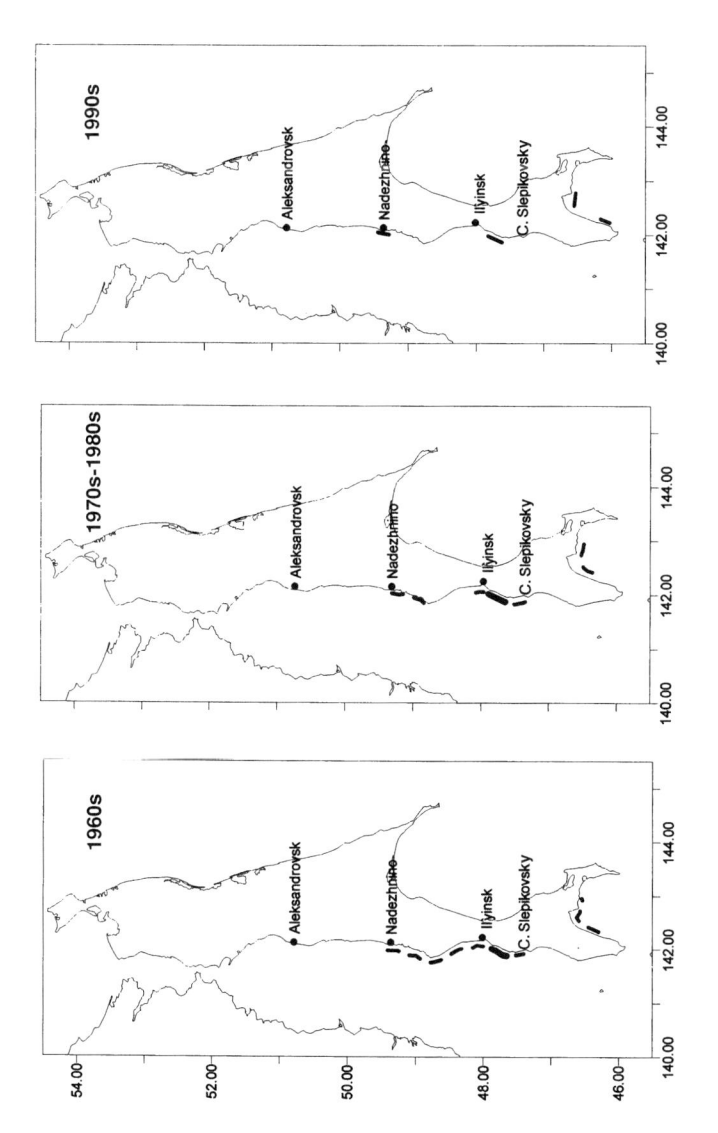

Figure 4. Spawning grounds of Sakhalin-Hokkaido herring in the 1960s to 1990s. The continuous line indicates sites of greatest spawning.

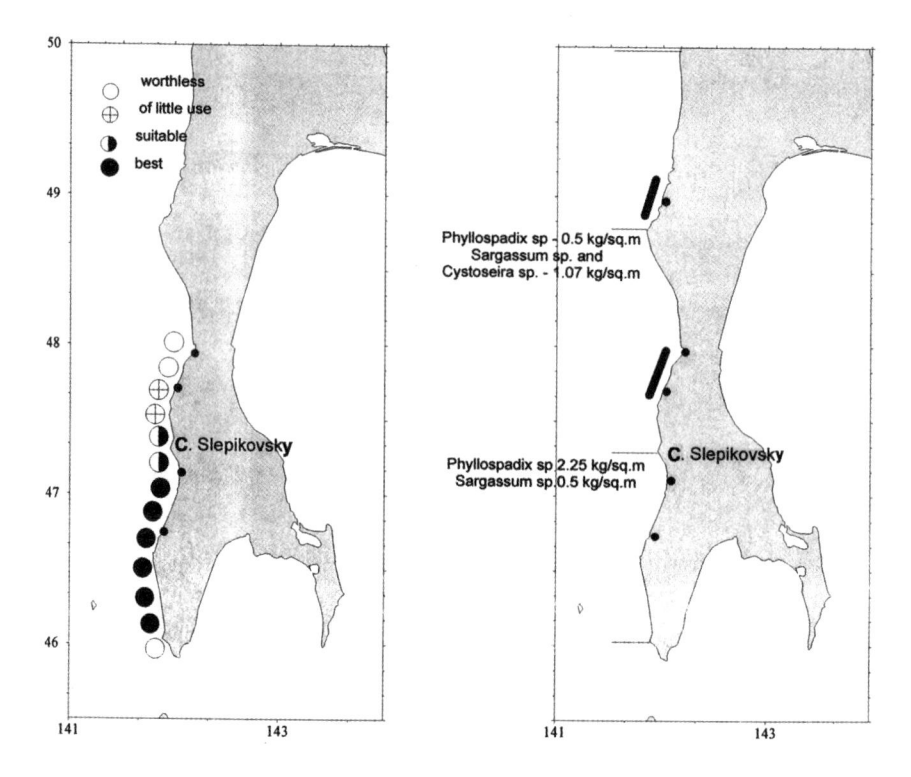

Figure 5. Classification of Sakhalin-Hokkaido herring spawning grounds (Fridland 1951, left) and spawning grounds in the 1990s and average biomass of algae (right).

Table 1. Density of Sakhalin-Hokkaido herring eggs (in millions per m²) on substratum.

Density of eggs	Period			
	1950s	1970s	1980s	1990-1998[a]
Average	2.90	1.25	0.80	0.40
Maximum	17.50	3.00	1.00	0.73

[a]Eggs were found only in 1994 and 1996.

18 m to 10 m, width from 500-700 m to 10-50 m, and to 2-3 m some years. Egg density varied from 10 eggs to 1 million eggs per m^2 (average 0.4 million eggs per m^2), unlike the 1950s, when the number of eggs varied from 1 to 2 million eggs per m^2 to 17.5 million eggs per m^2 (average 2.9 million eggs per m^2) (Table 1).

The number of layers in herring egg masses also decreased, from 40-51 to 1-3. In the last 10 years, it has been irrelevant to assess this index, because the number of eggs on the substratum has been very small. A catastrophic reduction of spawning areas has occurred in the most recent 10 years (since 1990). Only in 1994 and 1996, in May, were herring eggs observed. These were at the secondary spawning grounds only, and their abundance was tens and hundreds times less than in the 1950s. In the 1990s the Sakhalin-Hokkaido herring stock was at an extremely low level. Vast areas potentially good for spawning remained vacant. In the area southward from Cape Slepikovsky, which earlier had been the best for spawning in April-May, no run of spawners has been recorded for almost 30 years. The numerous features of population depression exist, and there is no evidence to suggest abundance will increase in the future.

Conclusion

Due to the catastrophic decline in numbers of Sakhalin-Hokkaido herring, the spawning grounds were reduced not only near the shores of Japan, but also near Sakhalin, where since the 1930s the main spawning grounds for Sakhalin-Hokkaido herring occurred. At one time spawning extended along the whole west coast. In the 1940s-1950s, the absence of spawning grounds was recorded near the northwestern coast and a reduction near the southwestern Sakhalin coast. In the 1960s, redistribution of spawning grounds took place. By this time, the spawning herring run had not been observed near the southwestern Sakhalin coast. During the following years those spawning grounds were reduced. By the end of the 20th century, herring spawning did not occur annually. Spawning at traditional places typical for the period of high abundance was completely absent on the whole. The large-scale spawning area reduction is the consequence of the decrease in abundance of Sakhalin-Hokkaido herring and it reflects the extremely depressed state of the population.

Acknowledgment

I would like to thank many SAKHNIRO pesonnel who contributed toward the collection and processing of herring data used in this analysis. In particular we would like to thank Sergey D. Bukin for his helpful comments.

References

Darda, M.A. 1969. Abundance and distribution of Sakhalin-Hokkaido herring in Sakhalin waters in 1969. Report SakhNIRO. Sci. Archives No. 1943. 66 pp. (In Russian.)

Druzhinin, A.D. 1964. Materials on biology of feeding herring (*Clupea harengus pallasi* Val.) in Sakhalin waters. Izv. TINRO 55:3-37. (In Russian.)

Fridland, I.G. 1951. Herring reproduction near southwestern Sakhalin coast. Izv. TINRO 35:105-145. (In Russian.)

Kondo, H. 1963. On the condition of herring (*Clupea pallasi*) in waters around Hokkaido and Sakhalin during recent years. Hokkaido Fisheries Experimental Station 3:1-18. (In Japanese.)

Motoda, S., and Y. Hirano. 1963. Review of Japanese herring investigations. Ext. du Rapp. et Proc. Verbaux 154:249-261.

Piskunov, I.A. 1952. Spring herring from the western coast of southern Sakhalin. Izv. TINRO 37:3-67. (In Russian.)

Probatov, A.N. 1954. Distribution of the Japan Sea herring. Izv. TINRO 39:21-57. (In Russian.)

Probatov, A.N. 1958. Fluctuations of Sakhalin-Hokkaido herring abundance due to oceanological conditions. Proceedings of Oceanographic Commission 3:124-124. (In Russian.)

Probatov, A.N. 1959. Runs of herring spawners to the western coast of southern Sakhalin. Rybn. Khoz. 2:2-29. (In Russian.)

Probatov, A.N., and M.A. Darda. 1957. Spawning herring of Kunashir Island. Izv. TINRO 44:3-11. (In Russian.)

Pushnikova, G.M. 1996. Fishery and stock condition of the Sakhalin waters herring. Scientific Proceedings of Far Eastern State Technical Fishery University 8:34-43. (In Russian.)

Pushnikova, G. 1997. Will the population of the Sakhalin-Hokkaido herring disappear from the industrial supply? Rus. Convers. News 3:14-17.

Rumyantsev A.I. 1958. Recourse condition of abundant Sakhalin-Hokkaido herring stock. Rybn. Khoz. 4:3-9. (In Russian.)

Rumyantsev A.I. 1963. Material by conditions herring stock at the Sakhalin-Hokkaido region. Report SakhNIRO. Sci. Archives No. 1326. 48 pp. (In Russian.)

Svetovidov, A.N. 1952. Fluctuations of the southern Sakhalin herring catches and their reasons. Zool. Zh. 31:831-842. (In Russian.)

Factors Influencing Location and Time of Spawning in Norwegian Spring-Spawning Herring: An Evaluation of Different Hypotheses

Aril Slotte
Institute of Marine Research, Bergen, Norway

Abstract

In January Norwegian spring-spawning herring (*Clupea harengus* L.) migrate from wintering grounds in fjords of northern Norway (68°N) toward spawning grounds distributed along approximately 1,500 km of the coast (58-70°N). Environmental factors may differ between spawning grounds, and the migrating herring may vary in abundance and be composed of several age and length groups in different physiological condition from a variety of hatching, retention, and nursery areas. The present paper evaluates different theoretical hypotheses describing the influence of such factors on spawning ground selection and spawning time. Regarding spawning ground selection, three hypotheses are evaluated: First, herring home to their natal spawning grounds. Second, independently of their origin herring repeatedly return to spawn at specific grounds through migration and homing patterns learned from random adult groups they joined the year preceding first spawning. Third, herring tend to select spawning grounds based on individual state (length, condition), the cost of migration, and prospects of larval survival. Regarding spawning time two hypotheses are evaluated: First, herring spawn in a decreasing order of body length due to a size-dependent maturation rate, which in turn may result in multiple spawning waves reflected by the number of age modes in the stock. Second, there will be only two major spawning waves regardless of age distribution, the repeat spawners first followed by the recruit spawners that have a delayed onset of maturation. According to this evaluation the observed spawning migration can best be explained by the third hypothesis regarding spawning ground selection and the second hypothesis regarding spawning time.

Introduction

Within herring populations the environmental preferences favoring the survival of eggs and larvae may be specific, leading to spawning at specific locations and times (Haegele and Schweigert 1985). In the case of Norwegian spring-spawning herring (*Clupea harengus* L.) (NSS herring) banks and shelf areas with stony or rocky bottom and depths less than 250 m are utilized for spawning (Runnstrøm 1941, Dragesund 1970). Such areas are distributed along approximately 1,500 km along the Norwegian coast between 58°N and 70°N. Traditionally the most important spawning grounds are found from north to south at Lofoten, Træna, Sklinnabanken, Haltenbanken, Frøyabanken, and along the districts of Møre, Sogn, and Rogaland (Fig. 1). In recent years the NSS herring stock has recovered from a state of severe depletion in the late 1960s. Simultaneously the stock has changed its feeding and wintering areas and hence the migration routes to the spawning grounds. Prior to the stock decline the herring spent the summer feeding season in the Norwegian Sea, and the wintering occurred to the east of Iceland. The herring had to cross the open waters of the Norwegian Sea before entering the Norwegian west coast to find the spawning grounds. After the stock decline oceanic feeding and wintering areas were abandoned, and instead the herring spent the entire year in Norwegian coastal waters and fjords (Dragesund et al. 1980, Hamre 1990, Røttingen 1990). When the large 1983 year class recruited in 1988, wintering commenced in Vestfjorden, Tysfjorden, and Ofotfjorden (Vestfjorden system), northern Norway. Furthermore, after the recruitment of the large 1991 and 1992 year classes the feeding area was once again extended westward into the Norwegian Sea (Dragesund et al. 1997, Misund et al. 1998). At present NSS herring (*Clupea harengus* L.) spend the feeding season, April-August, covering large areas of the Norwegian Sea, but in September they gather as a rather homogeneous mass in the Vestfjorden system. When the spawning migration begins in mid-January they split up again heading for spawning grounds located with different migration distances, from Lofoten in the north to Lista in the south.

It is this scenario that forms the basis of the present review (Fig. 1). The environmental factors may differ between spawning grounds. In addition, the population may vary in abundance and be composed of several age and length groups in different physiological condition from a variety of hatching, retention, and nursery areas. The question is whether these factors have any influence on spawning ground selection and spawning time, and the main objective of the paper is to evaluate different theoretical hypotheses already put forward to describe such influences. The present evaluation is divided into two main analyses—one regarding spawning ground selection and another regarding spawning time. Each analysis gives an overview of claimed models, their predictions, and the related observations and data, and ends with a conclusion.

Herring: Expectations for a New Millennium   257

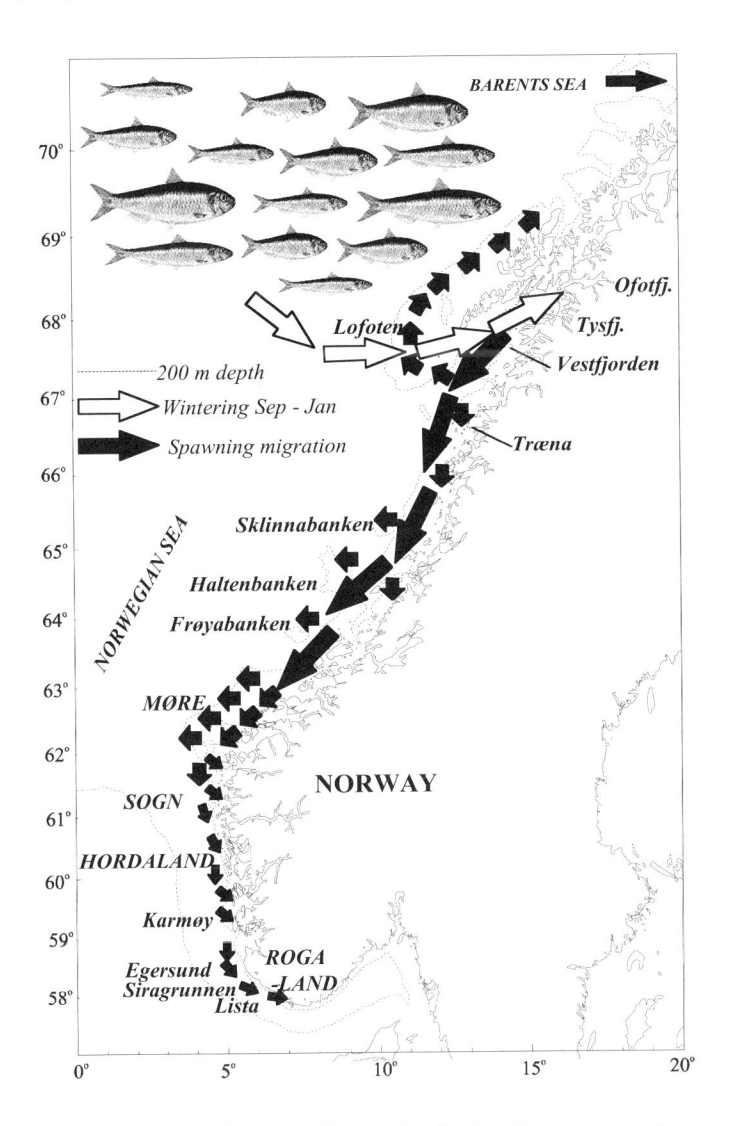

Figure 1. In April-August NSS herring feed in the Norwegian Sea. In September they gather in the Vestfjorden System. In mid-January they split up and head for spawning grounds. Areas and locations mentioned are specified on the map.

Spawning Ground Selection
Models of Spawning Ground Selection
Natal Homing
The first hypothesis claims that the selection of spawning grounds in herring is clearly related to their hatching, retention, and nursery areas; i.e., herring home to their natal spawning grounds. It claims that herring at different spawning grounds are discrete population units with independent life histories (Iles and Sinclair 1982, Sinclair 1988). This model of spawning ground selection is supported by the idea that each spawning ground is associated with a specific larval retention area, which due to hydrographical features provide a mechanism for imprinting and maintenance of reproductive isolation by natal homing. Sinclair (1988) proposed that larval retention would result in a number of survivors (members) within a population that will exceed the number of losses (vagrants), the member/vagrant hypothesis. From an evolutionary sense of view this model implies that straying herring will not successfully pass on their genes. Thus, in a sense they are evolutionary "losers."

Return Spawning
The second hypothesis also claims that herring repeatedly return to spawn at specific grounds, but this behavior is supposed to be independent of hatching, retention, and nursery areas. In this model the principal characteristics responsible for the structure, persistence, and integrity of a population are of behavioral rather than genetic nature; i.e., it is learning of spatial structures and not imprinting that is important with respect to homing (McQuinn 1997). Here local herring populations, which are formed through straying from existing populations (colonization), are perpetuated in geographic space through the social transmission of migration and homing patterns from adults to the recruiting juveniles in the year preceding first spawning. The recruiting herring, which could derive from various spawning grounds, must learn from older, more experienced repeat spawners how to find the respective feeding, wintering, and spawning grounds of the population, which they will keep on visiting in subsequent years.

State-Dependent Migration
The third hypothesis is, unlike the two other hypotheses, based on life history theories suggesting that animal behavior in general is influenced by the state of the animal, i.e., animals have state-dependent life history strategies (McNamara and Houston 1996). In addition it is specially directed toward NSS herring and the present migration pattern with spawning grounds located with different migration distances and in different environments. The hypothesis claims that natal homing or simply return spawning is neither a crucial nor a successful strategy on the long term, and that selection of spawning grounds in herring to a large degree is influenced by individual state (length, condition), the cost of migration

and prospects of larval survival; i.e., they tend to select the spawning grounds yielding the highest lifetime fitness (Slotte 1999a, Slotte and Fiksen 2000). In addition it claims that learning is another factor influencing spawning ground selection. Here the idea is that immature herring must learn migration routes from older repeat spawners, as also suggested by McQuinn (1997). In addition, learning may contribute to shifts in migration routes among repeat spawners too. This claim is based on the idea that individuals will tend to follow the mainstream of schools or school members. Another claim is that the herring will have some tendency toward return spawning at their first (or last) spawning ground—a tendency, which is not at all crucial from an evolutionary point of view, but rather an adaptation to secure the findings of spots with preferable spawning substrates and other important environmental conditions. In other words in the model of state-dependent migration herring may shift spawning grounds whenever this may be beneficial to their lifetime fitness.

Model Predictions of Spawning Ground Selection

Predicted Influence of Hatching, Retention, or Nursery Areas

One of the main differences between the three models is the predicted spawning ground in relation to hatching, retention, or nursery areas. The natal homing model predicts larval retention within each spawning area, and that imprinting and natal homing eventually will lead to genetic differences between spawning groups. The other two models predict that there is no clear relationship between selected spawning ground and the origin of the herring. On the other hand they do not neglect that a proportion of the stock in fact may spawn at their home grounds, but this is due to factors other than imprinting. In other words they predict that there will not be any significant genetic distance between herring from the different spawning grounds.

Predicted Influence of First Spawning Ground

All three models predict that the herring will have a tendency to return to the same spawning ground they spawned the first time (last time). However, they differ considerably in predictions of homing rates or straying. The natal homing hypothesis predicts 100% return rates; here the strays are genetically lost for the population. The other two hypotheses predict much more straying, and here they will not be genetically lost. It is the hypothesis of state-dependent migration that clearly predicts the highest straying rates or lowest homing rates.

Predicted Influence of Stock Size and Structure

The natal homing model does not predict that selections of spawning grounds may vary with stock size and structure. However, the return-spawning model predicts that the tendency to disperse (stray) is stronger within the recruits, and diminishes with reinforcement of traditional migration and spawning patterns (McQuinn 1997). Furthermore, it predicts that straying is more prevalent in unstable populations, i.e., following a population collapse or a recruitment boom. The idea is that recruiting herring, lacking older "teachers" to show them their way, will tend to find new spawning grounds. The state-dependent migration model also supports this idea, but it also predicts that spawning ground selection will vary with variations in year-class strength and stock size due to density-dependent recruitment; here the idea is that increasing stock size will lead to spreading of spawning products over larger areas to reduce density-dependent mortality.

The state-dependent migration model clearly contrasts with the return-spawning model in one main aspect; it predicts that herring, including repeat spawners, to a large degree also will shift spawning grounds in stable populations. More specifically, it predicts that learning in fact could be influenced by the length structure and condition of the stock, and thus be an important contributor to straying in stable populations. Due to size-dependent swimming speed (Ware 1975, 1978) and hydrodynamic advantages of swimming close to neighbors of similar size (Pitcher et al. 1985), herring may migrate from the wintering grounds toward the spawning grounds in schools segregated by size and condition (Pitcher and Parrish 1993), and may consequently be guided to a different spawning ground than the one used previously. This prediction is clearly different from that predicted by the other hypotheses, i.e., that schools will be composed of individuals homing to specific spawning grounds. It is based on the idea that staying in the school sometimes may be more important than any other factor for lifetime fitness. Clearly schooling is an anti-predator behavior (Pitcher and Parrish 1993), and thus splitting from a well-structured school may lead to increased probability of getting caught, which of course has dramatic effects on fitness. However, sometimes migration constraints make splitting of schools necessary. Another prediction from the state-dependent model is that recruits, small herring or herring in bad condition, could be unable to follow their "teachers" of larger size or with better condition. Instead these individuals with less migration potential may stop at suitable spawning grounds closer to the wintering area.

Predicted Influence of Environmental Conditions and Migration Constraints

Neither the natal homing model, nor the return-spawning model, take into account the migration constraints of a stock and the difference in environmental conditions that may be experienced between different spawning grounds. This may be due to the fact that these models are based on stocks that do not have the same large-scale migrations and significant differences in distances between wintering grounds and spawning grounds as found in NSS herring. Both models predict that herring for their first time spawn at a selected ground regardless of environmental conditions at other spawning grounds, regardless of migration distance, and regardless of individual body size or condition. Similarly they both predict that the herring will keep on spawning at these grounds regardless of changes in body size and condition, and regardless of environmental conditions at other spawning grounds. In summary these models predict that the age and length distributions should stay relatively stable within spawning grounds, whereas they should differ between spawning grounds, but this difference should not be related to the migration distance from wintering grounds. Perhaps the most important prediction from these models is that the condition of fish of the same body length should decrease with the migration distance from the wintering grounds, given the assumption that all herring have the same chance of feeding success regardless of their previous home spawning ground.

The migration constraints of a stock and the difference in environmental conditions at different spawning grounds are clearly taken into account by the state-dependent migration model (Slotte and Fiksen 2000). It assumes that NSS herring may have a state-dependent tradeoff between costs of migrating and benefits of spawning farther south where the larvae may encounter warmer waters resulting in increased survival. Here the idea is that the available energy could be traded off between metabolic and reproductive costs, which imply reduced fecundity given extensive migrations. The probability of survival until metamorphosis for an egg laid at each site is calculated by letting the larvae drift with currents and grow as a function of the temperature it is exposed to. With a size- or temperature-dependent mortality rate, the probability of surviving until metamorphosis is given by the time needed to reach this size. Roughly, the model predicts that larvae from the southern grounds drift through warmer temperatures than larvae hatched farther north, which makes southern areas more optimal for larval survival. However, optimum spawning site is the site yielding the highest product of fecundity and offspring survival probability. Here it is important to emphasize that NSS herring generally do not feed during the period from the arrival in the wintering area in September-October until the end of spawning in March-April (Slotte 1999b). Instead the energy expended on maturation, swimming, and maintenance metabolism is drawn from reserves stored during the summer feeding period. Slotte (1999b) has demonstrated that the southward spawning migration demands a lot of energy, and the relative energy loss (%) decreases with body length. Related to these data the state-dependent migration model clearly predicts that large fish migrate farther south than smaller fish, but due to the energetic costs of migration optimal spawning grounds are found farther north than the optimal hatching grounds. The model also predicts that variations in body condition may have considerable effects on optimal spawning site and fecundity. Perhaps most important, the model predicts that a strategy with 100% homing will result in major reductions in fitness if variations in condition are not taken into account. It is clearly optimal for herring with increasing body length and condition to migrate farther south and to increase fecundity at the same time, whereas reduced condition should confer reduced fecundity and migration.

Data on Spawning Ground Selection

Data on the Influence of Hatching, Retention, or Nursery Areas

Tagging experiments on Pacific herring (Hourston 1982) and Atlantic herring (Wheeler and Winters 1984) have demonstrated high return rates to specific spawning grounds. This has led to a general consensus that the herring maintains population integrity and persistence by repeatedly returning to spawn (i.e., home) with high precision at specific spawning grounds, but there is disagreement about the actual nature of this behavior. Stock discreteness and the member/vagrant hypothesis were subject for discussion at a workshop during the International Herring Symposium in Anchorage, Alaska, in 1990. It was apparent from the discussion at this workshop (Collie 1991) and from a review article (Stephenson 1991), that there were disagreements about the stock discreteness in herring and the validity of the member/vagrant hypothesis. Except for a genetic study by Stephenson and Kornfield (1990) evidence for the discrete population concept was lacking. The main uncertainty was related to the degree and mechanisms of natal homing and separation between neighboring groups.

The Pacific herring off British Columbia consist of several stocks of which each is connected to a number of spawning sites within a particular area (Hay and Kronlund 1987), and studies on larval distribution have indicated mixing within areas but not between areas (Hay and McCarter 1997). However, Hay and McCarter emphasized that these stocks mix during summer feeding migrations, and that larval imprinting could explain the observed geographical spawning distribution without demonstrating genetic differentiation. Thus, with exception of the genetic aspect, their data supported the larval retention hypothesis (Iles and Sinclair 1982, Sinclair 1988).

Similarly, Stephenson et al. (2001, this volume) conclude that the widespread movement and mixing of juvenile and adult herring outside of the spawning season indicate that population structure is related to discontinuities at the egg and larval stage. They further suggest that the body of literature and experience of Atlantic, Pacific, and Baltic herring indicate that herring sharing a common larval retention area belong to the same complex in which there may be several cells (spawning areas) with several spawning grounds. Here exchange would be expected to be highest within cell (among spawning grounds), less between cells, and low between complexes.

With regard to data on natal homing it is important to point out that the observed larval drift and distribution of NSS herring is quite unique compared to other stocks. The predictions of larval distribution, retention, and imprinting from the natal model does not fit well in this stock. One similarity between the wide range of spawning areas and grounds from 58°N to 70°N is the location within the northward running coastal current (Fig. 2). Regardless of hatching ground the great majority of the larvae drift northward and eventually end up in the Barents Sea, where they spend a nursery period of about 3-5 years before they join the mature population (Dragesund 1970, Holst and Slotte 1998). However, the main larval

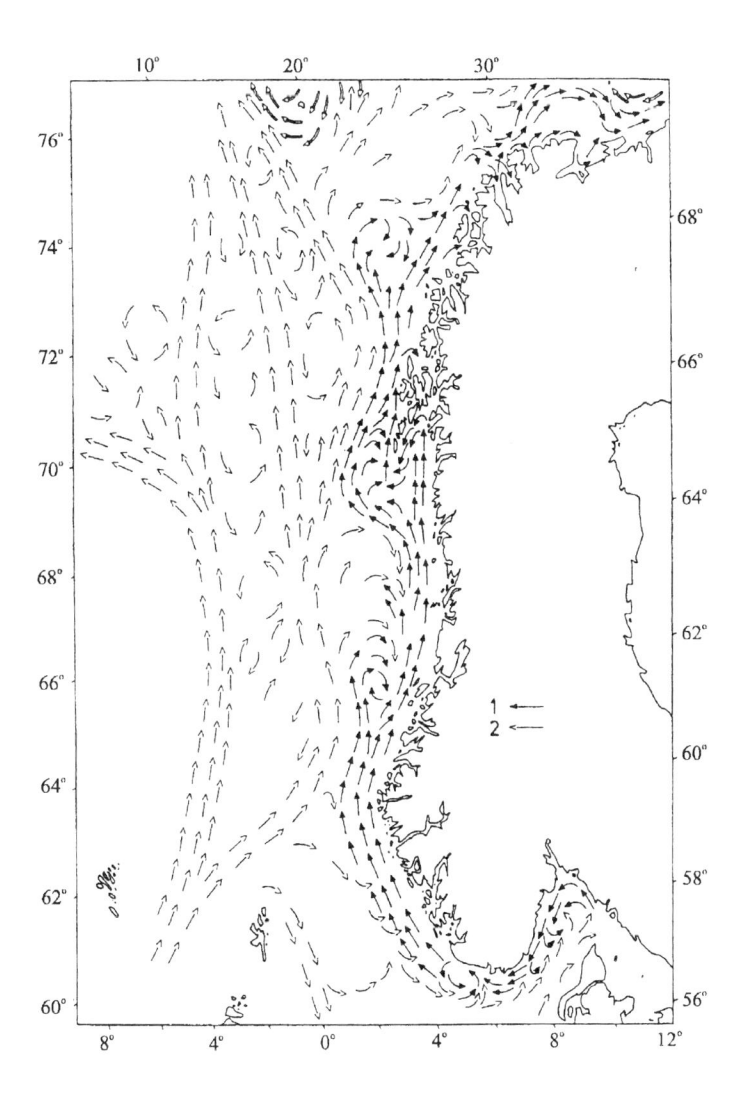

Figure 2. Map of coastal (1) and Atlantic (2) currents (adapted from Aure and Østensen 1993). The larval drift follows the coastal current at average speeds of 15-40 cm per second and maximum speeds at 100 cm per second.

drift toward the Barents Sea may be delayed due to retention at certain locations along the Norwegian coast, especially in connection with bank areas like Møre, Frøyabanken, Haltenbanken, Sklinnabanken, and Lofoten (Sætre 1999, Sætre et al. in press). In fact, a comparably smaller proportion of the larvae also end up in fjord systems along the coast (Dragesund 1970, Holst and Slotte 1998). The large range of temperature and ecological regimes experienced by the juveniles is reflected in differences in growth patterns throughout their geographic range. As a rule, the growth rate decreases northward, and consequently individuals from the coastal nurseries mature at a younger age than individuals from the Barents Sea.

The differential growth between coastal and Barents Sea nurseries has in fact been used to test the validity of natal homing in NSS herring (Holst and Slotte 1998). When spawning occurs farther south there is an increasing probability that larvae of NSS herring will drift into coastal areas and fjords (Dragesund 1970). Thus, if herring spawning at the southern grounds were dominated by the coastal component, this could provide some evidence for natal homing. Based on this fact Holst and Slotte (1998) studied the geographical, temporal, and age-dependent recruitment patterns of the coastal and the Barents Sea components of most year classes from 1930 to 1989. The relative importance of each component in the year classes varied with the fluctuating stock size observed during the period studied. In periods with a low spawning stock hardly any recruitment appeared from the Barents Sea juvenile areas, while at high stock levels the Barents Sea component dominated the spawning stock. The asynchronous maturation of the two components caused a recruitment pattern characterized by a high proportion of the coastal component at early ages and an increasing proportion of the Barents Sea component at older ages. Both components recruited along the entire spawning range studied, and once fully recruited, no latitudinal gradient in the spawning stock component composition was observed. Thus, in agreement with the predictions from the return-spawning model and the state-dependent migration model, this study provided no evidence that imprinting and natal homing drives the spawning ground selection in this stock.

Technically, the best evidence of natal homing would be if NSS herring from different spawning grounds differed genetically, but this has not yet been tested. However, it has been found that NSS herring differ both phenetically (Johannessen and Jørgensen 1991) and genetically (Jørstad and Pedersen 1986) from some local Norwegian fjord herring stocks. It differs phenetically (Ryman et al. 1984, Johannessen and Jørgensen 1991) from populations in the North Sea, Skagerak, Kattegat, and the Baltic Sea, whereas there appears to be little genetic difference between these stocks based on electrophoretic analyses (Ryman et al. 1984, Jørstad et al. 1991). However, Jørstad et al. suggest that genetic differences are likely to be found between these populations if more comprehensive studies with use of new techniques, as mitochondrial DNA analysis, are carried out in the future. Similarly, one may also be able to find differences between spawning grounds of NSS herring using DNA analysis. The question is what such differences will explain. At the plenary meeting after the genetic session at Herring 2000, Jørstad specifically warned about drawing early conclusions from genetic differences at the DNA level. Such projects should be carefully designed. A common way of doing genetic analysis of fish is simply to collect one or two samples from each area and test for between-area differences. This is, according to this author, simply not good enough, when working at a DNA level. I think it will be very difficult to draw any conclusions from such studies before it is proved that the same difference between spawning grounds is found in many samples throughout the spawning season, and for some years. This author has no knowledge of such studies in herring to this date.

Data on the Influence of First Spawning Ground

Although there does not appear to be any good evidence of natal homing in NSS herring, it is quite clear that this stock has a tendency to occupy the same spawning grounds year after year. Acoustic abundance data from recent years demonstrate that the relative distribution of year classes at spawning grounds from south to north differs, and that this difference is quite similar between years (Fig. 3), clearly indicating a tendency of return spawning. The Norwegian tagging experiments can also show some evidence of herring returning to spawn at the southern grounds. In 1999 an experimental fishery for recovery of tags was arranged south of 61°N. The recoveries showed that all individuals had been tagged south of 61°N, most in 1998 (Røttingen and Slotte 2001, this volume). On the other hand, these data do not disprove that a proportion of the herring tagged south of 61°N in fact have spawned farther north in subsequent years. Many tags from the experiments at southern grounds have also been recovered at the spawning areas north of 61°N. Since most of these tags were found in maturing and spent herring, it is difficult to conclude with certainty if they would spawn or had spawned in this area, or if they in fact were on their way to or from more southern grounds. There are similar problems with experiments conducted to the north of 61°N. However, there are good data on high homing rates to particular spawning grounds, from tagging experiments conducted at very low stock levels in the 1970s (Dragesund et al. 1980, Hamre 1990). Nevertheless, the same data and data presented in previous studies of NSS herring (Devold 1963, 1968; Dragesund 1970; Dragesund et al. 1980, 1997; Hamre 1990; Røttingen 1990) clearly demonstrate that this stock is highly dynamic and has a strong tendency to stray to new spawning grounds. This tendency to stray is not in agreement with the predictions from the natal homing model, but fit well with the predictions from the other two models.

Data on the Influence of Stock Size and Structure

The observed straying in NSS herring is clearly related to historic variations in stock size and structure. In accordance with predictions of the state migration model, increasing stock size has resulted in a wider spawning area. When the large 1983 year class was fully recruited in 1989, spawning occurred for the first time in 30 years at the traditional southern grounds off Karmøy (Johannessen et al. 1995). In 1997 when the 1991 and 1992 herring recruited and the spawning stock reached its highest biomass since the 1950s, the biomass at the southern grounds also increased significantly from estimates at 20,000 t to estimates at 200,000 t (Slotte and Dommasnes 1997). In 1999-2001 the spawning stock decreased. Simultaneously the biomass at the southern grounds tended to decrease (Slotte and Dommasnes 1997, 1998, 1999, 2000) and in 2001 the estimate was as low as 5,000 t.

Historic observations also indicate that straying related to variations in stock size and structure is associated with learning. The return-spawning model and state-dependent migration model predict that recruit spawners, in the absence of older repeat spawners to show them the way, may deviate from a particular migration pattern. This may explain the migration behavior observed in 1963, when a group of recruits mainly belonging to the strong 1959 year class commenced spawning at Lofoten, and feeding and wintering in the open sea to the north of Lofoten (Devold 1968). They were most likely herring leaving the Barents Sea nursery area without finding older "teachers" showing them the migration route. Their migration pattern was, however, only maintained until the summer in 1966, when the unit joined the main part of the stock wintering to the east of Iceland and spawning off Møre. This demonstrates a second aspect of learning, in which herring schools and school members will tend to follow the mainstream behavior.

A similar example of learning and straying is found after the stock collapse. Two different components of immature herring survived the heavy exploitation in the 1960s, one in the Barents Sea and one at the west coast of Norway (Dragesund et al. 1980, Hamre 1990). Both components spawned for the first time in 1973; the northern component spawned off Lofoten (as the 1959 year class in 1963) and the southern component spawned off Møre. The main thing here is that both components spawned in the vicinity of their nursery area. In the following years the two components developed as separate units with different spawning grounds, feeding areas, and wintering areas. The northern component spawned from northern Møre to Lofoten, whereas the southern component spawned at southern Møre earlier in the season. Both components fed off the coast, but the northern component distributed farther north and wintered in the Vestfjorden area, whereas the southern component wintered in fjords at Møre. It must be emphasized that the examples of high homing rates to particular spawning grounds in NSS herring (Dragesund et al. 1980, Hamre 1990), were obtained from tagging experiments on these two components. The low abundance and different migration patterns of the components probably reduced the mixing of schools, and thus prevented the herring from learning "new" migration routes. Local fjord herring stocks mixed to some extent with the two components in the wintering areas (Hamre 1990). Thus, it is possible that it was these local herring that guided the Norwegian spring spawners to their wintering areas. However, when the stock size increased, with the recruitment of the strong 1983 year class, the separation between the two different components disappeared. Wintering was established within the Vestfjorden area, whereas during the spawning season the stock was distributed all over the Møre shelf area and northward at Haltenbanken and Sklinnabanken and also south of 61°N (Hamre 1990, Røttingen 1990, Johannessen et al. 1995).

A third aspect of learning in relation to straying is that recruits, small herring or herring in bad condition, might be unable to follow their "teachers" of larger size and in better condition as predicted by the state-dependent migration model. This might be the reason why the large 1991 year class spawned farther north, for their first time, than the large 1992 year class (Fig. 3). In general year classes tend to leave the nursery areas and join the adult herring the summer preceding first spawning. This was also the case for the large 1991 year class, but at the same time the main bulk of the 1992 year class followed although it was not ready to spawn the year after. This is clearly another incidence indicating a tendency to follow the mainstream. As a result these year classes had a totally different basis for their first spawning migration. When the 1991 year class recruited there were few "teachers" of similar size. At that time the largest group of repeat spawners was the 8-years-older 1983 year class, which had much higher migration speed and migration potential than the 1991 year class. It seems likely that a large proportion of the 1991 year class therefore remained behind and instead spawned closer to the wintering area their first time. However, when the bulk of the 1992 year class recruited one year later, it had the possibility to school with similar sized "teachers" of the 1991 year class and was to a larger extent led southward by the fraction spawning off Møre.

Data on the Influence of Environmental Conditions and Migration Constraints

As mentioned above, the migration potential may be of importance for schooling herring with regard to learning, i.e., individuals may be taught new spawning grounds by following the main stream of similar sized school members, whereas individuals with lower size and condition may have to split from schools and undertake shorter migrations. One of the main reasons for undertaking the southward migrations seems to be the search for better environmental conditions for eggs and larvae. The body size and condition seems to be important as well. Since 1995 the distance of the southward spawning migration has been influenced by the body length and condition of the fish (Slotte and Dommasnes 1997, 1998, 1999; Slotte 1999a). More specifically these investigations have demonstrated that the mean body length in the overall spawning population and of given year classes tend to increase southward, and that the condition factor of given

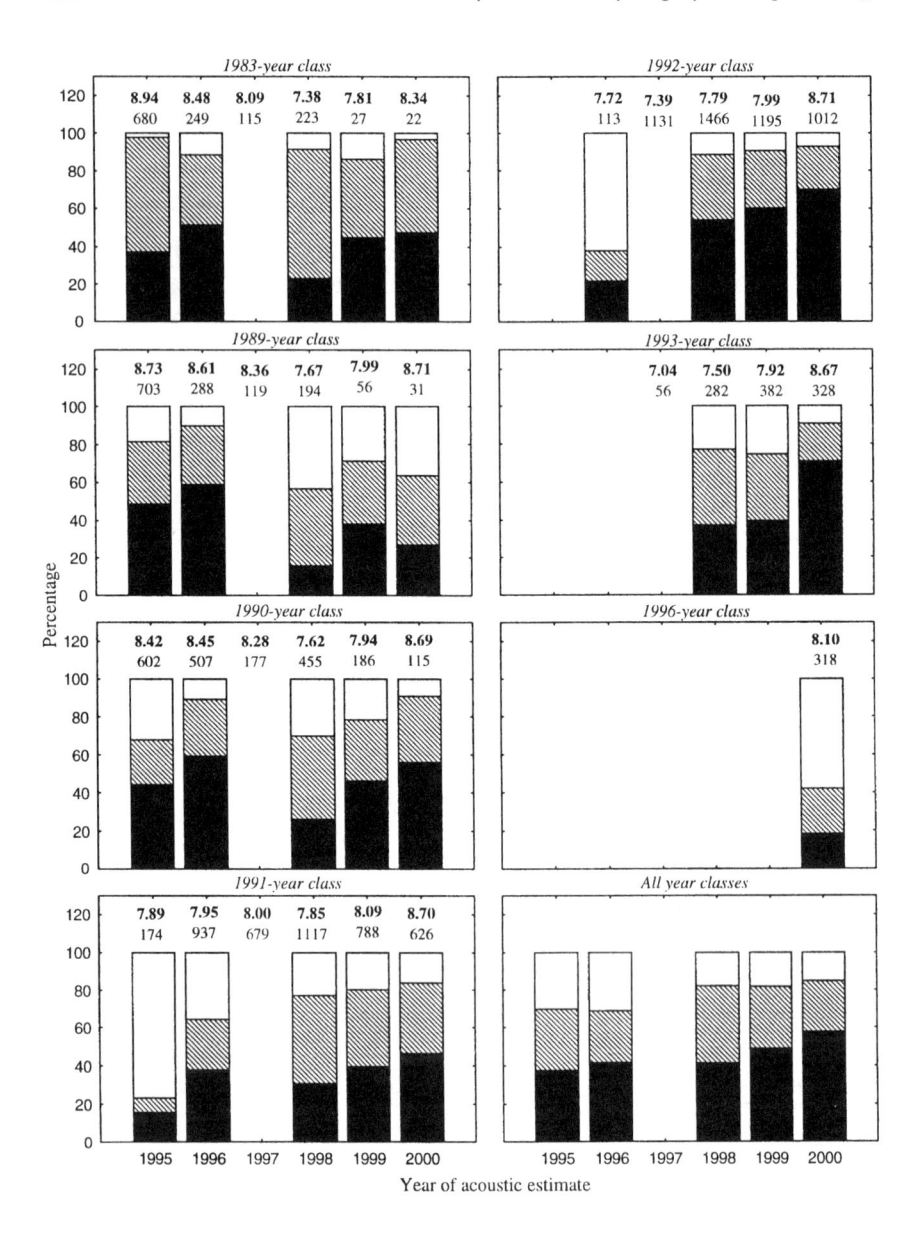

Figure 3. The relative abundance of year classes in NSS herring on a latitudinal scale from 1995 to 2000 as measured from acoustic surveys (in 1997 the total spawning area was not covered). Black bars = 62-64°N, patterned bars = 64-66°, and white bars = 66-70°N. The mean condition factor (outlined) in the wintering area in January prior to spawning migration and number of fish included in the analysis is given on the top of the bars. Acoustic data from Slotte (1999a), Slotte and Dommasnes (1998, 1999, 2000).

year classes, and of 1-cm length groups, tend to increase southward. The same picture is found when looking at the relative abundance of year classes along the coast (Fig. 3): clearly there is a tendency that a year class spawns farther south as it grows larger; at the same time there is a tendency that year classes may spawn farther to the north again if the condition is reduced. The gonad development and investment is also influenced by the condition of the fish, i.e., the stage of maturity and gonad condition factor also increases southward (Slotte 1999b). This relation between migration distance and gonad development is further supported by Oskarsson and others (Marine Research Institute, Reykjavik, Iceland, unpubl. data). They analyzed oozyte diameter, fecundity, and atresia from herring at various latitudes in 1997. The oozyte diameter and fecundity increased southward from 67°N to 62°N along with an increase in condition, and correspondingly the level of atresia decreased. At the main spawning area (62-64°N) the level of atresia was close to zero, whereas some fish sampled close to the wintering area (67°N) had up to 100%. All these findings are in accordance with predictions from the state-dependent migration model, and at the same time they do not fit with the predictions from the other two models.

The opportunities of Norwegian spring-spawning herring to spawn within the vicinity of the wintering area are numerous given the present migration pattern compared to periods prior to the stock decline, when the herring were wintering to the east of Iceland. One may therefore question how individuals approached migration constraints during previous periods. Data on individual condition and tagging experiments indicate similar behavior during the period with oceanic wintering. Coinciding with decreasing body condition (length specific weight) in the stock in the 1960s (Holst 1996), in excess of 90% of the herring spawning off the Faeroes were estimated to be Norwegian spring spawners, mainly contributed by the recruitment of the strong 1959 year class (Jakobsson 1970). In addition, individuals spawning some years (1960, 1963, 1964, 1965, 1966) off western Norway spawned off the Faeroes one or several years later (1967). It must also be emphasized that the herring stopped visiting the southernmost grounds (south of 61°N) in 1960. Perhaps the herring in poor condition migrating toward Norway to spawn from the oceanic wintering area east of Iceland instead stopped at the Faroes due to migration constraints. In fact, this reduced the migration distance by approximately 700 km. Similarly, the herring arriving within the vicinity of Møre after migration from Iceland reduced the migration distance by up to 500 km by spawning off Møre compared to at the southernmost grounds, which may explain why the herring stopped visiting the southern grounds after 1959. In accordance with the state-dependent migration model, some of these historical changes in the use of spawning grounds by NSS herring could also be due to changes in environmental conditions as well as migration constraints; but this is yet to be investigated.

Which Model Fits Best with the Observed Spawning Ground Selection?

The observed data on NSS herring larva distribution and spawning ground selection does not fit well with the predictions from the natal homing model. Such a model would only fit the Stephenson et al. (2001) model of complexes, cells, and spawning grounds, if one concluded that NSS herring is a large complex, with one single 1,500 km wide egg distribution area. Many of the predictions from the return-spawning migration model about straying with variations in stock size and structure fits quite well with the data. Nevertheless, this model fails to take into account the variations in environmental conditions between spawning grounds and the costs of reaching them. It fails to predict the observed selection of spawning grounds in relation to body length and condition. If homing, whether it is natal or simply return spawning, was of crucial importance in this stock, then one should expect that individuals would select spawning grounds along the coast independently of body length or condition, but they do not. In addition, due to the high costs of migration, one should expect decreasing condition southward, but the opposite relationship is observed. It seems that larger individuals and/or individuals in better condition choose both to migrate a longer distance southward and invest more energy into gonads. Individuals in very weak condition may spawn close to the wintering area with reduced fecundity or even skip a season through total absorption of the eggs (atresia). Thus, the predictions from the state-dependent migration model fit the data very well in every aspect. There appears to be no other plausible models that would yield the same fit between predictions and data, and therefore this hypothesis is found to be the true one (Table 1).

It seems likely that a state-dependent migration behavior could have been adapted in this highly migratory stock, since this behavior apparently is advantageous for individual fitness. However, although the state-dependent migration model fits well within the NSS herring dynamics, it might not fit in other stocks. The natal homing model and the return-spawning model has been developed from observations on less migratory stocks, and they may very well be true in many other stocks. Nevertheless, one weakness of these models is the missing link to evolutionary aspects like fitness. It seems that Darwin's words that evolution has developed by "survival of the fittest" have less meaning in these models compared to the state-dependent migration model.

Spawning Time

Observations at herring spawning grounds have indicated that they arrive and spawn in waves, and that this is related to the age and length structure of the population. However, there is disagreement about how spawning time is related to length or age and how many spawning waves there might be. The next analysis will evaluate two different hypotheses on this aspect in NSS herring.

Table 1. Spawning ground selection in NSS herring: an evaluation of different hypotheses.

| | Hypotheses | | | |
Factors	Natal homing	Return spawning	State dependent	Data
Hatching location	Yes	No	No	No
First spawning location	Yes	Yes	Yes	Yes
Stock size and structure	No	Yes	Yes	Yes
Environmental conditions	No	No	Yes	Yes
Migration potential	No	No	Yes	Yes
Hypothesis true/false	False	False	True	

The predicted influence of different factors (predicted = yes, not predicted = no) versus actual data. The state-dependent migration hypothesis is the only hypothesis where all predictions fit with observations, and it was therefore found to be true.

Models of Wave Spawning

Multiple Spawning Waves

The hypothesis of multiple spawning waves claims that herring spawn in a number of waves reflected by the number of age modes or length modes in the reproductive stock starting with the largest or oldest fish (Lambert 1987; Lambert and Messieh 1989; Ware and Tanasichuk 1989, 1990). The idea is that herring initiate maturation at the same time regardless of fish size, while the instantaneous maturation rate increases with fish size and thus results in earlier spawning in larger fish. From an evolutionary point of view size-dependent maturation rate will result in a progressively earlier spawning as a herring ages, which might enhance an individual's fitness over its reproductive life span, because of inter-annual uncertainties in food supply, predation pressure, and density-dependent interactions. This idea is analogous to the hypothesis that the recruitment success is likely to improve if the fish spawn in waves (Hay 1985, Ware and Tanasichuk 1989, Lambert 1990).

Two Spawning Waves

The hypothesis of two spawning waves is based on knowledge of NSS herring and claims recruit spawners spawn with a delay compared to the repeat spawners, due to a delayed onset of maturation rather than lower maturation rate (Slotte et al. 2000). Thus, regardless of the number of age or length modes in the population, the herring will always spawn in two waves. From an evolutionary point of view the delayed spawning in the recruits is not supposed to have any selective advantage. It is rather a phenomenon demonstrating that the recruits may have problems with timing their first spawning to environmental conditions in the same way as second-year spawners and older fish.

Predictions of Wave Spawning

The model of multiple spawning waves predicts that the state of maturity will increase with the body length, and that the difference in maturity between large and small fish will increase from the start of maturation until spawning. The model of two spawning waves predicts that the state of maturity at all times during the wintering and pre-spawning season will be equal among NSS herring of sizes 32 cm and upward, whereas it will increase with body length among herring at size 27-31 cm. This prediction is based on maturation data on NSS herring (Toresen 1990) indicating that the proportion of recruit spawners decreases with increasing fish lengths up to approximately 32 cm, after which roughly all herring are repeat spawners.

Data on Size-Dependent Maturation

Studies on both Norwegian spring-spawning herring and other herring stocks have demonstrated that the time of arrival at spawning grounds and spawning time are influenced by the population structure. Based on data of Norwegian spring-spawning herring collected prior to the severe stock decline in the late 1960s, the recruits were found to spawn progressively later in the season than the older repeat spawners (Rasmussen 1939, Runnstrøm 1941, Dragesund 1970). After reanalyzing data from Runnstrøm (1941), Lambert (1987) concluded that herring spawned in a decreasing order of age, suggesting that also among repeat spawners there was a difference in spawning time between age groups. Lambert and Messieh (1989) found a similar relationship between fish length and spawning time in Atlantic herring from Canadian waters. Studies of Ware and Tanasichuk (1989, 1990) gave more support to the hypothesis of herring spawning in a decreasing order of size. They claimed that Pacific herring (*Clupea pallasii*) initiate maturation at the same time regardless of fish size, while the instantaneous maturation rate increases with fish weight and thus results in earlier spawning in larger fish. Ware and Tanasichuk (1989) also concluded that wave spawning in Pacific herring is based upon the size-dependent maturation rate. Both Atlantic (Lambert 1987 and references therein; Lambert and Messieh 1989) and Pacific (Hay 1985, Ware and Tanasichuk 1989) herring are suggested to have spawned in up to four waves. Lambert (1987) and Lambert and Messieh (1989) found also that the number of spawning waves reflects the number of age modes or length modes in the reproductive stock.

Slotte et al. (2000) studied the relationship between spawning time and body length of NSS herring by use of the following index (I_G): gonad weight in percentage of expected gonad weight at full maturity for a given total length (L_T). The index was recorded for three wintering and spawning seasons. At all times I_G tended to increase with L_T in the range 27-31cm, whereas insignificant differences in I_G were found between L_T groups in the range 32-37 cm (Fig. 4). More support for the model of two spawn-

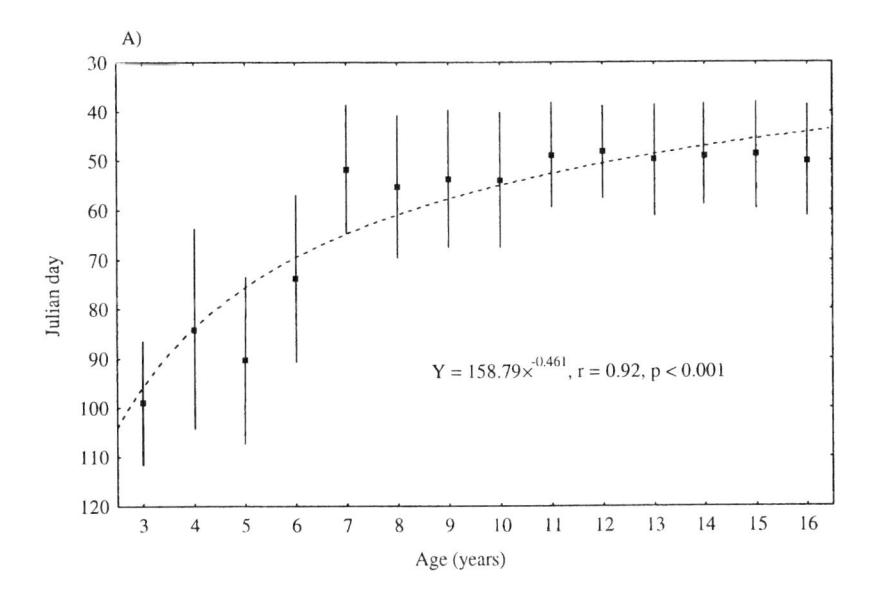

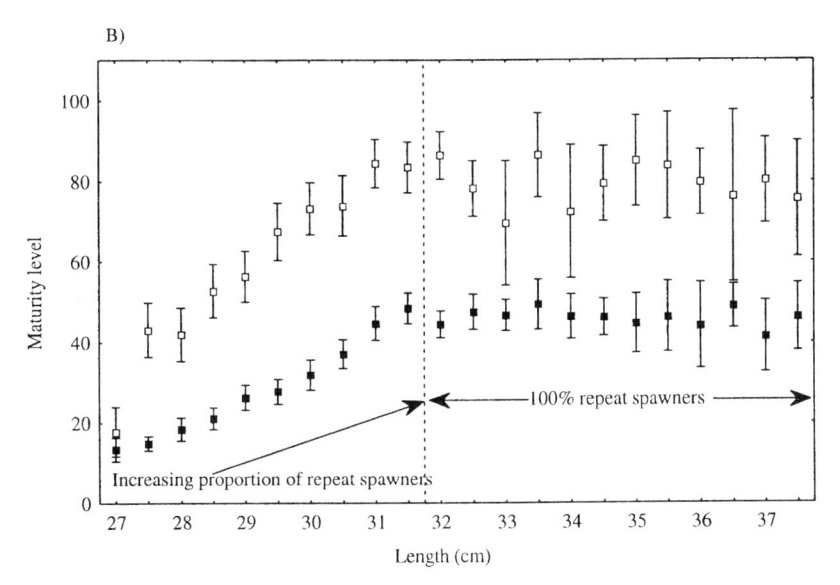

Figure 4. Upper: The relation between weighted mean date (±S.D.) for peak of maturity stage 6 (spawning) and age for NSS herring in 1934. Adapted from Lambert (1987), who did the calculations based on data from Runnstrøm (1941). Bottom: The relation between body length and maturity level (I_G) in male (n = 815) and female (n = 895) herring from the wintering area in December 1996. Mean values ±95% confidence limits (n ≥ 3). The length groups to the right of the dotted line are assumed to be repeat spawners. Adapted from Slotte et al. (2000).

ing waves is found in a recent study on oocyte diameter and histology (diameter of yolk granules, volume fraction of yolk granules, and chorion thickness) of pre-spawning NSS herring (Oskarsson et al. in press). These detailed studies of herring maturation also demonstrated that state of maturity did not differ with body length in 32 cm or larger fish. Slotte et al. (2000) also found that the relation between fish length and maturity level was maintained in the spawning stock throughout the wintering and spawning season. In addition the percentage of maturing (maturity stage > 2) herring tended to increase during the wintering season. This implies that the delayed spawning time in the smallest fish may be due to delayed onset of maturation (initiation of vitellogenesis, spermatogenesis) rather than a slower maturation rate.

A closer look at the results of Lambert (1987) suggests that his data in fact also support the model of two spawning waves, rather than the model of multiple spawning waves (Fig. 4). Lambert interprets that herring at age 7 on the average will spawn 20 days later than herring at age 16, but it seems that artificial effects of the curve-fitting procedure cause this difference. On the other hand, it seems appropriate that the mean date of peak occurrence of maturity stage 6 (running) increased in herring from age 7 toward age 3 years. In light of the recent data one may interpret Lambert's analysis of data from 1934 to reflect the results obtained for the same stock in 1994-1996; that recruits spawn with a delay compared to the repeat spawners. The data from 1934 indicate that the proportion of recruits decreased with age from age 3 to age 6 years, whereas herring at age 7 years or older were repeat spawners.

Which Model Fits Best with the Observations of Spawning Waves?

Both the recent and historic data on NSS herring are more in accordance with the predictions from the model of two spawning waves than the model of multiple spawning waves. Thus, the hypothesis of two spawning waves is accepted as the true one. However, this does not disprove that the hypothesis of multiple spawning waves may be true in other herring populations.

Acknowledgments

I thank the two anonymous referees and editor R.L. Stephenson for their valuable suggestions to improvement of the paper. The Norwegian Research Council funded the study.

References

Aure, J., and Ø. Østensen. 1993. Hydrographic normals and long-term variations in Norwegian coastal waters. Fisken og Havet 6. (In Norwegian.)

Collie, J. (moderator). 1991. Workshop on member/vagrant hypothesis and stock discreteness. In: Proceedings of the International Herring Symposium 1990. University of Alaska Sea Grant, AK-SG-91-01, Fairbanks, pp. 641-657.

Devold, F. 1963. The life history of the Atlanto-Scandian herring. Rapp. P.V. Reun. Cons. Int. Explor. Mer 154:98-108.

Devold, F. 1968. The formation and the disappearance of a stock unit of Norwegian herring. Fiskeridir. Skr. Ser. Havunders. 15:1-15.

Dragesund, O. 1970. Factors influencing year-class strength of Norwegian spring spawning herring (*Clupea harengus* L.). Fiskeridir. Skr. Ser. Havunders. 15:381-450.

Dragesund, O., J. Hamre, and Ø. Ulltang. 1980. Biology and population dynamics of the Norwegian spring spawning herring. Rapp. P.V. Reun. Cons. Int. Explor. Mer 177:43-71.

Dragesund, O., A. Johannessen, and Ø. Ulltang. 1997. Variation in migration and abundance of Norwegian spring spawning herring (*Clupea harengus* L.). Sarsia 82:97-105.

Haegele, C.W., and J.F. Schweigert. 1985. Distribution and characteristics of herring spawning behavior. Can. J. Fish. Aquat. Sci. 42(Suppl. 1):39-55.

Hamre, J. 1990. Life history and exploitation of the Norwegian spring-spawning herring. In: T. Monstad (ed.), Proceedings of the 4th Soviet-Norwegian Symposium, 12-16 June 1989. Institute of Marine Research, Bergen, Norway, pp. 5-39

Hay, D.E. 1985. Reproductive biology of Pacific herring (*Clupea harengus pallasi*). Can. J. Fish. Aquat. Sci. 42(Suppl. 1):111-126.

Hay, D.E., and A.R. Kronlund. 1987. Factors affecting the distribution, abundance and measurements of Pacific herring (*Clupea harengus pallasi*) spawn. Can. J. Fish. Aquat. Sci. 44:1181-1194.

Hay, D.E., and P.B. McCarter. 1997. Larval distribution, abundance, and stock structure of British Colombia herring. J. Fish. Biol. 51(Suppl. A):155-175.

Holst, J.C. 1996. Long term trends in the growth and recruitment pattern of the Norwegian spring-spawning herring (*Clupea harengus* Linnaeus 1758). Ph. D. thesis, University of Bergen, Bergen, Norway. ISBN 82-77444-032-4.

Holst, J.C., and A. Slotte. 1998. Effects of juvenile nursery on geographic spawning distribution in Norwegian spring spawning herring (*Clupea harengus* L.). ICES J. Mar. Sci. 55:987-996.

Hourston, A.S. 1982. Homing by Canada's west coast herring to management units and divisions as indicated by tag recoveries. Can. J. Fish. Aquat. Sci. 39:1414-1422.

Iles, T.D., and M. Sinclair. 1982. Atlantic herring: Stock discreteness and abundance. Science 215:627-633.

Jakobsson, J. 1970. The biological position of the 'Faeroese Bank' herring within the Atlanto-Scandian herring stocks. ICES C.M. 1970/H:12.

Johannessen, A., and T. Jørgensen. 1991. Stock structure and classification of herring (*Clupea harengus* L.) in the North Sea, Skagerrak/Kattegat and Western Baltic based on multivariate analysis of morphometric and meristic characters. In: Proceedings of the International Herring Symposium. University of Alaska Sea Grant, AK-SG-91-01, Fairbanks, pp. 223-244.

Johannessen, A., A. Slotte, O.A. Bergstad, O. Dragesund, and I. Røttingen. 1995. Reappearance of Norwegian spring spawning herring (*Clupea harengus* L.) at spawning grounds off south-western Norway. In: H.R. Skjoldal, C. Hopkins, K.E. Erikstad, and H.P. Leinaas (eds.), Ecology of fjords and coastal waters. Elsevier Science, pp. 347-363.

Jørstad, K.E., and S.A. Pedersen. 1986. Discrimination of herring populations in a northern Norwegian fjord: Genetic and biological aspects. ICES C.M. 1986/M:63.

Jørstad, K.E., D.P.F. King, and G. Nævdal. 1991. Population structure of Atlantic herring *Clupea harengus* L. J. Fish. Biol. 39(Suppl. A):43-52.

Lambert, T.C. 1987. Duration and intensity of spawning in herring (*Clupea harengus*) as related to the age structure of the mature population. Mar. Ecol. Prog. Ser. 39:209-220.

Lambert, T.C. 1990. The effect of population structure on recruitment in herring. J. Cons. Int. Explor. Mer 47:249-255.

Lambert, T.C., and S.N. Messieh. 1989. Spawning dynamics of Gulf of St. Lawrence herring (*Clupea harengus*). Can. J. Fish. Aquat. Sci. 46:2085-2094.

McNamara, J.M., and I. Houston. 1996. State-dependent life histories. Nature 380:215-221.

McQuinn, I.H. 1997. Metapopulations and the Atlantic herring. Rev. Fish Biol. Fish. 7:297-329.

Misund, O.A., H. Vilhjalmsson, S.H.I. Jakupsstovu, I. Røttingen, S. Belikov, O. Asthorsson, J. Blindheim, J. Jonsson, A. Krysov, S.A. Malmberg, and S. Sveinbjornsson. 1998. Distribution, migration and abundance of Norwegian spring spawning herring in relation to the temperature and zooplankton biomass in the Norwegian Sea as recorded by coordinated surveys in spring and summer 1996. Sarsia 83:117-127.

Pitcher, T.J., and J.K. Parrish. 1993. Functions of shoaling behaviour in teleosts. In: T.J. Pitcher (ed.), The behaviour of teleost fishes, 2nd edn. Croom Helm, London, pp. 364-439

Pitcher, T.J., A.E. Magurran, and J.I. Edvards. 1985. Schooling mackerel and herring choose neighbours of similar size. Mar. Biol. 86:319-322.

Rasmussen, T.H. 1939. Fluktuasjoner i vintersildens utvikling og gyting. Rep. Norweg. Fish. Invest. 6(1):70-73. (In Norwegian.)

Røttingen, I. 1990. The 1983 year class of Norwegian spring spawning herring as juveniles and recruit spawners. In: T. Monstad (ed.), Proceedings of the 4th Soviet-Norwegian Symposium. Institute of Marine Research, Bergen, Norway, pp. 165-203.

Røttingen, I., and A. Slotte. 2001. The relevance of a former important spawning area in the present life history and management of Norwegian spring spawning herring. In: F. Funk, J. Blackburn, D. Hay, A.J. Paul, R. Stephenson, R. Toresen, and D. Witherell (eds.), Herring: Expectations for a new millennium. University of Alaska Sea Grant, AK-SG-01-04, Fairbanks. (This volume.)

Runnstrøm, S. 1941. Quantitative investigations on herring spawning and its yearly fluctuations at the west coast of Norway. Fiskeridir. Skr. Ser. Havunders. 6(8):5-71.

Ryman, N.U., L. Lagercrantz, R. Andersson, R. Chakraborky, and R. Rosenberg. 1984. Lack of correspondence between genetic and morphologic variability patterns in Atlantic herring. Heredity 53:687-704.

Sætre, R. 1999. Features of the central Norwegian shelf circulation. Cont. Shelf. Res. 19(14):1809-1831.

Sætre, R., R. Toresen, H. Søiland, and P. Fossum. 2002. The Norwegian spring spawning herring: Spawning, larval drift, and retention. Sarsia 87. In press.

Sinclair, M. 1988. Marine populations: An essay on population regulation and speciation. Washington Sea Grant Program/ University of Washington Press, Seattle. 252 pp.

Slotte, A. 1999a. Effects of fish length and condition on spawning migration in Norwegian spring spawning herring (*Clupea harengus* L.). Sarsia 84:111-127.

Slotte, A. 1999b. Differential utilisation of energy during wintering and spawning migration in Norwegian spring spawning herring (*Clupea harengus* L.). J. Fish. Biol. 54:338-355.

Slotte, A., and A. Dommasnes, A. 1997. Abundance estimation of Norwegian spring spawning herring at spawning grounds 20 February-18 March 1997. Institute of Marine Research Internal Report No. 4, Bergen, Norway. 10 pp.

Slotte, A., and A. Dommasnes. 1998. Distribution and abundance of Norwegian spring spawning herring during the spawning season in 1998. Fisken og Havet 5. 10 pp.

Slotte, A., and A. Dommasnes. 1999. Distribution and abundance of Norwegian spring spawning herring during the spawning season in 1999. Fisken og Havet 12. 27 pp.

Slotte, A., and A. Dommasnes. 2000. Distribution and abundance of Norwegian spring spawning herring during the spawning season in 2000. Fisken og Havet 10. 18 pp.

Slotte, A., and Ø. Fiksen. 2000. State-dependent spawning migration in Norwegian spring spawning herring (*Clupea harengus* L.). J. Fish. Biol. 56:138-162.

Slotte, A., A. Johannessen, and O. Kjesbu. 2000. Effects of fish size on spawning time in Norwegian spring spawning herring. J. Fish. Biol. 56:295-310.

Stephenson, R.L. 1991. Stock discreteness in Atlantic herring: A review of arguments for and against. In: Proceedings of the International Herring Symposium. University of Alaska Sea Grant, AK-SG-91-01, Fairbanks, pp. 659-666.

Stephenson, R.L., and I. Kornfield. 1990. Reappearance of spawning Atlantic herring (*Clupea harengus harengus*) on Georges Bank: Population resurgence not recolonization. Can. J. Fish. Aquat. Sci. 47:1060-1064.

Stephenson, R.L., K.J. Clark, M.J. Power, F.J. Fife, and G.D. Melvin. 2001. Herring stock structure, stock discreteness, and biodiversity. In: F. Funk, J. Blackburn, D. Hay, A.J. Paul, R. Stephenson, R. Toresen, and D. Witherell (eds.), Herring: Expectations for a new millennium. University of Alaska Sea Grant, AK-SG-01-04, Fairbanks. (This volume.)

Toresen, R. 1990. Long term changes in growth and maturation in Norwegian spring-spawning herring. In: T. Monstad ed.), Proceedings of the 4th Soviet-Norwegian Symposium. Institute of Marine Research, Bergen, Norway, pp. 89-106.

Ware, D.M. 1975. Growth, metabolism, and optimal swimming speed of a pelagic fish. J. Fish. Res. Board Can. 32:33-41.

Ware, D.M. 1978. Bioenergetics of pelagic fish: Theoretical change in swimming speed and ration with body size. J. Fish. Res. Board Can. 35(2):220-228.

Ware, D.M., and R.W. Tanasichuk. 1989. Biological basis of maturation and spawning waves in Pacific herring (*Clupea harengus pallasi*). Can. J. Fish. Aquat. Sci. 46:1776-1784.

Ware, D.M., and R.W. Tanasichuk. 1990. A method for forecasting the timing of ripeness in Pacific herring (*Clupea harengus pallasi*) females. Can. J. Fish. Aquat. Sci. 47:2375-2379.

Wheeler, J.P., and G.H. Winters. 1984. Homing of Atlantic herring (*Clupea harengus harengus*) in Newfoundland waters as indicated by tagging data. Can. J. Fish. Aquat. Sci. 41:108-117.

Norwegian Spring-Spawning Herring (*Clupea harengus*) and Climate throughout the Twentieth Century

Reidar Toresen and Ole Johan Østvedt
Institute of Marine Research, Bergen, Norway

Abstract

A long-term virtual population analysis, VPA, (1907-1997) was made for Norwegian spring-spawning herring (NSSH). It shows that this herring stock has had large fluctuations during the last century. The spawning stock biomass (SSB) increased from a rather low level in the early years of the twentieth century and reached around 14 million metric tons by 1930. The SSB decreased to a level of around 10 million t by 1940, but increased again to a record high level of 16 million t by 1945. The stock then decreased over the next 20-year period, and fell to a level of less than 50,000 t by the late 1960s. Through the 1970s and 1980s the stock slowly recovered and after the recruitment of strong year classes in 1983, and in 1990, 1991, and 1992 the stock recovered to a spawning stock biomass about 10 million t.

We found that the long-term changes in spawning stock abundance are highly correlated with the long-term variations in the mean annual temperature of the inflowing Atlantic water masses (through the Kola section) into the northeast Atlantic region. The recruitment is positively correlated with the average temperature in the Kola section in the winter months, January-April, which indicates that environmental factors govern the large scale fluctuations in production for this herring stock.

This contribution is a summary of Toresen, R. and Østvedt, O.J. 2000. Variation in abundance of Norwegian spring-spawning herring (*Clupea harengus*, Clupeidae) throughout the 20th century and the influence of climatic fluctuations. Blackwell Science Ltd, Fish and Fisheries, 2000, 1(3):231-256.

Introduction

Marine fish populations vary in abundance. This was known long before it became a science to study fish abundances in the early years of this century, and has long caused challenges for both scientists and for fishery managers throughout the world. The Norwegian spring-spawning herring (NSSH) is a stock which has undergone large fluctuations (Devold 1963, Dragesund and Ulltang 1978). One of the first scientists to take note of the variations in fish abundance was Johan Hjort (1914) who described the variation in abundance of NSSH and a few other important fish stocks in the northeast Atlantic. Since then, it has been an everlasting challenge for scientists to find the answer to what governs the great changes in abundance. Bits and pieces of the answer fall into place now and then for different stocks, but the relation between physical oceanographic factors and biological parameters has been shown for only a few stocks (Myers 1998, Ottersen et al. 1998; Michalsen et al. 1998).

An analytical assessment of a fish stock requires reliable fishery statistics and representative sampling of biological data from the fisheries (including age determination). Both these elements were fulfilled for the NSSH from the very start of sampling in the early 1900s, and it is therefore possible to carry out a long-term analytical assessment of this stock. The paper on which this summary is based (Toresen and Østvedt 2000) contains details on a long-term VPA ranging from 1907 to 1998 and an evaluation of the stock fluctuations in relation to the long-term changes in ocean climate in the northeast Atlantic.

Material and Methods

The VPA

The data and the method used for the long-term virtual population analysis (VPA) are presented in Toresen and Østvedt (2000).

For the period 1950 until 1998, the assessment data as presented by ICES (1999) were used. Age distributions in the landings each year were used to convert annual landings to catch in numbers at age for the period 1907-1949. These numbers were used in a VPA-based back-calculation applying Pope's approximation (Pope 1972).

Stock Size and Recruitment in Relation to Climate

The water temperature in the Kola section is a long data series with quarterly temperature measurements from 1900 to present. Since 1921, the data series is of monthly measurements. The temperatures along the section give an indication of the temperature regime in the water masses where the herring larvae live the first months of life. The annual average for 0-200 m from 1900 to 1997 (Bochkov 1982) and the average values for the months January-April for the years 1921-1994 were used for the comparison with stock size and recruitment. Data from recent years have been provided by PINRO, Murmansk.

Results
Development of the Spawning Stock Biomass
As shown in Fig. 1, the spawning stock biomass increased from a rather low level of 2 million t in 1907 to a level of about 5 million t in 1911. This increase was due to the rather rich 1904 year class, which entered the spawning stock fully during 1911 and 1912. In the following years, the stock decreased to about 2 million t in 1920. It then increased again due to good recruitment, first the 1913 year class, and later more abundant year classes, such as 1918, 1922, 1923, and 1925. The stock reached a high level in 1930 with a spawning stock biomass of about 14 million t. The stock then decreased to around 9 million t in 1939, but increased rapidly during the early 1940s to a record high of 16 million t in 1945. This increase was caused by several abundant year classes in the 1930s. After 1946 the spawning stock decreased almost continuously until the stock collapsed in the late 1960s. The decrease was interrupted briefly due to recruitment of the rich 1950 and 1959 year classes. In the period 1970-1975, the SSB was estimated to be less than 100,000 t. During the 1970s and early 1980s the stock increased very slowly, and then accelerated due to the recruitment of the 1983 year class in the late 1980s. In the late 1990s, the spawning stock increased further because of the recruitment of the rich 1991 and 1992 year classes. Poor recruitment after 1993 has caused a decrease in SSB from 1998 onward (ICES 1999).

Correlation of SSB and Recruitment with Temperature
Eleven year classes produced more than 20 billion individuals as 3-year-olds in the period from 1904 to 1994. Ten of them were produced by SSBs larger than 4 million t and the three most abundant year classes were produced by SSBs larger than 10 million t. The estimated geometric mean of recruitment as 3-year-olds for all year classes in the period 1904-1994 is 2.3 billion individuals. However, the geometric mean recruitment in the years when the spawning stock biomass is above 2.5 million t is 5.1 billion individuals. The most abundant year class in the whole time range is that from 1950, estimated at 47 billion individuals (as 3-year-olds), while the 1992 year class is the second most abundant, estimated at 39 billion individuals.

The estimated time series of spawning stock biomass from the long-term VPA is highly correlated with the smoothed (19-year moving average) mean annual temperature in the Kola section (Fig. 1).

Further, the average temperatures (un-smoothed) during the winter months (January-April) in the Kola section were correlated with the annual estimated recruitment (0-group) for years when the estimated spawning

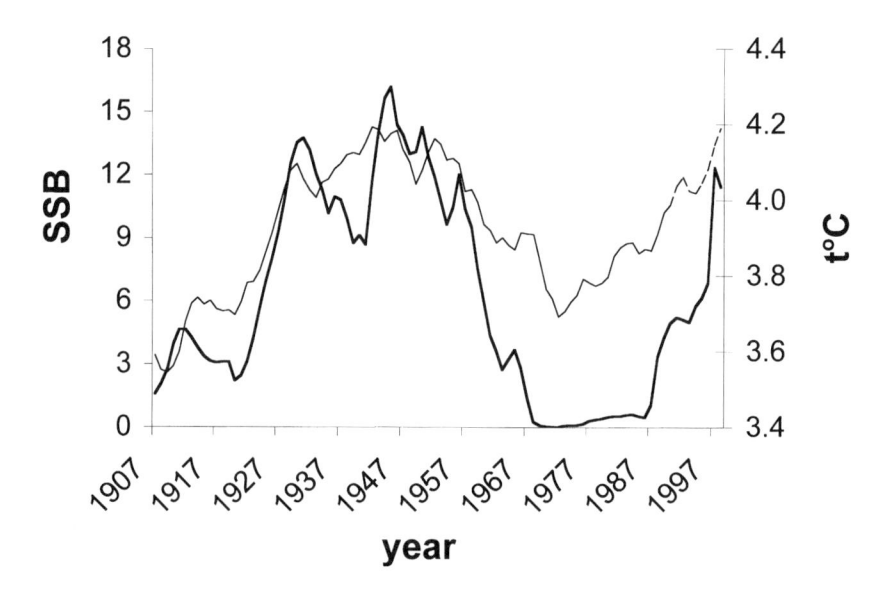

Figure 1. Estimated spawning stock biomass (metric tons × 10⁶) of Norwegian spring-spawning herring, 1907-1998 (heavy line) and long-term temperature fluctuations at the Kola section (Bochkov 1982).

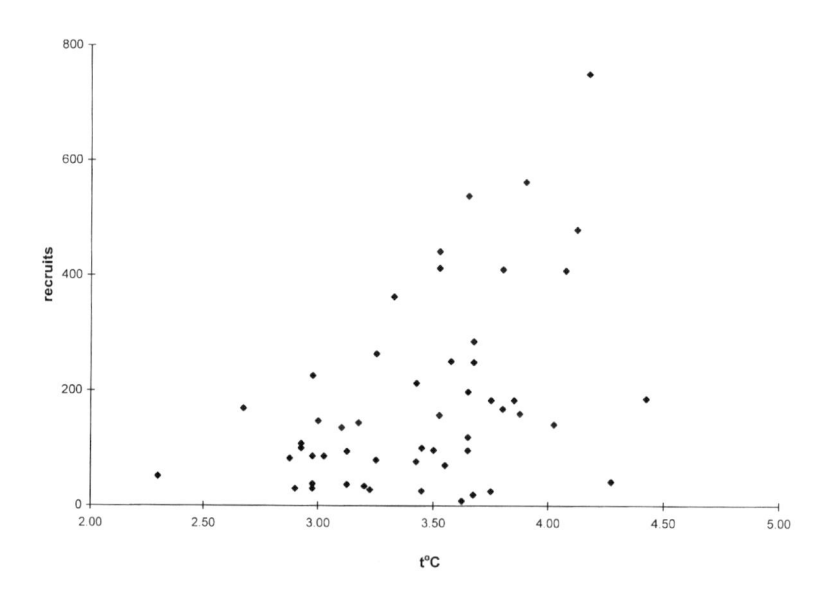

Figure 2. Estimated number of recruits as 0-group (n × 10⁶) versus temperature (°C) at the Kola section (Bochkov 1982).

stock was greater than 2.5 million t. The estimated correlation coefficient is 0.44 and is significant ($P < 0.01$) (Fig. 2).

The winter temperatures in the Kola section were correlated with the recruitment success (recruits per unit of SSB). This was done for the period 1970-1998. There is a significant relationship ($P < 0.05$) between the series.

Discussion

The maximal estimate of spawning stock size, 16 million t in 1945, is about double recent estimates of the stock (ICES 1999). Judging from what we know about the distribution area of this stock in the 1930s and 1940s (Runnstrøm 1936, Devold 1963), both the feeding grounds and the overwintering areas were substantially larger then than today. The stock used larger areas of the Norwegian Sea for feeding and spent the winters there (Devold 1963). Since the total distribution area was much larger than today, it is likely that the stock may have been much larger.

The correlation between the long-term development of the SSB and the temperature in the Kola section was significant. However, the temperature index series was smoothed considerably (19-year moving average) to show the long-term trend. The correlation would probably have been better if it were not for the total collapse of the spawning stock in the 20-year period from about 1965 until the mid-1980s. This collapse was caused at least in part by overfishing (Dragesund et al. 1980).

The estimated correlation between recruitment and temperature in the Kola section, although statistically significant, was not very strong. There is a substantial year-to-year variation in the temperature data. The correlation between temperature and recruitment shows that the temperature probably has a large effect on recruitment processes. This has been shown for other stocks in the northeast Atlantic (Sætersdal and Loeng 1987, Ottersen 1996, Sundby 1994). The plot (Fig. 2) of the recruitment versus temperature for Norwegian spring-spawning herring shows that high temperature is a necessary but not sufficient condition for good recruitment.

Concluding Remarks

The abundance of Norwegian spring-spawning herring has fluctuated during the twentieth century. The spawning stock increased in biomass during the first 30 years, from a low level of about 2 million t at the turn of the century to more than 16 million t in 1945. It fluctuated at a high level for about 15 years until about 1950, then it decreased steadily until its collapse in the late 1960s. The stock has increased in biomass during the last 10-year period and, in light of its history, seems still to be in a rebuilding state.

Long-term fluctuations in stock abundance seem to be governed by climatic fluctuations.

There is a positive correlation between stock development and long term fluctuations in the mean annual temperature in the Kola section in the Barents Sea. There is also a significant positive correlation between recruitment and the annual average temperature in the winter months in the Kola section.

References

Bochkov, Y.A. 1982. Water temperature in the 0-200 m layer in the Kola-Meridian in the Barents Sea, 1900-1981. Sb. Nauchn. Trud. PINRO Murmansk 46:113-122. (In Russian.)

Devold, F. 1963. The life history of the Atlanto-Scandian herring. Rapp. P.V. Reun. Cons. Int. Explor. Mer 154:98-108.

Dragesund, O., and Ø. Ulltang. 1978. Stock size fluctuations and rate of exploitation of the Norwegian spring spawning herring, 1950-1974. Fiskeridir. Skr. Ser. Havunders. 16(10):315-337.

Dragesund, O., J. Hamre, and Ø. Ulltang. 1980. Biology and population dynamics of the Norwegian spring-spawning herring. Rapp. P.V. Reun. Cons. Perm. Int. Explor. Mer 177:43-71.

Hjort, J. 1914. Fluctuations in the great fisheries of Northern Europe viewed in the light of biological research. Rapp. P.V. Reun. Cons. Int. Explor. Mer 20:1-228.

ICES. 1999. Report of the Northern Pelagic and Blue Whiting Fisheries Working Group. ICES Headquarters, Copenhagen, Denmark, 27 April-5 May 1999. ICES C.M. 1999/ACFM:18.

Michalsen, K., G. Ottersen, and O. Nakken. 1998. Growth of north-east Arctic cod (*Gadus morhua* L.) in relation to ambient temperature. ICES J. Mar. Sci. 55:863-877.

Myers, R.A. 1998. When do environment-recruitment correlations work? Rev. Fish Biol. Fish. 8:285-305.

Ottersen, G. 1996. Environmental impact on variability in recruitment, larval growth and distribution of Arcto-Norwegian cod. Dr. Scient. thesis, University of Bergen, Norway.

Ottersen, G., K. Michalsen, and O. Nakken. 1998. Ambient temperature and distribution of north-east Arctic cod. ICES J. Mar. Sci. 55:67-85.

Pope, J.G. 1972. An investigation of the accuracy of virtual population analysis using cohort analysis. Int. Comm. Northwest Atl. Fish. Res. Bull. 9:65-74.

Runnstrøm, S. 1936. A study on the life history and migrations of the Norwegian spring-herring based on the analysis of the winter rings and summer zones of the scale. Fiskeridir. Skr. Ser. Havunders. 5(2):5-102.

Sætersdal, G., and L. Loeng. 1987. Ecological adaptation of reproduction in northeast arctic cod. Fish. Res. 5:253-270.

Sundby, S. 1994. The influence of bio-physical processes on fish recruitment in an arctic-boreal ecosystem. Ph.D. thesis, University of Bergen, Norway. 190 pp.

Toresen, R., and O.J. Østvedt. 2000. Variation in abundance of Norwegian spring-spawning herring (*Clupea harengus*, Clupeidae) throughout the 20th century and the influence of climatic fluctuations. Fish and Fisheries 2000, 1(3):231-256.

Oocyte Degeneration in Female Recruits of Norwegian Spring-Spawning Herring (*Clupea harengus*)

G.P. Mazhirina and E.I. Seliverstova
Knipovich Polar Research Institute of Marine Fisheries and Oceanography (PINRO), Murmansk, Russia

Abstract

Intensive trophoplasmatic growth of vitellogenic oocytes and simultaneous spawning are typical of Norwegian herring. Spawning occurs on spawning grounds along the coast of Norway in February-March. In 1996-1997, samples of herring were collected for histological analysis on the spawning grounds along the coast of Norway (62°30′-65°30′N) from late February to mid-March. Catches consisted of herring at age 3-14 years of 26-39 cm total lengths. Specimens from the 1991 and 1992 abundant year classes made up the bulk of spawning aggregations. In 1996, the recruits comprised 36% of the samples and they were dominated by the Barents Sea migrants (88%). In 1997, the recruits comprised 50% of which the Barents Sea component dominated with 92%.

Resorption of oocytes was mainly noted in the Barents Sea recruits. In 1996, those were the specimens from the 1991 year class. They constituted 20% of the total amount of the recruits and 20.5% of the Barents Sea component. In 1997, atresia of oocytes was noted in herring from the 1992 year class. This fish made up 31% of the total amount of the recruits and 33% of the Barents Sea component. According to visual estimation the ovaries in such specimens were at maturity stages III and IV. Resorption of oocytes of all the stages of trophoplasmatic growth, from vacuolization to a termination of vitellogenesis, was histologically established to occur in ovaries of the Barents Sea recruits. Fairly high percentage of such specimens from the 1991-1992 year classes indicates a possible unfavorable ecological situation in the area of their occurrence.

Introduction

Breach in the process of reproduction, in particular, of normal development of gonads and gametes, is observed both in freshwater and saltwater fish. Degeneration of the oocytes results in a considerable reduction in fecundity, disturbances in periodicity of spawning, and extension of time of maturation. Degeneration of maturing reproductive products in saltwater fish is, in most cases, considered to be a usual phenomenon observed under the circumstances being unfavorable for spawning and which result in delay of spawning and overmaturity of reproductive products. In such cases, the degeneration of gametes mainly occurs in preovulatory and ovulatory periods (Galkina 1959, Koshelev 1984, Sacun and Svirsky 1992). The investigations carried out during recent years (Kurita et al. 2000) have shown that the resorption of oocytes was noted in repeated spawners of Norwegian spring-spawning herring mainly during wintering in October-November. In connection with the above, the degeneration of vitellogenic oocytes observed in spawning areas in recruits of Norwegian spring-spawning herring, which grew in the Barents Sea until the beginning of maturation (the Barents Sea component) and with the development of gonads migrated to the Norwegian Sea, is of undoubted interest.

Materials and Methods

Ovaries of Norwegian spring-spawning herring were collected in the spawning areas on banks from late February to mid-March 1996 (62°30′-65°30′N) and during the first half of March in the same area in 1997 (Table 1).

Herring specimens were aged by otoliths and scales. Specimens grown in the different areas: southern, northern, and Barents Sea, were distinguished by scale structure (Ottestad 1934; Runnstrøm 1936; Seliverstova 1968, 1970, 1977, 1990). A 7-point scale of gonad maturity was used.

In total, gonads from 191 females were fixed in Bouin's fluid. Histological treatment of gonads was done according to standard methods; i.e., setting into alcohol of increasing concentration, paraffin-xylol, and then putting in paraffin. Sections of 6-7 µm were stained with iron haematoxylin according to Heidenhein (Roskin and Levinson 1957).

Results

The histological samples gathered on the spawning grounds along the coast of Norway during 1996-1997 (Table 1) comprised herring at age 3-14 years (Table 2) of 26-39 cm total lengths. Specimens from the 1991 and 1992 abundant year classes constituted the bulk of the spawning aggregations. Age composition of herring from the histological samples agreed with an age structure of its prespawning and spawning aggregations during those years.

Table 1. Sites of collecting samples of herring for histological analysis, February-March 1996 and March 1997.

Year	Date	Position		Numbers, indiv.
1996	20 Feb	62°35'N	4°51'E	13
	24 Feb	62°42'N	5°01'E	12
	25 Feb	63°11'N	5°44'E	24
	28 Feb	65°31'N	10°12'E	22
Total				71
	1 Mar	64°21'N	8°09'E	6
	2 Mar	63°53'N	7°05'E	10
	3 Mar	63°45'N	7°06'E	12
	4 Mar	63°31'N	5°54'E	15
	5 Mar	63°15'N	5°54'E	11
	14 Mar	63°56'N	5°59'E	14
Total				68
1997	1 Mar	62°43'N	4°20'E	13
	2 Mar	62°55'N	5°09'E	11
	6 Mar	65°13'N	9°48'E	11
	10 Mar	65°11'N	9°54'E	8
	13 Mar	65°19'N	9°47'E	9
Total				52

Habitat conditions of herring during the first years of life influence scale structure (number of coastal, oceanic, spawning rings), growth rate, and age at first spawning. The characteristics mentioned give opportunity to distinguish the southern, northern, and Barents Sea components in spawning aggregations of Norwegian herring, to determine a ratio of recruits and repeated spawners (Ottestad 1934; Runnstrøm 1936; Seliverstova 1968, 1970, 1977, 1990).

The histological samples comprised specimens from the three components mentioned. Among the repeated spawners and recruits (Table 2) the specimens from the Barents Sea component were predominant which was typical of herring abundant year classes (Seliverstova 1990). In 1996, the amount of repeat spawners made up 64%, and 36% for recruits; the latter were represented by the Barents Sea migrants (88%). In 1997, the amount of recruits constituted 50% and the Barents Sea component among the recruits was 92% (Table 2).

Table 2. Age composition and amount of females with resorption of reproductive products of Norwegian herring spawning stock from histological samples taken on spawning grounds in February-March 1996-1997.

Year	Age	Number of females examined Indiv.	%	Repeat spawners Southern	Northern	Barents Sea	Total	Recruits Northern	Barents Sea	Total
1996	13	20	14.4	2	2	16	20	—	—	—
	11	1	0.7	—	—	1	1	—	—	—
	8	8	5.8	4	4	—	8	—	—	—
	7	11	7.9	1	6	4	11	—	—	—
	6	34	24.4	2	2	30	34	—	—	—
	5	60	43.2	—	4	8	12	5/0[a]	43/8	48/8
	4	5	3.6	3	—	—	3	1/1	1/1	2/2
Total	Indiv.	139		12	18	59	89	6/1	44/9	50/10
	%		100	13.5	20.2	66.3	100	12.0/16.7	88.0/20.5	100.0/20.0
1997	14	3	5.8	—	—	3	3	—	—	—
	12	1	1.9	—	1	—	1	—	—	—
	10	1	1.9	—	—	1	1	—	—	—
	8	2	3.9	—	1	1	2	—	—	—
	7	3	5.8	—	—	3	3	—	—	—
	6	18	34.6	—	3	11	14	—	4/0	4/0
	5	23	44.2	—	1	1	2	1	20/8	21/8
	3	1	1.9	—	—	—	—	1	—	1/0
Total	Indiv.	52		—	6	20	26	2	24/8	26/8
	%		100	—	23.1	76.9	100.0	7.7	92.3/33.3	100.0/30.8

[a] Total amount of the females examined is before slash; those with resorption of ova is after slash.

In 1996, the amount of the female recruits increased over the period of observations from 11.4% to 52.9%. From 1 to 6 March 1997, the amount of recruits in spawning aggregations increased from 11.9% to 58.6%. Adults participated in spawning and left spawning grounds earlier than other fish; the spawning was completed by recruits at age 3-6 years.

Common trophoplasmatic growth of oocytes and simultaneous spawning are typical of Norwegian herring. Coefficient of maturity for repeated spawners varied from 12.4% to 25.3%.

Prespawning fish contained ova at terminal stages of trophoplasmatic growth (oocyte filled with yolk and ripe oocyte); i.e., they were at maturity stages IV and IV-V (Fig. 1). Definitive diameter of ova varied from 1.25 to 1.39 mm. One of the features that indicated a beginning of spawning status was an appearance of swollen half-transparent oocytes; further, the number of transparent eggs increased and there was a transition of ovary to maturity stage V.

In female recruits the ovaries were at maturity stages III and IV. Anomalies in ovaries of recruits were observed. Gonads were asymmetric and deformed, with some cavities on their surface. The ovaries were mainly filled with vitellogenic oocytes 0.65-1.24 mm in diameter. According to the histological analysis, degeneration of oocytes at different stages of trophoplasmatic growth occurred in the ovaries of the fish examined. Different formation of oocytes by the beginning of atresia, i.e., from vacuolization to the oocyte filled with yolk, was the basis of morphological differences in resorption process. Degenerative variations manifested themselves in a breach of the membrane structure, disappearance of the nucleus, mixing of vitelline granules, and the shrinkage of the oocyte's volume (Fig. 2). Resorption of ripe gametes resulted, as a rule, in the reduction of the coefficient of maturity which varied from 3.9% to 10% in female recruits.

In 1996, resorption of oocytes was mainly observed in the Barents Sea recruits from the 1991 year class. Of the total amount of recruits, specimens with resorption of oocytes made up 20%, and 20.5% of the Barents Sea component. In 1997, atresia was noted only in the herring from the 1992 year class. Of the total amount of recruits such fish constituted 30.8% and 33.3% of the Barents Sea component (Table 2).

Discussion

Mass maturation of specimens at age 5-6 years and predominance of the Barents Sea component (Seliverstova 1970, 1977, 1990) are typical of the abundant year classes of herring. In connection with the above, it is natural that specimens from the 1991-1992 abundant year classes were predominant in the age composition of herring from the 1996-1997 year classes (ICES 1996) comprising the Barents Sea component (Table 2).

In 1996, herring from the 1991 year class were observed over the entire area investigated (Table 1). Herring specimens were 26-32 cm in total length (average 30.0 cm). Similar lengths were observed in February and March: 30.0 and 29.9 cm, respectively. Recruits from the 1991 year class with atresia of oocytes, observed on Sklinna Bank (1 individual of 29 cm total length) on 28 February and on 1-3 March on the Halten and Froya banks, were the smallest (26-29 cm total length, average 28.4 cm).

In 1997, recruits from the 1992 year class were observed on Buagrunnen and Sklinna banks, however, with atresia of oocytes being observed only on Sklinna Bank on 6 and 13 March. Lengths of the recruits were 26-31 cm (average 29.1 cm), and those of recruits with atresia of oocytes, 28-29 cm (average 28.7 cm).

Occurrence of herring recruits with atresia of oocytes on the northern banks proves the conclusion that the Norwegian herring female recruits, especially those with a retarded growth rate, do not migrate far, but move to the spawning grounds situated very close to their feeding area (Marty 1956) and wintering grounds (Slott 1999). The largest ripe and fat specimens migrate from the wintering grounds to the southern spawning grounds (Slotte 1999).

The studies performed earlier have shown a mass degeneration of vitellogenic oocytes to result in a gap in spawning (Galkina 1959, Koshelev 1984, Sacun and Svirsky 1992) and reduction in fecundity (Sacun and Svirsky 1992, Witthames and Walker 1995). It may be a result of a forced shortening of spawning migration distance and cutting of the spawning grounds area (Chepurnova 1966). Degeneration of oocytes occurs also under deterioration of feeding conditions and wintering of fish and can be a consequence of abnormal physiological status of reproductive system as a result of anthropogenic disturbance in hydrographic and hydrochemical regimes of water basins (Chepurnova 1966, Soin et al. 1984, Shatunovsky et al. 1996). The experimental works also showed that the deterioration of feeding conditions increases intensity of the atresia in oocytes (Ma et al. 1998).

Presently, it is difficult to establish factors provoking degeneration of vitellogenic oocytes in the Barents Sea recruits. For this, observations are necessary to be continued of the status of reproductive system in recruits from wintering and spawning areas. The rather high proportion of recruits with a resorption of reproductive products observed in the spawning areas of Norwegian spring-spawning herring indicates, in our opinion, a possible unfavorable ecological situation in the areas of their occurrence.

Figure 1. Sections of ovaries from Norwegian spring-spawning herring: (A) maturity stage IV, oocytes in the phase of filling with yolk; (B) maturity stage IV-V, ripe non-ovulated oocyte. Magnification 10 × 8.

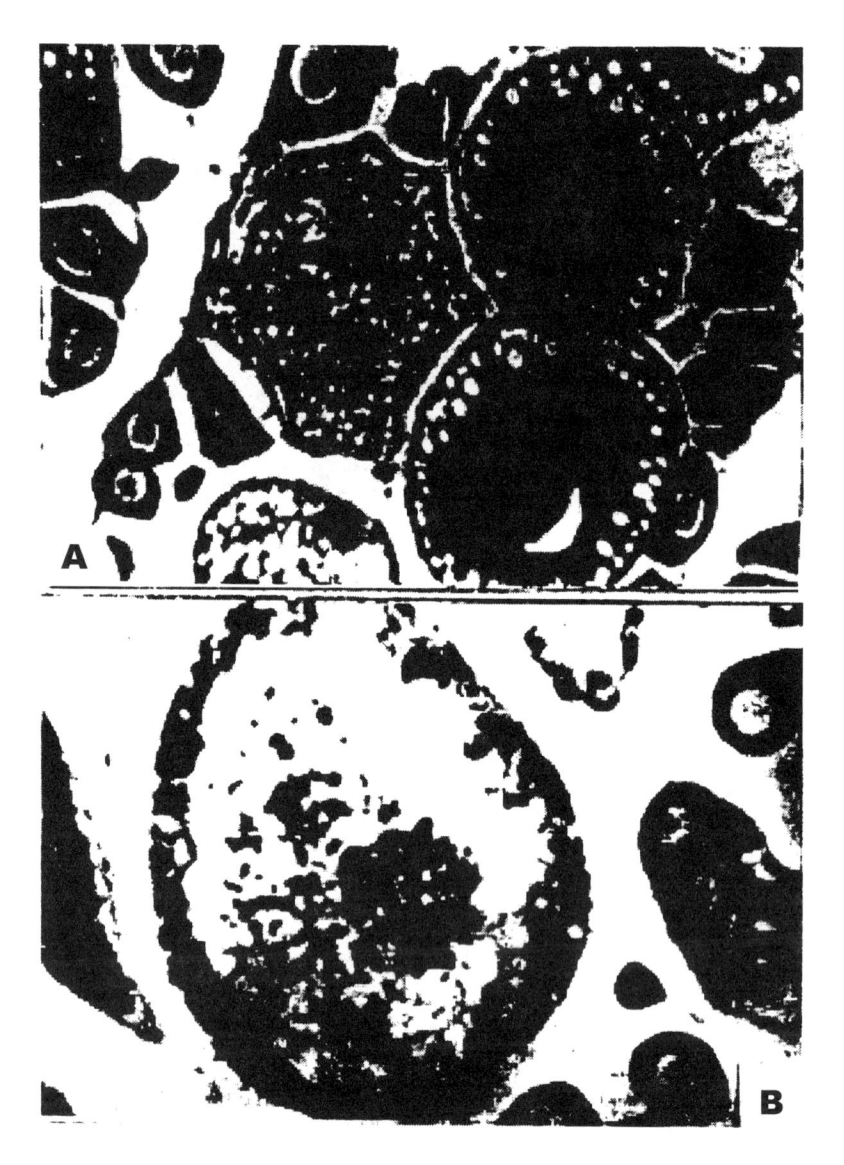

Figure 2. Degeneration of vitellogenic oocytes at different stages of trophoplasmatic growth in herring recruits at age 5: (A) maturity stage III, resorption of oocytes at the phase of vacuolization; total length 28.9 cm, Schmidt's length 26.3 cm. (B) maturity stage III-IV, degeneration of oocyte at the beginning of vitellogenesis; total length 29.0 cm, Schmidt's length 26.1 cm. (C) maturity stage IV, resorption of oocytes in phase of filling with yolk; total length 28.6 cm, Schmidt's length 25.8 cm. (D) total length 29.0 cm, Schmidt's length 26.2 cm. A and B, magnification 8×10; C and D, magnification 3.5×10.

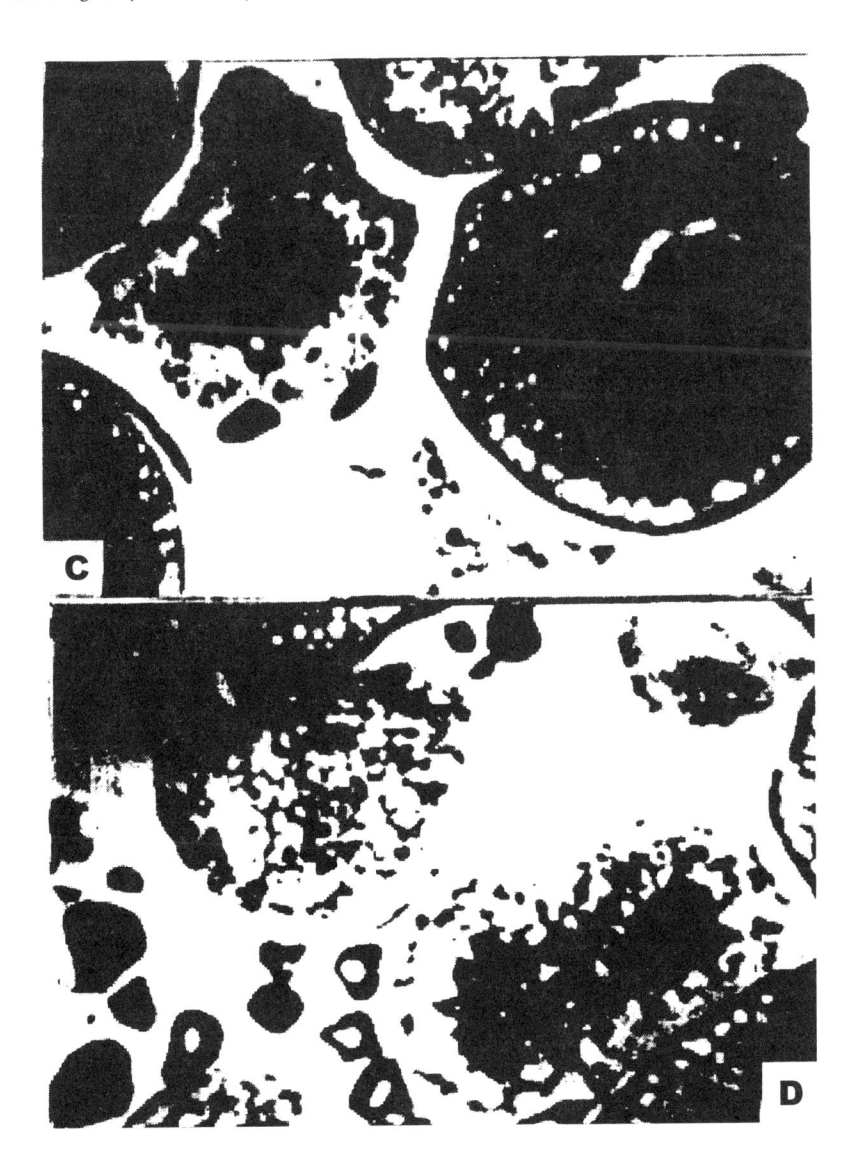

Figure 2. (Continued.)

References

Chepurnova, L.V. 1966. The breaks in ovogenesis of *Vimba vimba* natio *Carinata* (Pallas) due to the shortening of the route of spawning migration and to the diminished spawning area on the Dniester. Vopr. Ikhtiol. 6(1):51-58. (In Russian.)

Galkina, L.A. 1959. Degeneration phenomena in gametes of the Pacific herring. Papers of the USSR Academy of Sciences 126(2):404-405. (In Russian.)

ICES. 1996. Preliminary report of the international 0-group fish survey in the Barents Sea and adjacent waters in August-September 1996. ICES C.M. 1996/G:31. 37 pp.

Koshelev, B.V. 1984. Ecology of fish reproduction. Nauka Press, Moscow. 307 pp. (In Russian.)

Kurita, Y., A. Thoresen, M. Fonn, A. Svardal, and O.S. Kjesbu. 2000. Oocyte growth and fecundity regulation of Atlantic herring (*Clupea harengus*) in relation to declining body reserves during overwintering. In: Proceedings of the Sixth International Symposium on the Reproductive Physiology of Fish. Bergen, pp. 85-87.

Ma, Y., O.S. Kjesbu, and T. Jørgensen. 1998. Effects of ration on the maturation and fecundity in captive Atlantic herring (*Clupea harengus*). Can. J. Fish. Aquat. Sci. 55:900-908.

Marty, Yu.Yu. 1956. Main stages of the life-cycle of the Atlanto-Scandian herring. Tr. PINRO 9:5-61. (In Russian.)

Ottestad, P. 1934. Statistical analysis of the Norwegian herring population. Rapp. P.-V. Reun. Cons. Int. Explor. Mer 88:2-44.

Roskin, G.I., and L.V. Levinson. 1957. Microscopic instruments. Sovetskaya nauka, Moscow. 467 pp. (In Russian.)

Runnstrøm, S. 1936. A study on the life history and migration of the Norwegian spring herring based on the analysis of the winter rings and summer zones of the scale. Fiskeridir. Skr. 5(2). 103 pp.

Sacun, O.F., and V.G. Svirsky. 1992. Oocyte degeneration in the periods of perivitellogenesis and vitellogenesis during reproductive cycle in the Pacific sardine, *Sardinops sagax melanostica*. Vopr. Ikhtiol. 32(3):52-58. (In Russian.)

Seliverstova, E.I. 1968. The problem of the determination of abundance of the Barents Sea population of the 1950 year-class of the Atlanto-Scandian herring stock in the Norwegian Sea (according to data for 1954-1958). ICES C.M. 1968/H:12. 16 pp.

Seliverstova, E.I. 1970. Comparative characteristics of the Atlanto-Scandian herring of the 1950 and 1959 year-classes (ratio of types of growth: a rate of sexual maturity). ICES C.M. 1970/H:21. 10 pp.

Seliverstova, E.I. 1977. The relationship between the growth and maturity rates of the Atlanto-Scandian herring and the structure of their year-classes. ICES C.M. 1977/H:8. 39 pp.

Seliverstova, E.I. 1990. Recommendation on rational exploitation of the Atlanto-Scandian herring stock. PINRO, Murmansk. 83 pp. (In Russian.)

Shatunovsky, M.I., N.V. Akimova, and G.I. Ruban. 1996. Response of fish reproductive system on the anthropogenous influence Vopr. Ikhtiol. 36(2):229-238. (In Russian.)

Slotte, A. 1999. Effects of fish length and condition on spawning migration in Norwegian spring-spawning herring (*Clupea harengus* L.). Sarsia 84:111-127.

Soin, S.G., O.P. Danil'chenko, and A.S. Khandal'. 1984. Coaction of decreasing salinity and triethyl stannic chloride on the development of marine form of threespine stickleback *Gasterosteus aculeatus* L. (Gasterosteidae). Vopr. Ikhtiol. 24(3):481-489. (In Russian.)

Witthames, P.R., and M.G. Walker. 1995. Determinacy of fecundity and oocyte atresia in sole (*Solea solea*) from the Channel, the North Sea and the Irish Sea. Aquat. Living Resour. 8(1):91-109.

The Relevance of a Former Important Spawning Area in the Present Life History and Management of Norwegian Spring-Spawning Herring

Ingolf Røttingen and Aril Slotte
Institute of Marine Research, Bergen, Norway

Abstract

The main spawning and fishing areas for the Norwegian spring-spawning herring during the 19th century and the first half of the 20th century were located south of 60°N. From the late 1940s changes in the migration patterns of the herring led to a northward shift in the location of spawning activities, and by the beginning of the 1960s the spawning occurred only in areas located north of 62°N. Further, in the late 1960s this stock was depleted due to a large increase in fishing effort on all life stages of the herring. At the start of the rebuilding period the spawning occurred north of 62°N. However, in 1989, after a 30-year absence, the herring began spawning again in the historically important spawning areas south of 60°N. This paper discusses some underlying principles of this reappearance, describes the development of the spawning in these areas since 1989, and also evaluates the rationale behind the present regulation measures for this spawning habitat.

Spawning Distribution and Spawning Stock Abundance in Former Periods

Spawning of Norwegian spring-spawning herring has traditionally taken place along the Norwegian coast from Lofoten in the north to Siragrunnen in the south (Fig. 1), but the relative importance of the different spawning grounds has changed with time. Runnstrøm (1941a) regarded the grounds south of Bergen, particularly the ones to the west and south of the island of Karmøy, as the most significant in the 1930s and also in former periods

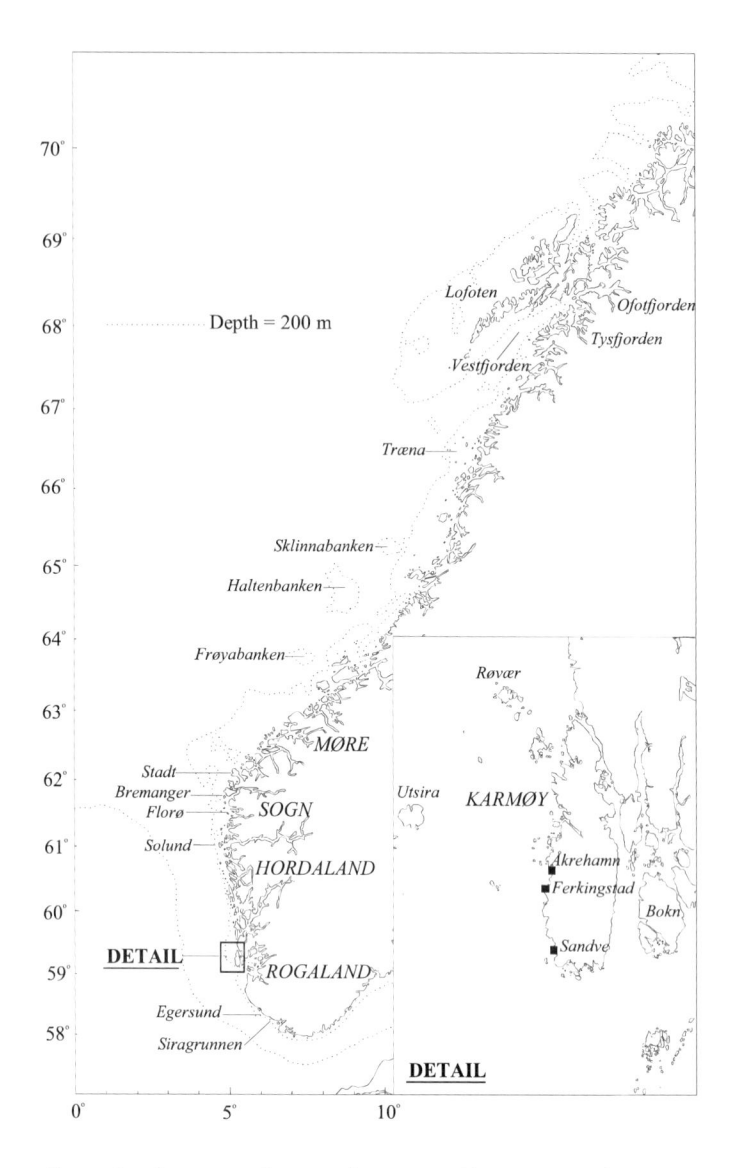

Figure 1. Important herring districts and locations on the Norwegian coast.

of rich herring fisheries. Between 70% and 80% of the landings during the first three decades of the 20th century were from these spawning areas. (Runnstrøm 1941b; Røttingen 1990a). In the present paper these areas are designated "southern spawning areas." Runnstrøm (1941b) mapped details of spawning areas off Karmøy and farther south along the coast by surveying the areas by use of echo sounder (one of the first echo surveys done on herring, Cushing 1973), and by grab to record the presence of herring roe (Fig. 2A and C). According to the catch statistics from the fishery the spawning areas off Møre were also used regularly during this period, whereas catches on spawning grounds north of Møre were sporadic, indicating that these spawning areas seemed to be used only a few seasons.

From the 1940s on several important structural changes regarding the spawning of the herring took place (Devold 1963). First, the herring appeared later and later (as judged from the first catches of pre-spawning herring) on the spawning grounds. Second, the spawning moved gradually more northward in the last years of the period 1946-1958. After 1959 the southern spawning areas were not utilized, whereas significant numbers of yolk sac larvae north of the main fishing areas at Møre indicated that banks north of Møre and even off the Lofoten islands were important spawning areas in the early 1960s (Dragesund 1970).

Information on abundance of herring in earlier "herring periods" can be inferred from reports of the fishery. The annual catch taken in southern spawning areas in the 1860s was on the order of 150,000 metric tons. At the time this was a considerable catch, taken by simple fishing gears available, such as land seine and small gillnets. This indicates that considerable amounts of spawning herring must have been present in these areas. Recent VPA analyses indicate very high spawning stock levels (12-15 million t) in the 1930s (Toresen and Østvedt 2000). In the 1950s, when the northward shift of the spawning areas took place, the estimates indicate a decreasing spawning stock from approximately 14 million t in 1950 to 6 million t in 1960 (ICES 2001). In the 1960s new technology, increasing fishing pressure on all life stages of the herring, and absence of regulation measures led to a depletion of the stock (Dragesund et al. 1980). The long-range migrations of the herring ceased, and the entire life cycle was spent in Norwegian coastal waters and fjords. The spawning of the reduced stock took place in the Møre area (Dragesund et al. 1980, Hamre 1990, Røttingen 1990a). After a long rebuilding period the spawning stock increased from 0.5 million t to several million t by 1988 with the recruitment of the 1983 year class (ICES 2001). This year class established a new migration pattern for the stock (Røttingen 1992). The main elements of the migration of the adult herring were spawning at the traditional areas in Møre, feeding in the Norwegian Sea, and wintering in the Vestfjorden area in northern Norway.

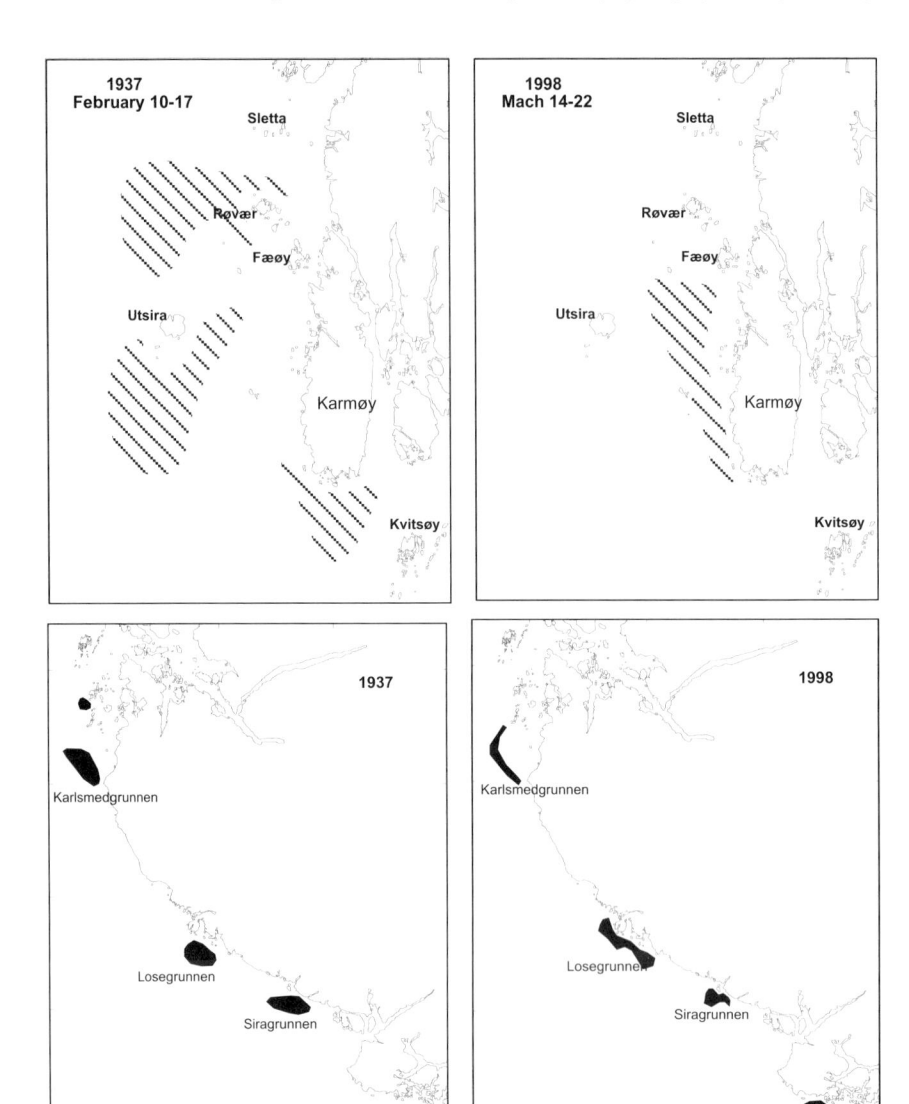

Figure 2. Distribution of spawning herring in 1937 (left) and 1998 (right). The upper figure for 1937 (modified from Runnstrøm 1941b and Cushing 1973) is based on acoustic surveys; the lower (modified from Runnstrøm 1941b) is based on sampling of herring roe. The figures for 1998 are based on acoustic surveys.

Present Management and Regulation in the Southern Area

In the spawning season of 1989 a new element was added to the migration pattern: Norwegian spring-spawning herring reappeared on the traditional spawning areas south of Bergen, after an absence of 30 years (Røttingen 1989). This was approximately 180 nautical miles south the southern border for the spawning in 1988.

By 1989 the stock was still regarded to be in a rebuilding state, and the primary objective for the management of the stock at that time was to obtain a further increase in the spawning stock. The fishery on the stock was regulated by a low fishing mortality (less than 5%). The total catch was divided into vessel quotas that could be taken anywhere the fisherman preferred. However, when the herring appeared in the southern spawning area, the question was asked if these areas should be given protection from the general fishery. Even though these were traditionally important spawning areas, the following scientific arguments were put forward in order to obtain a special regulation status for the southern areas:

1. A large geographical extension of the spawning area should be advantageous to the recruitment of the stock. A number of studies have emphasized the importance of dispersal in populations under high spatial and temporal environmental variability (Gadgil 1970, Roff 1975, Kuno 1981, Levin et al. 1984, Levin and Cohen 1991, Cohen 1993). Thus, by spreading the spawning products from north to south, the Norwegian spring-spawning herring should overcome the variations in environmental conditions and increase the overall recruitment to the stock (Dragesund 1970).

2. Another benefit of extending the spawning area could be a decreased density-dependent mortality of eggs and larvae. Recruitment has been related to stock density in herring by several authors (Anthony and Fogarty 1985, Stocker et al. 1985, Winters et al. 1986, Winters and Wheeler 1987). Development and survival of eggs are inversely correlated with egg density in herring (Taylor 1971, Galkina 1971, Hourston and Rosenthal 1981), and reduced density may also reduce competition for food among larvae (Kiørboe et al. 1988).

3. There are several indications that larvae hatched at the southern grounds have a higher chance of surviving until the age of recruitment. Slotte and Fiksen (2000) concluded from modeling that the southern larvae would drift northward with the coastal current in higher temperatures than larvae from more northern grounds, which ultimately would enhance the survival.

4. Larvae hatched at the southern grounds would be dispersed over a larger area, and a larger range of environments, than larvae hatched

farther north. For instance, larvae hatched off Karmøy will be spread northward with the coastal current to nursery areas (shore areas and fjords) along the entire coast and in the Barents Sea, whereas larvae hatched off Lofoten only have fjords in northern Norway and the Barents Sea as nursery areas.

5. There is an increasing probability that larvae from southern spawning will drift into coastal areas and fjords (Dragesund 1970), i.e., larvae will stay at higher temperatures through the juvenile stage. Warmer water would probably also increase the survival from metamorphosis until the stage of first maturation. In fact it has been suggested that temperature related effects are more important in the late larval–early juvenile phase than in the early larval period, due to the high predation rate in these stages (Sissenwine et al. 1984, Anthony and Fogarty 1985). There are in fact several studies showing that herring originating from coastal nursery areas grow much faster than individuals in the Barents Sea, and as a result they recruit to the spawning stock 1-2 years earlier (Lea 1929a,b; Ottestad 1934; Runnstrøm 1936; Holst 1996). The really large year classes are produced in the Barents Sea, while the nursery areas along the coast function as a buffer; i.e., individuals originating from these areas predominate in years with low recruitment (Holst and Slotte 1998).

Altogether these arguments indicated that if a large amount of herring were allowed to spawn in the southern area, the result could be a positive development of the stock. The management authorities responded to these arguments by giving the southern spawning areas special protection. In the national herring fishery regulation a maximum catch not exceeding 5% of the Norwegian TAC (total allowable catch) (this maximum catch level corresponds to less the 2% of the total TAC) is enforced in the southern areas. Further, only small coastal vessels are allowed to participate in this fishery.

The Southern Spawning Area Since 1989

The fraction of total spawning stock of Norwegian spring-spawning herring migrating to the southern grounds has been small. Table 1 gives estimates of herring biomass recorded by acoustic surveys in parts of the southern spawning areas since 1989. Although the precision of the estimates is low, the biomass estimates indicate that the fraction of the stock spawning south of 61°N is very small compared with the total spawning stock.

The recruitment of the 1983 year class was followed by a number of weak year classes, but strong year classes reappeared in 1991 and 1992. Herring were observed spawning in the southern area in these years, but judging from larval distributions (Fossum 1996) it is doubtful if any significant part of these strong year classes originates from spawning prod-

Table 1. Estimates of biomass (B, in 100,000 t) of Norwegian spring-spawning herring in the southern areas in 1990-2000, compared with the total spawning stock biomass (SSB).

	1990	1991	1992	1993	1994	1995
B	0.32	0.12	0.20	0.16	0.20	0.11
SSB	4.17	4.37	4.24	4.08	4.62	5.73

	1996	1997	1998	1999	2000	2001
B	0.17	2.15	0.80	0.76	1.10	0.10
SSB	7.67	9.18	8.42	7.77	6.72	6.11

Data on SSB (in million metric tons) is taken from VPA-runs (ICES 2001).

ucts produced in the southern areas. The recruitment of these strong year classes led to an increase in the spawning stock to approximately 9 million t, thus making the spawning stock comparable in size to the spawning stock in the 1950s, the last time the herring visited the southern spawning areas.

Thus, one may conclude that the southern spawning fraction has played a minor role in the recruitment and growth of Norwegian spring-spawning herring stock in the period after the reappearance, 1989-1999. The explanation to why such a small fraction has chosen to visit the southern spawning grounds after 1989 could be related to the present migration pattern.

Stock Identification and the Migration Pattern after 1989

The herring that reappeared on the spawning area were classified as Norwegian spring-spawning herring on the basis of vertebrae counts (>57 vertebrae) and scale markings (Røttingen 1989, Johanessen et al. 1995). The vertebrae counts were in agreement with the counts made in the same area in the 1930s by Runnstrøm (1941a). Further, the age distribution (almost entirely 1983 year class) of the herring that appeared in the southern areas was in exact correspondence with the age distribution in the spawning stock of the Norwegian spring spawners.

This stock identification was in accordance with other qualitative observations. The migrating herring had been detected by acoustic surveys and development of the fishery along the Norwegian coast on the spawning migration from the start in the Vestfjorden area (Røttingen 1989). Further, Karmøy is one of the larger Norwegian fishing communities, and it has a long tradition of herring fishing. Although there had been minor concentrations of herring in the area (mostly North Sea autumn spawners) since 1959, the herring, which appeared in 1989, were immediately identified as the "correct" type of herring. The appearance in 1989 was seen in connection with the former "herring periods." This type of herring had disappeared from the area in the 1760s, reappeared around 1810, disappeared in 1870, reappeared in the 1890s, and then spawned annually until 1959 when it again disappeared.

Later, tagging experiments supported this stock identification. Since 1990 more than 105,000 herring have been tagged on the southern spawning areas. Figure 3 shows the geographical distribution of the recoveries from these tagging experiments. These tag recovery data show that the herring spawning in the southern areas are found in the feeding area in the Norwegian Sea and in Vestfjorden, geographical areas that form parts of the present migration pattern of Norwegian spring-spawning herring (Røttingen 1992).

What was the mechanism behind this appearance on the southern area? The entire spawning of Norwegian spring-spawning herring in 1983 occurred north of 62°N (Røttingen 1990b). Thus the appearance of the 1983 year class in the southern area in 1989 was not a result of natal homing. Further, only very limited concentrations of local herring (North Sea autumn spawner type) were recorded in the southern area in 1983, and a rapid expansion in 1989 of a local population is therefore very unlikely. In this context the herring that reappeared on the southern spawning area was a discrete spawning stock unit (Norwegian spring-spawning herring) that expanded its spawning areas. The expansion included areas that had been used in former periods, and may have been related to the spawning stock increase from less than a half million t in 1986 to over 3 million t in 1989.

This is regarded as an example of recolonization of a former important spawning area, but it is important to note that the migration route to these southern areas is different from the last time the herring occupied these areas. The spawning migration routes are shown in Fig. 4. The pre-1970 spawning migration commenced in the open sea east of Iceland; at present it has started from Vestfjorden in northern Norway. Another element of the different spawning migration routes may be noted. The recorded spawning areas for this stock are along the Norwegian coast, and with the present situation the herring that migrate to the southern areas on their spawning migration route pass over areas where other herring of the same stock will spawn. This was not the case before 1970, where the major part of the spawning migration took part over deepwater areas. Further, the entire spawning migration is today carried out against the main coastal current system.

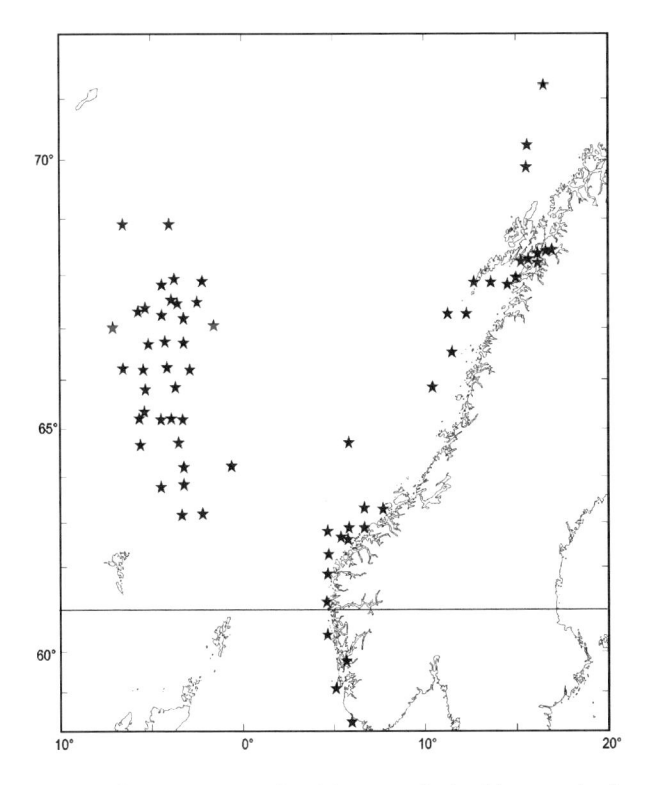

Figure 3. Recaptures (positions marked with a star) of herring tagged in the coastal areas of Norway south of 61°N in the period 1990-1998. (61°N is marked with a line.)

Repeat Homing to the Southern Areas?

As explained above the reappearance of the Norwegian spring-spawning herring in the southern area in 1989 was not an example of natal homing. But have these herring demonstrated repeat homing to the southern spawning area every year since reoccupation? Several authors argue that herring return year after year to the same spawning grounds as they spawned for the first time, regardless whether they were born there (Hourston 1982, Wheeler and Winters 1984). In 1999 an experimental fishery for recovery of tags was arranged south of 61°N (Fig. 5). The recoveries showed that all the recovered tagged herring (12 tags from an overall catch of 371 t) were tagged south of 61°N, indicating that the herring return to these southern spawning areas. On the other hand, these data do not disprove that a proportion of the herring tagged south of 61°N in fact have spawned farther north in subsequent years. Many tags from the experiments at the

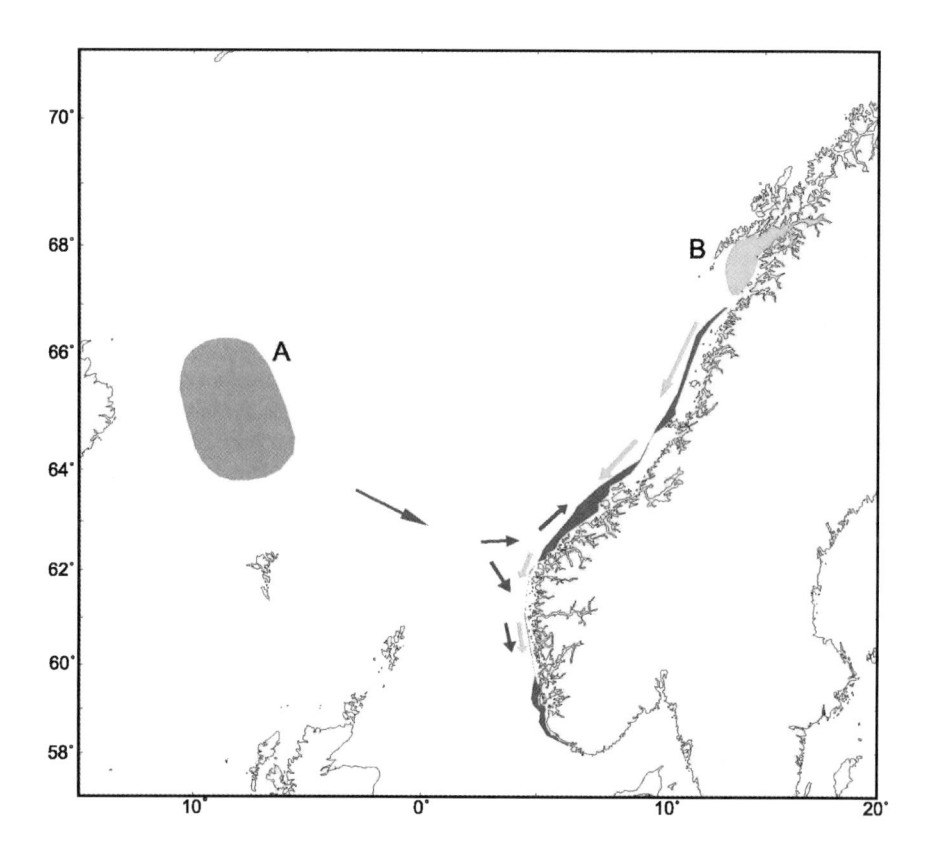

Figure 4. Spawning migration (dark arrows) and wintering area (A) before the stock collapse of the late 1960s. Present spawning migration (light arrows) and wintering areas (B). Spawning areas are indicated.

southern grounds have also been recovered at the spawning areas north of 61°N (Fig. 3). However, since most of these tags were found in maturing and spent herring, it is difficult to conclude with certainty if they would spawn or had spawned in this area, or if they in fact were on their way to or from more southern grounds. Only the few fish in a spawning stage can for certain prove such straying. An additional indication of return migration to the spawning areas along the Norwegian coast can be found in a stable age structure (relative distribution of the 1991 and 1992 year classes) of the spawning herring in the last years (Slotte and Dommasnes 1998, 1999, 2000). In the northern areas the 1991 year class dominates; in the southern areas 1992 is the most frequent year class in the spawning stock.

The above data give some evidence of the persistence of a local spawning population in the southern area. On the other hand some work, particularly

Herring: Expectations for a New Millennium 307

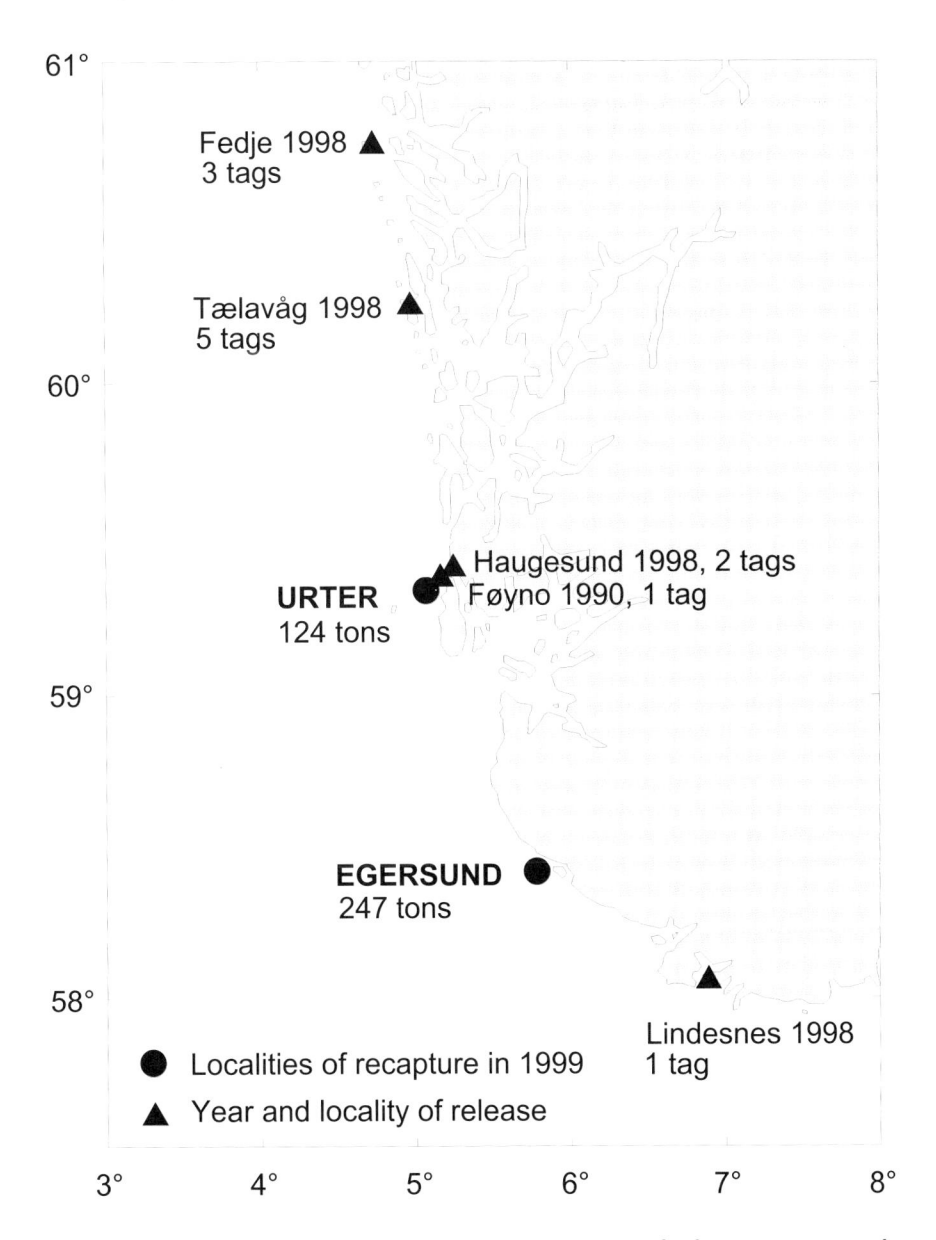

Figure 5. Release localities (triangles) and year of release for herring recaptured during an experimental fishery in March 1999. Recapture localities are marked with circle and size of catch stated.

within the perspective of energy loss during spawning migration, suggests that the same herring may not return to the southern spawning ground each year; they could choose not to migrate to the southernmost grounds due to migration constraints. Some data indicate that herring at the southern grounds comprise "elite" herring, being larger and/or in better condition than the average individuals of the stock. The herring migration to spawn at Egersund/Siragrunnen (Fig. 1) undertake a migration distance up to 1,500 km longer that the herring spawning off Lofoten. Data on energy loss show that the non-feeding herring migrating southward from the wintering area have 3-4 times higher relative weekly energy loss than during the wintering period, and the relative energy loss decreases with fish size (Slotte 1999a). Correspondingly, the herring body length and condition tend to increase with the migration distance southward, and also within year classes (Slotte and Dommasnes 1998,1999, Slotte 1999b). In this regard it has been demonstrated by modeling that only large individuals and/or individuals in very good condition may have advantage of increased fitness by migration to the more "profitable" southernmost grounds (Slotte and Fiksen 2000). This model further indicates that herring returning to the southern grounds regardless of the condition will not succeed. They will gain higher lifetime fitness by spawning farther north, and instead head southward in years with good condition.

Recently the acoustic estimates of Norwegian spring-spawning herring at the spawning grounds has been reviewed (Slotte 2001, this volume). This review agrees with McQuinn's theory (1997) that recruiting herring may learn the migration routes to their spawning grounds from repeat spawners, and that they have a tendency to return to these areas in subsequent years. McQuinn suggests that herring will have a tendency to stray in unstable populations, i.e., following a population collapse or a recruitment boom. Contrasting to this theory Slotte suggests that the herring also will have a tendency to stray in stable populations, a tendency which is related to environmental preferences and migration potential, i.e., the need to maximize lifetime fitness. Such straying is apparent from the acoustic estimates, which demonstrate that year classes have a tendency to spawn farther south as they grow to be larger/older, and a tendency to spawn farther north in years with poor conditions. Thus, the low spawning fraction at the southernmost grounds could possibly be due to not visiting the southern grounds before the year classes were fully recruited at the age of 5-6 years. This became clear in 1997, when the highest biomass estimates at the southern grounds were estimated. It was likely due to the increase in the total spawning biomass from approximately 6 to 9 million t (ICES 2001), which could induce density dependent mortality of progeny, and individuals with great migration potential chose to visit grounds at a farther distance. It should be mentioned here that in the year 2001 hardly any herring visited the southern grounds (stock estimates on the southern grounds were similar to 1990-1996), and at the same time the SSB (spawning stock biomass) was as its lowest since 1997.

Conclusions

Due to the tendency of herring populations to home to particular grounds during the spawning season, whereas they may occur in mixed aggregations with other populations during the rest of the year, it has been proposed that herring should be managed within each spawning area (Doubleday 1985, Sinclair et al. 1985), as in Pacific herring along the coast of Canada and the United States (Trumble and Humphreys 1985). Although the Norwegian spring-spawning herring occurs at a wide variety of spawning grounds along the Norwegian coast, such a management strategy was not considered in this stock until the return of herring at the southernmost grounds in 1989. The present management measures on the southern grounds are in force until 2004. Should they be prolonged? The main argument for introducing the protections in the 1990s was that these spawning areas would be a positive element in the development of the stock. That has not been fulfilled; the recruiting year classes to the Norwegian spring-spawning herring in the last years have originated from other spawning areas. This is mostly due to the fact that the biomass at the southern grounds never reached the amount hoped for. After a maximum level of 200,000 t in 1997 it decreased to 10,000 t again in 2001 despite the protection. It seems that the migration and homing pattern of Norwegian spring-spawning herring is quite unique and influenced to a larger degree by straying than in other stocks. One simply cannot be sure that a protection of herring at these grounds secures spawning in subsequent years. A real change in the abundance of spawning herring in the southern area can probably only take place if there is change in the overall migration pattern of this migratory stock. Such a change is linked to complex biological and environmental changes and not to protective measures on the southern spawning area.

One may therefore conclude that the expected increase in biomass at the southern grounds and a resulting increase in the overall recruitment have failed, but it may not be correct to say that the protection itself has been a failure. This management strategy has secured a certain spawning stock in these areas, which otherwise could have been fished down by an effective purse seine fleet. In addition, there has been an enrichment of the ecosystem after the herring returned, which has led to an increase in other fisheries. One may say that such a protection is in the spirit of the international agreement about species enrichment maintenance, which in fact could be good enough reason to prolong the protection.

References

Anthony, V.C., and M.J. Fogarty. 1985. Environmental effects on recruitment, growth, and vulnerability of Atlantic herring (*Clupea harengus harengus*) in the Gulf of Maine region. Can. J. Fish. Aquat. Sci. 42(Suppl. 1):158-173.

Cohen, D. 1993. Fitness in random environments. In: Y. Yoshimura and C.W. Clark (eds.), Adaptation in stochastic environments. Lecture Notes in Biomathematics 98. Springer-Verlag, Berlin, pp. 8-25.

Cushing, D. 1973. The detection of fish. Pergamon Press Ltd., Oxford. 200 pp.

Devold, F. 1963. The life history of the Atlanto-Scandian herring. Rapp. P.V. Reun. Cons. Int. Explor. Mer 154:98-108.

Dragesund, O. 1970. Factors influencing year-class strength of Norwegian spring spawning herring (*Clupea harengus* L.). Fiskeridir. Skr. Ser. Havunders. 15:381-450.

Dragesund, O., J. Hamre, and Ø. Ulltang. 1980. Biology and population dynamics of the Norwegian spring spawning herring. Rapp. P.V. Reun. Cons. Int. Explor. Mer 177:43-71.

Doubleday, W.G. 1985. Managing herring fisheries under uncertainty. Can. J. Fish. Aquat. Sci. 42(Suppl. 1):245-257.

Fossum, P. 1996. A study of the first-feeding herring (*Clupea harengus* L.) larvae during the period 1985-1993. ICES J. Mar. Sci. 49:51-59.

Gadgil, M. 1970. Dispersal: Population consequences and evolution. Ecology 52:253-261.

Galkina, L.A. 1971. Survival of spawn of Pacific herring (*Clupea harengus pallasi*) related to the spawning stock. Rapp. P.V. Reun. Cons. Int. Explor. Mer 160:30-33.

Hamre, J. 1990. Life history and exploitation of the Norwegian spring-spawning herring. In: T. Monstad (ed.), Proceedings of the 4th Soviet-Norwegian Symposium, 12-16 June 1989. Institute of Marine Research, Bergen, Norway, pp. 5-39.

Holst, J.C. 1996. Long term trends in the growth and recruitment pattern of the Norwegian spring-spawning herring (*Clupea harengus* Linnaeus 1758). Ph.D. thesis, University of Bergen, Bergen, Norway. ISBN 82-77444-032-4.

Holst, J.C., and A. Slotte. 1998. Effects of juvenile nursery on geographic spawning distribution in Norwegian spring spawning herring (*Clupea harengus* L.). ICES J. Mar. Sci. 55:987-996.

Hourston, A.S. 1982. Homing by Canada's west coast herring to management units and divisions as indicated by tag recoveries. Can. J. Fish. Aquat. Sci. 39:1414-1422.

Hourston, A.S., and H. Rosental. 1981. Data summaries for viable hatch from Pacific eggs deposited at different intensities on a variety of substrates. Can. Data. Rep. Fish. Aquat. Sci. 267. 56 pp.

ICES. 2001. Report of the Northern Pelagic and Blue Whiting Fisheries Working Group. ICES C.M. 2001/ACFM:17.

Johannessen, A., A. Slotte, O.A. Bergstad, O.Dragesund, and I. Røttingen. 1995. Reappearance of Norwegian spring spawning herring (*Clupea harengus* L.) at spawning grounds off southwestern Norway. In: H.R. Skjoldal, C. Hopkins, K.E. Erikstad, and H.P. Leinaas (eds.), Ecology of fjords and coastal waters. Elsevier B.V., pp. 347-363.

Kiørboe, T., P. Munk, K. Richardson, V. Christensen, and H. Paulsen. 1988. Plankton dynamics and larval herring growth, drift and survival in a frontal area. Mar. Ecol. Prog. Ser. 44:205-219.

Kuno, E. 1981. Dispersal and the persistence of populations in unstable habitats: A theoretical note. Oecologia 49:123-126.

Lea, E. 1929a. The herring scale as a certificate of origin, its applicability to race investigations. Rapp. P.V. Reun. Cons. Int. Explor. Mer 54:1-228.

Lea, E. 1929b. The oceanic stage in the life history of Norwegian herring. J. Cons. Cons. Int. Expl. Mer 4:3-42.

Levin, S.A., and D. Cohen. 1991. Dispersal in patchy environments: The effects of temporal and spatial structure. Theor. Popul. Biol. 39:63-99.

Levin, S.A., D. Cohen, and A. Hastings. 1984. Dispersal strategies in patchy environments. Theor. Popul. Biol. 26:165-191.

McQuinn, I.H. 1997. Metapopulations and the Atlantic herring. Rev. Fish Biol. Fish. 7:297-329.

Ottestad, P. 1934. Statistical analyses of the Norwegian herring population. Rapp. P.V. Reun. Cons. Int. Explor. Mer 88(3):1-45.

Roff, D.A. 1975. Population stability and the evolution of dispersal in a heterogeneous environment. Oecologia 19:217-223.

Runnstrøm, S. 1936. A study on the life history and migrations of the Norwegian spring herring based on analysis of the winter rings and summer zones of the scale. Fiskeridir. Skr. Ser. Havunders. 5(2):1-110.

Runnstrøm, S. 1941a. Racial analysis of the herring in Norwegian waters. Fiskeridir. Skr. Ser. Havunders. 6(7):5-10.

Runnstrøm, S. 1941b. Quantitative investigations on herring spawning and its yearly fluctuations at the west coast of Norway. Fiskeridir. Skr. Ser. Havunders. 6(8):5-71.

Røttingen, I. 1989. Reappearance of Norwegian spring spawning herring on spawning grounds south of 60°N. ICES C.M. 1989/H:22.

Røttingen, I. 1990a. A review of the variability in the distribution and abundance of Norwegian spring spawning herring and capelin. Polar Research 8:33-42.

Røttingen, I. 1990b. The 1983 year class of Norwegian spring spawning herring as juveniles and recruit spawners. In: T. Monstad (ed.), Proceedings of the 4th Soviet-Norwegian Symposium, 12-16 June 1989. Institute of Marine Research, Bergen, Norway, pp. 165-203.

Røttingen, I. 1992. Recent migration routes of Norwegian spring spawning herring. ICES 1992/H:18.

Sinclair, M., V.C. Anthony, T.D. Iles, and R.N. O'Boyle. 1985. Stock assessment problems in Atlantic herring (*Clupea harengus*) in the northwest Atlantic. Can. J. Fish. Aquat. Sci. 42:888-898.

Sissenwine, M.P., E.B. Cohen, and M.D. Grosslein. 1984. Structure of the Georges Bank ecosystem. Rapp. P.V. Reun. Cons. Int. Explor. Mer 183:243-254.

Slotte, A. 1996. Relations between seasonal migrations and fat content in Norwegian spring spawning herring (*Clupea harengus* L.). ICES C.M. 1996/H:11.

Slotte, A. 1999a. Differential utilisation of energy during winter and spawning migration in Norwegian spring spawning herring (*Clupea harengus* L.). J. Fish. Biol. 54:338-355.

Slotte, A. 1999b. Efffects of fish length and condition on spawning migration in Norwegian spring spawning herring (*Clupea harengus* L.). Sarsia 84:111-127.

Slotte, A. 2001. Factors influencing location and time of spawning in Norwegian spring spawning herring: An evaluation of different hypotheses. In: F. Funk, J. Blackburn, D. Hay, A.J. Paul, R. Stephenson, R. Toresen, and D. Witherell (eds.), Herring: Expectations for a new millennium. University of Alaska Sea Grant, AK-SG-01-04, Fairbanks. (This volume.)

Slotte, A., and A. Dommasnes. 1998. Distribution and abundance of Norwegian spring spawning herring during the spawning season in 1998. Fisken og Havet 5. 10 pp.

Slotte, A., and A. Dommasnes. 1999. Distribution and abundance of Norwegian spring spawning herring during the spawning season in 1999. Fisken og Havet 12. 27 pp.

Slotte, A., and A. Dommasnes. 2000. Distribution and abundance of Norwegian spring spawning herring during the spawning season in 2000. Fisken og Havet 10. 18 pp.

Slotte, A., and A. Johannessen. 1997a. Spawning of Norwegian spring spawning herring (*Clupea harengus* L.) related to geographical location and population structure. ICES C.M. 1997/cc:17.

Slotte, A., and A. Johannessen. 1997b. Exploitation of Norwegian spring spawning herring (*Clupea harengus* L.). before and after the stock decline: Towards a size selective fishery. In: D.A. Hancock, D.C. Smith, A. Grant, and J.P Beumer (eds.), Developing and sustaining world fisheries resources: The state of science and management. CISRO Publishing, Collingwood, Australia, pp. 103-109.

Slotte, A., and Ø. Fiksen. 2000. State dependent spawning migration of Norwegian spring spawning herring (*Clupea harengus* L). J. Fish. Biol. 56:138-162.

Stocker, M., V. Haist, and D. Fournier. 1985. Environmental variation and recruitment of Pacific herring (*Clupea harengus pallasi*) in the Strait of Georgia. Can. J. Fish. Aquat. Sci. 42(Suppl. 1):174-180.

Taylor, F.H.C. 1971. Variation in hatching success in Pacific herring eggs with water depth, temperature, salinity and egg mass thickness. Rapp. P.V. Reun. Cons. Int. Explor. Mer 160:34-41.

Toresen, R., and O.J. Østvedt. 2000. Variation in abundance of Norwegian spring spawning herring (*Clupea harengus*) throughout the 20th century and the influence of climatic fluctuations. Fish and Fisheries 1:231-256.

Trumble, R.J., and R.D. Humphreys. 1985. Management of Pacific herring (*Clupea harengus pallasi*) in the eastern Pacific Ocean. Can. J. Fish. Aquat. Sci. 42 (Suppl. 1):230-244.

Wheeler, J.P., and G.H. Winters. 1984. Homing of Atlantic herring (*Clupea harengus harengus*) in Newfoundland waters as indicated by tagging data. Can. J. Fish. Aquat. Sci. 41:108-117.

Winters, G.H., and J.P. Wheeler. 1987. Recruitment dynamics of spring-spawning herring in the northwest Atlantic. Can. J. Fish. Aquat. Sci. 44:882-900.

Winters, G.H., J.P. Wheeler, and E.L. Dalley. 1986. Survival of a herring stock subjected to a catastrophic event and fluctuating environmental conditions. J. Cons. Int. Explor. Mer 43:26-42.

Spawning Stock Fluctuations and Recruitment Variability Related to Temperature for Selected Herring (*Clupea harengus*) Stocks in the North Atlantic

Reidar Toresen
Institute of Marine Research, Bergen, Norway

Abstract

Fish stocks vary in abundance, and assessments show that exploitation has a large effect on stock size fluctuations. The effect of the physical environment on stock fluctuations is not well understood. In this paper, the main purpose is to shed light on the relationship between the temperature in surrounding water-masses and the fluctuations in spawning stock biomass and recruitment to herring stocks in the North Atlantic. A feature that may indicate the influence of external factors is the coherence or common pattern in the fluctuations in the abundance of herring stocks on both sides of the Atlantic Ocean. Time series of normalized spawning stock size were compared between five herring stocks in the North Atlantic Ocean. Common patterns in stock biomass were found between several of the stocks. A significant relationship was found between recruitment and temperature for Norwegian spring spawners only. North Sea herring, Icelandic summer spawners, and herring in the Bay of Fundy also had a pattern of enhanced recruitment at higher temperatures. Although it is found that herring stocks inhabiting warmer waters have a higher stock-recruitment relationship than the stocks inhabiting colder regions, it is believed that the temperature is a proxy for other environmental factors influencing the survival of larvae and hence the level of recruitment.

Introduction

As for other fish stocks, the adaptations of herring to various ecosystems may be traced in the biological estimates of the stocks. The most important of these, for fisheries assessment and management, are recruitment, growth, and mortality, which determine the stock's productivity. Productivity, stock size, and their fluctuations, vary considerably both interspecifically and intraspecifically between herring stocks (ICES 1999a,b; DFO 1995). Assessments and investigations show that for some herring stocks, the fisheries have played an important role in stock size fluctuations (Dragesund and Ulltang 1978; DFO 1995; ICES 1999a,b). During the 1980s and 1990s quite a lot of research was done in the field of climate change and fish populations, and there has been an increased effort to obtain and quantify relations between environmental variation and recruitment, growth, distribution, and migration of fish in the northern Atlantic regions: the Barents Sea (Gjøsæter and Loeng 1987, Loeng 1989, Ottersen et al. 1994, 1998; Sundby 1994, 2000; Loeng et al. 1995; Ottersen and Sundby 1995; Michalsen et al. 1998; Sundby 2000; Toresen and Østvedt 2000), the North Sea (Corten 1986, 1999; Nichols and Brander 1989; Maravelias and Reid 1995), the seas surrounding Iceland (Grainger 1979, Stefánsson and Jakobsson 1989, Astthorsson and Gislason 1994, Malmberg and Blindheim 1994) and eastern Canadian waters (Anthony and Fogarty 1985, deYoung and Rose 1993, Myers et al. 1993, Campana et al. 1995). Schweigert (1995) showed that Pacific herring recruitment is affected by changes in the physical environment. It is now accepted that stock size fluctuations are often affected by environmental factors. A greater understanding of how ocean climate affects the reproduction in herring stocks can be reached by studying the relationship between ocean climate indices and recruitment for several stocks. In this paper, time series of temperature in four regions in the North Atlantic are compared with characteristics of five herring stocks. The regions are: Gulf of Maine, Icelandic shelf, Norwegian Sea/Barents Sea, and the North Sea.

The intention of the present paper is to shed light on the following questions:

- Is there coherence between the time series of spawning stock biomass or recruitment between herring stocks in the North Atlantic?

- Is there any relationship between spawning stock development and temperature in these stocks?

- Is there a relationship between recruitment and temperature?

Materials and Methods

The herring stocks selected for the study are presented in Table 1. These stocks were chosen for several reasons: they have been important for fisheries for many years, and therefore biological data for these stocks have been regularly collected for decades; they represent a wide range of herring stock sizes and variability; and they are linked to a common environmental feature, the trans-Atlantic current system (Fig. 1). The stocks spawn

Table 1. The herring stocks selected for the study, and the source of the data.

Stock	Data source
North Sea herring	ICES 1999a
Norwegian spring-spawning herring	ICES 1999b, Jakobsson et al. 1993
Icelandic summer-spawning herring	ICES 1999b
Georges Bank herring	Anthony and Waring 1980, NEFSC 1996
Herring in the Bay of Fundy	DFO 1995

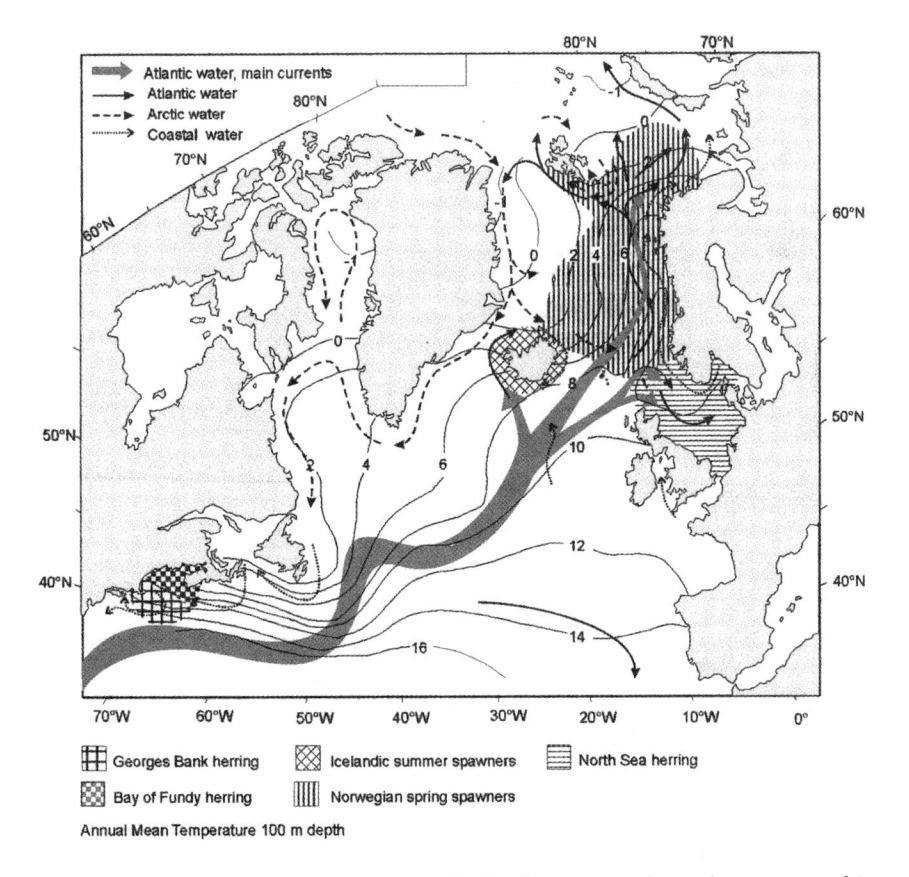

Figure 1. The North Atlantic Ocean with five herring stocks and oceanographic features.

Table 2. The source of the temperature data in the different regions.

Area	Station/section	Source
North Sea	9-11/Torungen Hirtshals	IMR, Bergen, Norway
Barents Sea	All/Kola	IMR, Bergen, Norway
Iceland	SI3	MRI, Reykjavik, Iceland
Georges Bank	Prince 5	FOC, St. Andrews Biol. St.
Bay of Fundy	Prince 5	FOC, St. Andrews Biol. St.

IMR = Institute of Marine Research, MRI = Marine Research Institute, FOC = Fisheries and Oceans Canada

in separate areas, and there is no mixing between them in any other stages. The stocks are not found to be genetically distinct, but isoenzyme analysis of tissues from adult Georges Bank herring revealed differences from other stocks in the Gulf of Maine (Stephenson and Kornfield 1990).

The biology of the different stocks is well described (Runnstrøm 1936; Devold 1963; Parrish and Saville 1965; Jakobsson 1973, 1980; Dragesund and Ulltang 1978; Anthony and Waring 1980; Iles and Sinclair 1982; Sinclair and Iles 1985; Grosslein 1987; Dragesund et al. 1997).

Assessment Data

The spawning stock biomass (SSB) and recruitment estimates were obtained from the assessments of the different stocks. Assessments for these stocks are determined either by working groups in the International Council for the Exploration of the Seas (ICES) and endorsed by the Advisory Committee for Fishery Management (ACFM) (North Sea herring, Norwegian spring spawners, and Icelandic summer spawners), or by working groups in the stock assessment systems in the United States and Canada (DFO 1995; NEFSC 1996; ICES 1999a,b).

Annual spawning stock biomass estimates for the five stocks are in Table 3. For presentation purposes, the annual SSB estimates are normalized to 1. The mean SSB for all stocks is set to 1 and the curve shows the annual stock estimates relative to this value.

Recruitment

Recruitment success (r/SSB), which is defined as the number of recruits per unit of SSB, indicate the year-to-year variability of how many recruits the different stocks have produced per unit of SSB. For exploratory purposes, the time series of r/SSB is shown in Fig. 2b.

The recruitment is here defined as the number of 1-ringers (winter annuli from otoliths) for autumn spawners and 2-ringers for spring spawners, as estimated as the number of fish at the start of the year in the assessment for the different stocks. In order to make the numbers comparable for

Table 3. Spawning stock biomass (SSB, thousand metric tons) and number of recruits (millions) for North Sea herring, Norwegian spring spawners, Icelandic summer spawners, Georges Bank herring, and herring in the Bay of Fundy.

	North sea SSB	North sea N recr.	Norw. spr. SSB	Norw. spr. N recr.	Icelandic SSB	Icelandic N recr.	Georges B. SSB	Georges B. N recr.	B. of Fundy SSB	B. of Fundy N recr.
1950			13,984	248,920	120	324				
1951			12,440	37,783	88	197				
1952			11,482	19,247	79	167				
1953			10,613	17,294	134	191				
1954			9,445	4,831	124	469				
1955			10,223	3,112	127	791				
1956			11,740	2,808	166	369				
1957			10,129	2,273	140	555				
1958			9,280	2,391	197	713				
1959			7,350	114,851	272	531				
1960			5,817	39,813	161	525				
1961			4,229	17,159	255	467				
1962	1,192	17,270	3,465	1,701	222	586				
1963	2,264	22,810	2,635	51,527	222	507				
1964	2,091	12,750	2,795	24,909	167	100				
1965	1,499	10,030	3,067	780	128	392			353	6,159
1966	1,310	14,440	2,595	2,370	80	178			473	1,298
1967	932	13,750	1,145	490	79	46	811	2,090	539	1,773
1968	417	7,880	219	139	24	34	559	1,413	545	2,321
1969	426	14,590	78	2,388	16	70	411	7,717	492	7,524
1970	376	11,500	31	101	21	90	365	1,184	459	1,149
1971	267	7,240	8	10	14	418	281	1,009	350	2,370
1972	289	3,570	2	139	12	132	114	1,663	311	1,646
1973	234	7,430	74	4,419	29	199	162	1,025	367	251
1974	163	940	85	2,935	46	554	227	1,290	538	732
1975	84	900	91	1,010	117	436	124	3,596	475	4,194
1976	81	1,470	146	3,487	129	196	65	2,764	368	1,380
1977	52	1,650	284	1,737	133	248	41	410	305	481
1978	71	3,610	354	2,131	176	254	25	2,357	213	1,647
1979	114	5,460	685	4,240	198	881	28	1,587	199	1,757
1980	139	8,660	468	533	213	238	24	1,602	345	2,580
1981	204	17,200	503	376	186	220	24	1,417	335	4,717
1982	288	15,330	501	797	193	503	21	3,423	272	5,592
1983	446	15,770	571	132,098	220	1,253	37	1,756	288	2,329
1984	721	27,370	594	5,526	233	703	73	2,483	345	1,393
1985	754	33,630	492	34,354	251	345	134	5,129	500	1,867
1986	771	26,920	414	1,874	264	495	186	7,160	750	2,351
1987	887	13,910	1,031	5,396	373	390	245	10,141	912	3,270
1988	1,144	13,340	3,268	12,898	444	1,004	278	14,164	826	3,185
1989	1,277	12,150	4,150	32,475	412	1,353	337	11,947	718	1,213
1990	1,169	11,600	4,848	59,234	371	738	479	7,047	704	3,317
1991	980	18,680	5,119	161,215	320	850	796	23,013	653	3,774
1992	716	15,940	5,016	204,160	381	327	1,204	29,845	694	1,222
1993	462	10,730	4,868	30,807	523	436	1,853		683	1,000
1994	513	13,200	5,604	15,223	532	1,221	2,352		653	
1995	501	17,020	5,948	8,198	509	471				
1996	488	18,440	6,652	26,996	415	973				

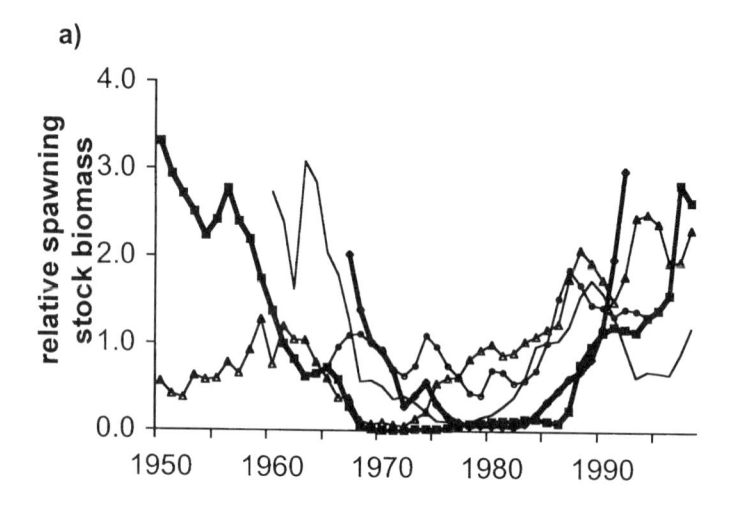

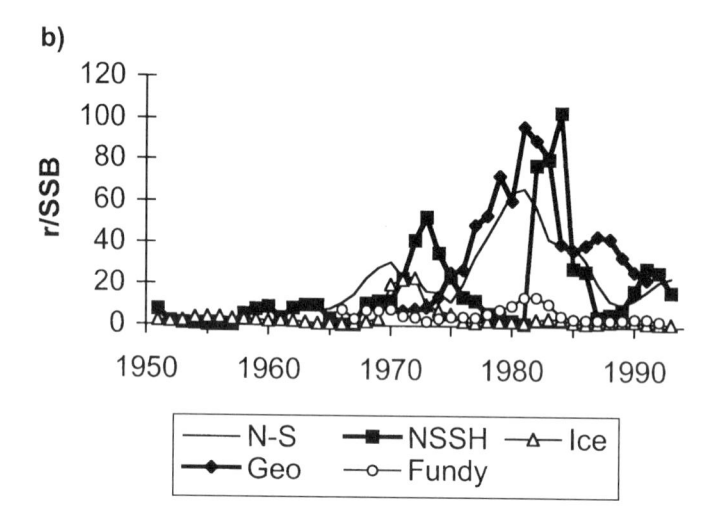

Figure 2. (a) Relative spawning stock biomass of five herring stocks, (normalized to 1). (b) Recruitment per unit of SSB (3-year moving average) for five herring stocks. N-S = North Sea herring, NSSH = Norwegian spring spawners, Ice = Icelandic summer spawners, Geo = Georges Bank herring, Fundy = herring in the Bay of Fundy.

spring and autumn spawners, the values were normalized to 7 months after metamorphosis. The autumn spawners in the North Sea spawn (on average) in October and metamorphose in May the following year. The first winter annulus is formed the following winter, and the fish are called 1-ringers on 1 January, 7 months after metamorphosis. Spring spawners, on the other hand, spawn in February and metamorphose in the summer (July) in the same year as they are hatched. The first annulus is formed the first winter. The comparable age (7 months after metamorphosis) for spring spawners will then be February in the year the otoliths have one winter ring. The number of spring spawners as 2-ringers are therefore adjusted positively for a natural mortality of 0.075 per month for 10 months to make the two groups "equal" in age.

Numbers of estimated recruits and SSB levels for the different stocks in a range of years are in Table 3. The SSB refers to estimates at the time of spawning the actual year, while the estimated recruits are the normalized estimated number of recruits produced in these years.

Coherence of Stock Development

Coherence in spawning stock biomass between fish stocks may indicate influence by external climatic factors. The spawning stock biomass was correlated between the five herring stocks applying a simple linear regression analysis for exploratory purposes.

Oceanographic Data

The oceanographic data used to examine the correlation between SSB, recruitment, and temperature in the different regions is given in Table 4.

The choice of temperature series for the different stocks is based on knowledge of the distribution of the youngest stages (larvae) of the stocks. In most cases the applied temperature time series will also be representative for the general hydrographic regime in each region.

For the autumn spawners, a time series of temperatures in the spring following the year they hatched was used. For North Sea herring, May-June temperatures in 20 m depth were used. For the Icelandic summer spawners the measured temperature at 50 m in June was used, while for the herring stocks in the Gulf of Maine region, the mean temperature in February at 25 m depth was used for both stocks. The herring distribute in shallow waters in the early phases of life, and the time series of temperatures with the most continuous data sets in the upper water column in the different areas were used. For Norwegian spring spawners, which was the only spring spawning stock in the investigation, the mean temperature at the Kola section, from 0 to 200 m in January, the same year as they were spawned, was used. The depths at the Kola section were chosen because there is a consistent time series of data from these depths, and it is believed that the average temperature at these depths are representative for the hydrographical regime in the waters where the larvae and young herring live.

Table 4. Temperatures (°C) used in correlation with spawning stock biomass and recruitment.

Year	Tor. 9-11	Kola	SI3	Prince 5	Year	Tor. 9-11	Kola	SI3	Prince 5
1950		4.6	6.1		1973	6.9	4.2	5.5	2.9
1951		4.6	6.7		1974	8.2	3.7	5.8	2.0
1952		4.0	3.2		1975	5.2	4.0	3.5	2.7
1953		3.6	3.0		1976	7.8	3.9	5.6	2.6
1954		4.9	6.2		1977	8.6	3.8	3.6	2.4
1955		4.3	5.4		1978	5.3	3.5	4.7	1.4
1956		3.5	5.2		1979	8.3	2.7	2.2	2.4
1957		3.7	6.6		1980	10.3	3.4	5.9	1.3
1958		3.7	4.7		1981	9.5	3.3	4.1	4.0
1959		3.9	5.2		1982	8.2	2.9	4.0	2.4
1960		4.3	6.4		1983	6.3	4.7	3.4	3.1
1961		3.9	6.5		1984	7.4	3.7	4.2	2.6
1962	8.0	4.2	5.3		1985	7.9	3.7	5.5	0.6
1963	7.7	3.6	2.6		1986	8.1	3.5	3.7	2.7
1964	7.6	3.5	5.6	1.1	1987	8.8	3.6	5.2	2.0
1965	7.2	4.4	4.0	1.2	1988	9.4	3.6	5.1	1.5
1966	10.0	3.0	2.9	3.1	1989	9.1	3.1	4.1	1.9
1967	9.3	2.8	0.7	0.6	1990	9.5	4.2	5.1	2.9
1968	6.7	3.7	1.9	3.0	1991	9.1	4.2	4.8	1.8
1969	8.3	3.5	1.7	3.1	1992	7.4	4.4	4.6	1.4
1970	7.1	4.3	1.4	0.9	1993	9.5	4.3	5.1	1.6
1971	7.6	3.8	3.4	1.4	1994	6.6	3.6	4.8	3.3
1972	6.4	3.7	5.1	1.8	1995	10.7	4.0		2.3

Tor.9-11 = The Torungen-Hirtshals (ref. IMR, Bergen, Norway), Kola = the kola section in the southern Barents Sea (IMR, Bergen, Norway), SI3 = hydrographic station at north coast of Iceland (MRI, Iceland), Prince 5 = hydrographical station in the Gulf of Maine (FOC, St. Andrews, Canada).

Results
Coherence in Long-Term Stock Size Fluctuations
The time series of the estimated spawning stock biomass (fluctuations around the average value normalized to 1) are shown in Fig. 2a. There is an overall common pattern of stock development through the last 40 years for the stocks in study. All stocks declined in the late 1960s and early 1970s. By 1970, the Norwegian spring-spawning herring and the Icelandic summer spawners had both reached a very low level of abundance. The Norwegian spring spawners collapsed and were at these low levels until the late 1980s, after which they recovered and increased in abundance. The Icelandic summer-spawners recovered sooner and started to recover again as early as the mid-1970s. The North Sea herring and the Georges Bank herring reached a low level later than the Norwegian spring spawners and Icelandic summer-spawners, by the late 1970s. The North Sea herring soon recovered and increased in abundance during the 1980s, while the Georges Bank herring stayed at a low level until about 1984. It then recovered and increased in abundance throughout the 1980s. The stocks all recovered to a higher level by the 1990s, but the North Sea herring had a small decline in the years 1994-1997.

For exploratory purposes, linear regressions between the stock trajectories of the different stocks were carried out. All stocks are significantly correlated except the relation between the Icelandic summer spawners and the two stocks, North Sea herring and the Norwegian spring spawners. Also, the Georges Bank herring did not correlate significantly with North Sea herring.

Recruitment Success—Coherence and Magnitudes
The time series of the recruitment success (r/SSB) is shown in Fig. 2b. Comparing Figs. 2a and 2b, there is a general trend of low r/SSB in the period when the stocks are in a decreasing SSB-trend, while the r/SSB for most stocks increases in the late phase of recession and in the early years of increase.

The Georges Bank herring has the largest mean number of recruits per unit of SSB, estimated at 33.3 individuals per kg SSB (C.V.= 1.04), while the herring in the Bay of Fundy has a mean of 6.0 per kg SSB (C.V.=0.84). The Icelandic summer spawners have the lowest number, estimated to a mere 3.5 (C.V.=1.3), while the Norwegian spring spawners also have a fairly low mean value of number of recruits per unit of SSB (kg), estimated to 16.2 (C.V.=2.2). The herring stock in the North Sea has a corresponding estimated number of 27.0 (C.V.=0.66).

Correlation between Time Series of SSB, Recruitment and Temperature

SSB and Temperature

The time series of spawning stock biomass and temperature for three stocks, with significant positive correlations (North Sea herring, Norwegian spring spawners, and Icelandic summer spawners) are shown in Fig. 3 a-c. The temperature data were smoothed by a 5-year moving average in order to follow the trends in temperature.

For herring on Georges Bank and in the Bay of Fundy, the correlation between the SSB and temperature was very weak.

Recruitment and Temperature

The time series of the estimated recruitment were correlated with the time series of temperatures in the respective areas. The correlations with significance levels are given in Table 5. There was a significant and positive correlation between recruitment and temperature for the Norwegian spring-spawning herring. The other stocks correlate poorly between temperature and recruitment. However, studying the recruitment-temperature plots of North Sea herring and Icelandic summer spawners (Figs. 4b and c), there is a pattern of higher recruitment at certain temperatures. For herring on Georges Bank and in the Bay of Fundy, no such pattern was found.

Discussion

Coherence in Spawning Stock Development

Natural external factors working on fish stocks are either biotic or abiotic, and in addition anthropogenic effects influence the stocks, of which the fisheries probably are the most important. There was a change in the way the herring fisheries were carried out in Norway and in Iceland in the late 1950s and early 1960s, which probably affected the fisheries quite a lot (Dragesund and Ulltang 1978, Jakobsson 1980, Bakken and Dragesund 1971). This change was due to the introduction of echo sounders and sonars as search tools onboard the fishing vessels. In addition the introduction of the power-block and fishing nets made of nylon on the purse seiners made the hauling of larger nets easier (Bakken and Dragesund 1971). These developments in gear and techniques made the vessels much more efficient, especially through the 1960s. The exploitation of the Norwegian spring spawners and Icelandic summer spawners increased drastically during this decade (Dragesund et al. 1997, Jakobsson 1980). After the collapse of these stocks by the end of the 1960s, the Norwegian herring fleet needed new resources of herring to exploit, and Norwegian exploitation of the North Sea herring developed during the late 1960s (ICES 1976). However, even though the fisheries may explain much of the development in the herring stocks in the northeast Atlantic in the 1950s and 1960s, scientists have advocated the importance of natural factors, and

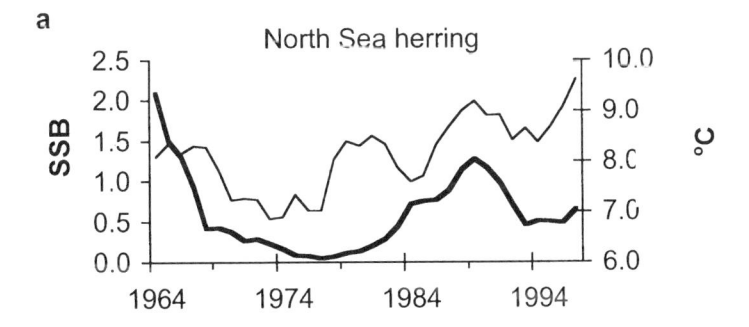

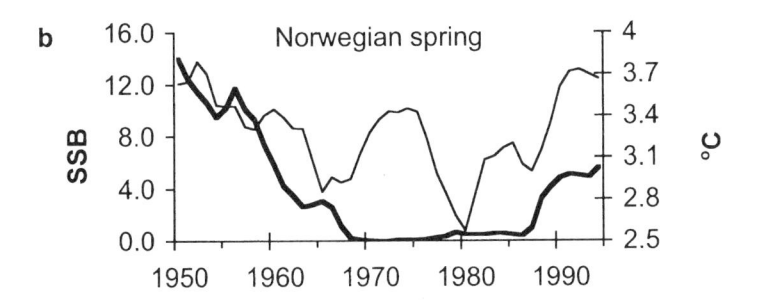

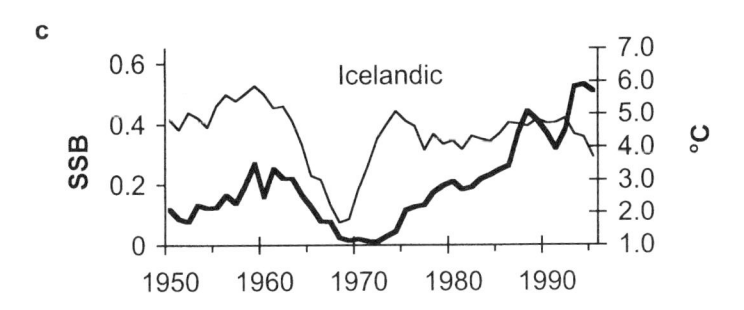

Figure 3. Time series of spawning stock biomass (thousand metric tons) and temperatures for (a) North Sea herring, (b) Norwegian spring-spawners, and (c) Icelandic summer spawners. Five-year moving average. Solid line = spawning stock, thin line = temperature.

Table 5. Correlation coefficients (r) in linear regressions between temperature (°C) and recruitment.

NS	NSSH	IC	GEO	BF
0.23	**0.43**	0.16	0.19	0.22

NS = North Sea herring, NSSH = Norwegian spring-spawning herring, IC = Icelandic summer-spawning herring, GEO = Georges Bank herring, and BF = herring in the Bay of Fundy. Bold denotes significance, $P<0.01$.

that there seemed to be a period of poor recruitment associated with the decline of the stocks (Bailey and Steele 1972; Dragesund and Ulltang 1972; Corten 1986, 1990, 1999; Jakobsson et al. 1993).

The two stocks living in the Gulf of Maine region also had a drop in abundance, and for the herring stock on Georges Bank the spawning stock biomass was estimated at a very low level in the period 1978-1987. The observed drop in abundance for the herring in the Bay of Fundy was not as drastic as that observed in the other stocks. Compared to the earlier period of the time series, the herring in the Bay of Fundy had a decreasing tendency in stock abundance through the years 1966 to 1980, after which it increased in abundance again as is observed for the other stocks. The minimum level coincided with the low level period for the Georges Bank herring.

Anthony and Waring (1980) describe how the herring fisheries developed in the region through the 1960s and 1970s. The authors point out several possible reasons for the Georges Bank herring decline, and conclude that the fisheries definitely had a large effect. It was, however, pointed out that recruitment during the 1970s was generally poor, which probably made the stock more vulnerable to exploitation.

The collapse of four of the stocks coincided with the huge change in the way the fisheries were carried out. However, the long time-series of climatic development through the twentieth century shows that there was a lower temperature period of the large-scale temperature fluctuations in the northeast Atlantic (Blindheim et al. 1981, Hansen-Bauer and Nordli 1998) during the late 1960s and 1970s. This cold period may have made the herring stocks less productive and made them more vulnerable to fishing (Jakobsson et al. 1993). Toresen and Østvedt (2000) show that the Norwegian spring-spawning herring was in a declining phase, through the late 1940s and the 1950s, long before the fishery really made an impact on the stock in the mid-1960s. It is therefore likely that there are both natural and anthropogenic effects on the herring stocks. It was probably not a coincidence that the development in the herring fishery took place

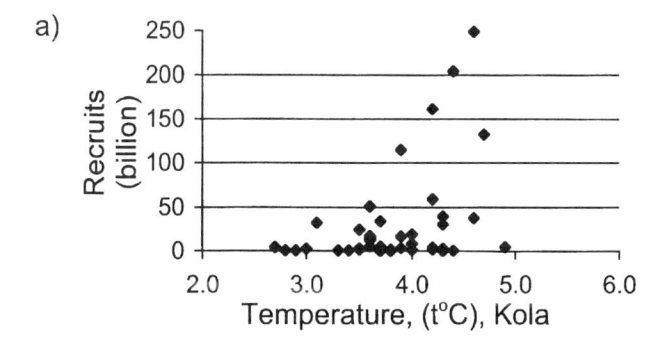

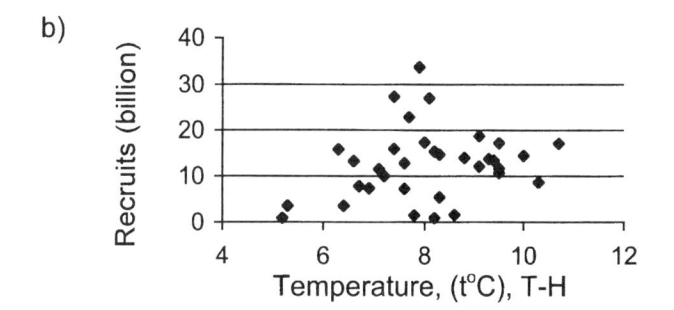

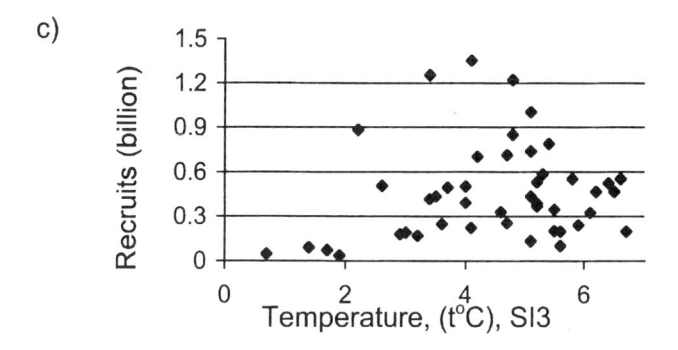

Figure 4. Recruits (estimated from stock assessments) versus temperature (°C) for (a) Norwegian spring spawners versus °C at Kola (Bochkov 1982), (b) North Sea herring versus t°C at the Torungen-Hirtshals section (IMR, Bergen, Norway), and (c) Icelandic summer spawners versus °C at SI3 (MRI, Reykjavik, Iceland).

at this time. Better fishing techniques are often developed when availability of fish is declining, and declining stock levels, originally caused by natural effects, may therefore in some cases be harvested harder than they should.

The Temperature Time Series

In the use of temperature time series for correlation with stock estimates, it is assumed that these indices represent the development in the climatic environment where the herring live. This may not be the case. There are many more data series on ocean climate in these regions and more work should be done in analyzing the relationship between ocean climate and fish stock estimates. For the North Sea herring and the herring stocks in the Gulf of Maine region, the mean temperatures at 20 m and 25 m were used. This is rather shallow, and the temperatures may have been influenced by local weather conditions. In such a depth, it is reasonable to assume that water temperature is influenced by atmospheric temperature conditions. However, temperatures at the same locations in other depths were also tested and were found to have weaker correlations than the chosen time series. The temperature used for correlation with the Norwegian spring spawners (the Kola section) was a calculated mean from surface to 200 meters and is known to correlate well to other fish stocks in the area (Sætersdal and Loeng 1987, Ottersen 1996). The temperature index off Iceland was chosen because it had the longest continuous data series and is situated in the middle of the distribution area of juvenile herring. The data series from the Gulf of Maine (Prince 5, Table 2) is situated at the entrance of the Bay of Fundy and it is assumed that the temperature at this location is representative for the ocean climate in the Gulf of Maine region.

SSB and Recruitment Versus Temperature

In the Norwegian and Barents seas, good year classes frequently occur simultaneously for several species, such as for herring, cod, and haddock (Ottersen and Loeng 2000, Toresen et al. 1999). Dragesund (1971) and Sætersdal and Loeng (1987) also found several cases of favorable recruitment coinciding for cod, haddock, and herring stocks in the Barents Sea. This is a clear indication of the presence of an environmental signal in the recruitment mechanisms. These features were related to increased inflow of Atlantic water masses to the Norwegian coast and to higher temperatures. Significant relations between temperature, growth, and year-class strength at the 0-group stage have been shown for the above-mentioned species (Ottersen and Loeng 2000).

The correlation between recruitment and temperature data used in this investigation indicate that temperature data gives a signal of recruitment for Norwegian spring-spawning herring, North Sea herring, and Icelandic summer-spawners. However, a simple linear regression analysis shows that the relationship is only significant for Norwegian spring spawners. In a complex environment, it is likely that many factors influence recruitment. Earlier investigations have shown that there is never a 100% correlation between recruitment and temperature (Ellertsen et al. 1987, Sætersdal and Loeng 1987, Ellertsen et al. 1989, Sundby 1994, Ottersen 1996, Helle and Pennington 1999), even with very long time series available (Toresen and Østvedt 2000), and the temperature is probably not directly linked to the recruitment level.

In the correlation between recruitment and temperature there is often a pattern of low probability of good recruitment at low temperatures and higher probability of good recruitment at higher temperatures. This feature has been shown for northeast arctic cod (Sundby 1994), and it is also present here, for the Norwegian spring-spawning herring (Fig. 4a). This means that for this herring stock, there is a lower probability of good recruitment when low local temperatures prevail, and a higher probability for good recruitment in a regime with higher temperatures. For the North Sea herring (Fig. 4b), there seems to be a higher probability of good recruitment around 8°C, with lower probability of good recruitment at both higher and lower temperatures. This feature also seems to apply for the herring in the Bay of Fundy (Fig. 4c), which has a higher probability of good recruitment at about 4°C.

Potential Mechanisms

Optimal temperatures do not ensure good recruitment. Poor recruitment also occurs at high temperatures, which show that the environmental factors, which are reflected in the temperature, are necessary but not sufficient conditions for good recruitment. The mechanism of the influence of the temperature on recruitment is not yet fully understood. Sætre (IMR) and others (unpubl.) showed that the year-class strength of Norwegian spring-spawning herring is determined during the first six months of life. Fiksen and Folkvord (1999) showed that the temperature has a direct effect on the growth of herring larvae if food is readily available. However, the large variation in recruitment level at high temperatures suggests other mechanisms than the direct influence. Fromentin and Plaque (1996) showed that there is a positive correlation between environmental factors and zooplankton abundance. In the northeast Arctic region the temperature is a proxy for inflow of Atlantic water masses (Loeng et al. 1997). These water masses contain suitable food particles (copepods) for herring larvae (Sakshaug et al. 1995, Melle and Skjoldal 1998). There is a higher probability of survival when growth increases (Fossum and Moksness 1993, 1995), and Sætre et al. 2001 showed that the larvae of rich year classes have better growth than the larvae of poor year classes. For North Sea herring, there are indications of similar mechanisms. Svendsen et al. 1995 showed that recruitment of North Sea herring was positively correlated with northerly wind during winter, which indicates that advection of near-surface water masses may be an important factor for successful recruitment.

Hamre and Hatlebakk (1998) and Barros et al. (1998) showed that predation on small herring by cod may reduce the abundance of rich year classes of herring during the adolescent phase in the southern Barents Sea. Although the effect in some years can be large, predation by cod merely adjusts the abundance of the year classes (Sætre [IMR] and others [unpubl.]).

Concluding Remarks

There is a significant coherence in spawning stock development through the last four decades between five herring stocks in the north Atlantic. This coherence can be explained by fishery, but temperature time series indicate that environmental factors also may have affected the stocks.

There is a positive significant relationship between temperature and recruitment for Norwegian spring-spawning herring. The other stocks did not have a significant positive correlation between temperature and recruitment. However, North Sea herring and Icelandic summer-spawners had a pattern of higher recruitment at specific temperatures.

References

Anthony, V.C., and M.J. Fogarty. 1985. Environmental effects on recruitment, growth, and vulnerability of Atlantic herring (*Clupea harengus harengus*) in the Gulf of Maine region. Can. J. Fish. Aquat. Sci. 42(Suppl. 1):158-173.

Anthony, V.C., and G.T. Waring. 1980. The assessment and management of the Georges Bank herring fishery. Rapp. P.V. Reun. Cons. Int. Expl. Mer 177:72-111.

Astthorsson, O.S., and A. Gislason. 1994. Distribution, abundance and length of pelagic juvenile cod in Icelandic waters in relation to environmental conditions. ICES Mar. Sci. Symp. 198:529-541.

Bailey R.S., and J.H. Steele. 1992. North Sea Herring Fluctuations. In: M.H. Glantz (ed.), Climate variability, climate change and fisheries. Cambridge Univ. Press, pp. 213-230.

Bakken, E., and O. Dragesund. 1971. Fluctuations of pelagic fish stocks in the northeast Atlantic and their technological and economic effects on the fisheries. OECD Fisheries Division. International Symposium on Fisheries Economics. Paris, 29 November-3 December 1971.

Barros, P., E.M. Tirasin, and R. Toresen. 1998. Relevance of cod (*Gadus morhua* L.) predation for inter-cohort variability in mortality of juvenile Norwegian spring-spawning herring (*Clupea harengus* L.). ICES J. Mar. Sci. 55:454-466.

Blindheim, J., H. Loeng, and R. Sætre. 1981. Long term temperature trends in Norwegian coastal waters. ICES C.M. 1981/C:19 13 pp.

Bochkov, Y.A. 1982. Water temperature in the 0-200 m layer in the Kola-Meridian in the Barents Sea, 1900-1981. Sb. Nauchn. Trud. PINRO Murmansk 46:113-122. (In Russian.)

Campana, S.E., R.K. Mohn, S.J. Smith, and G.A. Chouinard. 1995. Spatial implications of a temperature-based growth model for Atlantic cod (*Gadus morhua*) off the eastern coast of Canada. Can. J. Fish. Aquat. Sci. 52:2445-2456.

Corten, A. 1986. On the causes of the recruitment failure of herring in the central and northern North Sea in the years 1972-1978. J. Cons. Int. Expl. Mer 42:281-294.

Corten, A. 1990. Long-term trends in pelagic fish stocks of the North Sea and adjacent waters and their possible connection to hydrographic changes. Netherlands Journal of Sea Research. 25(1-2):227-235.

Corten, A. 1999. Evidence from plankton for multi-annual variations of Atlantic inflow into the northwestern North Sea. Netherlands Journal of Sea Research. 42:191-205.

deYoung, B., and G.A. Rose. 1993. On recruitment and distribution of Atlantic cod (*Gadus morhua*) off Newfoundland. Can. J. Fish. Aquat. Sci. 50:2729-2740.

Devold, F. 1963. The life history of the Atlanto-Scandian herring. Rapp. P.V. Reun. Cons. Int. Expl. Mer 154:98-108.

DFO. 1995. Evaluation of the stock status of 4WX herring. DFO Atlantic Fisheries Research Document 95/83.

Dragesund, O. 1971. Comparative analysis of year-class strength among fish stocks in the North Atlantic. Fiskeridir. Skr. Ser. Havunders. 16:49-64.

Dragesund, O., and Ø. Ulltang. 1972. The collapse of the Norwegian spring spawning herring stock. ICES C.M. 1972/H:11.

Dragesund, O., and Ø. Ulltang. 1978. Stock size fluctuations and rate of exploitation of the Norwegian spring spawning herring, 1950-1974. Fiskeridir. Skr. Ser. Havunders. 16(10):315-337.

Dragesund, O., A. Johannessen, and Ø. Ulltang. 1997. Variation in migration and abundance of Norwegian spring spawning herring (*Clupea harengus* L.). Sarsia 82:97-105.

Ellertsen, B., P. Fossum, P. Solemdal, S. Sundby, and S. Tilseth. 1987. The effect of biological and physical factors on the survival of Arcto-Norwegian cod and the influence on recruitment variability. In: H. Loeng (ed.), The effect of oceanographic conditions on distribution and population dynamics of commercial fish stocks in the Barents Sea. Proceedings of the Third Soviet-Norwegian Symposium, Murmansk, 26-28 May 1986. Institute of Marine Research, Bergen.

Ellertsen, B., P. Fossum, P. Solemdal, and S. Sundby. 1989. Relation between temperature and survival of eggs and first-feeding larvae of northeast Arctic cod (*Gadus morhua* L.). Rapp. P.V. Reun. Cons. Int. Expl. Mer 191:209-219.

Fiksen, Ø., and A. Folkvord. 1999. Modelling growth and ingestion processes in herring *Clupea harengus* larvae. Mar. Ecol. Prog. Ser. 184:273-289.

Fossum, P., and E. Moksness. 1993. A study of spring- and autumn-spawned herring (*Clupea harengus* L.) larvae in the Norwegian Coastal Current during spring 1990. Fish. Oceanogr. 2(2):73-81.

Fossum, P., and E. Moksness. 1995. Recruitment Processes of the 1991 year-class of spring spawning Norwegian herring (*Clupea harengus* L.) determined from otolith microstructure examination. In: D.H. Secor, J.M. Dean, and S.E. Campana (eds.), Recent developments in fish otolith research. Belle W. Baruch Library in Marine Science No. 19, University of South Carolina Press. ISBN 1-57003-011-1.

Fromentin, J.-M., and B. Planque. 1996. *Calanus* and environment in the eastern North Atlantic. I. Spatial and temporal patterns of *C. finmarchicus* and *C. helgolandicus*. Mar. Ecol. Prog. Ser. 134:101-109.

Gjøsæter, H., and H. Loeng. 1987. Growth of the Barents Sea capelin, *Mallotus villosus*, in relation to climate. Env. Biol. Fish. 20:293-300.

Grainger, R.J.R. 1979. Herring abundance off the west coast of Iceland in relation to oceanographic variation. J. Cons. Int. Expl. Mer 38(2):180-188.

Grosslein, M.D. 1987. Synopsis of knowledge of the recruitment process for Atlantic herring (*Clupea harengus*), with special reference to Georges Bank. NAFO. Scientific Council Studies 11:91-108.

Hamre, J., and E. Hatlebakk. 1998. System model (Systmod) for the Norwegian Sea and the Barents Sea. In: P. Rødseth (ed.), Models for multispecies management. contributions to economics. Physica-Verlag, Heidelberg-New York, pp. 93-116.

Hanssen-Bauer, I., and P.Ø. Nordli. 1998. Annual and seasonal temperature variations in Norway 1876-1997). DNMI (Det Norske Meteorologisk Institutt) KLIMA Report No. 25/98.

Helle, K., and M. Pennington. 1999. The relation of the spatial distribution of early juvenile cod (*Gadus morhua* L.) in the Barents Sea to zooplankton density and water flux during the period 1978-1984. ICES J. Mar. Sci. 56:15-27.

ICES. 1976. Report of the Herring Assessment Working Group for the Area South of 62°N. ICES C.M. 1976/H:2.

ICES. 1999a. Report of the Herring Assessment Working Group for the Area South of 62°N. ICES C.M. 1999/ACFM:12.

ICES. 1999b. Report of the Northern Pelagic and Blue Whiting Fisheries Working Group. ICES Headquarters, Copenhagen, Denmark, 27 April-5 May 1999. ICES C.M. 1999/ACFM:18.

Iles, T.D., and M. Sinclair. 1982. Atlantic herring: Stock discreteness and abundance. Science 215.

Jakobsson, J. 1973. Population studies on the Icelandic herring stocks. ICES C.M. 1973/H:4. 20 pp.

Jakobsson, J. 1980. Exploitation of the Icelandic spring- and summer-spawning herring in relation to fisheries management, 1947-1977. Rapp. P.V. Reun. Cons. Int. Expl. Mer 177:23-42.

Jakobsson, J., A. Gudmundsdottir, and G. Stefansson. 1993. Stock-related changes in biological parameters of the Icelandic summer-spawning herring. Fish. Oceanogr. 2(3/4):260-277.

Loeng, H. 1989. Features of the physical oceanographic conditions of the Barents Sea. Polar Research 10:5-18.

Loeng, H., H. Bjørke, and G. Ottersen. 1995. Larval fish growth in the Barents Sea. Can. Spec. Publ. Fish. Aquat. Sci. 121:691-698.

Loeng, H., V. Ozhigin, and B. Ådlandsvik. 1997. Water fluxes through the Barents Sea. ICES J. Mar. Sci. 54:310-317.

Malmberg, S.-A., and J. Blindheim. 1994. Climate, cod, and capelin in northern waters. ICES Mar. Sci. Symp. 198:297-310.

Maravelias, C.D., and D.G. Reid. 1995. Relationship between herring (*Clupea harengus*, L.) distribution and sea surface salinity and temperature in the northern North Sea. In: C. Bas, J.J. Castro, and J.M. Lorenzo (eds.), International Symposium on Middle-sized Pelagic Fish. Sci. Mar. 59(3-4):427-438).

Melle, W., and H.R. Skjoldal. 1998. Reproduction and development of *Calanus finmarchicus*, *C. glacialis* and *C. hyperboreus* in the Barents Sea. Mar. Ecol. Prog. Ser. 169:211-228.

Michalsen, K., G. Ottersen, and O. Nakken. 1998. Growth of north-east Arctic cod (*Gadus morhua*) L. in relation to ambient temperature. ICES J. Mar. Sci. 55:863-877.

Myers, R.A., K.F. Drinkwater, N.J. Barrowman, and J.W. Baird. 1993. Salinity and recruitment of Atlantic cod (*Gadus morhua*) in the Newfoundland region. Can. J. Fish. Aquat. Sci. 50:1599-1609.

NEFSC (Northeast Fisheries Science Center). 1996. Report of the 21st Northeast Regional Stock Assessment Workshop (21st SAW): Stock Assessment Review Committee (SARC) Consensus Summary of Assessments. NOAA/NMFS/NEFSC Ref. Doc.

Nichols, J.H., and K.M. Brander. 1989. Herring larval studies in the west-central North Sea. Rapp. P.V. Reun. Cons. Int. Expl. Mer 191:160-168.

Ottersen, G. 1996. Environmental impact on variability in recruitment, larval growth and distribution of Arcto-Norwegian cod. Dr. Scient. thesis, University of Bergen, Norway.

Ottersen, G., and Sundby, S. 1995. Effects of temperature, wind and spawning stock biomass on recruitment of Arcto-Norwegian cod. Fish. Oceanogr. 4:278-292.

Ottersen, G., and H. Loeng. 2000. Covariability in early growth and year-class strength of the Barents Sea cod, haddock and herring: The environmental link. Symposium on Recruitment Dynamics of Exploited Marine Populations: Physical-Biological Interactions. ICES J. Mar. Sci. 57:339-348.

Ottersen, G., H. Loeng, and A. Raknes. 1994. Influence of temperature variability on recruitment of cod in the Barents Sea. ICES Mar. Sci. Symp. 198:471-481.

Ottersen, G., K. Michalsen, and O. Nakken. 1998. Ambient temperature and distribution of north-east Arctic cod. ICES J. Mar. Sci. 55:67-85.

Ottersen, G., B. Ådlandsvik, and H. Loeng. 2000. Predicting the temperature of the Barents Sea. Fish. Oceanogr. 9(2):121-135.

Parrish, B.B., and A. Saville. 1965. The biology of the north-east Atlantic herring populations. Oceanogr. Mar. Biol. Ann. Rev. 3:323-373.

Runnstrøm, S. 1936. A study on the life history and migrations of the Norwegian spring-herring based on the analysis of the winter rings and summer zones of the scale. Fiskeridir. Skr. Ser. Havunders. 5(2):5-102.

Sakshaug, E., F. Rey, and D. Slagstad. 1995. Wind forcing of marine primary production in the northern atmospheric low-pressure belt. In: H.R. Skjoldal, C. Hopkins, K.E. Erikstad, and H.P. Leinaas (eds.), Ecology of fjords and coastal waters. Proceedings of the Mare Nor Symposium on the Ecology of Fjords and Coastal Waters, Tromsø, Norway. Elsevier Science B.V. ISBN 0-444-82096-5.

Schweigert, J.F. 1995. Environmental effects on long-term population dynamics and recruitment to Pacific herring (*Clupea pallasi*) populations in southern British Columbia. In: R.J. Beamish (ed.), Climate change and northern fish populations. Can. Spec. Publ. Fish. Aquat. Sci. 121:569-583.

Sinclair, M., and T.D. Iles. 1985. Atlantic herring (*Clupea harengus*) distributions in the Gulf of Maine-Scotian Shelf area in relation to oceanographic features. Can. J. Fish. Aquat. Sci. 42:880-887.

Stephenson, R.L., and I. Kornfield. 1990. Reappearance of spawning Atlantic herring (*Clupea harengus harengus*) on Georges Bank: Population resurgence, not recolonization. Can. J. Fish. Aquat. Sci. 47:1060-1064.

Stefánsson, U., and J. Jakobsson. 1989. Oceanographical variations in the Iceland Sea and their impact on biological conditions, a brief review. In: L. Rey and V. Alexander (eds.), Proceedings of the 6th Conference of the Comité Artique International. E.J. Brill, Leiden, pp. 427-455.

Sundby, S. 1994. The influence of bio-physical processes on fish recruitment in an arctic-boreal ecosystem. Ph.D. thesis, University in Bergen, Norway. 190 pp.

Sundby, S. 2000. Recruitment of Atlantic cod stocks in relation to temperature and advection of copepod populations. Sarsia 85:277-298.

Svendsen, E., A. Aglen, S.A. Iversen, D.W. Skagen, and O. Smedstad. 1995. Influence of climate on recruitment and migration of fish stocks in the North Sea. In: R.J. Beamish (ed.), Climate change and northern fish populations. Can. Spec. Publ. Fish. Aquat. Sci. 121:641-653.

Sætersdal, G., and H. Loeng. 1987. Ecological adaptation of reproduction in northeast arctic cod. Fisheries Research 5:253-270.

Toresen, R., and O.J. Østvedt. 2000. Variation in abundance of Norwegian spring-spawning herring (*Clupea harengus*, Clupeidae) throughout the 20th century and the influence of climatic fluctuations. Fish and Fisheries, Blackwell Science 1(3):231-256.

Toresen, R. et. al. 1999. Havets ressurser 1999, FiskenHav, Særnr. 1:1999. Institute of Marine Research, Bergen, Norway. (In Norwegian.)

Effect of Herring Egg Distribution and Environmental Factors on Year-Class Strength and Adult Distribution: Preliminary Results from Prince William Sound, Alaska

Evelyn D. Brown and Brenda L. Norcross
University of Alaska Fairbanks, School of Fisheries and Ocean Sciences, Institute of Marine Science, Fairbanks, Alaska

Introduction

Herring are important ecological and commercial species within Prince William Sound (PWS) that respond to long-term decadal scale factors and regime shifts. We wanted to refine our understanding of Pacific herring population structure and trends in Prince William Sound (PWS) by examining current and historic data on fish distribution within a spatially explicit ecological context. We also wanted to define temporal and spatial scales of processes affecting herring recruitment as well as the spatial extent of sub-population structure. We hypothesized that adult Pacific herring year-class strength and distribution was dependent on the initial distribution and density of herring embryos, modulated by ocean conditions within distinct spatial regions. We further hypothesized that all geographic units of herring spawning and recruitment within Prince William Sound respond to environmental forcing in unison, i.e., this is one stock or population. The null hypothesis was that population substructure (or metapopulations) exists as evidenced by a certain degree of regional independence in recruitment trends due to regional variability in key forcing variables. We defined sub- or metapopulation as a functionally operational spawning/recruitment unit (sector of the population) within a defined geographic region that can complete the life history cycle in the absence of other units and for which there exists a unique set of forcing variables (physical or biological) that may or may not overlap with other units. We found PWS to be an ideal location for the study of metapopulation structure because of its relative isolation from other stocks or populations.

The herring population crashed in 1993 due to a catastrophic disease outbreak (Marty et al. 1999) and restoration of the population could be impacted by current fisheries management practices. The Alaska Department of Fish and Game manages PWS as a single stock or population. If sub-population structure exists, there would be a risk of regional overfishing. A multitude of hypotheses addresses herring recruitment and factors controlling or limiting herring population restoration. This type of exploratory analysis is very useful in identifying key time and space scales of important processes and thus aids in focusing potential process or mechanistic studies. This is economical since process studies are often expensive.

Methods

The completion of this analysis required extensive data compilation from a variety of sources as well as a series of analytical steps. First, we defined temporal and spatial trends of early life history (eggs) distribution and recruitment; then we developed regional spawn predictor variables and two independent indices of recruitment (for testing of spurious variables). We then developed a set of regional and basin-wide (PWS) oceanographic and other environmental forcing variables and defined logical oceanographic regions. Using time-series methods, we tested coherence between regional spawning and recruitment trends and pooled regions with similar variability and coherence. Using scatterplots, cross-correlation, least-squares regression, and multiple regression techniques, we determined the interactions between the predictor variables (spawn and environmental factors), and defined existing functional relationships between the predictor and response variables of regional or pooled herring recruitment. Colinear predictor variables were eliminated and key variables identified. Spurious relationships were detected when the response function between regions or between independent recruitment indices changed in shape, sign, or significance. Finally we constructed generalized additive models (gam) (Hastie and Tibshirani 1990) using the key variables in a stepwise fashion selecting the most parsimonious model with the best fit explaining the highest degree of variation in recruitment. The gam approach was selected since several of the functions were nonparametric and the additional flexibility over generalized linear models was desired. The gam approach has been used in the past successfully in relating fish recruitment to stock biomass and environmental factors (Daskalov 1999).

Herring Predictor and Response Variables

Herring spawning and recruitment was highly variable over the 26-year period (1973-1999; Figs. 1 and 2) and between spawning regions. Overall recruitment was spiky and appeared to have a 4-year cycle (Fig. 2). Five spawn regions (Fig. 1) were developed, based on the Alaska Department of

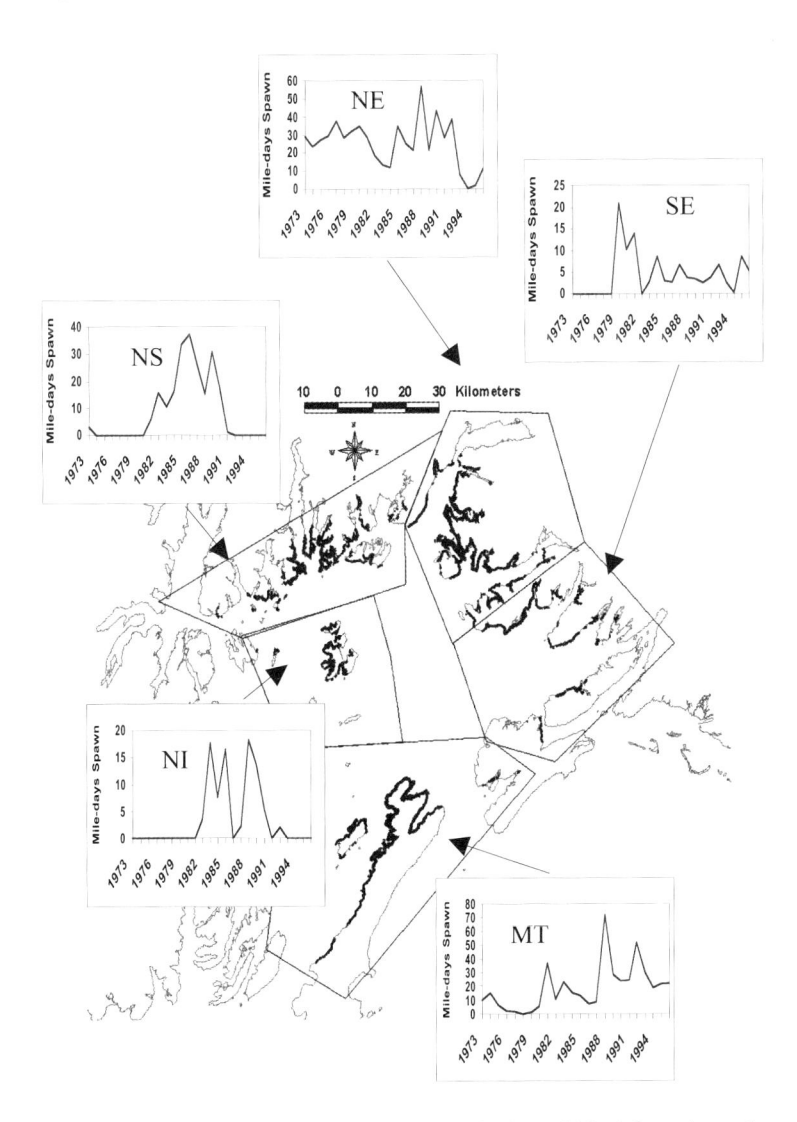

Figure 1. Composite spawning locations (thickened black line along the coast) in Prince William Sound from 1973 to 1999 and the divisions of spawn regions used by the Alaska Department of Fish and Game: Southeast (SE), Northeast (NE), North Shore (NS), Naked Island (NI), and Montague (MT). Inset graphs show regional trends in spawning magnitude recorded as mile-days, a cumulative measure of observed daily miles of spawn (from aerial surveys; Alaska Department of Fish and Game, Cordova Area Office, unpubl. data.)

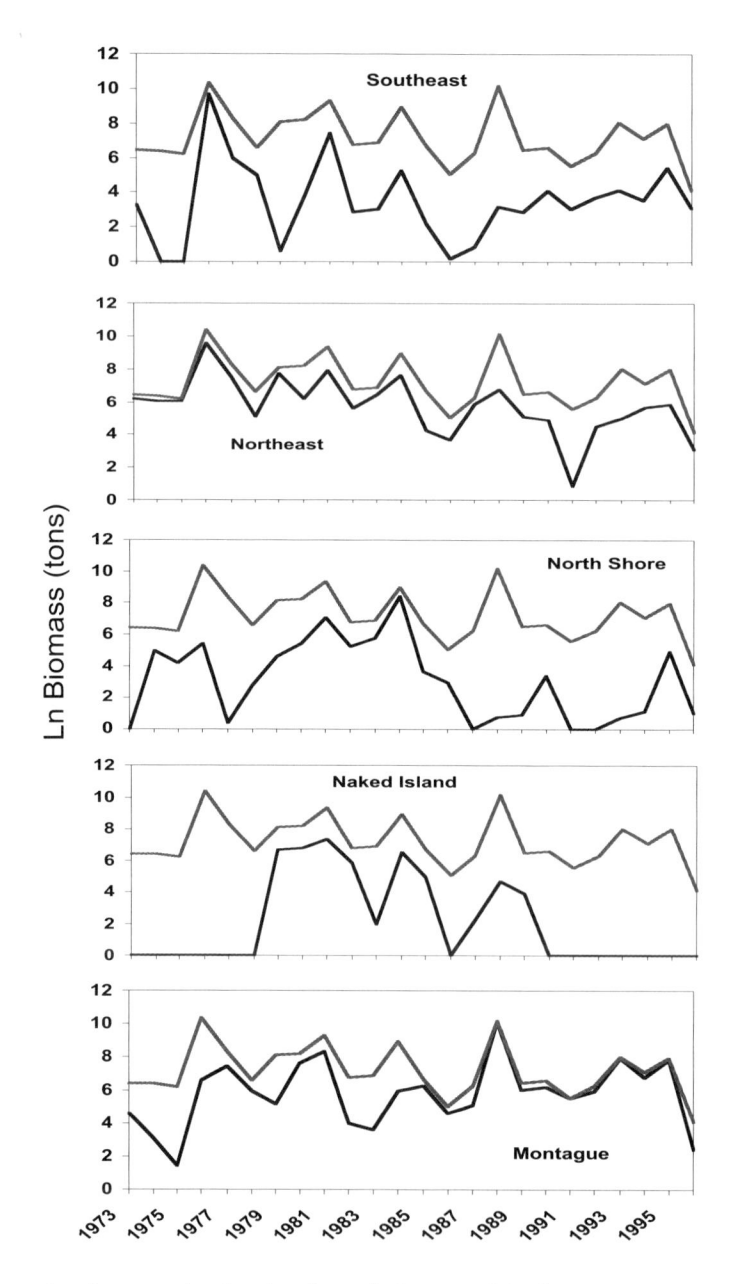

Figure 2. An example of regional trends for one of the four recruitment predictor variables: age-3, peak aerial survey (pas) (Brady 1987) index. The regional recruitment trends (log-transformed biomass in tons) are by year class and shown as black lines and compared to the total PWS recruitment trend (gray line) in each graph.

Fish and Games historic aerial survey records (Brady 1987), with a sixth variable representing all PWS spawn combined. The spawn variables were treated as predictors and included in a generalized form of a Ricker model introduced by Jacobson and MacCall (1995):

$$\ln(R) = \alpha + f(T) + g(S) + \varepsilon \quad \text{or} \quad \ln(\frac{R}{S}) = \alpha + h(T) + j(S) + \varepsilon$$

in the Ricker form where R is recruitment, T represents the environmental predictors, and S the spawning predictors.

The response variables consisted of two recruitment indices, peak aerial biomass (pas) (Brady 1987) and age-structured-analysis (asa) model output (Funk 1995). Peak aerial survey biomass is defined as the largest observed sum of estimated individual school biomass occurring over the range of spawn dates in a given geographic unit; a sighting tube is used to measure school size. Both pas and asa regional recruitment were developed using regional age composition information (Mark Willette, Alaska Department of Fish and Game, Cordova Area Office, unpubl. data and pers. comm.) for each of the regional spawning categories (5 regions, 1 total) for both age-3 and age-4 herring recruits resulting in a total of 24 response variables.

Environmental Predictor Variables

Four oceanographic regions (Fig. 3) were defined based on satellite imagery, composites of historic hydrography (1970s–early 1990s), and an ocean model of flow (Mooers and Wang 1998): (1) eastern PWS east of the central circulation gyre,(2) the northern quiescent region, (3) the western outflow region, and (4) the central dynamic section which is heavily influenced by the Gulf of Alaska (GOA). The western and northern regions were further subdivided resulting in a total of six oceanographic regions for more flexibility in analysis; regions or sub-regions were pooled if there was a lack of significant differences between variables. Regional oceanographic variables included mean temperature and salinity in the upper 20 m, wind, minimum and maximum air temperatures, and precipitation. Environmental variables applied sound-wide included freshwater discharge (Royer 1982), the Pacific Decadal Oscillation (PDO or PMDO from Enfield and Mestas-Nuñez 1999), upwelling at Hinchinbrook Entrance (opening to PWS), and an index of zooplankton biomass. Zooplankton was sampled using vertical tows as part of a plankton watch program operating at the Prince William Sound Aquaculture Association Armin F. Koernig salmon hatchery in southwest PWS.

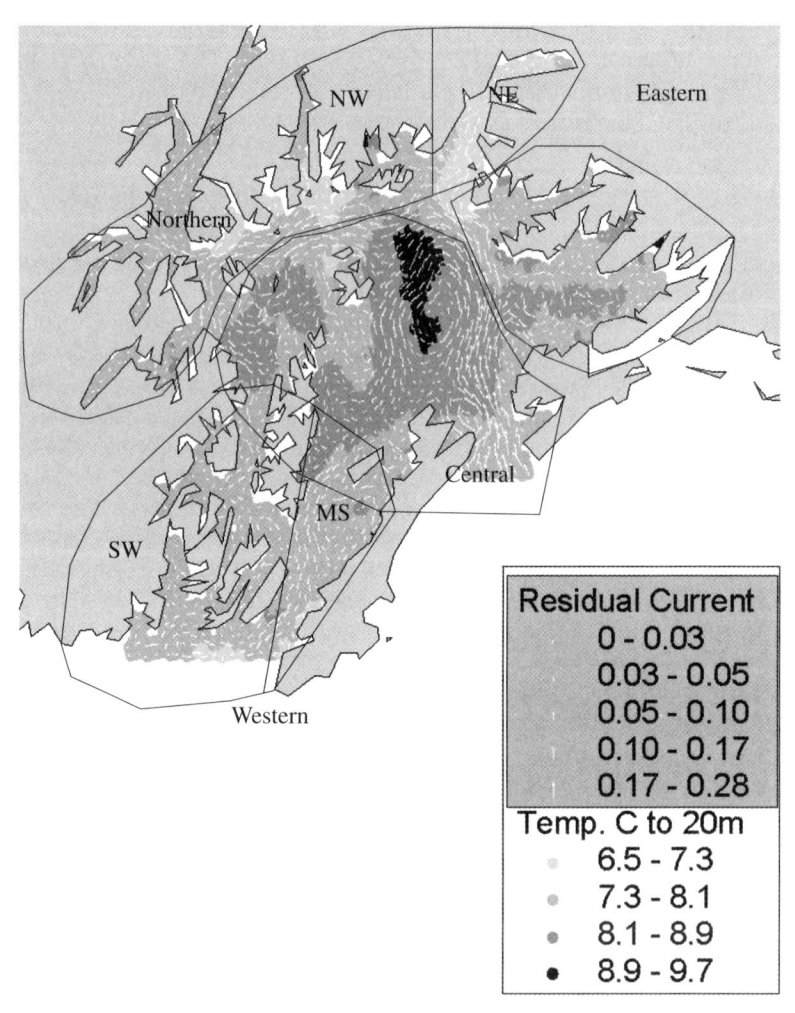

Figure 3. An example of an oceanographic composite used to delineate regions for development of environmental variables in Prince William Sound. Shown here is the May composite, with mean sea surface temperature (at 20 m depth) and modeled residual currents (Mooers and Wang 1998) from smoothed historical hydrographic data (University of Alaska Fairbanks, Institute of Marine Science oceanographic database). The four oceanographic regions and sub-regions are delineated: the Eastern area east of the central gyre, the Northern quiescent zone with the Northeast (NE) and Northwest (NW) sub-regions, the dynamic Central region, and the Western outflow area with the Montague Strait (MS) and Southwest (SW) sub-regions.

Results and Discussion

Prior to examining the effect of the environmental variables on recruitment, we examined the regional spawner-recruit relationships. In single linear comparisons of log-transformed spawning and recruitment, only age-3 regional Montague and all-area (total) age-3 recruitment were correlated significantly to spawning (for both recruitment indices). Using multiple regression, only age-3 Montague recruitment was related to spawn variables, but this time to both the northeast and Montague spawn abundance. Age-4 response variables were not related to any spawning variables. Using time series analysis, significant coherence was observed between recruitment in all northern and eastern (but not Montague) spawn regions. However, coherence in recruitment trends at age 4 was observed between Montague and the northeastern and southeastern regions. This was the first evidence of cross-regional migration in adult herring occurring between age 3 and age 4. Nearly every combination of recruitment regions was coherent at a lag of 4 providing statistical evidence of the 4-year recruiting cycle.

The relationships between the predictor and response variables were defined using gam. The initial environmental variable list was pared down from 117 to 29 and, not surprisingly, 50 were removed by colinearity alone. This is because the variables used were indications of environmental states that affected many of the factors simultaneously. Three predictor variables were key in the significant gam models. The average zooplankton biomass (from one site in PWS) lagged one year ahead of the herring cohort or birth year was critical in many models. This variable represents zooplankton biomass in April and May when the herring become age 1 and are still rearing in nursery bays. Summer temperature during July and August of the cohort year also had a positive relationship to recruitment. These temperatures occur during larval drift and recruitment of the newly metamorphosed age-0 herring into their nursery bays. Finally, regional or total spawn was important to some, but not all, of the regional recruitment variables and then only for age-3 herring. Spawn was never important in the gam models of age-4 herring. In all age-4 models, recruitment of age 3 from one or more of the neighboring regions was a critical predictor. Nonlinear and nonparametric smoothing functions (used to fit the predictor-response functional relationships) produced better model fits than parametric functions, and therefore we found gam modeling to be a powerful and useful tool for this type of ecological analysis.

Two examples of gam outputs are shown. At Montague Island, 76% of the variability in age-3 recruits was explained by a model including July and August mean sea surface temperatures in the Montague region (MSTJA, Fig. 4A), the average zooplankton biomass lagging a year ahead (AVEZL1), and the amount of spawn in both the Montague (MT) and Northeast (NE) regions. For the age-4 Montague recruit model (Fig. 4B), 70% of the variability in the log-transformed recruitment (pasmtb4) was explained by summer Montague sea surface temperatures (MSTJA), and log-transformed age-3 recruitment in the Montague region (pasmtb3) and in all regions combined (pastot3).

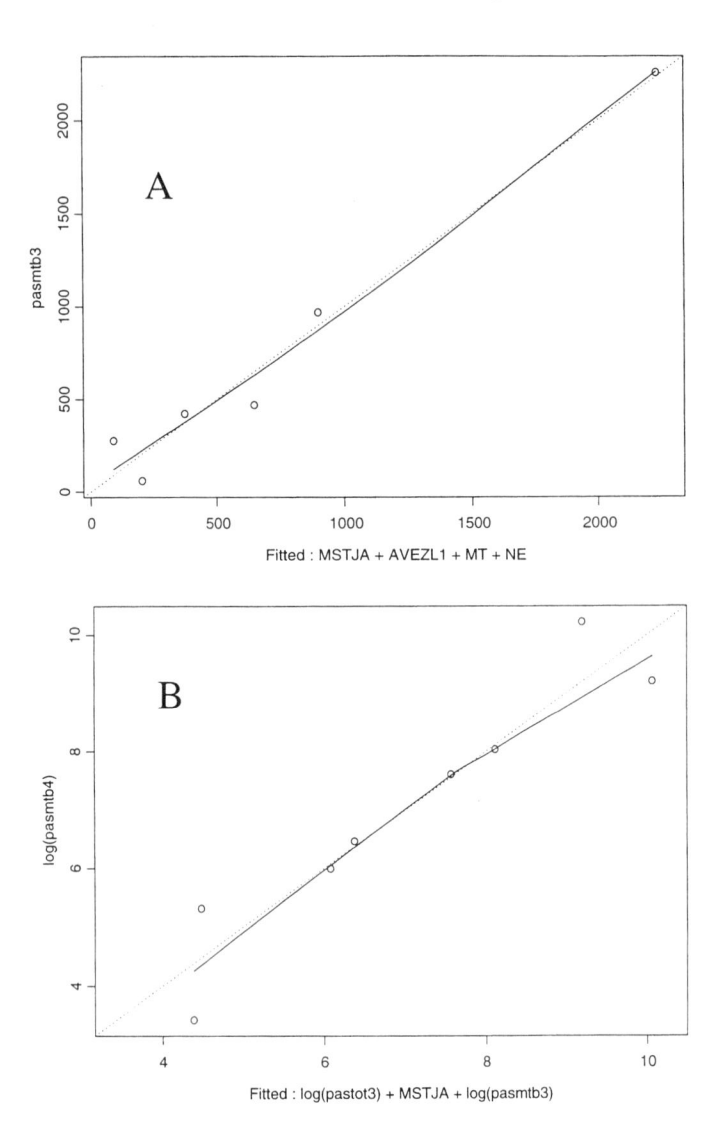

Figure 4. (A) The fitted gam for Montague age-3 recruitment (pasmtb3) with the Montague regional sea surface temperature for July and August (MSTJA) of the cohort year (no lag), the average zooplankton lagged one year forward (AVEZL1) of the cohort year, and spawn magnitude at the Montague (MT) and Northeast (NE) regions. The model explained 76% of the variability. (B) The fitted gam for log-transformed Montague age-4 recruitment (pasmtb4), summer sea surface temperature, and age-3 recruitment from the previous year for all regions combined (pastot3) and Montague (pasmtb3). The model explained 70% of the variability.

There were also interesting interactive effects between the key explanatory variables. The interactions could be further categorized into short-term or regional relationships and long-term decadal relationships. An example of significant short-term, regional interactions was between regional March and April precipitation and regional spawning variables; this relationship did not hold up on total spawn combined. Another example was the positive effect of March and April salinity on both northeast and southeast spawning as well as on average zooplankton biomass (short-term, but not regional). Other important short-term interactions were the effect of January and February upwelling on regional spawning (positive relationship with positive upwelling anomalies) and the negative impact of freshwater discharge in March switching to positive in May, on peak zooplankton biomass. An example of decadal scale interactions was the effect of the PDO on Montague, Northeast, and total spawn (Fig. 5A) in which the coherence between the two trends was significant. There was also evidence in all the regional hydrographic time series of a state or "regime" shift occurring at about 1989. Mean salinities and salinity anomalies, especially noticeable in the fall and spring, were significantly lower in the early 1990s compared to the late 1970s, early 1980s. There were also more bouts of negative temperature anomalies in the early 1990s as compared to the late 1970s, especially in the spring. There seemed to be a corresponding decadal shift in zooplankton biomass as well (Fig. 5A. The bloom, as measured in southwest PWS, was shorter and more intense in magnitude during the early 1980s as compared to the early 1990s. In addition, peaks in age-3 and age-4 recruitment corresponded with peaks in zooplankton biomass (Fig. 5B) explaining why zooplankton was important in the model. It is interesting to note that during the 1990s, when salinities were generally lower and zooplankton biomass was reduced, herring biomass declined.

The results generally supported our first hypothesis. Only models including environmental or a combination of environmental and spawning factors explained a significant amount of recruitment variability. In this case, environmental conditions during the winter and spring (March and April) appear to affect spawning magnitude. Summer temperatures (July and August) or related conditions during the late larval drift stage seem to be important in driving recruitment. We also know that zooplankton biomass in April and May, when the juvenile herring are age 1, affects recruitment.

The results led us to generally accept the null hypothesis that metapopulations exist in PWS. The evidence was based on the differences in parameterization of best-fit recruitment models between regions. The regional and area-wide environmental factors often explained more variability in regional recruitment than did spawning (regional or combined). However, there were large-scale environmental factors (e.g., PDO) that appeared to induce regional and area-wide environmental conditions and cause decadal shifts in herring abundance common to all regions. The results in this preliminary analysis were useful in identifying important temporal scale in key processes. We concluded that the large-scale, decadal

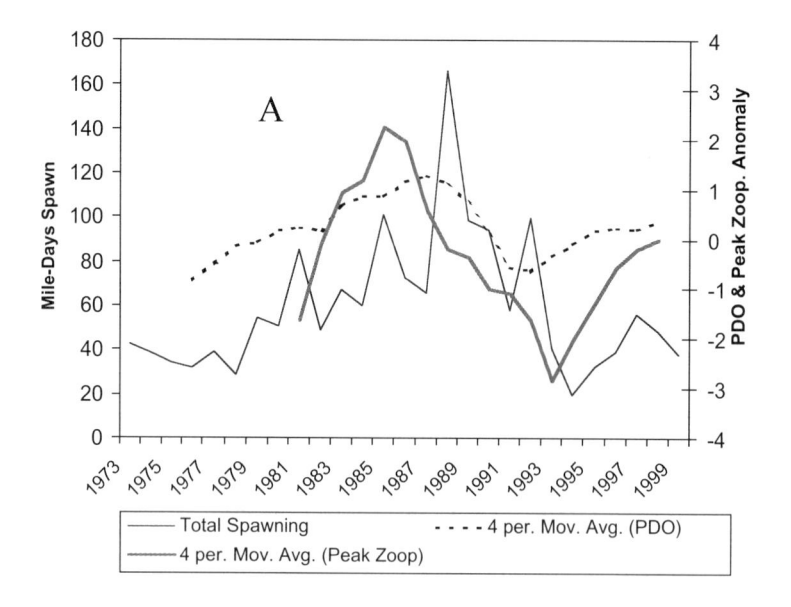

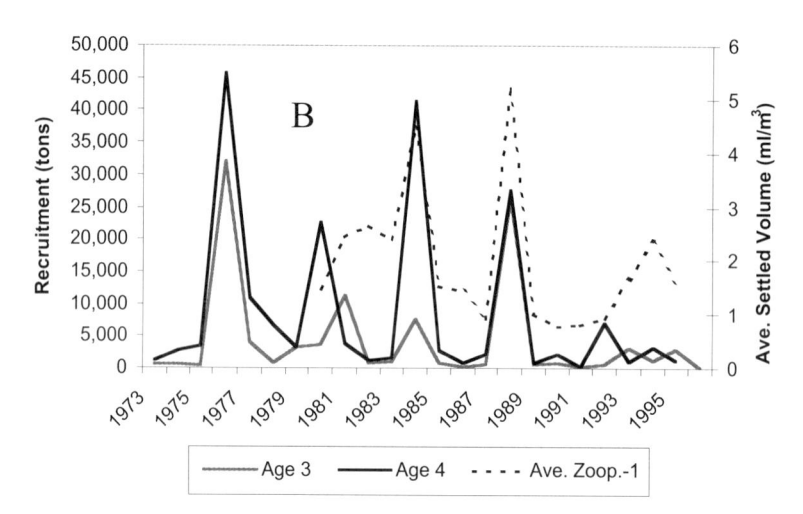

Figure 5. (A) The phasing of the Pacific decadal oscillation (PDO) anomaly smoothed as 4-year moving average, gray line); peak zooplankton biomass anomaly (smoothed as 4-year moving average, dashed line); and the total annual spawn magnitude in Prince William Sound (solid line). (B) Coherence between trends of herring recruitment (at age 3 and age 4 as biomass in tons) by year class and zooplankton biomass expressed as average settled volume lagged forward one year (when herring were age 1).

climate factors probably affect long-term trends in PWS herring abundance, but in an indirect fashion. The smaller-scale regional forcing factors likely directly affect herring processes, especially during early life history stages, and result in the short-term (4 years or less) noise observed in herring recruitment. The scale of regional forcing, based on the size of regions with distinct recruitment models, appears to occur between 4,000 and 8,000 km^2 in PWS.

Our results underscore the potential risk of treating PWS as a single management unit. If the eastern and Montague regions are distinct in terms of life history closure (spawning, rearing, and overwintering) or if the Montague region is dependent on eastern and northern regions for successful recruitment, then applying an area-wide fishing quota to a single region, especially in the east, would be unwise. We recommend further studies quantifying the inter-region exchange and dependency of herring between regions for recruitment. This information could help local managers, concerned resource users, and scientists identify protection measures that would best aid in recovery of the currently depressed PWS population.

References

Brady, J.A. 1987. Distribution, timing, and relative biomass indices for Pacific herring as determined by aerial surveys in Prince William Sound 1978-1987. Prince William Sound Data Report No. 87-14. Alaska Department of Fish and Game, Division of Commercial Fisheries, Juneau. 11 pp.

Daskalov, G. 1999. Relating fish recruitment to stock biomass and physical environment in the Black Sea using generalized additive models. Fish. Res. 41:1-23.

Enfield, D.B., and A.M. Mestas-Nuñez. 1999. Multiscale variabilities in global sea surface temperature and their relationships with tropospheric climate patterns. J. Climate 12:2719-2733

Funk, F. 1995. Age-structured assessment of Pacific herring in Prince William Sound, Alaska and forecast of abundance for 1994. Alaska Department of Fish and Game, Regional Information Report No. 5J95-00, Juneau. 40 pp.

Hastie, T., and R. Tibshirani. 1990. Generalized additive models. Chapman and Hall, London. 325 pp.

Jacobson, L.D., and A.D. MacCall. 1995. Stock-recruitment models for Pacific sardine. Can. J. Fish. Aquat. Sci. 52:566-572.

Marty, G.D., M.S. Okihiro, E.D. Brown, D. Hanes, and D.E. Hinton. 1999. Histopathology of adult Pacific herring in Prince William Sound, Alaska, after the *Exxon Valdez* oil spill. Can. J. Fish. Aquat. Sci. 56:419-426.

Mooers, C.N.K., and J. Wang. 1998. On the implementation of a three-dimensional circulation model for Prince William Sound, Alaska. Cont. Shelf Res. 18:253-277.

Royer, T.C. 1982. Coastal fresh water discharge in the northeast Pacific. J. Geophys. Res. 87:2017-2021.

Herring Occurrence in the Sound (ICES SD23) in Relation to Hydrographical Features

J. Rasmus Nielsen, Bo Lundgren, Torben F. Jensen, and Karl-Johan Stæhr
Danish Institute for Fisheries Research (DIFRES), North Sea Centre, Hirtshals, Denmark

Extended Abstract

Abundance and duration of the overwintering period of the Rügen spring-spawning herring (*Clupea harengus*) stock (RHS) in the central Sound between Denmark and Sweden (ICES Subdivison 23; Fig. 1) were investigated for hydrographical factors affecting relative distribution and triggering southward migration toward the spawning grounds. In the period 1993-1998 27 hydroacoustic monitoring surveys were carried out (Table 1, Fig. 2) covering the area with a standard set of acoustic transects. From concurrent biological sampling with scientific, multi-panel gillnets and with large pelagic trawls with fine meshed cod-ends, species composition and size and age were estimated. The results were compared with concurrently sampled hydrographical (CTD) data and data on water currents (Table 2, Fig. 3). The objectives of the study were to investigate

1. The biomass levels of RHS in the Sound over several years in all seasons of year;

2. The duration of the overwintering period for the RHS in the Sound based on the hypotheses that RHS overwinter in the Sound from September/November to December/January (h:1), and that the Sound is an important overwintering area for the stock (h:2), and that the main herring component in the Sound is RHS (h:3); and

3. The specific distribution and density patterns of herring in relation to hydrographical features and depth in different areas, seasons of the year, and between years in the Sound based on the hypotheses that during autumn and early winter the RHS concentrates in the deeper

Table 1. Overview of survey activities, and density per geographical stratum (t/nsm^2 and N/nm^2) of herring subdivided by survey and geographical strata. Furthermore, total biomass (BM) in tons (t) and abundance (N) in millions for the whole Sound area for each survey is presented.

OVERVIEW OF TYPES AND INTENSITY OF SAMPLING													
SURVEY	S0193	S0293	S0393	S0494	S0594	S0694	S0794	S1094	S1194	S1294	S0195	S0295	S0395
PERIOD	Sep. '93	Oct. '93	Dec. '93	Jan. '94	Feb. '94	Mar. '94	Apr. '94	Oct. '94	Nov. '94	Dec. '94	Jan. '95	Feb. '95	Mar. '95
Acoustic integration	17-22/9	25-30/10	29/11-4/12	10-15/1	14-19/2	14-19/3	11-16/4	17-20/10	21-27/11	12-16/12	9-16/1	6-10/2	27/2-4/3
Gillnet sampling	X	X	X	X	X	X	X	X	X	X	X	X	X
Trawl sampling		X						X					
Hydrogr. sampling								X	X	X	X	X	X
Exten. ind. sampling		X		X				X					

HERRING DENSITY IN NUMBER IN MILLIONS PER NM**2													
SURVEY	S0193	S0293	S0393	S0494	S0594	S0694	S0794	S1094	S1194	S1294	S0195	S0295	S0395
PERIOD	Sep. '93	Oct. '93	Dec. '93	Jan. '94	Feb. '94	Mar. '94	Apr. '94	Oct. '94	Nov. '94	Dec. '94	Jan. '95	Feb. '95	Mar. '95
STRATUM													
G01	6.48	3.71	-	-	-	-	0.17	2.20	0.33	2.22	0.61	0.20	1.79
G02	8.58	4.91	6.78	6.22	1.76	1.15	0.42	2.78	0.53	1.30	0.72	0.55	0.64
G03	7.54	5.54	6.57	9.94	13.78	1.47	0.86	2.79	0.49	2.81	1.39	0.52	0.84
G04	9.17	4.40	3.44	5.18	15.39	1.05	0.47	4.09	1.65	2.21	0.89	0.22	1.57
G05	3.05	3.50	1.78	2.36	3.78	0.62	0.15	2.30	1.24	1.09	1.12	0.23	0.38
G06	4.33	2.97	1.75	2.11	3.65	0.67	0.19	2.16	1.79	1.03	1.40	0.17	0.39
G07	3.71	1.91	2.16	1.60	0.71	0.78	0.10	1.49	1.31	0.83	0.62	0.21	0.42
G08	2.73	2.77	0.46	1.52	-	-	0.07	1.80	1.99	1.18	0.66	0.21	0.19
G09	3.32	3.05	3.06	0.83	0.82	-	0.04	2.54	1.09	0.96	1.18	0.31	0.14
G10	3.19	1.37	2.19	0.58	0.45	-	0.03	0.69	1.80	0.85	0.04	0.08	0.16
G11	1.50	2.32	1.85	0.91	0.03	-	0.02	0.23	0.30	0.50	0.03	0.03	0.35
G12	1.15	0.25	0.25	0.03	0.01	-	0.01	0.01	0.02	0.01	0.03	0.01	0.03
G13	-	-	-	-		-		0.01	0.01	0.01			
MEAN (by strata)	4.56	3.06	2.75	2.84	4.04	0.96	0.21	1.78	0.97	1.15	0.72	0.23	0.58
N: SOUND, TOTAL	1,151.44	792.08	680.47	674.29	835.97	132.02	51.30	513.32	320.07	314.88	205.17	61.42	127.22

HERRING DENSITY IN TONS PER NM**2													
SURVEY	S0193	S0293	S0393	S0494	S0594	S0694	S0794	S1094	S1194	S1294	S0195	S0295	S0395
PERIOD	Sep. '93	Oct. '93	Dec. '93	Jan. '94	Feb. '94	Mar. '94	Apr. '94	Oct. '94	Nov. '94	Dec. '94	Jan. '95	Feb. '95	Mar. '95
MEAN (by strata)	438.85	319.38	248.54	309.55	421.85	107.58	22.19	283.55	150.85	191.51	112.36	31.33	79.97
MIN (by strata)	148.38	30.56	30.63	5.09	2.22	90.93	1.44	0.98	1.92	2.63	5.68	0.65	3.60
MAX (by strata)	715.31	469.32	443.30	855.59	1,414.19	139.30	80.91	556.36	345.46	464.24	208.84	85.26	238.04
BM: SOUND, TOTAL	118,832.00	87,793.85	65,461.78	77,421.32	91,061.31	15,932.86	5,608.95	83,609.36	50,049.02	50,794.87	31,395.20	8,269.60	17,703.26

Herring: Expectations for a New Millennium 349

Table 1. (Continued.) Overview of survey activities, and density per geographical stratum (t/nm^2 and N/nm^2) of herring subdivided by survey and geographical strata. Furthermore, total biomass (BM) in tons (t) and abundance (N) in millions for the whole Sound area for each survey is presented.

OVERVIEW OF TYPES AND INTENSITY OF SAMPLING															
SURVEY	S0495	S0595	S0695	S0795	S0995	S0196	S0496	S1096	S1196	S0397	S0497	S1197	S0398	S0598
PERIOD	Apr. '95	May '95	Jul. '95	Aug. '95	Oct. '95	Mar. '96	Apr. '96	Oct. '96	Nov. '96	Mar. '97	Apr. '97	Nov. '97	Mar. '98	May '98
Acoustic integration	27/3-2/4	25, 30-31/5	10-11/7	6-11/8	9-13/10	18-22/3	10-14/4	30/9-6/10	11-17/11	3-9/3	5-10/4	11-15/11	30/3-3/4	17-19/5
Gillnet sampling	X			X	X			X	X	X	X	X	X	X
Trawl sampling			X	X	X	X	X	X	X			X	X	X
Hydrogr. sampling	X	X	X	X		X	X	X	X	X	X	X	X	X
Exten. ind. sampling	X					X		X				X	X	X

HERRING DENSITY IN NUMBER IN MILLIONS PER NM**2															
SURVEY	S0495	S0595	S0695	S0795	S0995	S0196	S0496	S1096	S1196	S0397	S0497	S1197	S0398	S0598
PERIOD	Apr. '95	May '95	Jul. '95	Aug. '95	Oct. '95	Mar. '96	Apr. '96	Oct. '96	Nov. '96	Mar. '97	Apr. '97	Nov. '97	Mar. '98	May '98
STRATUM														
G01	0.27	0.27	0.23	6.55	1.51	1.70	1.08	1.24	16.23	1.74	3.71	12.78	10.57	0.99
G02	0.24	0.36	0.20	2.19	1.45	0.99	0.23	5.64	7.24	2.43	4.51	11.84	2.56	0.50
G03	0.39	0.25	0.15	3.29	2.56	1.39	0.78	1.62	7.19	4.72	6.02	10.77	6.71	0.45
G04	0.36	0.32	0.13	2.80	1.52	1.85	0.88	9.12	8.52	3.36	2.77	6.84	6.72	0.66
G05	0.30	0.57	0.22	1.41	1.75	1.12	0.63	3.05	3.09	3.82	2.80	4.78	3.59	0.47
G06	0.58	0.42	0.11	0.77	0.57	0.77	0.99	3.08	2.70	6.11	2.13	3.47	1.53	0.61
G07	0.73	0.28	0.07	1.34	0.53	0.97	0.16	5.42	1.67	2.13	2.05	3.19	1.15	0.53
G08	0.19	0.28	0.04	0.54	2.10	0.32	0.34	3.08	1.58	0.01	1.00	2.34	1.07	0.44
G09	0.16	0.27	0.03	0.75	0.67	0.35	0.19	3.82	1.18	0.01	0.71	1.68	0.28	0.38
G10	0.11	0.46	0.03	0.57	0.31	0.29	0.18	1.82	0.57	0.01	0.20	1.32	0.52	0.23
G11	0.06	-	0.01	0.14	0.25	0.16	0.06	0.11	0.04	0.01	0.02	0.02	0.06	0.06
G12	0.03	0.01	0.01	0.16	0.02	0.01	0.01	0.07	0.02	0.01	0.01			
G13	-	0.01	0.01					0.01						
MEAN (by strata)	0.29	0.27	0.10	1.71	1.10	0.83	0.46	2.92	4.17	2.03	2.16	5.37	3.16	0.48
N: SOUND, TOTAL	86.91	82.67	24.80	370.40	284.93	207.56	113.07	839.50	857.73	553.13	537.45	1,125.67	608.93	116.48

HERRING DENSITY IN TONS PER NM**2															
SURVEY	S0495	S0595	S0695	S0795	S0995	S0196	S0496	S1096	S1196	S0397	S0497	S1197	S0398	S0598
PERIOD	Apr. '95	May '95	Jul. '95	Aug. '95	Oct. '95	Mar. '96	Apr. '96	Oct. '96	Nov. '96	Mar. '97	Apr. '97	Nov. '97	Mar. '98	May '98
MEAN (by strata)	38.17	35.38	6.92	297.55	176.42	138.20	76.24	295.85	390.04	211.36	223.33	733.89	287.48	29.42
MIN (by strata)	4.28	0.08	24.71	2.73	1.97	1.09	1.82	2.56	0.01	0.77	2.87	11.72	3.70	
MAX (by strata)	96.71	70.45	22.62	1,109.92	401.64	341.93	162.32	667.90	1,243.41	728.85	688.52	1,615.28	800.28	60.19
BM: SOUND, TOTAL	11,510.66	10,759.33	1,548.38	65,074.65	45,690.14	34,989.40	19,068.80	90,595.39	88,404.08	58,406.32	56,553.74	163,183.91	62,143.54	7,088.73

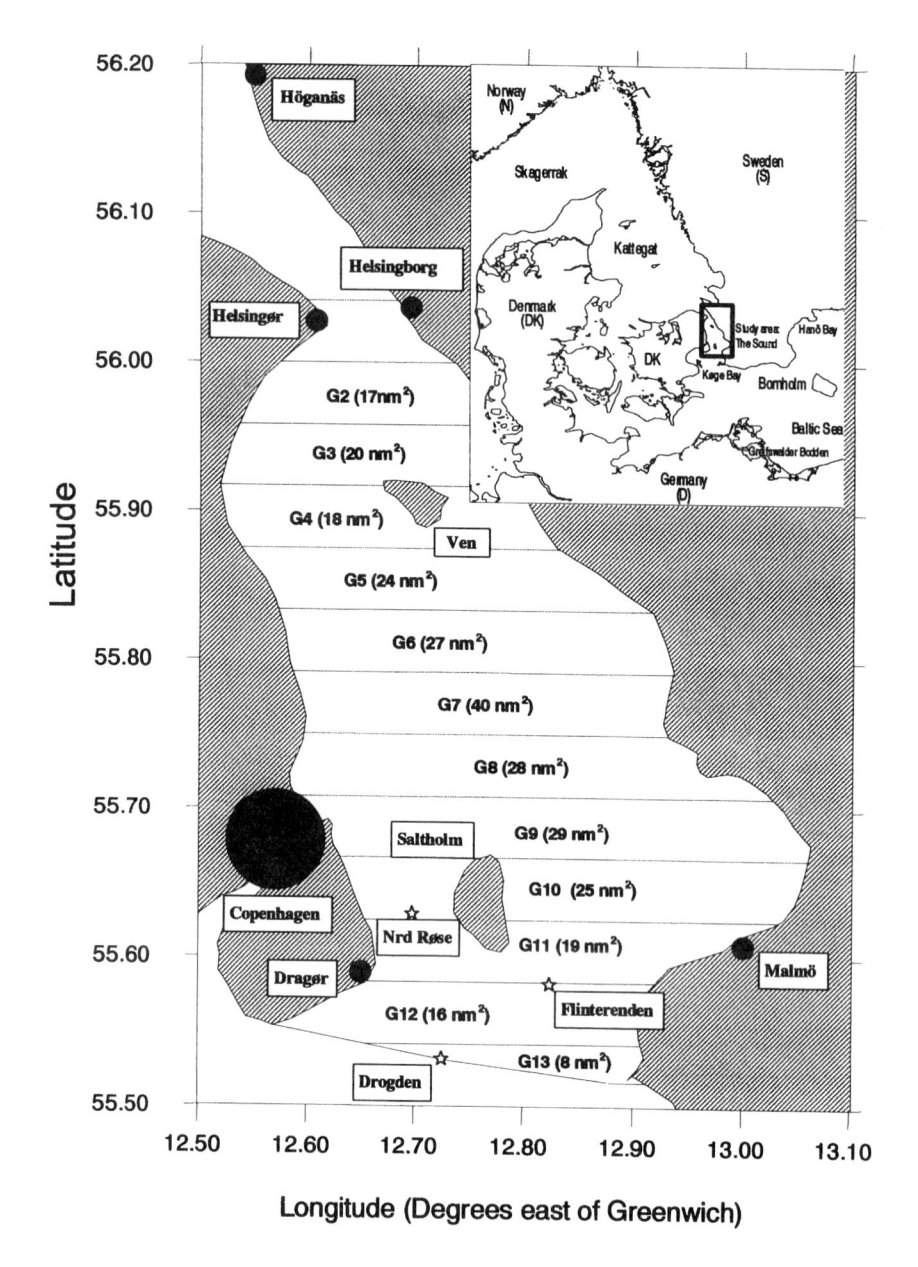

Figure 1. Map of the survey area in the Sound (ICES Subdivision 23) by geographical strata.

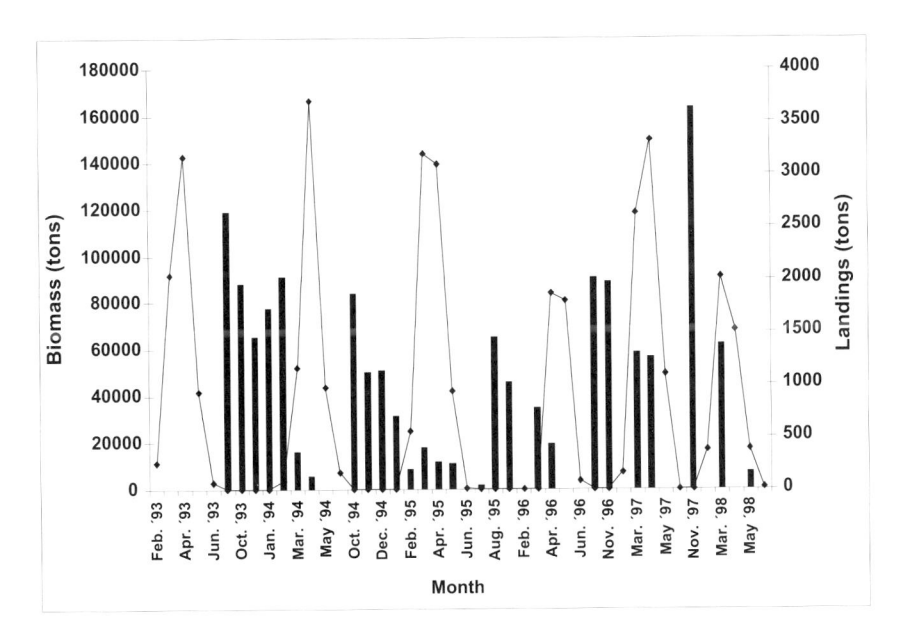

Figure 2. The total biomass in metric tons in the Sound (bars) and German landings in metric tons (solid lines) from commercial fishery at the spawning site at Greifswalder Bodden.

parts of the Sound below the pycnocline (h:4), and that in spring the RHS will concentrate in the southern part of the Sound close to the Drogden threshold (the shallow water area around Saltholm) before southward migration to the spawning grounds (h:5); whether water inflows to the Baltic may trigger southward migration of RHS from the Sound to the spawning grounds based on the hypothesis that southward migration indicated by declines in biomass levels of RHS in the Sound is directly correlated with longer periods of inflow to the Baltic (h:6); which size groups of herring migrate southward first based on the hypothesis that the largest herring start spawning migration first (h:7).

Herring abundance of 45,000-165,000 t in August-February, 5,000-60,000 t in March-May, and < 2,000 t in June-July were estimated (Table 1, Fig. 2). Relatively strong 1988, 1991, and 1994 year classes were found from age 2 in the autumn (Nielsen et al. 2001) and relatively high stock abundance in the autumn periods of 1993 and 1996-1997, respectively (Table 1, Fig. 2). The results suggest the Sound to be a main overwintering area for the RHS for the age 2+ group (Table 3), and indicate a longer overwintering period than hitherto assumed. This suggests rejection of

Table 2. Descriptive statistics of the GLM analyses of variance and estimated parameters from the additive GLM models as well as estimates of the different main effects from model 2. The sum of squared deviations (s. of sq.) for the various dependent effects are of type III sum of squares for the models, which for type III is independent of the order of the effects in the model. Finally, descriptive statistics of the test of normality for each model is shown.

Descriptive statistics: Source	Model	Degrees of freedom	Sum of squares	F	Probability > F
Model	1	125	2440,7	15.4	0.0001
Error	1	786	996,5		
Corrected total	1	911	3437,1		
R-square	1				$r^2 = 0.71$
Model	2	26	1984,3	46.49	0.0001
Error	2	885	1452,9		
Corrected total	2	911	3437,1		
R-square					$r^2 = 0.58$

Source		Degrees of freedom	Sum of squares	F	Probability > F
Year	1	5	160,42	25.31	0.0001
Season of year	1	1	82,23	64.86	0.0001
Geographical stratum	1	12	256,43	16.86	0.0001
Depth stratum	1	3	583,44	153.40	0.0001
Year * Season of year	1	3	28,12	7.39	0.0001
Year * Geographical stratum	1	55	132,44	1.90	0.0001
Year * Depth stratum	1	15	95,97	5.05	0.0001
Season of year * Depth stratum	1	3	53,83	14.15	0.0001
Geographical stratum * Depth stratum	1	28	122,69	3.46	0.0001
Year	2	5	267,76	32.57	0.0001
Season of year	2	1	88,26	53.68	0.0001
Geographical stratum	2	12	368,89	18.70	0.0001
Depth stratum	2	3	793,06	160.77	0.0001
Salinity range	2	3	21,08	4.27	0.0053

Logarithmic transformed estimates of parameters from main effect model (model 2):					
Parameter		Group	Estimate	T for H0; Par.=0	Pr > |T|
Intercept	2	Intercept	6.86	12.36	0.0001
Year	2	1993	1,02	4.46	0.0001
	2	1994	0,22	1.14	0.2552
	2	1995	-0,85	-4.73	0.0001
	2	1996	-0,41	-2.04	0.0413
	2	1997	-0,12	-0.59	0.5572
	2	1998	0		
Season of year	2	Autumn period	0,73	7.33	0.0001
	2	Spring period	0		
Geographical stratum	2	G01	2,69	5.89	0.0001
	2	G02	2,37	5.20	0.0001
	2	G03	2,38	5.25	0.0001
	2	G04	2,25	4.98	0.0001
	2	G05	2,05	4.53	0.0001
	2	G06	1,88	4.14	0.0001
	2	G07	1,07	2.34	0.0195
	2	G08	1,57	3.37	0.0008
	2	G09	1,48	3.18	0.0015
	2	G10	1,32	2.82	0.0049
	2	G11	0,87	1.86	0.0638
	2	G12	-0,03	-0.07	0.9409
	2	G13	0		
Depth stratum	2	6 m (5.0-7.5 m)	-0,90	-5.84	0.0001
	2	16 m (7.5-22.5 m)	1,35	8.39	0.0001
	2	26 m (22.5-32.5 m)	0,79	4.96	0.0001
	2	46 m (> 32.5 m)	0		
Salinity range (delta S)	2	1 (delta S < 2.0)	-0,61	-2.24	0.0251
	2	2 (2.0 <= delta S < 10.0)	-0,18	-0.63	0.5280
	2	3 (10.0 <= delta S < 20.0)	-0,10	-0.31	0.7589
	2	4 (delta S <= 20.0)	0		

Test of normality: Variable		N	W:Normal	Pr < W	
Residuals	1	912	0,98	0.05	
	2	912	0,99	0.86	

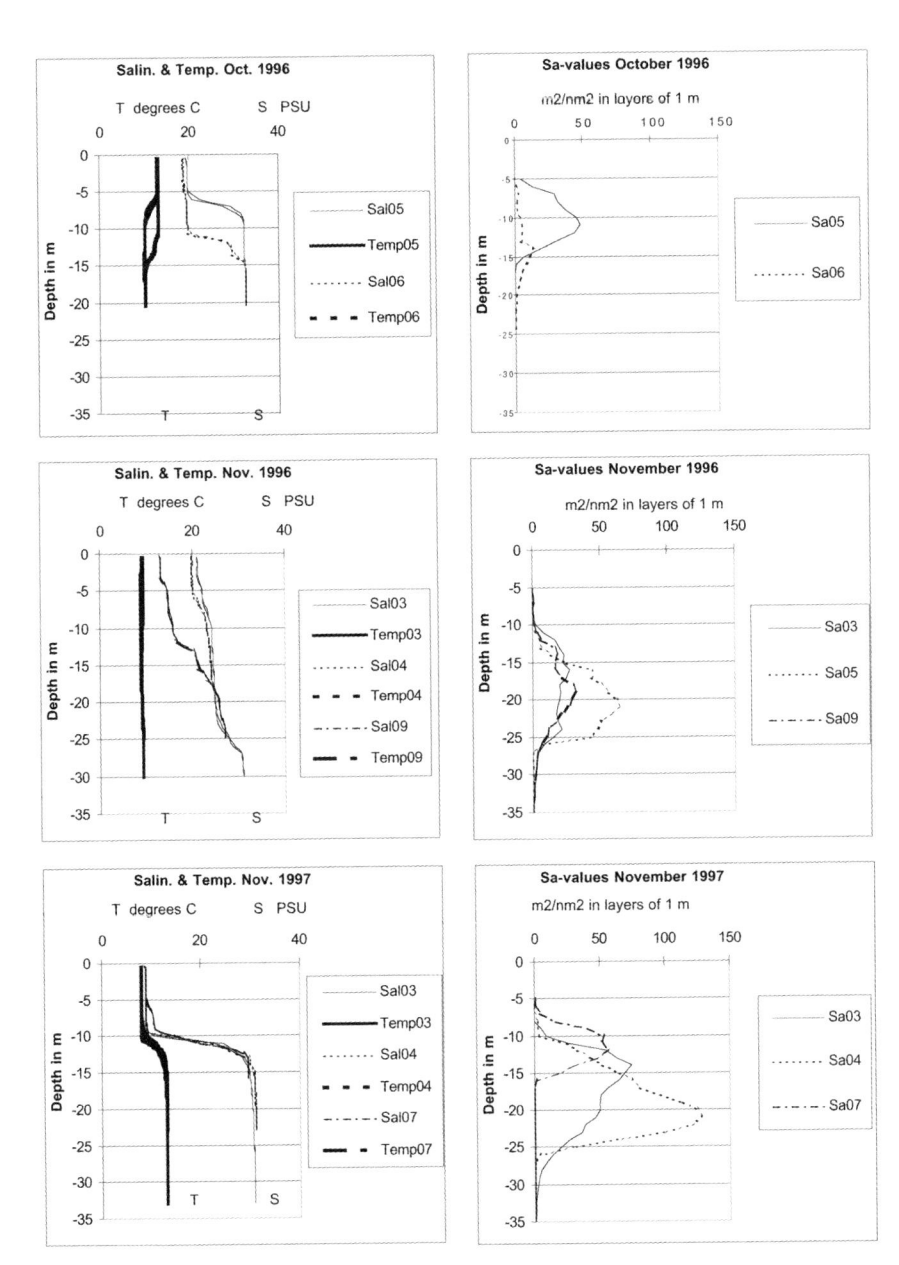

Figure 3. Comparison between typical salinity-temperature profiles (CTD) and typical acoustic density profiles (Sa values) at a selected position in the area north of the island of Ven in the Sound. The profiles cover the water column from sea surface to sea bottom at the respective localities, except for the surface layer from 0 to 5 m depth which is not covered in the acoustic density profile.

Table 3. Estimated biomass levels from hydroacoustic monitoring compared to landings and biomass levels of the RHS (as estimated by ICES, International Council for Exploration of the Sea 1999).

YEAR	BIOMASS SOUND Tons	ICES SSB Tons	LANDINGS Tons
1993	118.000	Not good quality	171.000
1994	84.000	Not good quality	163.000
1995	65.000	Not good quality	174.000
1996	91.000	162.000	132.000
1997	163.000	205.000	106.000

hypothesis h:1 and acceptance of hypothesis h:2. The year-class strengths found are in accordance with those obtained from larval abundance indices at the RHS spawning grounds (Table 4). Decreasing abundance in late spring was concordant with observed peaks in commercial landings (Fig. 2) and (subsequent) peak larval abundance indices at the RHS spawning grounds (Nielsen et al. 2001). This indicates that hypothesis h:3 should be accepted. The larger size groups seem to migrate southward to the spawning grounds in the spring before the smaller ones (Nielsen et al. 2001) leading to acceptance of hypothesis h:7. Herring density was analyzed statistically and described by a GLM-model (general linear models procedure in the SAS Statistical computer package where the GLM ANOVA can handle unbalanced data) in relation to year, season of year, geographical division, depth strata, temperature range, and salinity range (Table 2). The temperature and salinity ranges are the numerical differences between the lowest and the highest values in order to investigate the effect of the location of the thermocline and halocline, respectively. (Range classes are:

$\Delta T < 2.0 \Rightarrow \Delta T = 1$;
$2.0 \leq \Delta T < 10.0 \Rightarrow \Delta T = 2$;
$10.0 \geq \Delta T \Rightarrow \Delta T = 3$ (in degrees Celsius);
$\Delta S < 2.0 \Rightarrow \Delta S = 1$;
$2.0 \leq \Delta S < 10.0 \Rightarrow \Delta S = 2$;
$10.0 \leq \Delta S < 20.0 \Rightarrow \Delta S = 3$;
$20.0 \geq \Delta S \Rightarrow \Delta S = 4$ in parts per thousand).

Furthermore, density was analyzed in relation to major flows of water through the Sound to or from the Baltic Sea (Nielsen et al. 2001).

Highest densities were estimated in the northern areas in the autumn period, as well as in depths from 8 to 22 m below sea surface and just

Table 4. Larvae abundance indices at the spawning grounds for RHS (see Nielsen et al. 2001).

Year	Larvae number in millions
1986	1500
1987	1370
1988	1223
1989	63
1990	57
1991	236
1992	18
1993	199
1994	788
1995	171
1996	31
1997	54
1998	2202

below or in the halocline all year-round in areas with a stratified water column (Table 2, Fig. 3). Thus, the herring do not concentrate near the southern Drogden threshold in spring before southward migration. Consequently, hypothesis h:4 should be accepted and hypothesis h:5 rejected. Comparisons between the herring biomass estimates and the cumulative water flow in km past Nrd. Røse in different depth strata in the direction 0 degrees N in the Sound was made, where uphill water flow portions represents outflows from the Baltic and downhill portions inflows (see Nielsen et al. 2001). Indications of inflows to the Baltic triggering southward migration can be observed on some occasions; however, these results are not conclusive because of relatively few biomass estimates in 1996-1998 (Nielsen et al. 2001). This leads us to preliminary acceptance of hypothesis h:6. The identified patterns in herring occurrence agree with some conclusions on migration routes for RHS (between the main spawning and feeding grounds) from previously reported tagging studies, but the present investigations indicate prolonged overwintering in the Sound (Nielsen et al. 2001).

Acknowledgment

Modified from *Fisheries Research*, Volume number 50(2001):235-258. Authors: Nielsen, J.R., Lundgren, B., Jensen, T.F., and Stæhr, K.-J., title: "Distribution, density and abundance of the western Baltic herring (*Clupea harengus*) in the Sound (ICES Subdivision 23) in relation to hydrographical features," 24 pp. Copyright (2001), with permission from Elsevier Science.

Reference

Nielsen, J.R., B. Lundgren, T.F. Jensen, and K.J. Stæhr. 2001. Distribution, density and abundance of the western Baltic herring (*Clupea harengus*) in the Sound (ICES Subdivision 23) in relation to hydrographical features. Fish. Res. 50(2001):235-258.

The Norwegian Spring-Spawning Herring: Environmental Impact on Recruitment

R. Sætre and R. Toresen
Institute of Marine Research, Bergen, Norway

T. Anker-Nilssen
Norwegian Institute for Nature Research, Trondheim, Norway

P. Fossum
Institute of Marine Research, Bergen, Norway

Extended Abstract

The Norwegian continental shelf between 58° and 68°N is the spawning and hatching area for the Norwegian spring-spawning herring (Fig. 1). The herring spawn mainly at the southern-most coastal banks off central Norway in February-March. Since the start of the recovery of the Norwegian spring-spawning herring stock in the late 1970s, the main part of the stock spawned on a relatively limited shelf area at Møre and at the Sklinna Bank (Fig. 1). These locations have been stable spawning grounds since at least 1976. Parallel with the spawning stock increase, the herring stock reccurred at the traditional southern spawning sites, which had not been used since 1959, such as off Karmøy and the Sira Bank from 1989 onward. The Halten Bank was used as a spawning location from 1992 onward and the area off Lofoten Island from 1996. Since the early 1990s, first-feeding larvae have been found from the Sira Bank (Fig. 1) to about 69°N but the dominant spawning area is still at Møre between 62° and 63°30′N. The larvae hatch after approximately 3 weeks and are transported northward by the Norwegian Coastal Current into the nursery areas, mainly the Barents Sea. The hatching of the herring larvae usually starts during the first part of March with the main hatching during 20-30 March, and terminates around 20 April.

The Atlantic puffin (*Fratercula arctica*) is an important top predator of 0-group herring, and its colonies in Røst (Fig. 1), which constitute the largest aggregation of breeding seabirds in Norway, are strategically situ-

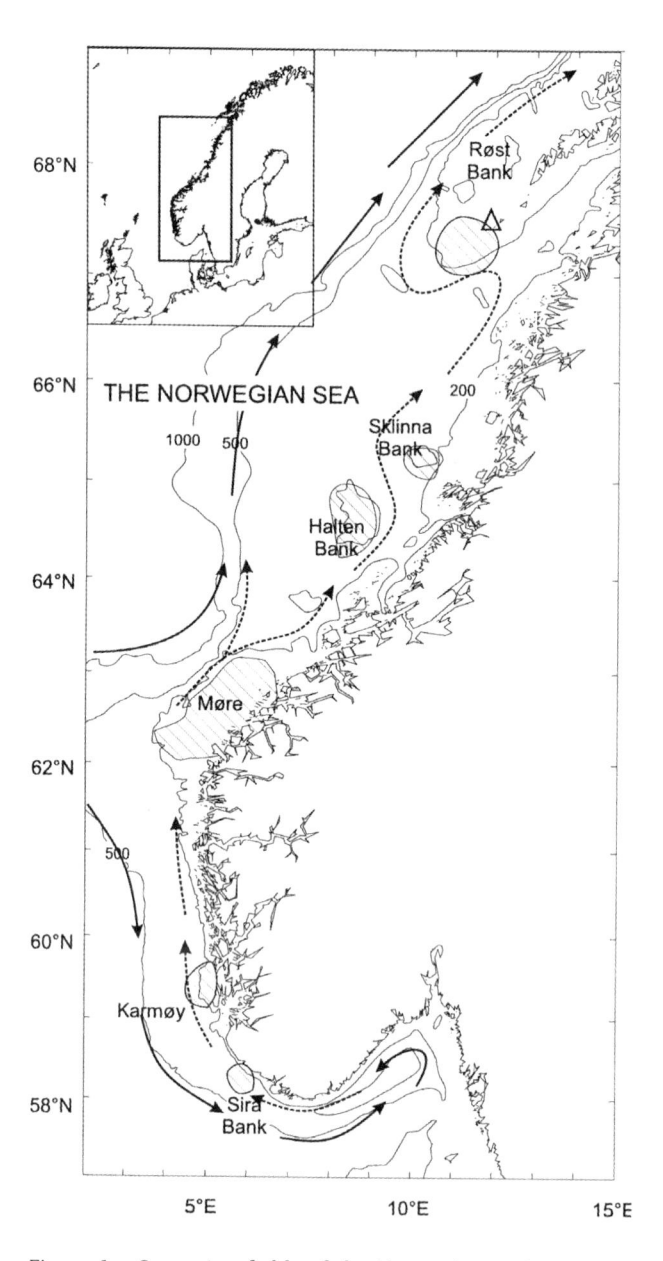

Figure 1. Spawning fields of the Norwegian spring-spawning herring and the persistent currents of Atlantic water (solid lines) and Coastal water (dotted lines). Triangle indicates the puffin colonies of the Røst archipelago.

ated at the northern part of the central Norwegian shelf (Sætre and others, unpubl.). During the last 20 years, the population of puffins breeding in Røst has dropped by 65% from almost 1.5 million pairs when monitoring started in 1979 to only 0.5 million in 1999. Most of this decline took place in the 1980s as a consequence of repeated reproductive failures attributed to the lack of 0-group herring, a staple food for nestling puffins in Røst. The fledging success of the puffins is positively correlated with the recruitment of the herring.

Our study is mainly focused on the period 1976-1999, during which there was a dramatic increase in the spawning stock size from approximately 100,000 t to more than 10 million t. The following questions were emphasized (Sætre and others, unpubl.):

- At what larval stage is the size of the new year class determined?
- What are the main driving mechanisms for regulating the recruitment?

Deduced by Lagrangian drifters, a number of retainment areas on the shelf between 62° and 68°N have been identified (Fig. 1). The mesoscale circulation within these locations includes both cyclonic and anticyclonic areas, and the residence time of the drifters within these may vary from 10 to 50 days (Sætre 1999). The anticyclonic retainment areas are usually associated with shallow banks. The trajectories of drifters, as well as the distribution pattern of the herring larvae and theoretical considerations, indicate that the banks may induce anticyclonic topographically trapped eddies or so-called Taylor columns. All the known spawning sites along the Norwegian coast are found in areas of topographically induced quasi-stationary eddies with increased residence time and with reduced vertical stability of the water. In such areas the newly hatched larvae have the possibility of being temporarily retained (retention areas). The term "retention area" is widely used for locations where different physical processes and larval behavior are mechanisms for retaining cohorts of fish larvae for a period of weeks or months.

The spawning area of the Norwegian spring-spawning herring consists of several separated spawning sites in the Norwegian Coastal Current, and each of them is associated with a retention area (Sætre et al. 2002). During their downstream drift, the herring larvae have the possibility of being temporarily entrained in several retention areas. These will most likely also entrain the prey of the larvae, such as copepod eggs and nauplii. Retention areas may favor growth rate and survival of the larvae by:

- increasing densities of food particles leading to higher growth rate and reduced mortality; and
- reducing vertical stability, resulting in a deeper-reaching wind-induced turbulence and thereby an increase in the predator-prey encounter rate.

High wind speeds in April seem to be favorable for recruitment. Good herring recruitment coincides with slower northerly larval drift than in years of poor recruitment. This indicates that interannual fluctuations in the degree of retention of the first-feeding larvae could be an important recruitment-regulating mechanism for the Norwegian spring-spawning herring (Sætre et al. 2002).

The high positive correlation between the fledging success of the puffins, and the recruitment of the herring at the 0-group stage, strongly indicate that year-class strength of the herring is predominantly determined before most of the herring larvae have passed Røst, i.e., in July. The extreme good year classes, such as 1983, 1991, and 1992, seem to be characterized by very low mortality after the larvae have passed Røst, i.e., during July-August. The natural mortality of the juvenile herring from the 0-group stage to the 3-year stage is only able to slightly modify this general picture by reducing some of the year classes from average at the 0-group stage to poor as 3-year-olds. Based on the mean temperature in the Barents Sea in January, the number of high wind speed events at Røst in April, and the fledging success of puffins at Røst, prediction of year-class strength of herring could be given in July.

A large-scale climatic signal, such as the North Atlantic Oscillation (NAO) winter index, is positively correlated to the southwesterly wind stress during the first quarter of the year as well as to the temperature condition both in the Barents Sea and along the coast of Norway. High values of these parameters favor good year classes of herring. Most likely, the variability in the circulation intensity and in the inflow of Atlantic water to the Barents Sea is a key factor in these relationships. During cold periods, the production and survival of the first-feeding larvae are low, always resulting in poor year classes. In warm periods, the survival of first-feeding larvae is good, and this is a necessary, but not sufficient, condition for a good year class at the 0-group stage (Sætre and others, unpubl.).

There seems to be a generic relationship between ocean climate parameters, such as temperature and wind stress, and recruitment for populations at the northern margin of their geographical distribution area. In the Norwegian and Barents seas system, high values of these are a necessary, but not sufficient, condition for successful recruitment. This implies increased probability for recruitment success with increasing values of ocean climate parameters. Large-scale variability in physical conditions thus sets the frame within which the other potential recruitment mechanisms, such as the availability of prey or predation, can operate. When controlling for the most important physical factors, the reproductive success of Atlantic puffins breeding in the Røst archipelago turned out to be the best biological predictor of herring year-class strength (Sætre and others, unpubl.).

References

Sætre, R. 1999. Features of the central Norwegian shelf circulation. Continental Shelf Res. 19:1809-1831.

Sætre, R., R. Toresen, H. Søiland, and P. Fossum. 2002. The Norwegian spring-spawning herring—spawning, larval drift and larval retention. Sarsia 87.

Disease and Population Assessment of Pacific Herring in Prince William Sound, Alaska

Terrance J. Quinn II
University of Alaska Fairbanks, School of Fisheries and Ocean Sciences, Juneau, Alaska

Gary D. Marty
University of California, School of Veterinary Medicine, Department of Anatomy, Physiology, and Cell Biology, Davis, California

John Wilcock
Alaska Department of Fish and Game, Juneau, Alaska

Mark Willette
Alaska Department of Fish and Game, Soldotna, Alaska

Abstract

Disease is rarely incorporated into fish stock assessment models because of a lack of information. A unique time series started in 1993-1994 on prevalence of two major diseases in Prince William Sound Pacific herring: viral hemorrhagic septicemia virus (VHSV) and the fungus-like organism *Ichthyophonus hoferi*. This research was prompted by a severe population decline in 1993 that shut down the fishery. Prevalence of *I. hoferi* appeared to be unrelated to the population decline. We modified the herring assessment model to let natural survival be linearly and negatively related to VHSV prevalence. Eight models allowed various natural mortality changes. There was a clear increase in natural mortality in 1992-1993, and VHSV information enhanced model fit compared to a constant natural mortality model. Because of higher VHSV prevalence in 1993 and 1997-1998, estimated natural mortality was higher in those years and estimated spawning biomass was reduced. However, the true effect of higher VHSV prevalence on natural mortality and spawning biomass cannot yet be unambiguously determined, because there is limited information for 1992-1993 and six of the eight mortality models produced nearly identical fits. Nevertheless, using VHSV prevalence for modeling and forecasting is more conservative than using constant natural mortality. Neither disease series was significantly correlated with recruitment, suggesting that disease has its main effect at the adult stage. However, the presence of a negative correlation between recruits and VHSV prevalence lagged 1 year suggests that disease may affect juveniles.

Introduction

Disease is a fundamental biological process that ultimately affects the longevity of individuals, and any processes that affect individuals may also impact populations. Studies have documented long-term population cycles in marine fish species (e.g., Lluch-Belda et al. 1992), and population decline has been associated with disease epizootics worldwide (e.g., Mellergaard and Spanggaard 1997, Whittington et al. 1997). However, the long-term ecological significance of enzootic disease in pelagic marine fish populations has never been examined. This study evaluates annual changes in disease prevalence within a schooling marine fish population: Pacific herring *Clupea pallasii* in Prince William Sound, Alaska. This information is integrated into a population stock assessment model to determine if the use of disease prevalence improves upon the status quo.

Pacific herring are schooling fish that range throughout coastal regions of the North Pacific. In Prince William Sound (PWS), Alaska, they first spawn when 3-5 years old, live up to 15 years, and weigh up to 300 g. Once mature, they spawn every year, but populations are usually dominated by a large year class about every 4 years. Pacific herring are an important component of the ecosystem as secondary consumers, and they also serve as high-quality prey for many marine mammals, birds, and fish. Commercial fisheries primarily harvest during the annual spawning season in April to obtain sac roe. Biomass estimates of prespawning aggregations of Pacific herring are commonly greater than 20,000 t. A single contiguous school can be several kilometers long and up to 2 km wide. At least one epidemic has been reported in Pacific herring (Tester 1942), and several extensive parasitologic surveys have been done (e.g., Moser and Hsieh 1992), but never has comprehensive medical pathology been applied to a fish population on a multiyear basis.

In 1989, the *Exxon Valdez* oil spill (EVOS) had significant oil-related effects in Pacific herring eggs and larvae (Brown et al. 1996). These effects were no longer apparent in 1990 or 1991. In 1992, a preliminary reproductive study was conducted on the 1988 year class, an abundant year class making its first important spawning contribution (Kocan et al. 1996). Because oil and oil-related effects were no longer demonstrable, and biologists were predicting record spawning biomass in 1993, the EVOS Trustee Council terminated all studies of Pacific herring after 1992.

During the winter of 1992-1993, the biomass of adult Pacific herring in PWS crashed. Over 110,000 t were predicted, but only an estimated 30,000 t actually returned to spawn. When the North American strain of viral hemorrhagic septicemia virus (VHSV) was isolated from several Pacific herring sampled from the population, it was hypothesized that a disease epidemic was responsible for the unexpectedly large natural mortality (Meyers et al. 1994). This was the first time VHSV was isolated from Pacific herring in Alaska, but isolation of VHSV from several other herring populations throughout the northeastern Pacific confounded interpretation of the significance of VHSV in population decline. Pacific herring fisheries were severely curtailed in 1993 and were closed in 1994, 1995, and most of 1996, resulting in significant loss to subsistence and commercial fishers. When the Pacific herring population continued to decline in 1994, a comprehensive disease study was initiated with the primary goal of evaluating the general health of the fish on a population scale. Results from the study in 1994 indicated that viral hemorrhagic septicemia virus (VHSV) and *Ichthyophonus hoferi* were the major pathogens contributing to disease (Marty et al. 1998), but the relative contribution of each pathogen was unknown. The study was repeated and greatly expanded in 1995 and continues to this day. Field study revealed that a decrease in prevalence of the two major pathogens was associated with population recovery, and a sharp increase in virus prevalence in 1997 was consistent with delayed recovery. In 1999, another population decline was observed in the surveys.

A rare opportunity exists to incorporate a time series of disease information with other knowledge about a population. Therefore, our objectives are to see if large changes in natural mortality occur in the stock assessment model for PWS herring by using disease information, to determine if the model is improved by using this information, and to examine whether recruitment (age 3) is related to this information.

Methods
Modeling Natural Mortality
Models to estimate fish populations commonly integrate many variables, including age, length, sex composition, fishing mortality, and indices of abundance such as catch per unit of effort and survey biomass (Quinn and Deriso 1999). For Pacific herring in Alaska, population biomass has been modeled and estimated since the early 1970s. Natural mortality M is usually treated as a constant in these models because of the difficulty in estimating mortality and the lack of comprehensive information about disease and predation (Vetter 1988). Evidence from a recent epizootic in the North Sea, however, provides strong evidence that use of a constant mortality in such models can lead to errors in stock assessment estimates (Patterson 1996).

Our work attempts to improve upon previous models by using actual disease information, as was done for North Sea herring by Patterson (1996). We concentrate on natural mortality models, because most of the data collected pertain to post-recruit fish. However, we also analyze recruitment information in relation to disease variables.

We assume that disease increases the natural mortality M for ages beyond the age of recruitment, or equivalently, that disease lowers the corresponding natural survival S. Presence of the pathogen does not necessarily equal disease, and fish that are not diseased today may be diseased tomorrow. The primary variable collected for disease studies is the pathogen sample prevalence x_t, the proportion of herring in the sample with the pathogen present.

We assume that natural mortality M_0 from sources other than disease is constant. Corresponding natural survival from these sources is $S_0 = \exp(-M_0)$. We then assume that natural survival decreases linearly as a function of disease prevalence, or $S = S_0 (1 - \beta x_t)$. Because $M = -\ln(S)$, the expression for natural mortality is

$$M = M_0 - \ln(1 - \beta x_t), \qquad (1)$$

where β represents the proportion of the infected population that dies from the infection. If $x_t = 0$ (no disease), then $S = S_0$ and $M = M_0$. If $x_t = 1$ and $\beta = 1$ (all fish are infected and die), then $S = 0$ and $M = \infty$. Parameter β can be greater than 1 if the incidence variable underestimates the mortality due to the disease. Other models, such as a proportional or regression relationship between M and x_t, could also be used (Patterson 1996).

Application to PWS Pacific Herring

Scientists with the Alaska Department of Fish and Game have assessed biomass of PWS Pacific herring with an age-structured assessment (ASA) model (Funk 1995; Wilcock 1998, unpubl.; Willette 1999, unpubl.). The ASA model provides an estimation framework to integrate the various sources of information about Pacific herring in PWS from 1980 to the present (1998 for this study).

The four main sources of information used to estimate population parameters in the stock assessment model are estimates of age composition from the purse-seine fishery ($n = 88$) and from annual spawning surveys ($n = 116$), annual estimates of egg production from egg surveys ($n = 10$), and annual observations of mile-days of milt from aerial surveys ($n = 19$). These observations are compared to comparable model quantities in a least squares setting to obtain parameter estimates. The egg and milt data sets are assumed to be lognormally distributed, so their residual sums of squares are on a logarithmic scale. The sum of squares for mile-days of milt is weighted by one-half relative to the other data sources because of conflicts with the other main sources. Other data include weight-at-age, fecundity-at-age, total seine catch, other fisheries' catch-at-age, and acoustic survey biomass (used as a lower bound). Estimated parameters in the model include recruitment, maturity, seine selectivity, and a milt calibration coefficient.

The pre-fishery population starts at the beginning of the spawning season and includes recruitment at age 3 for each year. The fisheries on PWS herring include purse-seine, gillnet, and pound fisheries in the spring (mainly for roe), and a food and bait fishery in the summer and fall. Total catch for the seine fishery and catch-at-age for the other fisheries are assumed measured without error in the model. A half-year convention for survival is applied, leading to the recursion relationship for age *a* and year *t*:

$$N_{a+1,t+1} = \left[\left(N_{a,t} - C_{seine} - C_{gill} - C_{pound}\right) \times S_t^{1/2} - C_{food}\right] \times S_{t+1}^{1/2} \quad (2)$$

where N = abundance, C = annual catch-at-age, and $S^{1/2}$ = half-year survival, calculated from natural mortality by the relationship $S^{1/2} = \exp(-M/2)$. The reason for the term $S_{t+1}^{1/2}$ in equation 2 is that we believe that the disease variable measured in year *t*+1 at the time of spawning affects the population in the last half-year of year *t* (before recruitment, spawning, and the spring fisheries) and the first half-year of year *t*+1 (after recruitment, spawning, and the spring fisheries). The corresponding number of spawners is the mature abundance after the spring fisheries, or

$$Spawners_t = \sum_a mat_a \left(N_{a,t} - C_{seine} - C_{gill} - C_{pound}\right)$$

in which *mat* is the proportion mature. Spawning biomass also uses weight-at-age in this equation.

Annual prevalence of VHSV and *Ichthyophonus* from sample(s) taken in the spring varied each year (Table 1). After 1993, 210-250 herring were individually examined for VHSV prevalence. In 1993, four batches of 5 fish, not distinguishable by individual fish, were examined for VHSV as an exploratory investigation. Researchers were surprised to find that three of the four batches were positive. The true prevalence could have been as low as 3/20 or as high as 15/20, depending on how many fish in a batch had the virus. We use three values (2/15, 6/15, and 10/15) for 1993 VHSV prevalence in different models to explore the sensitivity of the results. The time trend in *Ichthyophonus* prevalence shows little change from year to year (Table 1) and does not follow the observed pattern in PWS herring biomass. Prevalence was higher in 1994-1996 when no declines in spawning biomass were observed and lower in 1997-1998 just before the 1999 decline. Therefore, no models were constructed with *Ichthyophonus* prevalence.

We construct eight alternative models of natural mortality, summarized in Table 2. ADFG scientists used to estimate a constant natural mortality *M* for all years other than the 1992-1993 collapse period and a separate value for 1992-1993. They constrained the estimation so that natural mortality must be at least 0.25, which is thought to be the lowest realistic value of average natural mortality for Pacific herring. Usually the estimate converged to this constraint value of 0.25. Because our interests are in disease mortality above this average level, we fix $M_0 = 0.25$ and concentrate on estimating β, the disease-related natural mortality parameter.

Table 1. Prevalence of VHSV and *Ichthyophonus hoferi* in Prince William Sound Pacific herring sampled in March and April between 1993 and 1998. Sample size for 1994-1998 varies from 210 to 260 per year.

	VHSV Min	VHSV Med	VHSV Max	*Ichthyophonus*
1993	2/15=0.133	6/15=0.4	10/15=0.667	
1994		0.047		0.231
1995		0.023		0.192
1996		0.000		0.204
1997		0.146		0.162
1998		0.136		0.176

Note: In 1993, 3 out of 4 batches of 5 fish were VHSV+.

Table 2. Eight natural mortality models with respect to natural mortality in 1992 (M92), in 1993 (M93), and thereafter. Assumed 1993 VHSV prevalence is also shown. Natural mortality before 1992 in all models is 0.25.

Model	M92	M93	M after 1993	1993 VHSV prevalence
1	0.25	0.25	0.25	N/A
2	=M93	E	0.25	N/A
3	E	D	D	10/15
4	=M93	D	D	10/15
5	E	D	D	6/15
6	=M93	D	D	6/15
7	E	D	D	2/15
8	=M93	D	D	2/15

E, estimated; D, from disease model.
N/A, not applicable.

Model 1 is a base model in which natural mortality is constant and equal to 0.25 for all years 1980-1998. In model 2, disease information is not used, but a separate value for natural mortality is estimated for 1992-1993 (the ADFG approach prior to 1998), which assumes that the only change in natural mortality occurred in these 2 years. An attempt was made to run a modification of this model, in which a separate value for natural mortality after 1993 was estimated. However, the estimate converged to the constraint of 0.25 (the same as specified for model 2).

In models 3-8, natural mortality prior to 1992 is assumed to be equal to parameter $M_0 = 0.25$ and is determined from equation 1 using VHSV prevalence thereafter. In models 3 and 4, 1993 VHSV prevalence is assumed to be 10/15, the highest possible value. In model 3, natural mortality for the first half-year in 1992 after the time of spawning was estimated separately, and natural mortality thereafter was estimated from the disease model. ADFG used this model in 1998 and 1999. In model 4, natural mortality for the first half-year in 1992 is assumed equal to the value from the disease model in 1993. Models 5 and 6 are the same as models 3 and 4, except that 1993 VHSV prevalence is 6/15 (the median). Models 7 and 8 use a 1993 VHSV prevalence of 2/15, the lowest possible value.

We compare these models by examining their residual sums of squares and parameter estimates and by performing F-tests. Such comparisons show how much additional variability is explained by the disease data, whether the reduction in variability is statistically significant, and if the parameter estimates are credible.

A bootstrap procedure is used to illustrate the uncertainty in estimated natural mortality, spawning biomass, and recruitment. Natural mortality before 1992 was held fixed at 0.25, in accord with the estimation procedure, because information from the early period is so uncertain that it is likely that natural mortality is not estimable in this period. The bootstrap contained 1,000 replications of resampled data from the main data sources. For the egg and milt information, residuals (on a log scale) are resampled and added to predicted values from the original run. For seine and spawning age compositions, a parametric bootstrap is applied using the multinomial distribution with annual effective sample sizes derived from the original run. The disease information is not resampled, because it is an independent variable and the analysis is not set up as an errors-in-variables approach.

Disease could affect recruitment if it occurs at ages prior to the recruitment age. Correlations are computed between recruitment estimates in the stock assessment model and disease prevalence at lags 0 and 1 to study if this effect is important.

Table 3. Number of observations (*n*), number of parameters (*p*), residual sums of squares (RSS) and data source components, estimated natural mortality parameter β, 1999 estimated spawning biomass (SB99, in metric tons) for the eight models, along with selected tests of equality of models.

	Model							
	1	2	3	4	5	6	7	8
n	233	233	233	233	233	233	233	233
p	27	28	29	28	29	28	29	28
RSS	4.59	2.99	2.98	3.01	2.99	3.03	3.02	3.31
F-test	2 vs. 1	3 vs. 2	4 vs. 3		6 vs. 5		8 vs. 7	
F-statistic	109.4	0.4	1.4		2.2		19.1	
P-value	0.00	0.52	0.24		0.14		0.00	
β	N/A	N/A	0.76	1.01	1.13	1.68	1.28	4.47
SB99	39,491	41,277	36,702	35,705	34,689	32,467	33,498	19,821
RSS Components								
Seine agecomp.	0.32	0.26	0.28	0.29	0.28	0.31	0.28	0.42
Spawn agecomp.	0.39	0.27	0.26	0.26	0.26	0.26	0.26	0.30
Egg survey	2.98	1.47	1.44	1.46	1.43	1.47	1.43	1.61
Mile-days of milt	0.89	0.99	1.01	1.00	1.02	1.00	1.05	0.98

Results

Table 3 shows the number of observations, the number of parameters, sums of squares, estimated natural mortality parameter β, and estimated 1999 spawning biomass for the eight models. Six of the eight models (models 2-7) have comparable fits to the data. Model 1 (constant natural mortality) fits the data poorly, particularly in the period 1992-1993. The sum of squares for model 1 is significantly higher than for any other model. Of the other models, model 8, in which natural mortality in 1992-1993 comes from the disease model with the low value of 1993 VHSV prevalence, has a higher sum of squares. Its significant difference from model 7 suggests that actual mortality in 1992-1993 must be higher than that predicted by this model. For models 2-7, there are no significant differences among models. Thus, disease prevalence adequately accounts for mortality for 1992 and thereafter, and there is no need for a separate natural mortality parameter in 1992 in the disease models. Nevertheless, model 2 (which does not use disease information, estimates a separate mortality parameter for 1992-1993, and has natural mortality of 0.25 after 1993) fits the data as well as (even slightly better than) the disease models, suggesting that the impact of VHSV prevalence in 1997-1998 on natural mortality cannot be determined with certainty.

The estimate of parameter β ranges from 0.76 to 1.68 in models 3-7, which is near the value of 1 that would occur if disease prevalence measured natural mortality exactly. This result suggests that VHSV prevalence determined at a short interval in the spring is closely related to disease-related mortality and that there is no major underestimation of disease prevalence. Model 8 has an unrealistically high value of 4.47, suggesting that this model is less credible than the other disease models.

Estimated 1999 spawner biomass is lower in models 3-8 (when disease prevalence was used) than in models 1 and 2 (Table 3), because natural mortality in 1997 and 1998 is estimated higher than 0.25, as now shown. Estimated natural mortality for 1992-1998 is shown in Fig. 1 for the eight models classified by three values of 1993 disease prevalence. For the high 1993 value of 10/15, estimated natural mortality in 1992 and 1993 is substantially higher (roughly four-fold) than in other years, because the corresponding prevalence is much higher. When a 1993 value of 6/15 is used, the perception changes to a two-fold difference. When a 1993 value of 2/15 is used, estimated natural mortality in 1997 and 1998 is higher than in other years, because their prevalences are higher than 2/15. One other major feature of this figure is that estimated natural mortality in 1992 is much higher than in other years in models 3, 5, and 7, which estimate this parameter separately from the disease model. These high values are more indicative of model instability than of actual population changes, because there is either no or minor improvement in fit by comparing models 3 and 4, 5 and 6, or 7 and 8 (Table 3).

The models differ somewhat in their estimates of spawning biomass and recruitment over time (Fig. 2). Assuming constant natural mortality over all years (model 1) has the biggest effect. Spawning biomass declines sharply after 1989 in model 1 rather than being somewhat stable in other models until after 1992. All models show some rebuilding after 1994, but model 8 has a decline in 1997 and 1998. Recruitment values in 1990-1992 are lower for models 1 and 8 than for other models, and model 8 has much higher recruitment after 1994 (presumably to attempt to counteract the higher estimated natural mortality in 1997 and 1998). These results show that recruitment and natural mortality are interrelated in the parameter estimation, so that unanticipated results may occur when changing natural mortality assumptions in models.

Further understanding of the differences between model 1 and the other models comes from examining the sum of squares components for the four different data sources (Table 3). The two age composition components change little across the models. The fit to the egg survey information is degraded in model 1, and to a lesser extent model 8, compared to the other models, with a slightly better fit to the milt information. A graphical comparison of the fits from model 1 and model 4 (as a representative of the other models) to these two data sources (Fig. 3) shows that model 1 essentially ignores the increasing egg survey trend over the period 1988-1992, which is the opposite of the decreasing milt trend over the same period. The other models tend to average over the two sources of information, creating a slight peak in 1992. Consequently, each model has integrated the conflicting information, so that the predicted mile-days of milt and egg abundances in Fig. 3 are strongly correlated with estimated spawning

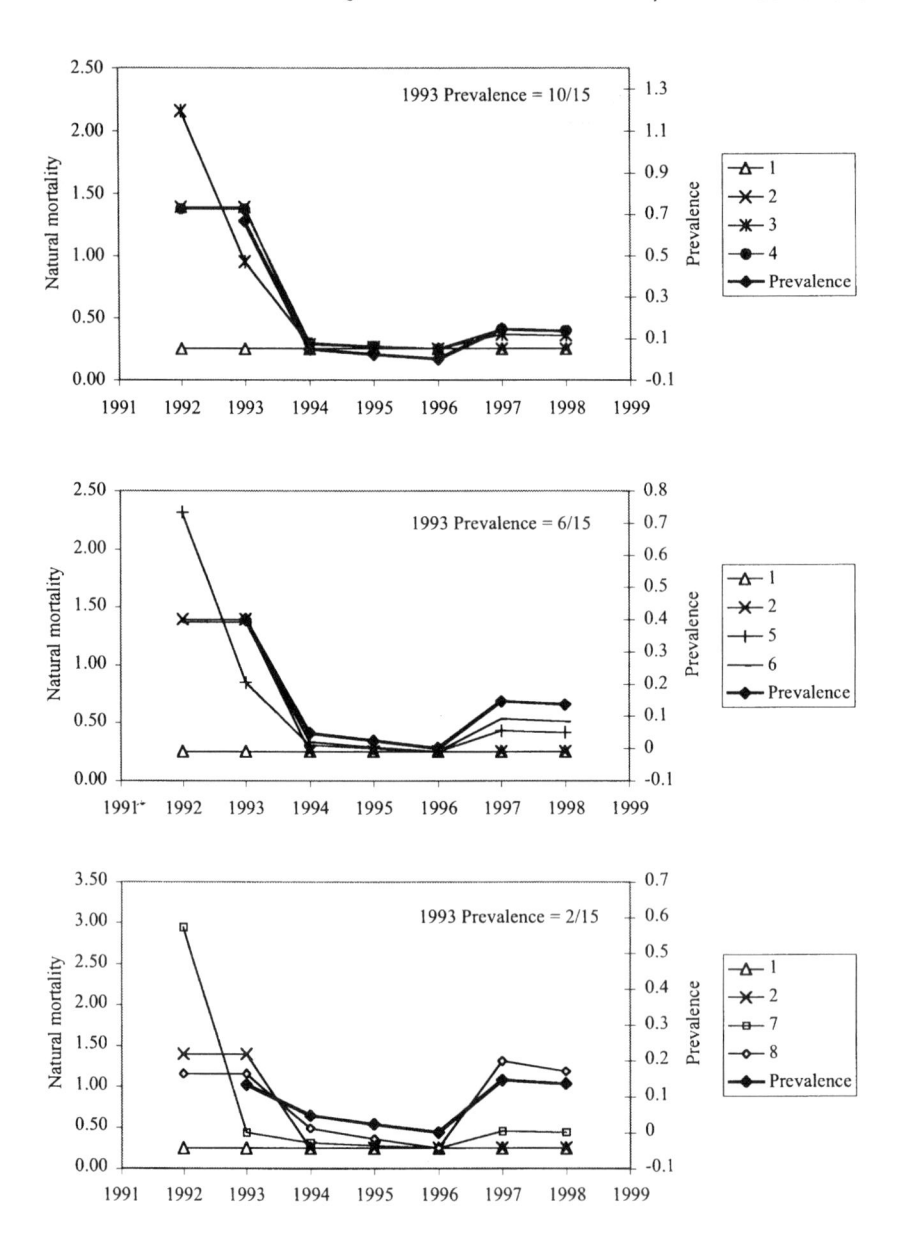

Figure 1. Estimated natural mortality over time from the eight models for three values of VHSV prevalence in 1993.

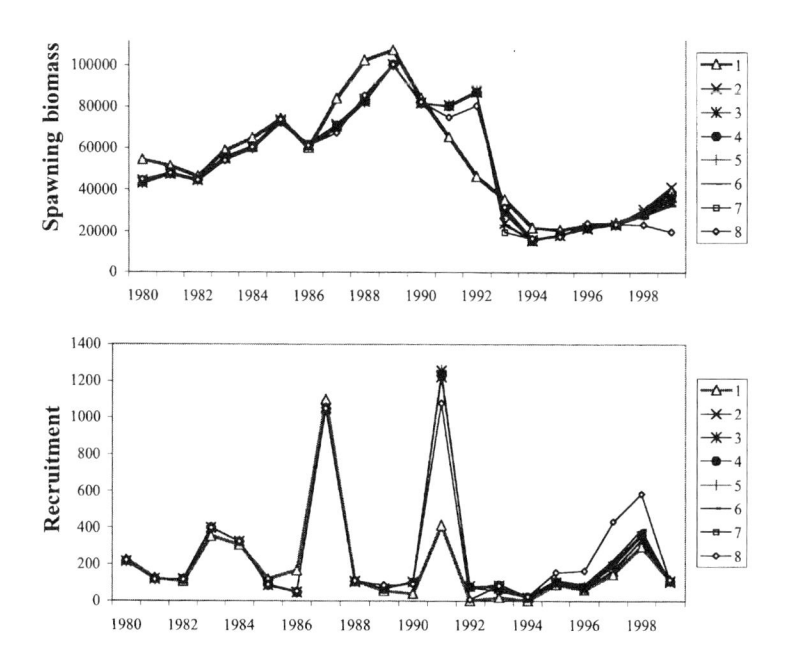

Figure 2. Estimated spawning biomass (t) and recruitment (millions) over time from the eight models.

biomass in Fig. 2. Thus, the use of constant natural mortality in model 1 forces adjustment of other population parameters in the estimation procedure to account for the steep drop in biomass between 1992 and 1993 apparent in the data sources. The primary adjustment in model 1 is a lowering of estimated recruitment over the period 1988-1992 (Fig. 2), which leads to a reduction in estimated spawning biomass with no peak in 1992 (Fig. 2).

Is it possible from these results to conclude that one of these models is best? Models 1 and 8 can be eliminated based on their poorer fits to the data. Models 7 and 8 based on the low 1993 prevalence value are less credible than others, because we suspect that the mortality event in 1992-1993 was larger than the recent 1999 event. Models 1 and 2 are less conservative in that they assume that natural mortality after 1993 is constant at 0.25, which ignores the possibility of an increase in natural mortality due to the 1997-1998 increase in VHSV prevalence. From the F-tests in Table 2, models 4 and 6 are more parsimonious than their counterparts 3 and 5, suggesting that it is not possible to estimate with confidence a separate mortality for the first half-year in 1992. Finally, the choice between models 4 and 6, with median and high 1993 prevalence, respectively, can only be resolved based on how strong the 1992-1993 mortality event is perceived in relation to that of 1999. This resolution awaits data

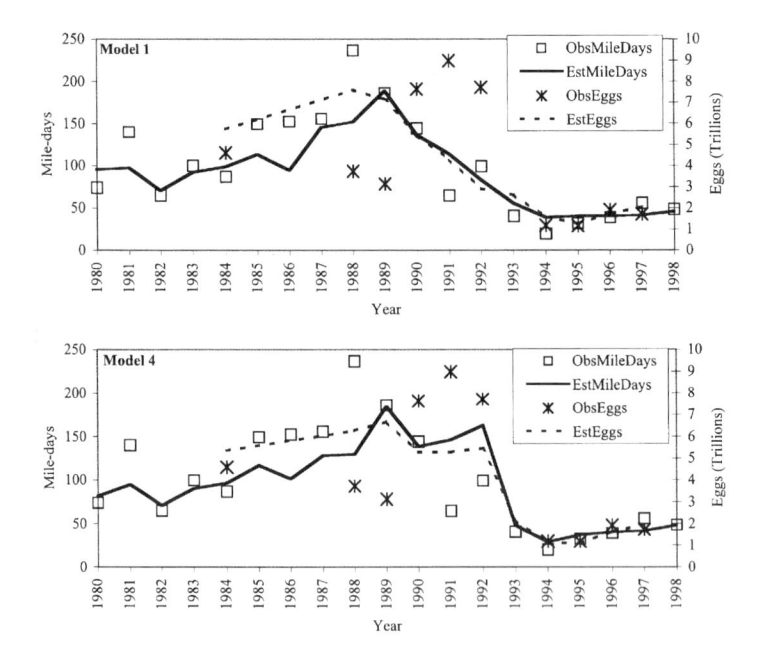

Figure 3. Observed and estimated values for mile-days of milt and egg abundance from model 1 (constant natural mortality) and model 4 (variable natural mortality related to disease prevalence).

collected after 1998, beyond the time period considered in this paper. We illustrate further model results using model 4 but do not assert that this model is better than model 6.

Bootstrap results on the parameter β show a left-skewed distribution and a 95% confidence interval of (0.62, 1.15) about the original estimate of 1.01 (Fig. 4). Therefore, this parameter is statistically significant from 0, in accord with the model testing results. Corresponding confidence intervals for natural mortality over time are narrow except for 1992-1993 (Fig. 5) and admittedly understate the uncertainty, being conditioned on the base level of 0.25. Their lower bounds are above 0.33 in 1992-1993, 1997, and 1998, the years with high VHSV prevalence.

Correlations between disease prevalence and the logarithm of recruitment at lags 0 and 1 are not significant for any of the three values of 1993 disease prevalence (Table 4). Nonsignificant results are also obtained when recruitment itself is used. Lag 1 was examined to see if there was a discernible prerecruit effect of disease for herring at age 2. While results are not significant, the strong negative correlations at lag 1 are notable, suggesting that a possible relationship could emerge with more years of data. The strongest relationship occurs with the high 1993 prevalence value.

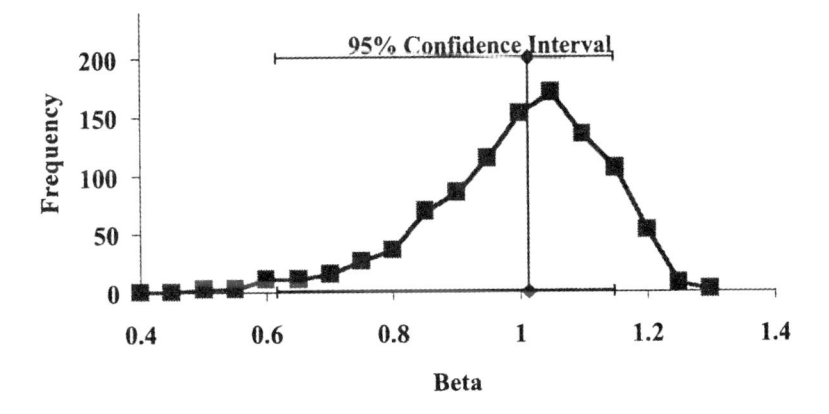

Figure 4. Bootstrap distribution of natural mortality parameter β from model 4. The point estimate and 95% confidence interval are also shown.

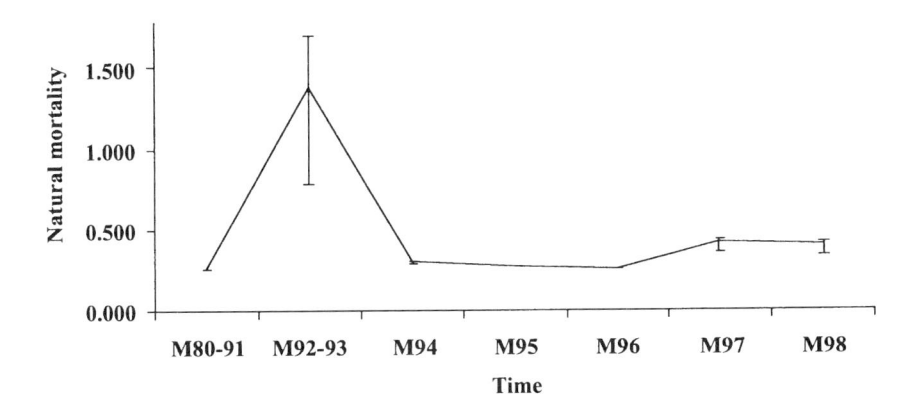

Figure 5. The 95% confidence intervals for natural mortality for 1992 and later from model 4 and the bootstrap procedure. The constant value 0.25 for natural mortality before 1992 is shown for comparison.

Table 4. Estimated correlations and *P*-values between logarithm of recruitment from model 4 and VHSV prevalence at lags 0 and 1 for the three values of 1993 VHSV prevalence.

Prevalence in 1993		Lag 0	Lag 1
10/15	Correlation	0.08	−0.74
	P-value	0.89	0.16
		NS	NS
6/15	Correlation	0.17	−0.67
	P-value	0.78	0.22
		NS	NS
2/15	Correlation	0.44	−0.27
	P-value	0.46	0.65
		NS	NS

NS, not significant.

Discussion

Disease information improved the Prince William Sound Pacific herring population model. The models with VHSV prevalence fitted the data better than model 1 with constant natural mortality over the entire time period and accounted for perceived mortality events in the population in 1992-1993 and 1999. Consequently, spawning biomass was reduced when these events occurred. Without disease information, such mortality events cannot be predicted ahead of time or accounted for in forecasts. Therefore, ongoing collection of disease information is necessary for accurate forecasting and careful management.

One caveat about our conclusions is the uncertainty in the strength of the relationship between VHSV prevalence and natural mortality. Natural mortality is frightfully difficult to estimate in assessment models, and the same is true here. The statistical significance of our results depends on prevalence of VHSV in 1992-1993, which is based on extremely limited information. It is clear that a major natural mortality event occurred in that period, leading to a population collapse, and that is consistent with VHSV being present. However, whether the increase in VHSV in 1997-1998 leads to a similar collapse in the future remains unclear. Model 2, based on constant natural mortality of 0.25 after 1993, fits the data as well as models using disease information (models 3-7). We do know that fewer fish returned to spawn in 1999 than were expected, which is consistent with the increase in VHSV in 1997-1998. Therefore, it would be a conservative assessment policy to use the disease model and refine it as new prevalence data are collected.

The lack of a relationship with the fungus *Ichthyophonus* differs from results in Patterson (1996) for North Sea herring. This does not mean that it does not affect PWS herring but rather that it has not expressed itself in terms of a large mortality event.

Our results suggest that disease primarily affects the adult stage. The disease event in 1992-1993 appeared to be linked to the condition of adults, because the decline in weight-at-age between fall 1992 and spring 1993 was the greatest observed (Pearson et al. 1999). Furthermore, the zooplankton bloom in PWS was also nearly nonexistent during the summer of 1992, and the adult herring population was very high, suggesting the possibilities of food limitation and subsequent stress on fish health.

The lack of significant correlation between estimated recruitment and disease might be due to confounding from other factors. Recruitment here is defined as the abundance of herring at age 3, as estimated from the catch-age model. Because recruitment is not measured directly, there may be a real disease effect on juveniles that was not discernible in the present analysis. The disease samples were from adult fish and these samples do not reflect disease incidence among juveniles, because juveniles and adults occupy different habitats and may experience different stressors. With Pacific herring and other fish species, the youngest and oldest individuals in any animal population are most susceptible to disease (MacLean et al. 1987, McVicar et al. 1988, Marty et al. 1998). We may not have seen a significant correlation between disease and recruitment, because disease is just one of many factors that determine recruitment, and we have a limited data set for examining effects of multiple factors.

Further improvements in understanding could be made by further investigating disease by size or age of herring. *Ichthyophonus* is more prevalent in older herring (G.D. Marty, unpubl.), and age-specific mortality- prevalence models may help to investigate whether there could be consequent delays in population recovery. VHSV affects younger herring more than older herring and we intend to examine this effect in future model development. The EVOS Trustee Council has funded a considerable effort to examine the relationship between juvenile condition, overwintering mortality, and recruitment (Foy and Norcross 2001). Although this work indicates that juveniles may just starve during the winter, disease or greater vulnerability to predation or both may be the cause of death. At any rate, extending our disease sampling to juveniles may shed more light on relationships to recruitment. This could be accomplished by stratifying disease sampling by age class along with measurements of condition (energy content). Further insight into how disease transfers through the population as it ages would also be revealing. It is possible that the prevalence of disease may have a density-dependent component that alters the temporal sustainability of the PWS herring population. Also, the addition of ulcer prevalence to the model may help, because it seems to correlate well with population mortality.

Acknowledgments

This work was funded by grants from the National Science Foundation (project numbers 9871982, 9901351) and by the *Exxon Valdez* Oil Spill Trustee Council through contracts with the Alaska Department of Fish and Game. The findings and conclusions presented by the authors are their own and do not necessarily reflect the views or position of the Trustee Council. The virus analysis was conducted by the ADFG Fish Pathology Laboratory, under the direction of Theodore Meyers. We thank editor Fritz Funk, referee Jeff Fujioka, and an anonymous referee for helpful comments that improved this paper.

References

Brown, E.D., T.T. Baker, J.E. Hose, R.M. Kocan, G.D. Marty, M.D. McGurk, B.L. Norcross, and J. Short. 1996. Injury to the early life history stages of Pacific herring in Prince William Sound after the *Exxon Valdez* oil spill. Am. Fish. Soc. Symp. 18:448-462.

Foy, R.J., and B.L. Norcross. 2001. Juvenile herring feeding ecology in Prince William Sound, Alaska. In: F. Funk, J. Blackburn, D. Hay, A.J. Paul, R. Stephenson, R. Toresen, and D. Witherell (eds.), Herring: Expectations for a new millennium. University of Alaska Sea Grant, AK-SG-01-04, Fairbanks. (This volume.)

Funk, F. 1995. Age-structured assessment of Pacific herring in Prince William Sound, Alaska and forecast of abundance for 1994. Alaska Department of Fish and Game, Juneau. Draft unpubl. manuscript.

Kocan, R.M., G.D. Marty, M.S. Okihiro, E.D. Brown, and T.T. Baker. 1996. Reproductive success and histopathology of individual Prince William Sound herring 3 years after the *Exxon Valdez* oil spill. Can. J. Fish. Aquat. Sci. 53:2388-2393.

Lluch-Belda, D., R.A. Schwartzlose, R. Serra, R. Parrish, T. Kawasaki, D. Hedgecock, and R.J.M. Crawford. 1992. Sardine and anchovy regime fluctuations of abundance in four regions of the world oceans: A workshop report. Fish. Oceanogr. 1:339-347.

MacLean, S.A., C.M. Morrison, R.A. Murchelano, S. Everline, and J.J. Evans. 1987. Cysts of unknown etiology in marine fishes of the Northwest Atlantic and Gulf of Mexico. Can. J. Zool. 65:296-303.

Marty, G.D., E.F. Freiberg, T.R. Meyers, J. Wilcock, T.B. Farver, and D.E. Hinton. 1998. Viral hemorrhagic septicemia virus, *Ichthyophonus hoferi*, and other causes of morbidity in Pacific herring *Clupea pallasi* spawning in Prince William Sound, Alaska, USA. Dis. Aquat. Org. 32:15-40.

McVicar, A.H., D.W. Bruno, and C.O. Fraser. 1988. Fish diseases in the North Sea in relation to sewage sludge dumping. Mar. Pollut. Bull. 22:169-173.

Mellergaard, S., and B. Spanggaard. 1997. An *Ichthyophonus hoferi* epizootic in herring in the North Sea, the Skagerrak, the Kattegat and the Baltic Sea. Dis. Aquat. Org. 28:191-199.

Meyers, T.R., S. Short, K. Lipson, W.N. Batts, J.R. Winton, J. Wilcock, and E. Brown. 1994. Association of viral hemorrhagic septicemia virus with epizootic hemorrhages of the skin in Pacific herring *Clupea harengus pallasi* from Prince William Sound and Kodiak Island, Alaska, USA. Dis. Aquat. Org. 19:27-37.

Moser, M., and J. Hsieh. 1992. Biological tags for stock separation in Pacific herring *Clupea harengus pallasi* in California. J. Parasitol. 78:54-60.

Patterson, K.R. 1996. Modelling the impact of disease-induced mortality in an exploited population: The outbreak of the fungal parasite *Ichthyophonus hoferi* in the North Sea herring (*Clupea harengus*). Can. J. Fish. Aquat. Sci. 53:2870-2887.

Pearson, W.R., R.A. Elston, R.W. Bienert, A.S. Drum, and L.D. Antrim. 1999. Why did the Prince William Sound, Alaska, Pacific herring (*Clupea pallasi*) fisheries collapse in 1993 and 1994? Review of the hypotheses. Can. J. Fish. Aquat. Sci. 56:711-737.

Quinn II, T.J., and R.B. Deriso. 1999. Quantitative fish dynamics. Oxford University Press. 542 pp.

Tester, A.L. 1942. Herring mortality along the south-east coast of Vancouver Island. Fish. Res. Board Can., Prog. Rep. Pac. Coast Stn. 52:11-15.

Vetter, E.F. 1988. Estimation of natural mortality in fish stocks: A review. Fish. Bull., U.S. 86:25-43.

Whittington, R.J., J.B. Jones, P.M. Hine, and A.D. Hyatt. 1997. Epizootic mortality in the pilchard *Sardinops sagax neopilchardus* in Australia and New Zealand in 1995. I. Pathology and epizootiology. Dis. Aquat. Org. 28:1-16.

Taking Stock: An Inventory and Review of World Herring Stocks in 2000

D.E. Hay,[1] R. Toresen,[2] R. Stephenson,[3] M. Thompson,[1] R. Claytor,[4] F. Funk,[5] E. Ivshina,[6] J. Jakobsson,[7] T. Kobayashi,[8] I. McQuinn,[9] G. Melvin,[3] J. Molloy,[10] N. Naumenko,[11] K.T. Oda,[12] R. Parmanne,[13] M. Power,[3] V. Radchenko,[6] J. Schweigert,[1] J. Simmonds,[14] B. Sjöstrand,[15] D.K. Stevenson,[16] R. Tanasichuk,[1] Q. Tang,[17] D.L. Watters,[12] and J. Wheeler[18]

[1]*Fisheries and Oceans Canada, Pacific Biological Station, Nanaimo, B.C., Canada*
[2]*Marine Research Institute, Bergen, Norway*
[3]*Fisheries and Oceans Canada, Biological Station, St. Andrews, N.B, Canada*
[4]*Fisheries and Oceans Canada, Bedford Institute of Oceanography, Dartmouth, Nova Scotia, Canada*
[5]*Alaska Department of Fish and Game, Juneau, Alaska*
[6]*Sakhalin Research Institute of Fisheries and Oceanography, Yuzhno-Sakhalinsk, Russia*
[7]*Marine Research Institute, Reykjavik, Iceland*
[8]*Hokkaido National Fisheries Research Institute, Hokkaido, Japan*
[9]*Institut Maurice Lamontagne, Mont-Joli, Québec, Canada*
[10]*The Marine Institute, Abbotstown, Dublin, Ireland*
[11]*KamchatNIRO, Petropavlovsk Kamchatsky, Russia*
[12]*California Department of Fish and Game, Belmont, California*
[13]*Finnish Game and Fisheries Research Institute, Helsinki, Finland*
[14]*FRS Marine Laboratory Aberdeen, Aberdeen, UK*
[15]*Institute of Marine Research, Lysekil, Sweden*
[16]*Maine Department of Marine Resources, West Boothbay Harbor, Maine*
[17]*Chinese Academy of Fisheries Science, Yellow Sea Fisheries Research Institute, Shandong, People's Republic of China*
[18]*Department of Fisheries and Oceans, Northwest Atlantic Fisheries Centre, St. John's, Nfld; Canada*

Present affiliation for T. Kobayashi is Tohoku National Fisheries Research Institute, Fisheries Research Agency, Shiogama, Japan. D.K. Stevenson is currently a contractor for NOAA/NMFS, Northeast Regional Office, Gloucester, MA.

Abstract

This paper provides an overview of the biology, catch history, and current biomass status of the world's major herring stocks. Herring occur throughout the northern coastal marine areas of the Atlantic and Pacific oceans and parts of the Arctic Ocean. Throughout their entire range, they have been used for food and commerce for millennia. Although several herring stocks have been fished commercially for more than 1,500 years, it is only within the last 100 years that nearly all stocks have been subjected to intense fisheries. During the last century, many stocks "collapsed" but most subsequently "recovered." A few experienced several collapses. Currently, some stocks are thriving but the present state of others is worrisome. This review provides a brief history and review of the present state of each stock provided by the biologists who are familiar with them. In most cases the stocks described in this paper represent aggregations of different biological subpopulations existing within "major stocks." In some instances, especially from multinational fisheries, contributions from different authors were integrated to describe certain stocks. Although our objective was not to explain the causes of variation in abundance, the effect of climatic variation on herring abundance is repeatedly mentioned by various contributors. The effect of climate on Baltic, North Sea, and Norwegian herring has been known for a long time but the effects of climate are reinforced in this report for other stocks. We also see that some changes appear to be synchronous among stocks within the same general areas of the Atlantic and Pacific. In particular, we see a pattern of increases in herring stocks in the southwest Atlantic (Georges Bank, Maine) but a decrease in some in the northwest Atlantic (Newfoundland). Another clear observation is that while there are many biological differences among herring stocks, almost all are subject to substantial fluctuations in abundance and the most severe declines are preceded or accompanied by intense fishing. The temporal duration of periods of crash and recovery vary, but in general it is about a decade or longer. This would be expected given the life span of herring, which ranges from about 6 or 7 years in extreme southern populations to 20 years or more in northern populations. In most instances, when fishing decreases the stocks "recover" but this has not occurred in the Hokkaido-Sakhalin stock, which was once one of the world's largest, with nearly 1 million t landed annually. There is no clear explanation for the failure of this stock to recover after all fishing has stopped, unless the population has been depleted beyond a point of no return. Although there is still hope that this stock may recover some day, after a decline of 50 years perhaps a "recovery" is not possible. There still are herring in the area but it is not clear if the small local stocks that still occupy those areas are part of the same biological entity that was once the great migratory Hokkaido-Sakhalin herring stock. The virtual disappearance of the Hokkaido-Sakhalin herring, when considered relative to the last century of herring fisheries in other areas of the world, indicates that herring stocks are remarkably resilient, but perhaps not indestructible.

Contents

Introduction
Part 1. Overview of Herring Distribution and Biology
Part 2. Eastern North Atlantic and Adjacent Seas
   Baltic Herring
   Norwegian Spring-Spawning Herring Fishery
   North Sea Herring
   Celtic Sea Herring
   West of Scotland Herring
   Icelandic Summer Spawning Herring
Part 3. Western North Atlantic
Part 3a. Northwestern North Atlantic
   East and Southeast Newfoundland
   West Coast of Newfoundland (4R) Herring
   Gulf of St. Lawrence
Part 3b. Southwestern North Atlantic
   A Note on Herring Stock Definitions in the Southern North Atlantic
   Scotia-Fundy
   Atlantic Coastal Stock Complex: Gulf of Maine and Georges Bank
Part 4. Eastern North Pacific
   Alaska
   British Columbia Herring
   California
Part 5. Western North Pacific
Part 5a. Northwestern North Pacific
   Notes on Different Forms of Herring in the Western North Pacific
   Russian Herring
Part 5b. Southwestern North Pacific
   Hokkaido-Sakhalin Herring
   Yellow Sea Herring
Part 6. Arctic Populations
   Greenland Sea
   Barents and Kara Seas
   White Sea
   Laptev and East Siberian Seas
   Chukchi Sea
   Beaufort Sea
   Other Waters: Hudson Bay, Baffin Bay, Davis Strait
Part 7. Synopsis, Summary, and Conclusions
References

Introduction

Throughout coastal regions of the Northern Hemisphere, herring (*Clupea pallasii* and *Clupea harengus*) have been fished for millennia as food. Herring have supported commercial fisheries in Europe for at least 1,500 years (Hodgson 1957). In the last 500 years, following technological advances in salt preservation, herring have guided the economies and politics of northern Europe, especially through policies of the Hanseatic League (Mollat du Jourdin 1993). In the western Atlantic, herring supported commercial fisheries when Europeans regularly fished in the Bay of Fundy about 500 years ago (Tibbo 1966). Pacific herring have been harvested commercially for over 500 years in northern Japan (Morita 1985) and during the last century, herring have supported intense Japanese and Russian fisheries (Motoda and Hirano 1963, Tyurnin 1973). On the Pacific coast of North America herring were widely used by indigenous peoples (Drucker 1955) and commercial herring fisheries began in California in the mid-1800s (Suer 1987), in British Columbia in the late 1870s (Taylor 1964), and in Alaska in the 1880s (Grosse and Hay 1988). This paper, prepared for an international symposium on herring at the end of the millennium, provides an inventory and brief review of the world's major herring stocks. The basic question addressed in the paper is: How have the world's herring fared after the last 100 years of commercial fisheries? The answers we provide are based on collaborative contributions from biologists representing nearly all major herring stocks of the world. Each provided a brief biological description of their stocks and, if available, time series data on spawning stock biomass and annual catches.

Although the objective of this paper may appear to be simple, in practice it is complex. Throughout the world, there are different types of agencies that assess and review herring. Some are multinational, such as the International Council for the Exploration of the Sea (ICES), others are managed by national governments, and still others by regional governments within countries. Among these different jurisdictions there are different practices for recording catch, with some included within calendar years and others within seasons. Further, there were several instances where there were minor discrepancies in estimates of catch when there was more than one person reporting on the same stocks. Aging conventions vary slightly among institutions so that the understanding of the real age of "age-1 herring" varies, by up to 1 year. This confounds comparisons of cohorts among different stocks. A difficult task was compiling historical catch data, because some regional catch data are regarded as confidential in some jurisdictions, including one in North America. Summaries of annual catch data, however, are usually published in one or more informal sources and large geographical summaries are presented in documents prepared by the Food and Agriculture Organization of the United Nations (FAO). In a few instances where unpublished data were not available we used published sources, some of which may have been slightly incomplete.

This paper represents a second attempt at such a global description of herring. The first was by Fedorov (1968), which is actually a small book (in Russian). Fedorov also presented biological summaries including maps showing migratory patterns and comments on different life history stages of major stocks. Although we had originally attempted to include about the same amount of detail as Fedorov, the resulting compilation was too long for these proceedings. Further, the inclusion of too much detail might detract from the main purpose of this paper, which is a commentary on the status of the world's herring stocks. Consequently, our comments on biological attributes of various herring stocks are brief.

The paper is organized into seven sections. Part 1 is a short description of the biology of herring with some comments on the significance of herring fisheries and research to the development of fisheries science. This is followed by five sections that describe the status of herring stocks in each of five regions: the eastern and western Atlantic, the eastern and western Pacific, and the Arctic. Part 2 describes the eastern Atlantic and adjacent seas including Iceland and the western Atlantic. Parts 3a and 3b describe stocks from Newfoundland to the Gulf of Maine. Part 4 describes the eastern Pacific from California to the eastern Bering Sea. Parts 5a and 5b describe the western Pacific from the western Bering Sea to the Yellow Sea. Part 6 briefly reviews herring in the Arctic Ocean and adjacent seas. The organization of each subsection, including headings, varies slightly, and reflects the emphases of the contributing authors. For the sake of brevity we did not attempt to provide maps showing all of the geographic names mentioned in the text. The paper concludes with Part 7, which provides some brief synthesis and summary comments for the whole paper.

Part 1. Brief Overview of Herring Distribution and Biology

Herring are members of the clupeid family. The family Clupeidae consists of about 330 species (Whitehead 1985), including Pacific and Atlantic herring. In the last century clupeids composed about one-third of total world catches (Blaxter and Hunter 1982), although this varies annually according to fluctuations of some of the largest stocks. Most clupeids are tropical, or subtropical, but Pacific and Atlantic herring are the most northern of clupeids, and the only ones in Arctic waters. The taxonomic usage has varied with time and location. Much of the literature has recognized Pacific and Atlantic herring and Baltic herring as separate subspecies: *Clupea harengus harengus* (Atlantic), *C. h. pallasii* (Pacific), and *C. h. membras* (Baltic). Baltic herring are coastal spawners living under conditions of reduced salinity. Recently usage has tended to favor recognition of Atlantic and Pacific herring as species (*C. harengus* and *C. pallasii*). Probably the most conspicuous difference is the occurrence of "fall-spawning" herring in the Atlantic, whereas Pacific herring are mainly late winter and spring spawners. Also, in general, Pacific herring tend to spawn closer to shore and often in intertidal waters, whereas Atlantic herring tend to spawn in deeper water, although some spring-spawning Atlantic herring also spawn in shallow water. For convenience, this report follows that practice of recognizing Atlantic and Pacific herring as species although we are uncertain about the best nomenclature for Arctic herring.

Herring spawn only once a year. All of the eggs are released within a single spawning period. The number of eggs increases with female size, with the smallest herring (~150 mm) having perhaps 10^4 eggs and the largest (~400 mm) having more (~10^5). The eggs are adhesive and deposited on gravel bottoms (particularly in Atlantic herring), or on vegetative substrate (Pacific herring and some spring-spawning Atlantic herring). Mean length and longevity vary with latitude, with smaller, shorter-lived herring (~5-6 years) in the south and larger, longer-lived populations in the north. Table 1 shows the age of maturity, age determination method (scales or otoliths), L_∞, approximate temperature at spawning, and maximal age for a number of different stocks. Tables 2 and 3 show the approximate age-specific length (centimeters) and weight (grams) from various stocks. In general, the northern herring grow larger and sometimes faster, in both the Atlantic and the Pacific. This simple generalization, however, is complicated by the observation that in some northern areas (such as some Icelandic populations) the larval stage may be prolonged and extend over the first winter, so age-specific growth rates during the early years may appear to be slower in such populations.

Most herring tend to migrate between summer feeding areas on shelf waters, to overwintering areas, which may be in nearshore protected waters, and then to spawning locations. In general, herring schools consist of individuals of similar size and/or age. Juveniles do not mix in their first year (often called age 0+). They school together and do not associate with older, larger juveniles (age 1+) or adult schools and in some areas may stay nearer to inside or sheltered waters. Herring are noted for their extensive vertical migrations, being deeper in the water by day and shallower at night. Herring are not herbivorous, but eat mainly zooplankton. Larvae consume copepod eggs and nauplii, plus young stages of other organisms, whereas juveniles tend to eat copepods and other zooplankton. Adults consume many different items but in many areas copepods and euphausiids dominate. Herring are both particulate feeders and filter feeders.

As a prey species, herring are consumed by nearly all animals large enough to eat them, and in some areas cod (*Gadus*) populations appear to rely on herring as a principal prey species. In the northern Pacific and other areas, herring populations support piscivorous salmonids as well as many marine bird and mammal species.

Herring have been the subject of biological investigations for several hundred years (Whitehead 1985) and information from herring has served

Table 1. Comparison of biological characteristics among herring stocks.

Location	Age at first maturity	Age determination method	Age units	L_∞ (cm)	Spawning temp. (°C)	T (years)
Norway	5-9	Scales	Ring	40	6	18-20
North Sea	2-3	Otoliths	Ring	35	8	12-16
Baltic	2-3	Otoliths	Annulus	19.1-28.4	4-18	20
Celtic	2-9	Otoliths	Annulus	10.2		
West of Scotland	2	Otoliths	Ring	34	14-16	
Iceland	4	Scales	Ring	36.2	10	16-18
Gulf of St. Lawrence, spring	5-9	Otoliths	Annulus	40.19		
Gulf of St. Lawrence, fall	5-9	Otoliths		43.84		
East-SE Newfoundland	3-5	Otoliths	Annulus	42		
West Newfoundland, spring	4-5	Otoliths	Annulus	38.8		20
West Newfoundland, fall	5-6	Otoliths	Annulus	41.5		20
Scotia Fundy	3-4	Otoliths	Annulus	40	8-12	16-20
Georges Bank, spring	3	Otoliths	Annulus	33.15	10-12	
Georges Bank, fall	3	Otoliths	Annulus	33.15	10-12	
Gulf of Maine	4			36.55	5-15	16
Beaufort Sea	6	Scales/Otoliths	Annulus	26.1	2.5-3.5	14-16
British Columbia	2-4	Scales	Annulus	25	8	10-12
Gulf of Alaska	3	Scales	Annulus	24	4	12-14
Bering Sea	4	Scales	Annulus	34	4	18-20
Western Bering Sea	3-7	Scales	Annulus	36.6	2-7	17
Hokkaido, Sakhalin	2-6	Scales	Annulus	44	4-6	18
Yellow Sea	2	Scales	Annulus	30.5	2-6	

These estimates are approximate and do not describe temporal or spatial variation. Fall and spring refer to fall and spring spawning stocks.

as a basis for investigations of other clupeids. Investigations on herring have influenced other aspects of fisheries science in most countries that have a history of herring fisheries. Russian fisheries literature, even back to the mid-nineteenth century, makes frequent reference to studies of Baltic and White Sea herring. Early research on herring has generated some of the most profound and enduring hypotheses in fisheries science, which have been applied to many other species. This includes the famous "Hjort" or "critical period" hypothesis (Hjort 1914) that attempts to explain the reason for strong year classes according to the survival of young herring at the larval period. Variation among herring populations within the Baltic and White Sea prompted much of what we know about marine fish physiology and much is derived from pioneering work by Hempel and Blaxter; e.g., see reviews by Blaxter and Holliday (1963) and Blaxter and Hunter (1982). Research on herring has provided the basis for several major developments in fisheries science and management (Stephenson 2001, this

Table 2. Comparison of length-at-age (centimeters) among stocks. The estimates are approximate and do not account for spatial or temporal variation.

Stock	Measurement	0+	1+	2+	3+	4+	5+	6+	7+	8+	9+	10+	11+	12+	13+	14+
Norway		12	17	22	25	27	29	31	33	35	36	37	38	39		
North Sea		10	20	25	27	28	29	30	30	31	31	32	33	34		
Baltic		10.4	14.5	16.2	17.9	19.1	20.1	20.9	21.6	22.2	22.6	23.3				
Celtic			23.2	24.8	26.4	27.1	27.6	28.3	28.8	29.3	30.1					
West of Scotland	SL		18.8	25	26.8	27.4	28	28.4	28.8	29.3	29.6					
Iceland			12	21	26	29	31	32	33	34	35	35	36	36	37	37
East-SE Newfoundland		11.5	18.5	23.6	26.8	28	30.5	31.4	31.8	32.3	33.3	33.9	35.9			
Scotia-Fundy		11.8	18.2	24.4	27.8	29.8	31.3	32.6	33.7	34.5	35.1	35.6	36.2	36.7	36.8	37.4
Georges Bank, spring			15.1	20.9	25	27.9	29.6	31	32.2	33.6	35.1					
Georges Bank, fall			14.6	21.8	25.3	27.7	29.1	30.1	31.4	32	33.3	33.2				
Georges Bank			15.9	21.6	25.5	28.2	30	31.3	32.2	32.8	33.2					
Gulf of Maine			19	22.9	25.7	26.2	28.3	29.4	30.3	31.5	32.4					
Beaufort	SL	13.9	18	20.2	21.6	22.6	23.4	24.2	24.7	25.3	25.7	26.1	26.4	26.7	26.8	

Table 2. (Continued.) Comparison of length-at-age (centimeters) among stocks. The estimates are approximate and do not account for spatial or temporal variation.

Stock	Measurement	0+	1+	2+	3+	4+	5+	6+	7+	8+	9+	10+	11+	12+	13+	14−
British Columbia	SL		16.2	18.3	19.5	20.5	21.2	21.7	22.3	22.8	23.2					
Gulf of Alaska			15	18	19	21	22	22	23	23	24					
Bering Sea					24	25	27	28	29	30	30	31	31	32	32	
Western Bering Sea	SL	11.3	17.9	22	24.5	26.1	27.5	28.5	29.4	30.1	30.7	31.5	32.2	33	33.7	34.6
Hokkaido-Sakhalin (1)	TL		15	22	26	29	30.5	32	33	34	34.5	35	35.3	35.6		
Hokkaido-Sakhalin (2)			13.3	19.8	22.8	25.8	28.5	29.8	30.9	31.8	32.7	33.5				
Hokkaido-Sakhalin (3)	SL	10.3	16.9	22.1	24.5	26.2	27.3	28.3	28.8	29.3	30	30.2	30.4	30.8	31.7	32
Lake Furen, Hokkaido			15.5	21	24	28	29	30								
Yellow Sea	FL		16.5	24.1	26.9	28.4	29.2	30.1								

Some measurements are in total length (TL) or fork length (FL). All others are assumed to be standard length (SL). The age of "0+" indicates fish in the first year of life but this varies according to stock, especially with some that overwinter as larvae (see Table 1). Several independent estimates are shown for some stocks. Fall and spring refer to fall- and spring-spawning stocks. The Lake Furen stock is from eastern Hokkaido (see Kobayashi 2001, this volume).

Table 3. Weight (grams) at age compared among different herring stocks.

Location	0+	1+	2+	3+	4+	5+	6+	7+	8+	9+	10+	11+	12+	13+	14+
Norway	10	85	155	250	300	370	400	425	440	470	490	500			
North Sea		60	130	180	210	220	230	240	250	270	300	330	350		
Baltic	19	27	37	46	54	61	67	73	78	85					
Celtic		93	126	153	169	185	198	208	218	228					
West of Scotland		62	142	172	183	194	204	212	222	230					
Iceland		10	71	140	200	244	284	305	330	353	367	382	398	400	
Gulf St. Lawrence, spring		71	144	166	197	230	259	288	300	319	297	319			
Gulf St. Lawrence, fall	10	96	166	222	258	289	315	336	356	366	402				
East-SE Newfoundland		50	108	156	175	221	240	261	265	306	320	367			
West Newfoundland, spring			102	155	200	230	283	305	345	384	420	449			
East Newfoundland, fall			81	126	165	201	225	263	295	324	344	410			

Table 3. (Continued.) Weight (grams) at age compared among different herring stocks.

Location	0+	1+	2+	3+	4+	5+	6+	7+	8+	9+	10+	11+	12+	13+	14+
Scotia-Fundy	11	46	118	182	229	264	298	331	357	371	384	399	416	427	419
Georges Bank, spring		23	63	108	146	174	202	223	245	287					
Georges Bank, fall	23	74	115	150	172	186	203	208	237	227					
Gulf of Maine		56	93	123	146	165	183	205	241	279					
Beaufort Sea	31	72	104	128	148	165	183	196	211	221	232	241	250	253	235
British Columbia		60	88	110	131	146	157	168	180	190					
Gulf of Alaska			93	109	130	148	165	190							
Bering Sea				171	213	247	270	299	325	341	363	375	381	401	418
Western Bering Sea	19	49	133	197	240	283	321	350	382	410	437	469	493	534	578
Hokkaido-Sakhalin (1)	20	80	140	200	230	260	290	320	340	360	370	380			
Hokkaido-Sakhalin (2)	20	80	120	180	240	270	300	330	360	400					
Hokkaido-Sakhalin (3)	24	72	140	191	242	279	317	360	374	398	416				
Yellow Sea	208	252	285	298											
45	146														

The estimates are approximate and do not account for spatial or temporal variation. The age of "0+" indicates fish in the first year of life but this varies according to stock, especially with some that overwinter as larvae (see Table 1). Several independent estimates are shown for some stocks, including 3 for the Hokkaido-Sakhalin stock. Fall and spring refer to fall and spring spawning stocks.

volume) including the early polar migration theory; development of the population/stock concept and contributions to the modern view of fish stock structure; tracking and quantification of year classes; and explanations for fluctuations in finfish abundance in the influential hypotheses of Hjort (1914) ("critical period"), Cushing (1972) ("match-mismatch"), and Iles and Sinclair ("larval retention"). Herring research also has served as the impetus for more recent development of hydroacoustic methods; explorations of the linkages between fisheries dynamics and hydrography (Heath 1989); new applications of theory to stock structure and metapopulations (McQuinn 1997b); and innovative approaches to fisheries regulation and management.

Part 2. Eastern Atlantic and Adjacent Seas

The section on the eastern Atlantic includes summaries from the Baltic Sea, North Sea, Celtic Sea, the Norwegian spring-spawning herring, and the Icelandic herring (Fig. 1). Probably several stocks, such as the Norwegian spring-spawning herring, or Icelandic herring, that we have called "Atlantic" herring, might also have been described as occurring in Arctic waters. It is customary, however, to view such stocks as "Atlantic" herring, and we follow that convention. Herring of the White Sea might also have been included here but we chose to include them as populations associated with the Arctic Ocean.

Baltic Herring

Herring are distributed throughout the Baltic Sea, including the Gulf of Bothnia where salinity is low (< 5 ppt). Herring spawn is patchy along the coast of the Baltic Sea in shallow water (depth usually 1-10 m). Most spawning is from April to June, but ranges from March to August. The preferred spawning substrate is vegetation. Small stocks of autumn-spawning herring spawn in deeper banks (10-20 m) offshore. Herring larvae and young herring are mainly found in shallow water close to the coast.

Adult herring conduct regular migrations from their spawning grounds to feeding areas in the open sea, mainly to the south (Aro 1989), and most are believed to return to the same spawning places. Herring spawning in the northern Baltic Sea, in the Gulf of Bothnia, stay mainly in that area also during feeding. Part of the herring spawning population in the southern Baltic Sea migrates to Skagerrak, northeastern North Sea, for feeding (Jönsson and Biester 1981).

Stock Structure

The intraspecific grouping of Baltic herring has been studied since the days of Heincke, who in 1882 described meristic variation in *Clupea harengus* and interpreted it as racial characteristics. He defined spring- and autumn-spawning races, based mainly on differences in number of vertebrae and seasonal development of gonads. Many others have continued to define a large number of intraspecific groups: races, stocks, and

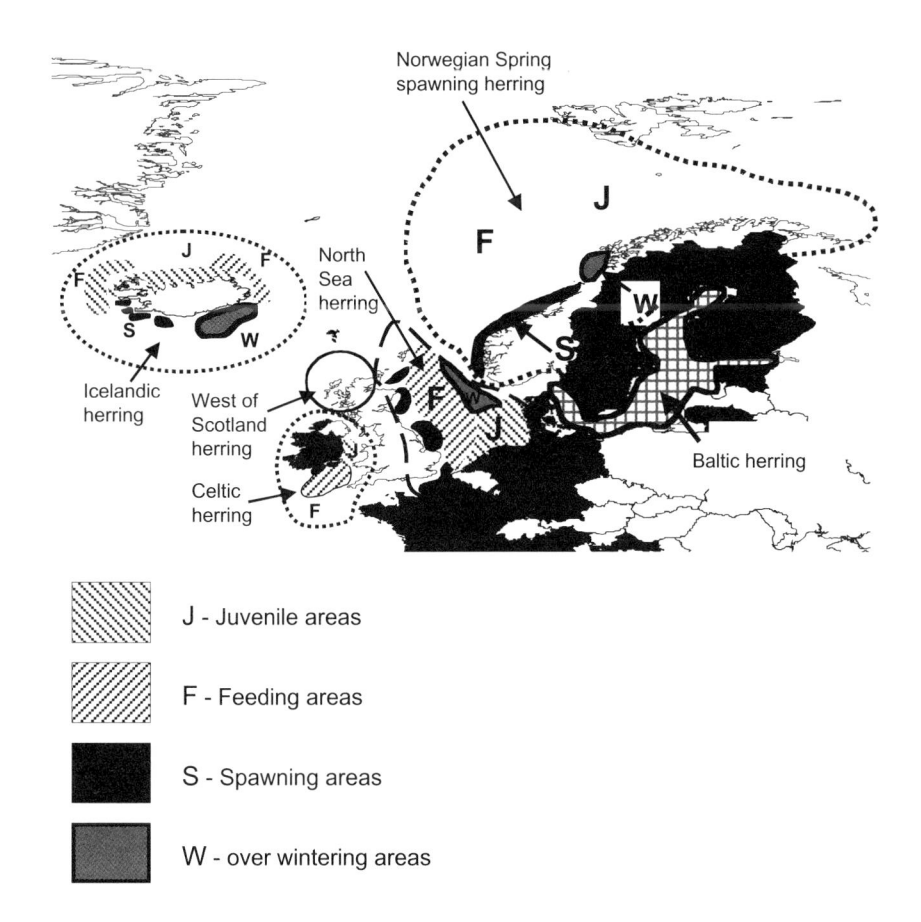

Figure 1. The eastern North Atlantic showing the major herring stocks. For most stocks, dark areas indicate spawning grounds (S) and shaded areas indicate either juvenile rearing areas (J), feeding areas (F), or overwintering areas.

populations (Hessle 1925; Kändler 1942; Popiel 1964; Rannak 1967; Kompowski 1971; Otterlind 1985; Ojaveer 1988; and overviews in Ojaveer 1989, Parmanne et al. 1994). It has, however, proved difficult to demonstrate similar genetic differentiation through studies of allozymes (Andersson et al. 1981, Ryman et al. 1984). The discreteness of Baltic herring stocks, like other stocks, remains an issue of debate and research (Smith and Jamieson 1986, Stephenson 1991).

Biology and Stock Assessment

The data used in the assessments (ICES 1999a, 2000) include catch statistics as officially reported, and, when necessary, supplemented by ICES working group estimates and biological data (age, weight, maturity) obtained by sampling the landings by national fisheries research institutes in all nine Baltic countries (Denmark, Estonia, Finland, Germany, Latvia, Lithuania, Poland, Russia, Sweden). The Baltic Fisheries Assessment Working Group has, since 1990, chosen to assess Baltic herring as four stocks (for units used earlier see Sjöstrand [1989]). Stock estimates from hydroacoustic surveys are used in tuning the sequential catch analysis (VPA). Surveys have been carried out in the Baltic since 1978 (Håkansson et al. 1979, ICES 1999b) in September and October, when the herring are dispersed in the Baltic proper on their feeding grounds. The youngest age groups stay closer to the coast and inside the archipelagos and are not properly covered.

Fishery and Management

The herring fishery remained a coastal fishery with nets, seines, and traps until the 1950s. The drastic increase of the fishery since then was made possible by the development of the trawl fishery and the practically unlimited market for fish in the eastern countries. The fear of continued increase in the catch capacity that could cause overexploitation and threaten the sustainability of the fishery led to the agreement of the Gdansk convention (Convention on Fishing and Conservation of the Living Resources in the Baltic Sea and the Belts). It was signed in 1973 by all countries around the Baltic and entered into force in 1974. Its aims, to achieve greater and closer cooperation between the parties in order to maintain the maximum stable productivity of the living resources of the region, are implemented by the International Baltic Sea Fishery Commission (IBSFC). The commission yearly requests advice on stock status and catch options from the ICES. There are no explicit management objectives agreed for the herring fisheries in the Baltic Sea. The IBSFC manages the fisheries through yearly catch limitations, total allowable catch (TAC), split into national quotas. Technical measures are applied additionally. Herring is managed in two units (subdivisions 22-29S, 32 and 29N-31) with separate TACs. Unfavorable market conditions for herring have been reflected in decreased landings for human consumption whereas the landings for industrial purposes have increased during the last few years.

Status by Stock (Assessment Unit)

In subdivisions 22-24 (western Baltic) and in division IIIa (Kattegat and Skagerrak) the spring-spawning stock spawns at several places in the Kattegat and southwest Baltic with its major spawning sites around the island of Rügen. The stock is migratory and mixes with the North Sea autumn-spawning herring in northeastern North Sea and Skagerrak during the feeding period. Landings increased from about 100,000 t in the late 1970s to around 200,000 t in the 1980s but have decreased since the 1980s. The state of the stock is uncertain due to problems in identifying spring and autumn spawners during the nonspawning time in the historical catch data and the lack of a coordinated comprehensive survey. In subdivisions 25-29 and 32 (central Baltic) herring are of a heterogeneous nature with a number of different spawning populations along the coasts. After spawning these are mixed in the open sea. Pronounced differences in size at age between herring from the southern and northeastern parts of the area contribute to the variability in stock estimates. Landings fluctuated around 300,000 t from the beginning of the 1970s to 1989 but have since decreased to around 200,000-250,000 t mainly due to a declining market for herring (Fig. 2a). **Although the stock size is uncertain, it appears certain that the biomass is decreasing as fishing mortality is increasing, and the stock is considered to be outside safe biological limits.** A marked decrease in mean weight at age for the herring has been registered since the beginning of the 1980s (Sparholt et al. 1994). Thus the spawning stock in numbers has, in contrast to the biomass development, been stable since the beginning of the 1980s or even increased. Simultaneously, the biomass of cod, the major herring predator, has decreased drastically and the frequency of influxes of saline water from the North Sea has been low. In subdivision 30 (Bothnian Sea) landings increased during the last decade from about 20,000 t to 60,000 t. The present state is difficult to judge due to low precision of assessment. There as an increased effort in the 1990s, believed to have increased fishing mortality. The spawning stock has declined after the peak in 1994. In subdivision 31 (Bothnian Bay) landings have never exceeded 10,000 t and have been around 5,000 t the last 5 years. Exploitation of the stock is considered to be within safe biological limits. The actual spawning-stock biomass (SSB) and fishing mortality are, however, not known. Production models do not indicate major changes in SSB and age compositions of catches are consistent with a lightly exploited stock.

Species Interaction–Multispecies Assessment

Interaction between the major species in the Baltic (cod, herring, and sprat) is important. Multispecies VPA (Helgason and Gislason 1979, Pope 1979) has been used to estimate total number of prey and predation mortalities. Data on food consumption and consumption rations of predators are needed in addition to data for the single species VPA. Results from multispecies assessments are presented (ICES 1999c) and summarized by Sparholt (1994), who also gives a comprehensive overview.

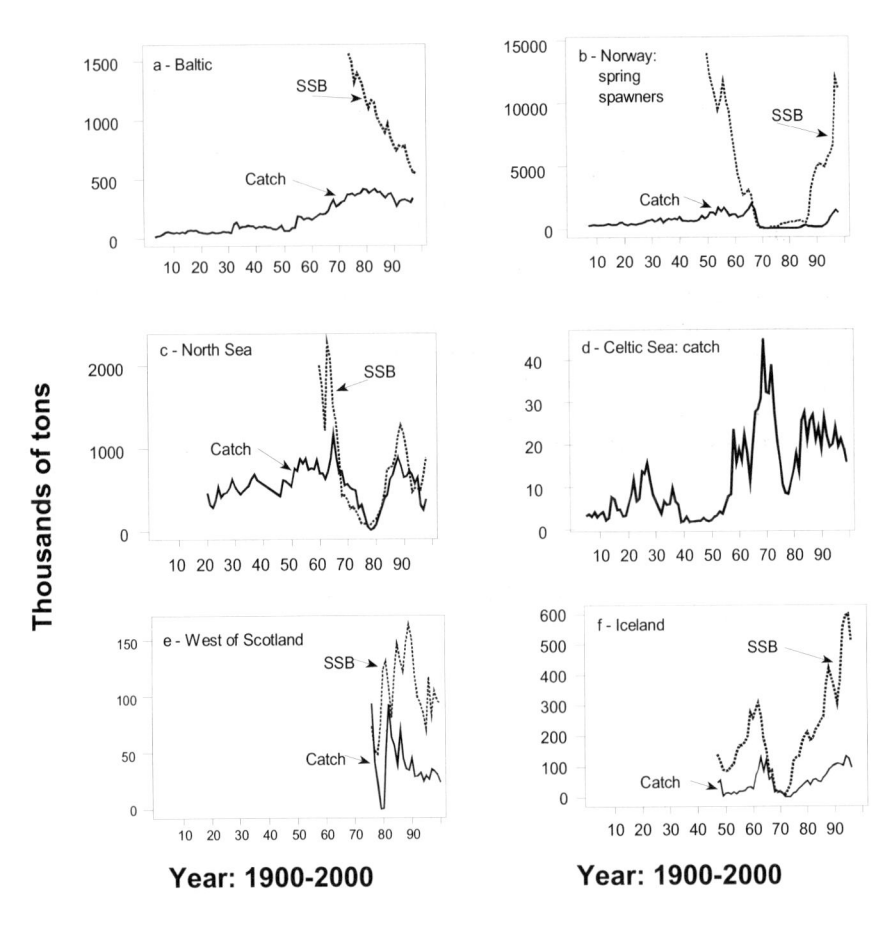

Figure 2. Catch (solid dark line) and spawning stock biomass (SSB, dashed line) for eastern North Atlantic herring stocks.

Norwegian Spring-Spawning Herring

Fishery

Annual catch statistics have been gathered by national authorities and reported to the ICES. These annual statistics have been published in Bulletin Statistique from the first edition in 1903 to the present. Also, fishing statistics are published in ICES reports (working group reports, and annual reports of the Advisory Committee for Fishery Management, ICES).

Biological Data and Assessments

Biological data are sampled by national fishery research institutes (in Norway, Iceland, and Russia). These are long-term data series. (In Bergen, there are records of biological data and age readings from 1907 to 1999.) Analytical assessments (VPAs) have been carried out annually since about 1980. These are based on catch statistics and abundance estimates of the stock. The abundance estimates are based on acoustic abundance estimation with echo sounders and tagging experiments (ICES 1999d). Published assessments (ICES 1999d) reports show that the stock fluctuated from high levels just after the Second World War to a stock collapse in the late 1960s. The stock has recovered again and is assessed to be at high levels again in recent years.

Biology of the Stock

The Norwegian spring-spawning herring is one of the stocks in the so-called Atlanto-Scandian group of herring stocks (Johansen 1919). The other two are the spring spawners and summer spawners in Iceland. These three stocks have certain similarities: they dwell in high latitudes, around Iceland and in the Norwegian Sea; grow to a large maximum size of about 40 cm; and the two spring spawners have the ability to migrate over large distances (Devold 1963).

The Norwegian spring spawners spawn in early spring (February-March) off the Norwegian coast in an area ranging from approximately 59°00′N to about 69°00′N (Devold 1963). The most important spawning grounds are found close to the coast at about 63°00′N. The larvae drift northward with the Norwegian coastal current and end up in the Barents Sea or in the fjords in northern Norway (Dragesund 1970a,b). By the end of summer, the larvae metamorphose and start shoaling. They remain in this nursery region until the spring in their third year of life and then migrate southward along the Norwegian coast where they mix with older age groups. At the age of 5-8 years (depending on growth), they mature and spawn (ICES 1999d). After spawning, the adult stock migrates from the Norwegian coast, in northwesterly direction, into the Norwegian Sea (Devold 1963). Here they feed during summer in May to September. Prior to the collapse of the stock in the late 1960s the feeding migration extended to the north coast of Iceland whence the herring migrated to the overwintering area east of Iceland and returned to the spawning grounds in January/February. However, presently the stock (then) migrates back to the Norwegian coast in September where it concentrates close to shore, in a smaller region in northern Norway (Slotte and Johannessen 1997). Here the spawning stock and young prespawning age groups overwinter in dense concentrations. In January, prior to spawning, the stock begins to migrate southward along the Norwegian coast as they complete the final stages of sexual maturation.

Fishery and Management

Historically, the stock was exploited at all stages: as juveniles, young, and adults. The record of fishery statistics, which is believed to be reliable, shows that there used to be a heavy exploitation of young individuals in the 1950s and early 1960s (Dragesund and Ulltang 1978). In the early 1960s, the total landings were about 2 million t of which three-quarters were juveniles. However, after the collapse of the stock, a minimum size regulation was enforced and after 1970 it has been illegal to catch small herring from this stock.

The fishery is carried out in spring, at the spawning grounds, and in the Norwegian Sea. Some fishing also takes place after the stock has returned to the Norwegian coast. The exploitation is shared in agreement between five parties: the European Union (EU), Iceland, the Faroe Islands, Russia, and Norway. On the basis of scientific advice from the ICES, the parties negotiate annually and agree on a total allowable catch and a share between them. Currently, Norway has a share of 57%, Iceland 16%, Russia 14%, EU 8%, and the Faroe Islands 5%. During the rebuilding phase of the stock, the landings have been restricted and it was a goal to keep the instantaneous fishing mortality rate below 0.05.

Historical Development and Status of the Stock

Assessment of the stock is based on annual landings statistics which are split by age groups based on age readings from samples of the catches. In addition, the abundance of the stock is estimated by acoustic abundance estimation and by tagging (ICES 1999d). In recent years the acoustic survey, covering the stock while it feeds in the Norwegian Sea, is done with participation of research vessels from several countries (Holst et al. 1998). The survey is coordinated by a planning group in the ICES. On the basis of the time series of the catch statistics, abundance estimates, and biological samples, an analytical assessment is carried out annually by an ICES working group (ICES 1999d). The assessments have been made annually since the mid-1970s, and recently a long-term VPA, from 1907 to 1998, was made for the stock (Toresen and Østvedt 2000). The assessments show that the spawning stock has probably fluctuated substantially through the last 50 years. In the early 1950s the spawning stock biomass was estimated to be about 10 million t (Fig. 2b). The biomass decreased drastically in the 1950s and 1960s and the stock collapsed in the late 1960s. In the early 1970s, the spawning stock was not detectable. Following strict regulations (i.e., minimum size regulations of 20 cm and moratorium on exploitation of adults in the years 1974-1977) the stock slowly recovered. In 1983, a very strong year class developed and matured in the late 1980s when the spawning stock increased to a level of about 4 million t. ***In the early 1990s several strong year classes emerged and by the late 1990s, the stock grew to a level of about 10 million t. However, recent assessments show that the stock probably will decrease in abundance in the near future because the most recent year classes (~1993) are poor.*** The current exploitation, with a total allowable catch of 1.3 million t, is estimated to be in excess of what will be produced, either as recruitment or individual growth, in the near future. In recent scientific advice, it is stressed that when a rich year class appears, it will still take at least 5 years before it matures to spawn. It was therefore agreed that the long-term management for the stock will be to limit the exploitation to avoid having the stock reduced to less than 2.5 million t. Historical data shows that below this level there is a lower probability of good recruitment.

North Sea Herring

According to Parrish and Saville (1965) three main centers and spawning times can be identified for herring in the North Sea: (1) the northwestern North Sea, from the Shetlands to the east coast of Scotland (Buchan), spawning from July to mid-September; (2) the English northeast coast and Dogger Bank, spawning from September to November; and (3) the southern Bight and eastern English Channel, spawning from November to January. Early investigations from these main spawning centers revealed consistent differences between them, especially between the northwestern and central North Sea spawners and those spawning in the southern North Sea and eastern English Channel. However, while feeding and overwintering, herring from the different spawning components mix. During the feeding period in summer (April-July) the major fisheries for adult herring takes place. Even though there are some morphological differences among stocks, the variance within each is substantial, and it is therefore not possible to tell the one from the others in catches. Because of the difficulties in allocating the amount of catch to the right stock component, it was decided in 1987 to pool the data for the different stocks together and make a common assessment of the three stock components.

Annually, landing statistics are reported to the ICES. On the basis of samples from the fishery, the catch is split in numbers per age group. In addition, various abundance estimates are carried out by North Sea coastal states. These estimates are based on different methods, as bottom trawl surveys, acoustic surveys, and larvae surveys. All surveys are coordinated by ICES planning groups. The time series of the estimates are of different length, but there are available abundance estimates based on each of the three methods, annually since 1979. The North Sea herring are assessed annually by the ICES. The current assessment is done by the integrated catch analysis (ICA) method. This method has been evolved from the theory described by Fournier and Archibald (1982) to the practical applications described by Deriso et al. (1985), and to new applications by Patterson in recent herring assessments (Patterson and Melvin 1996). The approach is fundamentally to build a model to describe the data. The intent of these integrated catch-at-age analyses is to provide the flexibility to incorporate all available types of information—ranging from catch-at-age data, acoustic survey, egg or larvae survey, catch-per-effort, length frequency distributions, stock recruitment relationships, to migration patterns—within one statistical framework that allows for parameter estimation by simultaneous fitting of the model to all data sources.

Biology

North Sea herring spawn in the western parts of the North Sea during late summer and autumn. The larvae drift southeastward into the German Bight and thereafter spread northward along the west coast of Denmark. Depending on the force of the currents and wind, various portions of the year class spread into the Skagerrak (which is the bight surrounded by Norway, Denmark, and Sweden). Skagerrak and the region off the west coast of Denmark are important nursery grounds for North Sea herring. The herring spend about 3 years in these eastern parts of the North Sea after which they migrate westward to join the spawning stock. After spawning, the adult stock concentrates in an area outside the southwest coast of Norway, and spends the winter in somewhat deeper waters there. In early spring, the herring spread, moving westward again, and start feeding in central parts of the North Sea.

Fishery

Historically, the herring in the North Sea have been exploited by several countries and herring fisheries have been especially important for the Netherlands, Scotland, Denmark, and Norway. The annual yield (Fig. 2c) has for many years been about 300,000 t of adult herring and some 100,000-200,000 t of small and juvenile herring (ICES 1999e). This exploitation, especially the part aiming for small fish, has not been sustainable. There has therefore been an international pressure to decrease the take of small herring, and in 1996 the managers in the EU and Norway agreed on a management policy to reduce that part substantially. Managers also agreed to regulating the fishery to keep fishing mortalities on adult herring at levels below 0.25.

Historical Development and Status of the Stock

The time series of the spawning stock biomass from 1960 to 1998, as estimated by the ICES Herring Assessment Working Group for the Area South of 62°N (ICES 1999e) shows that the SSB was probably at a level around 2 million t in the early 1960s. The spawning stock decreased steadily, and by the late 1960s it was estimated to be below 500,000 t. The stock continued to decrease, and by the mid-1970s the stock was at a level less than 100,000 t. From the late 1970s and on, the stock started to recover, and during the 1980s, increased to a high level of about 1.3 million t in 1989. **In the 1990s the stock decreased again, probably due to unsustainable exploitation, and reached a low level of less than 500,000 t in 1993. The stock remained at these low levels for several years before increasing again probably as a result of strict regulations limiting exploitation both of juveniles and adults. The SSB was estimated to be 870,000 t in 1998 and is expected to increase in the coming years if the managers agree to control exploitation.**

Celtic Sea Herring

The herring fishery in the Celtic Sea and ICES division VIIj is located off the south coast of Ireland, and covers ICES divisions VIIa South, VIIg, j, h, and k. The main fisheries now take place in the inshore waters along the Irish coast. The main spawning areas are located along the south and southwest coasts of Ireland. Within the spawning areas the main spawning grounds are well known and defined and individual spawning beds have been mapped. The main juvenile nursery areas are located in the northern part of the Irish Sea (division VIIa North) and in bays and estuaries along the south and southwest coast. In the winter, shoals congregate for spawning in inshore waters along the coastline and after spawning appear to migrate out to deeper water, although shoals of spent fish are also located in inshore waters but may not remain there for prolonged periods. The summer feeding areas are located in offshore waters (e.g., Labadie and Jones banks and the "Smalls" area). Adult migrations take place to and from the spawning and feeding grounds during spring and autumn. Juvenile migrations take place from the nursery areas when fish are in their second and third years to the spawning grounds for first-time spawning.

Fisheries Data

The assessment of this stock is now based on age analyses of the catches using a conventional integrated catch at age (ICA) model in which acoustic surveys are used as a tuning index. Two surveys are carried out each spawning season and data are available from 1990. Catches per effort of paired midwater trawlers were also successfully used from 1966 to 1979 and subsequently larval surveys were also successfully used as an index of spawning stock.

The Irish fleet now takes over 95% of the total catch, using paired midwater trawls. Most of the catch is taken by "dry hold" boats but some catches are also taken by vessels with refrigerated seawater tanks. Since the mid-1980s the fishery has been mainly dependent on the Japanese "roe" market. Catch-at-age data are available from 1958 and analytical assessments were carried out from 1958 to 1999. Data on total catch are available for the fishery since 1905 (Fig. 2d).

Annual catches prior to 1916 were less than 5,000 t. Subsequently they increased to about 10,000 t for a short time in the 1920s before decreasing again until the late 1950s. Catches increased rapidly in the 1960s and reached a peak of over 45,000 t in 1969. The fishery then collapsed and was closed for 5 years from 1977 to 1982. Following the reopening, catches have been stable at around 20,000 t, which is approximately the same as the TAC.

Status of Stock

The state of this stock at present is believed to be high. Recruitment has been good in the last number of years and fishing mortality has decreased to the lowest observed since the mid-1960s. Landings have been stable for a number of years at around the TAC of 20,000 t. The stock collapsed in the 1970s because of decreased recruitment and a high fishing mortality and the fishery was closed for 5 years. There are concerns for the stock because the historical data show that it is very sensitive to increased mortality. It is also clear that over the time period there have been two different periods of recruitment: a high and a low period. At present there is no method of estimating recruitment and it is therefore not known when recruitment may decrease and produce a decline in stock size. There are also concerns that because the market is highly dependent on the Japanese "roe" market there has been widespread discarding of herring unsuitable for roe production. These discards are difficult to estimate and may result in an underestimate of fishing mortality. Further concerns arise because of the proposals to remove gravel from some of the more important spawning grounds.

West of Scotland Herring

The West of Scotland herring is found distributed widely over the shelf area to the west and north of Scotland. The West of Scotland herring are assessed annually by ICES. However, because of the uncertainty in the total catch and the age structure the assessments have not been accepted. The assessment model used is the integrated catch analysis (ICA) method. This method has been evolved from the theory described by Fournier and Archibald (1982) to the practical applications described by Deriso et al. (1985), and to new applications by Patterson in recent herring assessments (Patterson and Melvin 1996). On the basis of samples from the fishery, the catch is split in numbers per age group. Samples are supplied from an international fishery with Scottish, Dutch, German, and English vessels. Currently sampling information is poor and there is uncertainty about the landings and the age distribution. In addition, abundance indices of 2-9+ herring are obtained by a Scottish acoustic survey. This survey is coordinated by an ICES planning group and carried out at the same time as the North Sea survey with overlapping boundaries. The time series for this survey, which is carried out in July, is from 1991 to 2000, excluding 1997 when it was conducted in June. From 1972 to 1991 a larvae abundance survey was conducted and used to provide an SSB index. This survey was halted in 1991 as the acoustic surveys provided age-disaggregated data and more precise indices. These assessment model outputs provide a guide for stock management. **Currently the stock is regarded as lightly exploited and recommended catches are set to the mean of the last 5 years.**

Biology

The West of Scotland herring stock is composed of two groups: spring and autumn spawners. Currently most of the population is made up of the latter group. Some herring mature and spawn with an age of "two winter rings" (wr) but most herring reach maturity by around 3 years (wr). Autumn spawning occurs from late August to October around the northwest of Ireland and to the west and north of the Outer Hebrides and off Cape Wrath, in depths up to 100 m (Rankine 1986). The spring-spawning component is currently of unknown extent, with spawning sites inshore particularly on Balantrae Bank and south of Arran on the Clyde (Parrish et al. 1959, Bailey et al. 1986). The period of incubation is temperature-dependent and about 3 weeks at ambient temperatures. Newly hatched larvae follow the current systems and drift to the north and east and then west across the north of Scotland (Dooley and McKay 1975, Heath and MachLachlan 1985). Some are retained on the west of Scotland but a large proportion, particularly those from the Hebrides and Cape Wrath spawning sites, are carried through the Fair Isle channel and travel well into the North Sea. In spring the larvae reach the nursery areas where they develop into juveniles. Young herring spend some time in the inshore areas and sea lochs and the Moray Firth in the North Sea before migrating offshore to join the adult population. There is some evidence to suggest, from tagging experiments and from using biological markers, that as herring mature, some of those that moved as juveniles to the east coast population make the return journey back to the West of Scotland spawning areas (MacKenzie 1985).

Fishery

Historically, several countries have exploited the herring in the West of Scotland and herring fisheries have been especially important for the Netherlands, Scotland, and Germany. Before 1970 the fishery was predominantly taken in the North Minch, yielding between 25,000 and 40,000 t annually (Baxter 1958, ICES 1979). In the 1950s these were driftnet and ring-net fisheries, converting to pair-trawlers and purse-seiners in the late 1960s. An offshore fishery by Dutch, German, and French trawlers started in the late 1960s, and for protection against overexploitation several spawning grounds were closed in September and October. Currently the bulk of the catch is taken by three main fleets: a Scottish inshore pair-trawl fleet which works around Barra and in the Minch; a Scottish purse-seine fleet which operates in the northern part of the area; and an offshore fleet, mainly Dutch and German freezer trawlers, which fish in the deeper waters at around 150 m depth near the edge of the continental shelf. Annual landings statistics are reported to ICES. The estimated SSB and catch are shown in Fig. 2e. The annual yield has for many years been about 40,000 t of adult herring age 2 wr and older (ICES 2001).

Historical Development and Status of the Stock

The time series of the spawning stock biomass from 1976 to 2000 as estimated by the ICES Herring Assessment Working Group for the Area South of 62°N (ICES 2001) and shows that the SSB was probably at a level around 120,000 t in the early 1980s. The stock may have been higher in the early 1970s but when the North Sea fishery was closed, fishing on the western stock increased. The western fishery was closed in 1979 and reopened along with the North Sea in the 1980s. Since then the fishery has been at around 30,000 t. The TAC from this area has been taken in the North Sea over the last 10 years. Recent changes in regulation have reduced this area misreporting and catches are thought to have been reduced. Currently catch advice is for the status quo catch, the average of the last 5 years.

Icelandic Summer-Spawning Herring

Fishery Data Records

Annual catch statistics on herring have been collected by national authorities and reported to the ICES since 1903 and subsequently published in Bulletin Statistique. However, it was not fully realized until about the middle of the twentieth century that the herring catches at Iceland consisted of three stocks; i.e., Icelandic summer and spring spawners as well as the Norwegian spring spawners all belonging to the Atlanto-Scandian herring (Fridriksson and Aasen 1950). When this was realized the herring catches in Iceland were split on the basis of intensive biological sampling. Therefore, reliable catch statistics on the Icelandic summer-spawning herring is available from 1948. In more recent years fishery statistics are also available in other ICES reports.

Biology

Biological data are collected by the Marine Research Institute in Reykjavík and are available since the late 1940s on this stock. Published assessment (e.g., ICES 1999f) reports show that the stock collapsed in the late 1960s. Since 1975 there has been a gradual recovery.

Summer-spawning herring off Iceland spawn in July, mainly south and southwest of the country. The larvae usually hatch in August, and they remain as larvae throughout the winter. They metamorphose the following spring. The nursery grounds are mainly in the fjords of north and northwest Iceland, and there the juveniles stay until they are about 2 years old. The adult component has two distinct feeding seasons: in the periods

April-June, i.e., prior to the spawning season; and August-October, i.e., following the spawning season (Jakobsson et al. 1969). The feeding areas are off the west and east coasts of Iceland, and the wintering areas of the adult stock have been variable in recent years. The adult herring overwintered in the east coast fjords in the 10 years from 1980 to 1989, but more recently most of the adults have overwintered off the east coast of Iceland, while the recruiting year classes have overwintered off southwest Iceland.

Development of the Fisheries, Exploitation, Management, and Stock Status

During the 1950s the fishery was exclusively made with driftnets and the catches varied from 15,000 t to 35,000 t (Fig. 2f). Sonar guided purse-seining started in 1960 and the catches increased to about 130,000 t in 1963. This was followed by a sharp decline and a fishing ban at the end of 1971. Since 1975 the catches increased gradually to about 100,000 t during the 1990s (Jakobsson 1980).

The fishing mortality was at a low level during the driftnet fishery in the 1950s but increased sharply during the purse-seine period in the 1960s. The SSB reached about 300,000 t in the beginning of the 1960s, fell to less than 20,000 t in the period 1968-1973, then gradually increased to about 500,000 t in the 1990s. During the recovery of the stock average recruitment increased with larger spawning stock (Jakobsson et al. 1993).

During the moratorium on fishing a new policy for future harvesting of the stock was formulated which included seasonal restrictions: (a) limiting the fishing season to the last 4 months of the year when the annual weight at age was at maximum; (b) minimum landing size of 27 cm and compulsory release of purse-seine catches that obviously consisted of small immature herring; and (c) acting on recommendations from the ICES, the target fishing mortality rate was set at $F_{0.1} = 0.22$ for this stock (Jakobsson 1973). **This policy has proved to be very successful and the SSB has been increased to about 500,000 t, which is almost twice as big as observed prior to the collapse during the 1960s.** Similarly, recruitment has been on a much higher level than previously observed. Since 1975 the fishery management has been based on individual quotas that have been freely transferable since 1990 (Jakobsson and Stefansson 1999). This ITQ system appears to work very well and is highly economical. Based on the experience of managing this stock during the period 1971-1999 the future prospects are optimistic as long as no unexpected environmental change takes place.

Part 3. Western North Atlantic

The western Atlantic includes summaries from two areas on the shores of Newfoundland, of which one (western Newfoundland) includes spring and fall spawners; the Gulf of St. Lawrence (with spring and fall spawners); the eastern shores of Nova Scotia and the Bay of Fundy; and the Gulf of Maine and Georges Bank (Fig. 3). The western Atlantic herring are divided into Part 3a (northwestern North Atlantic) and Part 3b (southwestern North Atlantic) because there is some overlap of stock configurations between the Bay of Fundy, Georges Bank, and the Gulf of Maine. This is explained briefly following the subheading "Southwestern North Atlantic."

Part 3a. Northwestern North Atlantic
East and Southeast Newfoundland
Biology and Migrations

East and southeast Newfoundland herring overwinter in deep waters within the coastal bays around the island. Spring spawners, which constitute the vast majority of the stocks, then migrate into shallow coastal waters (normally <10 m) where they spawn, normally from late April to early June. Subsequent to spawning, mature herring tend to disperse throughout the bays during the summer as they begin to actively feed. Along the northeast coast, the summer feeding migration tends to be northward of the overwintering grounds. During the fall, herring form schooling aggregations and begin their migration back into the bays.

Fisheries Data

For management purposes east and southeast Newfoundland herring comprise five stock complexes. The majority of fish in each of these stocks are spring spawners. Although the stocks are discrete during spawning, there is some overlap during the migratory phase (Wheeler and Winters 1984). Herring are harvested primarily by a fleet of small purse-seine vessels (< 20 m); they are also caught by fixed-gear fishers for lobster bait.

Prior to the 1970s, the fishery was executed by fixed-gear fishers and annual landings averaged less than 10,000 t (Fig. 4a). With the introduction of the mobile purse-seine fleet in the mid-1970s and the recruitment of two large year classes in the late 1960s, annual landings increased to a peak of approximately 30,000 t in the late 1970s. Quotas were first introduced in 1977 and were exceeded by approximately 50% from 1977 to 1979. This, combined with poor recruitment through the 1970s, led to the collapse of stocks. Consequently, the fishery was closed in most areas during the early 1980s. It reopened during the mid-1980s with the recruitment of a moderate-size year class in 1982. Landings peaked at approximately 20,000 during the 1980s and averaged approximately 6,000 t through the 1990s.

Analytical assessments are completed biannually for four of the five east and southeast Newfoundland herring stocks. Population sizes of spring spawners only are estimated for each stock using an integrated catch-at-age (ICA) analysis (Wheeler et al. 1999). Catch and weight-at-age matrices, by stock area, are available from 1970 to the present. Five series of abundance estimates were available for the most recent assessment, research

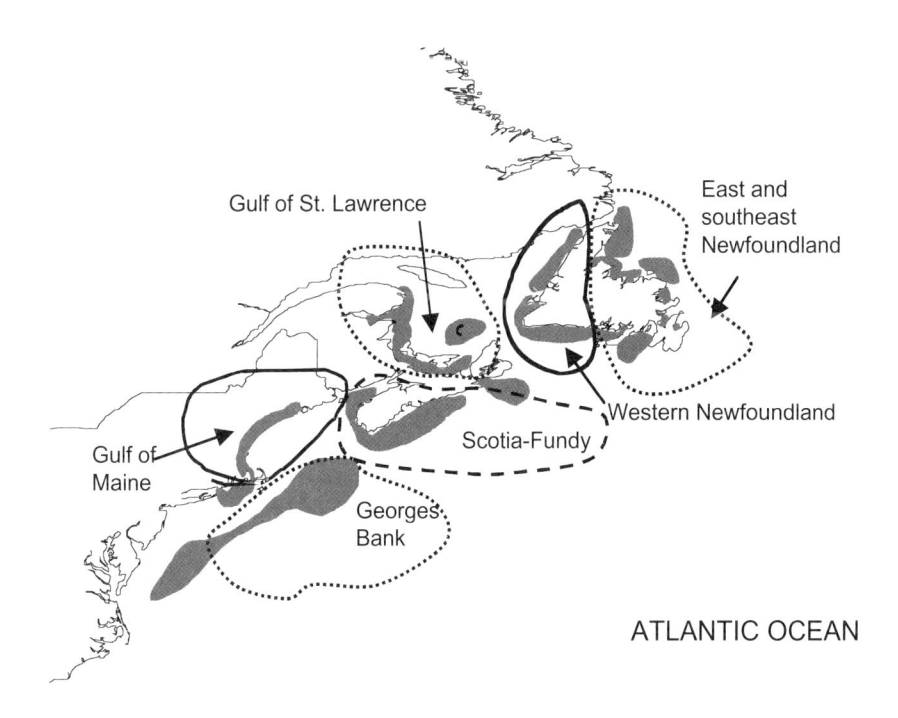

Figure 3. The western North Atlantic showing areas occupied by the major herring stocks separated by solid, dashed, or dotted lines. For most stocks, dark areas indicate the appropriate locations of spawning grounds.

gillnet catch rates (from index fishermen) and acoustic biomass estimates extending back to the 1980s, and commercial gillnet catch rates, gillnet fisher observations, and purse-seine fisher observations commencing in 1996. The status of each stock is defined by a stock status classification system based upon environmentally dependent stock-recruit relationships.

Commentary

Newfoundland herring stocks are at the northernmost range of herring in the northwest Atlantic. Strong recruitment of year classes tends to be very sporadic and is influenced by environmental conditions (Winters and Wheeler 1987). Below-normal water temperatures experienced during the early to mid-1990s were not conducive to strong recruitment. **Consequently, even with modest fisheries, most stocks have declined but stabilized at low levels compared to the 1970s.** Although there are not yet any indications, warming trends in water temperatures within the last few years may lead to increased recruitment.

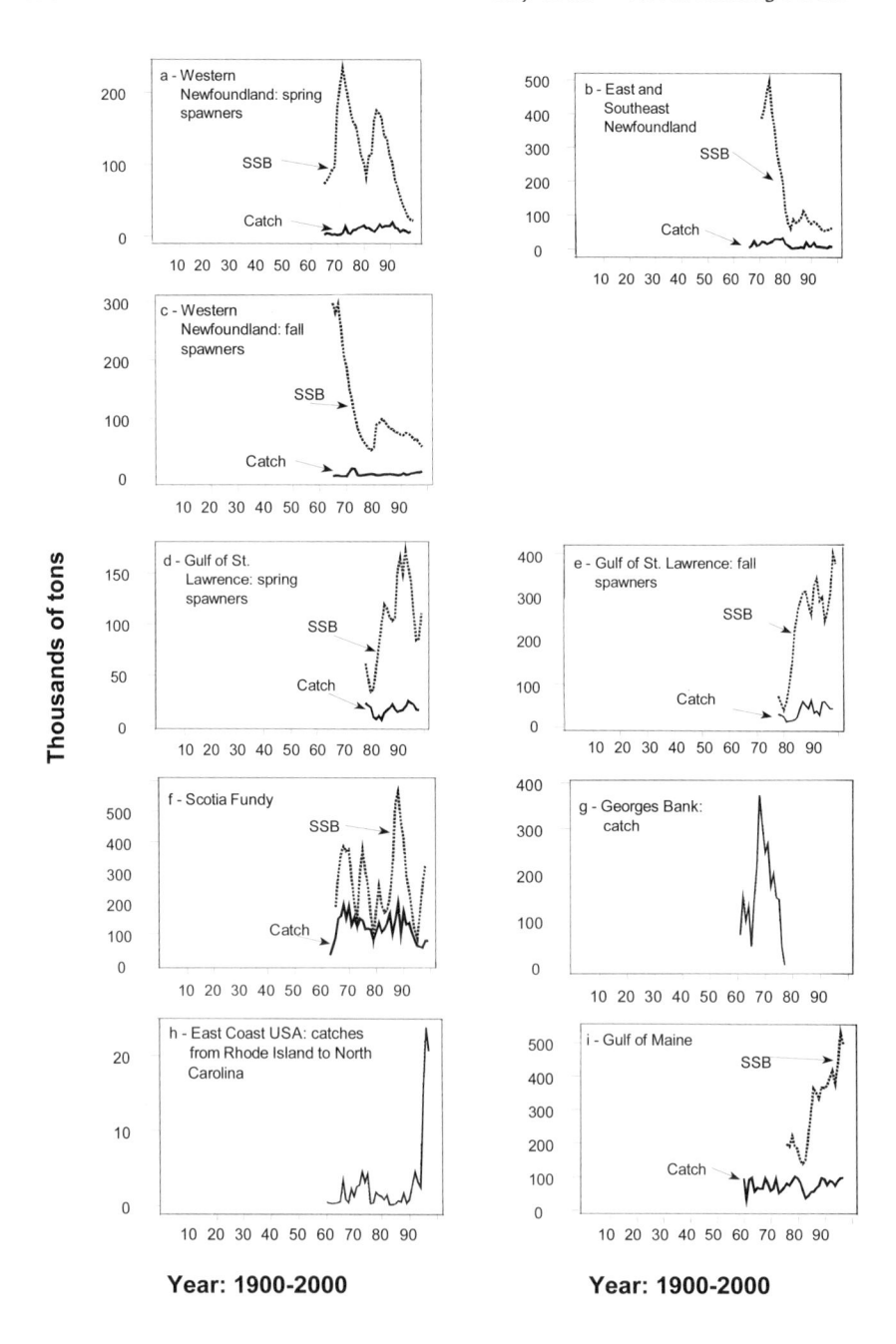

Figure 4. Catch (solid dark line) and spawning stock biomass (SSB, dashed line) for western North Atlantic herring stocks.

West Coast of Newfoundland (4R) Herring
Biology and Migrations
Within most of the geographic range of northwest Atlantic herring (*Clupea harengus* L.), including the west coast of Newfoundland (NAFO division 4R), some herring populations spawn in the spring (April to June) and others in the summer or autumn (July to October). Within each seasonal-spawning population (or stock), there are local spawning populations (or components) associated with specific spawning areas. Examples of spring-spawning components can be found in St. George's Bay, Port-au-Port Bay, and St. John Bay. These local components intermix throughout the range of the population, although most evidence suggests that once an individual fish spawns with a given local spawning component, it will return to spawn with that component year after year (Blaxter 1985). A local spawning component can therefore be considered as the basic biological unit to be protected from overexploitation. Local spawning components are, however, not independent of each other, as recruiting individuals may not spawn with their parental spawning component, but may be adopted by another local component, either with the same or a different spawning season (McQuinn 1997a). All the local components which together occupy a common geographic range, as delimited by their annual migration patterns, constitute the overall population (or metapopulation) which in turn defines the management area (McQuinn 1997b).

In the NAFO division 4R management area, individual fish cannot be attributed to their local population component if they are caught outside of the spawning season. Therefore the basic management unit has been defined as the seasonal-spawning stock, which can be determined from the stage of gonad development. The major spawning areas for the spring-spawning stock are located at the southern end of the coast in and around St. George's Bay and Port-au-Port Bay although several other spawning sites are known along the coast toward the north. Mature herring arrive and spawn in these areas from the end of April to the middle of June before dispersing. Autumn spawning is concentrated mainly north of Point Riche from mid-July to mid-September. At other times of the year, these two spawning stocks are mostly found in mixed schools in either feeding or overwintering areas. The major feeding areas, i.e., off St. George's Bay in the spring, north of Point Riche and in the Strait of Belle Isle in the summer, and off Bonne Bay in the fall, are associated with concentrations of copepods (red-feed) and/or euphausiids (krill) which are their main food items. Based on winter research survey data (McQuinn and Lefebvre 1995), they are believed to overwinter in the deeper waters of the Esquiman Channel.

Fisheries Data
Herring in western Newfoundland are exploited mainly from April to December by large (>25 m) and small (<20 m) purse-seiners and to a much lesser extent by fixed gillnetters. Since 1986, total herring landings from

the west coast of Newfoundland have averaged 17,300 t (from 12,400 t to 26,400 t) as compared to an average of 14,100 t for the previous decade. **The 1999 stock-status assessment (McQuinn et al. 1999) indicated that the spring-spawning stock is in danger of collapse. The autumn-spawning stock is declining gradually, while the exploitation rate has been slowly increasing.** Apart from the 1990 year class, recruitment to the spring-spawning stock has been below average since the 1987 year class recruited. The spring-spawner spawning-stock biomass (SSB) declined to a historical low of 14,000 t in 1999 (Fig. 4b). Recruitment to the autumn-spawning stock has been above average since the large 1979 year class, which has kept this stock at an intermediate level. The autumn-spawner SSB has been declining slowly, from 80,000 t in 1984 to 42,000 t in 1998 (Fig. 4c).

The latest assessment indicated that fishing mortality on these stocks has been increasing over the past 12-15 years and had been around $F_{0.1}$ for the spring spawners between 1991 and 1997. The closure of St. George's Bay and Port-au-Port Bay in 1995 had the desired effect of slowing the decline of this stock by concentrating fishing on the autumn spawners, of decreasing the quantity of spring spawners in the total catch, and of allowing these fish to spawn undisturbed. However, analyses have shown that the resumption of fishing in these southern bays in 1998 was premature, and that the concentrated harvesting of spring spawners in the spring fishery resulted in a sharp increase in fishing mortality, well above $F_{0.1}$. This is in agreement with comments received from inshore fishermen as well as the index-fisherman catch rates which suggest that **the stock has continued to decline since 1997 and has now reached a historical low.**

Commentary

These herring stocks, located near the northern end of the species distribution in the western Atlantic, are characterized by the occasional influx of very large year classes, up to 4 times the size of the standing stock which produced them, on roughly a 10- to 12-year cycle. For the spring-spawning stock, recruitment has been below average since the 1987 year class, and over 15 years have passed since the last large recruitment pulse (1980 and 1982 year classes). The production schedule of this stock over the past 30 years (McQuinn et al. 1999) shows that between 1987 and 1997, annual surplus production (recruitment + growth – natural mortality) rarely was positive, and annual net production (surplus production – fishing mortality) consistently was negative. This is mainly due to reduced average recruitment over this 11-year period, brought about by either reduced survival of young herring with less favorable environmental conditions, reduced spawning efficiency due to increased fishing pressure on spawning concentrations, and/or a possible increase in seal predation (although the consumption estimates are subject to large uncertainties). Regardless of the cause, the production of this stock (growth and recruitment) has not kept up with removals (catches and natural mortality), resulting in a declining SSB even though catches have been in line with the $F_{0.1}$ management strategy.

Gulf of St. Lawrence
Fisheries Data
Southern Gulf of St. Lawrence herring are harvested primarily by an inshore gillnet fleet fishing in 4T and a fleet of small purse-seine vessels (<20 m) in 4T and 4Vn. Two stocks of herring are harvested in these fisheries. The spring-spawning stock spawns before July 1 and the fall-spawning stock after July 1. During the spring and fall fishing seasons, larger seiners (>20 m) are prohibited from fishing in several areas set aside for exclusive fishing by the inshore fleet.

Prior to 1967, southern Gulf of St. Lawrence herring were exploited mainly by gillnets and average landings from 1935 to 1966 were 34,000 t. In the mid-1960s, a purse-seine fishery was introduced and average landings were 166,000 t from 1967 to 1972. Quotas were introduced in 1972 at 166,000 t and reduced to 40,000 t in 1973. Separate quotas for spring and fall spawners began in 1985. Catches of spring and fall spawners combined have been below the TAC since 1988. In the late 1970s and early 1980s spring and fall spawning stocks were about 10% of current stock sizes (Figs. 4d-e). The spring-spawner TAC was exceeded from 1994 to 1996 and was nearly caught in 1997 and 1998. The fall-spawner TAC has not been exceeded since 1986. Since 1981, the inshore fixed-gear component has had the majority of the catch of spring and fall spawners.

An ADAPT-VPA is the main assessment method. Separate assessments are provided for spring and fall spawners (Claytor et al. 1998, Claytor and LeBlanc 1999). Catch and weight-at-age matrices from 1978 to the present are estimated for each spawning group. Gillnet catch rates (1978-present) are the main abundance indices used to calibrate the VPA. For the spring spawners catch rates from the Escuminac, New Brunswick, and southeast New Brunswick fisheries are used. For the fall spawners, catch rates from all areas are used but the time series is split into two sections corresponding to the year in which there was a major shift in mesh size used in the gillnet fishery. The first is from 1978 to 1991, when 2 5/8" was the predominant mesh size (75-91%), and the second, after 1992 when the percentage using 2 5/8" dropped (54-67%) in favor of larger mesh sizes of 2 3/4" to 2 7/8". Most recently, biomass estimates from an annual acoustic survey (1994-present) have been used as auxiliary indices. It is expected that in time, the indices from the acoustic survey will become the main abundance index for spring and fall spawners. **Current stock levels estimated from the VPA for fall spawners are among the highest observed since 1978.** Uncertainties resulting from difficulties in estimating incoming recruitment and low abundance indices in the acoustic survey moderate this view. The abundance of fall spawners remains above those observed when the stock was very low in the early 1970s and late 1980s. Current stock levels estimated from the VPA for spring spawners indicate a stock size that has been about average since 1985 but above the low levels observed in the late 1970s and early 1980s.

Commentary

Estimates of 4+ spring spawner biomass peaked in 1995, when the 1991 year class, which was the largest on record, entered the fishery. This year class has been supporting the fishery since it first appeared in 1995. The 1992 year class was among the lowest since 1978 but the two most recently estimated year classes, 1993 and 1994, were above average. The result of these trends in year-class strength are that the biomass levels were relatively stable for the past 4 years. The $F_{0.1}$ fishing levels were between 16,000 t and 18,500 t from 1996 to 1999.

The history of recruitment since the spring-spawning stock started to rebuild in 1983 is that incoming 4-year-olds have ranged from 50 million to 150 million individuals. Two very strong year classes, 1988 and 1991, consisted of greater than 300 million individuals. It is only the influence of these year classes that increased 4+ biomass levels to above 80,000 t, and, unless year classes of this size appear again, no major increases in biomass or $F_{0.1}$ levels can be expected.

Prior to 1998, estimates of 4+ fall spawner biomass peaked in 1991, when the very large 1987 year class appeared in the fishery. The population declined until 1996, when the large 1992 year class appeared in the fishery. Since then, year classes have been above average and the population is growing. The $F_{0.1}$ fishing levels were between 50,000 t and 60,000 t from 1996 to 1999. The assessment of spring and fall spawning stocks is being improved by the development of an acoustic research survey that covers the majority of the stock area just after the gillnet fishery, and by increased involvement of fishing boats to collect acoustic data during surveys and regular fishing activity. These projects build on telephone surveys of the fleet which began in 1986 and workshops which began in 1994 to incorporate local fishers' knowledge as part of the assessment of these stocks.

Part 3b. Southwestern Atlantic

A Note on Herring Stock Definitions in the Southern North Atlantic

Atlantic herring from the eastern shore of Nova Scotia, the Bay of Fundy, and Gulf of Maine region are divided into three major spawning stocks: (1) a large-size stock that spawns in U.S. and Canadian waters on Georges Bank and in U.S. waters southeast of Cape Cod; (2) an intermediate-size stock that spawns in Canadian waters south of Nova Scotia; and (3) a smaller-size stock that spawns in U.S. coastal waters of the Gulf of Maine (Iles and Sinclair 1982). Each of these three stock complexes consists of several major and many minor spawning components. Prior to 1993 all three stock complexes were assessed separately. In 1993 the United States began using a spring bottom trawl survey as an index of abundance to tune VPA stock abundance models. Because adult herring from both the Georges Bank (5Z) and Gulf of Maine (5Y) spawning stocks occupy continental shelf waters south of Cape Cod in the spring and cannot be distinguished from each other, this resulted in the creation of the coastal stock "complex," an arbitrary assessment unit. On the other hand, since 1988 Canada has assessed and reported only on that part of 5Z which is east of the Great South Channel (longitude 69.0°W), thereby excluding a large segment (5Z west) of the stock complex.

Scotia-Fundy

Fisheries Data

Fisheries in recent years have been dominated by purse seine, weir, and gillnet with relatively minor landings by shutoff, trap, and midwater trawl. Most fishing takes place on dense summer feeding, spawning, and overwintering aggregations. Landings occur throughout the year, but occur mainly from July to October. In recent years landings have been measured (volumetric determination) by an independent dock-side monitoring program. Extensive biological sampling has been undertaken since 1986 (Power and Iles 2001, this volume). In recent years there has been considerable involvement by the fishing industry in recording size information and collecting biological samples in support of fishery evaluation.

Landings and biological information series are available for years since 1965 (Fig. 4f). Surveys of the abundance of distribution of herring larvae were undertaken annually from 1972 to 1998. Acoustic surveys of major overwintering aggregations were undertaken in the 1980s and early 1990s. Since 1997 there have been acoustic surveys of major aggregations and of spawning areas using commercial vessels (Melvin et al., 2001, this volume). **Annual landings have ranged from 30,000 t to 200,000 t and averaged 120,000 t for the period 1963-1999. The assessment indicates that the SSB fluctuated between 85,000 t and almost 600,000 t during the same period. The last decade saw a substantial drop in SSB (until 1996), followed by a substantial recent increase.**

Commentary

The 4VWX management unit contains a number of spawning areas separated by various degrees of space and time. Spawning units that are in close proximity, with similar spawning times, and which share a larval distribution area (e.g., Trinity Ledge and German Bank in southwestern Nova Scotia) are considered part of the same complex, and undoubtedly have much closer affinity than spawning units which are widely separated in space or time, and do not share a common larval distribution. Some spawning areas are large and offshore, whereas others are small, and more localized, sometimes very near shore or in small embayments. The situation is complicated further by the fact that some of these herring tend to migrate long distances, and to mix outside of the spawning period with members of other spawning groups. Some spawning areas are known from fishery sampling, tagging, etc. to have formed the basis for major historical fisheries, while others have not. For the purposes of evaluation and management, the 4VWX herring fisheries are divided into four components: (1) southwestern Nova Scotia/Bay of Fundy spawning component, (2) offshore Scotian Shelf Banks spawning component, (3) coastal Nova Scotia spawning component, and (4) southwestern New Brunswick migrant juveniles. Recognizing that each component has several spawning areas, and that mixing of fish occurs among components, industry and management have explored means of managing the complexity within each component (such as distributing fishing effort among spawning areas according to their relative size) and of taking appropriate account of interaction among components (such as restrictions on some areas of mixing).

Specific conservation objectives were reviewed and developed further during 1997. Three objectives and a number of targets within these objectives were defined as follows: Objective 1: Maintain reproductive capacity of herring in each management unit by (i) ensuring persistence of all spawning components in the management unit; (ii) maintaining biomass of each spawning component above a minimum threshold; (iii) maintaining a broad age composition for each spawning component; and (iv) maintaining a long spawning period for each spawning component. Objective 2: Prevent growth overfishing by continuing to strive for fishing mortality (F) below 0.1. Objective 3: Maintain ecosystem integrity ("ecosystem balance") or ecological relationships by implementation of a precautionary approach that requires further definition of target and limit reference points associated with these three objectives.

An "in-season" management process was implemented in the southwest Nova Scotia fishery during 1995 and has been extended to other areas and fisheries (Stephenson et al. 1999). This approach encourages surveying using the fishing fleet under scientific direction and control prior to fishing to ensure that fishing is distributed appropriately among various components of the stock (particularly among spawning components) according to the relative size and current state of each component (Melvin et al. 2001, this volume). It has improved data collection and enabled modifications to management decisions to be made with the involvement of participants and on the basis of up-to-date information.

Atlantic Coastal Stock Complex: Gulf of Maine and Georges Bank

Biology

Spawning occurs at more or less discrete locations on the northern edge and northeast peak of Georges Bank, on Nantucket Shoals, in the vicinity of Jeffreys Ledge, along the eastern Maine coast, and at several other poorly known sites along the Maine coast in the summer and fall (July-December). Eggs are deposited primarily on gravel substrate at water depths of 25-75 m. Larvae that hatch along the coast are transported in a southwesterly direction by the coastal Gulf of Maine current (Graham 1982, Chenoweth et al. 1989, Townsend 1992) and in a clockwise direction on Georges Bank. Larvae metamorphose into juveniles the following spring and remain in coastal waters in the Gulf of Maine, southern New England, and the mid-Atlantic states until the fall of their third year. Adults (age 3+) that spawn in the Gulf of Maine migrate south in the fall and mix with adults from Georges Bank and Nantucket Shoals to overwinter in southern New England and mid-Atlantic continental shelf waters, then return north in the spring and, following a summer feeding migration into the Gulf of Maine, reoccupy their respective spawning grounds in the late summer and fall.

History of Stock Assessments

Early assessments on Georges Bank and western Gulf of Maine herring were tuned using the catch of juvenile herring in the Maine fixed-gear fishery as an index of recruitment. These assessments did not rely on trawl survey data for tuning purposes and were applied to single stock components (ICNAF Redbook 1976; Anthony and Waring 1980a,b). Following the introduction of trawl survey abundance indices for VPA tuning purposes in the early 1990s, there were concerns about patchy spatial distribution of prespawning and spawning aggregations of adult herring in fall surveys. On the other hand, the use of spring survey data, collected when most of the adults from the two spawning stocks occupy shelf waters south of Cape Cod and cannot be differentiated, required the combination of the two stocks for assessment purposes. The most recent of these assessments, performed by the U.S. National Marine Fisheries Service in collaboration with state fishery scientists, was reviewed by the 27th Stock Assessment Review Committee in 1997 (NOAA 1998). Preliminary assessment information for the Gulf of Maine stock was also included in the 1998 stock assessment report. Additional information on the Gulf of Maine stock was presented at this meeting by D.K. Stevenson.

Georges Bank

Fisheries Data

Georges Bank once supported the largest herring fishery in the western Atlantic. The fishery began in 1961 when the former U.S.S.R. harvested 68,000 t of herring (Fig. 4g). From 1961 to 1965 the U.S.S.R. dominated the fishery with annual reported catches ranging between 38,000 and 151,000 t. The fishery expanded rapidly when Poland and the German Democratic Republic entered the fishery in the mid-1960s. Catches peaked at 374,000 t in 1968. Between 1967 and 1976 vessels from 12 countries, including Canada and the United States, participated in the fishery. The fishery collapsed in 1977 and between 1978 and 1993 there was no directed fishery for herring on the bank. The fishery reopened in 1994, when several Canadian herring vessels traveled to Georges Bank. Harvests were minimal until 1997 when 6,262 t were taken by U.S. vessels. Fishing activity by the U.S. fleet recently increased, with reported catches of 17,342 in 1998 and approximately 10,000 t in 1999.

Data Sources

There are several survey databases available in both Canada and the United States that can be, or have been, used to assess the status of Georges Bank herring. Repositories for the following survey data are the National Marine Fisheries Service (NMFS), Woods Hole, Massachusetts, and the Canadian Department of Fisheries and Oceans (DFO), St. Andrews, New Brunswick. The surveys include:

U.S. fall bottom trawl survey	1963-2000	NMFS
U.S. spring bottom trawl survey	1968-2000	NMFS
U.S. winter bottom trawl survey	1992-2000	NMFS
U.S. fall acoustic survey	1998-2000	NMFS
U.S. fall larval surveys (ICNAF, MARMAP, etc.)	1971-1994	NMFS
Canadian spring groundfish survey	1987-2000	DFO
Canadian adult/larval herring survey	1986-1995	DFO

Stock Assessment and Trends in Abundance

Since 1993 the United States has assessed Georges Bank herring as part of the coastal stock complex. Several of the databases listed above have been evaluated as possible tuning indices for VPAs. In the most recent U.S. assessment (June 1998), the spring bottom trawl and winter bottom trawl indices (number per tow for ages 2-8) were used to tune the VPA. In recent years Canada has not used a VPA to assess the status of Georges Bank herring. Instead, assessment advice has been based on biological characteristics, changes in spatial distribution, and comparisons of current larval abundance estimates with estimates from the early 1970s, prior to the collapse of the stock when SSB estimates for the Georges Bank stock were available (Melvin et al. 1996).

In its relatively short history, the Georges Bank herring stock has gone from one of the largest populations on the east coast of North America to virtually zero and back again. The extent of the decline in the 1970s was remarkable. Between 1979 and 1981 only 2 of more than 500 fishing sets in the fall bottom trawl survey produced any herring. The first signs of recovery occurred in 1984, when a large number of juveniles from the 1983 year class were captured. Between 1986 and 1993 there was a general increase in abundance, a protraction of spawning distribution, and positive signs of recruitment. In 1992, the spawning distribution (determined from the occurrence of larvae <10 mm) had returned to historical areas and included the northeast peak of the bank. **Since 1993, the abundance of herring on Georges Bank may have reached, or exceeded, precollapse levels.** In 1995 the U.S. estimate of SSB for the coastal stock complex was 787,000 t (of which approximately 80% was estimated to originate from Georges Bank). Canada estimated 100,000-200,000 t for Georges Bank east of the Great South Channel. **Biomass estimates derived from the most recent U.S. assessment (NOAA 1998) indicate a substantial increase in total stock size during the past few years, reaching about 3.5 million t in 1997.** The U.S. assessments are heavily influenced by spring trawl survey abundance indices, which increased significantly in recent years in southern New England and mid-Atlantic waters. Current stock size estimates are 3 times higher than estimates from the late 1960s when the fishery peaked on Georges Bank. Retrospective analysis has revealed that current stock sizes are overestimated in this assessment. However, preliminary biomass estimates obtained from prespawning aggregations of herring along the northern edge of Georges Bank in the fall of 1999 were similar to the VPA biomass estimates (William Overholtz, U.S. Department of Commerce, National Marine Fisheries Service, pers. comm.). A conditioned surplus production model produced a much more conservative stock size estimate of 1 million t and a maximum sustainable yield of 317,000 t (Stevenson et al. 1997, NOAA 1998).

Commentary

The Georges Bank herring stock provides a prime example of the detrimental effects of poor recruitment and overfishing. So effective were the combined effects of these two factors that the stock collapsed in 1977, just 15 years after fishing first started on the bank. From 1978 to 1984 there were so few herring on Georges Bank that many researchers, resource managers, and industry members believed the stock to be lost forever. Slowly during the mid-1980s research surveys began to document signs of recovery. However, it was not until 1992, when spawning was documented in most of the known historical areas, that investigators monitoring the stock's progress believed the stock could reach or exceed precollapse biomass levels. Details of the early stages (1983-1990) of recovery were documented by Stephenson and Power (1989), Smith and Morse (1990), and Stephenson and Kornfield (1990). The middle and late stages of the recovery were documented by Melvin et al. (1996) and most recently by NOAA (1998).

   The future outlook for Georges Bank herring is very positive. The most recent U.S. trawl survey (fall 1999) confirmed the continued expansion of the stock. Fishing mortality is at an extremely low level. However, as the fishery expands care must be taken to avoid a repeated overexploitation of the stock. Managers must consider the effects of rapid or large-scale expansion on the long-term sustainability of the stock. Geographical boundaries and the seasonal movement of this transboundary stock also need to be determined. Because of the transboundary nature of the stock, Canada and the United States have collaborated on recent assessments and are now planning to conduct a joint assessment of Georges Bank herring.

Finally we point out that in recent years, with the recovery of the Georges Bank stock, the abundance of herring on overwintering grounds in southern New England and the mid-Atlantic states also has increased dramatically, as have catches in the commercial fishery (Fig. 4h).

Gulf of Maine Stock

A fishery for juvenile herring developed in Maine and New Brunswick in the latter part of the nineteenth century, stimulated by the development of canning techniques. The last 15 years have seen the demise, in U.S. waters, of the nearshore fixed-gear juvenile fishery and a rise in importance of the offshore mobile gear (purse seines and midwater trawls) adult fishery. There is still a fixed-gear fishery in New Brunswick that harvests 15,000-20,000 t annually. The Gulf of Maine catch exceeded 75,000 t a year during the first decade of this century and again in the 1950s. The catch remained fairly steady at 80,000-100,000 t a year during the past 10 years as the fishery became less vulnerable to annual variations in juvenile abundance (Fig. 4i).

Biomass estimates are derived from a virtual population analysis of Gulf of Maine catch-at-age data (which include the New Brunswick fixed-gear fishery) and terminal fishing mortality estimates for individual year classes based on numbers caught at age (NOAA 1998, Stevenson 1998). **The results indicate that the Gulf of Maine stock doubled in size in a 3-year period during the mid-1980s, remained stable at 350,000-400,000 t between 1986 and 1994, then increased to about 500,000 t in 1996 and 1997 (Fig. 4g). A more recent VPA for this stock based on 1976-1998 catch-at-age data indicates that stock biomass in January 1998 declined to about 350,000 t** (presented at the Herring 2000: Expectations for a New Millennium symposium by D.K. Stevenson). The sharp increase in stock size that occurred between 1983 and 1986 was due to the recruitment of the large 1983 year class which, unlike previous year classes, was exploited very minimally at age 2. The development of the mobile gear fishery during the 1980s was based on the increased biomass of adults, in particular adults belonging to the 1983 year class.

Part 4. Eastern Pacific

In contrast to other parts of the world, herring populations in the eastern Pacific, from California to the Bering Sea, consist of a relatively large number (>20) of relatively small populations (most < 100,000 t) (Fig. 5). Most have limited migrations, with spawning and juvenile rearing areas in nearshore areas and summer feeding limited to the relatively narrow continental shelf.

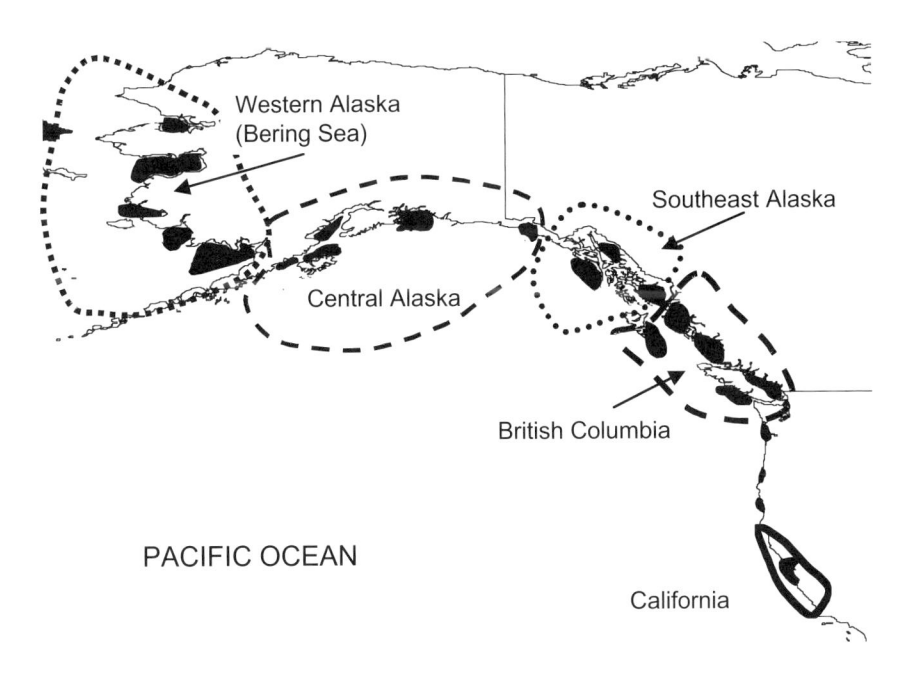

Figure 5. *The eastern North Pacific showing the major herring stocks separated by solid, dashed, or dotted lines. The dark areas indicate the approximate centers of spawning and overwintering.*

Virtually all eastern Pacific herring stocks use assessment methods that require some quantitative assessment of spawning. Pacific herring spawn mainly in nearshore, shallow inter- and subtidal waters. Over the years various techniques have developed to quantify herring spawning. This includes visual estimates from aerial surveys in remote areas of Alaska to detailed surveys using scuba divers, to make explicit counts of the total numbers of eggs on spawning grounds in California, Washington, and British Columbia.

Alaska

Biology

Within Alaskan waters Pacific herring spawn at discrete locations from Dixon Entrance in southeastern Alaska to Norton Sound. Like herring in other parts of the eastern Pacific, spawning occurs on intertidal and subtidal vegetation in spring. The timing of spawning is related to temperature, and progresses around the Alaska coast from March in southeastern Alaska, to June in Norton Sound. In warmer years, herring spawning occurs earlier throughout Alaska. However, there are some patterns in the time series of herring spawning that do not appear to be explained just by temperature variability. For example, since 1993 in Sitka Sound, spawning has occurred 3 weeks earlier than in all previous records. The changes in spawn timing can be biologically significant, affecting seabirds, shorebirds, marine mammals, and piscivores that are focused on herring spawning in the spring.

The life history strategy of herring in the Bering Sea is distinctly different from populations in the Gulf of Alaska and more southern regions in the eastern Pacific. Bering Sea herring attain large body size (to 500 g), whereas Gulf of Alaska herring reach only half that size. The eastern Bering Sea herring are long-distance migrators. The largest population spawns along the north shore of Bristol Bay, near the village of Togiak. Following spawning, these herring migrate in a clockwise direction down along the Alaska Peninsula, reaching the Unimak Pass area in early July (Funk 1990). They feed along the continental shelf edge, slowly moving northward to overwinter near the Pribilof Islands. The Bering Sea herring life history strategy appears to be an adaptation to take advantage of the distant rich feeding grounds and mild overwintering areas on the continental shelf edge while utilizing the protected inshore bays for summer larval nurseries. In contrast, Gulf of Alaska herring are smaller, have shorter life-spans, have more frequent recruitment events, and do not undergo long-distance migrations. In the Gulf of Alaska, recruitment events tend to occur synchronously over fairly broad areas that contain otherwise discrete spawning aggregations. Gulf of Alaska herring have some genetic distinction from Bering Sea herring (Grant and Utter 1984).

There appear to be strong, autocorrelated, and almost cyclic changes in size-at-age time series that date back to the reduction fisheries of the early 1920s. The cause of these apparent cycles in body size is not known, but the anomalies have been correlated to a time series of zooplankton abundance measured at Prince William Sound salmon hatcheries, and also to the Pacific Decadal Oscillation (Evelyn Brown, University of Alaska Fairbanks, pers. comm.). Only a mild effect of density dependence is seen in adult herring at studied locations in Alaska, such as Prince William Sound. However, density dependence could be an important mechanism affecting larval and juvenile herring growth, for which little time series data exist.

Recruitment also shows signs of periodic autocorrelated anomalies. In the Gulf of Alaska recruitment time series a 4-year cycle of strong year classes is apparent, although that pattern has changed recently. Recruitment events occur more frequently in the Gulf of Alaska (typically averaging every fourth year), whereas in the Bering Sea, strong recruitment events occur much less frequently, typically averaging every tenth year. Most areas experienced a positive response in recruitment associated with the 1977 regime shift. These recruitment indices were derived from routine agency stock assessments in support of fishery management.

Year-class abundance is quantified at a recruiting age of 3 in the Gulf of Alaska, and age 4 in the Bering Sea. Adult survival rate is usually treated as constant for stock assessment purposes. When this holds, the recruitment time series provide an excellent measure of abundance for comparison to long-term climate indices, particularly because herring early life history is fine-tuned to ocean processes with low tolerance for changing conditions. However, occasional adult herring epizootics have been observed in Prince William Sound, which can drastically alter adult survival rates. When adult survival rate changes substantially, the recruitment time series will not provide a good measure of adult abundance. Thus far, substantial changes in adult survival appear to be relatively rare, so that the recruitment indices typically provide a reliable index of abundance for both juveniles and adults.

Based on patterns in size at age and recruitment, Williams (1999) grouped Alaska herring into three categories: Bering Sea, Outer Gulf of Alaska, and Inner Gulf of Alaska. The spatial scale of these groupings reflects the spatial scale of oceanographic processes underlying herring productivity, as well as the different Bering Sea and Gulf of Alaska life history strategies. Fishery managers need to understand finer spatial scales of herring stock structure than these large groupings based on coherence in growth and recruitment anomalies. Because herring milt can be readily observed from aircraft and precisely defines spawning locations, fishery managers use maps of herring milt locations to define discrete groups of herring appropriately sized for management units.

Trends in Herring Abundance and Historical Catch in Alaska

In Alaska subsistence fisheries for Pacific herring predate recorded history. Traditional dried herring remains a major staple of the diet in Bering Sea villages near Nelson Island (Pete 1990), where salmon are not readily available. Alaska's commercial herring industry began in 1878 when 30,000 pounds of salt-cured product were prepared for human consumption. By 1882, a reduction plant at Killisnoo in Chatham Strait was producing 30,000 gallons of herring oil annually. The herring reduction industry expanded slowly through the early twentieth century, reaching a peak harvest of 142,000 t in 1934. Exploitation rates were quite high during the reduction fishery era, with large fluctuations in stock levels and annual harvests. As Peruvian anchovetta reduction fisheries developed, Alaska herring reduction fisheries declined, so that by 1967 herring were no longer harvested for reduction products in Alaska.

A Japanese and Russian trawl fishery for herring began in the central and eastern Bering Sea in the late 1950s, reaching a peak harvest of 146,000 t in 1970 (Fig. 6a). Substantial catches of herring for sac roe began throughout Alaska in the 1970s as market demand increased in Japan (Fig. 6b). Presently, herring are harvested primarily for sac roe destined for Japanese markets. Statewide herring harvests have averaged approximately 45,000 t in recent years. Much of the herring taken in these fisheries were from western Alaska coastal spawning stocks in the Bering Sea, which are the largest of Alaskan stocks (Fig. 6c).

Approximately 25 distinct fisheries for Pacific herring occur in Alaskan waters. Almost all of these herring fisheries are closely linked to a specific spawning population of herring. Most of the herring harvest currently occurs during sac roe fisheries, which harvest herring just before their spring spawning period. Both males and females are harvested, although the sac roe fisheries target the much higher-valued roe-bearing females. Alaska statutes require that the males also be retained and processed and not discarded as bycatch. Most sac roe fisheries occur during a series of short openings of a few hours each, spanning approximately 1 week. Fishing is not allowed between these short openings, to allow processors time to process the catch, and for managers to locate additional herring of marketable quality.

The present abundance in the Bering Sea, based mainly on the SSB estimates of the Togiak region, is much lower (~140,000 t) than the highest estimate of nearly 500,000 t in 1985. The abundances of the many smaller stocks in the same region show no clear trends in time.

Spawn-on-kelp fisheries harvest intertidal and subtidal macroalgae containing freshly deposited herring eggs. Both of these fisheries produce products for consumption primarily in Japanese domestic markets. Smaller amounts of herring are harvested from late July through February in herring food/bait fisheries. Most of the herring harvested in these fisheries are used for bait in Alaskan longline and pot fisheries for groundfish and shellfish. Smaller amounts are used for bait in salmon troll fisheries, with occasional utilization for human or zoo food.

Harvest policies used for herring in Alaska set the maximum exploitation rate at 20% of the exploitable or mature biomass, consistent with other herring fisheries on the west coast of North America. The 20% exploitation rate is lower than commonly used biological reference points (Funk 1991) for species with similar life history characteristics. In some areas, such as southeastern Alaska, a formal policy exists for reducing the exploitation rate as the biomass drops to low levels. In other areas, managers similarly reduce the exploitation rate as abundance drops, without the more formal exploitation rate framework. In addition to exploitation rate constraints, minimum threshold biomass levels are set for most Alaskan herring fisheries. If the spawning biomass is estimated to be below the threshold level, no commercial fishing is allowed. Threshold levels are generally set at 25% of the long-term average of unfished biomass (Funk and Rowell 1995).

British Columbia Herring
Stock Structure and Migrations
The herring population that occurs in British Columbia is believed to consist of five major migratory stocks as well as a large number of smaller

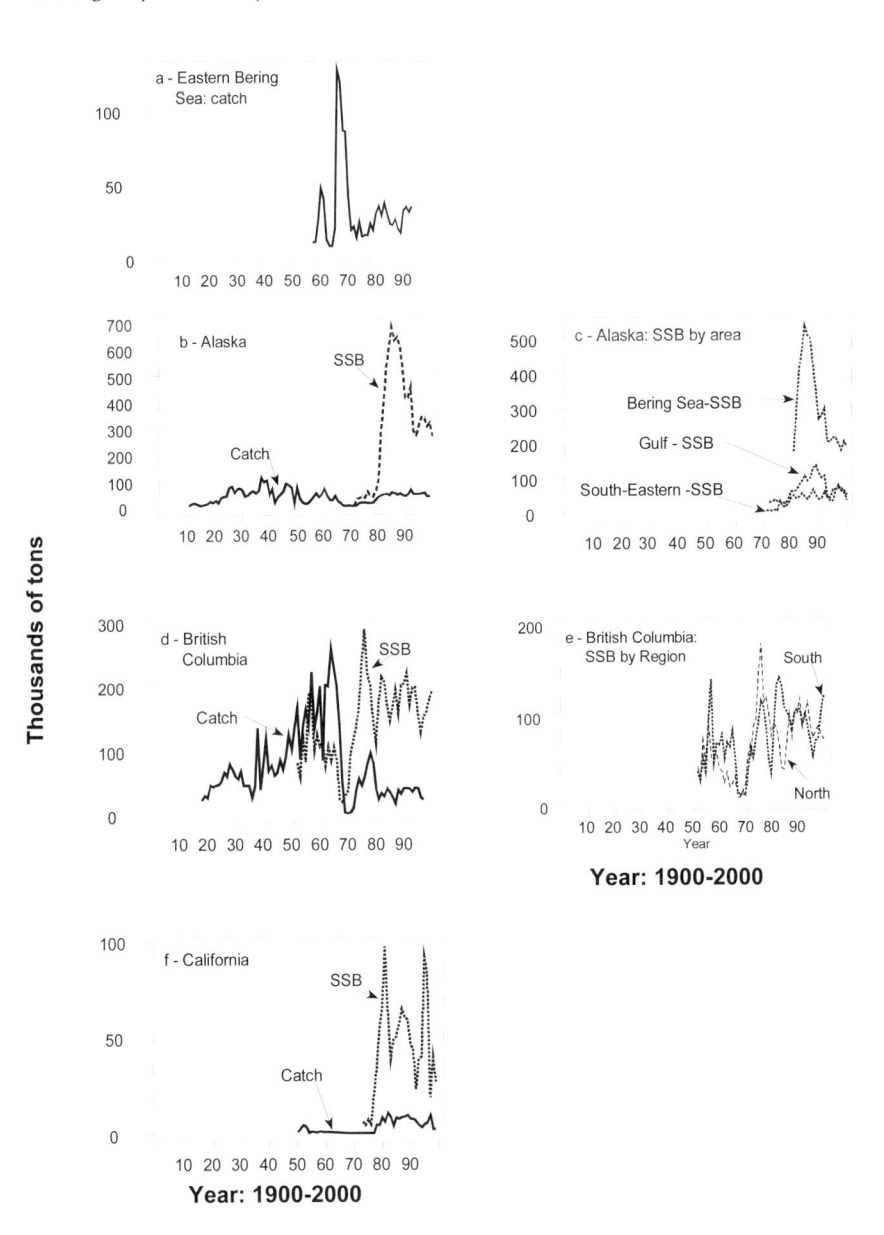

Figure 6. Catch (solid dark line) and spawning stock biomass (SSB, dashed line) for eastern North Pacific herring stocks. Figure 6b shows the sum of all Alaska stocks and Fig. 6c shows the SSB for each of the main subareas. Figure 6d shows the sum of all British Columbia stocks and Fig. 6e shows the SSB for northern and southern areas separately.

localized stocks (Hay and McCarter 1997, Schweigert et al. 1999). Based on their geographical spawning locations the five major stocks occur along the southeastern coast of the Queen Charlotte Islands, the north coast of British Columbia, the central coast of British Columbia, the west coast of Vancouver Island, and the Strait of Georgia. The localized stocks occur at the heads of many of the long inlets or fjords of the central and north coasts as well as in Johnstone Strait, the Strait of Georgia, and the west coast of the Queen Charlotte Islands. The spawning period for the major stocks begins in late February or early March in the south and progresses through mid-April as one moves northward (Haegele and Schweigert 1985). The spawning times for the localized stocks also occur in this period although there are some interesting exceptions from as early as late December or January in the Queen Charlotte Islands to as late as June or early July in the central coast. After spawning, the major migratory populations move offshore to the summer feeding areas, which are believed to be within Hecate Strait for the Queen Charlotte Islands and north coast stocks and likely some of the central coast stock. Fish that spawn within the Strait of Georgia migrate offshore through both the Strait of Juan de Fuca in the south and mix with west coast Vancouver Island fish on La Perouse Bank, as well as through Johnstone Strait into Queen Charlotte Sound or off the northwest coast of Vancouver Island where they could mix with either central coast or west coast of Vancouver Island stocks. Larval and juvenile distributions are not well known for other areas outside of the Strait of Georgia where they are ubiquitous (Haegele 1997, Hay and McCarter 1997). In general, surveys in all areas indicate diffusion of larvae away from the major spawning areas with time. Subsequently, juveniles appear to be found almost everywhere along the available shorelines.

Indications are that some juveniles begin to migrate offshore in the fall of their first year (Taylor 1964), although recent studies within the Strait of Georgia suggest that many juvenile herring overwinter there in their first year and do not migrate offshore until the following June-July, possibly following the adults as they return to offshore feeding grounds following spawning.

The Fishery

The first documented records of Pacific herring catches occurred in 1877 although herring were taken historically by aboriginal peoples. The earliest commercial fishery was for domestic consumption but larger quantities were exported to China as a salted product beginning in the early 1900s (Taylor 1964). The market for this product appears to have disappeared by the late 1920s, and in the early 1930s a reduction fishery began and expanded throughout the coast in the 1940s and 1950s as the sardine stocks collapsed off of California. By the early 1960s all herring stocks in British Columbia were being heavily exploited by this fishery, which collapsed and was closed in 1968 (Fig. 6d). The fisheries during this period were prosecuted initially with drag or beach seines and later with purse seines. The fishery occurred during the fall inshore migration of maturing stocks primarily from October through January. A limited amount of gillnetting occurred in the summer in the Strait of Georgia in the 1950s. After a 3-year closure an experimental roe herring fishery was opened in 1971 and it expanded rapidly through this decade. The roe fishery occurs on or near the spawning grounds during early March to early April. The roe fishery uses a combination of gillnets (45%) and seines (55%) to catch the quota. Concerns about collapsing stocks in the early 1980s led to the introduction of the currently used 20% fixed harvest rate policy to set quotas preseason. This policy was augmented with a fishing threshold or "cutoff" level in 1985 to ensure a spawning reserve for stock rebuilding in periods of natural stock decline (Stocker 1993). The harvesting policy has resulted in annual harvests of 30,000-40,000 t of herring coastwide during the past two decades although there have been fishing closures in each of the five stock areas during this period. In addition, there are small food and bait fisheries which harvest approximately 2,000 t annually. Another fishery that occurs in some localized areas of the coast is the spawn-on-kelp fishery, which is primarily undertaken by Native groups and lands about 360 t of product which includes the kelp and attached eggs.

Fisheries Data and Stock Assessment

Annual estimates of total landings have been collected since the early 1930s on a coastwide basis and since 1917 on a localized scale. These data are available in unpublished manuscripts at the Pacific Biological Station. Beginning in 1950, individual fishing companies were required to submit sales slips of their landings to the B.C. Department of Fisheries and these have been maintained in an electronic database to present. Similarly, much of the early data that was collected on the annual deposition of herring spawn and biological samples of the spawning runs is presented in summarized form in the annual B.C. Department of Fisheries reports. All available raw data for spawn deposition and biological sampling data such as length, weight, sex, and age are available in electronic form to present in a database. Other assorted biological data such as tagging and recovery information are also available in electronic databases.

The annual assessments of stock abundance rely on estimates from two analytical models. The first is a modification of the escapement model described by Schweigert and Stocker (1988) and relies on the data collected on spawn deposition. The second is a catch-age or age-structured model which is a modification of the model described by Fournier and Archibald (1982). Both models reconstruct mature stock abundance for the period since 1951 and forecast prespawning abundance for the next season. Forecasts of upcoming run size are based on the combination of estimates of surviving repeat spawners and newly recruited spawners which are presented as poor, average, and good, based on historical recruitment levels. The biological data required to annually monitor and assess stock abundance levels is obtained through a test fishing program which collects samples of herring from all the major spawning stocks and as many of the minor stocks as possible. In addition, scuba diver teams make annual surveys of the approximately 500 km or more of shoreline to assess herring spawn.

Biology

The age at first maturity appears to be the third year for most areas of British Columbia (Hay and McCarter 1999). A small proportion of the population in the southern stocks matures in the second year and in the northern areas in the fourth year. Maximal gonad weight at maturity is a function of fecundity and egg size but has been found to be about 25-30% of total body weight in B.C. stocks (Ware 1985). Similarly to gonad weight, fecundity is a conservative function of body weight and has been found to be constant at 200 eggs per gram of female weight (Hay 1985). Age determination is routinely done by scales although some trials with otoliths suggest that the oldest fish may be aged conservatively with scales. The largest herring observed in British Columbia was 310 mm long (SL), weighed 340 g, and was about 15 years old. Available data indicates that herring in British Columbia spawn at about 8°C (Hay 1985). Larval herring feed primarily on copepod nauplii as do juveniles which subsequently switch to euphausiids which are the major prey item for adult herring (Wailes 1936).

Stock Status

Pacific herring in British Columbia are short-lived, generally less than 8 years of age, which makes the stocks very dynamic and subject to marked fluctuations in abundance depending on environmental conditions during the prerecruit life stage. Indications for the west coast of Vancouver Island are that survival is inversely related to water temperature (Ware 1991), which has resulted in decreased survival in recent years as water temperatures have increased due to recurring El Niño events and possible global warming. However, all B.C. herring stocks have experienced marked fluctuations in abundance during the past century. Abundance declined for most stocks following the strong El Niños of 1940-1941 and 1958-1959 (Taylor 1964). Abundance of all stocks collapsed in the late 1960s following the large fisheries of the early 1960s combined with poor conditions for survival at this time. All stocks recovered rapidly in the early 1970s (Fig. 6e) following the reduction fishery closure. The two southern stocks declined in the mid-1980s following poor year classes in the early 1980s. Northern stocks were buoyed by a very strong 1977 year class. **Subsequently, strong 1985, 1989, and 1994 or 1995 year classes have maintained most stocks at healthy levels. The exception has been the west coast of Vancouver Island, which appears to have been adversely affected by warm water conditions since about 1976** (Schweigert et al. 1999). Prospects for the future of B.C. herring stocks will depend to a significant degree on the oceanic conditions over the next few decades. Should warm water continue to dominate the marine environment one may expect that survival for west coast of Vancouver Island and Queen Charlotte Islands stocks will continue to be poor. In addition, the impact of a rapidly increasing Pacific sardine stock on B.C. herring stock productivity is unknown.

California
Fishery Data
Commercial landings data dating back to 1916 are available in California Department of Fish and Game bulletins. The data are recorded by fish dealers and processors, who are required to fill out receipts for each landing and submit copies to the department. Landings statistics and other aspects of herring fisheries in California are also published in Department administrative reports and the journal, *California Fish and Game* (Watters and Oda 1997, State of California 2000).

Biology
Biological data are available from the California Department of Fish and Game and are published in administrative reports, a fishery bulletin, and the journal, *California Fish and Game* (e.g., Sprat 1981, 1987, 1992). Spawning population surveys were first conducted in California during 1954-1955 using acoustic and egg deposition surveys. Time series data exist from 1973 to the present for the San Francisco Bay and Tomales Bay spawning populations. These data include spawning biomass estimates, age composition, lengths, and weights. San Francisco Bay biomass estimates are derived from acoustic and egg deposition surveys, and fishery-independent samples of herring are collected from each school with midwater trawl gear for age determination. Tomales Bay biomass estimates are derived from egg deposition surveys and fishery-independent samples of herring are collected using variable-mesh gillnets.

California herring are at the southern end of the range for *Clupea pallasii* in the eastern North Pacific Ocean (Miller and Schmidkte 1956). Spawning occurs as far south as San Diego Bay but the largest spawning population utilizes San Francisco Bay, which is at 37°50′N. Spawning occurs from November through March in nearshore and estuarine environments. Most spawning areas are characterized as having reduced salinity, calm and protected waters, and spawning substrate such as marine vegetation or rocky intertidal areas; however, man-made structures such as pier pilings and rip rap are also frequently used spawning substrates in San Francisco Bay.

Throughout the spawning season, schools of herring enter bays and estuaries, where they may remain up to 3 weeks before spawning. School size varies but can be as large as tens of thousands of metric tons and kilometers in length in San Francisco Bay. Spawn depth distribution generally is shallower than 9 m, but has been found to a depth of 18 m in San

Francisco Bay. A large spawning run may last a week and can result in 30 km of shoreline covered by a 9-m-wide band of herring eggs.

Young-of-the-year herring remain in the bay until summer or early fall, when they migrate to the open ocean. Some herring reach sexual maturity at age 2 when they are about 155 mm (standard length) and all are sexually mature at age 3. California herring may live to be 9 years old and reach a maximum length of about 230 mm body length, although fish older than 7 are rare. Adults leave the bay immediately after spawning. Little is known about the distribution of herring in the open ocean (Reilly 1988).

Fishery and Management

In California, the Pacific herring (*Clupea pallasii*) fishery peaked three times during the past century in response to demand for herring. During the intervening years herring catches were low, when most herring were used as pet food, bait, or animal food at zoos. The herring reduction fishery peaked in 1918 at 3,630 t, but this fishery ended in 1919 when reduction of whole fish into fish meal was prohibited. From 1947 to 1954 herring were canned to supplement the declining supply of Pacific sardines; landings peaked in 1952 at 4,310 t. Canned herring, however, proved to be a poor substitute for sardines and limited demand led to the demise of this fishery by 1954.

In 1973, sac-roe fisheries developed in California and elsewhere along the west coast of North America to supply the demands of the Japanese market. Since then, the majority of herring landed in California have been for the roe market, with small amounts of whole herring marketed for human consumption, aquarium food, and bait. Herring sac roe from San Francisco Bay is typically smaller than sac roe from British Columbia and Alaska, but is highly valued for its unique golden coloration. California sac-roe herring landings peaked twice, at 10,433 t in 1982 and 10,705 t in 1997 (Fig. 6f). The lowest landings have occurred either during or just after El Niño events and corresponded with reduced population size.

The sac-roe fishery is limited to California's four largest herring spawning areas: San Francisco Bay, Tomales Bay, Humboldt Bay, and Crescent City Harbor. There also exists a small open-pound herring eggs-on-kelp fishery in San Francisco Bay. San Francisco Bay has the largest spawning population of herring and produces more than 90% of the state's herring catch. The four spawning areas are managed separately by the California Department of Fish and Game although it is not known whether adjacent spawning populations, such as San Francisco Bay and Tomales Bay, are the same stock.

For San Francisco and Tomales bays, catch quotas are based on the latest population estimates from acoustic surveys and spawning-ground surveys. Quotas are adjusted annually and are generally set at about 15% of the previous season's spawning biomass estimates. This percentage is adjusted lower if the population is at a low level and/or unfavorable conditions such as El Niño exist.

Since 1973, the herring fishery has been managed through a limited-entry system that has been carefully controlled. Until the 1997-1998 season, both round haul gear (purse seines and lampara nets) and set gillnet gear were allowed in the sac-roe fishery; since then only set gillnet gear is allowed in all of the sac-roe fisheries. The set gillnet gear has a minimum mesh size requirement, the purpose of which is to target age 4 and older herring.

Status of the Stock

Assessments are limited to the spawning portions of the San Francisco Bay and Tomales Bay stocks (the two largest in the state), and include fishery-independent estimates of biomass and age composition. **Since 1979, California's spawning biomass estimates ranged from a high of 96,841 t to a low of 18,675 t, with peaks occurring in 1982 (96,841 t), 1988 (64,375 t), and 1996 (91,725 t).** The lowest biomass estimates have occurred during or just after El Niño events: 38,174 t in 1984; 23,204 t in 1993; and 18,675 t in 1998. The lack of upwelling and associated warm water conditions that occur during El Niño events reduce the production of food for herring, which can affect their condition and survival. It also may displace herring to areas of colder water. **San Francisco Bay's population has not yet recovered from the effects of the 1997-1978 El Niño; spawning biomass was estimated at 24,853 t in 2000.**

Part 5. Western North Pacific

This section is dived into two subsections. The first is the northwestern North Pacific herring, which includes five major stocks in Russian waters (Fig. 7). The biological descriptions and explanations of trends in these five stocks are presented as a unit, so their presentation here under a single subheading was for convenience. The southwestern North Pacific includes the Hokkaido-Sakhalin herring and the presentation here represents an integration of information provided by several contributors. This section concludes with a description of Yellow Sea herring.

Part 5a. Northwestern North Pacific

Notes on Different Forms of Herring in the Western North Pacific

Several contributors to this symposium, and elsewhere, comment on different ecological forms of herring. Earlier it was noted that Bering Sea herring have different life history characteristics than more southerly herring in the eastern North Pacific. Similarly, in the northwestern Pacific there appear to be three distinct forms of herring: (1) a migratory "sea" herring, which is relatively long-lived and which undergoes considerable migrations; (2) a similar coastal form, which consists of a number of smaller populations and which spends most of its life close to shore with little or

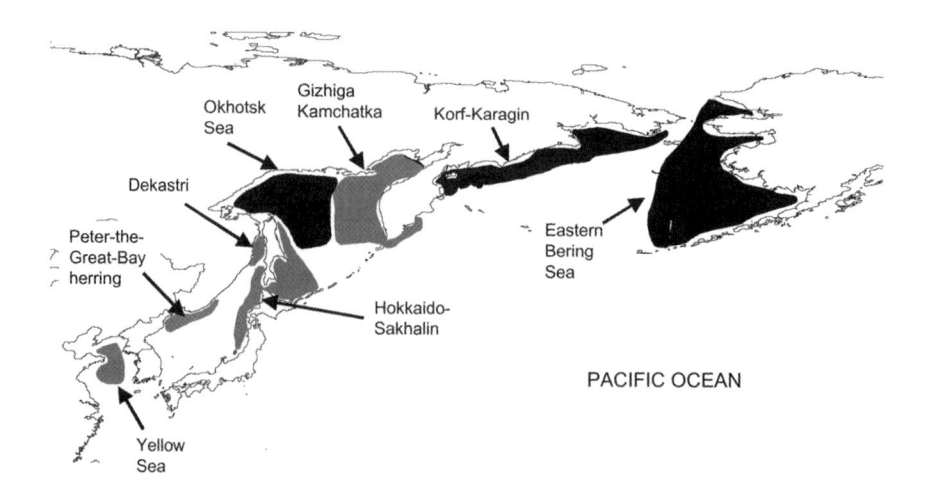

Figure 7. *The western North Pacific showing geographic ranges of the major herring stocks as black or shaded.*

no migrations; and (3) a unique "lagoon" form whose life history is connected with low-salinity lagoons or embayments. In general the marine forms comprise the largest populations and have supported commercial fisheries in the northwestern Pacific.

Much of the available review literature on the Russian Far East herring populations deals with several of the major populations in the same articles. For this reason, the following brief discussion of biology and fisheries also covers all of these groups, although we try to make clear any differences between the populations, from the western Bering Sea in the north to the Sea of Japan in the south. Two stocks, the Dekastri herring and Peter the Great Bay herring, might have been included with the next group (Part 5b, southwestern North Pacific herring) because they occur at the same latitudes as the Hokkaido-Sakhalin stock but their inclusion here is mainly for convenience. Because of their significance, the Hokkaido-Sakhalin stocks are described separately.

Russian Herring: Korf-Karagin, Gizhiga-Kamchatka, Okhotsk, Dekastri, and Peter the Great Bay Herring

Aside from the Hokkaido-Sakhalin herring (sometimes called the Sakhalin-Hokkaido herring) there are several major stocks in Russian waters of the western Pacific including the Korf-Karagin herring, Gizhiga-Kamchatka herring, Okhotsk herring, Dekastri herring, and Peter the Great Bay herring. (See Smirnov 2001 [this volume] for more detail on Gizhiga-Kamchatka stock.) These five stocks, as well as the Hokkaido-Sakhalin herring, constitute the "sea" form of western Pacific herring. In addition, Russian biologists recognize 20 or more smaller populations known as "coastal" or "lagoon" herring. These latter groups are not described in this review but we will point out that the "lagoon" form, which inhabits fresh and brackish water for part of its life cycle, may be unique to the western Pacific, and such forms have not been described for the eastern Pacific. There are a number of papers describing the genetics and biology of these lagoon herring (see Kobayashi 2001, this volume). In general it seems that lagoon herring (sometimes called "lake" herring) mature early (2-3 years) and grow rapidly, but have a short life-span with few older than 5 years of age (Ayushin 1963).

Biology

A small proportion of herring in the Peter the Great Bay population mature at age 1, although most mature at age 2 and 3 (Ayushin 1963). In most other populations sexual maturity and spawning begin at age 2 or 3. Okhotsk herring spawn from mid-May to Mid-June, with sea temperatures mainly between 2 and 8°C (Ayushin 1947, cited in Galkina 1961). Spawning grounds are mainly in the northwestern area of the Sea of Okhotsk between Tauiskaya Bay and Cape Ukoi. Spawning substrates include *Laminaria, Lessonia, Alaria, Cystoseira,* and red algae (Benko et al. 1987). The egg densities described for this area are extraordinarily high. In 1982 the average density was 10.9 million eggs/m^2 with some areas having more than 117 million eggs/m^2. This was substantially higher than the norms in other years in the same area, with means of 2-4 million/m^2. This density still is substantially higher than in most other parts of the world, by a factor of 3 or more (see list in Hay 1985 for interpopulation comparison).

Korf-Karagin herring spawn mainly in May in shallow bays and lagoons of Karagin and Korf bays (Naumenko 1996), mainly on eelgrass (*Zostera*) as the main substrate. These herring leave the area immediately after spawning and return to foraging areas. The range of foraging area inhabited depends on stock size and extends eastward during periods of high abundance, reaching longitude 178°E. During years of low abundance, foraging appears to be confined to Olyutorsk Bay (northeastern part of the Kamchatka Peninsula). Early winter migrations bring herring back to Olyutorsk Bay, where fish reside in moderately shallow depths (20-50 m) before moving to deeper depths, and continue moving westward in late November and December reaching traditional overwintering areas around Cape Goven (Naumenko 1996). Juvenile herring reside within Karagin Bay for the first 2 years of life and do not mix with adult schools. They move to Olyutorsk Bay in their third year. Herring appear to be fully recruited by age 4, at which time they are about 24 cm in length.

In general, the relative scale of variation in year-class strength differs between northern and southern populations: the difference between strong and weak year classes is less in the south and the frequency of stronger year classes is higher, with strong year classes appearing once every 3-6 years. In contrast, in northern populations strong year classes occur about once every 5 years. The causes of the variation in year-class strength have prompted much debate and research but without firm answers. It is clear, however, that there are some complex biotic interactions between some key species, and, in particular, it appears that in the western Bering Sea, herring abundance appears to be inversely related to pollock (*Theragra chalcogramma*).

Fishery Data and Stock Status
Catches for Russian stocks are indicated in Figs. 8a-e. Aside from the Hokkaido-Sakhalin herring, the commercial fisheries for Peter the Great Bay and Dekastri herring (Fig. 8a-b) were the earliest, with catches greater than 20,000 t made in the 1920s. Substantial commercial fisheries started in other stocks in later years: Korf-Karagin in 1939 (Fig. 8c) and Okhotsk (Fig. 8d) and Gizhiga-Kamchatka (Fig. 8e) in the mid- to late 1940s. Early fishing gear was passive, consisting of weirs set in the vicinity of spawning areas. In more recent years, catches were made with gillnets, trawls, and purse seines. Catch records of all of these stocks are characterized by substantial variation in annual catch, reflecting fluctuations in the availability of herring. In all stocks, the declines in catch rates and assessed biomass have led to prohibition of commercial fishing. The three northern populations supported large catches which began to increase rapidly in the late 1950s, with maximal total catches exceeding 500,000 t in the late 1960s. **This was followed by a rapid decline in all of the "marine" herring stocks, and a period of low abundance persisted until the late 1990s, when there were some indications of recovery of the Okhotsk and Korf-Karagin herring. The southern stocks, Dekastri and Peter the Great Bay herring, however, remain in a state of severe depression.**

Part 5b. Southwestern North Pacific
Hokkaido-Sakhalin Herring
Geography
Hokkaido-Sakhalin herring were the most abundant of all western Pacific herring populations. In the first half of this century, which seems to have been a favorable time for this population, this stock was very widely distributed and formed an industrially fished stock in the northern part of the Sea of Japan, the Sea of Okhotsk (along the shores of eastern Sakhalin to the Island of Iona, and along the Kuril Islands to the Island of Paramushir), and the Pacific Ocean (along the southern Kurils). Larval and juvenile herring inhabit the areas near south and southeastern Sakhalin and northwestern Hokkaido. After spawning, herring migrate farther to the marine sites which, at the present time, are mainly in the Tatar Strait (Sea of Japan). Herring spend the winter above 200-400 m depths, in water layers of 100-200 m in Tatar Strait (Sea of Japan) (Druzhinin 1963). Spawning occurs in

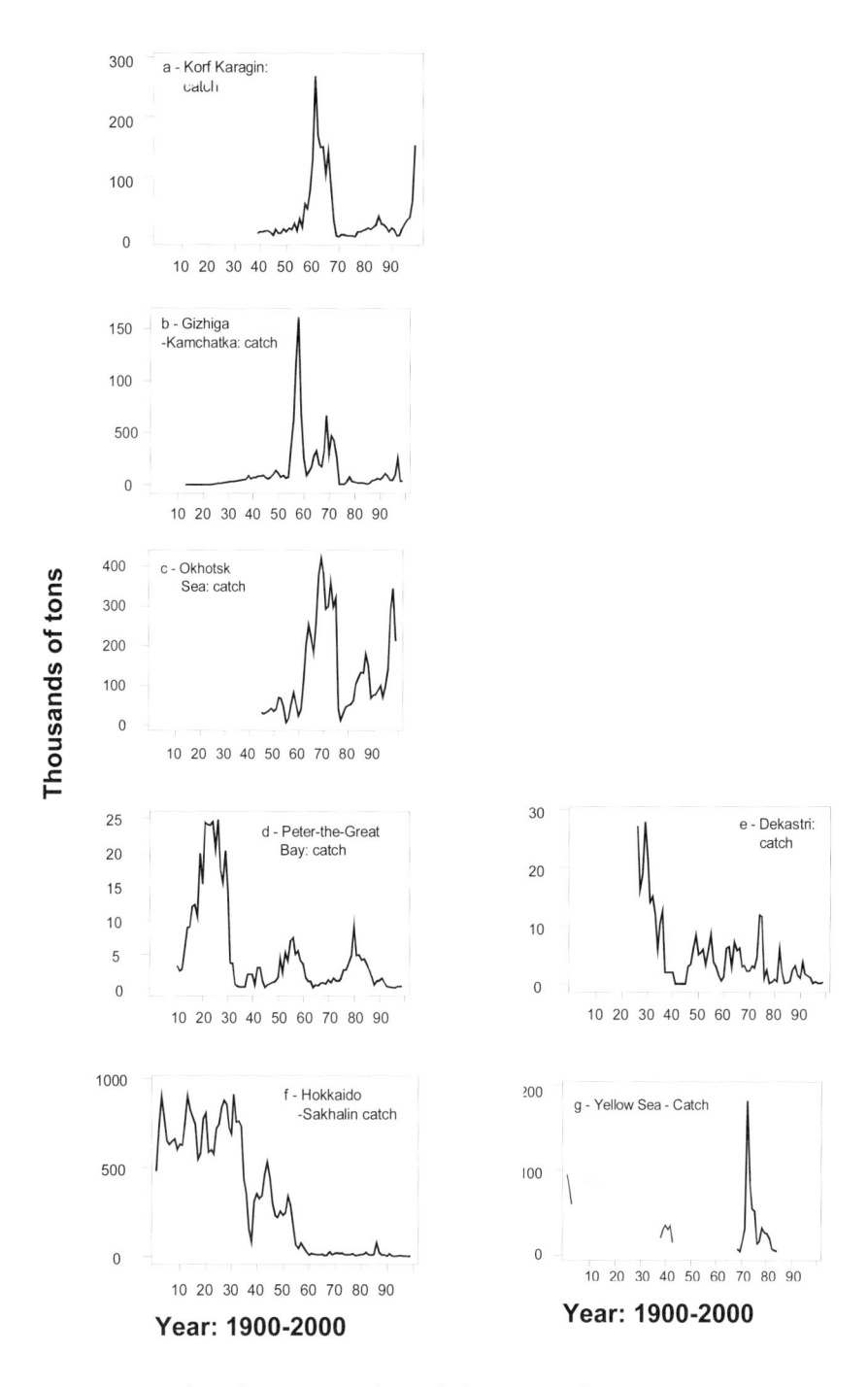

Figure 8. Catches of western North Pacific herring stocks.

May, although at the present time the distribution is very limited, mainly to a few sites on the western Sakhalin coast (Sea of Japan) at depths to 10 m. There are a number of small local stocks distributed throughout this area, but their relationship to the once-large Hokkaido-Sakhalin stock is unclear.

Annual catch statistics for northern Japan and southern Sakhalin have been gathered by national and provincial authorities in Japan and Russia since 1887 until the present time (Fig. 8f). These long-term series of catch and biological data have been recorded by staff of fisheries research institutes in Russia and Japan, including the national fisheries institutes and provincial fisheries experimental stations. Data on catch at age and body length composition at each age recorded from 1910 to 1954 for northern Japan are published in the reports of the Hokkaido Regional Fisheries Research Laboratory. Annual catch has been reported to the Russian-Japanese Fishery Commission. Various assessment methods include virtual population analyses and cohort analysis.

The historical peak in catch was recorded in 1897 as 973,000 t. After this period, total landing decreased annually and fishing grounds started moving north along the western coast of Hokkaido from the late nineteenth century. Since that time annual catch had gradually declined with continual fluctuation and the spawning herring had disappeared since 1955 from the Hokkaido coast. In those days herring was mainly caught by gillnet and set net. At present the Hokkaido-Sakhalin herring stock is at the lowest level of abundance for the whole history of the population's commercial exploitation (Motoda and Hirano 1963, Pushnikova 1996)

There appears to be three distinct periods of abundance in the last century, which can be appreciated by comparing estimated numbers of individuals:

Period 1: Late 1800s to 1950, period of high abundance. The stock varied annually from 15.2 billion (1948) to 228.2 billion (1940), and averaged 67.2 billion fish. The numbers of recruits for each generation in this period varied from 1.8 billion (1949) to 201.5 billion (1939), averaging 28.3 billion fish in each generation.

Period 2: From 1951 to 1960, period of population decline. The stock varied from 1.7 billion (1960) to 13.4 billion (1951), averaging 5.8 billion fish. Recruitment varied from 0.3 billion (1960) to 5.4 billion (1953), averaging 1.9 billion fish.

Period 3: 1961 to the present, period of population depression. The stock varied from 0.4 billion (1988) to 5.6 billion (1974), averaging 1.6 billion fish. Recruitment varied from 0.14 billion (1986) to 4.7 billion (1973), averaging 0.8 billion fish.

Commentary

After the collapse of this stock, only small fisheries have continued. These consist of gillnet fisheries in spring, near the spawning ground targeting small local populations, and trawl-net fisheries in the spring-autumn season in the feeding area that again target some small local populations in the Hokkaido-Sakhalin areas. The overall stock condition has been at quite a low level since 1955 and, except 1986, the annual catch has been less than 50,000 t. The 1983 year class, however, appeared to be an abundant year (from ages 2-5) and the 1988 year class also was noticeable, but much less than the 1983 year class. Year-class strength since then has been poor.

The reason that the stock has not recovered is not clear. One explanation is that oceanographic conditions have changed. Since late in the 1940s the sea-surface temperatures on the western coast of Hokkaido have been warmer during the winter-spring seasons but colder in the summer seasons. These oceanographic changes may affect the regional environment and the mechanism of recruitment of the Hokkaido-Sakhalin population. The major decline, however, coincided with the development of more intensive searching and fishing. This included the use of acoustic sounders to find fish and, in Russian waters, the use of airplanes for spotting their feeding areas. In the 1950s, trap nets continued to be used during the spawning period, but the industry also began using trawling nets to capture fish in the feeding areas. Immature and young fish were caught. The industrial pressure on young fish increased and this intense fishing pressure coincided with a period of natural decline in abundance.

Therefore, it is probable that the existing depression in abundance is caused for the most part by excessive fishing, which restricts recruitment. Historically, Russia and Japan both have harvested the same population of Hokkaido-Sakhalin herring. The result of an irrational fishing intensity during a period of natural decline has been a steady, long-term depression, to the point where some researchers have declared that in Japanese waters this population has disappeared in recent years (Kobayashi 1983).

Trend in Hokkaido-Sakhalin Herring Population

There are no indications of change in the extremely depressed status of Hokkaido-Sakhalin herring. This statement is based on the following observations: a trend toward decreasing productivity over successive generations; very infrequent formation of strong cohorts during the last 20 years; the absolute absence of spawning among the Hokkaido-Sakhalin herring in the coastal areas where they historically have spawned; the huge area of potential but currently empty spawning grounds; and the extraordinarily low numbers of reproductive fish.

Yellow Sea Herring
Biology
Yellow Sea herring mature early, with 99% mature at age 2. Minimum fork length and body weight of mature fish was 200 mm and 80 g for females and 168 mm and 46 g for males. Absolute fecundity ranged between 19,300 and 78,100 eggs and relative fecundity (number of eggs per gram) of net body weight was between 210 and 379 eggs. The estimate of eggs per unit length (E/L, or eggs per centimeter of fork length) was 93-269 eggs. Both E and E/L increased linearly with net body weight, exponentially with fork length, and with age. Relative fecundity did not vary with body weight, fork length, or age and was quite stable over time.

The growth rate of Yellow Sea herring varies seasonally, with rapid growth in summer and the period from late winter to the prespawning stage, but it is slow in the autumn and early winter. Growth rates were lowest during the spawning and postspawning period, but even then growth did not stop. This growth aspect was connected with seasonal variation of feeding. Growth rates varied with age and were greatest in young herring less than 3 years old, and slower at ages of 4 or more. There was no apparent relationship between rate of growth and population abundance.

Pacific herring in the Yellow Sea migrate between inshore and deeper water in response to seasonal changes. Yellow Sea herring have never been found in the area south of 34°N, and there are significant differences in the number of vertebrae, dorsal rays, anal rays, and scutes between Yellow Sea herring and Hokkaido herring in the Sea of Japan (Tang 1991). Therefore, the Yellow Sea herring is considered as a distinct population of Pacific herring.

The Yellow Sea herring has a long exploitation history. Its importance is demonstrated by the existence of villages and localities named for their association with it. In the last century, the commercial fishery for this species experienced three peaks (in about 1900, 1938, and 1972), each followed by a period of little or no catch (Fig. 8g). In 1967, a large number of 1-year-old herring appeared in bottom trawl catches. In 1972 a very strong 1970 year class was recruited to the fishery, and the stock reached its historical maximum abundance biomass, estimated at 200,000 t or 26.8×10^8 fish. **Since 1982, the stock has declined substantially because of a series of weak year classes. The catch decreased to below 1,000 t in 1989-1990, and there is no fishing today.**

Yellow Sea herring undergo rapid and extreme fluctuations in abundance. Tang (1981, 1987) reported that environmental conditions such as rainfall, wind, and daylight could strongly affect fluctuations in recruitment. Further, the long-term changes in biomass of Yellow Sea herring may be correlated with a 36-year cycle of dryness/wetness oscillation in eastern China (Fig. 9).

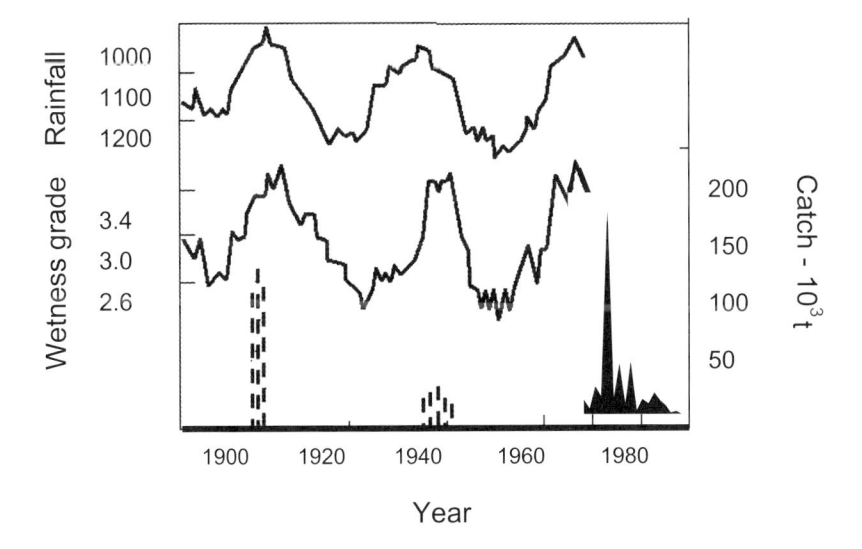

Figure 9. Fluctuations in herring abundance in the Yellow Sea and a 36-year cycle of wetness oscillations in eastern China (adapted from Tang 1981). The dashed vertical lines in lower left and middle show approximate years and tons of herring catches. Lower right shows catches during a period of herring abundance in the 1970s and early 1980s.

Part 6. Arctic Populations

Arctic populations of herring have not had a long history of exploitation, and therefore are literally and figuratively peripheral to one of the main objectives of this review, which is to review the status of herring stocks after a century of fishing. On the other hand, the review also is an "inventory," so for that reason we include some brief comments on Arctic herring populations. In general, the biological literature on Arctic herring populations, which is relatively sparse compared to more southern populations, tends to be associated with individual seas. We follow that approach.

There are a number of coastal seas within the Arctic Ocean (Fig. 10), and most have some records of herring. The Greenland Sea lies between northeastern Greenland and Spitsbergen. To the east of the Greenland Sea is the Barents Sea, on the north coast of Norway and northeastern Russia. The coastal seas on the north coast of Russia, from west to east, consist of four large seas: the Barents, Kara, Laptev, and East Siberian seas. The White Sea, in northeastern Russia, opens to the Barents Sea between the Kola and Kanin peninsulas. On the extreme northeastern coast of Russia, the Chukchi Sea separates western Alaska and the Chukchi Peninsula of Russia. On the north coast of North America, the Beaufort Sea spreads from northeastern Alaska to the western Canadian Arctic. The northeastern Canadian coast also includes the large Hudson Bay (mainly south of 65°N)

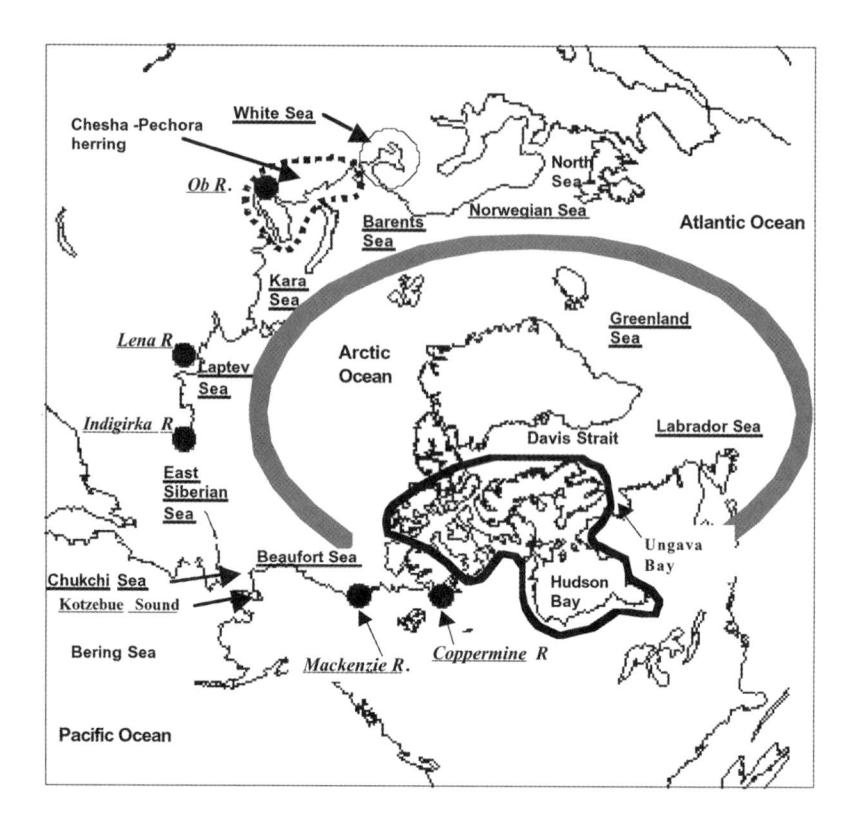

Figure 10. The Arctic Ocean and adjacent seas showing approximate herring distribution. Herring occur in each of the seas or estuaries of rivers with underlined names. The dashed line shows the distribution of the Chesha-Pechora herring, in the southern Barents and Kara seas. The dark circles indicate reports of herring found in river estuaries. The large semicircle indicates, figuratively, the zone across the North Atlantic and parts of the Arctic Ocean where herring distribution is nearly continuous. The dark area in the eastern Canadian Arctic indicates areas where herring do not occur, with the limits of their ranges being approximately the Coppermine River and Ungava Bay. [Author's note: subsequent to the preparation of this report, we have learned that there are indeed herring in parts of Hudson Bay. (J. Dodson, Biology Dept., Univ. of Laval, Québec, pers. comm.)]

as well as Baffin Bay (between Baffin Island and western Greenland). The large Davis Strait separates the Arctic Ocean and the Labrador Sea (part of the northwestern Atlantic Ocean).

Greenland Sea

The Greenland Sea is not documented as a spawning area for herring but some areas are occupied during summer feeding migrations of Norwegian spring-spawning herring. Also, juveniles from the Norwegian coast can extend to the southeastern areas of Spitsbergen.

Barents and Kara Seas

The Barents Sea has a long history of fisheries for herring, cod, and capelin (see Orlova et al. 2001, this volume). The main fisheries are on Norwegian spring-spawning herring as they are either feeding or overwintering. In addition, parts of the southern Barents Sea are occupied by a different form of herring, called by Svetovidov (1952) the "Chesha-Pechora" (low vertebral number herring). This form of herring extends from Mezan Bay, which separates the White Sea from the Barents Sea, to the Ob Inlet estuary, on the southwestern shores of the Kara Sea.

White Sea

The White Sea is an inlet of the Barents Sea (~95,000 km^2) in northwestern Russia opening into the Barents Sea between the Kola and Kanin peninsulas. Kandalashka Bay, in the southern section, is the deepest part of the sea (340 m). The Mezen, the Northern Dvina, and the Onega rivers empty into large bays of the White Sea. There are at least three distinct herring populations in the White Sea, each associated with different bays: the Onega Bay, Kandalaksha Bay, and Dvina Bay herring. They are exclusively spring spawners and there are a number of biological differences between these populations. Because of their nearshore spawning on macrophytes in the spring, as well as biological characteristics (especially vertebral number), Russian biologists have recognized White Sea herring as the same subspecies as Pacific herring (*Clupea harengus pallasii*). We mention this here, not as an endorsement or refutation of this taxonomic usage, but to point out that the White Sea herring are quite different from their neighboring Norwegian spring spawners. Commercial fisheries are small, landing about 1,200 t annually (Krixunov 1990). Each of the three populations in the White Sea declined in the early to mid-1960s, recovered in the 1970s, and then declined again (Krixunov 1990).

Laptev and East Siberian Seas

There is very little available information on herring in these areas. Svetovidov (1952) reported on herring taken in the delta of the Lena River, a large Siberian river that drains into the Laptev Sea. Spawning is reported to occur in Tiksi Cove. Less clear are reports of herring in the Indigirka

River estuary, which drains into the East Siberian Sea (Svetovidov 1952). Accounts of herring occurring (and spawning) in brackish estuarine Arctic waters is consistent, however, with reports from the North American Arctic (see below).

Chukchi Sea

Kotzebue Sound is on the extreme eastern part of the Chukchi Sea. The locations and dates of herring spawning in Kotzebue Sound (see Hay 2001, this volume). This area supports local subsistence fisheries and limited commercial fisheries. Although this area is part of the Chukchi Sea, it also can be viewed as the northern edge of the eastern Bering Sea, which supports large herring fisheries.

Beaufort Sea

The first record of Pacific herring (*Clupea pallasii*) caught in the Canadian Beaufort Sea was reported by Richardson (1823, cited in Riske 1960) in his description of fishes captured during Sir John Franklin's expeditions to the Polar Sea between 1819 and 1822. Riske (1960) lists several other early instances of herring being captured along the Beaufort Sea coast between the Mackenzie River Delta and Cape Bathurst. He collected 248 fish in 1958 to compare life history and morphometric and meristic characters with those of fish from the southern British Columbian coast. He concluded that Arctic and British Columbian herring constitute one taxonomic unit, *Clupea harengus pallasii*.

Recent collections of Pacific herring from the Beaufort Sea were made to address potential impacts of hydrocarbon exploration at Tuktoyaktuk and to evaluate the feasibility of a commercial roe fishery. Gillman and Kristofferson (1984) listed the sampling events that included catches of Pacific herring. These studies focused in the Mackenzie River Delta and adjacent areas. Sampling programs concentrated in the nearshore area. A perusal of published reports shows that all information on Pacific herring is based on about 2,300 fish, the vast majority being collected between 1981 and 1983 during the roe fishery feasibility study (Gillman and Kristofferson 1984).

Biology

According to Bond (1982), herring have been collected along the Arctic coast of North America eastward to the Yukon but appear to be more abundant east of the Mackenzie River Delta. No animals have been collected recently east of Paulatuk (Riske 1960). Seasonal variation in catch per unit of effort (Bond 1982, Lawrence et al. 1984) suggests that fish migrate from the shallow inshore spawning areas in June to offshore feeding areas and then return in August to overwinter and then spawn in the next spring. Beaufort Sea herring appear to tolerate low salinities. They have been reported as far as 100 km upstream from the Beaufort Sea, in the Mackenzie

River Delta (Hunter 1975). Lawrence et al. (1984) reported collecting herring from waters with salinities ranging between 13 and 30 ppt. Herring are also exposed to wide ranges in temperature. Lawrence et al. (1984) reported temperature ranges of 1.9-14.0, 7.3-13.9, and 2.1-8.5°C for June/July, August, and September, respectively. Gillman and Kristofferson (1984) measured sea surface temperatures ranging between 2.5 and 3.5°C during the spawning season. Nearshore waters warm rapidly. Tanasichuk et al. (1993) estimated that sea temperatures during the larval phase would be about 12.5°C.

Most of the available data are for age composition and size-at-age. Bond (1982) showed age-length data for a number of studies. His figure indicated that fish captured were between 2 and 16 years old. Growth in length appears to be linear and relatively rapid to age 8 with an asymptotic length of about 310 mm fork length. Bond's length-at-age estimates differ from those presented by Tanasichuk et al. (1993). It is not known whether this reflects a biological difference or differences in aging methods.

Tanasichuk et al. (1993) compared growth and reproductive characteristics of herring collected from the Beaufort Sea and British Columbia. Beaufort Sea herring became progressively longer-at-age after age 5 whereas mass-at-age was similar. Condition (length-specific mass) was greater for Strait of Georgia herring because size-specific ripe ovarian mass was 2.1 times heavier. These differences in growth may be related to differences in natural mortality which was almost twice as high for herring from British Columbia. Lawrence et al. (1984) and Bond (1982) reported that Beaufort Sea herring were not strict planktivores. Stomach contents included infaunal benthic organisms, epibenthos, and plankton. Stomach content weight of 0.12 and 0.62 g dry mass, respectively, suggests a very low daily ration.

Fisheries

Herring are not taken routinely and are only harvested domestically as incidental catch by gillnets. Attempts at developing commercial fisheries were unsuccessful, mainly because of high production and transportation costs (Corkum and McCart 1981, Gillman and Kristofferson 1984). Estimates of local spawning biomass are surprisingly low. Shields (1985) reported an estimated biomass of 8.2 t for one of the key spawning areas. His explanations for the low biomass estimates were that (1) the survey area might not be a major spawning location because of interannual variability in spawning locations, (2) a protracted spawning period meant that not all the spawn deposited in the study area would have been surveyed, and (3) it is uncertain if Beaufort Sea herring spawn every year.

Other Waters: Hudson Bay, Baffin Bay, Davis Strait

Although herring are known to occur in the southern Davis Strait area, along the coastal areas of Labrador and Greenland, there are no established spawning areas north of Newfoundland. Herring are documented as occurring in Bathhurst Inlet (Scattergood et al. 1959). Therefore there is a substantial geographical range where there are no records of herring occurrence (Fig. 10). This range extends from about Ungava Bay (adjacent to the Labrador Sea), as the approximate limit of their range to the west from the Atlantic, to Coppermine River (Riske 1960), as the approximate limit of their range to the east, from the Beaufort Sea, or Pacific Ocean.

The present herring distribution in the Arctic, even though not well known, probably represents remnants of a much more pronounced presence of herring in the Arctic when their distribution probably was circumpolar, probably during past periods of warmer climate. It will be interesting to see if climate changes forecasted for the remaining 99 years of this century will impact this herring distribution. One of the indicators could be the new presence of herring in new locations, such as Hudson Bay. Given the broad distribution of herring in other parts of the Arctic, the absence of herring in Hudson Bay is a puzzle.

Part 7. Synopsis, Summary, and Conclusions
What Have We Learned?

Has this inventory and review of herring stocks revealed anything that was previously unknown? From one perspective we suggest that the answer is no: the foregoing contributions have only served to amalgamate and assemble into a single document information that was well known for each stock. From another perspective we suggest that an inventory such as this does make some new contributions in the sense that the information derived from the "whole" may exceed the sum of the "parts." In this regard, we suggest that there may indeed be some conclusions that, while perhaps not entirely original, may be drawn more forcefully from this review than from the component contributions, or from some of the previous contributions cited in the references. For instance, one clear observation is the relative magnitude of the Norwegian spring-spawning herring. This single stock, during periods of high biomass (> 10 million t), is the equivalent, or greater, than all of the other herring stocks in the world combined! Another observation is that if "large" herring stocks are defined as those of 1 million t or greater, there are not many in the world—probably less than 10 worldwide.

A clear conclusion is that in the year 2000 there were still many herring stocks, and that 100 years of fishing between 1900 and 2000 has not resulted in their demise. It is probable, however, that there has been an elimination of spawning components within some stocks and the long-term effects of this are not clear. An equally clear observation, however, is that virtually all herring stocks have undergone substantial change in abundance in the past, and during periods of active fishing most have declined to the point where the stock has been described as "collapsed." From this we can make two other points. One is that we see that most stocks have recovered to the point where present, or recent, abundance has equalled or exceeded historical estimates. The other is that there is (or was) a very large stock, the Hokkaido-Sakhalin stock, that is seriously depleted, perhaps to the point of functional extinction. Similarly, there are grounds for concern about the Yellow Sea herring, Baltic herring, and two of the southern populations in the western Pacific: the Dekastri and Peter the Great Bay herring.

Elements Common to Different Contributions

A common theme running through many of the descriptions of herring stocks is the effect of environment or long-term climate change on herring abundance and distribution. Remarkably, the general impact of climate on herring has been understood for a long time; see, for instance, Rachel Carson's (1951) popularized discussion of climate and the changes on the Bohuslan herring of the middle ages in the Baltic, or more recent analyses of long-term climate change by Southward et al. (1988), Höglund (1978, and Øiestad (1994). The Russian scientific literature on herring makes frequent reference to effects of climate. Regardless, at the present time invoking climate change as an explanation for changes in fish stock abundance can be controversial. The controversy arises because it may be seen by some as a rationalization, or an excuse, to deflect criticism away from inadequate assessments or policies that resulted in the overfishing of herring.

There are grounds for scepticism about climate change as explanations for changes in stock abundance. As pointed out by Cushing (1971) 30 years ago, climate change is usually suggested as an explanation for most instances of collapse of fish stocks, but the arguments vary with the species and few explanations provide convincing biological mechanisms for the relationship between climate and fish abundance. In this review, many contributions point out that past fishing effort and practices were too high and not biologically sustainable but most also suggest that climate change is a very important factor affecting herring. There is now growing evidence that climate change is a key factor affecting abundance of many fish species; for instance, see Beamish (1995) and references therein. Therefore reference to climate as a factor affecting herring is not invoked here, or anywhere above, as an excuse for stock collapses. Rather, a useful observation from this review is that climate change was suggested by most contributors, independently, as an important factor affecting nearly all of the herring stocks described above. The problem is, however, that no one yet fully understands the biological mechanisms of climate change on herring abundance (Toresen and Østvedt 2000).

Stocks of Concern

We conclude by pointing out some of the most obvious and troubling observations in this review. One is that there has been a striking recovery of herring on the Georges Bank, and this appears to coincide with remarkable increases in herring abundance in the southern range of the western Atlantic, with herring catch extending south to North Carolina. Concurrently, the most northern western Atlantic herring stocks are declining (Newfoundland east and west coast spring spawners). The Atlanto-Scandian herring (Norwegian spring-spawning herring and Icelandic herring) appear to be in a period of high abundance, while North Sea and Baltic herring are in decline. The geographic and temporal patterns of abundance of herring in the Pacific are less clear. First, however, it should be noted that most stocks in the eastern Pacific (North American coast) are relatively small, with few exceeding 100,000 t, except for the Bering Sea. For the purposes of brevity, in this review we were forced to pool the abundance estimates of a number of smaller stocks. Regardless, the abundance of herring, from California to the Gulf of Alaska, indicates a mixed pattern, but generally most stocks had a higher spawning biomass in the late 1990s than they did in the previous five decades. The exception to this is Prince William Sound, and as pointed out in other papers in the Herring 2000 symposium, spawning grounds in this area were contaminated with oil in 1989 and the impact of that event may be responsible.

The general patterns of herring abundance in the southwestern North Pacific are not encouraging in the year 2000. The once great Hokkaido-Sakhalin stock has not recovered after a collapse that occurred more than 50 years ago and this is the only major herring stock in the world that has not recovered. Also worrisome is the Yellow Sea herring stock, which has not been abundant since the mid-1970s. Similarly, two other stocks in the southwestern North Pacific, the Peter the Great Bay and Dekastri herring, are severely depressed. Also the Gizhiga-Kamchatka stock appears to be low. On the other hand, the largest stock in the northwestern Pacific, the Okhotsk Sea stock, appears to be abundant at the present time, and the Korf-Karagin stock, low until recently, appears to be increasing.

Probably it is too simplistic to limit the discussion of "concern" only to those stocks that currently are at low levels of abundance. It is well understood that abundances of all herring stocks undergo major fluctuations of their own accord but we also understand that rapid stock reductions have been made worse by aggressive fishing strategies. Therefore another perspective is that any stock is a potential stock of concern where fishing management policies do not recognize the inherent variability of herring. There is a need for further development of management approaches that consider multiple factors and that can react quickly to complex changing conditions.

Limitations of This Review

Probably the available data and other biological information about herring equal or exceed that of any other marine species (Whitehead 1985) and we acknowledge that this report may not be as thorough or complete as we would have liked. The summary is incomplete on two counts. First, there was more information provided to us, or available, than we could include. Second, even with some deficiencies of data, more analyses would have been possible. From the perspective of understanding how fish stocks respond to fishing, and how climate affects fish and fisheries, it would appear to be wise to assemble the best information possible and make it as broadly available as possible.

Perhaps the greatest limitation of this review is the restricted description of the apparent complexity of stock structure of large herring stocks. The nature of the structure of marine fish stocks is the basis of active research and debate in the literature, and we could not do justice to this topic here. Similarly there is very little reference made in this review to the distribution and biology of the large number of "smaller stocks." There are many throughout nearly all regions, and their ecological and genetic relationship to larger stocks is not well understood.

References

Andersson, L., N. Ryman, R. Rosenberg, and G. Ståhl. 1981. Genetic variability in Atlantic herring (*Clupea harengus harengus*): Description of protein loci and population data. Hereditas 95:69-78.

Anthony, V.C., and G.T. Waring. 1980a. Assessment and management of the Georges Bank herring fishery. Rapp. P.-V. Reun. Cons. Int. Explor. Mer 177:72-111.

Anthony, V.C., and G. Waring. 1980b. Estimates of herring spawning stock biomass and egg production for Georges Bank-Gulf of Maine region. NAFO SCR Doc. 80/IX/135. 38 pp.

Aro, E. 1989. A review of fish migration patterns in the Baltic. Rapp. P.-V. Reun. Cons. Int. Explor. Mer 190:72-96.

Ayushin, B.N. 1963. Abundance dynamics of herring populations in the seas of the Far East, and reasons for the introduction of fishery regulations. Rapp. P.-V. Reun. Cons. Int. Explor. Mer 154:262-269.

Bailey, R.S., D.W. McKay, J.A. Morrison, and M. Walsh. 1986. The biology and management of herring and other pelagic fish species in the Firth of Clyde. Proceedings of the Royal Society of Edinburgh 90B:407-422.

Baxter, G.I. 1958. The composition of Minch herring stocks. Rapp. P.-V. Reun. Cons. Int. Explor. Mer 143(II).

Beamish, R.J. (ed.). 1995. Climate change and northern fish populations. 1995. Can. Spec. Publ. Fish. Aquat. Sci. 121. 739 pp.

Benko, Yu.K., Yu.N. Bogatkin, and R.K. Farkhutdinov. 1987. Biological basis of the use of artificial spawning grounds for the reproduction of Okhotsk herring. Translation by Plenum Publishing Co. from Biologiya Morya 1:56-61.

Blaxter, J.H.S. 1985. The herring: A successful species? Can. J. Fish. Aquat. Sci. 42(Suppl. 1):21-31.

Blaxter, J.H.S, and F.T.G Holliday. 1963. The behaviour and physiology of herring and other clupeids. Adv. Mar. Biol. 1:261-393.

Blaxter, J.H.S., and J.R. Hunter. 1982. The biology of clupeoid fishes. Adv. Mar. Biol. 20:1-223.

Bond, W.A. 1982. A study of the fishery resources of Tuktoyaktuk Harbour, southern Beaufort Sea coast, with special reference to the life histories of anadromous coregonids. Can. Tech. Rep. Fish. Aquat. Sci. 1119. 90 pp.

Carson, R. 1951. The sea around us. Oxford University Press, New York. 230 pp.

Chenoweth, S.B., D.A. Libby, R.L. Stephenson, and M.J. Power. 1989. Origin and dispersion of larval herring (*Clupea harengus*) in coastal waters of eastern Maine and southwest New Brunswick. Can. J. Fish. Aquat. Sci. 46(4):624-632.

Claytor, R.R., and C. LeBlanc. 1999. Assessment of the NAFO Division 4T southern Gulf of St. Lawrence herring stocks in 1998. Canadian Stock Assessment Secretariat Research Document 99/54.

Claytor, R., C. LeBlanc, C. MacDougall, and G. Poirier. 1998. Assessment of the NAFO Division 4T southern Gulf of St. Lawrence herring stock, 1997. Canadian Stock Assessment Secretariat Research Document 98/47.

Corkum, L.D., and P.J. McCart. 1981. A review of the fisheries of the Mackenzie Delta and nearshore Beaufort Sea. Can. Manuscr. Rep. Fish. Aquat. Sci. 1613. 55 pp.

Cushing, D. 1971. The dependence of recruitment on parent stock in different groups of fishes. J. Cons. Int. Explor. Mer 33:340-362.

Cushing, D.H. 1972. The production cycle and the numbers of marine fish. Symp. Zool. Soc. Lond. 29:213-232.

Deriso, R.B., T.J. Quinn, and P.R. Neal. 1985. Catch-at-age analysis with auxiliary information. Can. J. Fish. Aquat. Sci. 42:815-824.

Devold, F. 1963. The life history of the Atlanto-Scandian herring. Rapp. P.-V. Reun. Cons. Int. Perm. Explor. Mer 154:98-108.

Dooley, H.D., and D.W. McKay. 1975. Herring larvae and currents west of the Orkneys. ICES C.M. 1975/H:43

Dragesund, O. 1970a. Factors influencing year-class strength of Norwegian spring spawning herring (*Clupea harengus* Linne). Fiskeridir. Skr. Ser. Havunders. 15(4):381-450.

Dragesund, O. 1970b. Distribution, abundance and mortality of young and adolescent Norwegian spring spawning herring (*Clupea harengus* Linne) in relation to subsequent year-class strength. Fiskeridir. Skr. Ser. Havunders. 15(4):451-556.

Dragesund, O., and Ø. Ulltang. 1978. Stock size fluctuations and rate of exploitation of the Norwegian spring spawning herring, 1950-1974. Fiskeridir. Skr. Ser. Havunders. 16(10):315-337.

Drucker, P. 1955. Indians of the Northwest Coast. The American Museum of Natural History, The Natural History Press, Garden City, New York. 224 pp.

Druzhinin, A.D. 1963. The results of tagging the herring in Sakhalin waters in 1956-1960. Izv. TINRO 49:65-94. (In Russian.)

Fedorov, S.S. 1968. Biology and fisheries of marine herring. Moscow State University, Moscow. 140 pp. (In Russian.)

Fournier, D., and C.P. Archibald. 1982. A general theory for analyzing catch at age data. Can. J. Fish. Aquat. Sci. 39:1195-1207.

Fridriksson, A., and O. Aasen. 1950. The Norwegian-Icelandic herring tagging experiments. Report no. 1. Fiskeridir. Skr. Ser. Havunders. 9(11):1-34.

Funk, F. 1990. Migration of Pacific herring in the eastern Bering Sea as inferred from 1983-88 joint venture and foreign observer information. Alaska Department of Fish and Game, Regional Information Report 5J90-04, Juneau.

Funk, F. 1991. Harvest policy implications of yield per recruit models for Pacific herring in Alaska. In: Proceedings of the International Herring Symposium, University of Alaska Sea Grant, AK-SG-91-01, Fairbanks, pp. 453-462.

Funk, F., and K.A. Rowell. 1995. Population model suggests new threshold for managing Alaska's Togiak fishery for Pacific herring in Bristol Bay. Alaska Fish. Res. Bull. 2(2):125-136.

Galkina, L.A. 1961. Special features of the reproduction and early developmental stages in the northern regions of the Sea of Okhotsk. Problemy Severa 4:108-120.

Gillman, D.V., and A.H. Kristofferson. 1984. Biological data on Pacific herring (*Clupea harengus pallasi*) from Tuktoyaktuk Harbour and the Liverpool Bay area, Northwest Territories, 1981 to 1983. Can. Data Rep. Fish. Aquat. Sci. 485. 22 pp.

Graham, J.J. 1982. Production of larval herring, *Clupea harengus*, along the Maine coast. J. Northwest Atl. Fish. Sci. 3:63-85.

Grant, W.S., and F.M. Utter. 1984. Biochemical population genetics of Pacific herring (*Clupea pallasi*). Can. J. Fish. Aquat. Sci. 41:856-864.

Grosse, D.J., and D.E. Hay. 1988. Pacific herring, *Clupea harengus pallasi*, northeast Pacific and Bering Sea. In: N.J. Wilimovsky, L.S. Incze, and S.J. Westrheim (eds.), Species synopsis: Life histories of selected fish and shellfish in the northeast Pacific and Bering Sea. University of Washington, Seattle.

Haegele, C.W. 1997. The occurrence, abundance and food of juvenile herring and salmon in the Strait of Georgia, British Columbia in 1990 to 1994. Can. Manuscr. Rep. Fish. Aquat. Sci. 2390. 124 pp.

Haegele, C.W., and J.F. Schweigert. 1985. Distribution and characteristics of herring spawning grounds and description of spawning behaviour. Can. J. Fish. Aquat. Sci. 42(Suppl. 1):39-55.

Håkansson, N., S. Kollberg, U. Falk, E. Goetze, and O. Rechlin. 1979. A hydroacoustic and trawl survey of herring and sprat stocks of the Baltic proper in October 1978. Fisch.-Forsch. 2:7-23.

Hay, D.E. 1985. Reproductive biology of Pacific herring (*Clupea harengus pallasi*). Can. J. Fish. Aquat. Sci. 42(Suppl. 1):111-126.

Hay, D.E., and P.B. McCarter. 1997. Larval distribution, abundance, and stock structure of British Columbia herring. J. Fish Biol. 51(Suppl. A):155-175.

Hay, D.E., and P.B. McCarter. 1999. Age of sexual maturation and recruitment in Pacific herring. Can. Stock Assess. Secretariat Research Document 99/175. 39 pp.

Hay, D.E., M.J. Thompson, and P.B. McCarter. 2001. Anatomy of a strong year class: Analysis of the 1977 year class of Pacific herring in British Columbia and Alaska. In: F. Funk, J. Blackburn, D. Hay, A.J. Paul, R. Stephenson, R. Toresen, and D. Witherell (eds.), Herring: Expectations for a new millennium. University of Alaska Sea Grant, AK-SG-01-04, Fairbanks. (This volume.)

Heath, M. 1989. Transport of larval herring (*Clupea harengus* L.) by the Scottish coastal current. Rapp. P.-V. Reun. Cons. Int. Explor. Mer 191:85-91.

Heath, M.R., and P.M. MachLachlan. 1985. Growth and survival rates of yolk-sac herring larvae from a spawning ground to the west of the Outer Hebrides. ICES C.M. 1985/H:28.

Helgason, T., and H. Gislason. 1979. VPA-analysis with species interaction due to predation. ICES C.M. 1979/G:52.

Hessle, C. 1925. The herrings along the coast of Sweden. Publ. de Circonst. 89:1-57.

Hjort, J. 1914. Fluctuations in the great fisheries of northern Europe. Rapp. P.-V. Reun. Cons. Int. Explor. Mer 20:1-228.

Hodgson, W.C. 1957. The herring and its fishery. Routledge and Kegan Paul, London. 197 pp.

Höglund, H. 1978. Long-term variations in the Swedish herring fishery off Bohuslän and their relation to North Sea herring. Rapp. P.-V. Reun. Cons. Int. Explor. Mer 172:175-186.

Holst, J.C., F. Arrhenius, C. Hammer, N. Håkansson, J.A. Jacobsen, A. Krysov, W. Melle, and H. Vilhjalmsson. 1998. Report on the surveys of the distribution, abundance and migrations of the Norwegian spring-spawning herring, other pelagic fish and the environment of the Norwegian Sea and adjacent waters in late winter, spring and summer of 1998. ICES C.M. 1998/D:3. 61 pp.

Hunter, J.G. 1975. Fishery resources of the western Arctic. Fish. Res. Board Can. Manuscr. Rep. 1335. 32 pp.

ICES (International Council for the Exploration of the Sea). 1978. Report state of herring stocks around Ireland and the North West of Scotland. ICES C.M. 1979/H:2.

ICES. 1999a. Report of the Baltic Fisheries Assessment Working Group, ICES Headquarters, 14-23 April 1999. ICES C.M. 1999/ACFM:15. 555 pp.

ICES. 1999b. Report of the Study Group on Baltic Acoustic Data. ICES C.M. 1999/H:3.

ICES. 1999c. Report of the Study Group on Multispecies Model Implementation in the Baltic. ICES C.M. 1999/H:5.

ICES. 1999d. Report of the Northern Pelagic and Blue Whiting Fisheries Working Group. ICES Headquarters, Copenhagen, Denmark, 27 April-5 May 1999. ICES C.M. 1999/ACFM:18.

ICES. 1999e. Report of the Herring Assessment Working Group for the Area South of 62°N. ICES Headquarters, Copenhagen, Denmark, 15-24 March 1999. ICES C.M. 1999/ACFM:12.

ICES. 1999f. Report of the ICES Advisory Committee on Fishery Management, 1998. ICES Coop. Res. Rep. 229.

ICES. 2000. Report of the ICES Advisory Committee on Fishery Management, 1999. ICES Coop. Res. Rep. 236, Part 2. 405 pp.

ICES. 2001. Report of the Herring Assessment Working Group, for the Area South of 62°N. 13-22 March, 2001, Hamburg, Germany. ICES C.M. 2001/ACFM 12.

ICNAF Redbook. 1976. Report of the Standing Committee on Research and Statistics, App. II. Report of the Herring Working Group. Int. Comm. Northwest Atl. Fish., Dartmouth, Nova Scotia, August 1976.

Iles, T.D., and M. Sinclair. 1982. Atlantic herring stock discreteness and abundance. Science 215:627-632.

Jakobsson, J. 1973. Population studies on the Icelandic herring stocks. ICES C.M. 1973/H:4

Jakobsson, J. 1980. Exploitation of the Icelandic spring- and summer-spawning herring in relation to fisheries management, 1947-1977. Rapp. P.-V. Reun. Cons. Int. Explor. Mer 177:23-42.

Jakobsson, J., and G. Stefansson. 1999. Management of summer-spawning herring off Iceland. ICES J. Mar. Sci. 56:827-833.

Jakobsson, J., A. Gudmundsdóttir, and G. Stefánsson. 1993. Stock-related changes in biological parameters of the Icelandic summer-spawning herring. Fish. Oceanogr. 2:260-277.

Jakobsson, J., H. Vilhjálmsson, and S.A. Schopka. 1969. On the biology of the Icelandic herring stocks. Rit Fiskid. 4:6.

Johansen, A.C. 1919. On the large spring spawning herring in the northwest European waters. Medd. Komm. Havunders. 5.

Jönsson, N., and E. Biester. 1981. Wanderbewegungen des Rügenschen Frühjahrsherings in den Küsten—und Boddengewässern der DDR. Fisch.-Forsch. 19:47-51.

Kändler, R. 1942. Über die Erneuerung der Heringbestände und das Wachstum der Frühjahrs—und Herbstheringe in der westlichen Ostsee. Kiel. Meeresforsch. 6:73-89.

Kobayashi, T. 1983. Study of two genetically different spawning groups of Pacific herring (*Clupea harengus pallasi* Valenciennes) appearing in Ishikari Bay, Hokkaido. Bull. Hokkaido Reg. Fish. Res. Lab. 48:11-19.

Kobayashi, T. 2001. Biological characteristics and stock enhancement of Lake Furen herring distributed in northern Japan. In: F. Funk, J. Blackburn, D. Hay, A.J. Paul, R. Stephenson, R. Toresen, and D. Witherell (eds.), Herring: Expectations for a new millennium. University of Alaska Sea Grant, AK-SG-01-04, Fairbanks. (This volume.)

Kompowski, A. 1971. The types of otoliths in herring from the Southern Baltic. Pr. Morsk. Inst. Ryb. Gdyni 16(A):109-141.

Krixunov, E.A. 1990. Population dynamics of White Sea herring stocks. In: Proceedings of the International Herring Symposium, University of Alaska Sea Grant, AK-SG-91-01, Fairbanks, pp. 361-372.

Lawrence, M.J., G. Lacho, and S. Davies. 1984. A survey of the coastal fishes of the southeastern Beaufort Sea. Can. Tech. Rep. Fish. Aquat. Sci. 1220. 178 pp.

MacKenzie, K. 1985. The use of parasites as biological tags in population studies of herring in the North Sea and to the north and west of Scotland. J. Cons. Cons. Int. Explor. Mer 42:33-64.

McQuinn, I.H. 1997a. Year-class twinning in sympatric seasonal spawning populations of Atlantic herring *Clupea harengus*. Fish. Bull., U.S. 95:126-136.

McQuinn, I.H. 1997b. Metapopulations and the Atlantic herring. Rev. Fish Biol. Fish. 7:297-329.

McQuinn, I.H., and L. Lefebvre. 1995. Distribution, movements and size composition of spring-spawning herring in the northern Gulf of St. Lawrence. DFO Atl. Fish. Res. Doc. 95/57. 31 pp.

McQuinn, I.H., M. Hammill, and L. Lefebvre. 1999. An assessment and risk projections of the west coast of Newfoundland (NAFO division 4R) herring stocks (1965 to 2000). DFO Atl. Fish. Res. Doc. 99/119. 94 pp.

Melvin, G.D., F.J. Fife, M.J. Power, and R.L. Stephenson. 1996. The 1996 review of Georges Bank (5Z) herring stock. DFO Atl. Fish. Res. Doc. 96/29. 54 pp.

Melvin, G.D., R.L. Stephenson, M.J. Power, F.J. Fife, and K.J. Clark. 2001. Industry acoustic surveys as the basis for in-season decisions in a co-management regime. In: F. Funk, J. Blackburn, D. Hay, A.J. Paul, R. Stephenson, R. Toresen, and D. Witherell (eds.), Herring: Expectations for a new millennium. University of Alaska Sea Grant, AK-SG-01-04, Fairbanks. (This volume.)

Miller, D.J., and J. Schmidkte. 1956. Report on the distribution and abundance of Pacific herring, *Clupea pallasi*, along the coast of central and southern California. Calif. Fish Game 42:163-187.

Mollat du Jourdin, M. 1993. Europe and the sea. English translation copy. Blackwell Publishers, Oxford, U.K. 269 pp.

Morita, S. 1985. History of the herring fishery and review of artificial propagation techniques for herring in Japan. Can. J. Fish. Aquat. Sci. 42:222-229.

Motoda, S., and Y. Hirano. 1963. Review of Japanese herring investigations. Ext. du Rapp. et Proc.-Verbaux 154:249-261

Naumenko, N. 1996. Stock dynamics of the western Bering Sea. In: O.A. Mathisen and K.O. Coyle (eds.), Ecology of the Bering Sea. A review of Russian literature. University of Alaska Sea Grant, AK-SG-96-01, Fairbanks, pp. 169-175.

NOAA. 1998. Assessment of the coastal stock complex of Atlantic herring for 1998. Report of the 27th Northeast Regional Stock Assessment Workshop, Stock Assessment Review Committee (SARC). Northeast Fishery Science Center, Woods Hole, Massachusetts, NOAA/NFSC Ref. Doc. 98.

Øiestad, V. 1994. Historic changes in cod stocks and cod fisheries: Northeast arctic cod. ICES Mar. Sci. Symp. 198:17-30.

Ojaveer, E. 1988. Baltic herrings. Agropromisdat, Moscow, pp. 3-204. (In Russian.)

Ojaveer, E. 1989. Population structure of pelagic fishes in the Baltic. Rapp. P.-V. Reun. Cons. Int. Explor. Mer 190:17-21.

Orlova, E.L., E.I. Seliverstova, A.V. Dolgov, and V.N. Nesterova. 2001. Herring abundance, food supply, and distribution in the Barents Sea and their availability for cod. In: F. Funk, J. Blackburn, D. Hay, A.J. Paul, R. Stephenson, R. Toresen, and D. Witherell (eds.), Herring: Expectations for a new millennium. University of Alaska Sea Grant, AK-SG-01-04, Fairbanks. (This volume.)

Otterlind, G. 1985. The Rügen-herring in Swedish waters with remarks on herring population problems. Medd. Havsfiskelab. Lysekil 309:1-12.

Parmanne, R., O. Rechlin, and B. Sjöstrand. 1994. Status and future of herring and sprat stocks in the Baltic Sea. Dana 10:29-59.

Parrish, B.B., and A. Saville. 1965. The biology of the northeast Atlantic herring populations. Oceanogr. Mar. Biol. 3:323-373.

Parrish, B.B, A. Saville, R.E. Craig, G.I. Baxter, and R. Preistly. 1959. Observations on herring spawning and larval distribution in the Firth of Clyde in 1958. J. Mar. Biol. Assoc. U.K. 38:445-453

Patterson, K.R., and G.D. Melvin. 1996. Integrated Catch at Age Analysis Version 1.2. Scott. Fish. Res. Rep. 58.

Pete, M.C. 1990. Subsistence herring fishing in the Nelson Island and Nunivak Island districts, 1990. Alaska Department of Fish and Game, Division of Subsistence, Juneau, Tech. Pap. 196.

Pope, J.G. 1979. A modified cohort analysis in which constant natural mortality is replaced by estimates of predation levels. ICES C.M. 1979/H:16.

Popiel, J. 1964. Some remarks on the Baltic herring. ICES C.M. 1964/H:68.

Power, M.J., and T.D. Iles. 2001. Biological characteristics of Atlantic herring as described by a long term sampling program. In: F. Funk, J. Blackburn, D. Hay, A.J. Paul, R. Stephenson, R. Toresen, and D. Witherell (eds.), Herring: Expectations for a new millennium. University of Alaska Sea Grant, AK-SG-01-04, Fairbanks. (This volume.)

Pushnikova, G.M. 1996a. The state of supplies of Hokkaido-Sakhalin herring and the path toward stabilization of numbers. In: Fishery researches in the Sakhalin-Kuril region and adjacent areas. Yuzhno-Sakhalinsk, SakhTINRO 1:47-56. (In Russian.)

Pushnikova, G.M. 1996b. Fishery and stock condition of the Sakhalin waters herring. Scientific Proceedings of Far Eastern State Technical Fishery University 8:34-43. (In Russian.)

Rankine, P.W. 1986. Herring spawning grounds around the Scottish coast. ICES C.M. 1986/H:15.

Rannak, L. 1967. On biological groups of the spring spawning herring on the basis of otolith types. Eesti NSV Teaduste Akadeemia, Toimetised, 16/1:41-53. (In Estonian with English summary.)

Reilly, P.N. 1988. Growth of young-of-the-year and juvenile Pacific herring from San Francisco Bay, California. Calif. Fish Game 74:38-48.

Riske, M.E. 1960. A comparative study of north Pacific and Canadian Arctic herring (*Clupea*). M.S. thesis, University of Alberta, Edmonton, Alberta. 151 pp.

Ryman, N., U. Lagercranz, L. Andersson, R. Chakraborty, and R. Rosenberg. 1984. Lack of correspondence between genetic and morphologic variability patterns in Atlantic herring (*Clupea harengus*). Heredity 54(3):687-704.

Scattergood, L.W., C.J. Sindermann, and B.E. Skud. 1959. Spawning of North American herring. Trans. Am. Fish. Soc. 88:164-168.

Schweigert, J.F., and M. Stocker. 1988. Escapement model for estimating Pacific herring stock size from spawn survey data and its management implications. N. Am. J. Fish. Manage. 8:63-74.

Schweigert, J.F., C. Fort, and R. Tanasichuk. 1999. Stock assessment for British Columbia herring in 1998 and forecasts of the potential catch in 1999. Canadian Stock Assessment Secretariat Res. Doc. 99/21. 69 pp.

Shields, T. 1985. A diver survey in the Fingers area of Liverpool Bay, Northwest Territories. Archipelago Marine Research, Victoria, B.C. 87 pp.

Sjöstrand, B. 1989. Assessment review: Exploited pelagic stocks in the Baltic. Rapp. P.-V. Reun. Cons. Int. Explor. Mer 190:235-252.

Slotte, A., and A. Johannessen. 1997. Spawning of Norwegian spring spawning herring (*Clupea harengus* L.) related to geographical location and population structure. ICES C.M. 1997/CC:17.

Smith, P.J., and A. Jamieson. 1986. Stock discreteness in herrings: A conceptual revolution. Fish. Res. 4:223-234.

Smith, W.G., and W.W. Morse. 1990. Larval distribution patterns: Evidence for the collapse/recolonization of Atlantic herring on Georges Bank. ICES C.M., Pelagic Fish. Comm. 1990/H:87. 16 pp.

Southward, A.J., G.T. Boalch, and L. Maddock. 1988. Fluctuations in the herring and pilchard fisheries of Devon and Cornwall linked to change in climate since the 16th century. J. Mar. Biol. Assoc. U.K. 68:423-445.

Sparholt, H. 1994. Species interactions relevant for fish stock assessment in the central Baltic. Dana 10:131-162.

Sparholt, H., R. Aps, E. Aro, M. Fetter, J. Flinkman, G. Kornilovs, T. Raid, F. Shvetsov, B. Sjöstrand, and D. Uzars. 1994. Growth changes of herring in the Baltic. TemaNord 1994:532. 122 pp.

Spratt, J.D. 1981. The status of the Pacific herring, *Clupea harengus pallasii*, resource in California 1972 to 1980. Calif. Dept. Fish Game, Fish Bull. 171. 107 pp.

Spratt, J.D. 1987. Variation in the growth rate of Pacific herring from San Francisco Bay, California. Calif. Fish Game 73(3):132-138.

Spratt, J.D. 1992. The evolution of California's herring roe fishery: Catch allocation, limited entry, and conflict resolution. Calif. Fish Game. 78(1):20-44.

State of California. 2000. Final environmental document, Pacific herring commercial fishing regulations. Department of Fish and Game, Resources Agency.

Stephenson, R.L. 2001. The role of herring investigations in shaping fisheries science. In: F. Funk, J. Blackburn, D. Hay, A.J. Paul, R. Stephenson, R. Toresen, and D. Witherell (eds.), Herring: Expectations for a new millennium. University of Alaska Sea Grant, AK-SG-01-04, Fairbanks. (This volume.)

Stephenson, R.L. 1991. Stock discreteness in Atlantic herring: A review of arguments for and against. In: Proceedings of the International Herring Symposium. University of Alaska Sea Grant, AK-SG 91-01, Fairbanks, pp. 659-666.

Stephenson, R.L., and I. Kornfield. 1990. Reappearance of spawning Atlantic herring (*Clupea harengus harengus*) on Georges Bank: Population resurgence not recolonization. Can J. Fish Aquat Sci. 47:1060-1064.

Stephenson, R.L., and M.J. Power. 1989. Reappearance of Georges Bank herring: A biological update. Can. Atl. Sci. Adv. Comm. Res. Doc. 89/60. 14 pp.

Stephenson, R.L., K. Rodman, D.G. Aldous, and D.E. Lane. 1999. An in-season approach to management under uncertainty: The case of the SW Nova Scotia herring fishery. ICES J. Mar. Sci. 56:1005-1013.

Stevenson, D.K. 1998. Status of the coastal stock complex of Atlantic herring and a preliminary assessment of the Gulf of Maine stock. In: M.L. Mooney-Seuss, J.S. Goebel, H.C. Tausig, and M. Sweeney (eds.), Herring stock assessment research priorities. New England Aquarium Aquatic Forum Series Rep. 98-1:41-73.

Stevenson, D.K., K. Friedland, and M. Armstrong. 1997. Managing the U.S. Atlantic herring coastal stock complex for long-term sustainable yield. ICES C.M. 1997/DD:02.

Stocker, M. 1993. Recent management of the British Columbia herring fishery. In: L.S. Parsons and W.H. Lear (eds.), Perspectives on Canadian marine fisheries management. Can. Bull. Fish. Aquat. Sci. 226:267-293.

Suer, A.L. 1987. The herring of San Francisco and Tomales Bay. Ocean Research Institute, San Francisco. 64 pp.

Svetovidov, A.N. 1952. Fauna of the U.S.S.R.: Fishes: Clupeidae. Zool. Inst. Acad. Sci. U.S.S.R. New series 48(2), no.1. (English translation available from the Smithsonian Institution Libraries, Washington, D.C.)

Tanasichuk, R.W., A.H. Kristofferson, and D.V. Gillman. 1993. Comparison of some life history characteristics of Pacific herring (*Clupea pallasi*) from the Canadian Pacific Ocean and Beaufort Sea. Can. J. Fish. Aquat. Sci. 50:964-971.

Tang, Q. 1981. A preliminary study on the cause of fluctuation in year class size of Pacific herring in the Yellow Sea. Trans. Oceanol. Limnol. 2:37-45. (In Chinese with English abstract.)

Tang, Q. 1987. Estimation of fishing mortality and abundance of Pacific herring in the Yellow Sea by cohort analysis (VPA). Acta Oceanol. Sin. 6(1):132-141.

Tang, Q. 1991. Yellow Sea herring. In: J. Deng (ed.), Marine fishery biology. Agriculture Press, Beijing, pp. 296-356. (In Chinese.)

Taylor, F.C.H. 1964. Life history and present status of British Columbia herring stocks. Bull. Fish. Res. Board Can. 143. 81 pp.

Tibbo, S.N. 1966. The Canadian Atlantic herring fishery. Department of Fisheries, Can. Fish. Rep. 8:7-16.

Toresen, R., and O.J. Østvedt. 2000. Variation in the abundance of Norwegian spring spawning herring (*Clupea harengus*, Clupeidae) throughout the 20th century and the influence of climatic fluctuations. Fish Fish. 1:231-256.

Townsend, D.W. 1992. Ecology of herring larvae in relation to the oceanography of the Gulf of Maine. J. Plankton Res. 14(4):467-493.

Tyurnin, V.B. 1973. The spawning range of Okhotsk herring. Izv. Pac. Inst. Fish. Oceanogr. 86:12-21. (Transl. by Translation Bureau, Department of Secretary of State, Canada.)

Wailes, G.H. 1936. Food of *Clupea pallasii* in southern British Columbia waters. J. Biol. Board Can. 1(6):477-486.

Ware, D.W. 1985. Life history characteristics, reproductive value, and resilience of Pacific herring (*Clupea harengus pallasi*). Can. J. Fish. Aquat. Sci. 42(Suppl. 1):127-137.

Ware, D.W. 1991. Climate, predators, and prey: Behaviour of a linked oscillating system. In: T. Kawasaki et al. (eds.), Long-term variability of pelagic fish populations and their environment. Pergamon Press, Tokyo, pp. 279-291.

Watters, D.L., and K.T. Oda. 1997. Pacific herring, *Clupea pallasi*, spawning population assessment for San Francisco Bay, 1992-93. Calif. Dept. Fish Game, Mar. Res. Div. Admin. Rep. 97-3.

Wheeler, J.P., and G.H. Winters. 1984. Migrations and stock relationships of east and southeast Newfoundland herring (*Clupea harengus*) as indicated from tagging studies. J. Northwest Atl. Fish. Sci. 5:121-129.

Wheeler, J.P., B. Squires, and P. Williams. 1999. Newfoundland east and southeast coast herring: An assessment of stocks to the spring of 1998. Canadian Stock Assessment Secretariat Res. Doc. 99/13. 171 pp.

Whitehead, P.J.P. 1985. King herring: His place amongst the clupeoids. Can. J. Fish. Aquat. Sci. 42(Suppl. 1):3-20.

Williams, E.H. 1999. Interrelationships of Pacific herring, *Clupea pallasi*, populations and their relation to large-scale environmental and oceanographic variables. Ph.D. thesis, University of Alaska Fairbanks, Juneau.

Winters, G.H., and J.P. Wheeler. 1987. Recruitment dynamics of spring-spawning herring in the Northwest Atlantic. Can. J. Fish. Aquat. Sci. 44:882-900.

Survival of Pacific Herring Eggs on Giant Kelp in San Francisco Bay

Sara Peterson
San Francisco State University, Department of Biology–Larson Lab, San Francisco, California

Extended Abstract

An eggs-on-kelp fishery for Pacific herring (*Clupea pallasii*) roe operates in San Francisco Bay during the winter spawning period. During the past 10 years, eggs-on-kelp harvesting has developed into a substantial economic portion of the commercial herring roe fishery. Management of this fishing method is relatively new, and there are several biological questions about the effect of this fishery on the herring population. Giant kelp (*Macrocystis* sp.) is harvested from Monterey or southern California and suspended from open rafts immediately prior to the beginning of a spawn. If kelp is suspended too early, or the size of the spawning school is too small, the eggs-on-kelp product may be inadequate for market. In either of these cases, eggs-on-kelp is discarded with the assumption that the herring eggs will continue to develop and hatch. *Macrocystis* is not native to San Francisco Bay, and deteriorates at low salinities (Lobban et al.1985). The viability of kelp as a substrate for herring egg development is unknown. From a fishery management standpoint, the question of egg survival on discarded kelp is one of the unknown factors in deciding whether the amount of kelp used in the fishery should be regulated. The purpose of this study was to determine how the condition of unharvested and discarded kelp, along with other factors, affects the survival and hatching of herring eggs.

This study took place from January to March 1999, in Sausalito and San Francisco, California. I suspended kelp, obtained from fishery participants, on several small (1 m^2) rafts. In order to observe kelp at different stages of deterioration, the kelp was suspended both several days (2-5) prior to a spawning event and as soon as possible to the first day of spawning. Temperature and salinity were recorded on each sampling day. Once a spawning event ceased (fish were no longer observed in the area), several blades were separated from the raft to simulate what happens to the unmarketable eggs-on-kelp product that is discarded and left floating on the surface. These blades were placed in mesh boxes (collapsible crab pots covered with 0.25-inch mesh net) and submerged just under the surface (10 cm). Several blades were also attached outside the crab pots for comparison. Kelp condition was monitored every second day during embryo development for changes in color, turgidity, and integrity (i.e., still in one piece). A simple scale was used to assess kelp condition in the field (Table 1). Egg condition was assessed in the field according to eye and body pigmentation, movement within the egg casing, and number of body coils around the yolk sac (Stick 1990). When egg condition approached that of hatching, 3-cm^2 samples of eggs-on-kelp were collected and preserved in a solution of 70% ethanol and 30% seawater. Preserved embryos were analyzed and staged following development stages as described for Atlantic herring, *Clupea harengus* (Jones et al. 1978). Specifically, tail-free stage was identified by tail separation from the yolk and oblong yolk shape, and late embryo stage was identified by comparatively dark eye, head, and body pigmentation and the embryo body coiled around the yolk. Water temperature was relatively constant throughout the sampling period (Fig. 1)(spawning events 1-4). Salinity fluctuated within sampling periods and decreased throughout the season. Salinity remained relatively constant during spawning event 1, decreased during spawning events 2 and 3, and increased after the sixth day of spawning event 4 (Fig. 1).

In general, kelp deteriorated under these conditions found in San Francisco Bay, but how quickly kelp condition deteriorated appeared to depend on decreases in salinity. The more marked the decrease in salinity, the more accelerated the deterioration of the kelp. For example, during spawning event number 1, salinity remained consistent and the kelp remained in fair condition, deteriorating slowly over 20 days. In contrast, during spawning event number 4, salinity decreased sharply at day six, but increased afterward. The kelp continued to deteriorate and by day 20 the kelp was entirely decomposed (Fig. 1).

For all four spawning events, eye pigmentation and embryonic movement were observed in the field 10 days after the first day of spawning (marked by arrow points, Fig. 2). The condition of the kelp (at the arrow point) ranged from good to fair depending on the spawn sampled. For spawning events 1 and 2, both submerged and floating kelp were in at least fair condition or better 10 days after spawning began (Fig. 2). Kelp placed in the mesh box during spawning event 3 remained in fair condition longer than kelp left suspended. Kelp condition for spawning event 4 was less than fair when the eggs had reached the late embryonic stage (Fig. 2). It is important to note that for all spawning events, kelp remained in fair or better condition for at least 4 days after eggs were noted as having reached the late embryonic stage (Fig. 2).

One of the interesting results of this research was that for the more substantial spawns (those with several layers of eggs), kelp condition was not an impediment to egg survival but deterioration among the eggs themselves contributed to a large percentage of mortality. On blades where

Table 1. Kelp condition assessment scale. Kelp condition was assessed every second day during embryo development for changes in color, turgidity, and integrity (strength).

Scale	Condition
1	Kelp is dark green in color, blades and stipes are firm, stipe holds the weight of the blades, blades do not tear easily (can be "snapped").
2	Kelp begins to lose some color, becoming more dull green, stipes may be somewhat mushy in texture but still hold weight of blades, blades are firm and ragged on the edges.
3	Kelp is dull is color, stipes break when handled, blades firm, edges more ragged, water spots visible.
4	Kelp color dull and may be light green, stipes break easily and do not hold weight of blades when pulled from water, blades are easily punctured with finger, ragged on edges, but still hold together.
5	Kelp color same as above, stripes are mushy, falling apart, and do not hold blade weight, blade wilts when pulled from water, mushy and decomposing.

there were areas of higher egg deposition, normal egg development was observed until the late embryonic stage. In several instances, decay developed among some egg clusters and seemed to prevent the late-stage embryos from successfully hatching. From observing this on both kelp and pier pilings, I suspect that the underlying layers of eggs may die during development; then, as they begin to decay, contribute to the deterioration of surrounding embryos. Similarly, eggs that became covered with silt and particulate matter would reach the later embryonic stages but fail to successfully hatch. There are several strains of bacteria and/or water molds that are known to colonize fish eggs and organic matter in estuarine waters, and one or both of these processes probably occurs in San Francisco Bay herring eggs-on-kelp. Blades left outside the mesh boxes suffered almost immediate damage or complete loss due to predation. Suspended blades were not immune to this loss, because similar damage was seen on blades within 4 m of the water's surface. Several local species of birds were observed actively preying on floating and submerged eggs-on-kelp.

The results indicate that under the salinity and temperature conditions observed during this sampling period, giant kelp remained an adequate substrate for Pacific herring egg development. It should be noted however, that environmental conditions in San Francisco Bay are extremely variable from year to year (California Department of Fish and Game 1997).

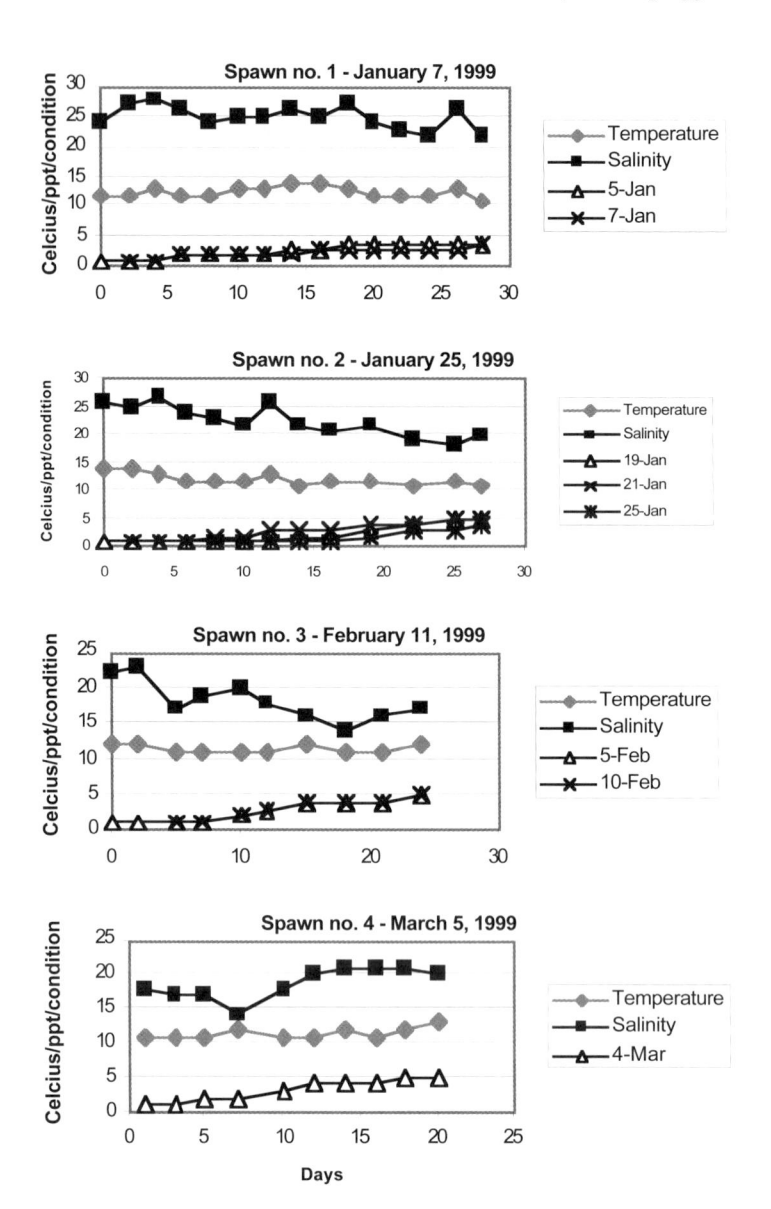

Figure 1. Temperature, salinity, and kelp condition. Kelp condition scored on a scale of 1 (good) to 5 (poor) (Table 1). Dates of spawns represent the first day of a spawning event. Dates within a spawning event refer to date of kelp suspension.

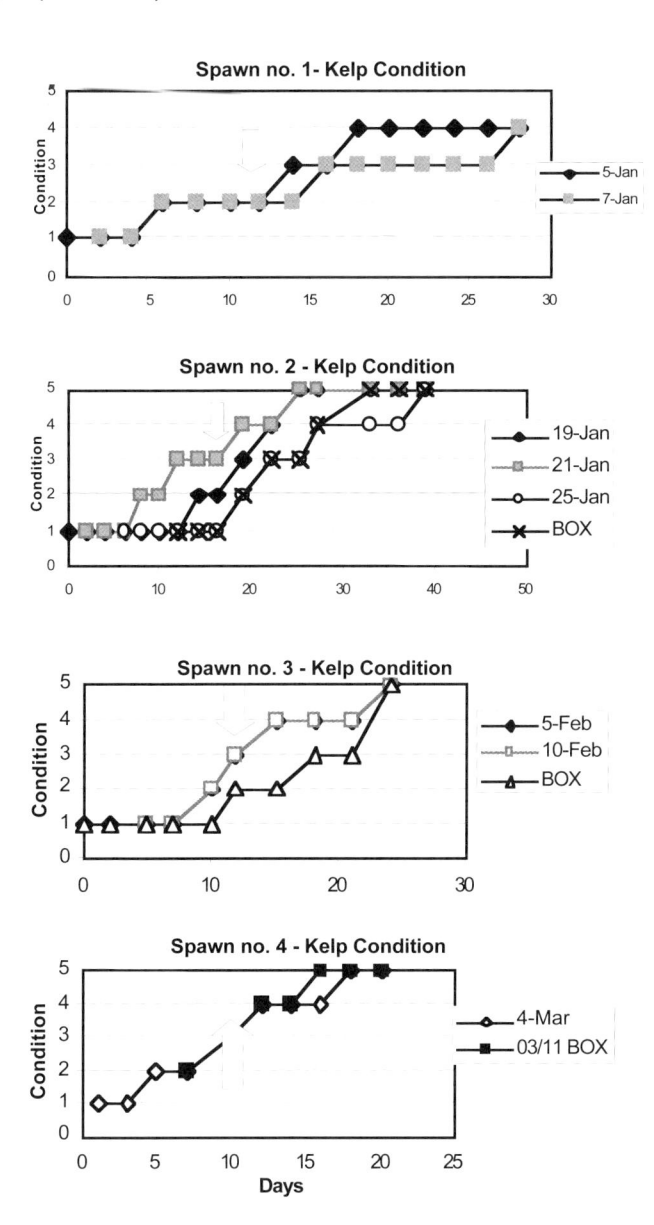

Figure 2. Kelp condition relative to egg stage. Arrow indicates day at which dark eye pigmentation and embryonic movement were observed. Kelp condition scored on a scale of 1 (good) to 5 (poor) (Table 1).

For instance, conditions during the 1997-1998 herring season were such that giant kelp would last no more than six days, which is insufficient time for embryonic development. Repeated studies over several seasons will be necessary to determine the larger implications of allowing fishery discard of eggs-on-kelp. Further research should also examine how bacterial growth and siltation affect successful hatching of eggs. Research on bacterial growth is not only important to eggs-on-kelp, but to any spawn where egg deposition is more than 2 or 3 layers. It was also observed that discarded eggs-on-kelp left floating on the surface or suspended are highly subject to predation. Predation of herring eggs by seabirds and invertebrates has been shown to contribute to egg loss (Haegele 1993), but determination of predation rates on eggs-on-kelp discard relative to eggs deposited on other substrates (pilings, boat bottoms, submerged and intertidal vegetation) requires further study.

References

California Department of Fish and Game. 1997. Environmental Document, Pacific Herring Commercial Fishing Regulations (Sections 163 and 164, Title 14, California Code of Regulations). State of California, The Resources Agency, Department of Fish and Game.

Haegele, C.W. 1993. Seabird predation of Pacific herring, *Clupea pallasi,* spawn in British Columbia. Can. Field-Nat. 107:73-82.

Jones, P.W., F.D. Martin, and J.D. Hardy Jr. 1978. Development of fishes of the Mid-Atlantic Bight. U.S. Fish and Wildlife Service, Office of Biological Services. Volume 1, pp. 112-131.

Lobban, C.S., P.J. Harrison, and M.J. Duncan. 1985. The physiological ecology of seaweeds. Cambridge University Press, New York.

Stick, K. 1990. Summary of 1989 Pacific herring spawning ground surveys in Washington State waters. Washington Department of Fisheries Progress Report No. 280. 49 pp.

Seasonal Variation in Herring Target Strength

E. Ona
Institute of Marine Research, Bergen, Norway

X. Zhao
Yellow Sea Fisheries Research Institute, Qingdao, China

I. Svellingen and J.E. Fosseidengen
Institute of Marine Research, Bergen, Norway

Abstract

The target strength of adult herring in captivity was monitored experimentally at three frequencies (18, 38, and 120 kHz) over a period of 2 years using carefully calibrated split-beam echo sounders. Periodic investigations of seasonal variations in fat content and gonad development show systematic changes in acoustic target strength of more than 3.0 dB, as well as a very high mean target strength as compared to expected values. The measurements, which are made close to the surface at depths of 5-20 m, indicate that the mean target strength of herring may be as much as 4-8 dB higher than the recommended target strength for this fish. The target strength variability and overall target strength level are well described at all three frequencies, and an improved target strength-to-size relation, also incorporating the effect of gonad development, is suggested.

Introduction

Acoustic abundance estimation of the Norwegian spring-spawning herring is now regarded as routine work at the Institute of Marine Research, Bergen (Røttingen 1990, Røttingen et al. 1994, Toresen and Barros 1995). In the application of acoustic fish abundance estimation, the target strength of the fish is a pivotal parameter for the conversion of integrated acoustic energy to absolute fish abundance (MacLennan 1990). The relation of target strength to fish length at 38 kHz currently applied in the acoustic assessment of the Norwegian spring-spawning herring is the one recommended by Foote (1987) for clupeoids:

$$TS = 20 \log L - 71.9 \qquad (1)$$

where the target strength, *TS*, is in decibels and length (*L*) is in centimeters, was adopted by the International Council for the Exploration of the Sea (ICES) (ICES 1988). The equation is the result of a simple linear regression with the slope preset to 20. The regression was based on four in situ estimates made in the mid-1980s. These are two data points gathered on North Sea herring by Foote et al. (1986), one from the Kattegat-Skagerrak herring by Degnbol et al. (1985), and one from the Baltic herring mixed with sprat (*Sprattus sprattus*) by Lassen and Stæhr (1985). The equation is now used for all acoustic surveys of the Norwegian spring-spawning herring. This includes the surveys on juveniles (Toresen and Barros 1995); the wintering adult component in the fjords of northern Norway (Røttingen et al. 1994); the spawning stocks of the coastal areas of west and southwest of Norway (Hamre and Dommasnes 1994, Johannessen et al. 1995); and on the feeding grounds in the Norwegian Sea (Slotte 1998).

The abundance estimates obtained from these surveys show somewhat different results. In particular, the estimates obtained during the spawning season differ from those obtained from the feeding area and the wintering grounds. At present acoustic estimates are only used as relative indexes in tuning the virtual population analysis (VPA) (ICES 1996). The conflicting results may be due to several factors, but one of the main reasons has been attributed to variations in the target strength of the fish. This may result either from changes in behavior, i.e., tilt angle (Huse and Ona 1996), or through systematic changes in the acoustic reflection properties of the fish as suggested by Ona (1990).

Using the pre-seining echo-integration technique, Hamre and Dommasnes (1994) found that the target strength of spent Norwegian spring-spawning herring was 4 dB lower than the one currently used. Measurements made in the past 10 years also show large differences in the target strength estimates obtained from different herring stocks (Degnbol et al. 1985, Lassen and Stæhr 1985, Foote et al. 1986, Rudstam et al. 1988, Kautsky et al. 1991, Reynisson 1993).

Due to the small specific acoustic impedance of gas compared to that of the fish flesh and bones as well as seawater, the gas-filled swim bladder is responsible for 90-95% of the acoustic energy reflected from the fish (Foote 1980a). Theoretical computations based on the exact form of the swim bladder (Foote 1985, Foote and Ona 1985) further demonstrated that the shape as well as the size of the swim bladder is important for the acoustic reflection.

Based on slicing and reconstructing methods, Ona (1990) found that both size and shape of the swim bladder could be significantly altered by changes in stomach contents, gonad development, and fat content of the fish. These in turn may cause diel, as well as seasonal, variations of the target strength of fish, suggesting that the acoustic fish abundance estimates may be significantly biased if a single, fixed *TS*-length relationship is to be used for all surveys. More detailed understanding of these effects, in particular in relation to potential seasonal variations of the target strength of fish, is therefore needed.

A European Union (EU) project (AIR3-CT94-2142) started taking measurements in 1995 to elucidate the variability of the target strength of fish to address the physiological factors causing the natural variations in the target strength of herring. The effects of gonad development and condition or fat content were studied experimentally by monitoring changes in target strength at three frequencies over 2 years on a captive population of adult herring. The main results from years one and two of these experiments are reported here.

Material and Methods
Experimental Design

The experiments were carried out at the fish farming plant of the Austevoll Aquaculture Station (Institute of Marine Research), situated at Austevoll, an island south of Bergen, where about 3 tons of herring were kept in captivity over more than 2 years. Herring from the Norwegian spring-spawning stock were captured in April 1995 at Langevåg, south of the experimental site. The fish were carefully transferred to the holding pen in Austevoll in a seawater tank on a transporting vessel, normally used to transport salmon. After an initial transport mortality of about 10% within the first 14 days, insignificant herring mortality was observed throughout the experimental period. The fish were fed with dry pellets (4.5-mm diameter) originally used as salmon smolt food. The daily supply varied from 0 to 25 kg depending on the water temperature and the response of the fish to the feed. Feeding was stopped 2 days prior to each measurement period, and the fish used for target strength measurements were not fed during measurements, usually conducted over 3-7 days.

The experiment was designed to make a series of target strength measurements following two full reproduction cycles of the captive herring population from summer 1995 to summer 1997, consisting of several discrete measurement periods each lasting approximately one week and covering distinct gonad maturation stages of the fish.

The experimental set-up used in this study is shown in Fig. 1. Captive herring were kept in a storage pen on the relatively "open" seaside of the floating pier, while the measurements were done in a separate pen on the inner side. The mobile instrument barrack, housing all the electronic equipment, was moved to the experimental site at the floating pier for each measurement period. The three split-beam transducers were mounted on a steel frame, which was in turn tied to the pier. Four buoys were attached

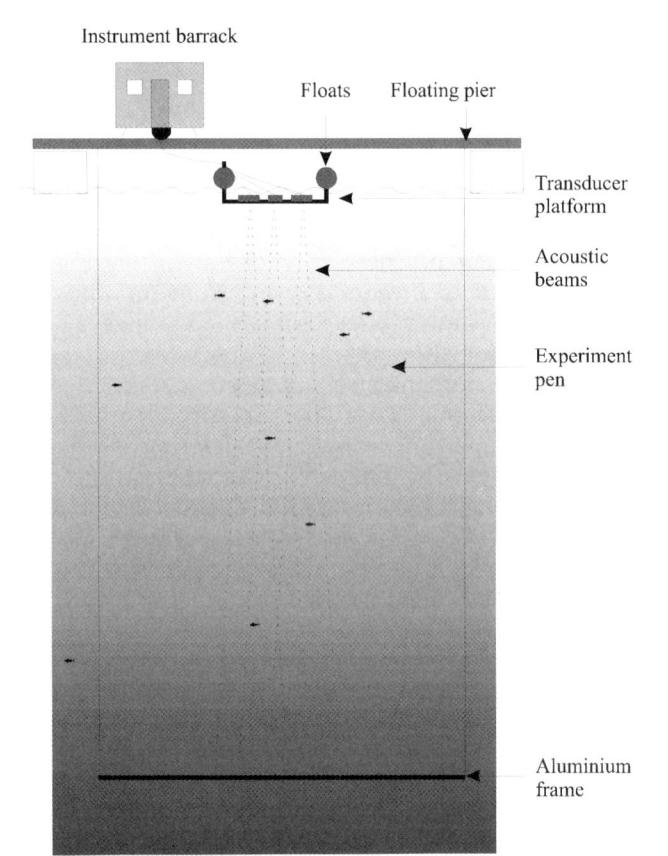

Figure 1. Simple sketch of the experimental set-up for target strength measurements: A, plan view. B, side view. (Not to scale.)

to the transducer frame to give the necessary lift, keeping the transducers horizontal at a depth of 0.65 m. The bottom depth beneath the measurement pen was about 37 m.

At the beginning of each measurement period, about 40-60 herring used for the target strength measurements were carefully transferred from the storage pen (12.5 × 12.5 × 7 m) to the measurement pen (12.5 × 12.5 × 21 m, mesh size 14 mm) using a scoop net. At least 4 hours were allowed for the fish to acclimate to the new pen before target strength data collection was started. During the measurements, the herring could freely move about in the 4,500 m³ net pen, frequently entering the acoustic beams, which only covered a small fraction of the central volume of the pen.

Biological Measurements

After target strength measurements, the bottom frame of the pen was slowly hoisted to a depth of about 1 m, and the herring were scooped over into a circular anaesthetizing bin in subgroups of 3-5 fish. The anaesthetizing solution was prepared by adding 25 ml Metomidate solution (1 gram of active agent) to roughly 40 liters of seawater. Swim bladder volume was measured according to Ona (1990), by repeated ventral massage from the pelvic fins toward the anal opening while holding the fish below the surface of the anaesthetizing bath. The swim bladder gas was collected in an inverted, submerged funnel, leading to a finely scaled volumetric cylinder with a measuring precision of 0.1 ml.

For each fish, the total length (L) was measured to the nearest centimeter, and the round fish weight (W) and gonad weight (W_g) were measured to the nearest gram. The condition factor (K) was calculated as

$$K = \frac{W}{L^3} \times 100 \qquad (2)$$

The gonadosomatic index (GSI) was calculated as

$$GSI = \frac{W_g}{W} \times 100 \qquad (3)$$

The swim bladder index (SBI) was defined as

$$SBI = \frac{V_{SB}}{L^3} \times 10,000 \qquad (4)$$

where V_{SB} is the volume of the swim bladder gas in milliliters (Ona 1990). The fat content of herring was determined through standard lab routine: Na_2SO_4 grinding with ethyl-ether extraction (Losnegard et al. 1979).

Acoustic Measurements

A SIMRAD EK500 scientific echo sounder (Bodholt et al. 1989) was used to collect single-fish-echo data. The echo sounder was configured to its maximum capacity, that is, with three frequencies working in parallel. The three frequencies used were 18, 38, and 120 kHz, with transducer nominal beam opening angles of 11, 12, and 9 degrees, respectively. All the three frequency transceiver channels are of the split-beam type. The echo sounder software versions used were EK500 V 5.00 and V 5.3. The target strength measurement methodology and the single target recognition criteria, as implemented in the echo sounder, were used as recommended in Ona (1999). An example of these settings, as well as the instrument parameter settings, is shown in Table 1. The actual sound speed was computed from temperature and salinity recorded, and applied during the calibration and measurements.

The filtered and accepted single-fish-echo data were logged via the serial port of the echo sounder and stored on a TCI 486 personal computer, using standard communication software PROCOMM (V 2.4, Datastorm Technologies Inc., 1986).

Calibrations of the overall acoustic system were performed prior to each measurement period applying the established standard reference target technique (MacLennan 1981, Foote 1982, Foote and MacLennan 1984). Calibration was conducted as described by Ona (1999), using the dedicated calibration software: LOBE V 5 (SIMRAD 1996). The two spheres used in this study and their nominal target strengths at different frequencies are listed in Table 2. Improved calibration accuracy at short range was obtained after installing the V 5.3 software in the EK-500 echo sounder (see Ona et al. 1996).

Data Analysis

The first step in the data analysis process was the selection of suitable data sets for target strength analysis, mainly based on the appearance of the echograms. Only recordings showing clear single fish traces and suitable fish densities were selected for further analysis. This procedure ensured an extremely low probability for falsely accepting multiple targets (see Ona 1999). Further, one frequency at a time, suitable range-limits were selected to be safely outside two times the transducer nearfield and for isolating false echoes from the net pen. Selected segments of the recordings were run through a target tracking software (Ona and Hansen 1991) isolating the target strength data from each herring track in separate data blocks. Finally, these data blocks were used to compute the swimming angles of the fish relative to the transducer, for data reformatting, and for further statistical analysis of target strength. A suitable cut-off beam angle of 5° off-axis was selected on the basis of statistical analysis of target distribution within the beam, ensuring a high signal-to-noise ratio on the targets detected inside this limit.

Table 1. Instrument technical specifications and parameter settings used for herring target strength measurements.

SIMRAD EK500	18 kHz	38 kHz	120 kHz	
Transceiver menu: Permanent settings				
Transducer type	ES18-11	ES38-12	ES120	
Absorption coefficient	3	10	38	dB/km
Pulse duration	Short	Medium	Medium	
	(0.7)	(1.0)	(0.3)	ms
Bandwidth	Auto	Auto	Auto	
	(1.8)	(3.8)	(12.0)	kHz
Maximum transmitting power	2,000	2,000	1,000	W
Two way beam angle	−17.3	−15.5	−18.5	dB
Alongship angle sensitivity	13.9	12.5	17.0	
Athwartship angle sensitivity	13.9	12.5	17.0	
Transceiver menu: Entered after calibration[a]				
TS transducer gain	24.4	21.2	22.9	dB
Alongship 3 dB beamwidth	11.2	12.0	8.6	dg
Athwartship 3 dB beamwidth	10.7	11.7	8.3	dg
Alongship offset angle	−0.12	−0.24	−0.08	dg
Athwartship offset angle	0.06	0.02	−0.17	dg
TS detection menu				
Minimum TS value	−60	−60	−60	dB
Minimum echo length	0.8	0.8	0.8	
Maximum echo length	1.3	1.3	1.5	
Maximum gain compensation	6.0 & 3.0	6.0	6.0	dB
Maximum phase deviation	3.0	3.0	5.0 & 3.0	

[a]Calibration results from 5 December 1995 are given here as an example.

Table 2. Target strength (dB) of two standard reference targets used for instrument calibrations.

Sphere type	18 kHz	38 kHz	120 kHz
Copper (60 mm)	−35.2	−33.6	
WC (38.1 mm)	−42.8	−42.3 & −42.2	−39.6 & −39.5

Refer to Table 1 for corresponding instrument settings.

Due to the schooling activity of the fish, only data collected during dark hours were included in the analysis. The average mean-track-target-strength within each time interval was computed in the intensity domain from all accepted, valid measurements made within the time interval. Depending on density, depth, and how the herring passed the transducer beam, usually several hundred accepted target strength measurements was recorded each hour. This will hereafter be referred to as the hourly mean target strength, and was later treated as an independent sample of the mean target strength of the fish. For each group of fish, the hourly mean target strengths were further averaged to provide the estimate of its overall mean target strength. All computations of the mean and associated statistics were performed on the acoustic cross section of the fish (i.e., in the linear domain, and later converted to the logarithmic domain). On occasions, where less than 50 tracks were detected within 1 hour, data from adjacent hours were pooled together. A minimum of 50 tracks was chosen to stabilize the variations of each hourly mean estimate.

The possible effect of the gonad development on the target strength of herring was studied using a general multiple linear regression analysis. In order to eliminate or reduce its size dependency, all the hourly mean target strengths ($\overline{TS}_H$) were normalized by the Root Mean Square length (RMS_L) of the fish. The result is the so-called b_{20}, as in the equation:

$$TS = 20\log \ell + b_{20} \qquad (5)$$

The specific formula used was

$$b_{20} = \overline{TS}_H - 20\log(RMS\_L + 0.5) \qquad (6)$$

The addition of 0.5 cm to the RMS length was due to the fact that the length of the fish was measured to the nearest centimeter below its total length. The length-normalized target strength or b_{20} was then selected as the dependent variable. The swim bladder index (SBI), gonadosomatic index (GSI), depth, mean tilt angle (Mean_Tilt), and the standard deviation of the tilt angle (SD_Tilt) were the candidates as predictor variables.

Following the usual TS to log fish size relation, a TS to Log(SBI) relation was assumed. For the depth effect, the practice of Halldórsson (1983), i.e., a log pressure dependence of the target strength, was followed. The pressure (P), expressed in units of atmospheric pressure (atm), was related to the range of the fish as follows:

$$P = \left(\overline{R} - R_d + 0.65 + 10\right)\big/10 \qquad (7)$$

where $\overline{R}$ is the mean range of the fish as measured by the echo sounder, R_d is the system range delay (Ona et al. 1996, Zhao 1996), 0.65 is the draft of the transducer, and 10 is the conversion factor from depth in meters to pressure in atmospheric units. The specific system delays used were 60 cm, 40 cm, and 0 cm for 18, 38, and 120 kHz echo sounders, when the EK-500 (V 5.00) was used. Since only echoes lying within 5° off-axis were selected for target strength analysis, the range, when added with the draft of the transducer, is an excellent approximation of depth within the range used. As no specific functional dependence of the fish target strength on the remaining three predictors was available, simple linear relationships were assumed. The starting model used was:

$$b_{20} = \beta_0 + \beta_1 \log(SBI) + \beta_2 \log P + \beta_3(GSI) + \beta_4(Mean\_Tilt) + \beta_5(SD\_Tilt) + \sum_{i=6}^{15} \beta_i(interaction) + \varepsilon \quad (8)$$

where β_0 is the constant term, $\beta_1...\beta_5$ are regression coefficients for the individual predictor variables, β_i is the regression coefficient for the interaction term and ε is a normally distributed error term.

Initially, all the interaction terms, limited to two variables for each term, were included in the model. A stepwise backward elimination procedure was performed using the General Linear Model option in SYSTAT (1992). The elimination procedure was started with the interaction terms, and the probability level for removal was set to 0.25. At each step, the interaction term that had the highest probability was removed and the revised model refitted. This procedure was repeated until no interaction term that had a probability level greater than 0.25 was left. The inclusion of the interaction terms in the final model was further constrained by a minimum tolerance of 0.1 to prevent the presence of strong intercorrelation between the predictor variables in the model (SYSTAT 1992). The final subset model was then tested at the 5% level.

Results

A summary of biological data for each group of fish associated with target strength measurements is shown in Table 3. A fairly stable mean length and weight was observed in the experimental groups, but with some variability in the range of lengths, resulting from the unselective transfer process between the holding pen and the experimental pen. A slight, gradual increase in weight and condition factor was, however, observed within the period from September 1995 to March 1996.

When the experiment initiated in September 1995, most of the fish were at the immature stage, and the gonads were too small to be accurately weighed (< 1 g). By October, the gonads gradually developed and the mean *GSI* increased steadily during the spring of 1996. On 26 March 1996, 50% of the fish had *GSI* of 15% or more.

Table 3. Biological data on the herring used for target strength measurements.

Parameter	n	Min.	Max.	Mean	S.D.
Second period, 2nd group. Measured on 26 October 1995.					
Length (cm)	68	27	37	33.0	2.9
Weight (g)	68	138	325	230.2	47.8
Condition factor	68	0.501	0.789	0.625	0.080
GSI (%)	68	< 0.4	8.3	2.2	1.8
Third period, 3rd group.[a] Measured on 8 December 1995.					
Length (cm)	60	25	36	32.1	2.8
Weight (g)	60	137	349	245.1	49.9
Fat content (%)	20	7.5	18.2	13.6	2.8
Condition factor	60	0.564	0.984	0.737	0.092
GSI (%)	60	0.4	16.8	5.4	4.0
V.B. (ml)	20	3.5	15.0	9.5	3.1
V.B./weight (%)	20	1.81	5.21	3.85	0.80
Swim bladder index	20	1.30	3.86	2.76	0.53
Third period, 4th group. Measured on 11 December 1995.					
Length (cm)	54	27	37	32.7	2.9
Weight (g)	54	156	352	240.0	50.1
Condition factor	54	0.541	0.894	0.685	0.083
GSI (%)	32	0.7	17.6	6.2	4.4
V.B. (ml)	17	4.2	13.7	8.4	2.9
V.B./weight (%)	17	1.64	4.68	3.48	0.82
Swim bladder index	17	1.03	2.94	2.34	0.53

[a]The swim bladder data were not included in the final analysis due to some mortality of the fish in the group.
V.B. stands for volume of the swim bladder.

Table 3. (Continued.) Biological data on the herring used for target strength measurements.

Parameter	n	Min.	Max.	Mean	S.D.
Fourth period, 5th group. Measured on 22 February 1996.					
Length (cm)	57	25	37	30.8	3.0
Weight (g)	57	122	359	228.4	57.1
Fat content (%)	29	3.3	19.3	11.1	4.2
Condition factor	57	0.584	1.052	0.773	0.105
GSI (%)	57	0.5	19.4	10.3	5.3
V.B. (ml)	26	3.3	14.7	7.7	2.7
V.B./weight (%)	26	1.58	5.06	3.37	0.84
Swim bladder index	26	1.18	3.40	2.52	0.62
Fifth period, 6th group. Measured on 26 March 1996.					
Length (cm)	59	25	37	31.2	3.2
Weight (g)	59	124	398	249.6	64.1
Fat content (%)	31	0.7	16.7	10.2	3.4
Condition factor	59	0.508	1.111	0.816	0.114
GSI (%)	58	0.8	32.3	14.7	7.8
V.B. (ml)	31	3.5	15.0	8.6	2.4
V.B./weight (%)	31	1.42	5.00	3.35	0.72
Swim bladder index	31	1.30	3.64	2.56	0.52

V.B. stands for volume of the swim bladder.

The mean fat content of the samples taken from the storage pen in September 1995 was 7.9%. This low fat content was probably due to limited food supply during the summer period. After the herring in the storage pen were feeding regularly on artificial food, the fat content of the fish increased accordingly. While the gonads developed, as a result of energy transfer, the fat content of the fish was reduced from a mean of 13.6% in December to 11.1% in February and further to 10.2% in March. The relative reduction, 3.4%, however, was much lower than the 10% observed in Norwegian spring-spawning herring (Slotte 1996). Throughout the experiment, the fat content of the herring varied greatly from fish to fish, with a minimum of only 0.7% in March 1996, and a maximum of 19.3% in February 1996.

The relationship between bladder volume and size of the fish is shown in Fig. 2, based on the data selected from September 1995 to June 1996. The correlation between swim bladder volume and fish length cubed is stronger than between swim bladder volume and fish weight. If the data from June 1996 are excluded, as the fish in this period represent the only cases where lipid deposition was apparent in the body cavity, the coefficient of determination (explained variation) between swim bladder size and fish size were further strengthened (linear regression: $r^2 = 0.511$ for volume vs. weight and 0.646 for volume vs. length cubed). The differences in the mean *SBI*, however, were not significant (ANOVA, $P > 0.3$) among the remaining groups.

The bladder volume per unit of fish body weight was inversely related to the fat content of the fish, but the correlation was weak. No significant relation was found between the swim bladder index and the fat content of the fish. Therefore, the intended search for a relation between the target strength and the fat content of the fish could not be made on this material. The main reason for this is due to the unexpected large within-group variability in fat content, and the unselective process of transferring fish from the holding pen to the experimental pen.

The mean target strength of each group of fish at each of the three frequencies is shown for six measurement periods in Table 4. For herring with mean lengths from 31 to 33 cm, the mean target strengths were found to be –38 to –34 dB at 38 kHz. At 18 and 120 kHz, the recorded mean target strengths were from –37 to –35 and –40 to –37 dB, respectively. The table also shows a systematic difference in the target strength of herring with frequency, most notably between 38 and 120 kHz. An increasing trend in the mean target strength over the period from December 1995 to March 1996 is also clearly observed.

The target strength of adult herring showed a broad range of variation. Differences from less than 1 dB to more than 25 dB were observed within single-fish tracks. This is due to the combined effect of acoustic directivity of the herring of this size at the measured frequencies and the tilt angle distribution. A detailed, long-term analysis of target strength ping-to-ping variability for one single herring showed that grouped fish and single fish measurements reproduced nearly identical target strength

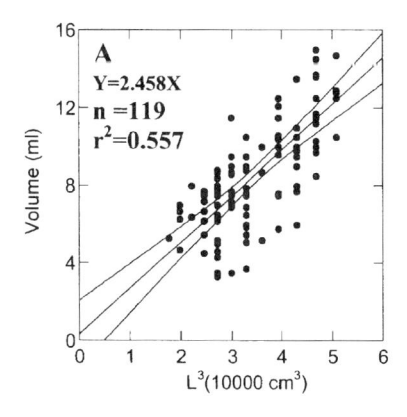

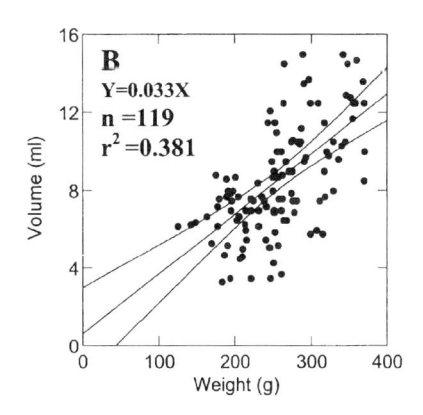

Figure 2. Relationship between the size of the swim bladder and the size of the fish. A, swim bladder volume vs. fish length cubed expressed in units of 10,000 cm³. B, swim bladder volume vs. total wet body weight. The lines shown are the best fits to the data and their 95% confidence intervals. However, since the constant terms are not significantly different from zero (P > 0.5 in both cases), the regression models given are without the constant term.

distributions (Zhao 1996). An example of the vertical swimming angle or the apparent tilt angle of the fish as estimated from target tracking on the 120-kHz transducer is shown in Fig. 3. Both methods used to compute the tilt angle (Fig. 3a and b) indicate that the fish were swimming almost horizontally on the average and the tilt angle distribution was essentially normal.

For the two frequencies commonly used in fisheries research (38 and 120 kHz), the mean target strength of herring increased considerably over the period of rapid gonad development from December to March (Table 4). This is also evident from the PDFs (probability density functions) of the mean-track target strength (Fig. 4) and from the summarized data on hourly mean target strengths (Fig. 5).

Figure 5 shows the box plots (McGill et al. 1978) of the hourly mean track target strength normalized by the RMS length of the fish, or b_{20} (equation 6). The notches around the median (the inner horizontal bar at the narrowest part of the notch) are clearly non-overlapping for the 38-kHz data and this is also roughly true for the 120-kHz data, therefore the median b_{20}s were different ($P < 0.05$) between any pair of groups of different measurement periods. The increasing trend of herring target strength over the three periods is thus significant.

In order to understand the role of the developing gonad on the observed changes of the target strength of herring, a general linear regression analysis was performed. Since some fishes in the second experimental group in December 1995 were dead at the end of the measurements, the data from this group were excluded from the regression analysis.

Table 4. Estimates of the target strength (*TS* in dB) of six groups of herring at three frequencies during September 1995–March 1996.

Group	Period	$\bar{L}$(cm)	S.D.	TS	C.I.	b_{20}	d.f.
		Fish length			18 kHz		
1	Sep. 1995	32.8	2.2	−35.2	(−36.4,−34.3)	−65.6	6
2	Oct. 1995	33.0	2.9	−34.4	(−34.7,−34.2)	−64.9	10
3	Dec. 1995	32.1	2.8				
4	Dec. 1995	32.7	2.9				
5	Feb. 1996	30.8	3.0	−37.4	(−38.5,−36.5)	−67.4	4
6	Mar. 1996	31.2	3.2	−34.6	(−35.0,−34.2)	−64.6	5

Group	Period	$\bar{L}$(cm)	S.D.	TS	C.I.	b_{20}	d.f.
		Fish length			38 kHz		
1	Sep. 1995	32.8	2.2				
2	Oct. 1995	33.0	2.9				
3	Dec. 1995	32.1	2.8	−37.3	(−37.6,−36.9)	−67.6	9
4	Dec. 1995	32.7	2.9	−38.5	(−39.0,−38.1)	−68.9	10
5	Feb. 1996	30.8	3.0	−36.1	(−36.5,−35.7)	−66.1	10
6	Mar. 1996	31.2	3.2	−33.6	(−33.9,−33.4)	−63.6	9

Group	Period	$\bar{L}$(cm)	S.D.	TS	C.I.	b_{20}	d.f.
		Fish length			120 kHz		
1	Sep. 1995	32.8	2.2				
2	Oct. 1995	33.0	2.9				
3	Dec. 1995	32.1	2.8	−40.0	(−41.3,−39.0)	−70.3	6
4	Dec. 1995	32.7	2.9	−40.5	(−41.0,−40.0)	−70.9	14
5	Feb. 1996	30.8	3.0	−38.3	(−39.1,−37.7)	−68.3	8
6	Mar. 1996	31.2	3.2	−37.1	(−37.5,−36.7)	−67.1	9

Note: The 38-kHz data may subject to a small correction due to the ping interval effect; see text.
The estimates are based on hourly mean track target strength and averaged in the linear domain. The 95% confidence intervals (C.I.) are converted from its counterparts in the linear domain assuming *t*-distribution with the degree of freedom (number of individual hourly means) indicated.

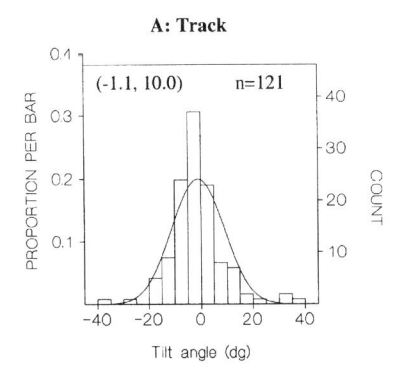

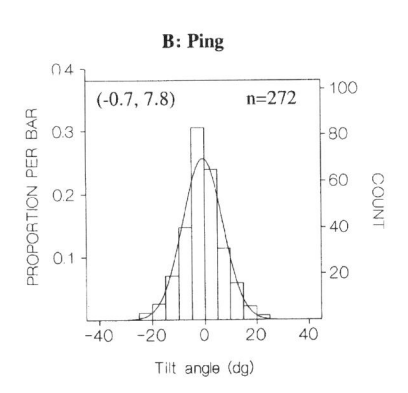

Figure 3. Apparent tilt angle distribution of a 34-cm herring observed by the 120-kHz transducer during a 30-hour period. The superimposed curves are the fitted truncated Gaussian (normal) distribution. Also shown in parentheses are the means and the standard deviations, respectively. A, mean track tilt angle. B, ping-to-ping tilt angle for swimming speed > 0.3 m per second.

The variations of the several candidate predictors over the three measurement periods analyzed here are shown in Fig. 6. Of the five factors considered, the swim bladder index (*SBI*) was previously found to be not significantly different among the three fish groups, and was therefore dropped from the candidate lists. The gonadosomatic index (*GSI*) was assumed to be constant within each measurement period, and only the three mean values were used in the regression analysis, one for each group of data. The rest of the predictor variables were calculated within the same time period as the hourly mean target strength data, and were thus attached to each dependent variable data (b_{20}) within each of three groups.

The results of the multiple linear regression for the data collected at 120 kHz are presented in Table 5. The b_{20}, hence the mean target strength of herring, was found to be significantly ($P < 0.05$) correlated with parameters of the tilt angle distribution as well as to the *GSI* of the fish. About 83% of the observed variance in the target strength of herring can thus be explained by the changes of the tilt angle distribution and the gonad development of the fish. None of the interaction terms were found to be significant at the 5% level and all of them were associated with extremely low tolerance (<0.05). They were thus excluded. The depth-related pressure dependence of the target strength of herring was also insignificant ($P > 0.05$) within the range studied. The final subset model for 120 kHz is:

$$b_{20} = -69.0 - 0.40(Mean\_Tilt) - 0.32(SD\_Tilt) + 0.24 GSI \qquad (9)$$

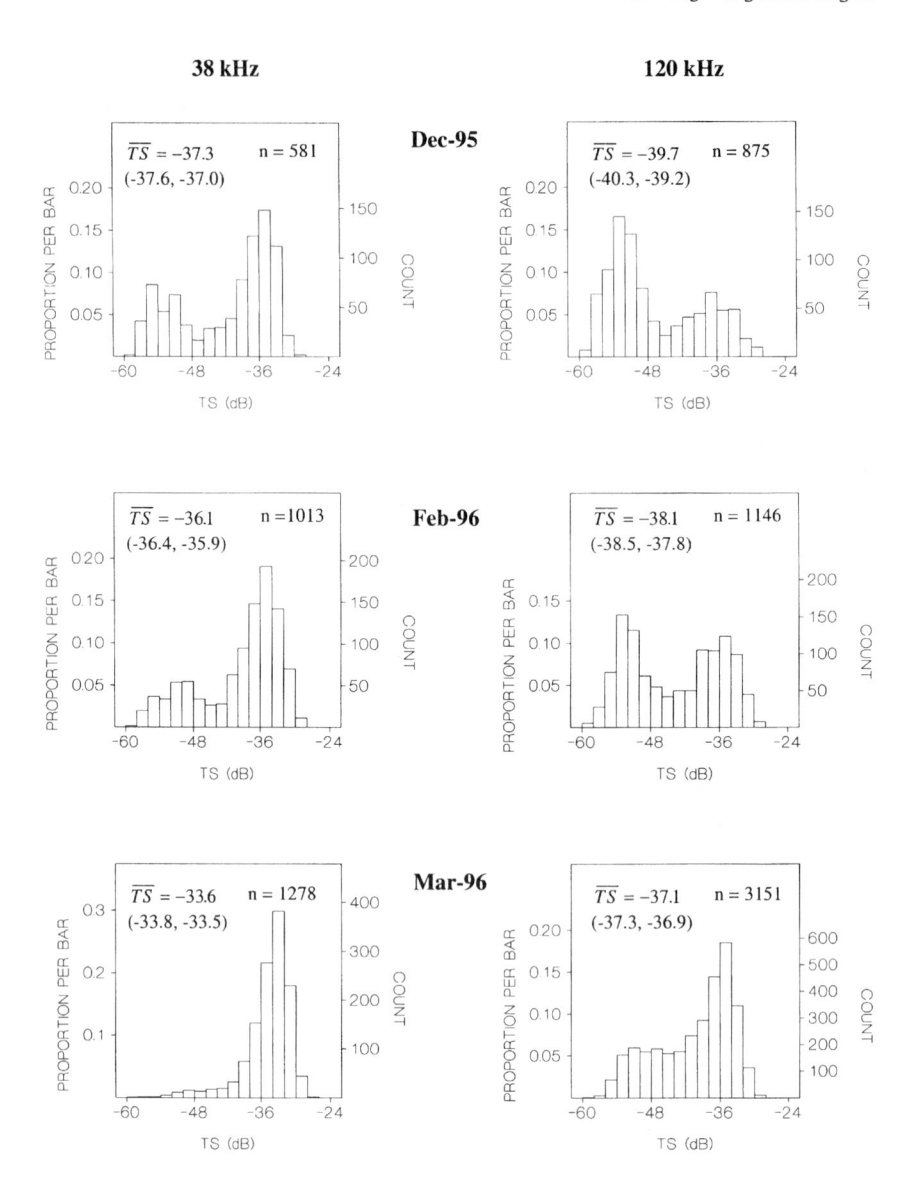

Figure 4. Progressive changes of the (mean track) target strength PDFs of herring at two frequencies over a 3-month period showing the plausible impact of gonad development on the target strength of herring. The mean target strengths are based on simple arithmetic operation on individual tracks performed in the linear domain. The 95% confidence intervals are converted from counterparts in the linear domain, which in turn were calculated using 2 standard errors of the mean.

A: 38 kHz

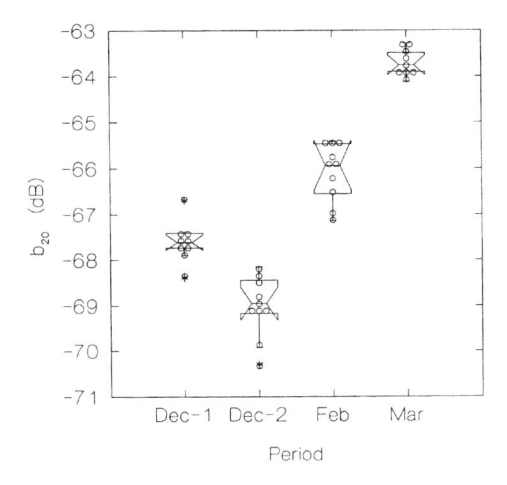

B: 120 kHz

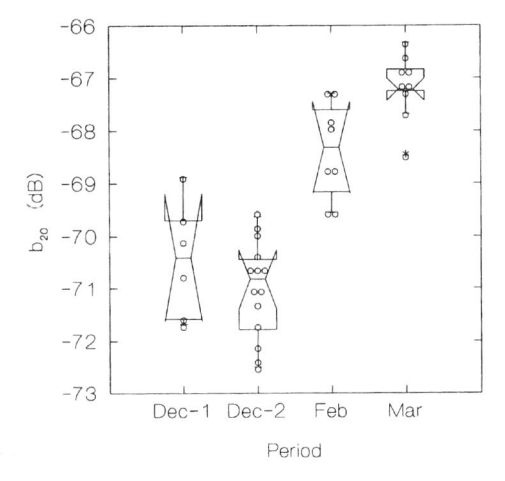

Figure 5. Notched box plot superimposed with actual data points (b_{20}) based on the hourly mean target strength of four groups of herring measured at three different sampling times. The non-overlapping notches among the three measurement periods indicate that there were significant increases in the target strength of herring over the period investigated.

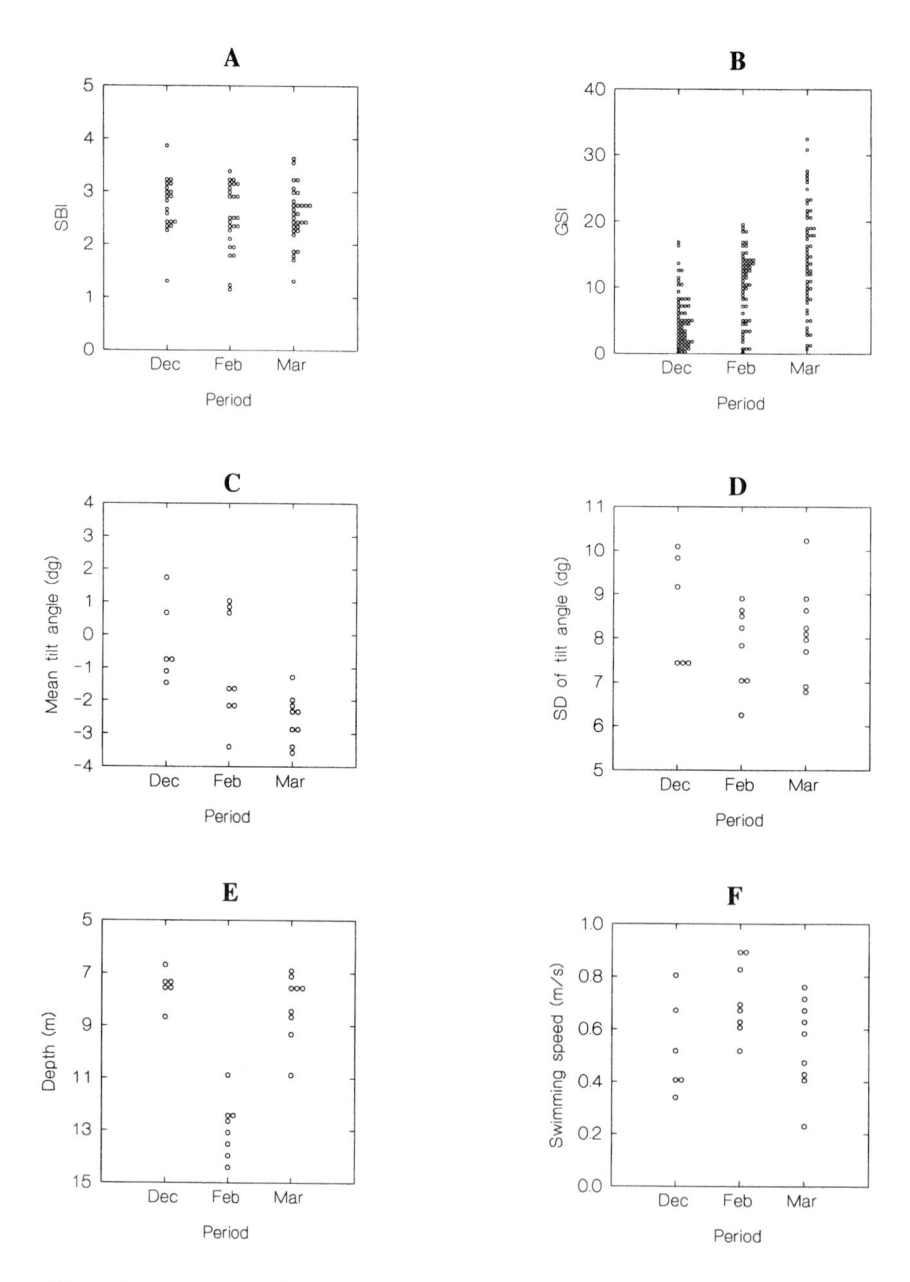

Figure 6. Variations of the five (A-E) potential factors that may cause the variation of the target strength of herring over the three measurement periods analyzed. The swimming speed of the fish is also shown (F). The swim bladder indexes (SBI, A) and the gonadosomatic indexes (GSI, B) were based on individual fish measurements, while the rest were the averages of many fish tracks. The depth (E) was calculated from the sea surface.

Table 5. Results and regression statistics for the analysis of target strength versus *GSI*, *Mean_tilt* and *SD_tilt*.

Variable	Coefficient (β)	S.E. (β)	Tolerance	$P(\beta=0)$
Constant	−69.0	1.3		<0.0005
GSI	0.24	0.05	0.60	<0.0005
Mean-tilt	−0.40	0.12	0.61	0.005
SD-tilt	−0.32	0.14	0.96	0.036

The subset model was in the form of $=\beta \mathbf{X} + \varepsilon$, where **X** denotes the predictor variable vector as listed in the first column, β denotes the regression coefficient vector, and ε is the normally distributed error term. Tolerance $= 1-r^2$, where r^2 is the squared multiple correlation between one predictor and other predictor variables included in the model. For the regression, $r^2 = 0.831$, S.E. (ε) = 0.69, $n = 23$, $P < 0.0005$.

with a standard deviation of 0.7 dB. The negative correlation between b_{20} and the two parameters of the tilt angle distribution is in accordance with the general knowledge on the effects of these two parameters (Foote 1980b). The positive correlation between b_{20} and *GSI* is thus substantiated, and about 2.2 dB out of the observed 3.2 dB increase in b_{20} at 120 kHz may thus be attributed to the development of the gonad over the 3-month period.

A similar regression exercise was also performed on the 38-kHz data, and a similar, positive correlation between b_{20} and *GSI* was also measured here. However, a small but possible discrepancy in the target strength data at this particular frequency due to a detected ping interval effect (see Ona et al. 1996, Zhao 1996) prevents us from suggesting a new equation here, before further analysis of the next 2 years of data. For target strength at 18 kHz, due to the lack of data in December 1995, the comparison cannot be made in the same periods as done at the other two frequencies. The target strength of herring at 18 kHz did also experience an increase from February to March 1996, but the trends seen at the higher frequencies are not as clear.

Discussion

One of the key questions when investigating target strength variability is the calibration accuracy, as compared to the expected and observed variations. A large effort was spent in the initial phase of the project to ensure that the highest possible accuracy was obtained throughout the measurements. The highest level of consistency between calibrations was obtained at the 120-kHz channel, where the system gain only dropped 0.2 dB (4.7%) over 1 year. For the 18- and 38-kHz channels, a larger variability in system gain was observed, 0.7 dB at 18 kHz, and 0.4 dB at 38 kHz, when calibrations at similar transmit powers were compared (Ona and Svellingen 1998). Calibration parameters may change as a result of temperature effects on the transducer and sound velocity. The measured variability during the study period is well within expected values. Temperature variation between the measurement periods ranged from 3.5° to 15.2°, and the sound velocity from 1,460 to 1,501 m per second, measured at a depth of 5 m. The actual calibration prior to measurement was used in each period, and periodic variation is therefore compensated for in the herring data. Measured deviations in the time varied gain (TVG) function at short range, due to inaccurate range measurement for the 38- and 18-kHz channels, were compensated for as described by Ona et al. (1996). Other possible error sources are related to the single target recognition system; the bias occurring when falsely accepting multiple targets; and to signal-to-noise-related problems which decrease the acceptance probability of small targets in the outer edge of the beam. Careful selection of recorded data sets, target tracking, and statistical analysis of detection probability over the acoustic beam was undertaken in order to minimize these errors. Detailed analysis and discussions of the errors in these measurements was reported by Zhao (1996) and in a more general manner in Ona (1999). From the error analysis it is concluded that these first measurement series have an overall measurement accuracy of about ±0.3 dB (7%) at the three frequencies used. For further improvement of this accuracy, later series were measured with a new narrow beam transducer (7°) at 38 kHz, a new, designated calibration sphere for 18 kHz, and improved digital resolution at all frequencies.

Since the measured herring target strength varied well outside the overall calibration accuracy the measured variability was assumed to be real. Several important findings are therefore discussed in relation to "established knowledge" for herring, as follows.

It is generally believed that the target strength of herring, or physostomous fish in general, should be lower than that of a physoclistous fish, such as cod. Abundant empirical evidence supports this (Foote 1987). Surprisingly perhaps, the target strength of herring measured during this experiment was very high. For herring of mean length about 31-33 cm, the calculated b_{20} ranges from –63.6 to –68.9 dB. These figures are considerably higher than the *TS*-length relationship currently applied in the acoustic abundance estimation of the Norwegian spring-spawning herring, or those actually recommended for physostomous and physoclistous fish in echo integration surveys (Foote 1987, Foote and Traynor 1988, Rose and Porter 1996).

Since the swim bladder is responsible for 90-95% of the total reflected energy from a fish (Foote 1980a), it may be interesting to compare the swim bladder index of herring with that of a gadoid fish. Taking the herring sampled from the storage pen as an example, the mean *SBI* was 2.64. One of the rare examples for gadoid fish is the reconstructed swim bladder of pollock (*Pollachius pollachius*) given in Foote (1985). The mean *SBI* here as 1.91. If the *SBI* can also be used to compare the intrinsic acoustic property between distinct species, the above comparison suggests that the target strength of our herring should not be less than that of a gadoid fish, at least at surface level. This was also partially supported by the findings of Nakken and Olsen (1977) in a classical tethered fish measurement where similar target strength was found for juvenile (6-12 cm) gadoids (cod and saithe) and clupeoids (herring and sprat). However, Nakken and Olsen also concluded that for bigger fish the target strength of clupeoids was lower than that of gadoids. This is probably due to the different swim bladder shapes of the two types of fish. For adult fish, the gadoid swim bladder generally has a lower aspect ratio (i.e., shorter but broader), with the effective reflecting dorsal area relatively larger, hence with higher target strength. Our findings at least suggest that the mean target strength of herring at surface levels is high, and at about the same level as published data for gadoids. A relatively long swim bladder seems to compensate for the narrower width, and a more organized orientation pattern (low spread) is likely to produce higher mean target strength.

Research aimed at investigating the effects of various physiological factors on the target strength of fish revealed the swim bladder of cod (*Gadus morhua*) to be drastically deformed by the gonads (Ona 1990). In particular, deformation was stronger by the ovaries, which are less flexible than the testes. The pressure from the gonads reduced the length and the posterior width of the swim bladder, making it a smaller acoustic target. If the gonads of herring affect the swim bladder in a similar way, a progressive reduction in the mean target strength of herring should be expected while the fish is maturing. The opposite, however, was observed during this study; and about 2.2 dB increase in the mean target strength can be attributed to the developing gonad during the period from December to March. This figure, however, was not based on morphological studies as it was for cod (Ona 1990); rather, it was the result of a regression analysis based on actually measured target strength data and other factors that may affect the target strength of herring simultaneously.

Since the swim bladder is the primary organ that is responsible for the reflected energy from fish (Foote 1980a, 1985; Furusawa 1988), any change in the swim bladder size should have an effect on the target strength. Although the length normalized target strength or b_{20} was used in the regression analysis, differences, if any, in the degree of inflation of the swim bladders were not accounted for by this normalization. In order to eliminate the size dependence of the target strength of herring, the relative size of the swim bladder should also be included in the analysis.

The most frequently used parameter is the percentage swim bladder volume per fish body volume, often expressed as the percentage swim bladder volume per unit of fish weight. However, although this parameter is most useful as a buoyancy index, it is a poor measure of the relative size of the swim bladder. This is because the fish body weight may be subject to large variation due to feeding or reproduction. Therefore, the swim bladder index (*SBI*) was used as a gauge for the relative size of the swim bladder. When the variable-weight parts of the fish, or the stomach content and maturation stage, are similar for different individuals, *SBI* can also be used as a buoyancy index. This is because *SBI* is simply the product of the swim bladder volume per unit of body weight and the condition factor of the fish.

The volume of the swim bladder, hence *SBI*, varied greatly among individuals. The largest individual *SBI* were about 3 times as high as the smallest *SBI* in each group (Table 5). Large variation in the swim bladder volume is commonly reported in herring (Blaxter and Batty 1984, Ona 1984), probably due to the variable fat content and the time of last replenishment of the swim bladder (Blaxter and Batty 1984). The final exclusion of *SBI* from the regression analysis, however, was because the mean *SBIs* were not significantly different among the three groups of fish analyzed, and therefore lacked the contrast needed as a predictor variable.

When reporting the dorsal aspect target strength of fish, the tilt angle is also very important (Nakken and Olsen 1977). In describing the tilt angle dependence of the mean target strength of an ensemble or a group of fish, the mean and the standard deviation of the tilt angle distribution of the fish are commonly used (Nakken and Olsen 1977; Foote 1980b, 1980c, 1985). This is because the target strength PDF is the convolution of the target strength directivity pattern and the tilt angle distribution of the fish.

In an attempt to eliminate tilt-angle-induced variation in the observed target strength, the mean and the standard deviation of the tilt angle distribution were included in the regression analysis. As expected, both were significantly correlated with the mean target strength of fish, and each took the correct sign (i.e., the correct direction of influence). However, the precision of the dependency, as reflected in the regression coefficients of these two variables in final model, is not known. The tilt angle distributions used in the regression analysis were based on the movements of the fish within the beam, as sensed by the split-beam transducer. A necessary assumption underlying this method was that the vertical swimming direction of the fish was the same as the longitudinal axis of the fish. Video analysis confirmed that the measured swimming angles corresponded well with the tilt angles in herring (Ona 2001). With our method, small deviation in each individual tilt measurement may further be expected from small wave movements of the floating transducer platform, and from the maximum digital resolution of the 120-kHz system, 3 cm. The general distribution pattern, however, shows a striking resemblance to photographically estimated tilt angle distributions in herring.

It is seen from Table 5 that the tolerance of the first two predictor variables, *GSI* and mean tilt, were relatively low. This was because these two variables were negatively correlated. The interaction term between these two variables was almost significant ($P = 0.052$), but its tolerance was extremely low (<0.05), therefore it was rejected for the final model.

However, the presence of a negative correlation between *GSI* and the tilt angle of the fish is an interesting phenomenon. Whether this happened by chance or there is a causal link between the two variables is not known. A shift in the center of gravity as well as a changed hydrodynamic flow over the fish body may occur in ripe herring, but such speculation requires corroboration by detailed analysis of swimming herring at different speeds.

Due to the absence of a gas-producing organ in the swim bladder of herring, the effect of pressure on target strength is expected to be large for herring (Ona 1984, 1990). Edwards and Armstrong (1983) found 2 to 3 dB reductions in the mean target strength when the encaged herring were lowered from 17.5 to 47.5 m of depth. Olsen and Ahlquist (1989) found similar depth effects in both cage and in situ experiments. How much of the reduction was due to swim bladder compression and how much was due to behavior differences is unknown. On the other hand, Reynisson (1993) was unable to detect any significant depth dependency in target strength of the Iceland herring within a large data set consisting of some 30 in situ measurements. The depth effect on the target strength of herring has been further analyzed experimentally and in situ during this experiment, but will be published later (Ona et al. 2001; Ona and others, submitted manuscript). Within the experimental pen at surface level, however, where herring was mostly distributed from 5 to 15 m of depth, the pressure term was not found to be significant ($P < 0.05$).

In a similar manner, it is believed that the swim bladder and the fat content are two important and complementary factors, which govern the buoyancy status of herring. Several authors (e.g., Brawn 1962; Ona 1984, 1990) have found that the fat content of herring bears a close negative correlation with the size of the swim bladder. Reynisson (1993) was also able to relate the fat content to the target strength of herring, and his tentative conclusion was that the target strength is lowered by about 0.2 dB for each 1% increase in fat content of the fish. However, no significant correlation between the swim bladder index and the fat content of the fish was found in this study. This may suggest that the intrinsic scattering property of our herring was not necessarily altered by their fat content, at least within the observed limits of this parameter (0.7-19.3%). It may also indicate that our methodology for retrieving the net for swim bladder volume measurements has been too rapid, or that the herring have adapted their swim bladder at different depths within the pen. Controlled acoustic measurements on surface-adapted individual herring may be needed to fully understand the indirect effect of fat on swim bladder volume and further on target strength.

When the effects of other factors, including the mean and standard deviation of the tilt angle distribution, were removed from the model, it was found, rather surprisingly, that the effect of the gonad development was to significantly increase the target strength of herring. It is thus hypothesized that, during gonad development, the swim bladder is steadily squeezed from its ventral size. If the swim bladder volume is constant, as was the case in this study, the apparent or effective dorsal area of the swim bladder, hence the target strength of the fish, may therefore increase during the gonad development period. This was observed for shock-frozen specimens with large gonads. The dissected section showed that the swim bladder was shallower and flatter than on herring without gonads. Shock-frozen samples were taken for both mature and immature fish to study this effect on swim bladder shape.

Whether we will find the same gonad effect in the field is, however, uncertain, and will depend on the in situ state of the swim bladder, which in turn may depend on the depth and time of observation during the gonad development period. New experimental measurements (not published) and in situ target strength measurements (Ona and others, submitted manuscript) show a clear depth dependency, but significantly lower than the one expected from free swim bladder compression theory. The magnitude of the observed gonad effect on target strength is in the same order of magnitude as the measured pressure effect over the depth range 10-300 m, and should therefore still be considered important in an expanded target strength relation for herring.

If acoustic surveys on herring in the future shall reflect the absolute abundance of herring, two important sources of bias must be removed or corrected for. These are the effect of vessel avoidance (Olsen 1990, Vabø 1999), and the effect of depth and seasonal variability in target strength. Results from repeated measurements of the Norwegian spring-spawning herring stock within 1 year indicate that the present estimates are biased when factors affecting target strength variability are not accounted for.

Acknowledgment

The financial support of European research project EU (AIR-CT94-2142) to perform these measurements is greatly appreciated.

References

Blaxter, J.H.S., and R.S. Batty. 1984. The herring swimbladder: Loss and gain of gas. J. Mar. Biol. Assoc. U.K. 64:441-459.

Bodholt, H., H. Nes, and H. Solli. 1989. A new echo-sounder system. Int. Conf. Progress in Fisheries Acoustics, Lowestoft. Proc. I. O. A., St Alban, U.K. 11(3):123-130.

Brawn, V.M. 1962. Physical properties and hydrostatic function of the swimbladder of herring (*Clupea harengus* L.). J. Fish. Res. Board Can. 19(4):635-655.

Degnbol, P., H. Lassen, and K.J. Stæhr. 1985. In-situ determination of target strength of herring and sprat at 38 and 120 kHz. Dana 5:45-54.

Edwards, J.I., and F. Armstrong. 1983. Measurement of the target strength of live herring and mackerel. FAO Fish. Rep. 300:69-77.

Foote, K.G. 1980a. Importance of the swimbladder in acoustic scattering by fish: A comparison of gadoid and mackerel target strengths. J. Acoust. Soc. Am. 67:2084-2089.

Foote, K.G. 1980b. Effect of fish behaviour on echo energy: The need for measurements of orientation distributions. J. Cons. Int. Explor. Mer 39(2):193-201.

Foote, K.G. 1980c. Averaging of fish target strength functions. J. Acoust. Soc. Am. 67(2):504-515.

Foote, K.G. 1982. Optimizing copper spheres for precision calibration of hydroacoustic equipment. J. Acoust. Soc. Am. 71:742-747.

Foote, K. G. 1985. Rather-high-frequency sound scattering by swimbladdered fish. J. Acoust. Soc. Am. 78:688-700.

Foote, K.G. 1987. Fish target strengths for use in echo integrator surveys. J. Acoust. Soc. Am. 82:981-987.

Foote, K.G., and D.N. MacLennan. 1984. Comparison of copper and tungsten carbide spheres. J. Acoust. Soc. Am. 75:612-616.

Foote, K.G., and E. Ona. 1985. Swimbladder cross sections and acoustic target strengths of 13 pollock and 2 saithe. Fiskeridir. Skr. Ser. Havunders. 18:1-57.

Foote, K.G., and T.J. Traynor. 1988. Comparison of walleye pollock target strength estimates determined from in situ measurements and calculations based on swimbladder form. J. Acoust. Soc. Am. 83:9-17.

Foote, K.G., A. Aglen, and O. Nakken. 1986. Measurement of fish target strength with a split-beam echo sounder. J. Acoust. Soc. Am. 80:612-621.

Furusawa, M. 1988. Prolate spheroidal models for predicting general trends of fish target strength. J. Acoust. Soc. Jpn. (E) 9:3-24.

Halldórsson, O. 1983. On the behaviour of the Icelandic summer spawning herring (*C. harengus* L.) during echo surveying and depth dependence of acoustic target strength in situ. ICES C.M. 1983/H:36.

Hamre, J., and A. Dommasnes. 1994. Test experiments of target strength of herring by comparing density indices obtained by acoustic method and purse seine catches. ICES C.M. 1994/B:17.

Huse, I., and E. Ona. 1996. Tilt angle distribution and swimming speed of overwintering Norwegian spring spawning herring. ICES J. Mar. Sci. 53:863-873.

ICES. 1988. Report of the Atlanto-Scandian herring and capelin working group. ICES C.M. 1988/Assess:10.

ICES. 1996. Report of the northern pelagic and blue whiting fisheries working group. ICES C.M. 1996/Assess:14.

Johannessen, A., A. Slotte, O.A. Bergstad, O. Dragesund, and I. Røttingen. 1995. Reappearance of Norwegian spring spawning herring (*Clupea harengus* L.) at spawning grounds off southwestern Norway. In: H.R. Skjoldal, C. Hopkins, K.E. Erikstad, and H.P. Leinaas (eds.), Ecology of fjords and coastal waters. Elsevier Science B.V., pp. 347-363.

Kautsky, G.A., N.A. Lemberg, and E. Ona. 1991. In situ target strength measurements of Pacific herring (*Clupea harengus pallasi*) in the eastern Strait of Georgia using dual-beam and split-beam sonar. In: Proceedings of the International Herring Symposium. University of Alaska Sea Grant, AK-SG-91-01, Fairbanks, pp. 163-183.

Lassen, H., and K.J. Stæhr. 1985. Target strength of Baltic herring and sprat measured in situ. ICES C.M. 1985/B:41.

Losnegard, N., B. Bøe, and T. Larsen. 1979. Undersøkelse av ekstraksjon-midler for bestemmelse av fett. Fiskeridirektoratet, Rapporter og Meldinger 1979(1). 4 pp. (In Norwegian.)

MacLennan, D.N. 1981. The theory of solid spheres as sonar calibration targets. Scott. Fish. Res. Rep. 22. 17 pp.

MacLennan, D.N. 1990. Acoustical measurement of fish abundance. J. Acoust. Soc. Am. 87:1-15.

McGill, R., J.W. Tukey, and W.A. Larsen. 1978. Variations of box plots. The American Statistician 32:12-16.

Nakken, O., and K. Olsen. 1977. Target strength measurements of fish. Rapp. P.-V. Reun. Cons. Perm. Int. Explor. Mer 170:52-69.

Olsen, K. 1990 Fish behaviour and acoustic sampling. Rapp. P.-V. Reun. Cons. Int. Explor. Mer 189:147-158.

Olsen, K., and I. Ahlquist. 1989. Target strength of fish at various depth, observed experimentally. ICES C.M. 1989/B:53.

Ona, E. 1984. In situ observations of swimbladder compression in herring. ICES C.M. 1984/B:18.

Ona, E. 1990. Physiological factors causing natural variations in acoustic target strength of fish. J. Mar. Biol. Assoc. U.K. 70:107-127.

Ona, E. (ed.). 1999. Methodology for target strength measurements. ICES Coop. Res. Rep. 235. 59 pp.

Ona, E. 2001. Herring tilt angles, measured through target tracking. In: F. Funk, J. Blackburn, D. Hay, A.J. Paul, R. Stephenson, R. Toresen, and D. Witherell (eds.), Herring: Expectations for a new millennium. University of Alaska Sea Grant, AK-SG-01-04, Fairbanks. (This volume.)

Ona, E., and D. Hansen. 1991. Software for target tracking with split-beam echo sounders: User manual. Institute of Marine Research, Bergen, Norway, Oct. 1991.

Ona, E., and I. Svellingen. 1998. Improved calibration of split beam echo sounders. ICES Fisheries and Acoustics, Science and Technology (FAST) Working Group, La Coruna, Spain, April 1998. 20 pp.

Ona, E., I. Svellingen, and J.E. Fosseidengen. 2001. Target strength of herring during vertical excursion. ICES Fisheries and Acoustics Science and Technology (FAST) Working group, Seattle, April 2001. 16 pp.

Ona, E., X. Zhao, I. Svellingen, and K.G. Foote. 1996. Some pitfalls of short-range standard-target calibration. ICES C.M. 1996/B:36.

Reynisson, P. 1993. In situ target strength measurements of Icelandic summer spawning herring in the period 1985-1992. ICES C.M. 1993/B:40.

Rose, G.A., and D.R. Porter. 1996. Target-strength studies on Atlantic cod (*Gadus morhua*) in Newfoundland waters. ICES J. Mar. Sci. 53:259-265.

Røttingen, I. 1990. The 1983 year class of Norwegian spring spawning herring as juveniles and recruit spawners. In: T. Monstad. (ed.), Biology and fisheries of the Norwegian spring spawning herring and blue whiting in the northeast Atlantic. Proceedings of the Fourth Soviet-Norwegian Symposium, Bergen, 12-16 June 1989, pp. 165-203.

Røttingen, I., K.G. Foote, I. Huse, and E. Ona. 1994. Acoustic abundance estimation of wintering Norwegian spring spawning herring with emphasis on methodological aspects. ICES C.M. 1994/(B+D+G+H):1.

Rudstam, L.G., T. Lindem, and S. Hansson. 1988. Density and in situ target strength of herring and sprat: A comparison between two methods of analysing single-beam sonar data. Fish. Res. 6:305-315.

Slotte, A. 1996. Relations between seasonal migrations and fat content in Norwegian spring spawning herring (*Clupea harengus* L.). ICES C.M. 1996/H:11.

Slotte, A. 1998. Spawning migration of Norwegian spring spawning herring (*Clupea harengus* L.) in relation to population structure. Dr. scient. thesis. University of Bergen, Norway.

SIMRAD Norge. 1996. SIMRAD EK500 Scientific Echo Sounder. Instruction manual, P2172E.

SYSTAT, Inc. 1992. SYSTAT for windows: Statistics, Version 5 edition. SYSTAT, Inc., Evanston, Illinois. 636 pp.

Toresen, R., and P. Barros. 1995. Acoustic estimates of abundance-at-age of juvenile Norwegian spring-spawning herring (*Clupea harengus* L.) in the Barents Sea from 1983 to 1993. In: P. Barros, Quantitative studies on recruitment variations in Norwegian spring- spawning herring (*Clupea harengus* Linnaeus 1758), with special emphasis on the juvenile stage. Dr. scient. thesis, University of Bergen, Norway, pp. 19-38.

Vabø, R. 1999. Measurement and correction models of behaviourally induced biases in acoustic estimates of wintering herring. Dr. Scient. thesis. University of Bergen, Norway.

Zhao, X. 1996. Target strength of herring (*Clupea harengus* L.) measured by the split-beam tracking method. M. phil. thesis, University of Bergen, Norway.

Estimates of Egg Loss in Pacific Herring Spawning Beds and Its Impact on Stock Assessments

Jake Schweigert and Carl Haegele
Fisheries and Oceans Canada, Pacific Biological Station, Nanaimo, British Columbia, Canada

Abstract

Pacific herring (*Clupea pallasii*) spawning grounds were assessed for egg loss during the incubation period in Barkley Sound on the west coast of Vancouver Island in 1988 and in Georgia Strait in 1989 and 1990. Estimates of the instantaneous egg loss rate based on counts of eggs in samples from 1988 were variable but data collected in 1989 and 1990 indicate an instantaneous rate of about 0.10 per day. General linear models were fitted to the data and indicated significant effects of vegetation type, depth, and bird abundance on observed egg numbers. However, bird abundance and sample depth did not appreciably affect the estimated variation accounted for by the models. Similar models were fitted to estimates of egg density using a statistical model to predict egg numbers from visual assessments of the number of egg layers, the proportion of the sampling quadrat filled with vegetation, and the dominant type of vegetation in the sampling plot. Results from this analysis were similar to those from the counted samples but suggested a more conservative egg loss rate of 0.05 per day. The estimated egg loss rates were used to investigate the impact of adjusting spawning biomass levels in southern British Columbia during 1999 for egg loss. Adjusted biomass estimates were not significantly different from the observed levels in Georgia Strait but did differ on the west coast of Vancouver Island. It was concluded that surveys of spawning grounds should be conducted as quickly as possible following the commencement of spawning to minimize bias in the estimates of stock biomass.

Introduction

The identification and quantification of the factors which affect the survival of Pacific herring eggs on the spawning grounds has a long history dating back to the late 1920s (Munro and Clemens 1931, Hart and Tester 1934; Cleaver and Franett 1946; Outram 1958; Haegele and Schweigert 1989, 1991; Rooper et al. 1999). Most of the earlier studies focused on the impacts of bird predation and speculation about the remedial effects on herring stocks of predator control. More recent work has tended to focus on the importance of abiotic factors such as exposure and storm effects on egg loss as well as the impacts of non-avian predation from invertebrates and whales (Hay and Miller 1982; Haegele, 1993a,b; Rooper et al. 1999). The objectives of much of the recent research in this area has been to investigate the linkages between survival during the early life history stages and subsequent recruitment to the adult stocks (Hay and Kronlund 1987, Schweigert 1995, Zebdi and Collie 1995).

In addition, there has been a very practical interest in the extent and magnitude of herring egg loss over the course of the incubation period as it relates to surveys of the spawning grounds used in annual assessments of spawning stock biomass (Haegele and Schweigert 1989,1991; Haegele 1991). Spawning ground surveys are used to estimate egg density which may be combined with information on spawning bed size to estimate total egg deposition and consequently infer spawning biomass for each stock area (Schweigert and Fournier 1982, Schweigert et al. 1985). If egg loss prior to these spawn surveys is sufficient to alter visual estimates of egg density, then this would result in underestimates of spawner biomass and possibly overly conservative harvesting practices. Secondarily, is the question of whether visual assessments of the egg deposition are accurate enough to detect a true decline in egg numbers over the incubation period and thereby require adjustment to improve the accuracy of the estimates of stock abundance.

Egg loss may be determined in two ways. Spawn can be sampled throughout the incubation period to determine if there is a decrease in egg density with time. Alternately, sources of egg loss can be identified and the degree of egg loss can be estimated. In this study, we report on the data collected from surveys over the course of the incubation period in two areas of southern British Columbia. These surveys were conducted over a relatively large geographical area so that results would have general applicability. Recently, Rooper et al. (1999) conducted similar surveys in Alaska and reported that average depth, cumulative time of exposure to air, location and exposure to weather, type of substrate, type of kelp, and bird abundance were all important factors in determining the rate of egg loss over the incubation period. In this paper we report on the effects of some of these factors in British Columbia and discuss the implications to the accuracy and precision of the resulting stock assessments of Pacific herring.

Materials and Methods

Surveys of Pacific herring spawn were conducted in 8 locations in 1988, 4 of which were on the Vancouver Island shoreline and 4 on smaller islands in northern Barkley Sound (Fig. 1). Similar surveys were conducted in the

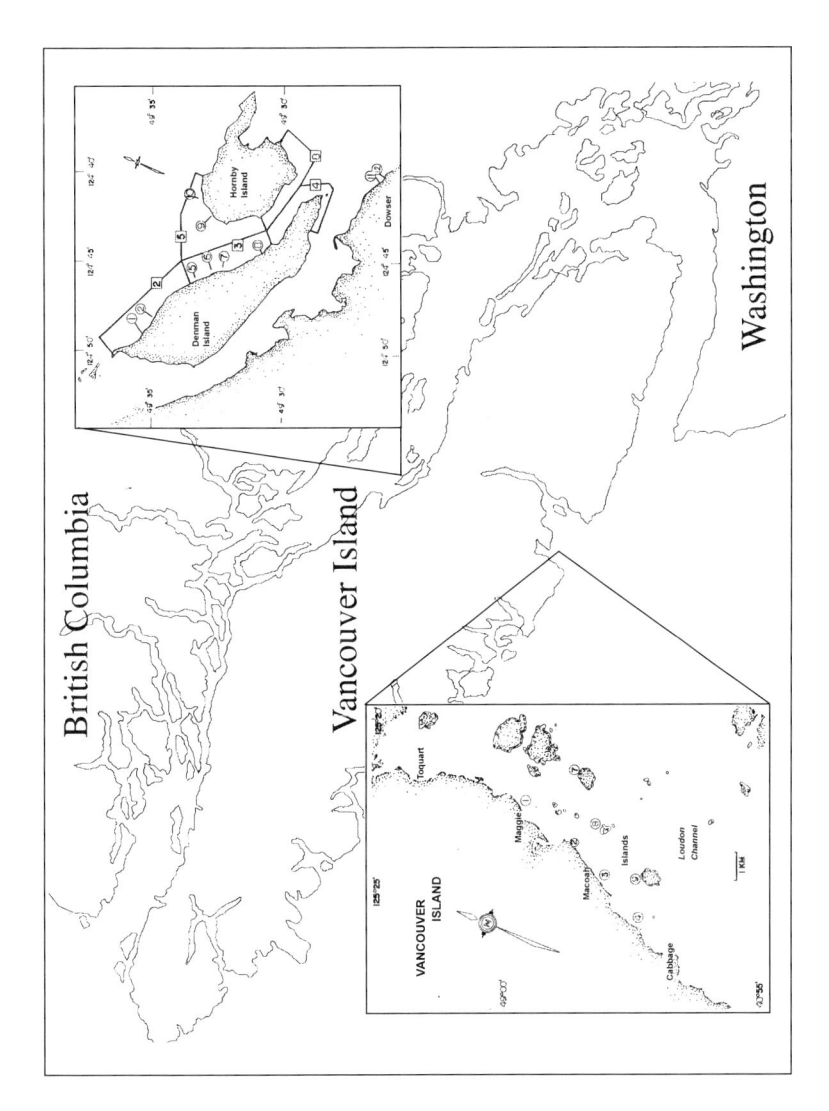

Figure 1. Map of Vancouver Island showing sampling locations in the Strait of Georgia and west coast of Vancouver Island (Barkley Sound) for the egg loss studies of 1988-1990.

Strait of Georgia (Fig.1) in 1989 (8 locations, 6 on Denman and 2 on Hornby Islands) and 1990 (6 locations, 2 on Denman, 2 on Hornby, and 2 on Vancouver Islands). Standard visual observations of the type used to estimate egg density from scuba surveys of herring spawn (Schweigert and Fournier 1982) were made and vegetation with attached eggs from 0.5 m^2 quadrats was sampled for egg enumeration. Predators of herring eggs were also identified and enumerated.

Barkley Sound

In 1988 a series of transects on historical herring spawning beds in northern Barkley Sound were identified using gillnet leadline, which was set perpendicular to the shore and deployed before spawning. About 5 sampling stations were equidistantly spaced along each transect over the depth range usually utilized by herring for spawning. Thus, stations were 10-40 m apart, depending on the slope of the bottom. Sampling sites were marked at these stations with 5 m long pennants attached to and perpendicular to the transect leadline. Pennants were flagged at 1 m intervals with surveyor's tape. Sample plots were 0.5 m^2 quadrats, a square with 0.7 m sides. One side of the quadrat was placed along the pennant with one corner at the meter mark and the other side extending away from the transect line. The quadrat was then flipped once shorewardly in an effort to make quadrat placement non-selective.

Transects were sampled on alternate days over 12 days of the incubation period after which some hatching may be expected to occur. On the first sampling day, visual observations were made at all sample sites and the rooted or attached vegetation in the quadrat at the first sample site at each station, and the adhering eggs, were removed for further analysis. On the second sampling day, visual observations were made at the remaining sample sites and the second plot at each station was harvested. This was repeated up to the last sampling day. Visual observations made by the divers were an estimate of the percent cover of vegetation (the proportion of the quadrat covered by vegetation), predominant vegetation type (seagrass, rockweed, kelp, sargassum, foliose red algae, filamentous red algae), plant height, number of egg layers on each vegetation type and on the bottom substrate, and depth. Observed depth was corrected to chart data to account for tidal fluctuation.

Harvested vegetation and eggs were separated by vegetation type into fractions, which were weighed, and weighed subsamples preserved in Gilson's fluid. The preserved samples were processed by immersion in alcoholic (25% by volume) 1N KOH at 40°C for approximately 1 hr to liberate the eggs. This solution dissolves the glue by which herring eggs are attached. Eggs were then preserved in 10% formalin for approximately 1 week. The preservative was vacuum extracted, the eggs weighed, and two aliquots of approximately 200 to 400 eggs removed and weighed, and the eggs counted. The number of eggs on each vegetation type was then calculated and eggs per m^2 was twice the sum of these estimates. Each sample was assigned to a dominant vegetation type based on which type had the most eggs attached. Eggs on bottom substrate were estimated from egg layer observations (1 layer = 340,000 eggs per m^2, Haegele et al. 1979).

Georgia Strait

In both 1989 and 1990, herring spawn was sampled on transects marked with gillnet leadline, laid perpendicular to shore. Problems with wave action on the placement of the pennants used in 1988 resulted in their discontinuance. In an attempt to increase precision on egg loss estimates at least 10 samples were collected on each transect and the maximum distance between sampling stations was 20 m. Hence, spacing between sampling stations and the number of stations per transect depended on spawn width; minimum distances between stations were 5 m and the maximum number of stations on a transect was 31. Each transect was resampled on every second day up to day 12 of the incubation period. In the 1990 surveys, the number of transects sampled was reduced to 4 in the Denman/Hornby Island areas and each transect was resampled every day up to day 14 of the incubation period. Sample plots in both years were 0.5 m^2 quadrats, the same as used in 1988. Quadrat placement was made non-selective by laying one side of the quadrat along the transect line and flipping the quadrat once to the left. Initially, we had planned to leave the transects in place over the course of the incubation period to standardize the sample collection and minimize the heterogeneity between sample collections at the "same" sites. However, theft of two transects lines early in the 1989 survey forced us to retrieve and reset the transect locations each sampling day. Transects were reset as close as possible to the same sampling location on each day but they were never close enough to be affected by previous sampling at the site.

As in 1988, divers made visual observations of the percent cover of vegetation (the proportion of the quadrat covered by vegetation), vegetation type, plant height, egg layers on vegetation and on bottom substrate, and depth at each station. Vegetation rooted within the quadrat and the attached eggs were removed for further analysis. Observed depth was corrected to chart data. Harvested vegetation and eggs were processed as described above to separate the eggs and aliquots counted to determine total number of eggs in each sample.

Statistical Analysis

In a recent study, Rooper et al. (1999) collected data on egg loss from a series of 0.1 m^2 sampling frames fixed in the spawning beds from Prince William Sound, Alaska. From these data they estimated the egg loss rate assuming a simple exponential decay and then conducted analysis of variance on the estimated egg loss rates and various environmental factors. Instead of assuming a function for the egg loss rate, we developed a series of general linear models to examine the effects of incubation time, water depth, substrate type, and bird abundance, on egg density in southern British Columbia herring spawning beds. Because our sampling procedures required resetting the transect lines each sampling day, it was not possible to compare data from the same sampling plot on consecutive days as was done in the Alaska study. Instead, we used the sampling transect or location as the sampling unit for monitoring egg density over time. Hence, we fit the following general model to the observations from each transect location for the data from 1988-1990 and to the combined data from each year:

$$\ln(Eggs/m^2) = \alpha - Z\ Incday + \beta\ Vtype + \gamma\ Depth + \delta\ Depth^2 + \lambda\ Birds + \varepsilon \quad (1)$$

where

Eggs per m² = estimated egg density per m² on transect *i* on day *j*,

Incday = incubation days or number of days following the beginning of spawning,

Depth = depth at which the sample was collected corrected to chart data,

Vtype = dominant type of vegetation in the sampling quadrat,

Birds = total number of birds observed in the sector surrounding the transect, and

Z = instantaneous rate of egg loss.

A series of versions of equation 1 were fitted by sequentially dropping a variable (*Birds, Depth²*, *Depth,* and *Vtype*) from the model and refitting (models 2-5). This provided a range of models from very simple to very complex to attempt to account for the variation in egg loss over the incubation period. Thus, model 1 includes all the variables, while model 2 includes model 1 minus birds, model 3 includes model 2 minus depth², and so forth. Stepwise regression also could have been applied to derive the statistically most significant model but this can result in spurious results given the limited number of data points in some instances so we opted to look at a priori determined models for each data set. The variables chosen for the analysis included those noted by previous authors as being important determinants of egg mortality. Thus, heavier spawn deposition appears to attract more birds, primarily gulls. In addition, Haegele et al. (1981) had demonstrated that egg density peaks near a depth of zero meters chart data and then decreases at greater or shallower depths so we fit both linear and quadratic terms to these data. Similarly, Schweigert (1993) demonstrated that there is a significant difference in the egg density found on different vegetation types. Another issue that arises is determining when the incubation period begins since the duration of spawning activity can vary among spawning locations from a few days to almost a week. We chose to use the date that spawning began as day 0 and enumerated incubation days subsequent to this date.

A question that arises from equation 1 is whether the visual observations collected by scuba divers provide similar estimates of egg loss rate to those determined from counts of eggs in sampled quadrats. In other words, does the visual diving survey provide an accurate assessment of the egg deposition or are adjustments to egg numbers necessary to provide unbiased estimates of egg deposition for assessing stock abundance. To examine this question, we refit the models (equation 1) to estimates of egg density calculated from the divers' visual assessments of the vegetation type, percent cover of vegetation, and egg layers in each sampling quadrat (equation 2) and compared the resulting egg loss rates with those from similar versions of equation 1 fit to the counted samples. For the visual observations, egg density was estimated from a general linear model fit to egg counts from vegetation samples collected over several years as described in Schweigert (1993). These estimates are referred to as the predicted egg density throughout and the equation which best described these data is:

$$Eggs_{ij} = 1033.6694 \; L_{ij}^{0.7137} \; P_{ij}^{1.5076} \; V_{ij} \; Q_j \qquad (2)$$

where
$Eggs_{ij}$ = estimated number of eggs in thousands per m² on vegetation type i in quadrat j,
L_{ij} = number of layers of eggs on algal substrate i in quadrat j,
P_{ij} = proportion of quadrat covered by algal substrate i in quadrat j,
V_{1j} = 0.9948 parameter for seagrasses in quadrat j,
V_{2j} = 1.2305 parameter for rockweed in quadrat j,
V_{3j} = 0.8378 parameter for flat kelp in quadrat j,
V_{4j} = 1.1583 parameter for other brown algae in quadrat j,
V_{5j} = 0.9824 parameter for leafy red and green algae in quadrat j,
V_{6j} = 1.0000 parameter for stringy red algae in quadrat j,
Q_1 = 0.5668 parameter for 1.00 m² quadrats,
Q_2 = 0.5020 parameter for 0.50 m² quadrats, and
Q_3 = 1.0000 parameter for 0.25 m² quadrats.

Finally, we applied the egg loss rates derived above to data collected during annual spawn deposition monitoring surveys in 1999 and present the results of adjusted and unadjusted mature biomass estimates as an example of the impact egg loss corrections could make on the estimate of total stock biomass. However, it is unknown how generally applicable these results may be since egg loss rate may change over time or between spawning areas. We used paired *t*-tests to compare the observed and adjusted biomass estimates for the Strait of Georgia and west coast of Vancouver Island herring sections.

Results

The estimated number of eggs per m² from counted samples for each transect in each year is plotted against the incubation day in Fig. 2 along

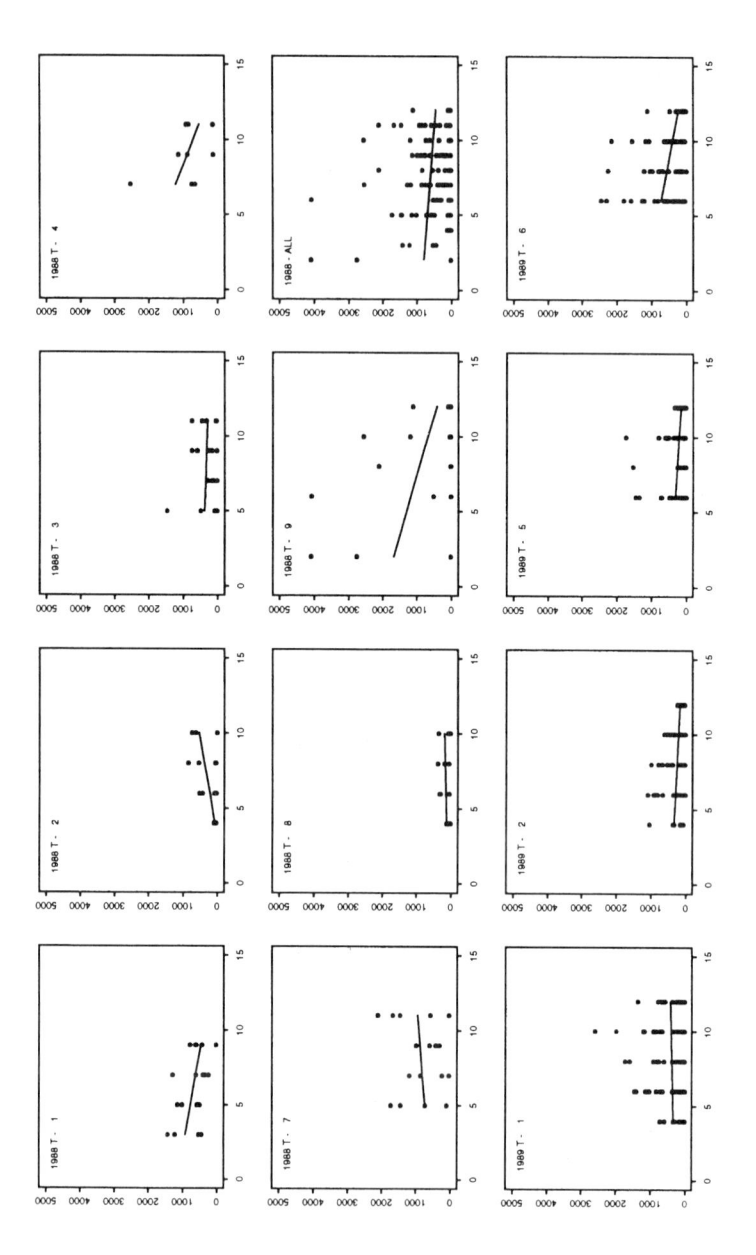

Figure 2. Egg density from egg counts observed over the course of the incubation period on transects from 1988 to 1990. Lines are the best linear fit to the data.

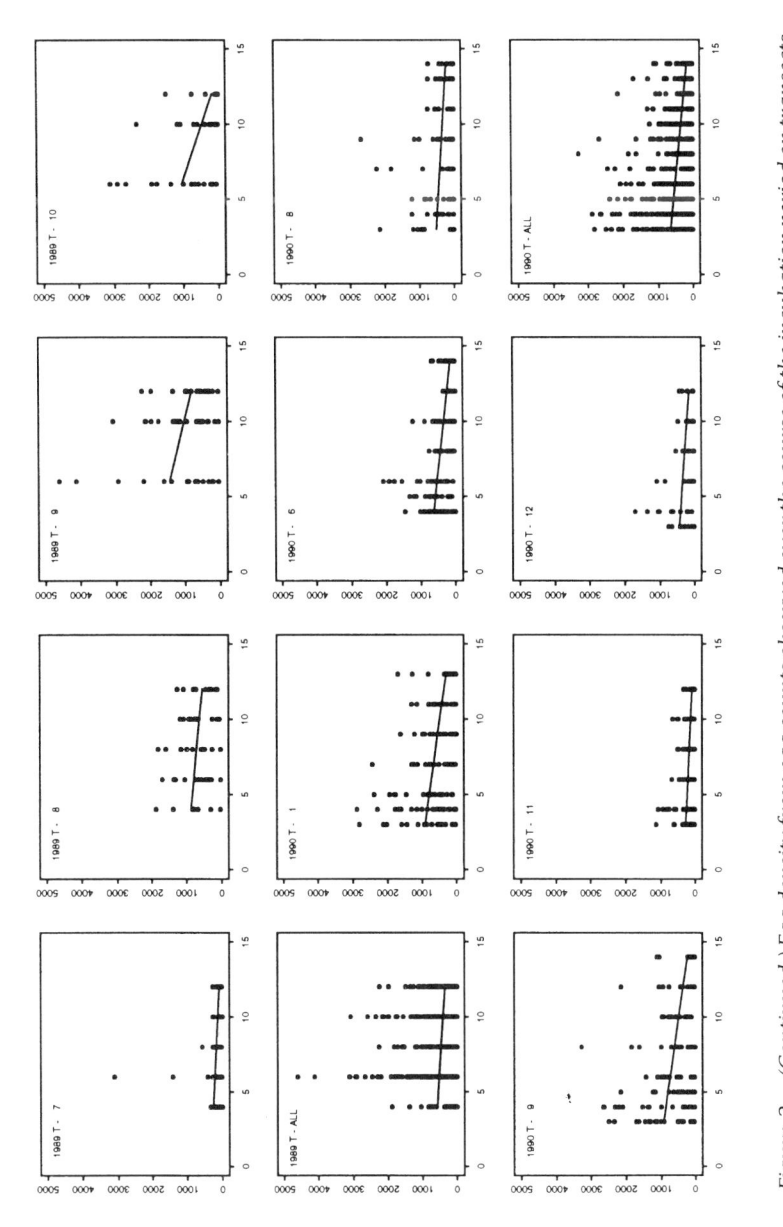

Herring: Expectations for a New Millennium 497

Figure 2. (Continued.) Egg density from egg counts observed over the course of the incubation period on transects from 1988 to 1990. Lines are the best linear fit to the data.

with the best fit linear regression line. It is evident from the plots that the data are quite variable and in a few instances show no evidence of a decline in egg numbers over time (1988: transects 2, 3, 7, and 8; 1989: transect 1). However, all three years do indicate similar rates of decline in egg numbers over time when all data for a year are combined together. The results for 1989 and 1990 also suggest similar rates of decline in egg numbers at each transect location (Fig. 2) although the egg density estimates from individual sampling plots are quite variable.

The logarithm of the estimated number of eggs per m^2 predicted from equation 2 is compared with the logarithm of the egg counts in sample quadrats from 1988-1990 in Fig. 3. The correspondence between the observed and predicted egg densities is quite good although there appears to be a slight tendency to overpredict at intermediate egg density. The model predictions of egg density for each transect in each year are plotted against the incubation day in Fig. 4 along with the best fit linear regression line. It is evident from these plots that the data are considerably more variable than for the estimates from egg counts. In fact, for some transects in all three years there are instances when the predicted egg numbers increase over time. For 1988, the combined egg estimates from all transects suggest an increase over time whereas for both 1989 and 1990 the amalgamated data suggest rates of decline in egg number similar to those determined from egg counts.

The results from fitting the five models described above to the egg loss data are presented in Table 1 and summarized for models 1-5 in Fig. 5. The results from the 1988 survey are very variable ranging from estimates of egg loss that are very highly negative (transect 4, model 1) to others which are positive. However, in very few cases are the results consistent between models and in most instances the results are not statistically significant. Even amalgamating all the data from 1988 together results in only model 1, indicating a decline in egg density over time whereas the other models all estimate a slight increase, yet all models are statistically significant. Apparently, there were insufficient samples collected to provide an accurate assessment of egg loss rate from this survey.

In 1989 the surveys included a similar number of transects as in 1988 but sampling began earlier in the incubation period, samples were collected more frequently on each transect, and the sampling of each transect occurred more often (every second day) over the course of the incubation period. The results in Table 2 and Fig. 5 indicate that estimates of the instantaneous egg loss are fairly consistent between models and transects at about 0.10 except for transects 6 and 10 where the egg loss rate is higher. The other interesting observation is that estimates of egg loss for model 1, which includes a term for bird abundance, are invariably higher than for the other models, which don't account for birds. In most instances, the estimated egg loss rate and the overall models for each transect are statistically significant. However, it appears that the addition of terms for depth and bird abundance do not account for markedly more of the variation in egg density than model 4 which includes only incubation day and vegetation substrate type.

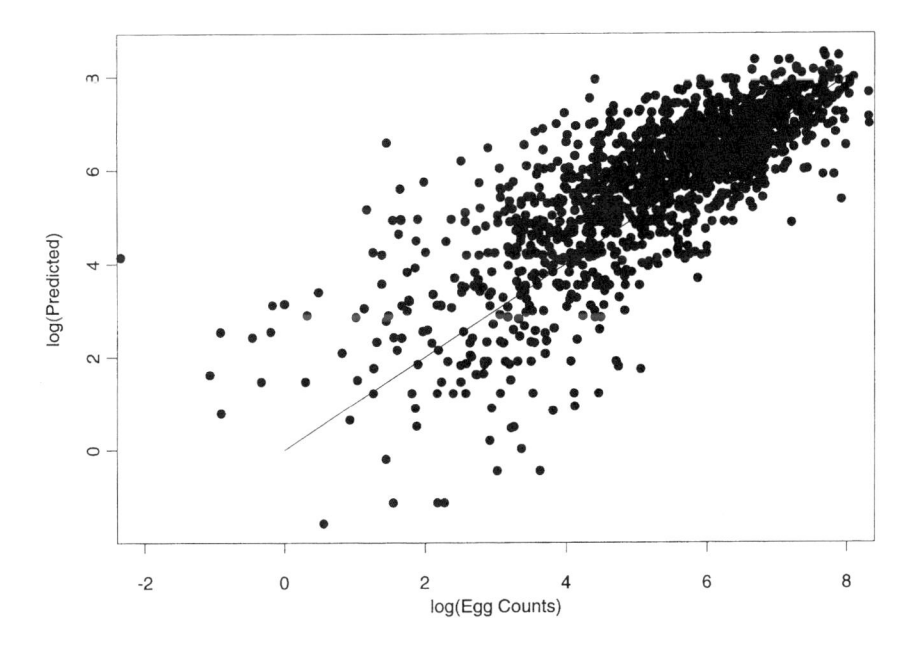

Figure 3. Comparison of logarithm of egg densities predicted from equation 2 and egg densities observed from counted samples taken for egg loss estimation from 1988-1990. Line is the 1:1 correspondence between the two variables.

In 1990, the survey was repeated in the Strait of Georgia but focused on only half the transects sampled in 1989. In addition, two transects, 11 and 12, were added on a later spawning bed outside the main study area. Sampling intensity was increased early in the incubation period when much of the egg loss appears to occur, and each transect was sampled daily immediately after spawning and then sampling was reduced to every second day. The results in Table 1 and Fig. 5 indicate instantaneous egg loss rates of about 0.10 (similar to those in 1989) for all transects except 6 and 11 where they are somewhat higher. Overall, results between transects appear slightly more variable than in 1989 and interestingly the egg loss rates estimated for model 1, which incorporates bird abundance, are lower than for the other models, the opposite result to 1989. Again, the estimated egg loss rates and model fits to the data are statistically significant in all instances but the addition of terms for depth and bird abundance do not markedly increase the total amount of variation in egg numbers explained. In both 1989 and 1990, the results for the combined data from all transects result in estimates of instantaneous egg loss rate of about 0.10.

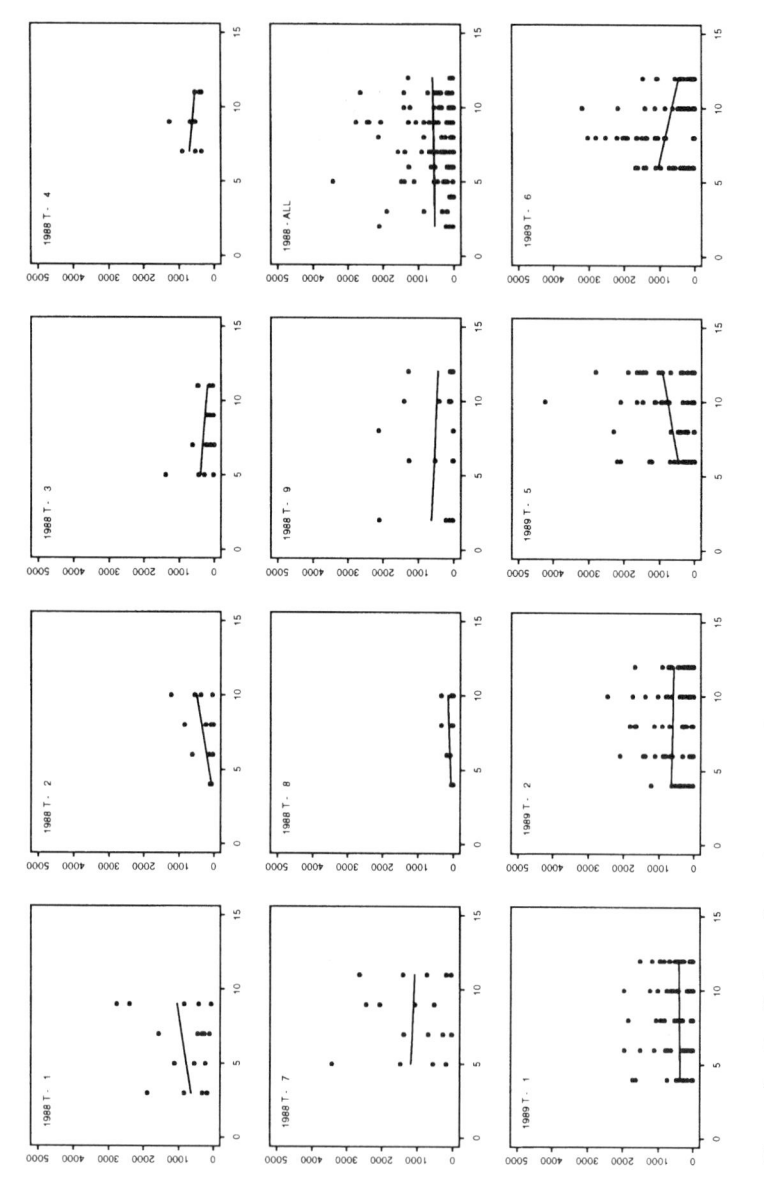

Figure 4. Predictions of egg density over the course of the incubation period at transects sampled from 1988 to 1990. Lines are the best linear fit to the data.

Herring: Expectations for a New Millennium 501

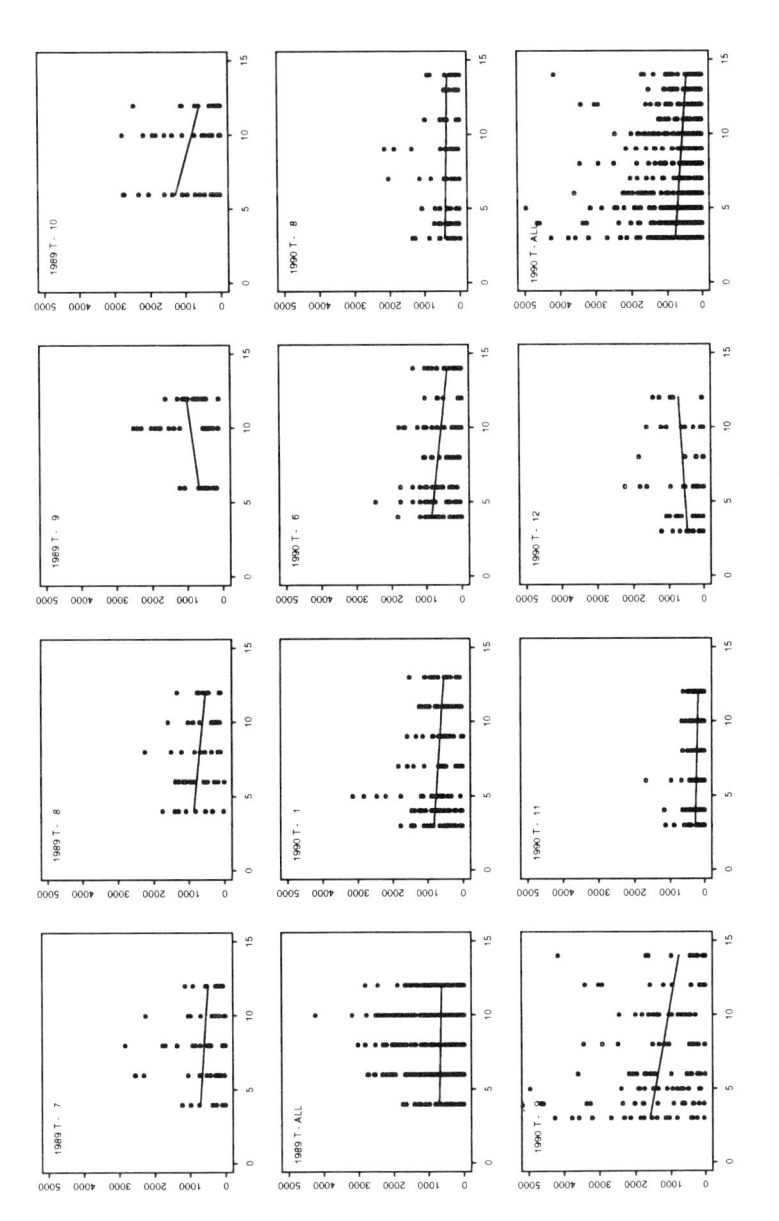

Figure 4. (Continued.) Predictions of egg density over the course of the incubation period at transects sampled from 1988 to 1990. Lines are the best linear fit to the data.

Table 1. Estimated instantaneous egg loss rates (–Z), variance explained (r^2), and significance levels for observed and predicted egg density data and five model formulations for each transect surveyed from 1988 to 1990 in southern British Columbia.

	Trans		Model 1[e] Obs	Model 1[e] Pred	Model 2 Obs	Model 2 Pred	Model 3 Obs	Model 3 Pred	Model 4 Obs	Model 4 Pred	Model 5 Obs	Model 5 Pred
1988	1	–Z	–0.28	–0.04	–0.10	0.13	–0.13	0.11	–0.15	0.04	–0.17[a]	0.01
	2	–Z	–0.27[b]	0.05	0.005	0.03	0.001	0.02	0.02	0.06	0.23	0.18
	3	–Z	0.18	0.31	–0.04	–0.02	–0.05	–0.02	–0.05	–0.09	0.07	0.01
	4[d]	–Z	–1.40	–0.49	–1.10[a]	–0.09	–1.12[b]	–0.09			–0.21	–0.06
	7	–Z	0.11	0.07	0.07	–0.03	0.10	0.03	0.01	–0.05	0.03	–0.02
	8[a]	–Z	–0.23	–0.23	–0.32	–0.26	–0.27	–0.21			0.09	0.08
	9	–Z	0.003	–0.02	0.11	0.07	0.12	0.08	0.01	0.05	–0.08	–0.01
	All	–Z	–0.05	0.04	0.07	0.11[b]	0.06	0.09[a]	0.001	0.03	0.002	0.03
	1	r^2	50[a]	41	42	36	39[a]	33	37[b]	16	17[a]	0
	2	r^2	69[b]	50	63[b]	50[a]	63[b]	49[b]	62[b]	39[b]	11	11
	3	r^2	58[b]	54[a]	53[b]	40	51[b]	37[a]	51[c]	34[b]	1	0
	4	r^2	61	41	58	15	55[a]	7			13	7
	7	r^2	65[a]	48	65[b]	45	63[b]	41	9	8	0	0
	8	r^2	59	46	55[a]	45	44[a]	33			2	2
	9	r^2	93[c]	78[c]	93[c]	77[c]	93[c]	77[c]	91[c]	77[c]	1	0
	All	r^2	41[c]	35[c]	39[c]	35[c]	38[c]	34[c]	34[c]	27[c]	0	0
1989	1	–Z	–0.10[a]	0.02	–0.10[b]	0.02	–0.09[b]	0.01	–0.11[b]	–0.02	–0.01	0.06
	2	–Z	–0.13[b]	–0.09	–0.07	–0.04	–0.08[a]	–0.04	–0.08[a]	–0.05	–0.06	–0.03
	5	–Z	–0.30[c]	–0.33[c]	–0.10	–0.07	–0.10	–0.07	–0.10	–0.07	–0.07	–0.07
	6	–Z	–0.38[c]	–0.43[c]	–0.25[c]	–0.34[c]	–0.25[c]	–0.34[c]	–0.25[c]	–0.32[c]	–0.22[c]	–0.30[c]
	7	–Z	–0.03	–0.02	–0.01	–0.01	0.004	0.02	–0.03	–0.01	–0.05	–0.04
	8	–Z	–0.08	–0.08	–0.08	–0.08	–0.07	–0.06	–0.08	–0.08	–0.05	–0.03
	9	–Z	–0.13[a]	–0.05	–0.11[a]	–0.04	–0.11[a]	–0.04	–0.13[b]	0.03	–0.05	0.08
	10	–Z	–0.41[c]	–0.20	–0.32[c]	–0.13	–0.32[c]	–0.13	–0.35[c]	–0.14	–0.40[c]	–0.20[b]
	All	–Z	–0.13[c]	–0.07[c]	–0.10[c]	–0.05[c]	–0.11[c]	–0.06[c]	–0.10[c]	–0.05[b]	–0.07[c]	–0.02
	1	r^2	62[c]	71[c]	62[c]	71[c]	61[c]	71[c]	61[c]	70[c]	0	1
	2	r^2	31[c]	16	26[c]	13	21[c]	12	19[c]	11	2	0
	5	r^2	35[c]	27[c]	28[c]	13[b]	28[c]	13[b]	25[c]	7	1	1
	6	r^2	61[c]	33[c]	55[c]	32[c]	52[c]	32[c]	52[c]	31[c]	11[c]	12
	7	r^2	28[c]	29[c]	27[c]	29[c]	26[c]	25[c]	12[a]	10	1	1
	8	r^2	18	32[b]	18	32[c]	15	27[c]	15	25[c]	16	1
	9	r^2	26[b]	39[c]	26[b]	38[c]	24[b]	34[c]	24[c]	22[b]	1	5[a]
	10	r^2	46[c]	28[b]	46[c]	27[b]	45[c]	27[b]	41[c]	27[c]	32[c]	10[b]
	All	r^2	19[c]	15[c]	18[c]	15[c]	18[c]	14[c]	16[c]	13[c]	1[c]	0
1990	1	–Z	–0.05	0.03	–0.09[c]	0.01	–0.09[c]	0.01	–0.10[c]	–0.002	–0.12[c]	–0.03
	6	–Z	–0.03	–0.07	–0.16[c]	–0.16[c]	–0.16[c]	–0.15[c]	–0.15[c]	–0.15[c]	–0.15[c]	–0.14[c]
	8	–Z	–0.002	–0.06[c]	–0.09[b]	–0.01	–0.09[b]	–0.01	–0.08[b]	–0.01	–0.08[a]	0.01
	9	–Z	–0.12[c]	–0.11[c]	–0.10[c]	–0.09[c]	–0.10[c]	–0.08[c]	–0.10[c]	–0.08[c]	–0.14[c]	–0.13[c]
	11	–Z	–0.14[c]	0.06	–0.12[c]	0.01	–0.14[c]	–0.01	–0.14[c]	–0.01	–0.16[c]	–0.04
	12	–Z	–0.04	0.06	–0.06	0.02	–0.04	0.04	–0.04	0.05	–0.08	0.01
	All	–Z	–0.11[c]	–0.08	–0.11[c]	–0.04[c]	–0.11[c]	–0.04[c]	–0.11[c]	–0.04[c]	–0.12[c]	–0.06[c]
	1	r^2	48[c]	39[c]	47[c]	39[c]	44[c]	35	41[c]	30[c]	9[c]	1
	6	r^2	57[c]	47[c]	55[c]	46[c]	20[c]	19[c]	19[c]	18[c]	17[c]	12[c]
	8	r^2	37[c]	26[c]	36[c]	26[c]	36[c]	25[c]	32[c]	23[c]	3[a]	0
	9	r^2	54[c]	66[c]	54[c]	66[c]	53[c]	64[c]	53[c]	64[c]	13[c]	9[c]
	11	r^2	23[c]	28[c]	23[c]	27[c]	18[c]	21[c]	17[c]	20[c]	13[c]	1
	12	r^2	41[c]	37[c]	41[c]	37[c]	31[c]	33[c]	30[c]	33[c]	3	0
	All	r^2	27[c]	31[c]	22[c]	25[c]	20[c]	24[c]	17[c]	18[c]	7[c]	2[c]

[a]Significant at the 0.10 level.
[b]Significant at the 0.05 level.
[c]Significant at the 0.01 level.
[d]Only one vegetation type observed on this transect.
[e]Model 1 includes incubation day, vegetation type, depth, depth^2, birds; model 2 includes incubation day, vegetation type, depth, and depth^2; model 3 includes incubation day, vegetation type, and depth; model 4 includes incubation day, and vegetation type; and model 5 only includes incubation day.

Herring: Expectations for a New Millennium 503

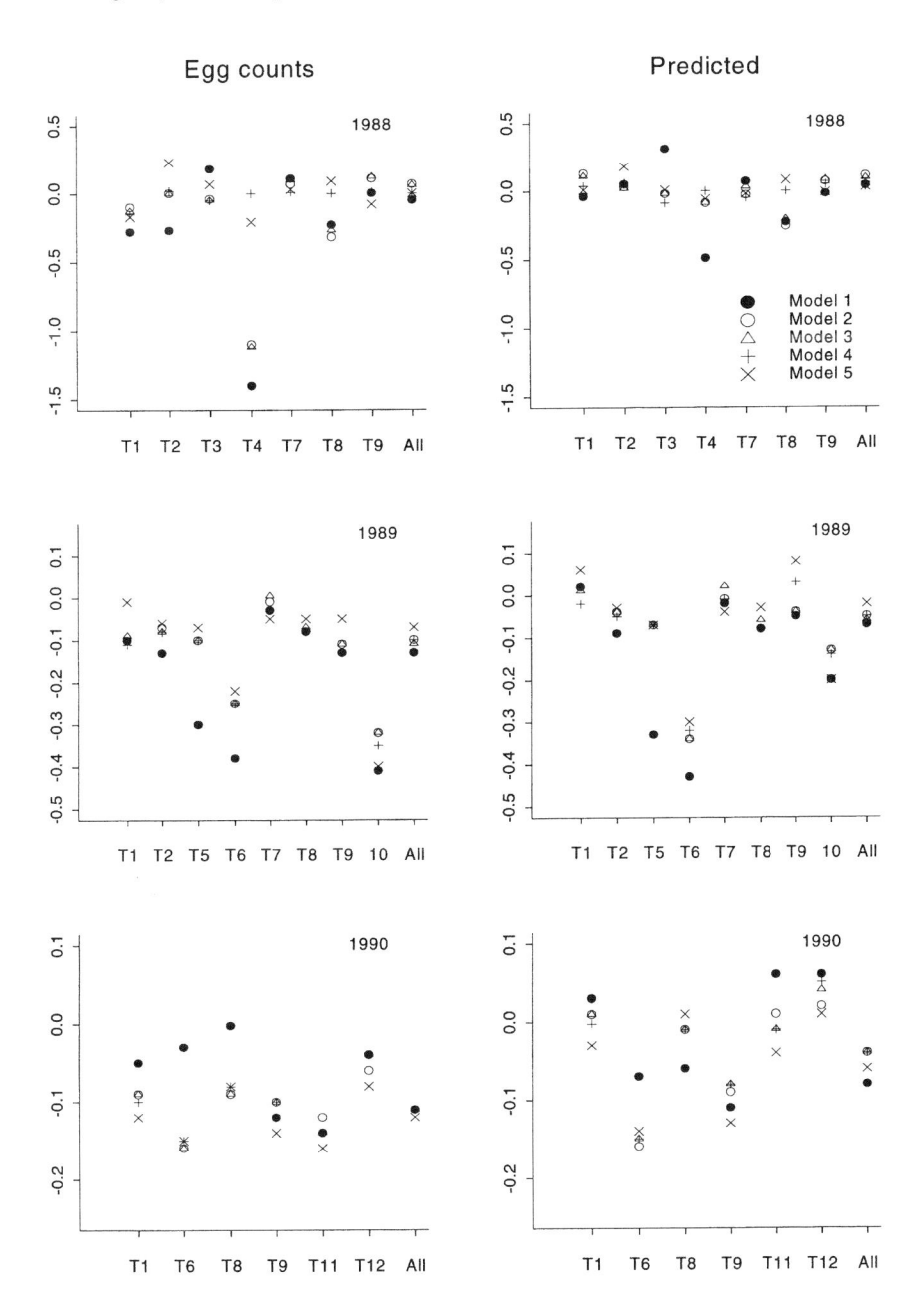

Figure 5. Estimated instantaneous egg loss rate from five general linear models at each transect and for the combined data from 1988-1990 surveys for both counted samples of egg density and for predicted egg density from equation 2.

Table 2. Estimated biomass for southern British Columbia herring spawning beds in 1999 surveys after adjusting for egg loss based on the mean number of incubation days between the beginning of spawning and the date of the survey.

Section	Mean Spawn day	Mean Survey day	Mean Incubation days	Observed Biomass	Adjusted biomass (t) $-Z = 0.10$	$-Z = 0.05$
141	77	79	2	8,122	8,705	8,408
142	64.5	75	10.5	41,079	51,715	46,091
143	73	79	6	21,550	25,779	23,570
172	75	86	11	4,530	5,758	5,107
173	81	89	8	4,898	3,978	5,435
Georgia St.				80,179	95,935	88,611
232	81	90	9	4,173	5,198	4,658
242	82	86	4	893	1,026	957
243	88	94	6	5,974	7,146	6,534
244	88	91.5	3.5	365	410	388
245	90	92	2	544	583	563
252	76.5	80	3.5	2014	2263	2,143
253	63	73	10	4,865	6,125	5,459
WCVI				18,828	22,751	20,701

The results for the fitting of these models to the egg predictions generally mirror those from the egg counts (Table 1, Fig. 5). The estimates for 1988 are less variable than for the counts but the estimated egg loss rates are all about zero or positive. Results for 1989 are very similar to those from the counts although estimated egg loss is slightly less than observed for the counted samples. The results for the 1990 data are more variable than for the egg count data and estimated egg loss rates are lower. Overall, fewer of the model fits to the predicted egg numbers were statistically significant and the proportion of the variation explained by the model fits to these data were usually lower with some notable exceptions in both 1989 and 1990. The estimated egg loss rate for the combined data from all transects in 1989 for model 4 was –0.05, about half the rate estimated from the egg counts. Similarly, in 1990 for model 4 the egg loss rate from the predicted data was –0.04 compared to –0.11 for the egg count data.

It is possible to provide an approximate estimate of the impact of egg loss on the total estimate of mature biomass from spawn surveys by determining the mean number of days that elapse between the beginning of herring spawning and the date on which surveys of the spawning beds occur. Table 2 presents the results of tabulating these data for herring surveys in 1999 of the spawnings in the herring sections (Haist and

Rosenfeld 1988) in the Strait of Georgia and the west coast of Vancouver Island. In general, the surveys occur within a week of the start of spawning activity in both areas. However, for the major spawning in the Strait of Georgia an average of 10.5 days elapsed prior to survey which results in the loss of about 20% of the total egg deposition. We calculated an adjusted biomass for these two areas by assuming instantaneous egg loss rates of 0.05 and 0.10 respectively, and adjusted the observed biomass based on the elapsed incubation days (Table 2). We calculated paired *t*-tests on the two sets of adjusted biomass estimates and the observed biomass for the two areas. Interestingly, the tests indicated no significant differences between the two adjusted values and the observed biomass in the Strait of Georgia where the elapsed days between spawning and survey were greater than for the west coast of Vancouver Island where both were significantly different ($P < 0.05$). This may result from the higher coefficient of variation in the Strait of Georgia estimates compared to the west coast of Vancouver Island biomass estimates.

Discussion

There have been several studies of egg loss from Pacific herring spawns but there is no consensus on its magnitude. Apparently, the two major causes of egg loss are predation and physical translocation through wave action. Munro and Clemens (1931) observed, inventoried, and sampled birds on herring spawning grounds near Nanaimo on the east coast of Vancouver Island over several years, concluding that while birds consumed a significant quantity of eggs it was not of a magnitude that would endanger the herring stocks. Outram (1958) conducted exclusion experiments on intertidal eelgrass *(Zostera marina),* on the west coast of Vancouver Island, and estimated that about 75% of eggs were lost during incubation and of these 39% were lost due to predation and 37% to wave action. Vermeer (1981) estimated that 75,000 surf scoters *(Melanitta perspicillata)* consumed the eggs from 1,030 t of herring along the west and east coast of Vancouver Island during two weeks in March 1978 representing about 1.4% of the total stock of herring. Palsson (1984) reported egg losses of 95-99% from very light density spawns in Puget Sound. Bird predation was the major cause of loss, followed by snail and gammarid predation. Haegele (1993b) also reported substantial egg losses due to invertebrate predation. Two studies on the east coast of Vancouver Island examined egg loss from wave action. Hart and Tester (1934) estimated that 40% of eggs in one spawn were washed ashore and that 70% of these eggs died. Hay and Miller (1982) found that 26% of eggs in a spawn in the Strait of Georgia were cast ashore in windrows. Both studies noted that adjacent spawns did not appear to experience this magnitude of egg loss.

Recent studies have focused on monitoring the egg loss over the course of the incubation period without specifically estimating the rates of predation (Haegele and Schweigert 1991, Rooper et al. 1999). The findings from both studies suggest that egg loss rate over the entire spawning bed is probably lower than previously reported. Both studies indicate instantaneous egg loss rates of 0.10 per day or less. In their study of Prince William Sound, Rooper et al. (1999) attribute much of the egg loss observed to bird predation which was more severe the longer the eggs were exposed and available to the predator. They also report that wave exposure was a significant factor determining egg loss rates. In this re-analysis of British Columbia data we determined that bird abundance did not significantly increase the egg loss rate although in 1989 the rates in the model were higher when birds were included. On the other hand, in 1990 rates were lower for the models that included bird abundance although bird abundance in 1990 appears to be about half what was observed in 1989 (Haegele 1991). Therefore birds may increase consumption rates in areas of high egg abundance, particularly in the intertidal areas, but such an increase may not appreciably affect overall egg numbers to an extent that we can detect. Although our study was intended to contrast protected areas with those exposed to wind and wave effects, we did not find any differences in egg loss among the transects in these two habitat types. There is a suggestion that egg loss rates were slightly elevated at transects 6 and 10 in 1989 and at transect 6 and 11 in 1990. These transects were in a more protected area and so might be subject to greater levels of bird predation, a finding also noted by Rooper et al. (1999).

Assessment of herring spawning biomass from egg surveys has always been assumed to be conservative because it is argued that not all spawning events are observed so some spawning beds will not be surveyed. Similarly, it was known that eggs were lost from the spawning beds prior to survey and although the magnitude of the loss was unknown it was assumed to be insignificant. Considering these factors alone, the estimated biomass was likely underestimated to an unknown degree. The results from this study indicate that instantaneous egg loss rates during 1989 and 1990 were approximately 0.10 per day. This equates to an overall loss of about 24% of the egg deposition by day 15 which corresponds roughly to the incubation period in southern British Columbia waters. Therefore, if assessment surveys are conducted shortly after spawning, i.e., within the first week following the beginning of spawning, egg losses are likely to be minimal (less than 18%) and bias in the total abundance estimates relatively small. However, results for the models based on the egg density predictions indicate that visual assessments of egg abundance tend to be conservative and that egg loss rates based on these assessments are likely to underestimate the true egg loss rate. On this basis, approximate adjustments for the apparent egg loss rate estimated here from visual assessments would still result in conservative estimates of spawning stock biomass (Table 2). Therefore, every effort should be made to conduct surveys of herring spawning sites as close to the initiation of spawning as is practical to minimize the effects of egg loss on the assessment of stock abundance. It may also be desirable to conduct periodic surveys to estimate the current egg loss rate which could then be used to adjust spawn survey estimates for any surveys that are conducted late in the incubation period and therefore would be expected to substantially underestimate the true spawning biomass.

References

Cleaver, F.C., and D.M. Franett. 1946. The predation by sea birds upon the eggs of the Pacific herring (*Clupea pallasi*) at Holmes Harbor during 1945. Wash. State. Dept. Fish. Biol. Rep. 46B.

Haegele, C.W. 1991. Spawn estimates and associated predator data for herring egg loss in Lambert Channel, Georgia Strait—1989 and 1990. Can. Data Rep. Fish. Aquat. Sci. 821. 103 pp.

Haegele, C.W. 1993a. Seabird predation of Pacific herring, *Clupea pallasi*, spawn in British Columbia. Can. Field-Nat. 107:73-82.

Haegele, C.W. 1993b. Epibenthic invertebrate predation of Pacific herring, *Clupea pallasi*, spawn in British Columbia. Can. Field-Nat. 107:83-91.

Haegele, C.W., and J.F. Schweigert. 1989. Egg loss from Pacific herring spawns in Barkley Sound in 1988. Can. Manuscr. Rep. Fish. Aquat. Sci. 2037. 37 pp.

Haegele, C.W., and J.F. Schweigert. 1991. Egg loss in Georgia Strait, British Columbia. Proceedings of the International Herring Symposium. University of Alaska Sea Grant, AK-SG-91-01, Fairbanks, pp. 309-322.

Haegele, C.W., R.D. Humphreys, and A.S. Hourston. 1981. Distribution of eggs by depth and vegetation type in Pacific herring (*Clupea harengus pallasi*) spawnings in southern British Columbia. Can. J. Fish. Aquat. Sci. 38:381-386.

Haegele, C.W., A.S. Hourston, R.D. Humphreys, and D.C. Miller. 1979. Eggs per unit area in British Columbia herring spawn depositions. Fish. Mar. Serv. Tech. Rep. 894. 30 pp.

Haist, V., and L. Rosenfeld. 1988. Definitions and codings of localities, sections, and assessment regions for British Columbia herring data. Can. Manuscr. Rep. Fish. Aquat. Sci. 1994. 123 pp.

Hart, J.L., and A.L. Tester. 1934. Quantitative studies on herring spawning. Trans. Am. Fish. Soc. 64:307-313.

Hay, D.E., and A.R. Kronlund. 1987. Factors affecting the distribution, abundance, and measurement of Pacific herring (*Clupea harengus pallasi*) spawn. Can. J. Fish. Aquat. Sci. 44:1181-1194.

Hay, D.E., and D.C. Miller. 1982. A quantitative assessment of herring spawn lost by storm action in French Creek, 1980. Can. Manuscr. Rep. Fish. Aquat. Sci. 1636. 9 pp.

Munro, J.A., and W.A. Clemens. 1931. Water fowl in relation to the spawning of herring in British Columbia. Biol. Board Can. Bull. 17. 46 pp.

Outram, D.N. 1958. The magnitude of herring spawn losses due to bird predation on the west coast of Vancouver Island. Fish. Res. Board. Can. Pac. Biol. Sta. Pro. Rep. 111:9-13.

Palsson, W.A. 1984. Egg mortality upon natural and artificial substrata within Washington state spawning grounds of Pacific herring *(Clupea harengus pallasi)*. M.S. thesis, University of Washington, Seattle. 191 pp.

Rooper, C.N., L.J. Haldorson, and T.J. Quinn II. 1999. Habitat factors controlling Pacific herring *(Clupea pallasi)* egg loss in Prince William Sound, Alaska. Can. J. Fish. Aquat. Sci. 56:1133-1142.

Schweigert, J.F. 1993. A review and evaluation of methodology for estimating Pacific herring egg deposition. Bull. Mar. Sci. 53(2):818-841.

Schweigert, J.F. 1995. Environmental effects on long-term population dynamics and recruitment to Pacific herring *(Clupea pallasi)* populations in southern British Columbia. In: R.J. Beamish (ed.), Climate change in northern fish populations. Can. Spec. Publ. Fish. Aquat. Sci. 121:569-583.

Schweigert, J.F., and D. Fournier. 1982. A model for predicting Pacific herring *(Clupea harengus pallasi)* spawn density from diver observations. Can. J. Fish. Aquat. Sci. 39:1361-1365.

Schweigert, J.F., C.W. Haegele, and M. Stocker. 1985. Optimizing sampling design for herring spawn surveys in the Strait of Georgia, B.C. Can. J. Fish. Aquat. Sci. 42:1806-1814.

Vermeer, K. 1981. Food and populations of surf scoters in British Columbia. Wildfowl 32:107-116.

Zebdi, A., and J.S. Collie. 1995. Effect of climate on herring *(Clupea pallasi)* population dynamics in the Northeast Pacific Ocean. In: R.J. Beamish (ed.), Climate change and northern fish populations. Can. Spec. Publ. Fish. Aquat. Sci. 121:277-290.

Herring Tilt Angles, Measured through Target Tracking

E. Ona
Institute of Marine Research, Bergen, Norway

Abstract

The target strength of herring in captivity was monitored experimentally at three frequencies (18, 38, and 120 kHz) over a period of 3 years using split-beam echo sounders. One of the fundamental parameters of a target strength measurement, when performed on free-swimming fish, is angular orientation relative to the horizontal. This is usually defined as the tilt angle, when vertical echo sounding is conducted. The swimming angle of the targets was measured directly through split-beam target tracking with specialized split-beam hardware and software. As the swimming angle and the tilt angle are not always identical, underwater video analysis was used to measure the relation between the swimming angle and the actual tilt angle for herring at different depths.

Introduction

At the commonly used frequencies for acoustic abundance estimation of herring (30-120 kHz), the herring is a quite directive target (Haslett 1977, Nakken and Olsen 1977). A small change in orientation of the fish relative to the transducer surface may drastically change the target strength at these frequencies. In fish roll aspect, the fish is more or less omnidirectional (Nakken and Olsen 1977). This is due to the circularity of the cross section of the swimbladder in this plane, but also due to the small dimension of the cross section as compared with the wavelength of the acoustic signal. In the pitch plane of the fish, however, fish with a swimbladder, comparable or longer than the wavelength will be acoustically directive, and the backscattering amplitude will be dependent on the pitch angle of the fish, relative to the transducer. In acoustic abundance estimation, the mean target strength or mean backscattering cross section is needed when converting integrated echo energy to biomass (MacLennan 1990). The preferred mean target strength to be used in this conversion are those collected in a direct measurement, in situ, on the surveyed population (Ona

1999). For many fishes, however, the target strength estimates obtained in a survey situation may not be representative for the actual target strength, as the typical fish density recorded cannot be resolved for single target strength analysis. Herring is a typical representative for these species, where most of the abundance is recorded as schools or as dense layers at variable depths. The available technology for direct target strength analysis, the single-, dual-, and split-beam echo sounders, all phase the same problems in resolving the most important densities recorded on a survey into single targets for "safe" target strength analysis. Therefore, the target strength data on herring are with a few exceptions (Huse and Ona 1996, Ona et al. 2000a) collected in very loose aggregations, often at night. As the mean target strength estimate is a convolution of the directivity pattern of the fish and its tilt angle distribution (Foote 1980), the estimate is quite sensitive to changes in tilt angle distribution. If the tilt angle distribution inside dense schools of herring is different than in loose aggregations, the mean target strength may be biased. In fact, the target strength estimate should ideally contain information on tilt angle distribution if this was possible. Photographically, it has been determined that the tilt angle distribution of herring will vary with herring behavior, and in particular between the feeding situation and the schooling situation (Beltestad 1973), but also with depth and light intensity in a wintering situation where no feeding occurs (Huse and Ona 1996).

During experimental measurements of seasonal variation in target strength, a control of the swimming behavior was needed in order to separate the effect of tilt angle distribution from the other effects studied (Ona and others, submitted manuscript). This was tried using a single fish target tracking system (Ona and Hansen 1991) utilizing the data from a split-beam echo sounder to compute swimming parameters on single fish passing the beam of a stationary acoustic transducer. As this system records the three-dimensional movement of the target through the beam, computation of the fish swimming speed, swimming direction, and other related parameters are straightforward (Brede et al. 1990, Ona and Hansen 1991, Ehrenberg and Torkelson 1996). The actual tilt angle of the fish is, however, not a direct derivative. An underlying assumption when trying to compute this from information on geometrical position is that the vertical swimming direction of the herring is exactly along the longitudinal axis of the fish. For fishes with normally high swimming speeds, like Atlantic mackerel, this is usually confirmed (He and Wardle 1986), but negatively buoyant fish swimming at low speed (<0.8 body lengths per second) often swim with a slight head-up posture. Since the gas-producing capacity of the herring swimbladder is still unclear (Blaxter and Batty 1984, Ona 1990), negative buoyancy is likely to occur at depth, and a similar swimming strategy as for mackerel is expected in herring if additional lift is needed. A few observations, however, indicate that herring use another, commonly used compensatory strategy for negatively buoyant animals, namely the "rise and glide" method (Huse and Ona 1996). In order to evaluate if the target tracking method could be used to determine the actual tilt angle of the fish, a video analysis of herring swimming behavior was conducted. The results from these, with examples of tilt angle distributions computed by the tracking method, are reported here.

Material and Methods

During vertical excursion experiments with captive herring for target strength analysis (Fig. 1), the experimental pen ($12.5 \times 12.5 \times 21$ m; 4,500 m^3), was rearranged with a top frame and net into a cage. After careful transfer of 50 herring from the holding pen (see Ona et al. 2001 [this volume]), the cage was stepwise lowered to 50 m in one experiment in August 1996, and further to 100 m in April 1997. During the depth excursions, an underwater, 360-degree pan video camera was used to monitor the behavior of the herring inside the cage. The camera could be lowered from the top frame of the cage to the bottom by pulling the recorder cable, which was guided through a 50-mm nylon ring attached to the top panel. The image of a plumb line, attached to the camera, was used for vertical reference during video analysis of herring swimming behavior. Video recordings and target strength measurements were made sequentially, since the camera and cable interfered with the acoustic recordings. Based on the visual inspection of the video data, there seemed to be little difference between the swimming angle and the actual tilt angle of herring at any depth. To further quantify this, some of the recorded data were selected for a more detailed analysis. If a fish is swimming along a straight line through the acoustic beam, without changing depth, the recorded tilt angle between consecutive detections will be zero. If the fish is swimming along this line with a slight head-down or head-up posture (Fig. 2), however, the tilt angle and the swimming angle will be different. The tilt angle of the fish is measured as defined by Olsen (1971): The angle between the fish center line, or imaginary line running from the root of the tail to the tip of the upper jaw, and the true horizontal. To analyze the data, the recorder was connected to standard, commercial hardware and software for image analysis, and a videotape editing system. The movement of herring could here be displayed frame by frame on a PC display with an internal, two-dimensional pixel-based reference system.

When a single herring passed the vertical reference, oriented within ±10° of the plane of the photographic axis, the video was stopped, slowly reversed, and further displayed one frame at a time. On the first frame, the exact position of the snout of the herring was tagged with a marker on the screen (Fig. 3). The film was then started, frame by frame, until the tail passed the marker and stopped. The distance from the marker to the imaginary line between the snout and the root of the tail was then measured digitally with the cursor. In order to make this measurement independent of distance from the camera, the units used were parts of tail-root width. For herring of this size, 32 cm, the unit corresponds to an angular resolu-

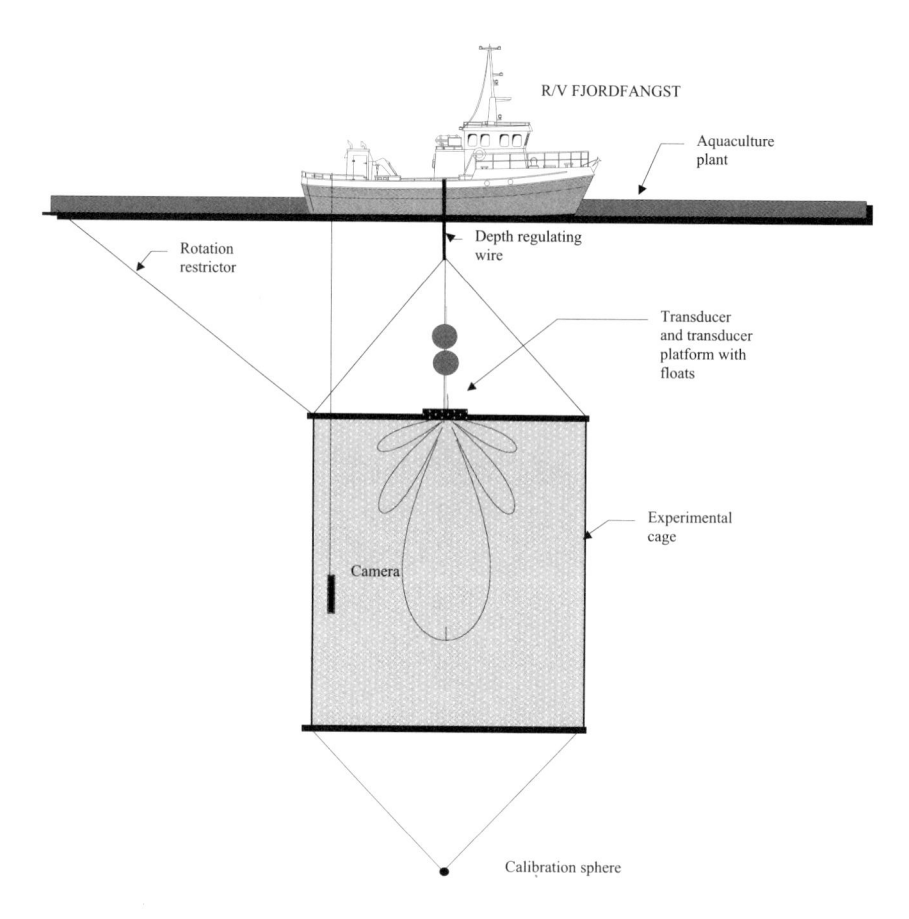

Figure. 1. Simple sketch of the experimental set-up for target strength measurements and video analysis of herring swimming behavior. Cage dimensions are $12.5 \times 12.5 \times 21$ m.

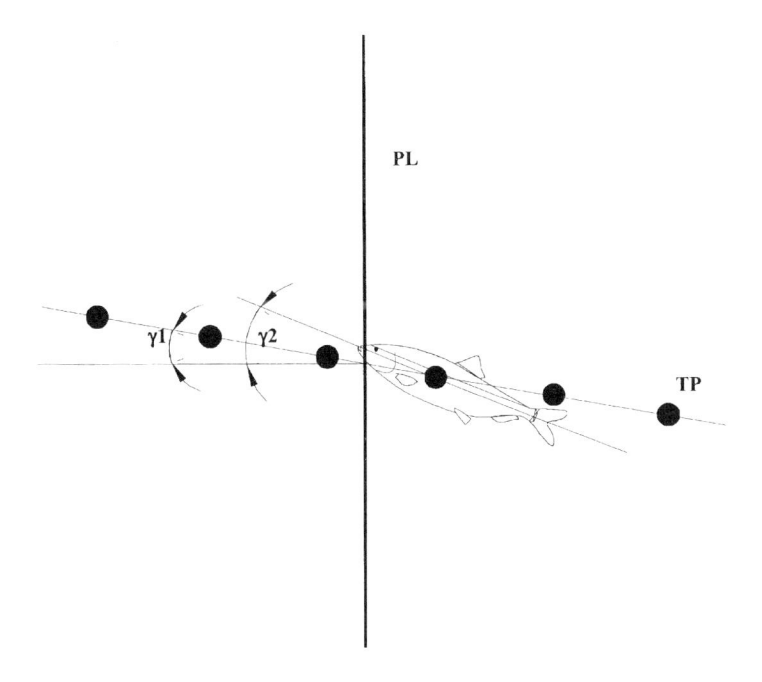

Figure 2. Specification of the difference between the swimming angle, γ1, and the actual tilt angle, γ2. The detected target positions are marked, TP, and the vertical reference plumb line, PL, is shown.

tion of about 1° or, more exactly, 1.16°, as determined from a series of measurements of the ratio between root-of-tail width and the length of herring, as when displayed on the screen.

Examples of estimated swimming angles for herring are made using the Simrad EK-500 split-beam echo sounder, 120-kHz system with a digital resolution of 3 cm in the range measurements (Bodholt et al. 1989). The transducer, with a full beam width of 8.6°, circular, has an angular resolution of 0.13°. The standard target strength and angular data for the echo sounder were used for target tracking (Ona and Hansen 1991) and further to compute the swimming angles of the fish.

Results and Discussion

The results from the video analysis are summarized in Table 1. The carefully selected data from 5 depth intervals, from 5 to 40 m, show that the herring generally swim straight along its imaginary center line, and that the difference between actual tilt angle and swimming angle are small at

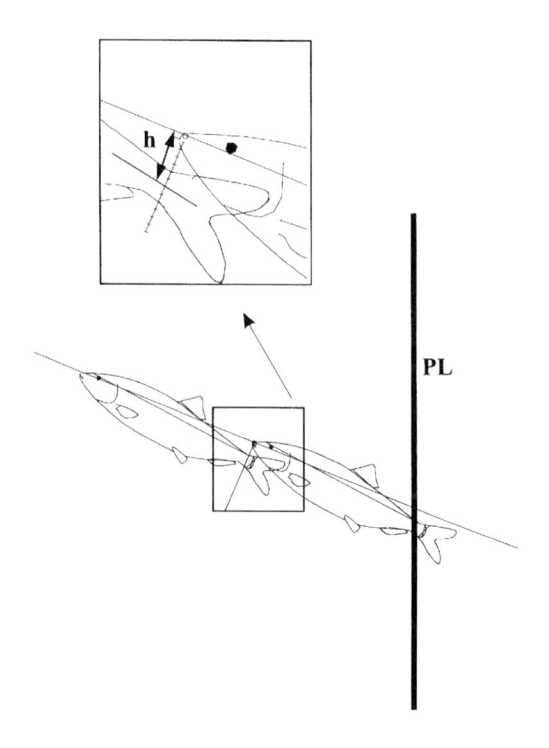

Figure 3. Measurement of the difference between swimming angle and the tilt angle, made from video recordings. PL, plumb line; h, distance. The measurement procedure is described in the text.

all depths. The recordings were made under natural light conditions with no artificial light. The exact swimming speeds were not recorded in the video analysis, but the split-beam tracking system recorded a mean speed of 35 cm per second, or about 1.1 body lengths per second. A slightly negative mean head-down posture, –1.61°, was recorded at the 5-m depth, while the mean values at 20, 25, 30, and 40 m were either close to zero, or slightly head up. As the demand for angular measurements is strict with respect to orientation relative to the photographic axis, only 362 measurements were accepted. If all data are grouped (Fig. 4), the recorded deviation between mean swimming angle and mean tilt angle is close to zero (–0.06°, S.E. = 0.10°).

The herring is a constant swimmer, and is seldom observed swimming below 0.8 body lengths per second, even when hibernating in the wintering areas of the stock (Huse and Ona 1996). When schooling and feeding, individual swimming speeds are presumably larger (Harden-Jones 1968). Hydrodynamically, the herring is also shaped like a fast swimmer,

Table 1. Results from the analysis of difference between swimming angle, $\gamma 1$, and actual tilt angle, $\gamma 2$, from video analysis.

Date	Time	Camera depth (m)	Mean $\Delta\gamma$ (deg.)	N	S.E.
18 Aug. 1996	1200-1300	5	−1.61	90	0.18
23 Aug. 1996	1200-1400	25	+0.49	24	0.39
27 Aug. 1996	1000-1400	40	+1.12	91	0.16
28 Aug. 1996	1100-1200	30	+0.41	62	0.22
29 Aug. 1996	1600-1700	20	−0.19	95	0.10
All data			−0.06	362	0.10

The mean difference, $\Delta\gamma$, number of measurements, and standard error of the mean are given for each series. Negative values indicate that the center of the root of the tail is passing below the snout marker. The actual tilt angle is then larger than the swimming angle; i.e., $\gamma 2 = \gamma 1 - \Delta\gamma$.

and only at very low speeds, or at large negative buoyancy, are deviations between swimming angle and actual tilt expected.

From our analysis, we may conclude that this difference is so small that its effect on estimates of actual tilt angle measurements from target tracking is comparable to the accuracy in photographically estimated tilt angles. If these are made with single-frame cameras, the precision of an individual tilt angle is about 1° when only selecting fish perpendicular to the photographic axis (Olsen 1971). The use of stereo-camera methods may increase this accuracy.

Examples of tilt angle distributions obtained by target tracking are shown in Figs. 5 and 6. The first two distributions are recorded on a single individual herring, 34 cm, freely swimming inside the experimental pen for a period of 30 hours. Two estimates of the overall tilt angle distribution may be made from the tracking data. If the herring passes the beam along a fairly straight line, the mean tilt angle may be computed from the entire passage (Fig. 5a). Depending on target range, swimming speed, and pulse repetition rate, the fish is detected several times during the passage, and tilt angle between successive detections may be computed (Fig. 5b). Both methods show that this herring, mainly swimming at shallow depths in the pen, passes the transducer beam at a nearly horizontal angle (−1.1° and −0.7°) with a standard deviation of about 10°. These distributions are close to photographically estimated tilt angle distributions of herring under similar conditions. Ona (1984) measured the tilt angle distributions of herring in net pens at several adaptation depths. His photographic measurements showed that for herring at depths of 1.5 and 4 m, the tilt angle distributions were −3.9° (S.D. = 12.8°) and 0.2° (S.D. = 11.9°), respectively. Beltestad (1973) recorded distributions of 3.8° (S.D. = 6.0°) during the day and −3.2° (S.D. = 13.6°) at night. An example of a larger sample (Fig. 6) recorded at night in March 1997, where

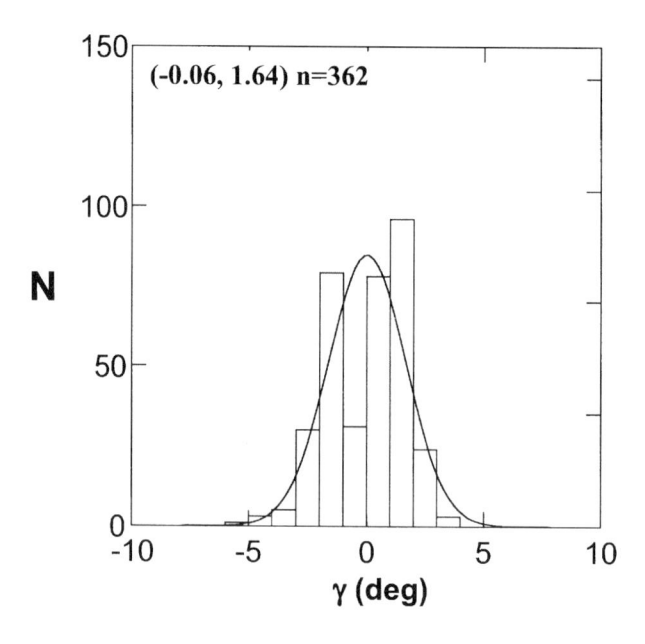

Figure 4. Overall distribution of the difference between swimming angle and tilt angle of herring. Mean = −0.06°, S.D. = 1.94°.

the mean tilt is −3.1° and the standard deviation is 14.2°, is also very close to photographically determined tilt angle distributions. The accuracy of the tracked angle is dependent on the range resolution of the split-beam system, its angular resolution, and the horizontal movement of the fish. If the movement of the fish is small compared to the range resolution, as is the case at the lower frequencies of the split-beam system used, sudden steps in angular estimates are seen, connected to the steps in range bins. With the 120-kHz system, however, the accuracy of each angular measurement on herring swimming at 35 cm per second is estimated to be about 2°. In most cases, this is sufficient accuracy to describe the main properties of the tilt angle distribution. Technical improvements of the method have already been made at the lower echo sounder frequencies, and parallel estimates of tilt angle distribution at three frequencies are now possible. In particular this improvement was needed when establishing a methodology for in situ target strength measurements on herring at 38 kHz in deep water or inside layers with high volume density, as described by Ona and others (submitted manuscript). The method described may only be used when the conditions for single target recognition by split-beam echo sounder systems are fulfilled, i.e., in low-density situations or at short range. When herring are schooling at high density during daytime, tilt angles may not be measured by this method.

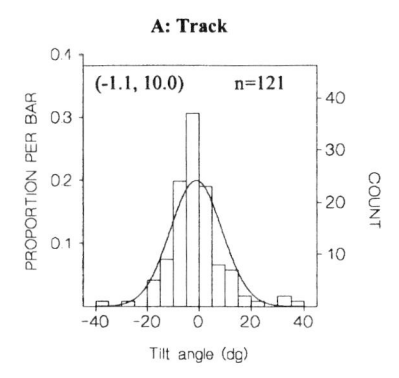

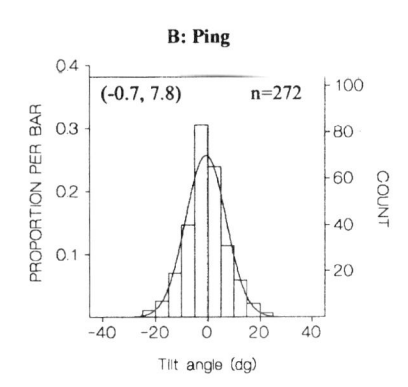

Figure 5. The apparent tilt angle distribution of a 34-cm herring observed by the 120-kHz transducer during a 30-hour period. The superimposed are the fitted truncated Gaussian (normal) distribution. Also shown in the parentheses are the means and the standard deviations, respectively. A, mean track tilt angle; B, ping-to-ping tilt angle.

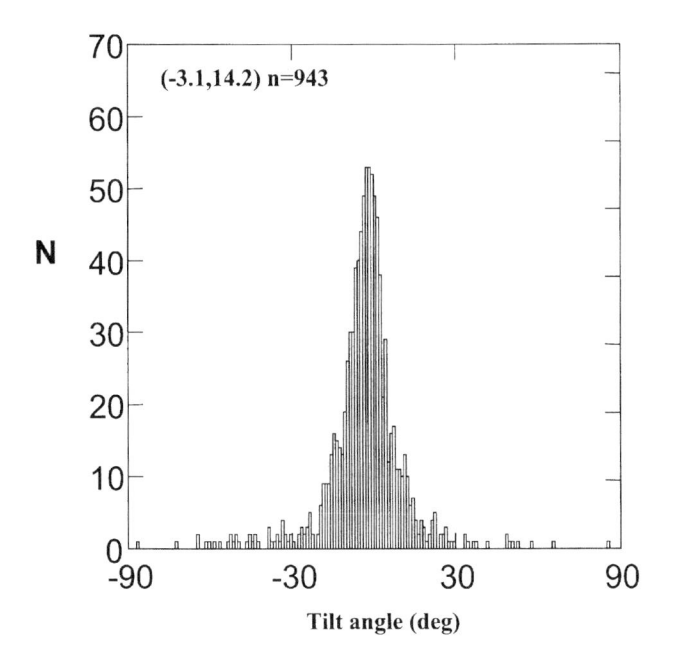

Figure 6. Estimated tilt angle distribution from target tracking at 120 kHz in March 1997, recorded over 5 hours at night. The parameters for the normal distribution are indicated.

Acknowledgment

Partial financial support from European Union research project EU (AIR-CT94-2142) for this investigation is acknowledged.

References

Beltestad, A.K. 1973. Feeding behavior and vertical migration in 0-group herring (*Clupea harengus L.*) in relation to light intensity. Cand. Real. thesis, University of Bergen. (In Norwegian.)

Blaxter, J.H.S., and R.S. Batty. 1984. The herring swimbladder: Loss and gain of gas. J. Mar. Biol. Assoc. U.K. 64:441-459.

Bodholt, H., H. Nes, and H. Solli. 1989. A new echo-sounder system. Int. Conf. Progress in Fisheries Acoustics, Lowestoft. Proc. I. O. A., St. Alban, U.K. 11(3):123-130.

Brede, R., F.H. Kristensen, H. Solli, and E. Ona. 1990. Target tracking with a split-beam echo sounder. Rapp. P.-V. Reun. Cons. Int. Explor. Mer 189:254-263.

Ehrenberg, J.E., and T.C. Torkelson. 1996. Application of dual-beam and split-beam target tracking in fisheries acoustics. ICES J. Mar. Sci. 53:329-334.

Foote, K.G. 1980. Averaging of fish target strength functions. J. Acoust. Soc. Am. 67(2):504-515.

Harden-Jones, F.R. 1968. Fish migration. The Camelot Press Ltd., London. 325 pp.

Haslett, R.W.G. 1977. Automatic plotting of polar diagrams of target strength of fish in roll, pitch and yaw. Rapp. P.-V. Reun. Cons. Int. Explor. Mer 170:74-82.

He, P., and C.S. Wardle. 1986. Tilting behavior of the Atlantic mackerel, *Scomber scombrus*, at low swimming speeds. J. Fish Biol. 29(Suppl. A):223-232.

Huse, I., and E. Ona. 1996. Tilt angle distribution and swimming speed of overwintering Norwegian spring spawning herring. ICES J. Mar. Sci. 53:863-873.

MacLennan, D.N. 1990. Acoustical measurement of fish abundance. J. Acoust. Soc. Am. 87:1-15.

Nakken, O., and K. Olsen. 1977. Target strength measurements of fish. Rapp. P.-V. Reun. Cons. Perm. Int. Explor. Mer 170:52-69.

Olsen, K. 1971. Orientation measurements of cod in Lofoten obtained from underwater photographs and their relation to target strength. ICES C.M.1971/B:17. 8 pp.

Ona, E. 1984. Tilt angle measurements on herring. ICES C.M. 1984/B:19.

Ona, E. 1990. Physiological factors causing natural variations in acoustic target strength of fish. J. Mar. Biol. Assoc. U.K. 70:107-27.

Ona, E. (ed.). 1999. Methodology for target strength measurements. ICES Coop. Res. Rep. No. 235. 59 pp.

Ona, E., and D. Hansen. 1991. Software for target tracking with split-beam echo sounders: User manual. Institute of Marine Research, Bergen, Norway.

Ona, E., X. Zhao, I. Svellingen, and K.G. Foote. 1996. Some pitfalls of short-range standard-target calibration. ICES C.M. 1996/B:36.

Ona, E., X. Zhao, I. Svellingen, and J.E. Fosseidengen. 2001b. Seasonal variations in herring target strength. In: F. Funk, J. Blackburn, D. Hay, A.J. Paul, R. Stephenson, R. Toresen, and D. Witherell (eds.), Herring: Expectations for a new millennium. University of Alaska Sea Grant, AK-SG-01-04. Fairbanks. (This volume.)

Zhao, X. 1996. Target strength of herring (*Clupea harengus* L.) measured by the split-beam tracking method. M. phil. thesis, University of Bergen, Norway.

Latitudinal Difference in Recruitment Dynamics of Clupeid Fishes: Variable to the North, Stable to the South

Yoshiro Watanabe, Norio Shirahuji, and Masayuki Chimura
Ocean Research Institute, University of Tokyo, Tokyo, Japan

Abstract

Subarctic *Clupea pallasii*, temperate *Sardinops melanostictus*, subtropical *Etrumeus teres*, and tropical *Spratelloides gracilis* are important commercial clupeid fish species in Japan. The cumulative catch in number of *Clupea pallasii* in each year-class fluctuated by 3.1 orders of magnitude during the period 1907-1955. The estimated numbers of age 1 *Sardinops melanostictus* recruits fluctuated by more than 3 orders of magnitude between 1980 and 1991, when the population was at its maximum. During this period, a large proportion of young-of-the-year *Sardinops melanostictus* migrated north in summer to feed in the subarctic Oyashio waters. When the population was at a sub-maximum level, the juveniles remained in the warm Kuroshio Current and its associated waters, and fluctuations in *Sardinops melanostictus* recruitment was relatively small. The CPUE of subtropical *Etrumeus teres* in western Japan ranged from 17.6 to 87.7 t per boat per year in the 30-year period 1969-1998. Since a large part of this catch was composed of fish younger than 1 year, recruitment variability in *E. teres* seems to be less than 1 order of magnitude. The CPUE of tropical *Spratelloides gracilis* by the local purse seine fishery fluctuated from 70 to 850 kg per day; its life span is believed to be less than a year, leading to recruitment variability of approximately 1 order of magnitude. The latitudinal differences in the recruitment dynamics of clupeid fishes appear to be associated with the variability in their physical and biological environments. This paper examines and compares biological traits responsible for the differences in recruitment dynamics of the four clupeid species.

Introduction

The recruitment variability of commercially exploited fish populations has been a key subject of study in fisheries science since it was recognized that fluctuations in the fish catch are due largely to variable recruitment rather than to changes in fish migration (Hjort 1917). The critical period hypothesis proposed by Hjort emphasized the importance of the early life stages of fish in the determination of year-class strength, but was not supported by substantial field survey data. Mechanistic hypotheses of fish recruitment variability in relation to marine environments appeared in the 1970s. Cushing (1975) identified the temporal match/mismatch between the spring bloom and the first feeding period of fish larvae. Lasker (1975) found that the subsurface chlorophyll maximum layer in a vertically stratified water column was crucial for successful feeding in early larval *Engraulis mordax*, emphasizing the importance of spatial ocean stability on larval survival.

These two pioneering hypotheses in fish recruitment variability attempted to substantiate the critical period hypothesis proposed by Hjort (1917) in relation to variable ocean environments, and suggested the first feeding stage as a critical period for fish recruitment determination. However, recruitment studies based on field observations have shown that mortality in the early larval stage may not be a determining factor in recruitment (Peterman et al. 1989, Butler 1991, Watanabe et al. 1995). Houde (1989) demonstrated that episodic events such as mass mortality at the first feeding stage may not affect recruitment more than do subtle changes in growth and mortality rates during early life. Campana (1996) found that growth rates during the larval and early juvenile stages, estimated by otolith radius at 90 days of age, could predict year-class strength in Atlantic cod, *Gadus morhua*. Therefore, environmental changes could have cumulative effects through the larval and juvenile stages in determining recruitment.

Comparisons of closely related temperate species from different waters, such as *Sardinops* from the California, Humboldt, Benguela, and Kuroshio currents provide a potential approach to investigating the dynamics of fish populations' response to ocean and climate changes (Hunter and Alheit 1995). The marine ecosystems in northern and tropical/subtropical waters are very different. Therefore comparisons of the reproductive biology in adult and early life stages among fishes from latitudinally different waters offer another potential approach to the investigation of fish recruitment dynamics.

There are 15 species of clupeids distributed in the waters around Japan (Masuda et al. 1984), but only three are major commercial target species: subarctic *Clupea pallasii*, temperate *Sardinops melanostictus*, and subtropical *Etrumeus teres* (Fig. 1). Tropical *Spratelloides gracilis* are fished in local waters in southern Japan. The distributional range of *Clupea pallasii* extends from the Yellow Sea to the western Bering Sea (Whitehead 1985a), while *Sardinops melanostictus* is found in the northwestern Pacific and adjacent seas (Kuroda 1991) (Fig. 2). The range of *E. teres* extends from western

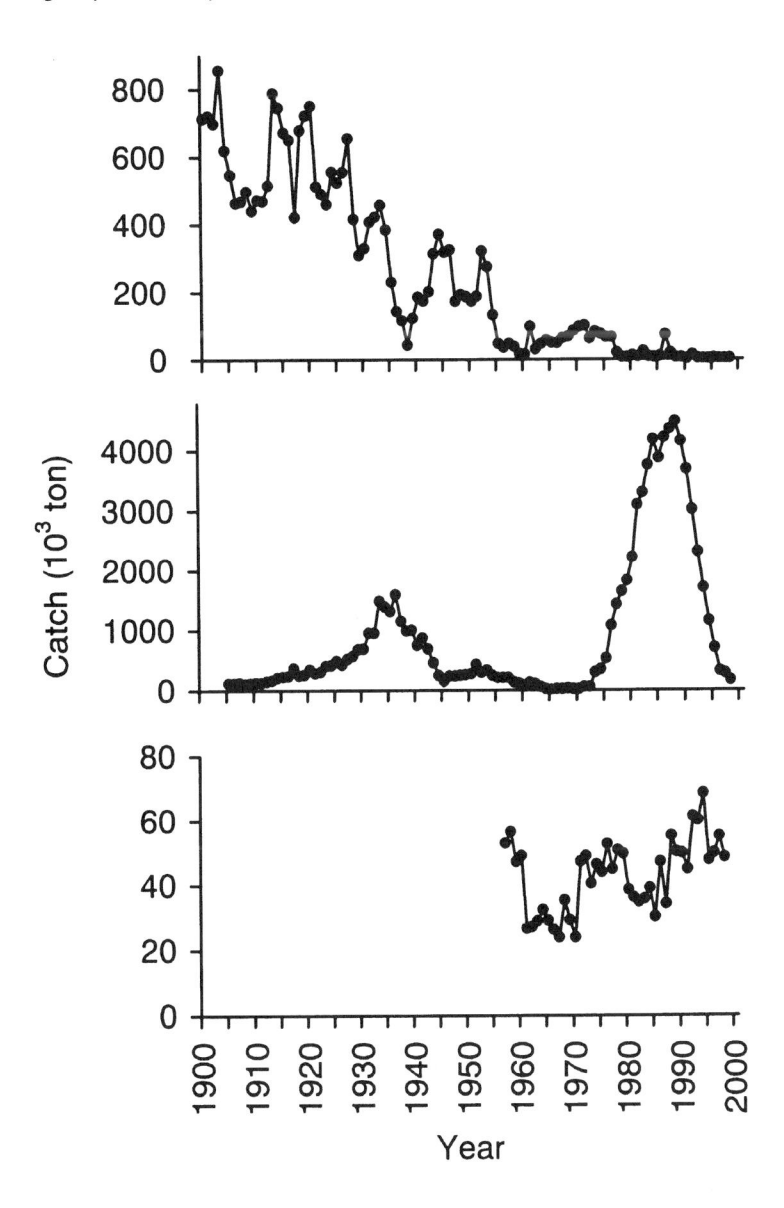

Figure 1. Time series of annual total catch of Clupea pallasii (top), Sardinops melanostictus (middle), and Etrumeus teres (bottom) in Japan. Data of Clupea pallasii during 1900-1959 are after Hanamura (1963) and 1960-1998 from the Fisheries Statistics (Statistics and Information Department 2000). Data on Sardinops melanostictus during 1905-1989 are after Kuroda (1991), and 1990-1998 are from Fisheries Statistics (Statistics and Information Department 2000). Data on Etrumeus teres are from Fisheries Statistics (Statistics and Information Department 2000).

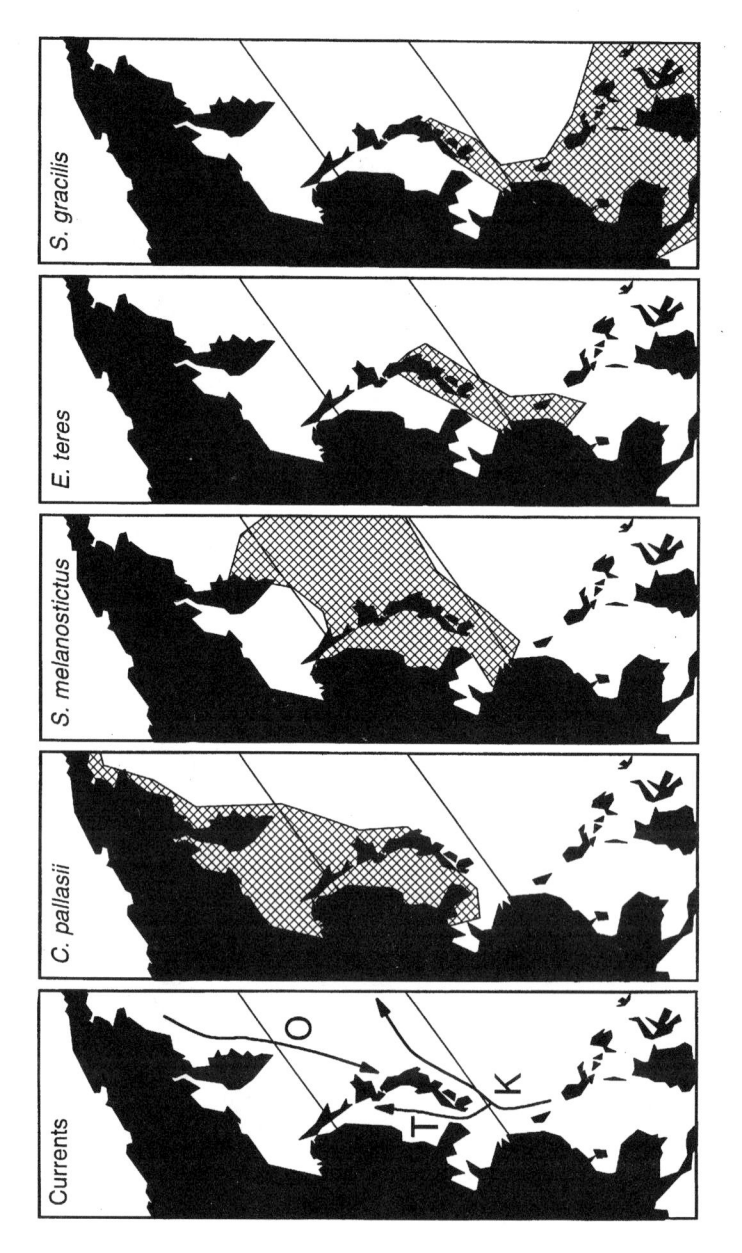

Figure 2. Geographical distributions (meshed) of the four clupeid species. The Kuroshio (K), Tsushima (T), and Oyashio (O) current paths are shown in solid lines in the left panel. Thin lines in each panel indicate 30°N and 50°N latitude.

Japan to the South China Sea (Yamada et al. 1986), and *Spratelloides gracilis* ranges from southern Japan to southeast Asia (Whitehead 1985a). The patterns of fluctuation in the annual catches of the three species are quite different (Fig. 1, data for *Spratelloides gracilis* not available). The catch of *Clupea pallasii* was 870×10^3 t in 1903, and then tended to decline with sharp peaks and deep troughs until the 1950s, and then remained low during the second half of the century. Catch of *Sardinops melanostictus* peaked in the 1930s and 1980s, but precipitously declined after the peak at 4.5 million t in 1988. Catches of *E. teres* remained at levels of $20\text{-}70 \times 10^3$ t for 42 years after 1957. The population dynamics of these three clupeid species therefore differed considerably. In this study, we compare and contrast the recruitment variability of these four clupeid species including *Spratelloides gracilis*, and consider the latitudinal differences in their recruitment dynamics.

Methods

The cumulative number of Hokkaido spring herring caught was calculated for each of the 1907-1955 year classes, based on the catch-at-age data reported by Hanamura (1963). This herring stock constituted the Hokkaido-Sakhalin stock of the Pacific herring (*Clupea pallasii*), which virtually disappeared from the spawning grounds along the west coast of Hokkaido in the mid-1950s.

The number of age-1 recruits of Japanese sardine (*Sardinops melanostictus*) was estimated by Wada (1985, 1998), based on the number of immigrant age-1 fish caught in the Doto fishing region, which is delimited by the coastline of Hokkaido Island, the 42°40′N line, 143°10′E line, and the southeast line extending from Cape Nosappu. This region is in the northern-most part of the migration range of the sardine. The estimated biomass here increased from 0.7×10^6 t in 1976 to 3.6×10^6 t in 1984, then declined to 0.3×10^6 t in 1992 (Wada 1998). *Sardinops melanostictus* was not found in these fishing grounds after 1994 (Mihara 1998), but the grounds off northern Honshu Island continued to produce substantial sardine catches. Since the migration range of *S. melanostictus* changed with the population decline, we used the data from the 1975-1991 year classes in this study.

For *Etrumeus teres*, the time series (1957-1998) of total annual catch by different size categories of purse seining boats in Japan (Statistics and Information Department 2000) was obtained. The CPUE (catch-per-unit-effort in tons per boat per year) was calculated based on the number of purse seine boats smaller than 50 t, which are principal purse seiners for the round herring fishery, and the total annual catch by the boats in this size category for the period 1969-1998 (Statistics and Information Department 2000). The size frequency distribution for *Etrumeus teres* caught by set nets (large trap nets) in the Kumano-nada Sea, east of the Kii Peninsula on the Pacific coast of central Japan was analyzed (Yamada 1994). The modal body length increased from 6.5 cm in April to 18.5 cm in February of the following year. The largest fish caught were approximately 23 cm long, and estimated to be 2-3 years old, but a large proportion of the catch was composed of fish smaller than 20 cm (≤1 year old). In *E. teres*, fluctuations in the catch and CPUE seem to more or less directly reflect recruitment variability.

The CPUE (kg per boat per day) of the blue sprat *Spratelloides gracilis* by small purse seiners was calculated, based on the data for the catch in mass and total days fishing in the local waters around Cape Shionomisaki at the tip of Kii Peninsula, from 1989 to 1998 (Wakayama Prefectural Fisheries Experimental Station, unpubl.). These local waters are at the northern distributional limits of *S. gracilis*, and no reported data on the generation time or life span for these populations were available. However, in tropical Indo-Pacific waters, *S. gracilis* lives for less than 4 months (Milton et al. 1991). Since *S. gracilis* is a short-lived species, we used CPUE as an index of recruitment.

Results and Discussion
Clupea pallasii

The cumulative catch in numbers of *Clupea pallasii* fluctuated widely over the 49 years represented in our data, ranging from 5.3×10^9 in 1915 to 0.004×10^9 in the 1952 year class (Fig. 3). Recruitment was very weak after the appearance of the large 1948 year class (2.1×10^9), and in 1959 there was no catch in the coastal waters of Hokkaido, northern island of Japan (Hanamura 1963). The magnitude of recruitment variability in *Clupea pallasii* calculated from these data was as large as $10^{3.1}$ over the 49-year period. Looking at shorter-term variability, in the 1930s the cumulative catch ranged from 5.2×10^9 in 1939 to 0.03×10^9 in the 1933 and 1934 year classes, representing 2.2 orders of magnitude.

Clupea pallasii were 4-5 years old and 26-28 cm long at first maturation in the first half of this century (Kobayashi 1993). The maximum age in the catch each year ranged from 10 to 15 years (Hanamura 1963). On average, adults continued spawning for 5 or more consecutive years. Dominant year classes ($\geq 2.0 \times 10^9$) appeared 10 times over the 49-year period (Fig. 3), or once every 5 years on average. This suggests that favorable environmental conditions after first maturation gave adult herring one or two opportunities to produce many progeny, establishing large year classes. Under variable environmental conditions found in subarctic waters, spawning for many years after first maturation seems to be a crucial ecological trait for herring to produce substantial numbers of offspring.

Sardinops melanostictus

The estimated number of age-1 recruits of *Sardinops melanostictus* fluctuated widely over the 17 years in our data set. Numbers of *S. melanostictus* varied from 16.8×10^9 in the 1983 year class to 0.03×10^9 in the 1990 year class, representing a difference of 2.7 orders of magnitude (Fig. 4). Although

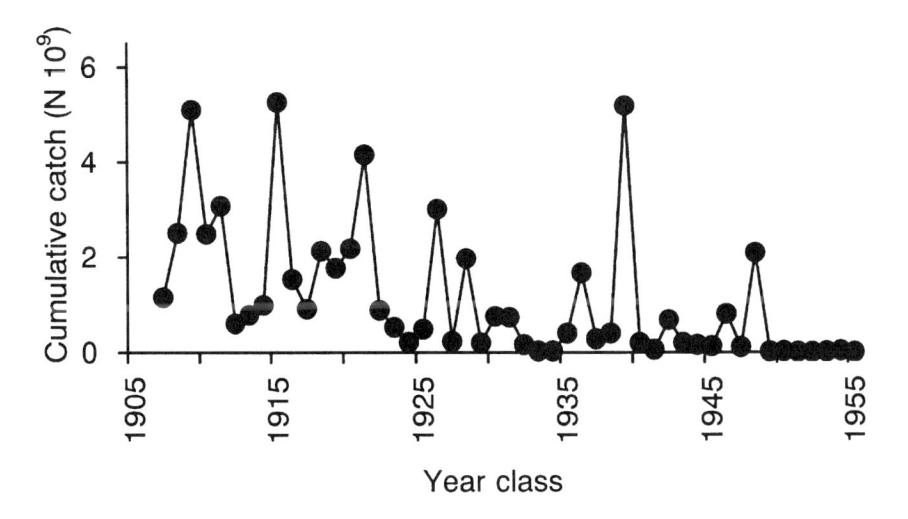

Figure 3. Cumulative catch in number of Clupea pallasii in Hokkaido from the 1907 to 1955 year classes (Hanamura 1963).

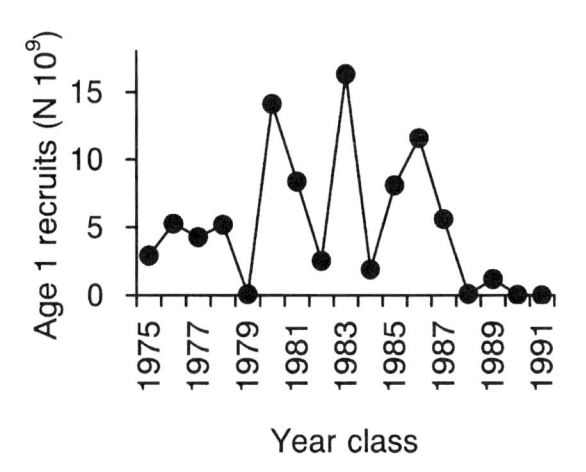

Figure 4. Number of age-1 recruits of Sardinops melanostictus in 1975-1991 year classes (Wada 1998).

the catch had declined after 1989, the estimated egg production of the sardine along the Pacific coast of Japan was $6,600 \times 10^{12}$ in 1990 (Ishida and Kikuchi 1992), which was much larger than the estimated 850×10^{12} eggs produced in 1983. The recruitment failure of the 1990 year class was due to high natural mortality after the first feeding stage (Watanabe et al. 1995). The estimated number of age 1 recruits in the 1991 year class was virtually zero, even though egg production in 1991 was as high as $3,900 \times 10^{12}$ (Zenitani et al. 1995). Overall magnitude of recruitment variability seems to have been larger than 3 orders of magnitude.

Recruitment of *Sardinops melanostictus* varied considerably from 1980 to 1991 (Fig. 4). The estimated latitudinal center of the spawning grounds of *S. melanostictus* in the waters off western and central Japan tended to move from the coastal area toward the offshore Kuroshio Current (Watanabe et al. 1996). When this occurred, a large proportion of eggs and larvae were transported by the current to the waters off Boso Peninsula, and to the eastern waters by the Kuroshio Extension (Heath et al. 1998). Larvae and juveniles transported to the Kuroshio Extension then migrated north to the subarctic Oyashio and the Doto fishing region. The migration of young-of-the-year (YOY) sardines to the subarctic waters was confirmed by catches of YOY fish in the Doto fishing region. The YOY catch of the 1980, 1984, 1985, and 1989 year classes was substantial (Mihara 1998). This implies that YOY sardines depended on the subarctic waters for feeding in the 1980s, when recruitment variability increased. The YOY catch was negligible from 1973 to 1979 (Mihara 1998), and there seems to have been relatively little variability from 1975 to 1978 (Fig. 4), when YOY sardines did not migrate far north to subarctic waters.

In *Sardinops melanostictus*, age at first maturity varies with fluctuations in the population. In 1953 when the population was small, 59% of age-1 females collected in the Sea of Japan had oocytes larger than 0.8 mm in diameter (Ito 1961). In the dominant 1980 year class, large parts of the cohort did not become matured at age 2 (Watanabe 1987). In 1995, 40% of age-1 fish in Tosa Bay had hydrated oocytes (Morimoto 1998). A large proportion of the sardines caught by fishing boats are 1- to 4-year-old fish (Wada 1998). These fish spawned for 2-3 consecutive years after first maturity in the years shown in Fig. 3. The sardine population did not decrease because of one year of recruitment failure in 1979; but the decline was inevitable because recruitment failed for the 4 consecutive years from 1988 to 1991 (Figs. 1, 4).

Etrumeus teres

The total catch in mass of *Etrumeus teres* ranged from 24 to 68,000 t in the 42 years between 1957 and 1998 (Fig. 1). Although the spawning area of this species largely overlaps that of the Japanese sardine in the waters off western Japan (Zenitani et al. 1995), the total catch of this species remained stable during the period when *Sardinops melanostictus* experienced a 500-fold increase in the 23 years from 1965 to 1988 and a $1/_{27}$ decline in 1990.

Cohorts of *Etrumeus teres* hatched in February, began increasing their gonad index in October at a body length of 16-17 cm, and matured in February of the following year (Yamada 1994). Since the maximum body length is approximately 23 cm, the life span of *E. teres* is estimated to be 2-3 years. A large part of the *E. teres* catch in Japan is composed of 0+ and 1+ fish (Yamada 1994). The CPUE of this species by medium- and small-size purse seining boats in Japan tended to increase after the second half of the 1980s. The CPUE ranged in 0.7 order of magnitude from 17.6 to 87.7 t per boat per year during the 30 years from 1969 to 1998 (Fig. 5). The stable CPUE in the last 30 years indicated that recruitment variability was not as large in *E. teres* as in *Clupea pallasii* and *Sardinops melanostictus*.

Spratelloides gracilis

The CPUE of *Spratelloides gracilis* in the waters around Cape Shionomisaki fluctuated (70-850 kg per day) by about 1 order of magnitude from 70 to 850 kg per day (Fig. 6), but implications for recruitment variability may be limited because the data originated from the local purse seine fishery. A distinct population peak such as seen in *Clupea pallasii* and *Sardinops melanostictus* was not observed in *E. teres*. The recruitment variability in short-lived *Spratelloides gracilis* (Milton et al. 1991) seems to be more like *E. teres* than that of *Clupea pallasii* or *Sardinops melanostictus*.

Recruitment Variability

In subarctic *Clupea pallasii* and in temperate *Sardinops melanostictus* during the 1980s, when YOY migrated north to the subarctic waters for feeding, recruitment variability was greater than 2-3 orders of magnitude, while in the warm water species *Etrumeus teres* and *Spratelloides gracilis* recruitment variability seems to have been within 1 order of magnitude. In general, primary production in boreal-temperate waters is seasonally cyclic and can be inter-annually variable (Levinton 1995). Inter-annual variability depends on the nutrient supply from deeper layers, which is affected by vertical mixing of the water column in winter. In the low latitudinal waters, on the other hand, the nutrient supply from deeper layers is limited, and primary production occurs at low levels almost year-round, using nutrients regenerated within the euphotic zone. Primary production is stable both seasonally and inter-annually. The latitudinal differences in the recruitment dynamics of clupeid fishes appear to be associated with the variability in the primary productivity and other physical and biological environments.

Among the 15 clupeid species distributed in the northwestern Pacific, *Clupea pallasii* is solely in subarctic waters, while *Sardinops melanostictus* and *Sardinella zunasi* are in temperate waters. The other 12 species are distributed mainly in subtropical and tropical regions, which is consistent with the observation that most clupeoid species are tropical or subtropical (Whitehead 1985b). Whitehead stated that "Quite possibly it is the high-latitude single-species stocks that are the later variants on the clupeoid

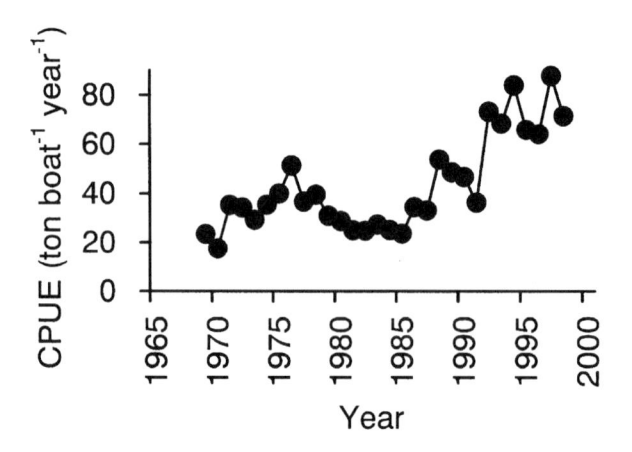

Figure 5. CPUE of Etrumeus teres calculated for purse seining boats <50 t. Total catch by the boats in this size category constituted 39-57% of total annual catch of this species in Japan.

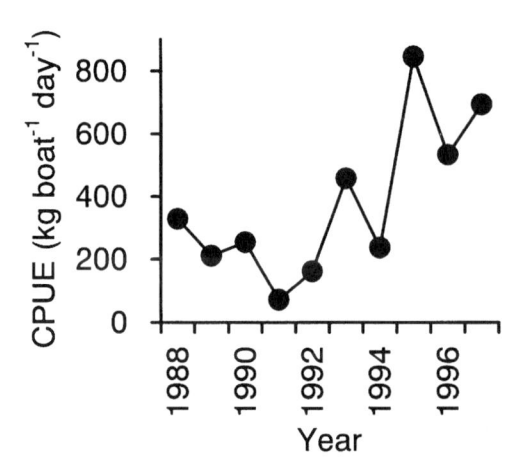

Figure 6. CPUE of Spratelloides gracilis in the waters around Kushimoto at the tip of the Kii Peninsula on the Pacific coast of central Japan (Wakayama Prefecture, Japan, pers. comm.).

Table 1. Life history traits of clupeid fishes in the northwestern Pacific.

	Larval TL at hatching (mm)	Maximum BL (mm)	Life span (year)	Age at first maturation (year)
Clupea pallasii[a]	7.2	360	15	4-5[b]
Sardinops melanostictus	3.0	240	8	2
Etrumeus teres	3.8[c]	230[d]	2[e]	1[e]
Spratelloides gracilis	4.6[c]	90[d]	<2[f]	<1[f]

[a]Masuda et al. (1984)
[b]Kobayashi (1993)
[c]Uchida et al. (1958)
[d]Yamada et al. (1986)
[e]Yamada (1994)
[f] N. Shirahuji (Ocean Res. Inst., Univ. Tokyo, unpubl.)

theme." *Clupea pallasii* is considered to have developed biological traits adaptive to northern subarctic waters, allowing it to take advantage of the region's high productivity. *What biological traits allowed Clupea pallasii to adapt to the productive but variable subarctic waters? What biological parameters in reproduction and early life history are responsible for the different recruitment dynamics observed in these species?*

Biological traits of the four species show that the cold water species tend to be larger in body size and longer in life span than warm water species (Table 1). The cold water species continue spawning for many years after first maturation, and therefore an adult fish can reproduce its progeny if it encounters one or two favorable years for reproduction under variable environment in the high latitudinal waters. The warm water species have only 1 or 2 years for spawning, and an adult fish needs to succeed in reproduction in each year. Such life history strategy is possible under stable environments in subtropical and tropical waters. We need to answer the questions above by conducting further comparative study in the reproductive biology of these four clupeid species.

References

Butler, J.L. 1991. Mortality and recruitment of Pacific sardine, *Sardinops sagax caerulea*, larvae in the California Current. Can. J. Fish. Aquat. Sci. 48:1713-1723.

Campana, S.E. 1996. Year-class strength and growth rate in young Atlantic cod *Gadus morhua*. Mar. Ecol. Prog. Ser. 135:21-26.

Cushing, D.H. 1975. Marine ecology and fisheries. Cambridge University Press, Cambridge. 278 pp.

Hanamura, N. 1963. A study on the method of prediction of the Hokkaido spring herring resources. Bull. Hokkaido Reg. Fish. Res. Lab. 26:1-66.

Heath, M., H. Zenitani, Y. Watanabe, R. Kimura, and M. Ishida. 1998. Modelling the dispersal of larval Japanese sardine, *Sardinops melanostictus*, by the Kuroshio Current in 1993 and 1994. Fish. Oceanogr. 7:335-346.

Hjort, J. 1917. Fluctuations in the great fisheries of northern Europe viewed in the light of biological research. Rapp. P.V. Reun. Cons. Int. Exp. Mer 20:1-228.

Houde, E.D. 1989. Subtleties and episodes in the early life of fishes. J. Fish Biol. 35(Suppl. A):29-38.

Hunter, J.R., and J. Alheit. 1995. International GLOBEC Small Pelagic Fishes and Climate Change Program. Report of the first planning meeting, La Paz, Mexico, June 20-24, 1994. GLOBEC Report No. 8, Plymouth, England. 72 pp.

Ishida, M., and H. Kikuchi (ed.). 1992. Monthly egg production of the Japanese sardine, anchovy, and mackerels off the southern coast of Japan by egg censuses: January 1989 through December 1990. Nansei Reg. Fish. Lab., Kochi. 86 pp.

Ito, S. 1961, Fishery biology of the sardine, *Sardinops melanosticta* (T. & S.), in the waters around Japan. Bull. Japan Sea Reg. Fish. Lab. 9:1-201

Kobayashi, T. 1993. Biochemical analyses of genetic variability and divergence of populations in Pacific herring. Bull. Natl. Res. Inst. Far Seas Fish. 30:1-77.

Kuroda, K. 1991. Studies on the recruitment process focusing on the early life history of the Japanese sardine, *Sardinops melanostictus* (Schlegel). Bull. Natl. Res. Inst. Fish. Sci. 3:25-278.

Lasker, R. 1975. Field criteria for survival of anchovy larvae: The relation between inshore chlorophyll maximum layers and successful first feeding. Fish. Bull., U.S. 73:453-462.

Levinton, J.S. 1995. Marine biology. Oxford University Press, New York. 420 pp.

Masuda, H., K. Amaoka, C. Araga, T. Ueno, and T. Yoshino (eds.). 1984. The fishes of the Japanese Archipelago. Tokai Univ. Press, Tokyo. 448 pp.

Mihara, Y. 1998. Immature and adult stocks. In: Y. Watanabe and T. Wada (eds.). Stock fluctuations and ecological changes of the Japanese sardine. Koseisha-Koseikaku, Tokyo, pp. 9-18.

Milton, D.A., S.J.M. Blaber, and N.J.F. Rawlinson. 1991. Age and growth of three species of tuna baitfish (genus: *Spratelloides*) in the tropical Indo-Pacific. J. Fish Biol. 39:849-866.

Morimoto, H. 1998. Maturation. In: Y. Watanabe and T. Wada (eds.). Stock fluctuations and ecological changes of the Japanese sardine. Koseisha Koseikaku, Tokyo, pp. 45-53.

Peterman, R., M.I. Bradford, N.C.H. Lo, and R.D. Methot. 1989. Contribution of early life stages to interannual variability in recruitment of northern anchovy (*Engraulis mordax*). Can. J. Fish. Aquat. Sci. 45:8-16.

Statistics and Information Department. 2000. Ministry of Agriculture, Forestry, and Fisheries. Annual report of fisheries and aquaculture productions in 1998. Norin-Tokei-Kyokai, Tokyo. 327 pp.

Uchida, K., S. Imai, S. Mito, S. Fujita, M. Ueno, Y. Shojima, T. Senta, M. Tahuku, and Y. Dotu. 1958. Studies on the eggs, larvae and juveniles of Japanese fishes. Faculty of Agriculture, Kyushu Univ., Fukuoka, Japan. 89 pp.

Wada, T. 1985. Population dynamics of Japanese sardine, *Sardinops melanostictus,* caught by the domestic purse seine fishery in the waters off the coast of southeastern Hokkaido. Bull. Hokkaido Reg. Fish. Res. Lab. 52:1-38.

Wada, T. 1998. Migration range and growth rate in the Oyashio area. In: Y. Watanabe and T. Wada (eds.), Stock fluctuations and ecological changes of the Japanese sardine. Koseisha-Koseikaku, Tokyo, pp. 27-34.

Watanabe, T. 1987. Appearance of the dominant 1980 year-class of Japanese sardine. Bull. Japan. Soc. Fish. Oceanogr. 51:34-39.

Watanabe, Y., H. Zenitani, and R. Kimura. 1995. Population decline of the Japanese sardine, *Sardinops melanostictus,* owing to recruitment failures. Can. J. Fish. Aquat. Sci. 52:1609-1616.

Watanabe, Y., H. Zenitani, and R. Kimura. 1996. Offshore expansion of spawning of the Japanese sardine *Sardinops melanostictus* and its implication to egg and larval survival. Can. J. Fish. Aquat. Sci. 53:55-61.

Whitehead, P.J.P. 1985a. FAO species catalog. Vol. 7. Clupeoid fishes of the world. An annotated and illustrated catalogue of the herrings, sardines, pilchards, sprats, anchovies and wolf herrings. Part I: Chirocentridae, Clupeidae and Pristigasteridae. FAO Fish. Synop. 125, Vol. 7(1). 303 pp.

Whitehead, P.J.P. 1985b. King herring: His place amongst the clupeoids. Can. J. Fish. Aquat. Sci. 42(Suppl. 1):3-20.

Yamada, H. 1994. Ecology of round herring *Etrumeus teres* in Kumano-nada Sea. Bull. Japanese Soc. Fish. Oceanogr. 58:286-292.

Yamada, U., M. Tagawa, S. Kishida, and Y. Honjo. 1986. Fishes of the East China Sea and the Yellow Sea. Seikai Reg. Fish. Res. Lab., Nagasaki, Japan. 493 pp.

Zenitani, H., M. Ishida, Y. Konishi, T. Goto, Y. Watanabe, and R. Kimura (eds.). 1995. Distribution of eggs and larvae of Japanese sardine, Japanese anchovy, mackerels, round herring, Japanese horse mackerel, and Japanese common squid in the waters around Japan, 1991 through 1993. Fisheries Agency Resources Management Research Report, Series A-1. Natl. Res. Inst. Fish. Sci., Yokohama. 368 pp.

Estimation of First-Year Survival of Pacific Herring from a Review of Recent Stage-Specific Studies

Brenda L. Norcross and Evelyn D. Brown
University of Alaska Fairbanks, School of Fisheries and Ocean Sciences, Institute of Marine Science, Fairbanks, Alaska

Abstract

We extracted stage-specific estimates of survival of Pacific herring (*Clupea pallasii*) from spawning through age-0 in Prince William Sound (PWS), Alaska, from recent studies. We used those estimates to calculate potential cumulative upper and lower limits of survival through the first year of juvenile life. Spawn timing, duration, amount, and location, as well as water temperatures during spawning may affect survival year-class strength, though no estimates of the magnitude of the effect on survival were available. Egg survival was affected primarily by wave action and predation with an estimated survival rate of 24-45%. Estimated survival for posthatch herring larvae was 50-100%. Starvation, predation, and transport to suitable nursery areas contribute to the mortality of later-stage larvae, but estimates of mortality varied among studies. Simulation modeling of larval drift in PWS, using a generalized daily mortality rate for herring in Alaska and British Columbia, which estimated mortality together with transport rates out of PWS, resulted in an overall estimated survival of 1-7%. During residence in nursery areas in summer through autumn, food availability, competition, predation, and disease may affect survival of juvenile herring. Acoustic surveys in PWS estimated relative density of juvenile herring in summer and autumn, resulting in an estimated survival rate of age-0 herring in October of 2-21%. During the winter (November through March), juvenile herring may run out of stored energy and starve. Conditions varied among nursery areas so modeled estimates of overwinter survival were 5-99%. Cumulating these stage-specific survival estimates yielded a range in survival estimates of 1 to 6,500 juveniles per 1 million eggs; i.e., three orders of magnitude. This compilation identifies the late larval and early juvenile stages as having the lowest survival estimates. Furthermore, the wide range of survival estimates for the winter juveniles was related to nursery areas.

We suggest field monitoring studies that could increase the precision of these estimates of juvenile herring survival.

Introduction

Pacific herring (*Clupea pallasii*) occur along the western coast of North America from Baja California to the Beaufort Sea, which encompasses the coastline of Alaska (Blaxter 1985, Tanasichuk et al. 1993). Herring have been important to the people of Prince William Sound (PWS) in southcentral Alaska (Fig. 1) for centuries. The indigenous peoples harvest herring and herring spawn-on-kelp as an important subsistence food source (J. Fall, Alaska Department of Fish and Game [(ADFG]), Subsistence Division, pers. comm.). Presently, herring are fished commercially for roe and spawn-on-kelp, human consumption, and bait. In PWS, commercial herring harvest peaked in 1938 at 50,870 t but total catch has varied greatly during its recorded history (Fig. 2). Since the late 1970s and the inception of the roe fishery, the value of the fishery has ranged from $187,000 US in 1996 to $12.2 million US in 1992 (Morstad et al. 1998). Though herring may be caught at an earlier age, they are fully recruited to these fisheries at age 4.

The biomass of herring in PWS was high and increasing prior to 1993 (Fig. 2). On 24 March 1989, during the period of high biomass, the tanker vessel *Exxon Valdez* ran aground on Bligh Reef in northeastern PWS (Fig. 1) and spilled 42 million liters of crude oil. Immediately following the oil spill, from 1 to 20 April 1989, herring spawned in PWS. At the time of the oil spill, little was known about early life stages (Brown et al. 1996), making injury to the PWS Pacific herring population from the *Exxon Valdez* oil spill difficult to determine. However in 1989, herring embryos and larvae had morphological and genetic damage and low survival, and herring larvae had slow growth rates (Kocan and Hose 1995, Hose et al. 1996, Kocan et al. 1996a, Norcross et al. 1996).

Estimates of spring spawning herring biomass from 1989 through 2000 ranged from 102,481 t in 1992 to 14,378 t in 1994 (Morstad et al. 1998; J. Wilcock and F. Funk, ADFG, pers. comm.). Only 20-25% of the forecasted return of the 1989 year class was found in PWS in 1993 (Funk 1995, Marty et al. 1998), apparently due to viral hemorrhagic septicemia virus (VHSV) (Marty et al. 1999). The disease causes external and internal hemorrhaging (Wolf 1988) and is either immediately fatal or the fish become immune to it. VHSV is known to be more common in fishes, especially age-0 (Kocan et al. 1997), that have compromised immune systems due to exposure to oil (Carls et al. 1998). Thus, herring catches were reduced in 1993 and the fisheries were closed from 1994 to 1996. The harvest prior to the collapse was 48,317 t in 1992; the highest catch since 1993 was 10,979 t in 1997 (Fig. 2). The 1998 adult herring biomass in PWS was estimated as approximately 38,640 t of which approximately 4,184 t were harvested by commercial fisheries (Morstad et al. 1999). The population appeared to increase with a forecasted 1999 abundance of 39,500 m (Morstad et al. 1999). How-

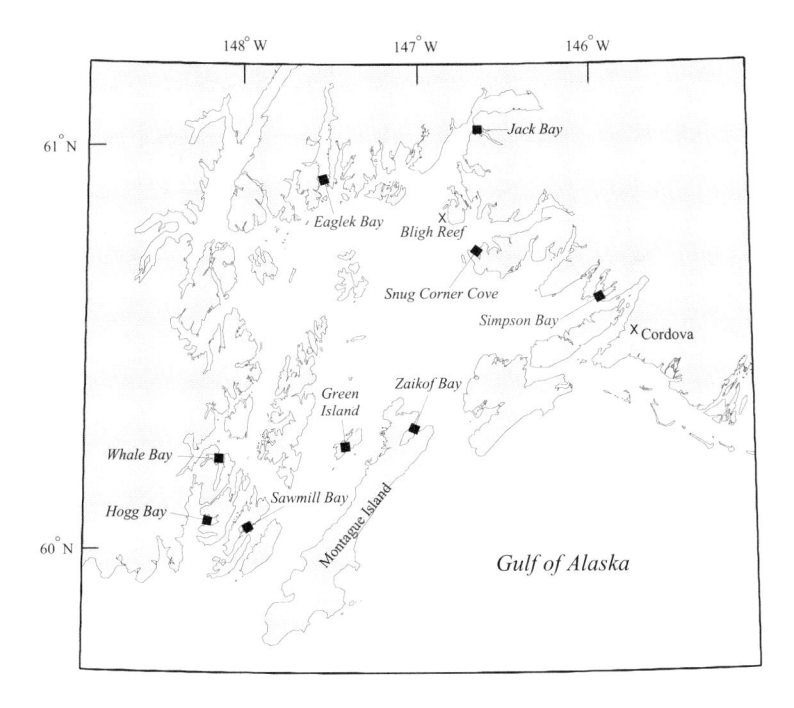

Figure 1. Study area in Prince William Sound, Alaska, including bays in which samples of juvenile herring were collected.

ever, the population of herring in PWS again collapsed in 1999. The fishery did not open and VHSV is suspected (G. Marty, University of California Davis, pers. comm.).

In response to the *Exxon Valdez* oil spill and the outbreak of VHSV, many studies were undertaken investigating a variety of factors that affect survival of embryos, larvae, and juvenile herring in PWS. Unlike papers that consider sources of mortality separately, in this paper we compile the results of multiple studies and evaluate their combined impact on the early life stages of herring. Lasker (1985) explored factors affecting egg (predation, variable production) and larval mortality (starvation, predation, transport, and competition) of clupeoids to determine the limitations of production. Herring studies conducted in PWS since 1989 allow us to examine more variables affecting egg and larval survival. The juvenile life stage of herring in PWS was studied from 1995 through early 1998 by the Sound Ecosystem Assessment (SEA) program (Norcross et al. 2002), allowing us to include survival of that stage as well. The objective of this paper is to review recent studies of mortality during the first-year life stages of herring with the focus in PWS. We assimilate estimated ranges of

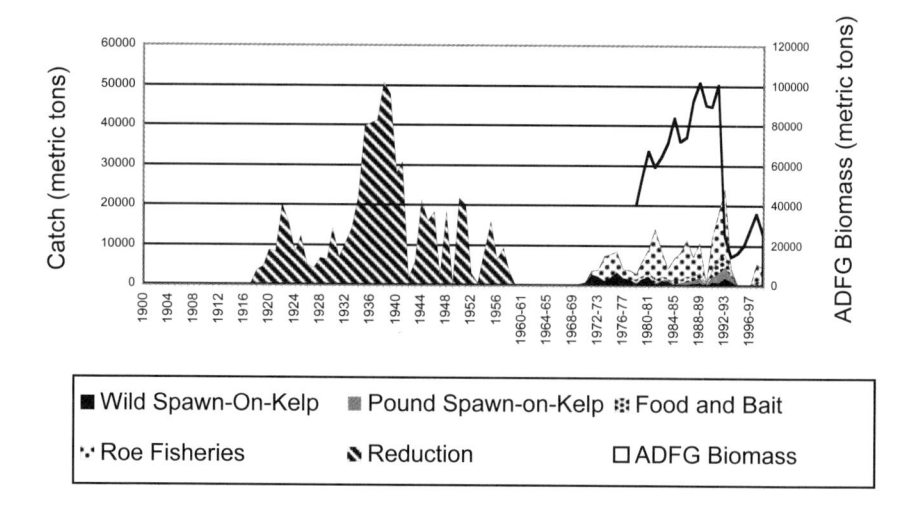

Figure 2. Historical commercial catch of herring in Prince William Sound, separated by fishery (Reid 1971, Morstad et al. 1998), and recent biomass estimates of adult herring by the Alaska Department of Fish and Game used to manage the fisheries in Prince William Sound, Alaska.

survival for egg, larval, and juvenile stages and calculate the potential cumulative upper and lower limits of survival through the first year. Comparison of the wide ranges of stage-specific survival estimates indicates the most vulnerable life stages of herring.

Spawning Effects

No prior studies in PWS estimated survival based on date of spawn, water temperature at or prior to spawning, or amount and location of spawn. In addition to behavioral and physiological processes, ecological factors such as water temperature and tidal cycle operate to control timing of spawning and location of spawning grounds (Hay 1985, 1990). Factors affecting spawn distribution and abundance are not well known, though local amount of spawn deposition may be directly related to stock abundance (Hay and Kronlund 1987).

Generally the warmer the water or the lower the latitude, the earlier date the maturation and spawning time of Pacific herring (Hay 1985, Ware and Tanasichuk 1989). Spawning in southeast Alaska occurs over 2-3 weeks in late April and early May when water temperatures are 5-6°C (Carlson 1980). PWS herring spawn at water temperatures of 4°C in calm seas (Biggs and Baker 1993). From 1973 to 1998 (Fig. 3) the mean date of spawn was 20 April ± 12 days (95% C.I.) (Biggs et al. 1992; J. Wilcock, ADFG, unpubl.

Herring: Expectations for a New Millennium 539

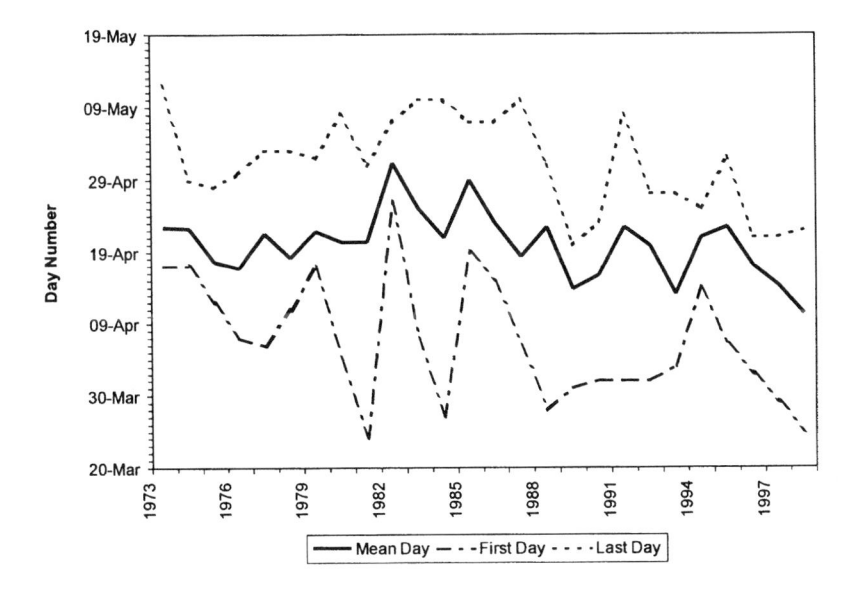

Figure 3. Mean, first, and last date of spawning of herring in Prince William Sound 1973-1998 (Biggs et al. 1992; J. Wilcock, ADFG, unpubl. data).

data). Each year of the SEA study, 1995-1998, spawning occurred earlier than the overall mean (Fig. 3), and spring water temperatures increased over the previous year (Gay and Vaughan 2002).

Based on ADFG aerial surveys, the total shoreline to receive herring spawn from 1973 to 1999 encompasses most of the coastline of the five ADFG sectors (North Shore, Naked Island, Northeast Shore, Southeast Shore, and Montague Island; Fig. 4a) and varied by an order of magnitude during that time (Table 1). Herring within PWS use a consistent range of spawning locations, especially the Northeast and Montague areas (Table 1). Fidelity to these two spawning sectors is evident even in the least (23 km in 1994) and greatest (268 km in 1988) spawn-extent years (Fig. 4b), as well as immediately following the oil spill (157 km in 1989; Fig. 4c). During the 1995-1998 SEA study, the spawn extent was 33-69 km mainly in Montague, and either Northeast or Southeast sectors (Fig. 4d). In years of low herring spawning biomass such as 1994, spawning sites are contracted and distribution of larvae may be affected.

Distribution of eggs is dependent upon the type of vegetation on which eggs are laid and the slope of the beach (Haegele et al. 1981). Diver estimates of egg numbers are expanded to the kilometers of shoreline spawned in each ADFG sector; and all sector estimates are summed to produce an estimate of total eggs deposited (Willette et al 1998). Estimates of annual

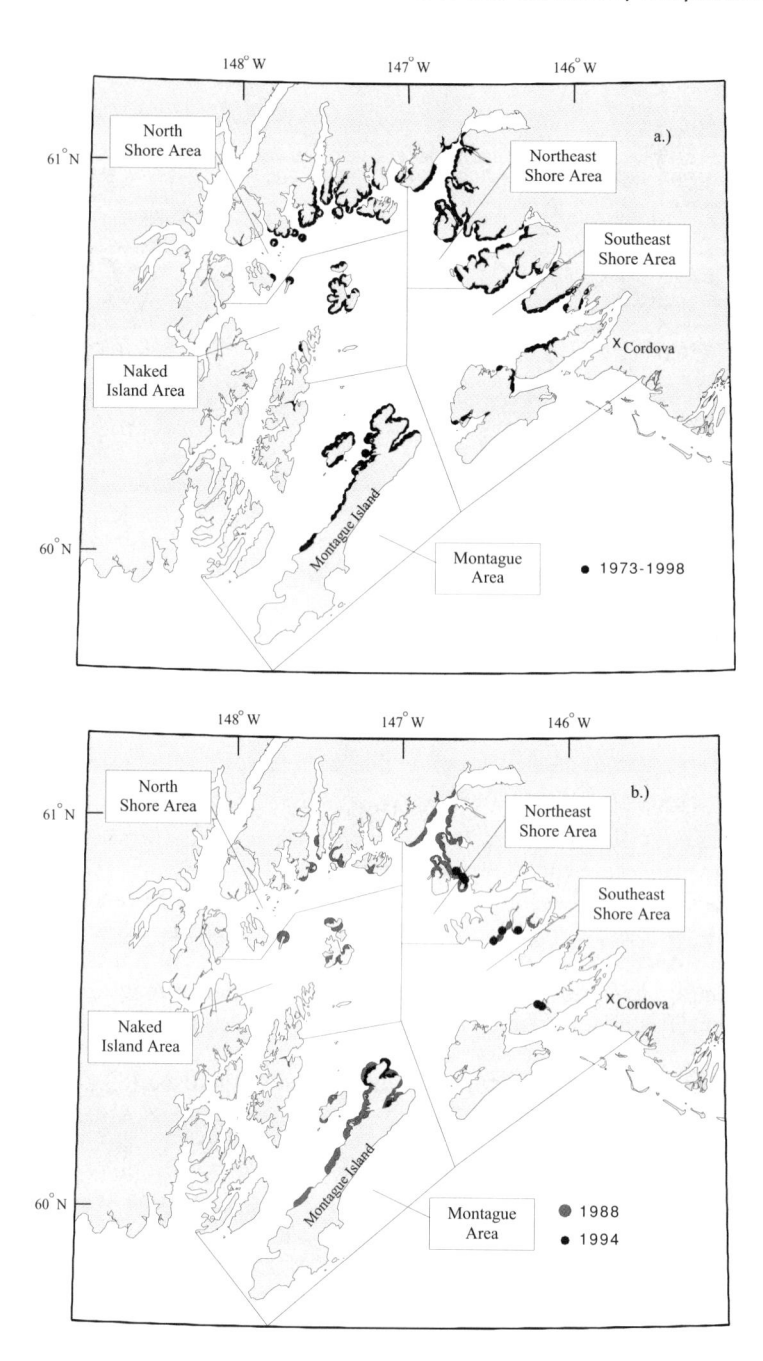

Figure 4. Pacific herring spawning locations by ADFG sector in Prince William Sound (a) combined 1973-1999; (b) in 1994, the year with the least, and 1988, the year with the most kilometers of spawn.

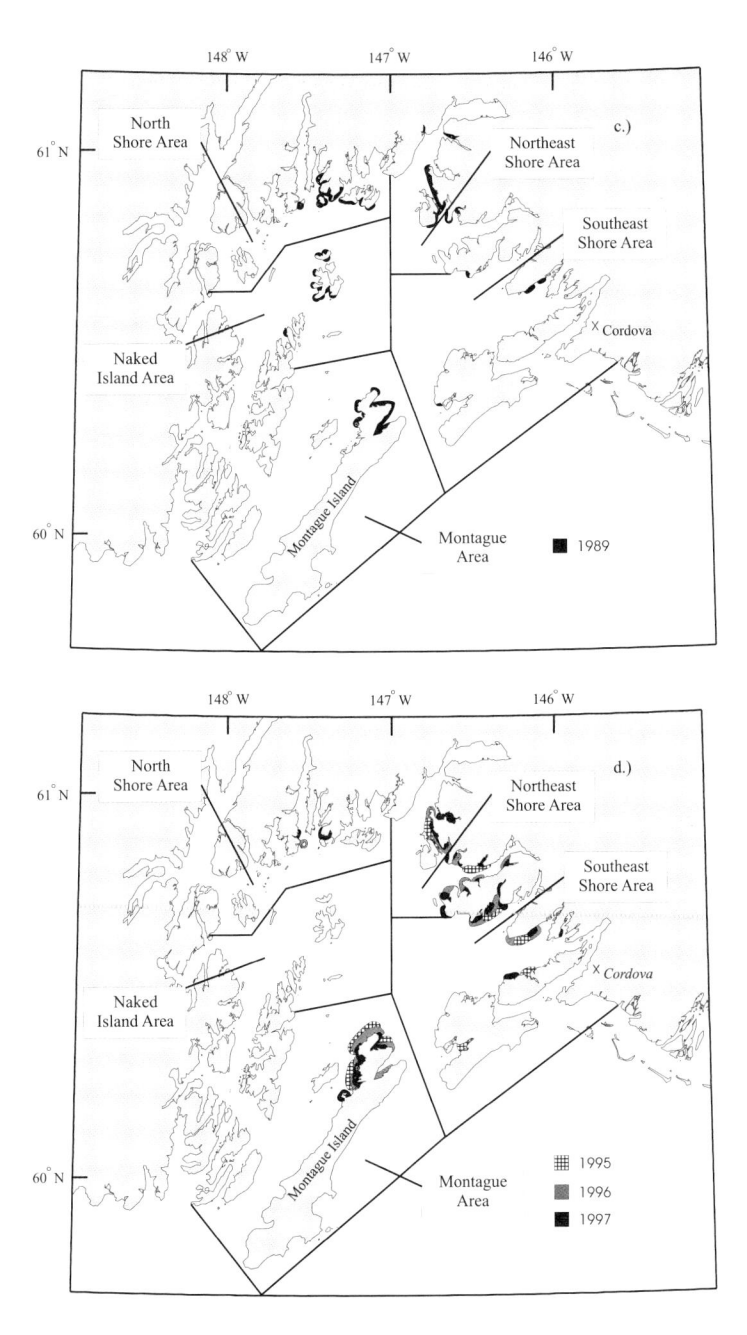

Figure 4. (Continued.) Pacific herring spawning locations by ADFG sector in Prince William Sound (c) in 1989, the year of the Exxon Valdez oil spill; and (d) 1995, 1996, and 1997 during the SEA project.

Table 1. Cumulative kilometers of shoreline on which herring spawned in PWS, 1973-1999, apportioned by ADFG survey sector (Biggs et al. 1992, Donaldson et al. 1993, Donaldson et al. 1995, Willette et al. 1998).

Year	Southeast	Northeast	North Shore	Naked Island	Montague	Total
1973	0.0	47.5	4.8	0.0	10.0	62.3
1974	0.0	37.8	0.0	0.0	24.2	62.0
1975	0.0	44.6	0.0	0.0	10.5	55.1
1976	0.0	47.7	0.0	0.0	3.7	51.4
1977	0.0	60.9	0.0	0.0	2.4	63.3
1978	0.0	45.9	0.0	0.0	0.3	46.2
1979	33.8	52.3	0.0	0.0	1.6	87.7
1980	16.6	56.0	0.0	0.0	8.9	81.5
1981	22.4	46.7	9.3	0.0	59.2	137.6
1982	0.5	30.4	25.4	5.6	16.9	78.8
1983	4.5	21.4	16.9	28.5	37.2	108.5
1984	14.0	19.5	26.2	12.2	25.0	96.9
1985	5.2	56.4	53.9	26.4	21.3	163.2
1986	4.5	41.1	59.7	0.0	11.3	116.6
1987	11.1	34.3	42.3	3.7	13.7	105.1
1988	6.1	91.6	24.8	29.5	115.9	267.9
1989	5.6	34.8	48.4	22.1	46.5	157.4
1990	4.2	70.4	29.3	8.7	39	151.6
1991	6.3	45.9	1.9	0.0	39.0	93.1
1992	11.6	51.8	0.0	0.5	56.4	120.3
1993	2.7	8.9	0.0	0.0	21.3	32.9
1994	0.3	0.5	0.0	0.0	22.5	23.3
1995	9.3	3.2	0.0	0.0	20.2	32.7
1996	3.9	16.1	0.3	0.0	23.5	43.8
1997	11.2	25.1	3.3	0.0	28.9	68.5
1998	27.5	11.1	3.4	0.0	25.8	67.8
1999	23.7	7.4	0.0	0.0	9.8	40.9
Total km	225.0	1,009.3	349.9	137.2	695.0	2,416.4
Average km	8.3	37.4	13.0	5.1	25.7	89.5
% total	9.3	41.8	14.5	5.7	28.8	

total egg deposition from 1984 to 1997 varied from approximately 9×10^{12} in 1991 to 1.4×10^{12} in 1997 (Willette et al. 1998; J. Wilcock and F. Funk, ADFG, unpubl. data). The number of kilometers of spawn does not directly relate to egg deposition. In 1988, 268 km of spawn yielded a total egg abundance of 3.7×10^{12}, while 23 km yielded 2.1×10^{12} eggs in 1994. In 1990-1992, when the herring population was large, the mean number of eggs deposited in PWS was $8.1 \pm 2.0 \times 10^{12}$ (95% C.I.) (Willette et al. 1998; J. Wilcock and F. Funk, ADFG, unpubl. data). During the SEA study, egg abundance was $1.9 \pm 0.8 \times 10^{12}$ (Willette et al. 1998). A ± 5-18% variance is associated with the egg estimates (Willette et al. 1998; F. Funk, pers. comm.).

Egg Survival

Herring eggs may be lost due to wave action (Hart and Tester 1934, Hay and Miller 1982); bird, fish, invertebrate, and marine mammal predation (Haegele and Schweigert 1989, 1991, Haegele 1993, Rooper 1996); and density of egg deposition (Hay 1985). The largest source of herring embryo mortality in PWS occurs where eggs are physically removed from their substrate by predation or wave forces (Rooper 1996). The effect of predators and wave exposure is governed by the depth of egg deposition. Seventy-eight percent of the variation in egg loss could be explained by bird abundance and water depth where eggs were deposited, versus only 54% variation explained using depth alone (Rooper 1996). Bishop and Green (2002) estimated that in 1994 birds consumed 27% of the herring eggs deposited.

Estimating egg survival required a combination of several studies. Cumulative time of air exposure was the most effective predictor of egg loss (Rooper et al. 1998). Herring embryos incubate in intertidal and shallow subtidal areas for 22 to 24 days in PWS (Biggs et al. 1992, Brown et al. 1996). Based on an air-exposure model, 76% of eggs are lost over the entire incubation period (Rooper 1996), providing 24% egg survival. That may be a low survival estimate, as a refinement of the model shows 69% of eggs lost per day over the first 5 days (Rooper et al. 1998). Of the eggs remaining attached, embryos die from dehydration, asphyxiation, fungus, and unknown causes at a rate of 2.5% per day (Biggs and Baker 1993). Combining these rates of loss and applying them over an incubation period of 22 to 24 days results in 43-45% egg survival. Thus, the estimated range of egg survival is between 24% and 45%.

Posthatch Survival

Hatching is a stressful transitional period when genetically imperfect or damaged larvae hatch but cannot survive (Purcell et al. 1990). Exposure to unusually warm air temperatures caused contorted bodies and reduced jaws and pectoral fins in 2-25% of yolk-sac larvae. An additional 4% to 68% of post-yolk-sac larvae had underdeveloped jaws, which may impair feeding and lead to starvation-induced mortality (Purcell et al. 1990). When air temperatures were average, these abnormalities were less prevalent. Reduced or absent jaws were found in May 1989 following the *Exxon Valdez* oil spill in 60-100% of herring larvae at nearshore sites in PWS (Hose et al. 1996) and 88% of larvae at offshore sites (Norcross et al. 1996), resulting in an overall mortality of 52% (Brown et al. 1996). The high incidence of abnormalities in 1989, as compared to 1995, was compounded by effects of the oil spill, including genetic mutations (Brown et al. 1996, Hose et al. 1996, Norcross et al. 1996). Craniofacial and mandibular malformations naturally occur in up to 25% in yolk-sac larval herring and 58% in post-yolk-sac larvae (Struhsaker 1977, Nanke 1981). In 1990 and 1991 no morphologic abnormalities of larvae were found in PWS (Hose et al. 1996) but in May 1995, the occurrence of jaw abnormalities was 61% in nearshore areas. The variation in posthatch mortality due to abnormal development of jaws is high and appears to have geographic and interannual differences. Survival ranges from 50% to100%.

Larval Survival

Clupeoid larvae may be transported into either favorable or unfavorable nursery areas (Lasker 1985). Larval herring are transported horizontally by water currents although they can adjust their vertical position in the water column (Graham 1972, Sinclair and Iles 1985, Stephenson and Power 1991). A three-dimensional numerical ocean current model for PWS (3D-PWS model) was used to study the potential transport of herring larvae from spawning sites to nursery areas (B.L. Norcross and others, unpubl. data). Circulation parameters based on 1996 conditions (Wang et al. 1997, 2002) were used to simulate physical dispersal of larvae as passive tracers (Norcross et al. 2002). Starting parameters included spawning locations (Fig. 4d) and estimated number of herring eggs deposited in 1996 (Willette et al. 1998) with a hatch duration, i.e., particle release, of 1-15 May (McGurk and Brown 1996).

Starvation (McGurk 1984) and predation (Purcell et al. 1987, Purcell and Grover 1990, Wespestad and Moksness 1990, McGurk et al. 1993) are the major sources of loss of herring larvae. In laboratory experiments, McGurk (1984) showed 18-36% of first-feeding herring larvae starve. In British Columbia mortality estimates range from 0.013 per day offshore to 0.064 per day inshore (Alderdice and Hourston 1985), and from 0.02 to 0.16 per day between cohorts (McGurk 1989). A geometric mean mortality of Pacific herring larvae off B.C. was calculated as 0.083 per day (McGurk 1993). Though mortality is thought to be greatest within the first few weeks (Wespestad and Moksness 1990), only average values over the length of the larval period were available. Therefore, in the larval herring transport model, mortality was held constant at a value of 0.05 per day over the entire simulated drift period of 1 May through 1 August.

The dispersal of simulated herring larvae (SHL) shows that larvae originating in different locations can reach the same nursery destinations

(B.L. Norcross and others, unpubl. data); i.e., progeny from different spawning populations mix within a nursery area. This result is similar to the distribution of larval Pacific herring found in British Columbia (Hay and McCarter 1997), but unlike that of Atlantic herring (*Clupea harengus*) in the North Atlantic Ocean (Iles and Sinclair 1982). A wide dispersal of larvae to numerous nursery areas insures some will encounter conditions that promote survival and minimize density-dependence effects within a single nursery area.

The model showed most SHL being advected into bays and retained in PWS while 1.3% were transported out of PWS (Norcross et al. 2002, unpubl. data). We assumed that all larvae that exited PWS were lost. By August the larvae metamorphose (Paul and Paul 1999) to juveniles in nursery areas (Stokesbury et al. 2000). The model showed that primarily because of mortality, only 7% of the SHL survived in PWS to 1 July and 1% of SHL survived in PWS to 1 August (B.L. Norcross and others, unpubl. data).

Autumn Juvenile Survival

As part of the SEA project, four bays in the north, south, east, and west regions of PWS were surveyed nine times from June 1996 to March 1998 (Stokesbury et al. 1999, 2000, 2002). Fish schools were surveyed acoustically and sampled with nets to determine species composition and size structure. A length-dependent scaling constant was used to convert the reflected acoustic energy into estimates of relative density (Stokesbury et al. 2000, 2002). A standardized transect length was used to yield relative density-per-unit-effort (number of fish per square kilometer) of juvenile herring for comparison among years and bays.

The relative abundance of age-0 herring declined significantly between August and October in both 1996 and 1997 (Table 2), averaging 15% seasonal survival (Stokesbury et al. 2002). At the same time, average length increased sharply for both year classes. Between August and October zooplankton was abundant, and juvenile herring were feeding and increasing their energy stores (Norcross et al. 2002). Though many juvenile pollock were present in the same bays, they were physically separate from the herring by being lower in the water column (Stokesbury et al. 2000). Thus interspecific competition for food may not have been the cause of autumn juvenile mortality. However, intraspecific competition for food may have occurred in bays with the densest populations of age-0 herring (Norcross et al. 2002). Salmon may eat juvenile herring in the autumn (Stokesbury et al. 2002). Pacific cod (*Gadus macrocephalus*) are known to consume juvenile herring in British Columbia (Walters et al. 1986) but were not noted in high abundance in PWS. Diseases such as VHSV, which occurs in PWS (Meyers et al. 1994) and is linked to the collapse of the herring fishery in 1993 (Marty et al. 1998), are rapidly spread by waterborne exposure and affect survival of juvenile herring (Kocan et al. 1997). The SEA study did not investigate the mechanism of mortality for juvenile herring in autumn. The estimated autumn survival of juvenile herring in PWS, based on acoustic estimates, over 2 years (Table 2) was 2% to 21%.

Table 2. Relative densities of age-0 herring (number of fish/km^2) along transects in four bays in PWS.

Date	Eaglek	Simpson	Whale	Zaikof
Jun. 1996	24,195	66,473	47,692	
Aug. 1996		524,123	105,707	
Oct. 1996		11,665	2,536	
May 1997		16,429		
Jul. 1997			150,528	
Aug. 1997	10,149	296,572	28,240	
Oct. 1997	15,518	62,989		9,017
Mar. 1998		5,064		1,513

Blanks indicate bay not sampled that month or no age-0 herring collected.

Winter Juvenile Survival

The timing of metamorphosis from larval to juvenile stage allows a very short growing season in which age-0 herring prepare for their first winter. Lab, field, and modeling experiments determined that energy content of juvenile herring in fall is critical to overwinter survival (Paul and Paul 1998a, Norcross et al. 2002), making that season one of the most vulnerable times for natural mortality, similar to Atlantic herring (Wilkins 1967). Summer food availability and possible feeding competition within nursery areas affect the autumn condition of age-0 herring (Paul et al. 1998). The limited amount of food consumed in the winter is not sufficient to meet metabolic needs of all juvenile herring (Foy and Norcross 1999). Field collections verify a decline in energy stores of age-0 herring from December to March (Foy and Paul 1999). The smallest age-0 fish were at risk of starvation during winter because their energy intake was less than their winter metabolic requirements (Paul and Paul 1999). In addition, herring may be more vulnerable to predators during the winter if hunger suppresses their antipredator behavior (Robinson and Pitcher 1989, Pitcher and Parrish 1993, Robinson 1995).

Overwinter mortality of age-0 herring was modeled using October energy content of herring and winter water temperature (Patrick 1999; Patrick et al. 2002). The model assumed no feeding during the winter, although modest amounts of feeding are known to occur (Foy and Norcross 1999, Foy and Paul 1999). The overwinter model demonstrated that mortality due to starvation may regulate recruitment (Patrick et al. 2002). Predicted size and energy density of fish surviving the winter showed patterns similar to spring field collections (Patrick et al. 2002). Furthermore, the low survival predicted by the model for Simpson Bay is consistent with expected effects of competition for limited food resources (Norcross et al.

2002). The estimates of survival from the model (Table 3) are of the same order of magnitude as acoustic measures of density of juvenile herring in winter 1997-1998. In Simpson Bay (Fig. 1), 5% of SHL were predicted to survive; 8% survival was measured (Table 2). In Zaikof Bay, 56% were expected to survive and 17% survival was measured. Warmer winter temperatures in 1997-1998 increased metabolic costs, making energy content of age-0 herring at the start of winter even more important for survival (Norcross et al. 2002). Model simulations indicated that the probability of herring survival varied among bays in PWS, possibly due to different physical and biological conditions within bays, resulting in a range of overwinter survival of 5% to 99% (Table 3). The acoustic estimate of average age-0 winter survival for 1995, 1996, and 1997 year classes was 51% for the bays sampled (Stokesbury et al. 2002).

Year Class Survival

Survival during each successive life stage governs the size of a year class of Pacific herring. Basic components are found in the early life history of all herring year classes: spawning amount, location, and timing; loss and predation of eggs; direction of larval dispersion and timing of larval entry into bays and fjords; food, predation, and temperature in nursery areas; condition and size of juveniles; and winter duration and water temperature. During the course of the first year of life of herring, positive or negative conditions may affect any of these components. The interactions of these effects, which are multiplicative, determine survival (S) and thus size of a year class as follows:

$$S_{age-1} = (S_{egg})(S_{larvae})(S_{autumn\ juveniles})(S_{winter\ juveniles})$$

We calculated survival for the potential extremes, i.e., cumulative lowest and highest estimates possible at each life stage. The result is a three order of magnitude range of survival estimates. Based on our calculations, from each 1 million eggs deposited, 1-6,500 juvenile herring may survive the first year (Table 4). This agrees with the variability in herring catch of three orders of magnitude found in Japan (Watanabe et al. 2001). When this general range of survival is applied to specific spawning events, the implications of potential variability to recruitment are enormous. The same years used to illustrate distribution of spawning around PWS (Fig. 4) were estimated (Table 5). The result is a difference between millions and tens of billions of herring surviving through their first year.

Discussion

Lasker's (1985) answer to his question "what limits clupeoids?" was "almost everything"; he then went on to ask "when in the life cycle?" As Lasker did, we conclude that many factors affect survival of herring, and we pro-

Table 3. Survival estimates from overwinter fasting model (Patrick 1999).

Year	Bay	Survival
1995-1996	Green Island	91%
1995-1996	Hogg Bay	7%
1995-1996	Jack Bay	92%
1995-1996	Sawmill Bay	99%
1995-1996	Snug Corner Cove	64%
1995-1996	Simpson Bay	64%
1995-1996	Whale Bay	86%
1995-1996	Zaikof Bay	39%
1996-1997	Eaglek Bay	76%
1996-1997	Simpson Bay	18%
1996-1997	Whale Bay	86%
1996-1997	Zaikof Bay	56%
1997-1998	Eaglek Bay	95%
1997-1998	Simpson Bay	5%
1997-1998	Zaikof Bay	56%

See Fig. 1 for locations.

Table 4. Estimation of year-class survival based on approximate ranges of early life-stage survival for each 10 billion eggs deposited.

Life stage	Percent survival	Number of herring eggs ($\times 10,000,000$)
Eggs	24-45%	2,400,000-4,500,000
Posthatch	50-100%	1,200,000-4,500,000
Larval drift	1-7%	12,000-315,000
Fall juveniles	2-21%	240-66,150
Winter juveniles	5-99%	12-65,489

Table 5. Estimated potential range of first-year survival for herring in PWS, Alaska, for the given years.

Year	Estimated number of eggs deposited	Estimated range of juvenile survival
1988	3.73×10^{12}	4.5×10^6 to 2.4×10^{10}
1989	3.13×10^{12}	3.8×10^6 to 2.1×10^{10}
1994	2.06×10^{12}	2.5×10^6 to 1.4×10^{10}
1995	2.00×10^{12}	2.4×10^6 to 1.3×10^{10}
1996	2.68×10^{12}	3.2×10^6 to 1.8×10^{10}
1997	1.97×10^{12}	2.4×10^6 to 1.3×10^{10}

Estimated eggs deposited from F. Funk (ADFG, Juneau, pers. comm.) and Willette et al. (1998).

vide answers to his question about life stage. Lasker's (1985) review of clupeoids included only egg and larval stages. Our analysis shows that in Pacific herring vulnerability is not equal among early life history stages, and that similar to results for Atlantic herring (de Bartos and Toresen 1998), juveniles can experience greater mortality than eggs or early larvae. Estimated survival is lower during the larval drift (<7%) and fall juvenile (<21%) stages than during egg (>24%) or posthatch larval (>50%) stages (Table 4).

Like clupeoids in general (Lasker 1985) and Pacific herring in PWS in particular, recruitment of Atlantic herring in Maine results from a complex interaction of processes in which no single factor is dominant (Campbell and Graham 1991). While the study in Maine noted that recruitment is usually determined in the larval stage, it further noted that in some years, recruitment might be determined at the juvenile stage. Comparisons between spring-spawning Pacific herring and autumn-spawning Atlantic herring are tenuous. It is interesting to note that food supply in winter and autumn (larval stage) was a primary limiting factor in Maine (Campbell and Graham 1991). In PWS, food availability in summer and autumn (Foy and Norcross 1999) affects condition of juvenile herring as they enter the winter fasting period (Paul et al. 1998, Foy and Paul 1999, Paul and Paul 1999, Patrick, in review) and is a key factor governing survival in fall and winter juveniles.

Due to differences among bays, overwinter survival of juvenile herring has the most variability (5% to 99%) (Patrick 1999; Patrick et al. 2002) of the life stages identified in this study (Table 4). Habitat conditions in various bays and fjords of PWS, as measured by temperature, abundance, energy value and source of zooplankton, size and energy content of juvenile herring, are not uniform in space and time (Norcross et al. 2002). This inequality in habitat is reflected in inequality in survival and may be linked to interannual variation in local climate and oceanographic conditions affecting PWS (S.M. Gay, unpubl.; Gay and Vaughan 2002). Additionally, the uncertain nature of larval drift patterns affects the density of juvenile herring in nursery areas (B.L. Norcross and others, unpubl. data), compounding the effects of local variation in habitat quality on observed survival through density-dependent competition (Norcross et al. 2002). Furthermore, it is possible that when spawning is more evenly distributed around PWS, the likelihood of herring larvae being transported to low-density or high-quality nursery areas would increase, thus increasing the likelihood of juvenile herring survival.

In this review, we calculated only the extremes of survival at each life stage based on the best and worst case estimates; however, any combination of survival estimates for each life stage is possible. The results of this exercise outline a mechanism to "predict" juvenile survival through monitoring of specific life stages, especially the most variable and vulnerable ones that are likely to have the largest influence on year-class strength. The time and space scales used for measurements will affect the accuracy of the estimation of survival. For example, the overall variability in the survival estimate (Table 4) could be reduced an order of magnitude through annual monitoring of autumn condition of juveniles in representative nurseries and modeling overwinter survival to provide regional estimates of survival (Norcross et al. 2002). Another way to increase precision by an order of magnitude of the overall survival estimate would be to couple additional monitoring of regional density of juvenile herring with regional estimates of survival. The precision of estimated survival as presented here (Table 5) might not be meaningful for management, but with refinement could provide estimates of increasing or decreasing trends of herring year classes 2-3 years before they enter the fishery. Meteorologists and economists do not predict daily weather or precise economic conditions a year ahead. They use indicators to predict trends. The outcome of this analysis is not intended to be on par with daily weather predictions or, for fish, preseason adult biomass estimates, but rather to suggest a mechanism to examine long-term trends in year-class strength.

No single life stage or variable is the "key" to predicting year-class strength of herring in PWS. Herring populations are affected by the environment (Cushing 1982) and experience variable recruitment (Zheng 1996). Strong year classes of herring occurred every 4 years from 1976 to 1988 in PWS (Zheng et al. 1993, Zheng 1996). It was assumed that the environment shifted to a favorable set of conditions every fourth year (Zheng et al. 1993), but it has since been conceded that the cause for the cycle has not been found (Zheng 1996). Nothing in this review indicated a factor affecting a life stage of herring with a periodicity of 4 years.

When strong year classes first appear in the commercial fishery, they are age 3 or 4; i.e., 2-3 years beyond our estimates of survival. While mortality at older ages is generally low compared to that of larvae and juveniles, some factors, such as disease, may be significant. VHSV causes an acute disease, especially in juvenile fish (Kocan et al. 1997; Marty et al. 1998; G.D. Marty, University of California School of Veterinary Medicine, pers. comm.). It may cause the collapse of a fishery, or more commonly may significantly limit recruitment (G.D. Marty, pers. comm.). The disease was prevalent in PWS in 1997 and 1998 (G.D. Marty, pers. comm.), thus its effect could be combined with estimates of juvenile survival presented here. However, VHSV also affects older age classes (Marty et al. 1998, Hershberger et al. 1999), an effect not accounted for here. Another disease found in PWS (Marty et al. 1998), caused by the fungus *Ichthyophonus hoferi*, is chronic and decreases the life expectancy of the fish (G.D. Marty, pers. comm.). A recent modeling study shows no correlations between these two diseases and recruitment, perhaps indicating that their main effect is in the adult stage (Quinn et al. 2001).

Survival of all life stages of herring and the effects of these diseases, especially VHSV, also were affected by a nonpredictable disastrous event, the *Exxon Valdez* oil spill in 1989. The added stress of exposure to oil just prior to spawning led to the development of hepatic necrosis caused by VHSV (Marty et al. 1999) as activation of VHSV occurs upon exposure to crude oil (Carls et al. 1998). Impaired reproductive success in 1992 could not be directly related to oil exposure in 1989 (Kocan et al. 1996b) or 1995 (Johnson et al. 1997). Thus the *Exxon Valdez* oil spill directly affected survival of the 1989 year class of herring eggs (McGurk and Brown 1996) and larvae (Norcross et al. 1996, Marty et al. 1997, Hose and Brown 1998), and contributed to the collapse of the fishery in 1993 (Marty et al. 1999). There is no evidence to link exposure to oil, the susceptibility to VHSV of herring and the collapse of the herring fishery in 1993 to a similar VHSV-induced collapse of the herring fishery in 1999 (G.D. Marty, pers. comm.; Quinn et al. 2001).

Acknowledgments

We thank the many scientists and technicians from UAF, ADFG, and PWSSC and fishermen from Cordova, along with Cordova Air, who participated in the field collections. We especially want to thank Ted Cooney and Jennifer Allen for their help and encouragement, and Tom Kline, A.J. Paul, and Kevin Stokesbury for ideas that contributed to this paper. Thanks to Doug Hay, Sue Hills, and Brenda Holladay for comments on this manuscript. The Exxon Valdez Oil Spill Trustee Council through the Sound Ecosystem Assessment project funded this project. The findings presented by the authors are their own and not necessarily the position of the Trustee Council.

References

Alderdice, D.F., and A.S. Hourston. 1985. Factors influencing development and survival of Pacific herring (*Clupea harengus pallasi*) eggs and larvae to beginning of exogenous feeding. Can. J. Fish. Aquat. Sci. 42(Suppl. 1):56-68.

Biggs, E.D., and T.T. Baker. 1993. Studies on Pacific herring, *Clupea pallasi*, spawning in Prince William Sound following the 1989 *Exxon Valdez* oil spill, 1989-1992. Annual Report for Natural Resource Damage Assessment Fish/Shellfish Study Number 11, *Exxon Valdez* Oil Spill Trustee Council, Anchorage, Alaska. 52 pp.

Biggs, E.D., B.E. Haley, and J.M Gilman. 1992. Historic database for Pacific herring in Prince William Sound, Alaska, 1973-1991. Alaska Department of Fish and Game, Division of Commercial Fisheries, Regional Informational Report 2C91-11, Anchorage. 335 pp.

Bishop, M.A., and S.P. Green. 2002. Predation on Pacific herring (*Clupea pallasi*) spawn by birds in Prince William Sound, Alaska. Fish. Oceanogr. In press.

Blaxter, J.H.S. 1985 The herring: A successful species? Can. J. Fish. Aquat. Sci. 42(Suppl. 1):21-30.

Brown, E.D., B.L. Norcross, and J.W. Short. 1996. An introduction to studies on the effects of the *Exxon Valdez* oil spill on early life history stages of Pacific herring, *Clupea pallasi*, in Prince William Sound, Alaska. Can. J. Fish. Aquat. Sci. 53(10):2337-2342.

Campbell, D.E., and J.J. Graham. 1991. Herring recruitment in Maine coastal waters: An ecological model. Can. J. Fish. Aquat. Sci. 48:448-471.

Carls, M.G., G.D. Marty, T.R. Meyers, R.E., Thomas, and S.D. Rice. 1998. Expression of viral hemorrhagic septicemia virus in prespawning Pacific herring (*Clupea pallasi*) exposed to weathered crude oil. Can. J. Fish. Aquat. Sci. 55:1-10.

Carlson, H.R. 1980. Seasonal distribution and environment of Pacific herring near Auke Bay, Lynn Canal, southeastern Alaska. Trans. Am. Fish. Soc. 109:71-78.

Cushing, D.H. 1982. Climate and fisheries. Academic Press, London. 373 pp.

de Bartos, P., and R. Toresen. 1998. Variable natural mortality rate of juvenile Norwegian spring-spawning herring (*Clupea harengus* L.) in the Barents Sea. ICES J. Mar. Sci. 55:430-442.

Donaldson, W., S. Morstad, E. Simpson, J. Wilcock, and S. Sharr. 1993. Prince William Sound management area, 1992 annual finfish management report. Alaska Department of Fish and Game, Regional Information Report No. 2A93-12, Anchorage. 148 pp.

Donaldson, W., S. Morstad, E. Simpson, J. Wilcock, and S. Sharr. 1995. Prince William Sound Management Area, 1993 annual finfish management report. Alaska Department of Fish and Game, Regional Information Report No. 2A95-20, Anchorage. 148 pp.

Foy, R.J., and B.L. Norcross. 1999. Spatial and temporal differences in the diet of juvenile Pacific herring (*Clupea pallasi*) in Prince William Sound, Alaska. Can. J. Zool. 77:697-706.

Foy, R.J., and A.J. Paul 1999. Winter feeding and changes in somatic energy content for age 0 Pacific herring in Prince William Sound, Alaska. Trans. Am. Fish. Soc. 128:1193-1200.

Funk, F. 1995. Age-structured assessment of Pacific herring in Prince William Sound, Alaska and forecast of abundance for 1994. Alaska Department of Fish and Game, Regional Information Report No. 5J95-00, Juneau. 40 pp.

Gay, S.M., and S.L. Vaughan. 2002. Seasonal hydrography and tidal currents of bays and fjords in Prince William Sound. Fish. Oceanogr. In press.

Graham, J.J. 1972. Retention of larval herring within Sheepscott Estuary, Maine. Fish. Bull., U.S. 70:299-305.

Haegele, C.W. 1993. Seabird predation of Pacific herring, *Clupea pallasi*, spawn in British Columbia. Can. Field-Nat. 107:73-82.

Haegele, C.W., and J.F. Schweigert. 1991. Egg loss in herring spawns in Georgia Strait, British Columbia. In: Proceedings of the International Herring Symposium. University of Alaska Sea Grant, AK-SG-91-01, Fairbanks, pp. 309-322.

Haegele, C.W., and J.F. Schweigert. 1989. Egg loss from Pacific herring spawns in Barkeley Sound in 1988. Can. Manuscr. Rep. Fish Aquat. Sci. 2037.

Haegele, C.W., R.D. Humphreys, and A.S. Hourston. 1981. Distribution of eggs by depth and vegetation type in Pacific herring (*Clupea harengus pallasi*) spawnings in southern British Columbia. Can. J. Fish. Aquat. Sci. 38:381-386

Hart, J.L., and A.L. Tester. 1934. Quantitative studies on herring spawning. Trans. Am. Fish. Soc. 64:307-313.

Hay, D.E. 1985. Reproductive biology of Pacific herring (*Clupea harengus pallasi*). Can. J. Fish. Aquat. Sci. 42 (Suppl. 1):111-126.

Hay, D.E. 1990. Tidal influence on spawning time of Pacific herring (*Clupea harengus pallasi*). Can. J. Fish. Aquat. Sci. 47:2390-2401.

Hay, D.E., and A.R. Kronlund. 1987. Factors affecting the distribution, abundance, and measurement of Pacific herring (*Clupea harengus pallasi*) spawn. Can. J. Fish. Aquat. Sci. 44:1181-1194.

Hay, D.E., and P.B. McCarter. 1997. Larval retention and stock abundance structure of British Columbia herring. J. Fish Biol. 51(Suppl. A):155-175.

Hay, D.E., and C.C. Miller. 1982. A quantitative assessment of herring spawn lost by storm action in French Creek, 1980. Can. Manuscr. Rep. Fish. Aquat. Sci. 1636.

Hershberger, P.K., R.M. Kocan, N.E. Elder, T.R. Meyers, and J.R. Winton. 1999. Epizootiology of viral hemorrhagic septicemia virus in Pacific herring from the spawn-on-kelp fishery in Prince William Sound, Alaska, USA. Dis. Aquat. Org. 37:23-31.

Hose, J.E., and E.D. Brown. 1998. Field applications of the piscine anaphase aberration test: Lessons from the *Exxon Valdez* oil spill. Mutation Res. 399:167-178.

Hose, J.E., M.D. McGurk, G.D. Marty, D.E. Hinton, E.D. Brown, and T.T. Baker. 1996. Sublethal effects of the *Exxon Valdez* oil spill on herring embryos and larvae: Morphologic, cytogenetic, and histopathological assessments, 1989-1991. Can. J. Fish. Aquat. Sci. 53(10):2355-2365.

Iles, T.D., and M. Sinclair. 1982. Atlantic herring: Stock discreteness and abundance. Science 215:627-633.

Johnson, S.W., MG. Carls, R.P. Stone, C.C. Brodersen, and S.D. Rice. 1997. Reproductive success of Pacific herring, *Clupea pallasi*, in Prince William Sound, Alaska, six years after the *Exxon Valdez* oil spill. Fish. Bull., U.S. 95:748-761.

Kocan, R.M., and J.E. Hose. 1995. Laboratory and field observations of sublethal damage in marine fish larvae: Lessons from the *Exxon Valdez* oil spill. In: R.M. Rolland, M. Gilbertson, and R.E. Peterson (eds.), Chemically induced alterations in functional development and reproduction of fishes. SETAC (Society of Environmental Toxicology and Chemistry) Tech. Publ., pp. 167-176.

Kocan, R.M., J.E. Hose, E.D. Brown, and T.T. Baker. 1996a. Pacific herring (*Clupea pallasi*) embryo sensitivity to Prudhoe Bay petroleum hydrocarbons: Laboratory evaluation and in situ exposure at oiled and unoiled sites in Prince William Sound. Can. J. Fish. Aquat. Sci. 53:2366-2375.

Kocan, R.M., G.D. Marty, M.S. Okihiro, E.D. Brown, and T.T. Baker. 1996b. Reproductive success and histopathology of individual Prince William Sound Pacific herring 3 years after the *Exxon Valdez* oil spill. Can. J. Fish. Aquat. Sci. 53:2388-2393.

Kocan, R.M., M. Bradley, N. Elder, T. Meyers, W. Batts, and J. Winton. 1997. North American strain of viral hemorrhagic septicemia virus is highly pathogenic for laboratory-reared Pacific herring. J. Aquat. Animal Health. 9:279-290.

Lasker, R. 1985. What limits clupeoid production? Can. J. Fish. Aquat. Sci. 42:31-37.

Marty, G.D., M.S. Okihiro, E.D. Brown, and D. Hanes. 1999. Histopathology of adult Pacific herring in Prince William Sound, Alaska, after the *Exxon Valdez* oil spill. Can. J. Fish. Aquat. Sci. 77:1-8.

Marty, G.D., J.E. Hose, M.D. McGurk, E.D. Brown, and D.E. Hinton. 1997. Histopathology and cytogenetic evaluation of Pacific herring larvae exposed to petroleum hydrocarbons in the laboratory or in Prince William Sound, Alaska, after the *Exxon Valdez* oil spill. Can. J. Fish. Aquat. Sci. 54:1846-1857.

Marty, G.D., E.F. Freiburg, T.R. Meyers, J. Wilcock, T.B. Farver, and D.E. Hinton. 1998. Viral hemorrhagic septicemia virus, *Ichtyophonus hoferi*, and other causes of the morbidity in Pacific herring *Clupea pallasi* spawning in Prince William Sound, Alaska, USA. Dis. Aquat. Org. 32:15-40.

McGurk, M.D. 1984. Effects of delayed feeding and temperature on the age of irreversible starvation and on the ranges of growth and mortality of Pacific herring larvae. Mar. Biol. 84:13-26.

McGurk, M.D. 1989 Advection, diffusion and mortality of Pacific herring larvae *Clupea harengus pallasi* in Bamfield Inlet, British Columbia. Mar. Ecol. Prog. Ser. 51:1-18.

McGurk, M.D. 1993. Allometry of herring mortality. Trans. Am. Fish. Soc. 122:1035-1042.

McGurk, M.D., and E.D. Brown. 1996. Egg-larval mortality of Pacific herring in Prince William Sound, Alaska, after the *Exxon Valdez* oil spill. Can. J. Fish. Aquat. Sci. 53(10):2343-2354.

McGurk, M.D., A.J. Paul, K.O. Coyle, D.A. Ziemann, and L.J. Haldorson. 1993. Relationships between prey concentration and growth, condition, and mortality of Pacific herring, *Clupea pallasi*, larvae in an Alaskan subarctic embayment. Can. J. Fish. Aquat. Sci. 50:163-180.

Meyers, T.R., S Short, K. Lipson, W.N. Batts, J.R. Winton, J. Wilcock, and E. Brown. 1994. Association of viral hemorrhagic septicemia virus with epizootic hemorrhages of the skin in Pacific herring *Clupea pallasi* from Prince William Sound and Kodiak Island, Alaska, USA. Dis. Aquat. Org. 19:27-37.

Morstad S., D. Sharp, J. Wilcock, T. Joyce, and J. Johnson. 1998. Prince William Sound management area, 1997 annual finfish management report. Alaska Department of Fish and Game, Regional Information Report No. 2A98-05, Anchorage. 163 pp.

Morstad S., D. Sharp, J. Wilcock, T. Joyce, and J. Johnson. 1999. Prince William Sound management area, 1998 annual finfish management report. Alaska Department of Fish and Game, Regional Information Report No. 2A99-20, Anchorage. 161 pp.

Nanke, R.L. 1981. Observations on deformed fish larvae in Long Island Sound and Niantic Bay, Connecticut. Rapp. P.V. Reun. Cons. Int. Explor. Mer 178:355:356.

Norcross, B.L., J.E. Hose, M. Frandsen, and E.D. Brown. 1996. Distribution, abundance, morphological condition, and cytogenetic abnormalities of larval herring in Prince William Sound, Alaska, following the *Exxon Valdez* oil spill. Can. J. Fish. Aquat. Sci. 53(10):2376-2387.

Norcross, B.L., E.D. Brown, R.J. Foy, M. Frandsen, S. Gay, T.C. Kline Jr., D.M. Mason, E.V. Patrick, A.J. Paul, and K.D.E. Stokesbury. 2002. A synthesis of the early life history and ecology of juvenile Pacific herring in Prince William Sound, Alaska. Fish. Oceanogr. In press.

Patrick, E.V. 1999. Validation reference for the models for physiological response and survival of age-0 Pacific herring during winter fasting. *Exxon Valdez* Oil Spill Trustees Council Final Report, Anchorage, Alaska.

Patrick, E.V., D.M. Mason, R.J. Foy, B.L. Norcross, A.J. Paul, K.D.E. Stokesbury, and T.C. Kline Jr. 2002. Model of the winter physiology and survival of age-0 Pacific herring (*Clupea pallasi*) in Prince William Sound, Alaska. Fish. Oceanogr. In Press.

Paul, A.J., and J.M. Paul. 1998a. Comparisons of whole body energy content of captive fasting age zero Alaskan Pacific herring (*Clupea pallasi* Valenciennes) and cohorts overwintering in nature. J. Exp. Mar. Biol. Ecol. 226:75-86.

Paul, A.J., and J.M. Paul. 1998b. Spring and summer whole body energy content of Alaskan juvenile Pacific herring. Alaska Fish. Res. Bull. 5(2):131-136.

Paul, A.J., and Paul, J.M. 1999. Interannual and regional variations in body length, weight and energy content of age-0 Pacific herring from Prince William Sound, Alaska. J. Fish Biol. 54:996-1001.

Paul, A.J., J.M. Paul, and E.D. Brown. 1998. Fall and spring somatic energy content for Alaskan Pacific herring (*Clupea pallasi* Valenciennes 1847) relative to age, size and sex. J. Exp. Mar. Biol. Ecol. 223:133-142.

Pitcher, T.J., and J.K. Parrish. 1993. Functions of shoaling behaviour in teleosts. In: T.J. Pitcher (ed.), Behaviour of teleost fishes. Chapman & Hall, New York, pp. 363-439.

Purcell, J.E., and J.J. Grover. 1990. Predation and food limitation as causes of mortality in larval herring at a spawning ground in British Columbia. Mar. Ecol. Prog. Ser. 59:55-61.

Purcell, J.E., D. Grosse, and J.J. Grover. 1990. Mass abundances of abnormal Pacific herring larvae at a spawning ground in British Columbia. Trans. Am. Fish. Soc. 119:463-469.

Purcell, J.E., T.D. Siferd, and J.B. Marliave. 1987. Vulnerability of larval herring (*Clupea harengus pallasi*) to capture by the jellyfish *Aequorea victoria*. Mar. Biol. 94:157-162.

Quinn II, T.J., G.D. Marty, J. Wilcock, and M. Willette. 2001. Disease and population assessment of Prince William Sound Pacific herring. In: F. Funk, J. Blackburn, D. Hay, A.J. Paul, R. Stephenson, R. Toresen, and D. Witherell (eds.), Herring: Expectations for a new millennium. University of Alaska Sea Grant, AK-SG-01-04, Fairbanks. (This volume.)

Reid, G.M. 1971. Age composition, weight, length, and sex of herring, *Clupea pallasi*, used for reduction in Alaska, 1929-66. NOAA Tech. Rep. NMFS SSRF 634. 25 pp.

Robinson, C.J. 1995. Food competition in a shoal of herring: The role of hunger. Mar. Behav. Physiol. 24:237-242.

Robinson, C.J., and T.J. Pitcher. 1989. Hunger motivation as a promoter of different behaviors within a shoal of herring: Selection for homogeneity in fish shoal? J. Fish. Biol. 35:459-460.

Rooper, C.N. 1996. Physical and biological factors affecting Pacific herring egg loss in Prince William Sound, Alaska. M.S. thesis, University of Alaska Fairbanks. 198 pp.

Rooper, C.N., L.J. Haldorson, and T.J. Quinn II. 1998. An egg-loss correction for estimating spawning biomass of Pacific herring in Prince William Sound, Alaska. Alaska Fish. Res. Bull. 5:137-142.

Rooper, C.N., L.J. Haldorson, and T.J. Quinn II. 1999. Habitat factors controlling Pacific herring (*Clupea pallasi*) egg loss in Prince William Sound, Alaska. Can. J. Fish. Aquat. Sci. 56:1133-1142.

Sinclair, M., and T.D. Iles. 1985. Atlantic herring (*Clupea harengus*) distributions in the Gulf of Maine-Scotian Shelf area in relation to oceanographic features. Can. J. Fish. Aquat. Sci. 42:880-887.

Stephenson, R.L., and M.J. Power. 1991. Diel vertical movement of Atlantic herring in relation to food availability and abundance. Proceedings of the International Herring Symposium. University of Alaska Sea Grant, AK-SG-91-01, Fairbanks, pp. 73-83.

Stokesbury, K.D.E., R.J. Foy, and B.L. Norcross. 1999. Spatial and temporal variability in juvenile Pacific herring (*Clupea pallasi*) growth in Prince William Sound, Alaska. Environ. Biol. Fishes 56:409-418.

Stokesbury, K.D.E., J. Kirsch, and B.L. Norcross. 2002. Natural mortality estimates of juvenile Pacific herring (*Clupea pallasi*) in Prince William Sound, Alaska. Can. J. Fish. Aquat. Sci. In Press.

Stokesbury, K.D.E., J. Kirsch, E.D. Brown, G.L. Thomas, and B.L Norcross. 2000. Seasonal variability in Pacific herring (*Clupea pallasi*) and walleye pollock (*Theragra chalcogramma*) spatial distributions in Prince William Sound, Alaska. Fish. Bull., U.S. 98:400-409.

Struhsaker, J.W. 1977. Effects of benzene (a toxic component of petroleum) on spawning Pacific herring, *Clupea harengus pallasi*. Fish. Bull., U.S. 75:43-49.

Tanasichuk, R.W., A.H. Kristofferson, and D.V. Gillman. 1993. Comparison of some life history characteristics of Pacific herring (*Clupea pallasi*) from the Canadian Pacific Ocean and Beaufort Sea. Can. J. Fish. Aquat. Sci. 50:964-971.

Walters, C.J., M. Stocker, A.V. Tyler, and S.J. Westerheim. 1986. Interaction between Pacific cod (*Gadus macrocephalus*) and herring (*Clupea harengus pallasi*) in the Hecate Strait, British Columbia. Can. J. Fish. Aquat. Sci. 43:830-837.

Wang, J., C.N.K Mooers, and E.V. Patrick. 1997. A three dimensional tidal model for Prince William Sound, Alaska. In: J.R. Acinas and C.A. Brebbia (eds.), Computer modelling of seas and coastal regions III. Computational Mechanics Publications, Southampton, pp. 95-104.

Wang, J., M. Jin, E.V. Patrick, J. Allen, S.L. Vaughan, D. Eslinger, and R.T. Cooney. 2002. Numerical simulations of the seasonal circulation patterns and thermohaline structures of Prince William Sound, Alaska. Fish. Oceanogr. In press.

Ware, D.M., and R.W. Tanasichuk. 1989. Biological basis of maturation and spawning waves in Pacific herring (*Clupea harengus pallasi*). Can. J. Fish. Aquat. Sci. 46:1776-1784.

Watanabe, Y., N. Shirahuji, and M. Chimura. 2001. Latitudinal difference in recruitment dynamics of clupeid fishes: Variable to the north, stable to the south. In: F. Funk, J. Blackburn, D. Hay, A.J. Paul, R. Stephenson, R. Toresen, and D. Witherell (eds.), Herring: Expectations for a new millennium. University of Alaska Sea Grant, AK-SG-01-04, Fairbanks. (This volume.)

Wespestad, V.G., and E. Moksness. 1990. Observations on growth and survival during the early life history of Pacific herring *Clupea pallasi* from Bristol Bay, Alaska, in a marine mesocosm. Fish. Bull., U.S. 88:191-200.

Wilkins, N.P. 1967. Starvation of the herring, *Clupea harengus* L.: Survival and some gross biochemical changes. Comp. Biochem. Physiol. 23:503-518.

Willette, M., G. Carpenter, K. Hyer, and J. Wilcock. 1998. Herring natal habitats. Final Report for Restoration Project 97166 to the *Exxon Valdez* Oil Spill Trustee Council, Anchorage, Alaska. 67 pp.

Wolf, K. 1988. Viral hemorrhagic septicemia. Chapter 18. In: Fish viruses and fish viral diseases. Cornell University Press, Ithaca, pp. 217-348.

Zheng, J. 1996. Herring stock-recruitment relationships and recruitment patterns in the North Atlantic and Northeast Pacific oceans. Fish. Res. 26:257-277.

Zheng, J., F.C. Funk, and G.H. Kruse. 1993. Evaluation of threshold management strategies for Pacific herring in Alaska. In: G. Kruse, D.M. Eggers, R.J. Marasco, C. Pautzke, and T.J. Quinn II (eds.), Proceedings of the International Symposium on Management Strategies for Exploited Fish Populations. University of Alaska Sea Grant, AK-SG-93-02, Fairbanks, pp. 141-165.

Herring Stock Structure, Stock Discreteness, and Biodiversity

R.L. Stephenson, K.J. Clark, M.J. Power, F.J. Fife, and G.D. Melvin
Department of Fisheries and Oceans, St. Andrews, New Brunswick, Canada

Abstract

Herring exhibit "complex" stock structure, with many spawning areas. Unfortunately, the degree of affinity of herring to individual spawning grounds, the degree of interchange among spawning grounds, and therefore the ecological significance of individual spawning grounds are not known. This paper reviews the state of understanding of herring stock structure as it relates to the significance of individual spawning grounds.

A conceptual model of herring stock structure divides populations into "complexes" defined primarily by discontinuities at the larval stages, with additional structure ("cells") resulting from homing tendencies to particular spawning grounds. Most management units are at the scale of the stock complex, and there has been little attention to the significance of maintaining individual spawning locations. Individual spawning grounds are prone to erosion by overfishing, and it is suggested that maintenance of diverse spawning locations (cells) is important to conserving intraspecific biodiversity and evolutionary or adaptive potential.

The evaluation and management of 4WX (Southwest Nova Scotia) herring provides a case study in which management of stock components has been attempted. Recognizing the possibility of progressive erosion of spawning grounds within an apparently appropriate overall total allowable catch (TAC) for the stock complex, a number of additional strategic targets have been established to maintain reproductive capacity. An in-season management scheme, which includes industry surveys of spawning grounds prior to fishing, has been implemented to document the status of spawning grounds and to spread fishing effort appropriately.

Introduction

There is a long history of biological investigation of Atlantic herring, resulting largely from the historical importance of herring to the economies of European nations. Historical observations and research on herring were central to the evolution of a number of important concepts in fisheries science, including the development of the population concept in marine fisheries about a century ago (reviewed by Sinclair and Solemdal 1988, Sinclair and Smith 2001, Stephenson 2001 [this volume]). In spite of this long history the stock or population structure of herring remains an unresolved issue. Herring are known to exhibit a high level of stock complexity or population richness (Sinclair 1988), with many spawning areas, but there is debate concerning the integrity and significance of these.

A highlight of the previous Alaska International Herring Symposium held in 1990 was a lively discussion on herring stock structure and the concepts that had been put forward regarding marine finfish stock structure in the few years preceding that symposium. In the discussion (Collie 1991) conference participants debated particularly the "member/vagrant" hypothesis (Sinclair 1988, Sinclair and Iles 1988) and the issue of stock discreteness. The final paper in the 1990 symposium proceedings (Stephenson 1991) summarized the arguments for and against "discrete stocks" of herring, and provided a conceptual view of stock structure (Fig. 1).

In the subsequent decade, discussion and debate concerning stock structure in marine finfish has continued. In fact, the topic has become more relevant with the recognition that stock/fishery failures have continued to occur in spite of current fisheries management, and that a new approach to management is required (Stephenson 1999). There has also been increasing interest in recent years arising from developments in population modeling approaches (including the metapopulation concept McQuinn 1997, Hanski 1999); increasing concern regarding a broader spectrum of the effects of fishing (Gislason et al. 2000); and improved techniques to demonstrate stock differences (e.g., O'Connell and Wright 1997, Mustafa 1999). However, the major impetus for continued interest in stock structure has been the movement to a more cautious ("precautionary") approach to fisheries management which includes greater emphasis on preserving all aspects of biodiversity, including within-species diversity (genetic, phenotypic, spatial, etc.) (Stephenson and Kenchington 2000, Smedbol and Stephenson 2001).

We suggest that there are four important questions that need to be addressed concerning herring stock structure:

1. Do populations contain subunits that are discrete on time scales of relevance to assessment and management?

2. How are these best described and modeled (e.g., are they "metapopulations," stock complexes, mixed stocks, sub-stocks)?

Herring: Expectations for a New Millennium 561

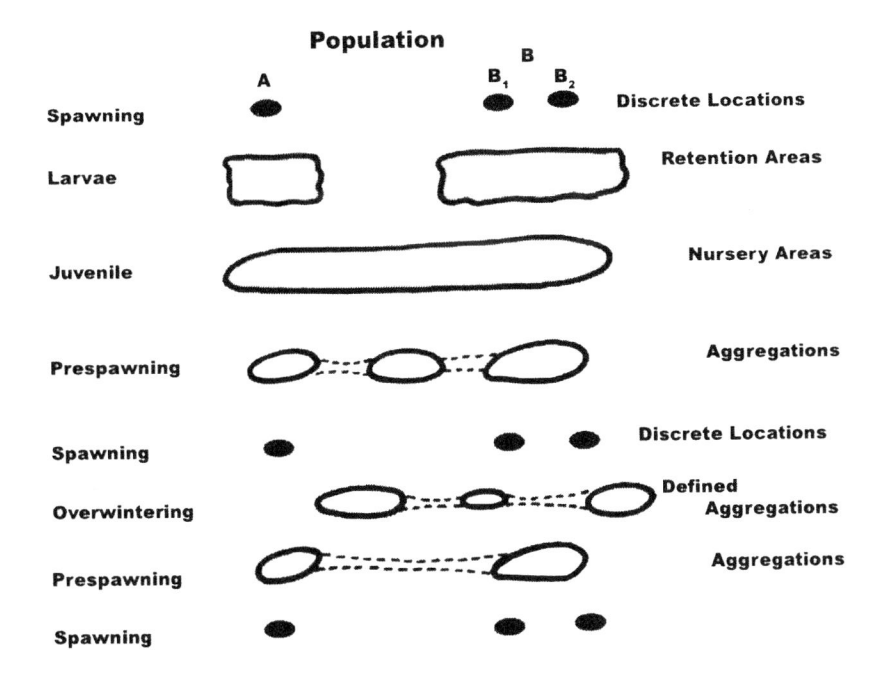

Figure 1. Conceptual model of the potential for mixing at life history stages of Atlantic herring populations put forward by Stephenson (1991). In this example, three groups of herring were separate at spawning, but overlapped to different degrees at other life history stages. Mixing between A and B occurred only at the juvenile and adult stages, whereas the two spawning groups marked B overlapped also at the larval stage. It was proposed that the two spawning groups marked B would be part of the same population complex separate from A, and that separation at the larval stage seemed to be a critical factor in maintaining population discreteness.

3. What is the ecological significance of individual spawning grounds (e.g., does removal of a spawning area represent extirpation or extinction)?

4. How should individual spawning grounds be treated in management?

In this paper, we present a case study on Southwest Nova Scotia (4WX) Atlantic herring, a group for which the importance of individual spawning components has recently been realized, and where management has been changed in an attempt to protect the diversity of spawning. We then review and extend the general model of herring stock structure, referring to emerging concepts of "metapopulation" thinking and of intraspecific biodiversity.

Investigation of Stock Structure in 4VWX Herring: A Case Study

Management units for herring of the Scotian Shelf, Bay of Fundy, and Gulf of Maine have evolved as the understanding of herring stock structure and movement has improved. Until very recently, the prevailing view of herring stock structure had large management units corresponding to statistical areas (NAFO units 4VWX, 5Y, and 5Z+6). This is exemplified by figures of stock structure that accompanied the developments in management (including TACs) under the auspices of the International Commission for Northwest Atlantic Fisheries (ICNAF) (Stephenson and Gordon 1991). According to that view fisheries were supported by a few important spawning areas. While the existence of other spawning areas was known from the literature and from historical fisheries (e.g., spring spawners south of Grand Manan), these were not considered in assessment and management.

Over the past decade or so, herring spawning areas of the Scotian Shelf, Bay of Fundy, and Gulf of Maine areas have been examined more closely. Information has been compiled from a variety of sources including scientific publications, fishery information, surveys of the distribution of larvae, and a questionnaire survey of coastal fishers of Nova Scotia (Denny et al. 1998, Clark et al. 1999, Harris and Stephenson 1999). Figure 2 shows that there are many more locations where herring are now known to spawn, or where herring have spawned within the past two decades than were recognized previously in management. Assessment and management of the 4VWX area has evolved to consider separate southwest Nova Scotia/Bay of Fundy, Coastal Nova Scotia and Offshore Scotian Shelf components (Fig. 3) with assessment and management of individual spawning grounds within each (Stephenson et al. 1997, 1999b).

The large (approximately 100 kt per year) southwest Nova Scotia/Bay of Fundy herring fishery has relied primarily on herring that aggregate and spawn off southwest Nova Scotia and in the Bay of Fundy. Much of the fishery has taken place either on prespawning aggregations or on the spawning grounds. Several years ago members of the scientific community and industry observed that the fishery was focusing each year on a single spawning area, and that over time the fishery was having to change locations due to decreasing abundance (Stephenson et al. 1994). It seemed that the fishery was progressively eroding herring spawning grounds. The best recent example is Trinity Ledge, a spawning ground that was the subject of intense fishing during the 1980s. Herring from Trinity Ledge comprised up to 40% of the total summer landings in 1985, but the spawning group apparently collapsed by 1989 (Chang et al. 1995). Spawning season closures of the area have led to recovery over the past decade, but the spawning is not yet at the level that it was in the mid-1980s (Stephenson et al. 2000).

The ecological significance of erosion of spawning groups is not known, but consistent with a precautionary approach, an attempt has been made in recent years to protect herring spawning grounds in the Scotia/Fundy

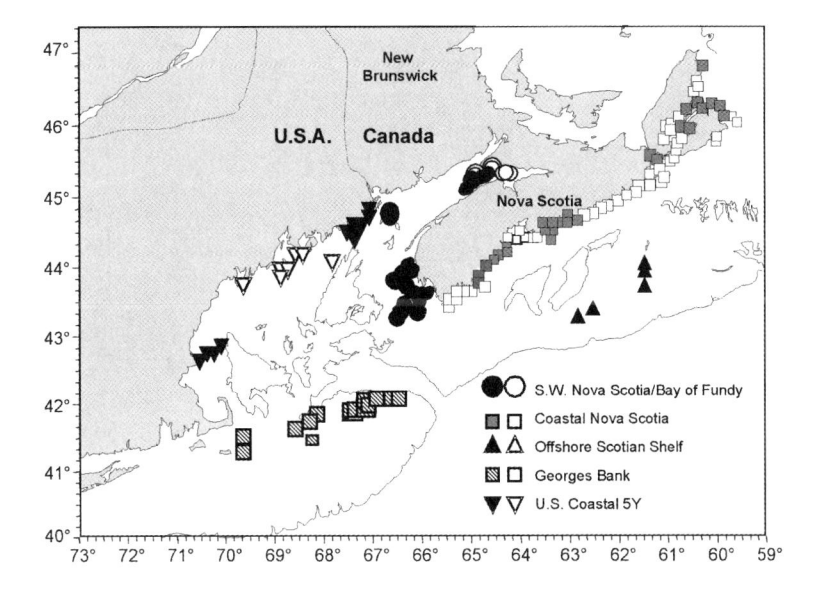

Figure 2. Atlantic herring spawning locations of the stock complexes used in managing Scotian Shelf, Bay of Fundy, and Gulf of Maine herring. Solid symbols indicate current spawning; open symbols indicate locations of spawning within the last two decades.

area herring fisheries. Conservation objectives for herring in this area were modified to refer explicitly to the maintenance of reproductive capacity along with prevention of growth overfishing, and maintenance of ecosystem integrity and ecological relationships (Sinclair 1997, Stephenson et al. 1998). The following strategic targets were established to "maintain reproductive capacity":

- Maintain all spawning grounds.
- Maintain a minimum spawning stock biomass for each spawning ground.
- Maintain a broad age distribution.
- Maintain full time period of spawning "runs."
- Document unstudied spawning areas.

In recognition that the management area contained several spawning grounds, and that these were subject to erosion through disproportionate fishing effort, management of this fishery adopted additional protection of individual spawning grounds. In order to maintain the viability of each

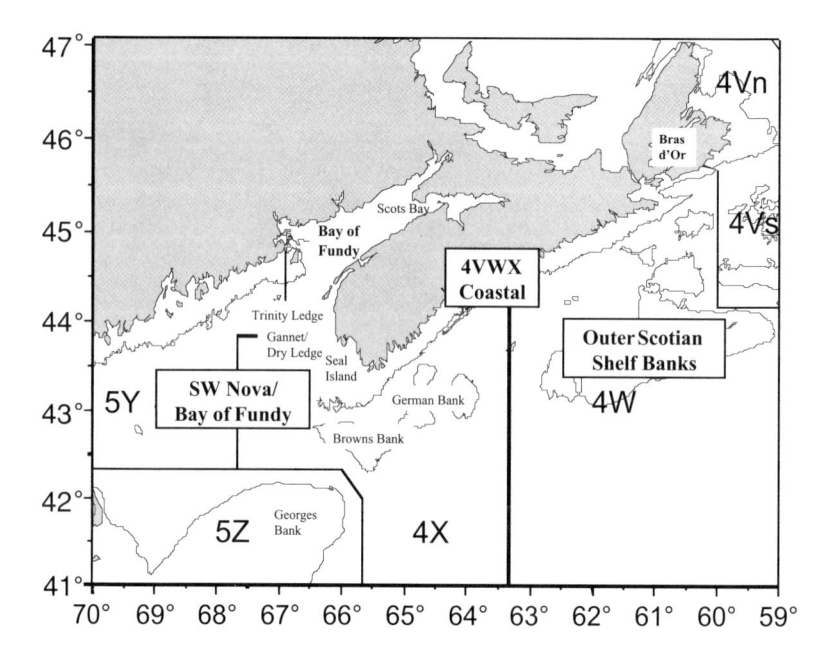

Figure 3. Management areas identified for herring of the Scotian Shelf and Bay of Fundy, and major locations mentioned in the text.

spawning component, it was considered important, in addition to a restriction on total allowable catch, to spread the fishing effort among spawning grounds according to their relative size. This was accomplished by introducing pre-fishing surveys of spawning grounds including Scots Bay, Trinity Ledge, and German Bank. Under this "survey, assess, then fish" protocol, it has become standard practice to protect individual spawning grounds by undertaking pre-fishing surveys, and fishing only a portion (less than 20% in areas with a TAC; less than 10% in other areas) of what had been documented (Stephenson et al. 1999a, Melvin et al. 2001 [this volume]).

This was facilitated by the introduction of an in-season management scheme (Stephenson et al. 1999) in which decisions on the timing and distribution of fishing within the overall plan were delegated to an industry-government team, so that decisions could be made during the season on the basis of best available information. Decisions regarding the spatial and temporal distribution of quota within the season are based on surveys undertaken almost exclusively by industry vessels. These surveys first provided relative estimates (derived from mapping fish schools), but now provide quantitative information from sounders and sonars of commercial vessels (Melvin et al. 1998, Melvin and Power 1999, Melvin et al. 2001). The information is used to match the portion of quota that can be taken in a particular time period with the amount of spawning stock biomass that has been documented in a particular area and time, and in this way to prevent disproportional effort in any area. The "survey, assess, then fish" protocol has resulted in area closures, delayed fishing until sufficient quantities of fish were observed, and increases in the allocations for individual spawning areas where large aggregations were documented. Herring taken at times other than spawning are presumed to be from mixtures, but maturity and age composition are monitored to test this assumption. Within-season monitoring of fish size and maturity in some mixed fisheries is designed to prevent disproportionate erosion of small stocks when they are mixed with larger ones, particularly in overwintering areas.

This case study provides one of the first examples of considering stock subunits (individual spawning grounds) in management. This has been achieved by recognizing the existence and importance of spawning groups within a management area, by establishing objectives relating to the protection of stock diversity, and by adoption of fishing practices that assure preservation of stock structure.

Model of Herring Stock Structure

Herring of the Scotia-Fundy region off Canada's east coast are typical of herring in most areas (Sinclair 1988). They aggregate to spawn in a number of discrete locations (Fig. 2) which differ in timing, but have a characteristic spawning period (of a few weeks) each year. Some spawn as early in the spring as April (e.g., Bras d'Or Lakes), others as late as November (e.g., Georges Bank). Demersal eggs (stuck to the bottom) hatch in about ten days into planktonic larvae which have been shown to remain in "larval retention areas" for several months (Iles and Sinclair 1982). After metamorphosis juvenile herring move widely, often it seems near shore, and mix with juveniles from other spawning areas. Herring first mature and spawn at three or four years of age, then begin an annual pattern of spawning, overwintering, and summer feeding. Tagging studies have shown considerable migration and intermixing of herring outside of the spawning areas. However, spawners return to the same spawning location year after year, and herring are presumed to "home" to the spawning areas they themselves were from.

While there are certainly some unique features among populations of Atlantic, Pacific, and Baltic herring (compare for example, Hay 1985, Sinclair 1988, Parmanne 1990, McQuinn 1997), we suggest that there are a number of common features of herring stock structure. Spawning areas are discrete. Most spawning areas are occupied each year, although some may be ephemeral. Spawning areas can usually be grouped according to geographical features (bays, banks, etc.) into discontinuous units (cells) in which there is persistent spawning. Larvae are pelagic, and have been shown to be associated with predictable hydrographic features such as a region of well-mixed water (Iles and Sinclair 1982, Stephenson and Power

1988, Chenoweth et al. 1989, Hay and McCarter 1997). Juveniles mix and migrate widely, but usually not in association with adults. Adults mix in overwintering, feeding, and prespawning aggregations, but separate out to spawn. Spawning occurs with a high degree of consistency in location and timing. Tagging evidence for repetitive spawning at the same site, stock-specific population dynamics (regularity of spawning, growth and age characteristics, and population abundance over time), discrete larval distributions, and occasional differences in biological or biochemical features (stock identification studies) imply a high degree of homing (Stephenson 1991).

We have attempted to illustrate the general pattern of herring stock structure conceptually in Fig. 4. Spawning takes place at discrete sites, most of which are occupied annually but some of which may be ephemeral. Often a number of spawning sites are in close proximity on recognized spawning grounds. Spawning grounds have different degrees of separation both spatially (different bays or banks) and temporally (in extreme cases spring vs. autumn). Larvae from one or more spawning grounds spend weeks or months in a distribution that is often associated with specific hydrographic features (larval retention area). Juvenile herring migrate widely and mix (in nursery areas, often near shore) with juveniles from other spawning grounds. Adults, usually distributed differently from juveniles, also mix and migrate, but separate out to spawn where they have spawned previously, apparently with a high degree of homing to the area from which they originated.

Given the widespread movement and mixing of juvenile and adult herring outside of the spawning season, it seems reasonable to conclude that population structure would be related to discontinuities at the egg and larval stage. It is hypothesized that major break points (discontinuities) in spawning and larvae distribution are critical in defining herring stock structure. Herring that share a common larval retention area are considered to belong to the same *complex* (A or B in Fig. 4). Within the complex, there is strong fidelity to spawning locations, which are separate in time and space. Spawning grounds can be readily divided on the basis of geographical or temporal separation into *cells* (A1, A2, B1 etc.) in which there is at least one persistent spawning area (for example spawners in the same bay or bank in the same season). Explicit recognition of cells (made up of one or more spawning grounds, some of which may be ephemeral) is a major advancement in this model compared with that put forward previously (Stephenson 1991; Fig. 1). It accounts for the observation that most spawning grounds are occupied every year while some are not, and it implies a gradient of affinity among spawning grounds. Herring spawning in close proximity (temporally and spatially) are likely more closely related than those that spawn in a different location or time. Exchange is expected to be highest within the cell (e.g., among spawning grounds within A1), less between cells within the complex (between A1 and A2 for example), and low between complexes (A vs. B).

Herring populations exist at a range of scales, from very small and localized or isolated (e.g., Bras d'Or Lakes in Eastern Canada) to very large and widespread (Norwegian spring-spawning herring). Synthesis and con-

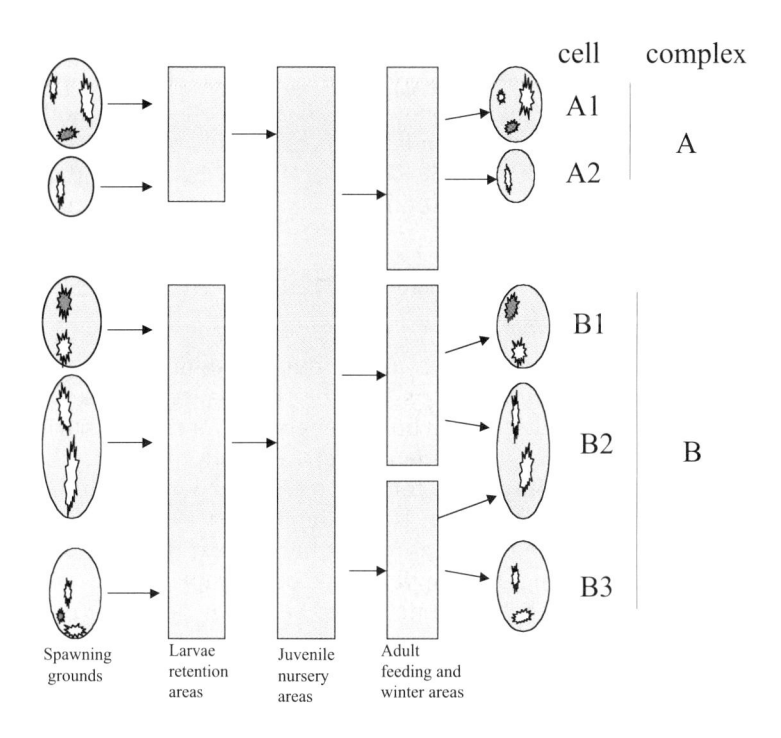

Figure 4. Conceptual model of herring population structure. Persistent (white) and ephemeral (shaded) spawning areas (irregular symbols) are grouped on the basis of geographical or temporal separation into "cells" (ovals) in which there is at least one persistent spawning area. Spawning cells are grouped into a "complex" on the basis of the distribution of larvae (larval retention areas). Juvenile and adult herring mix widely, but separate out to spawn with a high degree of fidelity to the complex, and to the cell. Mixing is expected to be highest within the cell, and lowest between complexes.

sensus regarding stock structure are further complicated by the use of different or ambiguous terminology (for example stock, homing, larval drift and retention). We suggest, however, that the conceptual model put forward here (and summarized in Fig. 4) best represents the body of literature and experience of Atlantic, Pacific, and Baltic herring, and offers a useful template for study, evaluation, and management of herring.

Implications for Assessment and Management of Herring Stocks

Herring stocks are characterized by numerous spawning groups with different spawning times and locations. Most herring management areas contain subunits which are discrete on time scales of relevance to management. It has been suggested that the metapopulation concept, which has become popular in describing some complex terrestrial populations, may be appropriate for herring (McQuinn 1997, Stephenson 1999, Smedbol and Stephenson 2001). While this requires further evaluation, it is clear that most herring stocks must be thought of as stock complexes of some sort. In that context, the conceptual model posed here provides a useful framework for management.

The ecological significance of eliminating spawning groups is not known. We do not know how removal of herring from a spawning ground will impact the ability of the complex to respond to natural and anthropogenic perturbation. We are uncertain as to the value of individual spawning grounds, if any, to preserving the intraspecific biodiversity and evolutionary or adaptive potential of herring. Given this uncertainty, we suggest that under a precautionary approach the spatial and temporal structure of herring spawning should be preserved to maintain within-species biodiversity (Stephenson 1999, Stephenson and Kenchington 2000, Smedbol and Stephenson 2001). Maintenance of the full diversity of spawning groups should be the default approach in management. In the conceptual model (Fig. 4) this corresponds to preservation of at least all cells, and perhaps all major spawning grounds within cells.

The experience of the past few decades has shown that herring stocks are prone to collapse. The abundance of herring stocks fluctuates naturally. Fishing pressure can hasten decline. Much has been written about the collapse of large stocks (such as Georges Bank and the Norwegian spring-spawning herring; Anthony and Waring 1980, Jakobsson 1985). Further, fisheries are known to have eroded and eliminated spawning on some grounds within stock complexes, including the Dogger Bank spawning grounds of the North Sea complex in the 1970s (Burd 1985, Nichols and Brander 1989), and the Trinity Ledge spawning ground in southwest Nova Scotia stock complex (Chang et al. 1995). Also there have been instances of elimination of some spawning areas in the absence of (or with very low) fishing, although these have been less well documented. In all cases, recovery of a stock following collapse has been slow (Stephenson 1997). We suggest that recovery time would be a function of the "proximity" of spawning in the conceptual model described above. Recovery from a remnant of the same cell might be slow—but replacement from neighboring parts of the complex and especially from another complex would be even slower. The relative difference in the time to reappearance of spawning after collapse of herring on Trinity Ledge (within three years) and on Georges Bank (about 10 years) is consistent with this hypothesis. We suggest that in addition to the usual conservation of overall stock biomass, effective management of herring requires preservation of spatial and temporal distribution of spawning.

References

Anthony, V.C., and G. Waring. 1980. The assessment and management of the Georges Bank herring fishery. Rapp. P-V. Reun. Cons. Int. Explor. Mer 177:72-111.

Burd, A.C. 1985. Recent changes in the central and southern North Sea herring stocks. Can J. Fish. Aquat. Sci. 42(Suppl. 1):192-206.

Chang, B.D., R.L. Stephenson, D.J. Wildish, and W.M. Watson-Wright. 1995. Protecting regionally significant marine habitats in the Gulf of Maine: A Canadian perspective. In: Improving interactions between coastal science and policy. Proceedings of the Gulf of Maine Symposium, Kennebunkport, 1-3 November 1994. National Academy Press, Washington, DC, pp. 121-146.

Chenoweth, S.B., D.A. Libby, R.L. Stephenson, and M.J. Power. 1989. Origin, and dispersion of larval herring (*Clupea harengus*) in coastal waters of eastern Maine and southwestern New Brunswick. Can. J. Fish. Aquat. Sci. 46:624-632.

Clark, K.J., D. Rogers, H. Boyd, and R.L. Stephenson. 1999. Questionnaire survey of the coastal Nova Scotia herring fishery, 1998. DFO Canadian Stock Assessment Secretariat Res. Doc. 99/137. 54 pp.

Collie, J. (moderator). 1991. Workshop on member/vagrant hypothesis and stock discreteness. In: Proceedings of the International Herring Symposium. University of Alaska Sea Grant, AK-SG-91-01, Fairbanks, pp. 641-657.

Denny, S., K.J. Clark, M.J. Power, and R.L. Stephenson. 1998. The status of the herring in the Bras d'Or Lakes in 1996-1997. DFO Atlantic Fish. Res. Doc. 98/80. 32 pp.

Gislason, H., M. Sinclair, K. Sainsbury, and R. O'Boyle. 2000. Symposium overview: Incorporating ecosystem objectives within fisheries management. ICES J. Mar. Sci. 57:468-475.

Hanski, I. 1999. Metapopulation ecology. Oxford University Press. 313 pp.

Harris, L.E., and R.L. Stephenson. 1999. Compilation of available information regarding the Scotian Shelf herring spawning component. DFO Canadian Stock Assessment Secretariat Res. Doc. 99/181. 30 pp.

Hay, D.E. 1985. Reproductive biology of Pacific herring (*Clupea harengus pallasi*). Can. J. Fish. Aquat. Sci. 42(Suppl. 1):111-126

Hay, D.E., and P.B. McCarter. 1997. Larval distribution, abundance, and stock structure of British Columbia herring. J. Fish Biol. 51(Suppl. A):155-175.

Iles, T.D,. and M. Sinclair. 1982. Atlantic herring: stock discreteness and abundance. Science 215:627-633.

Jakobsson, J. 1985. Monitoring and management of the Northeast Atlantic herring stocks. Can. J. Fish. Aquat. Sci. 42(Suppl. 1):207-221.

McQuinn, I.H. 1997. Metapopulations and the Atlantic herring. Rev. Fish Biol. Fish. 7:297-329.

Melvin, G.D., and M.J. Power. 1999. A proposed acoustic survey design for 4WX herring spawning components. DFO Canadian Stock Assessment Secretariat Res. Doc. 99/63. 15 pp.

Melvin, G.D., R.L. Stephenson, M.J. Power, F.J. Fife, and K.J. Clark. 2001. Industry acoustic surveys as the basis for in-season decisions in a comanagement regime. In: F. Funk, J. Blackburn, D. Hay, A.J. Paul, R. Stephenson, R. Toresen, and D. Witherell (eds.), Herring: Expectations for a new millennium. University of Alaska Sea Grant, AK-SG-01-04, Fairbanks. (This volume.)

Melvin, G.D., K.J. Clark, F.J. Fife, M.J. Power, S.D. Paul, and R.L. Stephenson. 1998. Quantitative acoustic surveys of 4WX herring in 1997. DFO Canadian Stock Assessment Secretariat Res. Doc. 98/81. 28 pp.

Mustafa, S. (ed.). 1999. Genetics and sustainable fisheries management. Blackwell Science Ltd., Oxford.

Nichols, J.H., and K.M. Brander. 1989. Herring larval studies in the west-central North Sea. Rapp. P-V. Reun. Cons. Int. Explor. Mer 191:160-168.

O'Connell, M., and J.M. Wright. 1997. Microsatellite DNA in fishes. Rev. Fish Biol. Fish. 7:331-363

Parmanne, R. 1990. Growth, morphological variation and migrations of herring (*Clupea harengus* L.) in the northern Baltic Sea. Finn. Fish. Res. 10:1-48.

Sinclair, M. 1988. Marine populations. An essay on population regulation and speciation. University of Washington Press, Seattle. 252 pp.

Sinclair, M. (chair). 1997. Report of the Maritimes Region Herring Workshop, 18-19 February 1997. Canadian Stock Assessment Proceedings Series 97/12. 58 pp.

Sinclair, M., and T.D. Iles. 1988. Population richness of marine fish species. Aquat. Living Resour. 1:71-83.

Sinclair, M., and P. Solemdal. 1988. The development of 'population thinking' in fisheries biology between 1878 and 1930. Aquat. Living Resour. 1:189-213.

Sinclair, M.M., and T.D. Smith. 2001. The notion that fish form stocks. ICES Mar. Sci. Symp. 215.

Smedbol, R.K., and R.L. Stephenson. 2001. The importance of managing within-species diversity in modern fisheries, with examples from the northwestern Atlantic. J. Fish Biol. 59(Suppl. A).

Stephenson, R. L. 1991. Stock discreteness of Atlantic herring: A review of arguments for and against. In: Proceedings of the International Herring Symposium. University of Alaska Sea Grant, AK-SG-91-01, Fairbanks, pp. 659-666.

Stephenson R.L. 1997. Successes and failures in the management of Atlantic herring fisheries: Do we know why some have collapsed and other survived? In: D.A. Hancock, D.C. Smith, A. Grant. and J.P. Beumer. Developing and sustaining world fisheries resources: The state of science and management. 2nd World Fisheries Congress proceedings. CSIRO, Australia, pp. 49-54.

Stephenson, R.L. 1999. Stock complexity in fisheries management: A perspective of emerging issues related to population subunits. Fisheries Research 43:247-249.

Stephenson, R.L. 2001. The role of herring investigations in shaping fisheries science. In: F. Funk, J. Blackburn, D. Hay, A.J. Paul, R. Stephenson, R. Toresen, and D. Witherell (eds.), Herring: Expectations for a new millennium. University of Alaska Sea Grant, AK-SG-01-04, Fairbanks. (This volume.)

Stephenson, R.L., and D. J. Gordon. 1991. Comparison of meristic and morphometric characters among spawning aggregations of northwest Atlantic herring, *Clupea harengus* L. In: Proceedings of the International Herring Symposium. University of Alaska Sea Grant, AK-SG-91-01, Fairbanks, pp. 279-297.

Stephenson, R.L., and E. Kenchington. 2000. Conserving fish stock structure is a critical aspect of preserving biodiversity. ICES C.M. 2000/Mini:07.

Stephenson, R.L, and M.J. Power. 1988. Semidiel vertical movements in Atlantic herring *Clupea harengus* larvae: A mechanism for larval retention? Mar. Ecol. Progr. Ser. 50:3-11.

Stephenson, R.L., K. Rodman, D.G. Aldous, and D.E. Lane. 1999a. An in-season approach to management under uncertainty: The case of the SW Nova Scotia herring fishery. ICES J. Mar. Sci. 56:1005-1013.

Stephenson, R.L., M.J. Power, J.B. Sochasky, F.J. Fife, and G.D. Melvin. 1994. Evaluation of the 1993 4WX herring fishery. DFO Atl. Fish. Res. Doc. 94/88. 50 pp.

Stephenson, R.L., M.J. Power, F.J. Fife, G.D. Melvin, and S.D. Paul. 1997. 1997 evaluation of the stock status of 4WX herring. DFO Canadian Stock Assessment Secretariat Res. Doc. 97/61. 28 pp.

Stephenson, R.L., M.J. Power, K.J. Clark, G.D. Melvin, F.J. Fife, and S.D. Paul. 1998. 1998 evaluation of 4VWX herring. DFO Canadian Stock Assessment Secretariat Res. Doc. 98/52. 58 pp.

Stephenson, R.L., M.J. Power, K.J. Clark, G.D. Melvin, F.J. Fife, S.D. Paul, L.E. Harris, and S. Boates. 1999b. 1999 evaluation of 4VWX herring. DFO Canadian Stock Assessment Secretariat Res. Doc. 99/64. 85 pp.

Stephenson, R.L., M.J. Power, K.J. Clark, G.D. Melvin, F.J. Fife, T. Scheidl, C.L. Waters, and S. Arsenault. 2000. 2000 evaluation of 4VWX herring. DFO Canadian Stock Assessment Secretariat Res. Doc. 2000/65. 114 pp.

Biological Characteristics and Stock Enhancement of Lake Furen Herring Distributed in Northern Japan

Tokimasa Kobayashi
Tohoku National Fisheries Research Institute, Miyagi, Japan

Abstract

Most Lake Furen herring underwent maturation at age 2 with a mean fork length for ages 1-7 of 15.5, 21.0, 24.0, 26.5, 28.0, 29.0, and 30 cm, respectively. To clarify the migration area of this population more than 300,000 juveniles were tagged (mean TL = 6-8 cm). Every year since 1993 tagged herring were released after they were reared in the lake for 2-3 weeks. There were 6,473 herring captured in Lake Furen during spawning seasons and 328 were tagged fish. In addition 83 tagged herring were discovered among 11,747 herring captured outside of Lake Furen during their feeding seasons. No tagged herring have been discovered in other spawning grounds. This tagging study showed that Lake Furen herring exhibited a homing characteristic and their migration area was from Cape Erimo (about 300 km westward) to Cape Shiretoko (about 100 km northward). The percentage of recaptured fish relative to the number of juveniles released for each year class ranged from 1.1% for the 1997 year class to 12.5% for the 1995 year class. The ratio of recaptured fish to the weight of the total catch rose to about 4% in 1998 and later years but it was less than 1% before that. However, it has not been demonstrated yet that this attempt at stock enhancement was fully successful.

Morphometric Variation among Spawning Groups of the Gulf of Maine–Georges Bank Herring Complex

Michael P. Armstrong
Massachusetts Division of Marine Fisheries, Annisquam River Marine Fisheries Station, Gloucester, Massachusetts

Steven X. Cadrin
National Marine Fisheries Service, Northeast Fisheries Science Center, Woods Hole, Massachusetts

Abstract

The purpose of this study was to characterize morphometric variation between the two major spawning components of Atlantic herring, *Clupea harengus*, in the Gulf of Maine–Georges Bank stock complex and to evaluate the use of morphometric differences for stock discrimination. Morphometric characters, including both traditional and truss network distances, were measured on herring from pre- and postspawning aggregations on Jeffreys Ledge (inshore Gulf of Maine) and Georges Bank. Prespawning herring were morphometrically distinct from postspawning herring on the same spawning ground, principally due to differences in abdominal size. Many truss measurements were affected by spawning condition while most of the traditional measurements were not. The Jeffreys Ledge and Georges Bank stocks could not be effectively discriminated using morphometrics based on prespawning samples due to the confounding effects of spawning condition on morphometry. Extrinsic samples of postspawning herring were classified into their respective spawning groups using discriminant analysis of morphometric characters with 88% accuracy. This study indicates that morphometric characters can be used to distinguish spawning stocks of Atlantic herring in the northwest Atlantic with moderate accuracy. However, due to the confounding effects of spawning condition, these analyses can only be accomplished on postspawning fish.

Introduction

Atlantic herring, *Clupea harengus*, that spawn off southwest Nova Scotia, on Georges Bank and Nantucket Shoals, and in coastal waters of the Gulf of Maine have historically been recognized as distinct stocks (Anthony 1972, Stephenson and Gordon 1990, NEFSC 1998), although the discreteness of these spawning grounds remains controversial (Stephenson 1990, Safford and Booke 1992). Assessments performed by the United States prior to 1991 were specific to either Georges Bank–Nantucket Shoals (GB-NS) or the Gulf of Maine (GOM) components. Since 1991, the stock complex has been assessed as a whole, because specific spawning components cannot be distinguished from survey samples and the mixed stock fisheries (NEFSC 1992). The Gulf of Maine–Georges Bank stock complex has been defined to include Atlantic herring from the southern extent of their northwest Atlantic range to the western shore of the Bay of Fundy, including Georges Bank (NEFSC 1998). The current stock assessment assumes all individuals are part of a single highly migratory population composed of distinct spawning stocks.

Geographic patterns in fishing effort and trends in abundance make assessment and management of the stock complex difficult. The Georges Bank spawning component collapsed in the 1970s due to intense fishing by distant water fleets (Anthony and Waring 1980), but low levels of exploitation following the collapse have allowed the resource to rebuild (Stephenson and Kornfield 1990; Smith and Morse 1993; Overholtz and Friedland 2002). As most catches of herring currently occur in the inshore waters of the Gulf of Maine, this portion of the stock complex likely has a higher rate of exploitation. If it is indeed a separate stock, distinct from the Georges Bank stock, then a reduction in the abundance of this population could be masked by the continued growth and larger size of the Georges Bank population. An alternative approach that would be more sensitive to changes in components is to conduct stock assessments on individual components of the stock complex. This requires the ability to discriminate between fish from the various stocks and to be able to assign removals and survey samples to individual stocks.

A number of methods have been used to discriminate herring stocks including genetic techniques (Kornfield et al. 1982, Grant 1984, King 1984, Kornfield and Bogdanowicz 1987, Safford and Booke 1992), parasite fauna (McGladdery and Burt 1985, Chenoweth et al. 1986, Moser 1990, Stephenson 1990), tagging (Harden-Jones 1968, Wheeler and Winters 1984), and meristics and morphometrics (Anthony 1972, Parsons 1975, Meng and Stocker 1984, King 1985, Schweigert 1990, Stephenson 1990, Safford and Booke 1992), with varying degrees of success. Of these methods, meristics and morphometrics have shown the most utility for stock discrimination. Morphometric analyses can likely be accomplished more quickly than meristic analyses, especially using image analysis equipment. Speed and ease of accomplishment is essential if stock discrimination needs to be performed between several stocks at several locations and times of year, as would be necessary to characterize mixed stock fisheries.

The purpose of this study was to characterize the morphometric variation within and between Atlantic herring from the two major spawning grounds in the stock complex (Jeffreys Ledge and Georges Bank) and further, to evaluate if morphometric differences could be used to discriminate these two stocks. Additionally, the effect of spawning condition and age on morphometric characters was examined.

Materials and Methods

Data

Atlantic herring were collected in September 1998, on Jeffreys Ledge and Georges Bank from a commercial midwater trawler fishing on prespawning aggregations (all fish were ripe or running ripe). Samples were taken at the dock and catch locations were obtained from the captain of the vessel. Each sample represented a single 2- to 3-hour tow. Samples of postspawning herring (spent or resting) were obtained during the National Marine Fisheries Service (NMFS) Autumn Bottom Trawl Survey in October 1998 from stations on Jeffreys Ledge and northeastern Georges Bank. Locations at which samples were obtained are presented in Fig. 1. An additional sample of unknown spawning affinity was obtained from the winter fishery near Block Island in southern New England during January 1999.

A digital image of the sagittal view of each fish was recorded and saved. A suite of morphometric characters was measured on the two-dimensional image using image analysis software (UTHSCSA ImageTool). Thirty straight-line measurements were made, including both traditional and truss network measurements (Strauss and Bookstein 1982; Table 1, Fig. 2). Sex and developmental stage were also recorded for each fish. Otoliths were removed and imbedded in Permount resin in black plastic trays and aged whole using standard methods (Dery 1988). The sample sizes were as follows: prespawning Jeffreys Ledge, 373; prespawning Georges Bank, 416; postspawn Jeffreys Ledge, 122; postspawn Georges Bank, 103.

Statistical Analysis

All variables were natural log transformed prior to analysis. Data were screened for outliers by examination of bivariate scatter plots of individual variables on total length and by examination of Studentized residuals from linear regressions of each variable on total length. Additionally, a principal component analysis (PCA) (SAS PRINCOMP, SAS Institute Inc. 1989) was run and bi-plots of the first four principal components were examined for outliers. Ten fish with grossly aberrant measurements for one or more variables were identified through these methods and excluded from further analysis.

Several data sets were assembled to examine different facets of morphometric variation within and between the two spawning groups. The

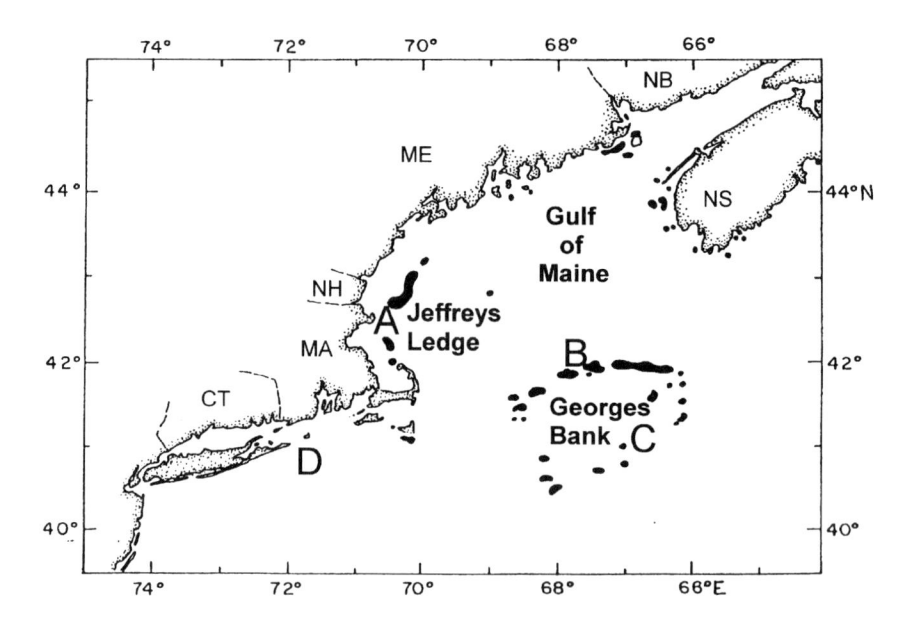

Figure 1. Location of sampling sites and historic spawning areas. A = Jeffreys Ledge, both pre- and postspawn samples; B = Georges Bank, prespawn sample; C = Georges Bank, postspawn sample; D = sample from the winter commercial fishery, unknown spawning affinities.

data sets were as follows: prespawning fish collected from the commercial fishery, postspawning fish collected from the NMFS Bottom Trawl Survey, postspawning fish from Georges Bank only, and pre- and postspawning fish from Jeffreys Ledge. Reduced data sets using only the most common age in all samples were also created.

Patterns of morphometric variation were initially examined using principal component analysis. Group discrimination was accomplished using discriminant analysis on size-adjusted data. The size adjustment was accomplished using multigroup principal component analysis (MGPCA; Thorpe 1988). Size components were removed from morphometric distances by setting first component scores to zero and transforming the adjusted score matrices back to the original variable space to derive size-adjusted data matrices (Burnaby 1966, Rohlf and Bookstein 1987). Discriminant analysis with equal prior probability was performed on the size-adjusted data with SAS DISCRIM using jack-knifed classification. Stepwise discriminant analysis (SAS STEPDISC) was used to select variables to use in the group discrimination. Multivariate analysis of variance (MANOVA) was used to examine shape differences among spent, resting, and immature herring in the Georges Bank postspawning sample.

Table 1. List of morphometric characters measured (asterisk indicates characters used in the discriminant analyses).

Name	Description
TL	Total length
A*	Snout to front of eye
B	Snout to posterior edge of circumorbital bone
C	Snout to posterior of opercular bone
D	Snout to anterior dorsal fin
E*	Snout to pectoral fin
F*	Snout to pelvic fin
G	Length of pectoral fin
H*	Length of pelvic fin
I	Pectoral fin to anterior dorsal fin
J	Pectoral fin to pelvic fin
K*	Head height
L	Base of dorsal fin
M	Anterior dorsal fin to pelvic fin
N	Posterior dorsal fin to pelvic fin
O	Anterior dorsal fin to anterior anal fin
P*	Anterior dorsal fin to posterior anal fin
Q*	Caudal peduncle height
R	Anterior dorsal fin to dorsal peduncle
S*	Anterior dorsal fin to ventral peduncle
T	Base of anal fin
U*	Pelvic fin to anterior anal fin
V*	Pelvic fin to dorsal peduncle
W	Pelvic fin to ventral peduncle
X*	Posterior dorsal fin to anterior anal fin
Y	Posterior dorsal fin to posterior anal fin
Z*	Posterior dorsal fin to dorsal peduncle
AA*	Posterior dorsal fin to ventral peduncle
BB	Anterior anal fin to dorsal peduncle
CC*	Anterior anal fin to ventral peduncle

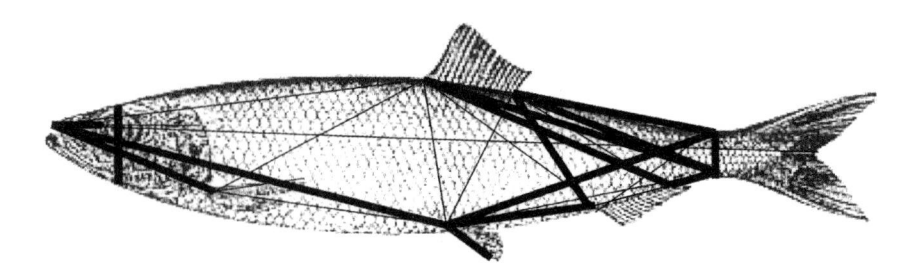

Figure 2. Morphometric features of Atlantic herring. Bold lines indicate distances used in discriminant analysis.

Results

All samples were a mix of ages 3-7, but were dominated by age 4. Prespawning fish from both areas were in various stages of ripening, although the Jeffreys Ledge fish were, in general, more well developed. PCA of pre- and postspawning herring from Jeffreys Ledge showed clear separation along the second component (PC2; Fig. 3). Correlations between all variables and first principal component (PC1) scores were positive, indicating that PC1 can be interpreted as a size component (Table 2). Correlations between variables and PC2 scores were greatest for the truss measurements that captured the abdomen height (M, N). As a precaution, we eliminated M and N from further analyses because they were so greatly affected by spawning condition.

PCA of postspawning fish from Georges Bank and Jeffreys Ledge showed moderate separation along the PC1 axis (accounting for size differences between the samples) but little separation along the PC2 axis (accounting for shape variation; Fig. 4). Stepwise discriminant analysis resulted in the inclusion of 15 characters, and accurately classified 88% of extrinsic samples into their correct spawning group (Table 3). Characters important in the discrimination included E, H, K, Q, U, V, X, Z, and AA (Table 4). Inclusion of only age-4 fish decreased the classification success to 79%. A multivariate analysis of variance (MANOVA) found no significant difference between spent, resting and immature fish in the postspawn sample from Georges Bank (Wilk's Lambda = 0.0334, $P > 0.05$). All univariate comparisons between these fish were nonsignificant for all 15 characters. This analysis indicated that using postspawning fish effectively eliminated the confounding effect of differences in spawning condition.

Prespawning fish from Georges Bank and Jeffreys Ledge showed no clear separation based on PCA (Fig. 5). Discriminant analysis successfully classified 64% of individuals into their correct spawning group (Table 5). Characters important in the discrimination included E, I, P, S, U, V, AA, and CC (Table 6). Classification success did not improve when only age-4 fish were used.

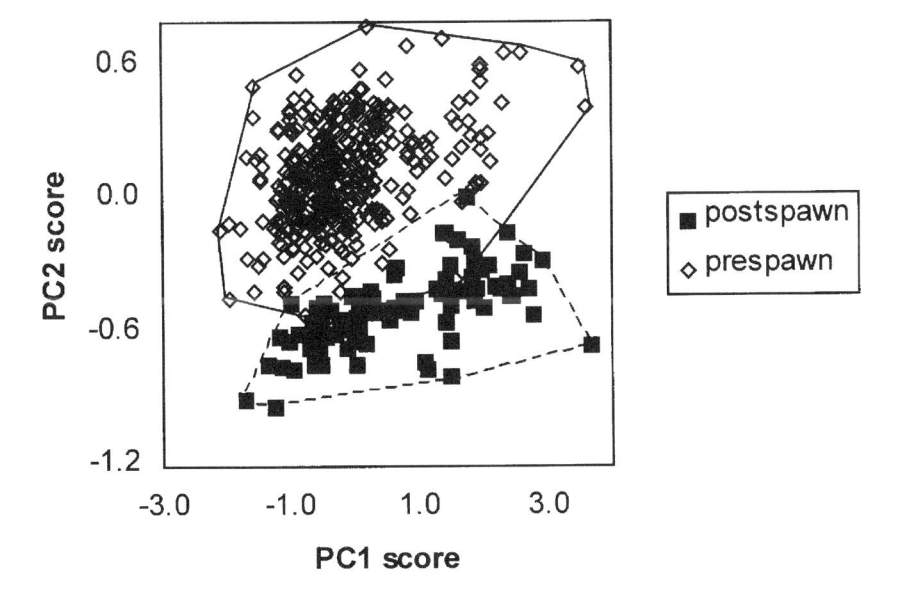

Figure 3. PC scores of Atlantic herring from pre- and postspawning aggregations on Jeffreys Ledge.

The sample of herring from the winter fishery in southern New England was classified using discriminant function based on the postspawning data set. The sample was classified as being 70% from the Georges Bank stock and 30% from the Jeffreys Ledge stock.

Discussion

Atlantic herring showed clear morphometric differences between prespawning and postspawning individuals. Development of the gonads results in an expansion of the abdomen which was captured by the truss components and clearly seen in principal component or discriminant analyses. This led us to conclude that discrimination of stocks based on truss distances that measured abdominal size would be confounded by differences in spawning times. For instance, two putative stocks could be discriminated if their times of spawning, and consequently their gonadal development, were slightly different, even if these morphometric differences did not exist outside of the spawning period. Alternatively, if differences in abdomen height did exist between stocks, the difference may be hidden by the expansion due to developing gonads. This presents difficulties when trying to discriminate stocks, since spawning times are seldom synchronized among stocks. In fact, this is one characteristic that

Table 2. Eigenvectors of the PCA using the pre- and postspawning data set from Jeffreys Ledge. Characters with high loadings are those that are most affected by spawning condition.

	PC1	PC2
A	0.209	−0.011
B	0.189	−0.062
C	0.196	−0.082
D	0.187	−0.002
E	0.205	−0.141
F	0.188	0.062
G	0.201	−0.118
H	0.198	−0.111
I	0.178	0.155
J	0.184	0.184
K	0.179	0.002
L	0.207	−0.192
M	0.129	0.538
N	0.126	0.522
O	0.198	0.057
P	0.196	−0.052
Q	0.178	−0.073
R	0.192	−0.049
S	0.193	−0.065
T	0.192	−0.305
U	0.191	0.203
V	0.172	0.095
W	0.188	−0.012
X	0.178	0.188
Y	0.185	0.034
Z	0.185	0.019
AA	0.182	0.001
BB	0.165	−0.092
CC	0.187	−0.276
Proportion of variance explained	0.665	0.079

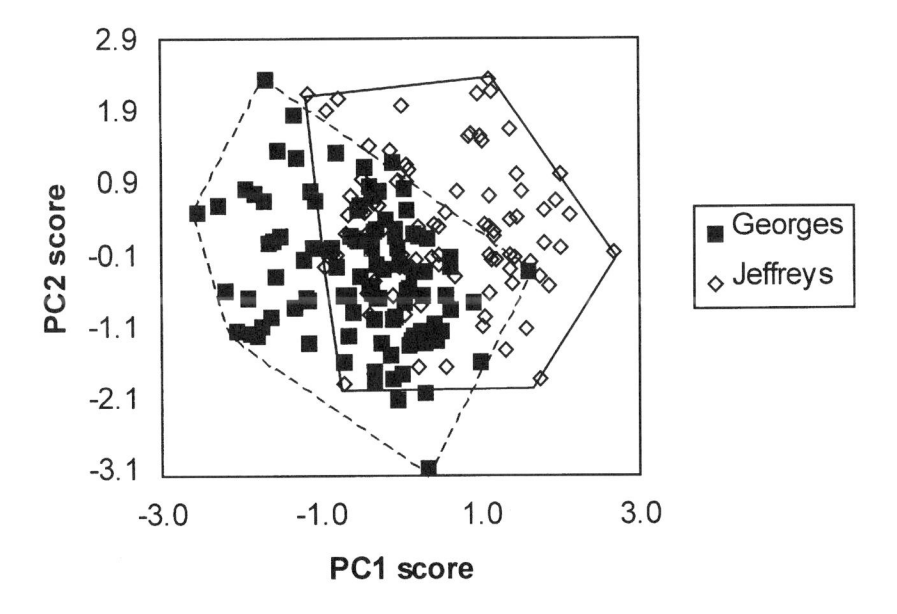

Figure 4. PC scores of Atlantic herring from postspawning aggregations on Jeffreys Ledge and Georges Bank.

Table 3. Number of fish classified to spawning group as extrinsic samples and associated error rate using the size-adjusted discriminant function of the postspawning samples.

Sample area	Classified as Georges Bank	Classified as Jeffreys Ledge	n	Error rate
Georges Bank	85	18	103	17.5%
Jeffreys Ledge	10	112	122	8.2%
Overall error rate				12.4%

Table 4. Univariate test statistics (d.f. = 1, 787) for the 15 variables used in the discriminant analysis using the postspawning data set.

	F value	Test probability
A	1.53	0.2177
E	120.08	0.0001
F	0.02	0.8871
H	38.44	0.0001
I	1.60	0.2067
K	5.73	0.0175
P	2.51	0.1143
Q	25.47	0.0001
S	1.90	0.1692
U	5.91	0.0159
V	34.78	0.0001
X	23.42	0.0001
Z	27.96	0.0001
AA	24.05	0.0001
CC	0.003	0.9574

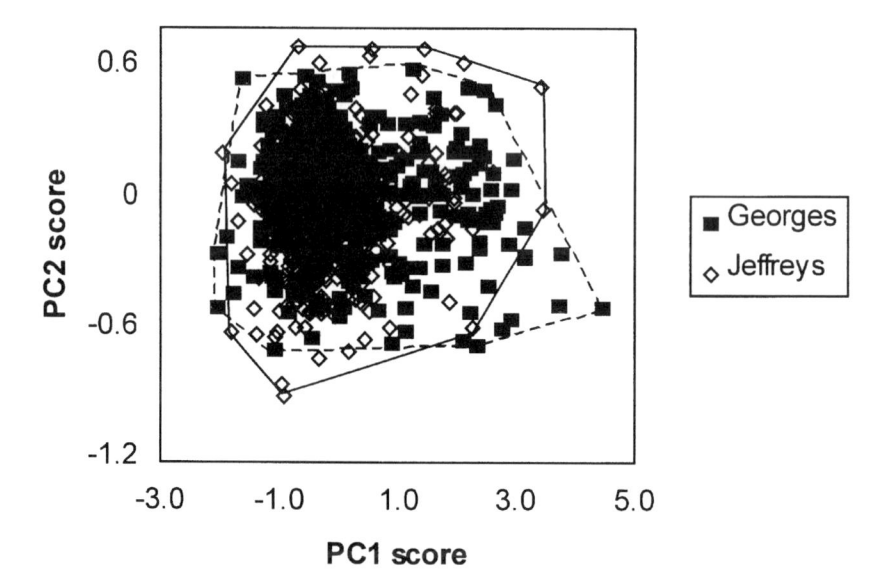

Figure 5. PC scores of Atlantic herring from prespawning aggregations on Jeffreys Ledge and Georges Bank.

Table 5. Number of fish classified to spawning group as extrinsic samples and associated error rate using the size-adjusted discriminant function of the prespawning samples.

Sample area	Classified as Georges Bank	Classified as Jeffreys Ledge	n	Error rate
Georges Bank	295	121	416	29.1%
Jeffreys Ledge	166	207	373	44.5%
Overall error rate				36.4%

Table 6. Univariate test statistics (d.f. = 1, 787) for the 15 variables used in the discriminant analysis using the prespawning data set.

	F value	Test probability
A	0.99	0.3209
E	22.74	0.0001
F	0.02	0.8955
H	0.01	0.9379
I	21.41	0.0001
K	8.06	0.0046
P	39.05	0.0001
Q	2.29	0.1309
S	10.62	0.0012
U	4.10	0.0431
V	22.99	0.0001
X	2.59	0.1081
Z	0.07	0.7951
AA	4.31	0.0383
CC	6.06	0.0141

has often been used to indicate discreteness of herring stocks. Because of the intermixing of stocks during nonspawning times, the only way to assure that individuals are members of a specific spawning group is to collect samples during spawning periods, preferably using only ripe or running ripe fish. However, these analyses show that abdominal measures from ripe herring cannot be used to discriminate stocks in a manner that will be applicable to nonspawning fish. In order to use fish in spawning condition, morphometric characters that are not affected by spawning condition must be used.

Discrimination of prespawning fish from Jeffreys Ledge and Georges Bank was poor (classification success 65%), likely because this analysis was confounded by differences in spawning condition between the two areas. Herring on Georges Bank spawn about 2-3 weeks later than the herring inshore on Jeffreys Ledge (M.P.A., unpublished data). This was evident from the samples where, even though the samples were taken only a few days apart, fish in the Jeffreys Ledge postspawn sample were all resting, while fish in the Georges Bank postspawn sample were a mix of resting and spent, indicating Jeffreys Ledge fish had completed spawning sooner, assuming no movement between areas. Although the two stocks can be discriminated based on morphometric characters as seen in the analysis with postspawn fish, the differences are confounded by the effects of spawning condition. Both Georges Bank and Jeffreys Ledge prespawn samples contained fish in various stages of ripeness. Overall, Jeffreys Ledge fish were more developed. However, because the classification was being driven by spawning condition rather than stock differences, less-developed fish from Jeffreys Ledge were classified into Georges Bank and more developed Georges Bank fish were classified as Jeffreys Ledge, resulting in low classification success.

Classification success was high using the postspawning data set. Many of the characters that showed differences between pre- and postspawners were also important in discriminating the stocks. In general, Georges Bank fish had a shorter pelvic fin (H), shorter but taller head (E and K), and a wider and longer area of the body (U, V, X, Z, AA) between the posterior edge of the dorsal to the caudal peduncle, although the peduncle itself (Q) was narrower than in Jeffreys Ledge fish. Overall, the Georges Bank fish could be described as "stockier" than the Jeffreys Ledge fish, especially in the head and caudal regions.

Using postspawned fish presents difficulties because there is less certainty that the fish are members of a specific spawning group. However, because our samples were taken shortly after spawning, this is less of a concern. It is likely that individuals from a specific spawning group maintain themselves as a school after spawning and remain in the area of spawning for at least a short time. As evidence of this, commercial purse-seine fishermen, who are prohibited by law from taking ripe and running ripe fish during the spawning season, target postspawning fish adjacent to spawning aggregations (M.P.A., personal observation). Nonetheless, it is possible that fish from other spawning groups contaminated our postspawning samples. Cross contamination between Jeffreys Ledge and Georges Bank would be less likely owing to their separation by greater than 150 miles. Any presence of herring from other spawning groups would violate our assumption that each sample comes from a single spawning group, and may have contributed to decreasing our classification success.

This study indicates that it is possible to discriminate Georges Bank spawners from Jeffreys Ledge spawners, with the caveat that the samples must be taken shortly after spawning rather than before spawning. Spawning condition greatly affects truss measurements and, unfortunately, traditional measurements that are unaffected by spawning condition such as head height, snout length, etc., show little discriminatory power by themselves.

Confining analyses to specific ages did not improve discriminatory power, suggesting that the differences between stocks maintain themselves over time and among cohorts. This will be an important consideration for future research when it may be necessary to process large numbers of fish. Simply taking morphometric measurements without removing, processing, and reading otoliths will reduce processing time.

The mix of stocks found in the sample from the winter fishery (30% Jeffreys Ledge, 70% Georges Bank) is in agreement with the relative size of these stocks (NEFSC 1998). This implies that both stocks winter in southern New England and form a well-mixed group. However, this analysis is purely exploratory. This classification does not consider that there may be fish from other spawning areas (e.g., eastern Maine, western Georges Bank, Nantucket Shoals) that may have morphologically distinct forms.

These results confirm morphological patterns in the Gulf of Maine-Georges Bank complex found by other researchers. Anthony (1972) examined counts of vertebrae and pectoral fin rays of herring from several locations around the Gulf of Maine including eastern and western Maine, Jeffreys Ledge-Stellwagen Bank, Nova Scotia, and Georges Bank. In general, he found counts to be highest in herring from Nova Scotia and lowest in Georges Bank herring. There were statistically significant differences in counts for both vertebrae and pectoral fin rays among the groups but the differences did not maintain themselves over the years for which he had samples. Anthony (1972) concluded that consistent differences in mean counts of meristic characters indicated limited mixing of groups of herring. Such separation means that, although the groups may not be genetic stocks, they can function as stock units in a herring management program.

Safford and Booke (1992) examined morphometric differences between herring from Trinity Ledge, Nova Scotia, and Jeffreys ledge. They found significant differences between the two spawning grounds but the discriminant function only performed well in one of two years. They also performed traditional starch gel electrophoresis of enzymes and found no stock structure.

The suggestion of stock structure indicated by the present study is not supported by previous genetic studies (Grant 1984, Kornfield and Bogdanowicz 1987, Safford and Booke 1992). The genetic studies indicate that there is enough mixing among spawning groups to prevent fixation of distinct alleles; thus, the morphometric differences seen between Jeffreys Ledge and Georges Bank spawners may be environmentally induced. Morphometric differences may result from differences in life history between the two groups, including perhaps migration and spawning patterns, trophic differences, and exposure to different environmental cues during important developmental periods. Although the spawning groups do not appear to be true genetic stocks, the presence of significant differences in morphometric characteristics indicates limited mixing of the stocks at spawning time, and so the spawning groups can be treated as unit stocks for management purposes.

Although we have successfully documented significant morphometric differences between two spawning groups in the Gulf of Maine, this is the first step in the stock identification process (Cadrin 2000) culminating in the ability to assign fishery catches and survey samples to individual stocks. There are many more discrete spawning areas in the Gulf of Maine and on Georges Bank and Nantucket Shoals that we have not yet sampled, and the degree of morphometric variation within and between all these spawning herring has yet to be determined.

Acknowledgments

We thank Genevieve Casey for her major contribution to imaging the fish and to Vaughn Silva of the National Marine Fisheries Service, for his help in acquiring samples and in aging. We also thank Fred Bailey, owner of the fishing vessel *K & J II*, for supplying commercial herring samples. This is Massachusetts Division of Marine Fisheries Contribution No. 1.

References

Anthony, V.C. 1972. Population dynamics of the Atlantic herring in the Gulf of Maine. Ph.D. thesis, University of Washington, Seattle. 266 pp.

Anthony, V.C., and G. Waring. 1980. Assessment and management of the Georges Bank herring fishery. Rapp. P.-V. Reun. Cons. Int. Explor. Mer 177:72-111.

Burnaby, T.P. 1966. Growth-invariant discriminant functions and generalized distances. Biometrics 22:96-110.

Cadrin, S.X. 2000. Advances in morphometric analysis of fish stock structure. Rev. Fish Biol. Fish. 10:91-112.

Chenoweth, J.F., S.E. McGladdery, C.J. Sindermann, T.K. Sawyer, and J.W. Bier. 1986. An investigation into the usefullness of parasites as tags for herring (*Clupea harengus*) stocks in the western north Atlantic, with emphasis on the use of the larval nematode *Anisakis simplex*. J. Northwest Atl. Fish. Sci. 7:25-33.

Dery, L.M. 1988. Atlantic herring, *Clupea harengus*. In: J. Pentilla and L.M. Dery (eds.), Age determination methods for Northwest Atlantic species. NOAA Tech. Rep. NMFS 72:17-22.

Grant, W.S. 1984. Biochemical population genetics of Atlantic herring, *Clupea harengus*. Copeia 1984(2):357-364.

Harden-Jones, F.R. 1968. Fish migration. Edward Arnold, London.

King, D.P.F. 1984. Genetic analysis of enzyme polymorphisms in herring. Heredity 52:121-131.

King, D.P.F. 1985. Morphological and meristic differences among spawning aggregations of north-east Atlantic herring, *Clupea harengus* L. J. Fish Biol. 26:591-607.

Kornfield, I., and S.M. Bogdanowicz. 1987. Differentiation of mitochondrial DNA in Atlantic herring, *Clupea harengus*. Fish. Bull., U.S. 85(3):561-568.

Kornfield, I., B. Sidell, and P. Gagnon. 1982. Stock definition in Atlantic herring: Genetic evidence for discrete fall and spring spawning populations. Can. J. Fish. Aquat. Sci. 39:1610-1621.

McGladdery, S.E., and M.D.B. Burt. 1985. Potential of parasites for use as biological indicators of migration, feeding, and spawning behavior of northwestern Atlantic herring (*Clupea harengus*). Can. J. Fish. Aquat. Sci. 42:1957-1968.

Meng, H.J., and M. Stocker. 1984. An evaluation of morphometrics and meristics for stock separation of Pacific herring (*Clupea harengus pallasi*). Can. J. Fish. Aquat. Sci. 41:414-422.

Moser, M. 1990. Biological tags for stock separation in Pacific herring (*Clupea harengus pallasi* Valenciennes) and the possible effect of "El Niño" currents on parasitism. In: Proceedings of the International Herring Symposium. University of Alaska Sea Grant, AK-SG-91-01, Fairbanks, pp. 245-254.

NEFSC (Northeast Fisheries Science Center). 1992. Atlantic herring advisory report. In: Report of the 13th Northeast Regional Stock Assessment Workshop (13th SAW): Public review workshop. Northeast Fish. Sci. Cent. Ref. Doc. 95-02. Available from: National Marine Fisheries Service, 166 Water St., Woods Hole, MA USA 02543-1026.

NEFSC. 1998. Atlantic herring advisory report. In: Report of the 27th Northeast Regional Stock Assessment Workshop (27th SAW): Public review workshop. Northeast Fish. Sci. Cent. Ref. Doc. 98-14. Available from: National Marine Fisheries Service, 166 Water St., Woods Hole, MA USA 02543-1026.

Overholtz, W.J., and K.D. Friedland. 2002. Recovery of the Gulf of Maine-Georges Bank herring complex: Perspectives from bottom trawl survey data. Fish. Bull., U.S. 100. (In Press)

Parsons, L.S. 1975. Morphometric variation in Atlantic herring from Newfoundland and adjacent waters. Int. Comm. Northwest Atl. Fish. Res. Bull. 11:73-92.

Rohlf, F.J., and F.L. Bookstein. 1987. A comment on shearing as a method for "size correction." Syst. Zool. 36:356-367.

Safford, S.E., and H. Booke. 1992. Lack of biochemical, genetic, and morphometric evidence for discrete stock of Northwest Atlantic herring, *Clupea harengus harengus*. Fish. Bull., U.S. 90:203-210.

SAS Institute, Inc. 1989. SAS/STAT user's guide, version 6, fourth edition. Volumes 1 and 2. SAS Institute Inc., Cary, North Carolina. 943 pp.

Schweigert, J.F. 1990. Comparison of morphometric and meristic data against truss networks for describing Pacific herring stocks. Am. Fish. Soc. Symp. 7:47-62.

Smith, W.G., and W.W. Morse. 1993. Larval distribution patterns: Early signals for the collapse/recovery of Atlantic herring *Clupea harengus* in the Georges Bank area. Fish. Bull., U.S. 91:338-347.

Stephenson, R.L. 1990. Stock discreteness in Atlantic herring: A review of arguments for and against. In: Proceedings of the International Herring Symposium. University of Alaska Sea Grant, AK-SG-91-01, Fairbanks, pp. 659-666.

Stephenson, R.L., and D.J. Gordon. 1990. Comparison of meristic and morphometric characters among spawning aggregations of northwest Atlantic herring, *Clupea harengus* L. In: Proceedings of the International Herring Symposium. University of Alaska Sea Grant, AK-SG-91-01, Fairbanks, pp. 279-296.

Stephenson, R.L., and I. Kornfield. 1990. Reappearance of spawning herring on Georges Bank: Population resurgence not recolonization. Can. J. Fish. Aquat. Sci. 47:1060-1064.

Strauss, R.E., and F.L. Bookstein. 1982. The truss: Body form reconstruction in morphometrics. Syst. Zool. 31:113-135.

Thorpe, R.S. 1988. Multiple group principal components analysis and population differentiation. J. Zool. (Lond.) 216:37-40.

Wheeler, J.P., and G.H. Winters. 1984. Homing of Atlantic herring in Newfoundland waters as indicated by tagging data. Can. J. Fish. Aquat. Sci. 41:108-117.

Herring of the White Sea

G.G. Novikov, A.K. Karpov, A.P. Andreeva, and A.V. Semenova
Moscow State University, Biology Department, Moscow, Russia

Extended Abstract

The herring of the White Sea belong to a low-vertebrae group of the genus *Clupea*. The taxonomic status of the White Sea herring is a subject of long-term discussions. In the monograph *Clupeidae*, Svetovidov (1952) categorically objected to the division of the species *Clupea harengus* into two species, *C. harengus* and *C. pallasii*. Both Svetovidov (1952) and Berg (1923) argued for the occurrence of two subspecies of *C. harengus* with various ecomorphs (Table 1), using vertebral numbers as the basic morphological criterion (besides geographical) for the distinction of the subspecies. Nikolskiy (1950) gave subspecies status to these basic ecological forms based on the geographical criterion; he did not consider vertebral number to be an important character. However, Rass (1985) maintained that *Clupea* contains two species (Table 1), and this classification has generally been adopted. A summary of the taxonomic alternatives is given below:

Svetovidov 1952, *Clupea harengus* L.:
C. harengus harengus L.   C. harengus pallasii Val.
(high vertebrae)   (low vertebrae)
n. *harengus*   s. *suworowi*
n. *membras*   n. *maris-albi*

Nikolskiy 1950, *Clupea harengus* L.:
C. harengus harengus L.
C. harengus membras L.
C. harengus maris-albi Berg
C. harengus suworowi Rab.
C. harengus pallasii Val.

Rass 1985:
Clupea harengus L.   Clupea pallasii Val.
C. harengus harengus L.   C. pallasii pallasii Val.
C. harengus membras L.   C. pallasii maris-albi Berg
                             C. pallasii suworowi Rab.

Table 1. Some meristic characters of low- and high-vertebrae herring (after Rass 1985).

Taxon of genus in Northern Hemisphere	Number of vertebrae	Teeth on vomer	Abdominal scales Between P and V	Abdominal scales Between V and A
Clupea harengus L. (high vertebrae)	(54)56-59(60)	Well developed	Developed	
C. harengus harengus L. (Atlantic herring)	(55)56-59(60), Mean 57.06	Well developed	Developed	Developed
C. harengus membras L. (Baltic herring)	54-57, Mean 54.7	Well developed	Developed	Developed
Clupea pallasii Val. (low vertebrae)	52-55	Less developed	Less developed or absent	More or less developed
C. pallasii pallasii Val. (eastern Pacific herring)	53-55, Mean 54.1	Less developed than for *C.h. membras* L.	Less developed or absent	More or less developed
C. pallasii marisalbi Berg (White sea herring)	52-55, Mean 53.2	Less developed	Less developed	More or less developed
C. pallasii suworowi Rabinerson (Tzosha-Pechora herring)	53-55, Mean 53.5	Less developed	Less developed	More or less developed

P = pectoral fin base; V = ventral fin base; A = anal fin base.

Ideas about microevolutionary processes and migration of herring in the North Atlantic in the postglacial period depend on opinions regarding the taxonomic status of these groups. The basic disagreements are connected with the origin and status of the low-vertebrae herring in the White Sea and Czech-Pechora region. The center of origin of herring from the genus *Clupea* is considered to be the Atlantic. About 3 million years ago, in the Pliocene, herring penetrated through Bering Strait to the Pacific Ocean (Berg 1918, 1934; Andriashev 1939; Svetovidov 1952; Grant 1984). Then, in the postglacial period, about 5,000 years ago, low-vertebrae herring had a backward migration, and they were distributed along the estuaries of the northern ocean, reaching the White Sea and even some of the northern Norway fjords (Derjugin 1929, Jorstad et al. 1984). Svetovidov (1952) believed that the herring came to the Pacific Ocean from the Atlantic along northern Asia. In the postglacial period, the Atlantic herring penetrated to the Baltic, White, and Barents seas, forming a range of groups distinguished by biological features. Thus, according to Svetovidov, in the northern parts of the range both low- and high-vertebrae herring spawned (Iceland, northern Norway, etc.), mainly in the spring, whereas in the southern parts of the range (North Sea, English Channel etc.) high-vertebrae herring spawned, mainly in the autumn. The main features of low-vertebrae herring compared to high-vertebrae herring are the low spawning temperature, and spawning in comparatively shallow and brackish waters (Table 2). The life cycles of low-vertebrae herring are not connected with the main water currents, like the Gulf Stream, but they are associated with the local current systems. These systems determine the spatial distribution and differentiation of herring stocks.

It is known that herring can form a set of morphologically and ecologically distinguishable groups within one large basin. These groups represent local stocks with different growth rates, fecundity, vertebrae numbers, and abdominal scale numbers. They spawn in different locations at different times and have different nursery areas (Nikolskiy 1950). The White Sea herring is no exception.

Hydrological conditions (current systems, temperatures, and salinities) in the White Sea promote the formation of ecologically different local groups in the local bays. They are characterized by life cycle features, growth rate, duration of life, spawning time, and special hydrological conditions on the spawning grounds. According to general opinion, there is a local group in each bay of the White Sea. Some authors (Averintsev 1927, 1928; Dmitriev 1946; Tambovzev 1957; Lajus 1966) believe that these groups are genetically isolated.

However, according to Lapin (1971, 1974, 1978), White Sea herring represent a single population. According to this hypothesis all herring at the age of 1-2 years live in Onega and Dvina bays, and they are then distributed in different bays where they later can be characterized by different growth rates. It is supposed that the effluent currents from the distant parts of these bays (observed by Nadejin 1963) are directed to the central part of the White Sea, and the spring-spawning herring juveniles drift from the inner areas of the bays. The individuals with the highest growth rate reach the inner parts of Kandalaksha Bay. This hypothesis does not explain why this drift is repeated annually and why the individuals from the same bay grow with different intensity.

The results from genetic analysis of the spawning groups of spring-spawning herring in various bays of the White Sea are presented in this study. To investigate genetic differentiation between groups of White Sea herring, four polymorphic systems where used: two lactate dehydrogenase loci (*LDH-1*; LDH-2**), maleate dehydrogenase (*MDH-4**), and glucosophosphate isomerase (*GPI-1**). Samples of white muscle were collected from herring on spawning grounds, frozen, and transported to Moscow for analysis. At the *LDH-1** locus, two alleles (of three known for this system) were found, where the fastest allele (*200*) is the most common for low-vertebrae herring. Three alleles are registered at the *LDH-2**

Table 2. Biological features of herring *Clupea harengus* (after Svetovidov 1952).

High-vertebrae group	Low-vertebrae group
Vertebral counts	
51-60 (55-57)	47-57 (52-55)
Abdominal scales between V and A	
12-16	10-14
Time of spawning	
Spring spawning[a]	Spring spawning
Summer spawning	
Autumn spawning	
Winter spawning	
Temperature at time of spawning	
Always higher than 0°C	Near 0°C
On the average 6-12 (16°C)	−1.8 to +6 (15°C)
Water salinity on spawning ground	
Usually 29-35 ‰	Usually freshwater
Sometimes 4-10 ‰	to >1-2 ‰
(Baltic Sea, spring spawning)	Sometimes 29-32 ‰
Location of spawning area	
Far from shore, on bank	Coastal zone
Max. depth 200 m	Depth to 15 m
Substratum for spawning (predominantly)	
Different kind of bottom	Underwater sea plants
Different kind of bottom	
Migrations	
Extensive migrations[b]	Less extensive migrations
Less apparent locality[b]	Apparent locality

[a] In north (Iceland, Norway, etc.) spring spawning only; in south (English Channel, North Sea) autumn spawning dominant.
[b] Besides Baltic Sea and North Sea.

Table 3. Allele frequencies for three polymorphic loci in pooled samples of spawning herring from different regions in the White Sea.

	Average allele frequency	LDH-2* 120	LDH-2* 100	MDH-4* 100	MDH-4* 70	GPI-1* 200	GPI-1* 150
Mezen Bay	96	0.970	0.020	0.980	0.020	0.310	0.680
Dvina Bay	226	0.923	0.077	0.944	0.056	0.211	0.703
Onega Bay	332	0.900	0.100	0.896	0.104	0.252	0.710
Kandalaksha Bay	313	0.908	0.092	0.896	0.104	0.241	0.700

Rare alleles are not included *(LDH-2* and GPI-1*)*.

locus, and the *120* allele is found at a higher frequency. A similar situation was observed for the *MDH-4** locus, where one (*MDH-4*100*) of three alleles has the highest frequency and *MDH-4*70* is found in individuals as heterozygotes. A more complex system with five alleles was found in *GPI-1**, and here only the data from the two most common alleles are given (Table 3).

The phenotype distributions observed at the different loci did not differ from the distribution expected according to Hardy-Weinberg. Therefore, it is assumed that the interpretation of the electrophoretic pattern observed is correct, and the groups can be characterized by the allele frequencies. Analysis of the polymorphic protein loci of herring samples collected annually in the same area over several years revealed that the allele frequencies are similar (data not shown). Comparison of samples taken in the same area, using Fishers criterion (Urbach 1963), show no significant differences between years. Apparently, this stability of genetic characteristics in the spawning groups over years can support the occurrence of homing and similar origin of the herring for the same spawning groups. Therefore, it is possible to pool the data obtained from the herring samples over several years, and the herring groups from each area can be characterized by the average values of allele frequencies as shown in Table 3.

Pairwise comparison of the groups using the allele frequencies at the *LDH-2** and *MDH-4** loci demonstrated significant differences between the herring samples from Mezen Bay and from Kandalaksha and Onega bays. However, according to allele frequency estimated for the *MDH-4** loci, the herring from Dvina Bay also differ significantly from the herring from Onega and Kandalaksha bays. As mentioned above, no significant annual variation of allele frequencies was found in any of the herring spawning bays investigated. Thus, the reasons for annual allele frequency stability in the bays and the relationships between the various herring stocks are now the subject of further investigations.

Acknowledgments

This study was supported by the Russian Foundation for Basic Research, the Foundation "Universities of Russia"; Russian Research Institute "SevPINRO" in Arkhangelsk, and Institute of Marine Research in Bergen, Norway.

References

Andriashev, A.P. 1939. An outline of the zoogeography and origin of the fish fauna of the Bering Sea and adjacent waters. Izdanie Leningradskogo Gosudarstvennogo Universiteta, Leningrad. 187 pp. (In Russian.)

Averintsev, S.V. 1927. The White Sea herring. Part 1. Tr. Nauchno-Issled. Inst. Rybn. Khoz. 2(1):41-77. (In Russian.)

Averintsev, S.V. 1928. The White Sea herring. Part 2. Tr. Nauchno-Issled. Inst. Rybn. Khoz. 3-4:73-141. (In Russian.)

Berg, L.S. 1918. Motive of resemblance of the northern part of Atlantic and Pacific Ocean fauna. Izv. Akad. Nauk. 6(12):1835-1842. (In Russian.)

Berg, L.S. 1923. Freshwater fishes of Russia. Izd. 2. Ch.1. M. L., 501 pp. (In Russian.)

Berg, L.S. 1934. About amphiboreal distribution of marine fauna in the Northern Hemisphere. Izv. Geographizeskogo obschestva 16:71-79. (In Russian.)

Derjugin, K.M. 1929. Fauna of the White Sea and the conditions of its origin. Priroda 9:17-28.

Dmitriev, N.A. 1946. Biology and fishery of the White Sea herring. Pischepromizdat 5. 88 pp. (In Russian.)

Grant, W.S. 1984. Biochemical genetic divergence between Atlantic *Clupea harengus* and Pacific *Clupea pallasii* herring. Copeia 1983(3):714-719.

Jørstad, K., G. Dahle, and O.I. Paulsen. 1984. Genetic comparison between Pacific herring (*Clupea pallasi*) and Norwegian fiord stock of Atlantic herring (*Clupea harengus*). Can. J. Fish. Aquat. Sci. 51(Suppl.1):233-239.

Lajus, D.L. 1996. What is the White Sea herring *Clupea pallasi marisalbi* Berg 1923? A new concept of the population structure. Publ. Espec. Inst. Esp. Oceanogr. 21:21-230.

Lapin, J.E. 1971. Regularity of the dynamics of fish population in connect with duration of its vital cycle. Nauka. 173 pp. (In Russian.)

Lapin, J.E. 1974. Variability of the vertebrae number and larvae monomers of White Sea herring *Clupea harengus pallasi* n. *marisalbi* Berg in connection with ecology. Vopr. Ikhtiol. 14, 4(87):546-557. (In Russian.)

Lapin, J.E. 1978. General characteristics of White Sea herring. Nauka, Moscow, pp. 37-52. (In Russian.)

Nadjozin, G.V. 1963. Meaning of currents in White Sea herring life. Vopr. Ikhtiol. 3, 4(29):616-624. (In Russian.)

Nikolskiy, G.V. 1950. Special ichthyology. Sovetskaya Nauka, Moscow. 436 pp. (In Russian.)

Rass, T.S. 1985. On the systematics of marine herring of genus *Clupea*. Ekologicheskie issledovanija perspektivnich objektov marikulturi v Belom more. Leningrad, pp. 3-56. (In Russian.)

Svetovidov, A.N. 1952. Fauna of U.S.S.R. Fishes. Vol. 2. Clupeidae. Izdatel'stvo Akademii Nauk SSSR, Moscow and Leningrad. 365 pp. (In Russian.)

Tambovzev, B.M. 1957. Low-vertebrae herring of the Barents and White seas. Tr. Murmanskoi biologicheskoi stanzii 3:169-174. (In Russian.)

Urbach, V.U. 1963. Mathematical statistics for biologists and physicians. Moscow. 323 pp. (In Russian.)

A Tagging Experiment on Spring-Spawning Baltic Herring (*Clupea harengus membras*) in Southwest Finland in 1990-1998

J. Kääriä
University of Turku, Archipelago Research Institute, Turku, Finland

M. Naarminen
Finnish Game and Fisheries Research Institute, Helsinki, Finland

J. Eklund
Vaasa Administrative Court, Vaasa, Finland

N. Jönsson
University of Rostock, Fachbereich Biologie, Rostock, Germany

G. Aneer
County Administrative Board of Stockholm, Environment and Planning Department, Stockholm, Sweden.

M. Rajasilta
University of Turku, Archipelago Research Institute, Turku, Finland

Abstract

Altogether 26,476 prespawning, mature herring from commercial trap nets were tagged during spawning time in May and June 1990-1998 in southwest Finland. The total number of recoveries is 94 (0.4%). The results indicate possible homing to a relatively large spawning area but exact homing to the tagging site was not observed. Of the recoveries made over 2 months after tagging, the majority was recorded within the Archipelago Sea, at moderate distances from the tagging site. Recoveries outside the Archipelago Sea were made mainly in the Bothnian Sea, indicating a north-south migration between the Archipelago Sea and the Bothnian Sea. Two tagging methods were compared and in the more successful method, where herring were kept in water during the treatment, some fish survived more than 2 years.

Introduction

In the Archipelago Sea (northern Baltic Sea, southwest Finland; Fig. 1), Baltic herring (*Clupea harengus membras* L.) spawn mainly in April/May-July. During the spawning period, the herring shoals arrive at the spawning grounds in the inner parts of the archipelago, where the spawning fish are exploited by the trap-net fishery. Herring spawning beds (an area where spawn has been deposited more or less homogeneously during one spawning act) are found in the same locations year after year (Rajasilta et al. 1993). Among the shores potentially suitable for spawning, herring seem to prefer certain shores to others (Kääriä et al. 1997).

It is known that spawning herring choose their spawning substrate by testing it (Stacey and Hourston 1982, Aneer et al. 1983). As herring do not spawn on all potentially suitable shores, it might be that the fish choose the spawning beds according to some criteria.

In general, the selection of spawning beds (on the scale of a few hundred meters) by the herring is random. Normally, in the Pacific, the Atlantic, and the Baltic herring, it is not possible from year to year to predict the exact spawning beds (Runnström 1941, Tyurnin 1973, Hourston and Rosenthal 1976, Johannessen 1986, Messieh 1987, Aneer 1989, Hay and McCarter 1997).

Our own diving investigations have shown that exactly the same spawning beds are used yearly for at least 10-19 years and, on some of the spawning beds, trap nets have been used for decades. The spawning beds in the Airisto Inlet led us to hypothesize that herring once having spawned on a certain bed imprint to the location and return to the same location for subsequent spawnings. Thus, a large part of the repeat spawning herring would return to the same spawning bed for later spawnings. To test this hypothesis, we tagged herring on the same spawning bed over a period of 7 years.

Herring tagging experiments have mostly been made to plot large-scale migrations (Jönsson and Biester 1981, Hourston 1982, Wheeler and Winters 1984, Aro 1988, Parmanne 1990), while reproduction ecology and spawning bed selection seldom have been the main objective of mapping studies. The recovery percentage, especially among Baltic herring (Parmanne 1990), is low and with modest resources it is not possible to obtain extensive results. However, every single recovery provides important information about herring behavior.

In this paper, we present our tagging experiment data from a 7-year period during which two different tagging methods were used, and we also compare the results of the two methods.

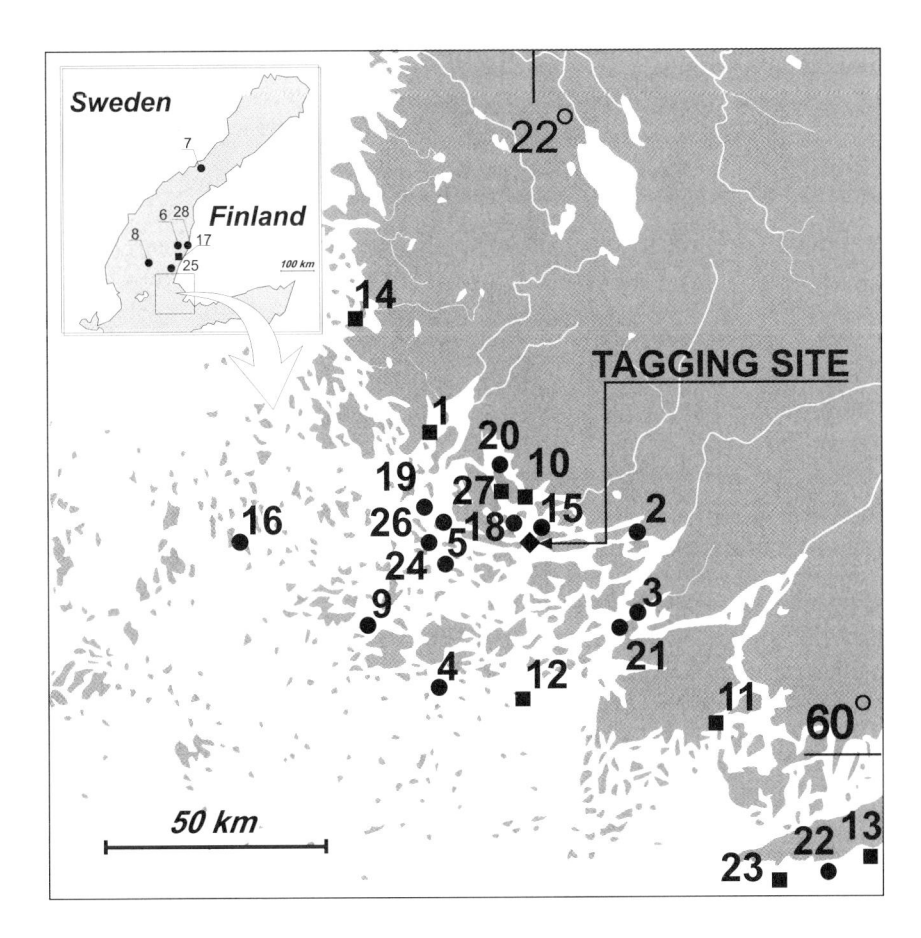

Figure 1. Study area in the Baltic Sea off southwest Finland. Tagging site off Järvistensaari Island in the Airisto Inlet (♦) and recoveries in spawning time (1 May-31 July) (■) or in feeding and overwintering time (●). For more details see Table 2.

Material and Methods

The tagging was done in the Airisto Inlet in the Archipelago Sea, southwest Finland (Fig. 1). The mean depth in the area is about 20 m, but in the north-south direction, deeper fracture lines (50-60 m) form channels on the sea bottom. Three small rivers discharge into the area, the largest being the Aurajoki River (mean flow = 8.5 m^3 per second; Pitkänen 1994). Salinity varies from 4.5 to 7.0 psu throughout the water column, with very little seasonal difference, except for the river mouth and surface water areas which are affected by the rivers (Vuorinen and Ranta 1987, Hietaranta 1990, Viitasalo et al. 1990).

Fish were caught with a trap net on the same spawning bed every year. In the net, herring swim freely in the rear end of the gear, in a net enclosure about 5 m deep and 12 m by 6 m wide (Fig. 2). Tagging was mainly done when a new shoal had arrived. To remove the fish from the trap net, the herring were led into a corner of the net by lifting the bottom of the net. During tagging and for about 3 days afterward the trap-net entrance was closed. The length of the fish tagged varied between 14 and 22 cm, which is the normal size range of the spawning shoals at the tagging site (Rajasilta et al. 1993). The trap-net catch consists almost exclusively of mature fish (gonad stages from 4 to 6). In the spawning shoals, small numbers of immature fish (<13 cm) are found but these fish were not tagged.

Two different tagging methods were used. The first method (group A, see Fig. 3 and Table 1) is essentially that used by Jönsson and Biester (1981). The fish (about 30-50 at a time) were lifted from the trap net with a hand net (without water) into a 60-liter plastic container with water. The container was transported about 100 m to the shore where the fish were tagged. Fish were lifted out of the water by hand (with plastic gloves) and a Reverdin needle was passed through the dorsal muscle under the dorsal fin (Fig. 3). The nylon line of the tag was hooked onto the needle, and pulled through. A yellow (22 × 5 × 0.1 mm) tag with nylon line was secured with a knot some distance from the fish's back, leaving the dorsal fin free. The tagging operation took about 30 seconds. Tagged herring were collected into another container and released within 20 minutes after tagging in small schools (about 10-20 fish). Altogether 20,646 herring were tagged in this manner between 1990 and 1994.

In the second method (group B in Fig. 3 and Table 1), the fish were handled and tagged under water. About 10 fish at a time were lifted from the trap net along with some water in a special hoop net and put into a small net cage. The cage with about 100 herring was towed very slowly (about one herring body length per second) to a suitable sheltered spot nearby. About 10 fish were lifted with the hoop net into a dark plastic container that was covered with a black plastic bag. For tagging, each herring was captured by hand under water, and put into the black tagging chamber. A plastic plug was passed through the dorsal muscle under the dorsal fin with a hollow needle (length 40 mm, outer diameter 1.2 mm; type 18 G). The red plastic tag (22 × 5 × 0.1 mm) with plug (diameter 1 mm) was secured with plastic fastening. The tagged fish were released immediately. Altogether 5,830 herring were tagged in this manner between 1994 and 1998.

Every year newspaper notices informed the public, and the herring fishermen in the area were contacted personally, to return the tags to the Game and Fisheries Research Institute when found. The institute recorded the returns.

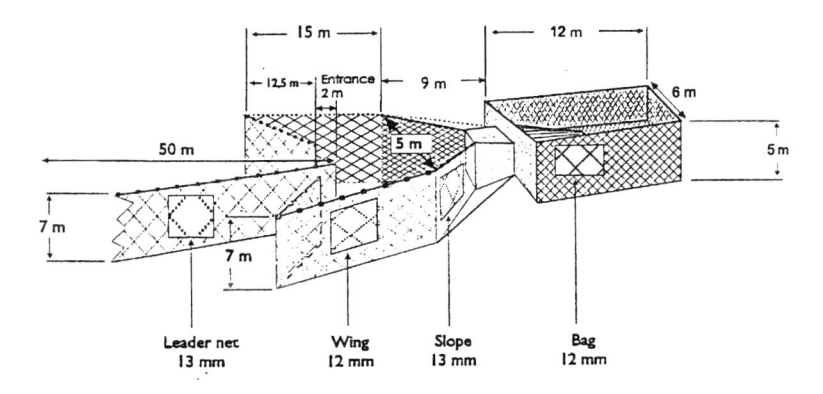

Figure 2. Structure of trap net.

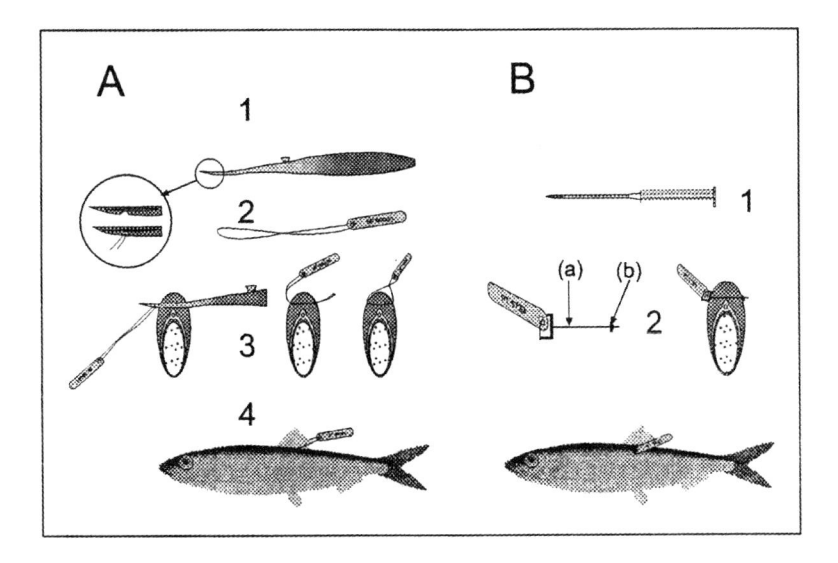

Figure 3. Tagging methods. Method A: (1) Reverdin needle; (2) yellow tag with nylon line; (3) Reverdin needle passed through the dorsal muscle under the dorsal fin. The nylon line was hooked onto the needle and pulled through; (4) the line was secured with a knot, leaving the dorsal fin clear. Method B: (1) hollow needle, length 40 mm, outer diameter 1.2 mm; (2) red plastic tag (22 × 5 × 0.1 mm) with plug (a) (diameter 1 mm) secured with plastic fastening (b).

Table 1. Recoveries of herring tagged at Järvistensaari Island (see Fig.1) in 1990-1998. Grouping (A/B) according to methods described in material and methods.

Date	Number of tagged herring	Group A/B	Total recoveries	Total ‰ of recoveries	Recoveries After 2 months	Recoveries After 2 years
22-28 May 1990	1,956	A	5	2.6	1	0
21-26 June 1991	5,029	A	9	1.8	7	0
14-15 May 1992	1,780	A	0	0.0	0	0
11-24 May 1993	5,850	A	7	1.2	3	0
24 May-4 June 1994	6,031	A	6	1.0	1	0
26 May 1994	640	B	6	9.4	5	3
1-2 June 1995	1,857	B	12	6.5	3	1
26 May-10 June 1998	3,333	B	49	14.7	10	1
Total	26,476		94	3.6	30	5

Results and Discussion

Altogether 26,476 herring were tagged between 1990 and 1998 (Table 1) and the total number of recoveries is 94 (0.4%). In the present study, we consider those 28 recoveries which were made at least 2 months after tagging and for which the site of the recovery is known (Table 2). Unfortunately, data on maturity were not available.

Nine fish were recaptured within the two following years during spawning time in May-July (Fig. 1, Table 2). Exact homing to the tagging site was not observed. The distance of the recoveries from the tagging site varied from 7 to 140 km, suggesting that homing is not very precise, but fish return to a relatively large spawning area. It should be noted that the tag recovered closest to the tagging site (number 10 in Fig. 1) was found loose on the shore. Most recoveries were made in the innermost parts of the Archipelago Sea, but two were made at the entrance of the Gulf of Finland and one in the Bothnian Sea (Fig. 1). In view of this distribution, homing to the spawning bed used in the previous spawning season is not probable. Although few in number, the recoveries indicate a more general homing to the spawning area (about 200 km in the innermost part in the Archipelago

Table 2. Recoveries (at least two months from tagging date and recovery place known) of herring tagged in 1990-1995.

No	Tagging date	Group	Tagging length, if measured (mm)	Recovery date	Gear, if known	Notes
1	22-28 May 1990	1	180	27 May 1991	Trap net	
2	21-26 June 1991	2	160	13 Sept. 1991		
3	21-26 June 1991	2		15 Sept. 1991	Gillnet	
4	21-26 June 1991	2		15 Oct. 1991		
5	21-26 June 1991	2		17 Nov. 1991	Gillnet	
6	21-26 June 1991	2		24 Nov. 1991		
7	21-26 June 1991	2		27 Feb. 1992	Trawl	
8	11-24 May 1993	2	180	4 Oct. 1993		
9	11-24 May 1993	2	180	14 Jan. 1994	Trawl	
10	11-24 May 1993	1	180	8 May 1994		Loose on shore
11	26 May 1994	1	156	18 July 1995	Gillnet	
12	26 May 1994	1	150	29 May 1996	Gillnet	Stuck to gillnet
13	26 May 1994	1	153	21 June 1996	Gillnet	Stuck to gillnet
14	24 May-4 June 1994	1	180	11 June 1995	Trawl	
15	1-2 June 1995	2	169	17 Aug. 1996	Gillnet	Stuck to gillnet
16	1-2 June 1995	2	180	17 Aug. 1995	Gillnet	
17	1-2 June 1995	1	169	31 July 1997		
18	26 May 1998	2	150	2 Sept. 1998	Gillnet	
19	10 June 1998	2	171	9 Oct. 1998	Gillnet	
20	10 June 1998	2	170	12 Oct. 1998	Gillnet	
21	26 May 1994	2	220	1998	Gillnet	Stuck to gillnet
22	10 June 1998	2	151	20 Dec. 1998	Gillnet	
23	26 May 1998	1	160	14 May 1999	Gillnet	Starving
24	10 June 1998	2	142	5 Oct. 1999	Gillnet	
25	26 May 1998	2	177	27 Aug. 1999	Gillnet	Stuck to gillnet
26	26 May 1998	2	165	3 Feb. 2000	Gillnet	
27	26 May 1998	1	185	31 May 2000	Gillnet	
28	10 June 1998	2	211	21 April 2000	Gillnet	

Grouping according to time of recovery. Group 1 recovered during spawning time (May 1-July 31). Group 2 recovered during the rest of the year. See Fig.1.

Sea in Fig. 1), comparable to homing by Pacific and Atlantic herring (Hourston 1982, Wheeler and Winters 1984). In general, our results seem to be similar to those of the Pacific herring and the Atlantic herring.

Homing to a spawning area was observed earlier in the Baltic herring by, for example, Otterlind (1957), Weber (1975), Jönsson and Biester (1981), and Parmanne and Sjöblom (1986). Homing behavior in the Archipelago Sea herring was also demonstrated by the tagging experiments of Parmanne (1990) in 1975-1988. Parmanne even found recoveries from the same trap net from which the fish had been tagged in the previous year. However, most of his recoveries in the following spawning period were spread throughout the Archipelago Sea, although some recoveries were made outside the Archipelago Sea, as far as 300 km from the tagging site.

After spawning, herring shoals migrate out to the feeding and overwintering areas, which are situated in the outer archipelago and the open sea. Nineteen herring were recaptured during the feeding and overwintering period (Fig. 1, Table 2). The recoveries were made in the Archipelago Sea, in the Bothnian Sea, or at the entrance of the Gulf of Finland. The pattern of recoveries indicates that as the herring migrate outside the Archipelago Sea, they seem to migrate mainly north into the Bothnian Sea. No recoveries were made in the outer archipelago or in the Baltic proper.

Parmanne (1990) reported a similar migration pattern, but recorded recoveries also from the central and southern Baltic. His tagging was done in the northwest part of the Archipelago Sea, which is approximately 35 km northwest of our tagging site. Herring spawning in that area may have a different migration pattern from herring of the Airisto area, but this is not likely because there are no physical or other barriers between the areas. The shift in the fishery may explain the difference between Parmanne's (1990) and our results. The main offshore fishing area during Parmanne's tagging in the 1980s was in the central and northern Baltic. During our study period in the 1990s the fishery shifted to the Bothnian Sea, which is now the most important trawl fishery area in Finland (Parmanne 1993). A possible shift in the migration pattern of the fish would be masked by the fishery shift, as the chances of tag recovery in the areas concerned depend on fishing intensity. It should be mentioned that the fishery in the Baltic proper underwent enormous changes during the 1990s. Changes in the former Eastern Europe bloc led to reductions in the fishery and these changes may have affected the results of our study.

With tagging method A (Fig. 3), the recovery rate was low (0.13%) compared to the recovery rates of Jönsson and Biester (1981) with the same tagging method (9.2%) in the southern Baltic Sea. In method B, the recovery rate was somewhat higher (1.15%) compared to method A, but still very low. Of the 67 recoveries with method B, 5 fish had survived more than 2 years from the tagging. With method A, only one fish had survived 1 year (number 14 in Table 2 and Fig. 1). One obvious reason for the low recovery rate is the small size of herring in the northern Baltic. This makes the fish more vulnerable to tagging injuries; for example, the loss of scales.

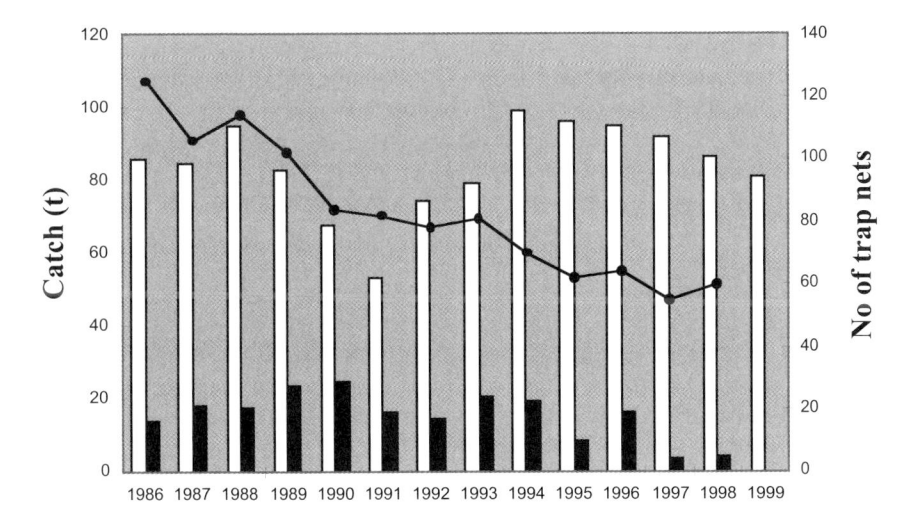

Figure 4. Total catch of herring in Finland between 1986 and 1999 (white column; 1,000 ×t) (Finnish Game and Fisheries Research Institute 2000). Total number of trap nets in the Airisto Inlet (within a distance of 15 km from the tagging site) (line) and total catch of these trap nets in 1986-1998 (black column; 100 × t) (Niinimäki 1999).

Our tagging also coincided with a recession in the market and the number of trap nets for herring in the study area (Fig. 4), leading to a reduction in fishing effort. Thus, in 1997, for example, the number of trap nets in operation in the tagging area (about 15 km from the tagging site) was only 55, compared to 125 in 1986. In trap-net and trawl catches the tags are easily lost. The trap net is emptied by means of a pump, giving the fisherman little time to notice the tags.

Our results indicate possible homing to a relatively large spawning area but imprecise homing to the exact spawning bed in where taggings have been done.

Acknowledgments

We wish to thank Ilppo Vuorinen and two anonymous referees for comments on the manuscript and fisherman Antero Eloranta for letting us collect herring from his trap net. We also thank the staff and students of the Finnish Fisheries and Environmental Institute and numerous other people for assistance during the field work. The Environmental Office of the City of Turku, the TOP Foundation, and the Academy of Finland financially supported the study.

References

Aneer, G. 1989. Herring (*Clupea harengus* L.) spawning and spawning ground characteristics in the Baltic Sea. Fish. Res. (Amst.) 8:169-195.

Aneer, G., G. Florell, U. Kautsky, S. Nellbring, and L. Sjöstedt. 1983. In-situ observations of Baltic herring (*Clupea harengus membras*) spawning behaviour in Askö-Landsort area, northern Baltic proper. Mar. Biol. (Berl.) 74:105-110.

Aro, E. 1988. A review of fish migration patterns in the Baltic Sea. ICES Symposium Baltic 13:1-57.

Finnish Game and Fisheries Research Institute. 2000. Professional Marine Fishery: Official statistics of Finland 1986-1999. Agriculture, Forestry, and Fishery 7:1-39.

Hay, D.E., and P.B. McCarter. 1997. Larval distribution, abundance, and stock structure of British Columbia herring. J. Fish Biol. 51:155-175.

Hietaranta, R. 1990. Saaristomeren eläinplanktonin vertikaalijakauma kesällä 1987 ja miniminäytemäärä. M.Sc. thesis, University of Turku, Department of Biology, pp. 1-40. (In Finnish.)

Hourston, A.S. 1982. Homing by Canada's west coast herring to management units and divisions as indicated by tag recoveries. Can. J. Fish. Aquat. Sci. 39:1414-1422.

Hourston, A.S., and H. Rosenthal. 1976. Sperm density during active spawning of Pacific herring (*Clupea harengus pallasi*). J. Fish. Res. Board Can. 33:1788-1790.

Johannessen, A. 1986. Recruitment studies of herring (*Clupea harengus* L.) in Lindaaspollene, western Norway. Fiskeridir. Skr. Ser. Havunders. 18:139-240.

Jönsson, N., and E. Biester. 1981. Wanderbewegungen des Rügenschen Frühjahrsherings in den Küsten- und Boddengewässern der DDR. Fischerei-Forschung Wissenschaftliche Schriftenreihe 19:47-51. (In German.)

Kääriä, J., M. Rajasilta, M. Kurkilahti, and M. Soikkeli. 1997. Spawning bed selection by the Baltic herring (*Clupea harengus membras*) in the archipelago of SW Finland. ICES J. Mar. Sci. 54:917-923.

Messieh, S.N. 1987. Some characteristics of Atlantic herring (*Clupea harengus*) spawning in the southern Gulf of St. Lawrence. NAFO (Northwest Atl. Fish. Organ.) Sci. Counc. Stud. 11:53-61.

Niinimäki, J. 1999. Turun ja Naantalin edustan merialueen ammatti- ja kirjanpitokalastus vuonna 1998. Lounais-Suomen vesiensuojeluyhdistys r.y. 147:1-20. (In Finnish.)

Otterlind, G. 1957. Från strömmingsmärkningarna på ostkusten. Ostkusten 3:24-30. (In Swedish.)

Parmanne, R. 1990. Growth, morphological variation and migrations of herring (*Clupea harengus* L.) in the northern Baltic Sea. Finn. Fish. Res. 10:1-48.

Parmanne, R. 1993. Larval abundance and catch composition of Baltic herring in the Gulf of Bothnia in 1974-1991. Aqua Fenn. 23:75-84.

Parmanne, R., and V. Sjöblom. 1986. Recaptures of Baltic herring tagged off the coast of Finland in 1982-85. ICES C.M. J.28.

Pitkänen, H. 1994. Eutrophication of the Finnish coastal waters: Origin, fate and effects of riverine nutrient fluxes. Publications of the Water and Environment Research Institute, National Board of Waters and the Environment, Finland 18:1-45.

Rajasilta, M., J. Eklund, J. Hänninen, M. Kurkilahti, J. Kääriä, P. Rannikko, and M. Soikkeli. 1993. Spawning of herring *(Clupea harengus membras* L.) in the Archipelago Sea. ICES J. Mar. Sci. 50:233-246.

Runnström, S. 1941. Quantitative investigations on herring spawning and its yearly fluctuations at the west coast of Norway. Fiskeridir. Skr. Ser. Havunders. 8:1-71.

Stacey, N.E., and A.S. Hourston. 1982. Spawning and feeding behavior of captive Pacific herring, *Clupea harengus pallasi.* Can. J. Fish. Aquat. Sci. 39:489-498.

Tyurnin, B.V. 1973. The spawning range of Okhotsk herring. Proc. Pacific Inst. Fish. Oceanogr. 86:12-21. (Transl. by Transl. Bur., Dep. Secretary of State, Canada.)

Viitasalo, M., I. Vuorinen, and E. Ranta. 1990. Changes in crustacean mesozooplankton and some environmental parameters in the Archipelago Sea (northern Baltic) in 1976-1984. Ophelia 31:207-217.

Vuorinen, I., and E. Ranta. 1987. Dynamics of marine meso-zooplankton at Seili, northern Baltic Sea, in 1967-1975. Ophelia 28:31-48.

Weber, W. 1975. A tagging experiment on spring-spawning herring of the Kiel Bay. Ber. Dtsch. Wiss. Komm. Meeresforsch. 24:184-188.

Wheeler, J.P., and G.H. Winters. 1984. Homing of Atlantic herring *(Clupea harengus harengus)* in Newfoundland waters as indicated by tagging data. Can. J. Fish. Aquat. Sci. 41:108-117.

Microsatellite Polymorphism and Population Genetic Structure of Atlantic Herring in the Baltic and Adjacent Seas

Carl André
Tjärnö Marine Biological Laboratory, Strömstad, Sweden

Fredrik Arrhenius
National Board of Fisheries, Institute of Marine Research, Lysekil, Sweden

Mats Envall and Per Sundberg
Göteborg University, Department of Zoology, Göteborg, Sweden

Extended Abstract

Herring, *Clupea harengus*, in the Kattegatt and southwest Baltic originates from several spawning stocks. The relative contribution of these stocks is, however, uncertain, due to limited knowledge about spatial distributions and migration patterns. Earlier efforts to separate stocks using various genetic techniques have shown little or no genetic differentiation among herring stocks in the area (Ryman et al. 1984, Dahle and Eriksson 1990). Recently, Shaw et al. (1999) showed large-scale genetic population structure in Atlantic herring using microsatellite markers. Here, we used microsatellite polymorphism in five loci to examine the genetic structure in and among three herring spawning populations in the Baltic (Bothnian Bay, Hanö Bight, and Rüegen); one location off Ålesund, Norway; and one off the Hebrides, Scotland ($n = 50$).

Overall F_{ST} values calculated for the five populations ranged from 0 to 0.0063 for individual loci, and F_{ST} for all loci combined, 0.0024 (Table 1), was statistically different from zero, which suggest limited gene flow among populations.

Pairwise comparisons of F_{ST} indicate a separation between Baltic populations vs. the adjacent populations (Table 2). This is supported in an exact G test of population differentiation between all Baltic populations

Table 1. Overall F_{ST} values for five herring microsatellite loci.

Locus	Cha 17	Cha 20	Cha 63	Cha 113	Cha 123	All loci
F_{ST}	0.0035	0.0011	−0.005	0.0063	0.0056	0.0024
P-value	**0.018**	0.36	n.a.	**0.016**	**<0.001***	**0.007***

F_{ST} P-values <0.05 are in bold; comparisons considered statistically significant after correction for multiple tests are denoted with *.

Table 2. Pairwise estimates of F_{ST} among herring population samples, all five loci combined.

	Bothnian Bay	Hanö Bight	Rüegen	Hebrides	Ålesund
Bothnian Bay	—	−0.0008	0.0021	0.0029	0.0045
Hanö Bight	0.680	—	0.0020	0.0010	−0.0002
Rüegen	0.100	0.160	—	0.0034	0.0079*
Hebrides	**0.047**	0.330	**0.034**	—	0.0009
Ålesund	**0.0087**	0.540	**0.0002***	0.300	—

Below the diagonal (lower left) are shown P-values for the pairwise estimates of F_{ST}. P-values <0.05 are in bold; comparisons considered statistically significant after correction for multiple tests are denoted with *.

pooled and the two other populations (Hebrides and Ålesund) pooled (P = 0.0002). However, the geographical distance alone does not explain the observed pattern of genetic differentiation (Mantel test: r = 0.3; P = 0.27). It should be noted that the results regarding the sample from Hanö Bight should be interpreted with caution since this sample showed both significant heterozygote deficiency and evidence for linkage disequilibrium (between Cha 20 and each of the other loci).

In conclusion, these findings are consistent with morphological evidence of a distinct Baltic population (Rosenberg and Palmén 1982) and support the separate management of Baltic stocks.

References

Dahle, G., and A.G. Eriksen. 1990. Spring and autumn spawners of herring (*Clupea harengus*) in the North Sea, Skagerrak and Kattegat: Population genetic analysis. Fish. Res. (Amst.) 9:131-141.

Rosenberg, R., and L.-E. Palmén. 1982. Composition of herring stocks in the Skagerrak-Kattegatt and the relations of these stocks with those of the North Sea and adjacent waters. Fish. Res. (Amst.) 1:83-104.

Ryman, N.U., L. Lagercrantz, R. Andersson, R. Chakraborty, and R. Rosenberg. 1984. Lack of correspondence between genetic and morphologic variability patterns in Atlantic herring (*Clupea harengus*). Heredity 53:687-704.

Shaw, P.W., C. Turan, J.M. Wright, M. O'Connell, and G.R. Carvalho. 1999. Microsatellite DNA analysis of population structure in Atlantic herring (*Clupea harengus*), with direct comparison to allozyme and mtDNA RFLP analyses. Heredity 83:490-499.

Microsatellite Population Structure in Herring at Three Spatial Scales

Arran A. McPherson
Dalhousie University, Department of Oceanography, Halifax, Nova Scotia, Canada

Christopher T. Taggart
Dalhousie University, Department of Oceanography and Marine Gene Probe Lab, Halifax, Nova Scotia, Canada

Paul W. Shaw
University of Hull, Department of Biological Sciences, Hull, United Kingdom

Patrick T. O'Reilly and Doug Cook
Dalhousie University, Marine Gene Probe Lab, Halifax, Nova Scotia, Canada

Abstract

We quantify population structure in herring at three spatial scales using 4-5 microsatellite DNA markers. Significant genetic structure was detected between putative species (Pacific vs. Atlantic herring), as well as among populations at ocean basin scales (northeast vs. northwest Atlantic) and at regional management scales (NAFO Division 4X). This study is the first that reports application of microsatellite markers to northwest Atlantic herring, where genetic support for population structure has been scant. Although the results are for 1 year only, and should therefore be interpreted with caution, these data imply that microsatellite analyses may enable detection of population structuring in Atlantic herring. The utility of the method and analyses are discussed and suggestions (related to sampling and interpretation) are provided.

Introduction

An objective of fisheries management is to ensure the integrity of a stock while maintaining (and even maximizing) harvests. Central to this concept is the identification of "populations" that are assumed to share a number of phenotypic traits (e.g., growth rate, fecundity, morphometric measurements) as well as, in the case of marine fish as reviewed by Carvalho and Hauser (1994), identifiable patterns of stock and recruitment. Identifying population structure within and among stocks therefore has the potential to greatly increase the ability to anticipate the effects of harvest levels among stock components.

Population structure can be delineated using a variety of techniques. Meristic, (e.g., variation in the number of vertebrae), morphometric, (e.g., variation in dimensions of body parts and/or rates of change), parasitic load (e.g., variations in number and type of parasites), and demographic (e.g., age structure, fecundity, and mortality) methods have long been used in the classification of groups of marine fish when the means and variance of these measurements have been observed to differ among populations (e.g., Messieh 1975). However, there are limitations associated with these techniques as the physical environment is known to influence the expression of traits over the life of the individual, therefore potentially converging nonheritable traits. This environmental influence is of greater consequence for a migratory species whose range could encompass a variety of environments, hence contributing to an increased variability in the nonheritable traits among populations. Therefore population-specific differences become difficult to interpret.

Neutral genetic variation can be used to differentiate among populations, and it is not thought to be subject to confounding environmental influences. However, the ability to reject the hypothesis of no population structure depends on a variety of factors, including the type and number of genetic markers used and the time-scale of divergence between populations. The successful application of molecular genetic markers in identifying population structure in freshwater and anadromous fish, that can be shown to be physically isolated (to some extent), has prompted marine fishery managers to consider genetic data in assessing population structure. Unlike the case for freshwater and anadromous fish, however, the environment of marine species may lack obvious physical barriers to dispersal and migration (Waples 1998). Therefore, delineation of marine populations with a "potentially" high capacity for gene flow (known, assumed, or otherwise) is problematic as complex oceanographic processes coupled with the migratory capabilities of the animals may create complex spatial patterns of population structure.

Given that relatively few successfully reproducing migrants are needed to reduce genetic differentiation to very low levels and that marine populations may be too large to markedly diverge by genetic drift (but note that variation in reproductive success may reduce the effective population size and therefore increase susceptibility to drift; Hedgecock 1994), the genetic "signal" indicating population differences of marine fish may be small relative to the sampling "noise." Therefore, measures must be taken to optimize the probability of detecting differences when they occur (i.e., by increasing the signal-to-noise ratio).

Herring are an ideal model species for exploring the genetic basis of population structure because they are one of the most studied marine teleosts (e.g., there is more literature on herring than any other fish species as indexed by the Food and Agriculture Organization; FAO 1999) and several herring-specific population structuring mechanisms have been hypothesized (Iles and Sinclair 1982, Smith and Jamieson 1986, McQuinn 1997) which have provided a framework in which we may better articulate questions related to population structure in this species. From a more pragmatic perspective, Atlantic herring (*Clupea harengus*) are a valuable resource in the North Atlantic Ocean (> 2 million metric tons catch in 1995; FAO 1999), and recent fluctuations in catch on historical fishing grounds (e.g., Stephenson 1997) and the requirements for better conservation practices (e.g., Stephenson 1999) demand a clarification of the population structure of this species.

Genetic analyses of Atlantic herring based on allozyme and mtDNA markers (e.g., Ridgeway et al. 1971, Grant 1984, Ryman et al. 1984, King et al. 1987, Kornfield and Bogdanowicz 1987, Dahle and Eriksen 1990, Safford and Booke 1992, Jørstad et al. 1994) have been largely unable to reject the null hypothesis of no genetic differentiation among populations at management unit scales, nor in many cases at ocean basin scales. In contrast, evidence for homing from tagging studies (e.g., Wheeler and Winters 1984), for different population metrics among neighboring groups (e.g., Messieh 1975), and for predictable variations in spawning times and locations among groups (e.g., Sinclair and Tremblay 1984) is consistent with the existence of distinct populations. Thus, quantitative studies that focus on measuring the degree of genetic isolation among herring populations, and then construct testable hypotheses concerning the mechanisms that may maintain it, are essential to resolve the above inconsistencies. The conflicting evidence for herring population structure is not unique; it is shared with a variety of marine fish species.

Genetic analyses based on variation at a suite of microsatellite loci developed for Pacific herring (*Clupea pallasii;* O'Connell et al. 1998a) have provided evidence for significant population structure in Alaskan Pacific herring (O'Connell et al. 1998b) and northeast Atlantic herring (Shaw et al. 1999). Grant and Utter (1984) found allozyme-based differences in the northern Pacific herring that were not apparent in O'Connell et al. (1998b). However, the results of Shaw et al. (1999), that demonstrated subbasin and interocean scale differences among herring, suggest that microsatellite loci may be key in quantifying population structure in northwest Atlantic herring. Microsatellite markers may detect structure at finer spatial and temporal scales than many other genetic methods due to higher levels of variability thought to be a result of high rates of mutation (Bentzen 1997).

Here we report an attempt to quantify population structure in herring at three spatial scales. Northeast (NE) and northwest (NW) Atlantic herring are compared to Pacific herring to quantify the degree of differentiation between groups that are clearly isolated geographically but whose life history similarities suggest a subspecies relationship (Svetovidov 1963 as outlined in Jørstad et al. 1994). Atlantic herring collected from the NE and NW Atlantic are then used to assess differentiation at the Atlantic basin scale, where the potential for genetic exchange might be presumed to be negligible. Finally, genetic variation among herring collected from three locations (NAFO Division 4X) in the NW Atlantic are then assessed at the management scale, where the potential for genetic exchange might be presumed to be greater than at the ocean basin scale.

Materials and Methods

Spawning-stage Atlantic herring (blood and/or muscle samples) were collected (Fig. 1) from the NW Atlantic at Spectacle Buoy (SB, 43.618°N, 66.124°W) and Scot's Bay (ScB, 45.224°N, 64.976°W). Mature fish (nonspawning) were also collected in the vicinity of Emerald Basin (EB, 44.294°N, 62.376°W). DNA extraction procedures for NW Atlantic samples follow Ruzzante et al. (1996). Five microsatellite loci, Cha 17, Cha 20, Cha 63, Cha 113 and Cha 123 were amplified (annealing temperatures were modified from those in O'Connell et al. 1998a) and scored. Four microsatellites (17, 20, 63, 113; allelic data were extracted from Shaw et al. 1999) were assayed in the NE Atlantic herring, represented by Barents Sea (BS, 70.53°N, 31.583°E) Norwegian spring-spawning herring (spawning stage), and the Pacific herring (nonspawning stage) collected seaward of Vancouver Island (PC). Single locus statistics (allele sizes, number of alleles, observed heterozygosity) were calculated for all herring samples and conformation to Hardy-Weinberg equilibrium (HWE) was tested for all loci in all collections (Miller 1997) to assess random allele assortment (nonassortative mating) within populations.

F_{ST} estimates (Wright 1951 as amended by Weir and Cockerham 1984) of population structure were calculated using F-STAT (Goudet 1996) that assumes an Infinite Allele Model of microsatellite mutation and R_{ST} (Slatkin 1995) estimates were calculated using R_{ST}-CALC (Goodman 1997), that assumes a Stepwise Mutation Model. Allele size data were expressed in terms of standard deviations from the mean for R_{ST} analyses, as recommended by Goodman (1997) to minimize effects of unequal variance among loci and unequal sample sizes. Permutation tests (1,000 resampling trails per comparison) were used to determine significance values for all tests. We also employed an Exact test (Raymond and Rousset 1995) to assess the statistical significance of locus-specific allele frequency differences be-

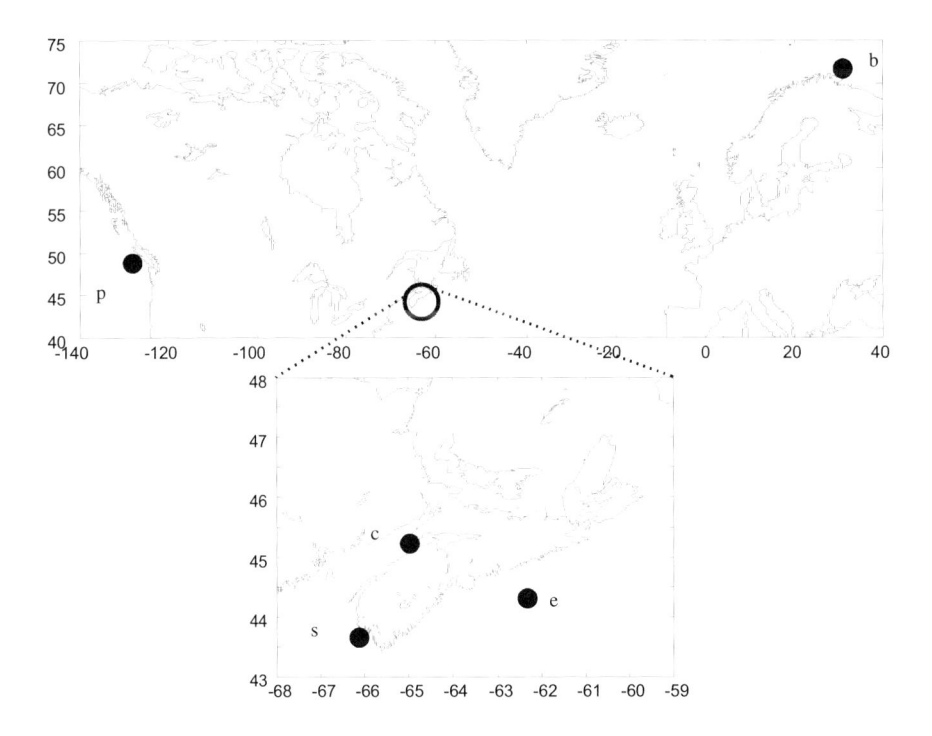

Figure 1. Sample locations of Atlantic and Pacific herring. Abbreviations are as follows: (p) Pacific herring collected seaward of Vancouver Island; (b) Atlantic herring from the Barents Sea; (e) Atlantic herring collected from Emerald Basin; (s) Atlantic herring collected from Spectacle Buoy; and (c) Atlantic herring collected from Scot's Bay.

tween pairs of population samples (Miller 1997) allowing an assessment of discriminatory utility at each locus.

Allele classes (base pairs; bp) were binned at each locus prior to parameter estimation in all pairwise comparisons to facilitate interlab comparisons of microsatellite data. The number of bins and bin-widths were determined at each locus by progressively increasing the number of bins and then comparing the binned frequency distribution to the original nonbinned distribution of all samples pooled for each locus respectively. The least number of bins (and thus bin width) that provided a frequency distribution not significantly different from the original distribution (Kolmogorov-Smirnov $P > 0.05$) was chosen to establish an objectively based binning criterion. In all cases, bin-widths were found to meet this criterion at 4 bp. Consequently, allele classes differ by 4 bp for each locus in all comparisons.

Results

Single locus statistics (allele sizes, number of alleles, observed heterozygosity) are similar across all samples (Table 1) and results for the NW Atlantic (SB, ScB, and EM) are consistent with those reported for Pacific and NE Atlantic herring (O'Connell et al. 1998b, Shaw et al. 1999). With the exception of one locus (Cha 20) in the Barents Sea population, no population deviated significantly from Hardy-Weinberg Equilibrium; i.e., consistent with samples being drawn from within randomly mating populations.

All pairwise R_{ST} and F_{ST} estimates between the Pacific (PC) and each of the four Atlantic (BS, SB, ScB, and EB) sites were significant (Table 2). Significant population structuring, as inferred by both F_{ST} and R_{ST} analyses, was also observed between the NE and NW Atlantic (see Table 2; pairwise comparisons between the NE and each of the NW samples). At the smallest spatial scale investigated (within the NAFO Division 4X region) significant structuring, using both F_{ST} and R_{ST}, was observed between herring from Spectacle Buoy and Emerald Basin, though the magnitude of differences were half those observed at the basin scale. No significant differences were observed between Scot's Bay herring and either Emerald Basin or Spectacle Buoy herring.

An Exact test showed locus-specific significant differences among populations (Table 3). All loci were useful (i.e., reported significant differences) in discriminating among Atlantic and Pacific herring. However, as the geographic scale of the comparison decreased from ocean scale to management unit scale, fewer loci were of discriminatory value. In fact, two to three loci showed significant comparisons including Cha 123, a locus not used by Shaw et al. (1999). When pairwise comparison results were pooled over all loci, the same population pairs showed significant differences as were found using F_{ST} and R_{ST}.

Discussion

This study reports significant differences in herring at three spatial scales ranging from Atlantic vs. Pacific to those within a NW Atlantic management unit. The significant differences among Atlantic and Pacific herring are consistent with Svetovidov (1963 as outlined in Jørstad et al. 1994) who argued a subspecies relationship based on an analysis of biological traits, Grant (1984) who argued a distinct species relationship from genetic (allozyme) information, and with Domanico et al. (1996) who estimated a 3.1 million year divergence between Atlantic and Pacific herring when assessed using ribosomal DNA sequence variation.

The significant differences detected between NE and NW Atlantic herring are inconsistent with Grant (1984), who was unable to reject the hypothesis of no structure at basin scales using 40 allozyme loci. Our results suggest that microsatellites are valuable in detecting population structure at ocean basin scales. At the finest geographic scale considered (NAFO

Table 1. Descriptive statistics for microsatellite DNA analysis of herring samples showing sample size (n), the number of alleles per locus, the range of allele sizes in base pairs (bp), observed heterozygosity, and the χ^2 value/probability associated with Hardy-Weinberg equilibrium (HWE) estimates (Miller 1997).

	Pacific Vancouver $n = 30$	NE Atlantic Barents Sea $n = 50$	NW Atlantic Emerald Basin $n = 40$ Scot's Bay $n = 50$ Spectacle Buoy $n = 49$
Cha 17			
No. of alleles	21	22	23
Allele size (bp)	96-154	90-152	96-170
Heterozygosity	0.93	0.94	0.89
HWE	0.01/0.99	1.44/0.23	0.94/0.33 EM
			0.02/0.64 ScB
			1.65/0.19 SB
Cha 20			
No. of alleles	15	27	25
Allele size (bp)	108-180	96-158	96-186
Heterozygosity	0.77	0.72	0.93
HWE	0.020/0.66	8.62/**0.003**	0.16/0.69 EM
			0.11/0.65 ScB
			0.49/0.48 SB
Cha 63			
No. of alleles	17	13	13
Allele size (bp)	128-166	126-156	130-174
Heterozygosity	0.77	0.84	0.88
HWE	0.01/0.99	0.50/0.50	0.70/0.4 EM
			0.64/ 0.46 ScB
			1.23/0.27 SB
Cha 113			
No. of alleles	18	16	18
Allele size (bp)	100-150	104-134	94-132
Heterozygosity	0.87	0.77	0.93
HWE	0.54/0.46	3.26/0.06	0.01/0.94 EM
			0.5/0.5 ScB
			0.02/0.90 SB
Cha 123			
No. of alleles			23
Allele size (bp)			152-222
Heterozygosity			0.93
HWE			0.5/0.5 EM
			0.45/0.49 ScB
			0.30/0.58 SB

Table 2. Pairwise F_{ST} above the diagonal (θ/P) and R_{ST} below diagonal (ρ/P) estimates of population structure in herring collected from the northeast (Barents Sea), northwest (Spectacle Buoy, Emerald Basin, and Scot's Bay) Atlantic and the Pacific.

Location	Spectacle Buoy 5 loci	Emerald Basin 5 loci	Scot's Bay 5 loci	Pacific 4 loci	Barents Sea 4 loci
Spectacle Buoy		0.008/ $P < 0.01$	0.006/ $P = 0.11$	0.047/ $P < 0.01$	0.039/ $P < 0.01$
Emerald Basin	0.039/ $P = 0.002$		0.005/ $P = 0.16$	0.046/ $P < 0.01$	0.06/ $P < 0.01$
Scot's Bay	0.025/ $P = 0.15$	0.013/ $P = 0.21$		0.042/ $P < 0.001$	0.028/ $P < 0.01$
Pacific	0.098/ $P < 0.0001$	0.068/ $P = 0.001$	0.073/ $P < 0.001$		0.06/ $P < 0.01$
Barents Sea	0.193/ $P < 0.0001$	0.290/ $P < 0.0001$	0.180/ $P < 0.001$	0.185/ $P < 0.0001$	

All significant results remain significant at $P < 0.05$ after Bonferroni correction (Manly 1985).

Table 3. Pairwise Exact test results for allele frequency differentiation (Raymond and Rousset 1995) at each locus.

Locations	Cha 17	Cha 20	Cha 63	Cha 113	Cha123
*PC-BS	$P = 0.0020$	$P < 0.001$	$P < 0.001$	$P < 0.001$	N/A
*PC-EB	$P = 0.055$	$P < 0.001$	$P < 0.001$	$P = 0.002$	N/A
*PC-ScB	$P = 0.008$	$P < 0.001$	$P < 0.001$	$P < 0.001$	N/A
*PC-SB	$P = 0.007$	$P < 0.001$	$P < 0.001$	$P < 0.001$	N/A
*BS-EB	$P = 0.56$	$P = 0.037$	$P < 0.001$	$P < 0.001$	N/A
*BS-ScB	$P = 0.31$	$P = 0.01$	$P < 0.001$	$P < 0.001$	N/A
*BS-SB	$P = 0.053$	$P < 0.001$	$P < 0.001$	$P < 0.001$	N/A
EB-ScB	$P = 0.65$	$P = 0.43$	$P = 0.100$	$P = 0.041$	$P = 0.049$
*EB-SB	$P = 0.88$	$P = 0.24$	$P < 0.001$	$P = 0.071$	$P = 0.050$
ScB-SB	$P = 0.65$	$P = 0.12$	$P = 0.14$	$P = 0.012$	$P = 0.030$

PC = Pacific; BS = Barents Sea; EB = Emerald Basin; SB = Spectacle Buoy; ScB = Scot's Bay. Shaded areas indicate significant differences ($P < 0.05$) between samples for that locus. * indicates combined Exact test probability over all loci is significant at $P < 0.001$. N/A: no data available for one of the populations at this locus.

Division 4X in the NW Atlantic), significant differences were observed between herring from Emerald Basin and herring caught at Spectacle Buoy. While it is necessary to test this pattern in subsequent years for it to be of direct use for management purposes, these results suggest microsatellite-based genetic evidence for management scale population structure of herring in the NW Atlantic, which is compelling as three different tests, based on different models, were used.

The implications of management-scale population differences are in some ways unique for 4X herring as Scotia-Fundy herring are currently managed using an "in-season management approach" as described by Stephenson et al. (1999). Each spawning ground is assessed (in season) and spawning ground-specific quotas are established accordingly. Therefore, the results presented here support the precautionary management currently employed. However, for most other marine fish management scenarios, there has not been this degree of success. The failure to recognize discrete populations within a stock complex may explain both the collapse and recovery failure in many marine fish populations (Frank and Brickman 2000). In fact, those authors suggest that when biological reference points are developed from aggregate (stock) data representing distinct (unit) populations and employed in conventional assessment models, the results are likely to be inaccurate and nonconservative.

However, the results of the within-management-unit comparisons should be interpreted with caution. As herring are thought to exhibit spawning ground fidelity (if not natal), and are known to mix at other seasonal stages (e.g., feeding), spawning stage herring collected from their spawning ground should be used to characterize population structure. While we can be reasonably confident that the Pacific herring have limited opportunity to mix with either the NE or the NW Atlantic populations, the assumption of no mixing is more problematic within the 4X management unit comparison, due to the close geographic proximity of sampling locations. This is particularly important when sampling different spawning groups that may overlap in time and space outside the spawning period or location, as is typically the case for comparisons at management-unit scales. Therefore, the state of the Emerald Basin herring sample (not in spawning condition) limits the utility of these results beyond providing evidence for small-scale population structure in the Nova Scotia and Bay of Fundy management unit. In addition, we recognize that the issue of temporal stability must be addressed as the next step in avoiding the sampling artifacts possible with such highly migratory fish (Waples 1998). However, we note that migration and mixing among adults does not necessarily imply reproductive mixing (i.e., gene flow) among putative populations, especially when considering those that exhibit spawning-ground fidelity.

Although differences were detected at all spatial scales observed, the magnitude of differences observed at the species/subspecies comparison ($F_{ST} < 0.05$) of Atlantic and Pacific herring suggest limited differentiation. This may be due to the high rates of mutation expected at microsatellite loci and the potential for convergence of allele sizes. F_{ST} was designed for characters that are considerably less polymorphic than microsatellites. Nevertheless, it is routinely applied to allele frequency data generated by any of several molecular markers (including mtDNA and microsatellites). The statistic can be interpreted as a ratio of the expected heterozygosity of an individual in an equivalent random mating total population minus the expected heterozygosity of an individual in an equivalent random mating subpopulation, to the expected heterozygosity of an individual in an equivalent random mating total population. Thus, F_{ST} estimates inbreeding in subpopulations relative to the total population. In doing so, the magnitude of the statistic is influenced by high levels of heterozygosity. Consider an extreme example involving two populations, each with an expected heterozygosity of 95% but with no alleles in common. The maximum pairwise F_{ST} value possible would be 0.05 (i.e., less than the homozygosity). Therefore, as Hendrick (1999) illustrated, the actual upper limit of the F_{ST} statistic is limited by the homozygosity (1–heterozygosity). Therefore, should we consider standardizing the F_{ST} value reported against the maximum F_{ST} possible given the observed homozygosities? If shown to be robust, this may prevent misleading interpretations of small F_{ST} values, that may correspond to very significant differences among groups of marine fish as responsible managers faced with difficult decisions may not fully appreciate the subtleties of the statistic that is easily and mistakenly interpreted (in its most frequently used form) as the proportion of the maximum possible differentiation.

When applying F_{ST} to address marine fish population structure, the magnitude of the error associated with each estimate is of significance. For potentially high gene flow marine species, the expected F_{ST} error estimates (due to nonrandom sampling) can be of a similar magnitude to the estimate itself when small samples sizes are used (<50). Therefore, following Waples (1998), expected random sampling error ($1/2n$) should be estimated for a given sample size (n) and be used when interpreting results. For example, the significant ($P < 0.05$) Emerald Basin to Spectacle Buoy comparison had an estimated F_{ST} value of 0.008 relative to the error associated with that estimate at 0.01. Therefore, our result falls within what may be attributed to sampling error or unexplained fluctuations in allele frequencies. Thus, temporal stability, which is required to substantiate the results reported here, is essential for confidence that differences are biologically meaningful in the context of stock identification and can be used to interpret spatial patterns in marine fish populations.

Because of the complexities involved with interpreting results averaged over loci (i.e., the error is dependent on sample size and independent of the significance of F_{ST}), we advocate considering the pairwise population comparisons at each locus (e.g., Exact test; Table 3). Significant differences at two to four loci were found in each pairwise comparison suggesting that all populations (including those within management unit 4X) are at least partially reproductively isolated and should therefore be considered distinct if differences can be shown to be reproducible and temporally stable. Bentzen (1997) advocates that if even one of several loci yield a significant result, it may be biologically meaningful given the obstacles (huge population sizes, sampling biases) faced in detecting legitimate population differences in the marine environment.

An additional, and somewhat more intuitive, hazard when comparing two populations for discriminatory purposes, is the influence that the number of "uninformative loci" (relative to "informative" loci) have on population structuring estimates when averaged over all loci. We do not advocate arbitrarily abandoning uninformative loci for a variety of reasons. However, one could argue that an "informative" locus is being maintained by selection, and if indeed the population differences did accrue through selection, there remains evidence of limited gene flow and thus of population structure at time scales that are relevant to management. Remembering that selection can result from differential fishing practices, the consequence of such a process in the short-term (several generations) evolution of population structure and dynamics is unclear. Alternatively, the differentiation might equally be a biologically meaningful result of the frequently assumed neutral variation that is detectable using at least the one locus that has been identified by chance. Thus, given the clearly debatable quandary, we argue, regardless of the driving mechanism, resolving variation that arises from sampling error and unexplained temporal fluctuations in allele frequencies is deserving from a tractable research perspective when we are increasingly faced with collapsing populations, the resultant concerns about genetic biodiversity, and the demand for precautionary fishing practices.

Acknowledgments

We are grateful to R. Stephenson, M. Power, J. Fife, and their colleagues at the St. Andrew's Biological Station (Canadian Department of Fisheries and Oceans), Pelagics Research Council, and many commercial fishers for their efforts in securing herring samples. We thank C. Reiss, A. Pickle, T. McParland, and J. Jeffrey for assistance and advice. This manuscript benefited from the comments of two anonymous reviewers whose suggestions improved the clarity and focus of the text. Funding was provided to MGPL and the Oceanography Department at Dalhousie through a research partnership with the Pelagics Research Council, to C.T.T. by the Natural Sciences and Engineering Research Council of Canada (NSERC). A.A.M. was supported by an NSERC scholarship.

References

Bentzen, P. 1997. Seeking evidence of local stock structure using molecular genetic methods. In: I. Hunt von Herbig, I. Kornfield, M. Tupper, and J. Wilson (eds.), The implications of localized fishery stocks. Proceeding from the Implications of Localized Fishery Stocks Symposium. Natural Resource, Agricultre, and Engineering Services, Portland, Maine, pp. 20-30.

Carvalho, G., and L. Hauser. 1994. Molecular genetics and the stock concept in fisheries. Rev. Fish Biol. Fish. 4:326-350.

Dahle, G., and A. Eriksen. 1990. Spring and autumn spawners of herring (*Clupea harengus*) in the North Sea, Skagerrak and Kattegat: Population genetic analysis. Fish. Res. (Amst.) 9:131-141.

Domanico, M., R. Phillips, and J. Schweigert. 1996. Sequence variation in the ribosomal DNA of Pacific (*Clupea pallasi*) and Atlantic herring (*Clupea harengus*). Can. J. Fish. Aquat. Sci. 53:2418-2423.

Food and Agriculture Organization. 1999. Species information sheet, *clupea Harengus*. [http://www.fao.org/fi/sidp/htmls/species/cl_ha_ht.htm]

Frank, K., and D. Brickman. 2000. Allele effects and compensatory population dynamics within a stock complex. Can. J. Fish. Aquat. Sci. 57:513-529.

Goodman, S. 1997. R_{ST} CALC: A collection of computer programs for calculating unbiased estimates of genetic differentiation and determining their significance for microsatellite data. Mol. Ecol. 6:881-885.

Goudet, J. 1996. F-STAT 1.2, a program for IBM PC compatible to calculate Weir and Cockerham's 1984 estimators of *F*-statistics. J. Hered. 86:485-486.

Grant, W. 1984. Biochemical population genetics of Atlantic herring, *Clupea harengus*. Copeia 1984(2):357-364.

Grant, W., and F. Utter. 1984. Biochemical population genetics of Pacific herring (*Clupea pallasi*). Can. J. Fish. Aquat. Sci. 41:856-864.

Hedgecock, D. 1994. Does variance in reproductive success limit effective population sizes of marine organisms? In: A. Beaumont (ed.), Genetics and evolution of aquatic organisms. Chapman and Hall, London, pp. 122-134.

Hendrick, P. 1999. Highly variable loci and their interpretation in evolution and conservation. Evolution 53:313-318

Iles, T., and M. Sinclair. 1982. Atlantic herring: Stock discreteness and abundance. Science 215:627-633.

Jørstad, K., G. Dahle, and O. Paulsen. 1994. Genetic comparison between Pacific herring (*Clupea pallasi*) and a Norwegian fjord stock of Atlantic herring (*Clupea harengus*). Can. J. Fish. Aquat. Sci. 51(Suppl. 1):233-239.

King, D., A. Ferguson, and I. Moffett. 1987. Aspects of the population genetics of herring, *Clupea harengus*, around the British Isles and in the Baltic Sea. Fish. Res. (Amst.) 6:35-52.

Kornfield, I., and S. Bogdanowicz. 1987. Differentiation of mitochondrial DNA in Atlantic herring, *Clupea harengus*. Fish. Bull., U.S. 85:561-568.

Manly, B. 1985. The statistics of natural selection. Chapman and Hall, London.

McQuinn, I. 1997. Metapopulations and the Atlantic herring. Rev. Fish Biol. Fish. 7:297-329.

Messieh, S. 1975. Delineating spring and autumn herring populations in the southern Gulf of St. Lawrence by discriminant function analysis. J. Fish. Res. Board Can. 32:471-477.

Miller, M. 1997. Tools for population genetic analyses (TFPGA) v.1.3. Department of Biological Sciences, Northern Arizona University, Flagstaff, AZ 86011-5460.

O'Connell, M., M. Dillon, and J. Wright. 1998a. Development of primers for polymorphic microsatellite loci in the Pacific herring (*Clupea harengus pallasi*). Mol. Ecol. 7:358-259.

O'Connell, M., M. Dillon, J. Wright, P. Bentzen, S. Merkouris, and J. Seeb. 1998b. Genetic structuring among Alaskan Pacific herring (*Clupea pallasi*) populations identified using microsatellite variability. J. Fish Biol. 53:150-163.

Raymond, M., and F. Rousset. 1995. An exact test for population differentiation. Evolution 49:1280-1283.

Ridgeway, G., D. Lewis, and S. Sherburne. 1971. Serological and biochemical studies of herring populations in the Gulf of Maine. Rapp. P.-V. Reun. Cons. Int. Explor. Mer 161:21-25.

Ruzzante, D., C.T. Taggart, and D. Cook. 1996. Spatial and temporal variation in the genetic composition of a larval cod (*Gadus morhua*) aggregation: Cohort contribution and genetic stability. Can. J. Fish. Aquat. Sci. 53:2695-2705.

Ryman, N., U. Lagercrantz, L. Andersson, R. Chakraborty, and R. Rosenberg. 1984. Lack of correspondence between genetic and morphologic variability patterns in Atlantic herring (*Clupea harengus*). Heredity 53:687-704.

Safford, S., and H. Booke. 1992. Lack of biochemical genetic and morphometric evidence for discrete stocks of northwest Atlantic herring *Clupea harengus harengus*. Fish. Bull., U.S. 90:203-210.

Shaw, P., C. Turan, J. Wright, M. O'Connell, and G. Carvalho. 1999. Microsatellite DNA analysis of population structure in Atlantic herring (*Clupea harengus*) with direct comparison to allozyme and mtDNA data. Heredity 83:490-499.

Sinclair, M., and M. Tremblay. 1984. Timing of spawning on Atlantic herring (*Clupea harengus harengus*) populations and the match-mismatch theory. Can. J. Fish. Aquat. Sci. 14:1055-1065.

Slatkin, M. 1995. A measure of population subdivision based on microsatellite allele frequencies. Genetics 139:457-462.

Smith, P., and A. Jamieson. 1986. Stock discreteness in herring: A conceptual revolution. Fish. Res. 4:223-234.

Stephenson, R. 1997. Successes and failures in the management of Atlantic herring: Do we know why some have collapsed and others survived? In: D.A. Hancock, D.C. Smith, A. Grant, and J.P. Beumer (eds.), Developing and sustaining world fisheries resources: The state of science and management. Proceedings of the 2nd World Fisheries Congress. CSIRO, Australia.

Stephenson, R. 1999. Stock complexity in fisheries management: A perspective of emerging issues related to population subunits. Fish. Res. (Amst.) 43:247-249.

Stephenson, R., K. Rodman, D. Aldous, and D. Lane. 1999. An in-season approach to management under uncertainty: The case of the SW Nova Scotia herring fishery. ICES J. Mar. Sci. 56:1005-1013.

Svetovidov, A. 1963. Fauna of U.S.S.R. Fishes. Vol. 2. Clupeidae. 428 pp. (Transl. from Russian by Israel Prog. Sci. Transl., Jerusalem.)

Waples, R. 1998. Separating the wheat from the chaff: Patterns of genetic differentiation in high gene flow species. J. Hered. 89:438-450.

Weir, B., and C. Cockerham. 1984. Estimating *F*-statistics for the analysis of population structure. Evolution 38:1358-1370.

Wheeler, J., and G. Winters. 1984. Homing of Atlantic herring (*Clupea harengus*) in Newfoundland waters as indicated by tagging data. Can. J. Fish. Aquat. Sci. 41:108-177.

Wright, S. 1951. The genetical structure of populations. Ann. Eugenics 15:323-354.

Intermingling of Herring Stocks in the Barents Sea Area

K.E. Jørstad
Institute of Marine Research, Nordnes-Bergen, Norway

G.G. Novikov
Moscow State University, Department of Ichthyology, Moscow, Russia

N.J. Stasenkova
SevPINRO, Arkhangelsk, Russia

I. Røttingen
Institute of Marine Research, Nordnes-Bergen, Norway

V.A. Stasenkov
SevPINRO, Arkhangelsk, Russia

V. Wennevik
Institute of Marine Research, Nordnes-Bergen, Norway

A.N. Golubev
Moscow State University, Department of Ichthyology, Moscow, Russia

O.I. Paulsen
Institute of Marine Research, Nordnes-Bergen, Norway

A.K. Karpov
Moscow State University, Department of Ichthyology, Moscow, Russia

L.A. Telitsina
SevPINRO, Arkhangelsk, Russia

A. Andreeva and A.N. Stroganov
Moscow State University, Department of Ichthyology, Moscow, Russia

Extended Abstract

The most commercially important herring stock in the northeast Atlantic is the Norwegian spring-spawning herring. Spawning takes place along the greater part of the Norwegian coast, and the juveniles use Norwegian fjords and the Barents Sea as nursery areas. In addition to the large oceanic herring stock, which performs large-scale feeding migrations (Marty 1959, Parrish and Saville 1967, Dragesund et al. 1980, Røttingen 1990), a number of genetically distinct fjord populations are also found in Norwegian seawaters (Jørstad et al. 1991). Of these minor populations the Trondheimfjord herring has been managed separately since 1980. The genetically peculiar Balsfjord herring (Jørstad and Pedersen 1986, Kjørsvik et al. 1990) spawns in shallow waters, and genetic analyses have suggested that this herring is more closely related to Pacific herring (Jørstad et al. 1994) than to Atlantic herring.

Some of the fjord populations of herring in northern Norway (Balsfjord, Lake Rossfjord; see Fig. 1) are also characterized by a low number of vertebrae, which suggests a relationship with the low-vertebrae herring (sometimes called Arctic herring) populations in Russian coastal areas and the White Sea. The taxonomy of these herring groups is quite controversial, but they are usually treated as subgroups within the Pacific herring, *Clupea pallasii* (Svedovidov 1963). The most commercially important population within the group spawns in shallow waters close to the southern Barents Sea with main spawning sites in Tshoscha Bay south of Kolguev Island. This herring population also migrates into the Barents Sea, overwinters in the Goose Bank area, and can be caught in bottom trawls together with juveniles of Norwegian spring-spawning herring (see Fig. 2). In some years the 0-group and juveniles of Norwegian spring spawners can be found as far east as around Kolguev Island and Pechora Sea (Nilsen et al. 1994).

The overlap of different groups of herring in the Barents Sea region was demonstrated during the survey with R/V *Johan Hjort* in the Barents Sea in February 1993. Genetic analyses using starch gel electrophoresis and selective staining of enzymes (Harris and Hopkinson 1976, Jørstad et al. 1991) revealed that two genetically different groups of herring existed in the region, and these could be found in the same trawl catch (Fig. 2). This situation caused several problems, and during the joint Russian-Norwegian herring surveys in autumn 1994, abundance estimates of the juvenile herring from the different groups could not be made because no formal separation criteria existed. The problem was further discussed at the ICES Working Group of Atlanto-Scandian Herring and Capelin in Copenhagen, October 1994, and the working group recommended developing commonly accepted separation criteria.

Based on this recommendation, the Institute of Marine Research, Bergen, Norway, established cooperation with Moscow State University, which has carried out genetic investigations on fish stocks in the White Sea for many years. Also included in the cooperation was SevPINRO in

Herring: Expectations for a New Millennium 631

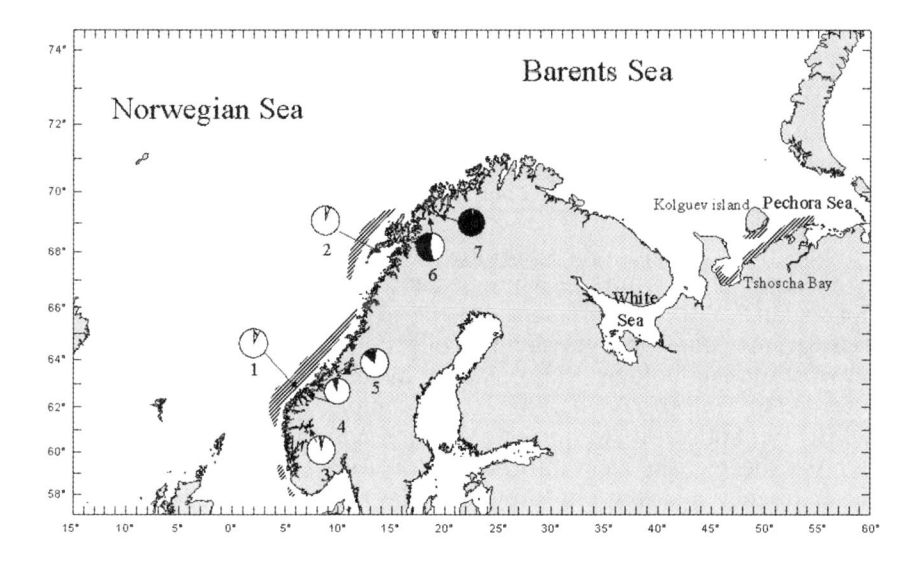

Figure 1. Pie diagrams of allele frequencies of LDH-2* in Norwegian herring populations and main spawning areas (shadow) of Norwegian spring spawners and low-vertebrae herring populations in Russian coastal areas. The data are from Jørstad et al. (1991): 1, Møre; 2, Lofoten; 3, Sognefjord; 4, Romsdalsfjord; 5, Trondheimfjord; 6, Lake Rossfjord; 7, Balsfjord.

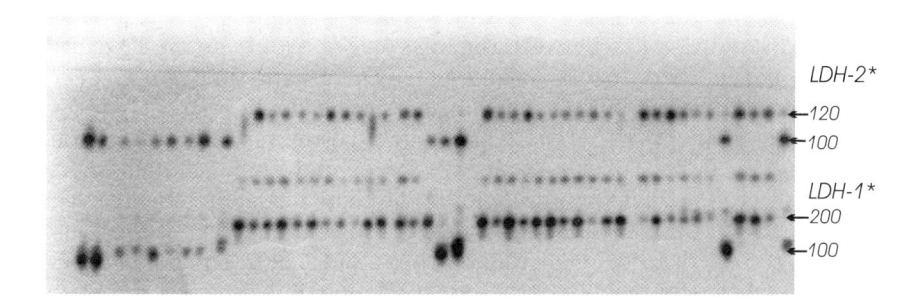

Figure 2. Banding pattern of lactate dehydrogenase from white muscle tissue. The two different loci and alleles are indicated. Samples 1-9 (from left) represent the typical banding seen in Norwegian spring spawners, while samples 10-23 are individuals from the low-vertebrae herring. The samples to the right consist of mixture of fish from both herring populations taken in the same trawl catch. All herring samples were collected in the Goose Bank area in the Barents Sea by R/V Johan Hjort during February 1993.

Arkhangelsk, a research institute associated with PINRO in Murmansk responsible for management of resources in the Russian northeast coastal area and the White Sea. During 1994 and 1995 a joint Russian-Norwegian herring investigation program was developed and the actual fieldwork was carried out in spring and early summer 1996. The Norwegian R/V *G.O. Sars* covered the southeast Barents Sea outside the Russian economic zone, while the Russian research vessel *Ivan Petrov* collected herring samples from the different spawning sites in shallow waters in the Russian coastal areas from the White Sea and east to Pechora. In 1997 the Norwegian R/V *Michael Sars* repeated the 1996 investigation in the southeast Barents Sea. The total sample collection consisted of 28 localities (trawl stations). Biological information, including vertebral counts, was recorded from all specimens. White tissue samples were analyzed by starch gel electrophoresis at sea immediately after collection.

As expected, the most informative loci were *LDH-1** and *LDH2**, and the frequency distributions supported the results obtained for herring samples collected on Goose Bank in the Barents Sea in 1993 (see banding pattern, Fig. 2). For the reference samples of low-vertebrae herring collected at the Russian coastal spawning sites the frequency of *LDH-1*200* allele was nearly 1.0. In contrast, the juvenile herring of the Norwegian spring-spawning population were nearly fixed for a different allele, *LDH-1*100,* and the individuals could be diagnostically identified to the two respective herring groups. A similar pattern was found for the *LDH-2**: the low-vertebrae herring were almost fixed for *LDH-2*120* and Norwegian spring spawners nearly fixed for *LDH-2*100*. Most surprisingly, the majority of the bottom trawl samples collected by *G.O. Sars* in the area around Kolguev Island consisted of a mixture of herring from the two groups. In the trawl samples taken between the Russian coast (outside the 12 nautical mile border) and Kolguev Island, between 15% and 20% of the specimens were Norwegian spring-spawning herring as identified by diagnostic genotypes at the *LDH-1** locus. This contrasted the results obtained for the herring samples collected on the different spawning localities where only low-vertebrae herring were found. The same intermingling area was also investigated in 1997, and the results confirmed the earlier observations.

The results clearly demonstrate the importance of international scientific cooperation for resolving stock identification problems in marine fish resource management. The samples of spawning herring in Russian and Norwegian coastal areas represent important references, laying the foundation to evaluate the extent of intermingling of the two populations in the Barents Sea during feeding migrations. The investigation also confirmed the discriminatory power of the genetic methods used to distinguish the two main groups of herring. In such areas where intermingling of different herring populations occurs, the correct population identification is fundamental for population abundance estimates and the development of future harvest prognosis.

Acknowledgment

In addition to internal resources from the different institutes, the herring investigation received funding from the Norwegian Ministry of Foreign Affairs.

References

Dragesund, O., J. Hamre, and Ø. Ulltang. 1980. Biology and population dynamics of the Norwegian spring-spawning herring. Rapp. P.-V. Reun. Cons. Int. Explor. Mer 177:43-71.

Harris, H., and D.A. Hopkinson. 1976. Handbook of enzyme electrophoresis in human genetics. North-Holland, Amsterdam.

Jørstad, K.E., and S.A. Pedersen. 1986. Discrimination of herring populations in a northern Norwegian fjord: Genetic and biological aspects. ICES C.M. 1986/H:63.

Jørstad, K.E., G. Dahle, and O.I. Paulsen. 1994. Genetic comparison between Pacific herring *Clupea pallasi* and a Norwegian fjord stock of Atlantic herring *Clupea harengus*. Can. J. Fish. Aquat. Sci. 51(Suppl. 1):233-239.

Jørstad, K.E., D.P.F. King, and G. Nævdal. 1991. Population structure of Atlantic herring, *Clupea harengus* L. J. Fish. Biol. 39(Suppl. A):43-52.

Kjørsvik, E., I.J. Lurås, C.C.E. Hopkins, and E.M. Nilssen. 1990. On the intertidal spawning of Balsfjord herring (*Clupea harengus* L.). ICES C.M. 1990/H:30. 27 pp.

Marty, Yu.Yu. 1959. The fundamental stages of the life cycle of Atlantic-Scandinavian herring. Fish. Res. Board Can. Transl. Ser. 167:5-68a.

Nilssen, K.T., I. Ahlquist, J.E. Eliassen, T. Haug, and L. Lindholm. 1994. Studies of food availability and diet of harp seals (*Phoca groenlandica*) in south-eastern Barents Sea in February 1993. ICES C.M. 1994/N:12. 24 pp.

Parrish, B.B., and A. Saville. 1967. The biology of the north-east Atlantic herring populations. Oceanogr. Mar. Biol. Ann. Rev. 5:404-447.

Røttingen, I. 1990. A review of variability in the distribution and abundance of Norwegian spring spawning herring and Barents Sea capelin. Polar Res. 8:33-42.

Svetovidov, A.N. 1963. Fauna of U.S.S.R. Fishes. Vol. 2. Clupeidae. 428 pp. (Transl. from Russian by Israel Prog. Sci. Transl., Jerusalem.)

Gizhiga-Kamchatka Herring Stock Level and Catch Potential

Andrey A. Smirnov
Pacific Research Institute of Fisheries and Oceanography, Magadan Branch (MoTINRO), Magadan, Russia

Abstract

The Gizhiga-Kamchatka herring is the second largest population inhabiting the Sea of Okhotsk, smaller only than the Okhotsk herring. In the years of its high numbers (the 1950s), its spawning grounds covered the coastal areas of Shelikof Bay and northwestern Kamchatka. The feeding grounds of the Gizhiga-Kamchatka herring are in the eastern Sea of Okhotsk.

The Gizhiga-Kamchatka herring fishery began in 1925. The catches increased to 161,000 t in 1958. By the early 1970s the stock level had declined, due to overfishing and lack of recruitment. After a ban on fishing was issued in 1974, the stock recovered. Fishing resumed in 1988, with catches set at 10-12% of the spawning stock. By the early 1990s the stock level had increased by several times, but it hasn't reached the level of the 1950s.

The increase in the number of herring is likely connected with conservation measures and with changes in climate and oceanographic conditions in the North Pacific Ocean.

The reduced catches of Gizhiga-Kamchatka herring in recent years may have contributed to high stock levels and reduced growth through density-dependent effects. Dorsal and pectoral fin ray counts appear to be a useful tool to separate Gizhiga-Kamchatka herring from Okhotsk herring when the stocks are mixed on the foraging grounds.

We recommended increasing the total fishery catch to 20.7% of the spawning stock biomass per year, with 6.5% of the catch taken during the spawning period.

Introduction

Gizhiga-Kamchatka herring is the second largest population in the Sea of Okhotsk, smaller only than the Okhotsk herring. In the years of its high numbers (the 1950s), its spawning grounds covered the coastal areas of

Shelikof Bay and northwestern Kamchatka (to 57°N). The herring feeding grounds are in the eastern Sea of Okhotsk and, in some years, the Pacific Ocean. Feeding ground locations were determined from fish marked near the northern coasts of the Shelikof Bay during the spawning season of the 1950s-1960s; the fish were then caught near southeastern Kamchatka and the Pacific coasts of the northern Kurils (Fig. 1).

The commercial catch of Gizhiga-Kamchatka herring began in 1925 near western Kamchatka, and in 1937 in Shelikof Bay. Bottom gillnets and throw nets were used to catch herring as they entered the coastal area for spawning. Since 1955, small purse seines have been used from boats to catch the spawning herring. These nets, when fished close to shore, harmed spawning areas by mechanically removing kelp with roe. Each year, there was a regular growth in catch, from 600 t in 1937 to 161,000 t in 1958. Following the peak catch in 1958 catches declined (Pravotorova 1965).

By the 1970s, due to overfishing and several successive poor year classes, the stock had decreased in size and the spawning and foraging areas shrank to the Shelikof Bay area. In 1974 commercial harvest of herring was banned, and a limited catch of only 1-2% of the stock was permitted during the spawning season, in order to gather data about the population.

The closure of the herring fishery favored a gradual recovery of the stock. Commercial fishing on both spawning and foraging herring resumed in 1988, with harvest levels set at 10-12% of the biomass.

In this paper we review growth and spawning and foraging distributions during periods of high and low abundance, investigate methods of separating Gizhiga-Kamchatka herring from other stocks on the foraging grounds, link the age composition of the stock to harvest policy, and speculate on mechanisms that determine year class size.

Methods

Location and timing of spawning is determined from research fishing activities, diving investigations on the spawning grounds, and aerial surveys of spawning herring accumulations. Spawning ground research activities are also used to estimate age composition and biomass and to determine allowable harvest.

To study the feasibility of using fin rays to separate Okhotsk and Gizhiga-Kamchatka herring we collected herring from the spawning grounds of each stock in 1999. Sixty herring were taken on the border of the Okhotsk spawning area near Magadan, and 90 specimens were collected in the center part of the Gizhiga-Kamchatka herring spawning area near Evensk. We enumerated the number of rays in dorsal, pectoral, and anal fins of each specimen for comparisons of the two areas.

Annual catch quotas are determined from population condition, numbers, biomass, and the strength of recruiting year classes. Usually the exploitation rate is approximately equal to the natural mortality rate. The allowable catch volume was determined by the method offered by Malkin (1995).

Abundance of spawning stock generations was determined by aerial survey, average weight of specimens of each age, and age composition of the population. The numbers obtained during the entire period of the study were averaged and divided into three categories: strong, average, and poor.

Results

Gizhiga-Kamchatka herring move to the spawning grounds in the coastal area of Shelikov Bay in the middle of May. The exact time of this process varies appreciably. The earliest approaches were observed on May 7, with the latest on June 6. Usually the approach to the central spawning area occurs May 15 through May 25.

The Gizhiga-Kamchatka herring spawn somewhat later than other populations of Okhotsk herring. Normally, spawning progresses from west to east.

According to the annual aerial survey results, the traditional spawning grounds near the northern coasts of the bay have not become larger during the last few years. But there is an increase in the density of roe deposited on the spawning substrate. According to the underwater survey data, the roe density deposited on the main spawning grounds was 1,642,000 eggs per m^2 in 1988, and 2,469,000 eggs per m^2 1999 (Vyshegorodtsev 1994). In 1999, in addition to spawning near the northern coast of Shelikof Bay, herring also returned to spawning grounds near western Kamchatka which may also indicate a growth of the herring population (Fig. 2).

The growth in the number of fish in the herring population is due to climatic and oceanic conditions in the North Pacific, which result in structural changes in the fish community as follows: the numbers of herring become higher and the numbers of pollock become lower. These processes are also influenced by a weakening West Kamchatka Current, which carries the ocean waters along Kamchatka to the north. A decrease in an incoming flow of cold water favored the development of shelf conditions, which are good for herring (Shuntov 1998). Conservation measures, such as a prohibition of using throw nets from boats in the main spawning areas, are also important.

We believe that the abundance of the herring population is determined by climatic and oceanic conditions in the North Pacific, which result in structural changes in the fish community where herring and pollock abundance are inversely proportional. These species are influenced by a weakening West Kamchatka Current, which carries the oceanic waters along Kamchatka to the north. A decrease in the incoming flow of cold waters favors the development of shelf conditions which are good for herring (Shuntov 1998). The conservation measures, such as a prohibition of us-

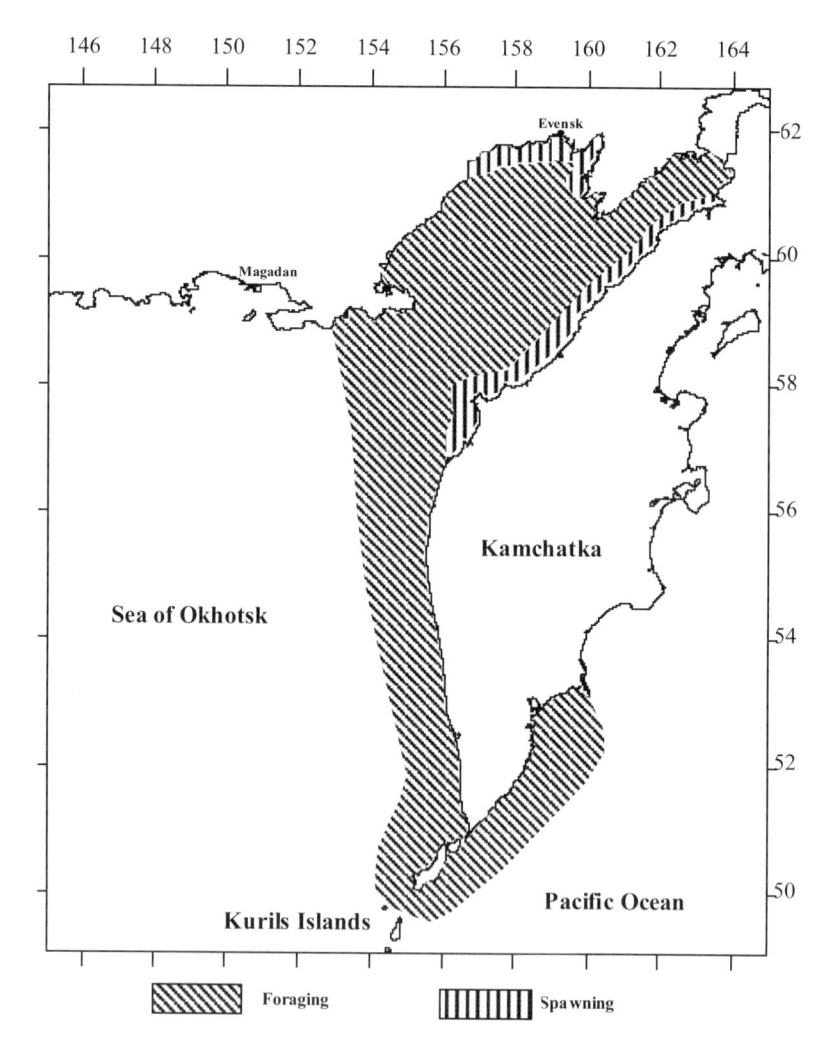

Figure 1. Spawning and foraging areas of Gizhiga-Kamchatka herring during the 1950s.

Herring: Expectations for a New Millennium 639

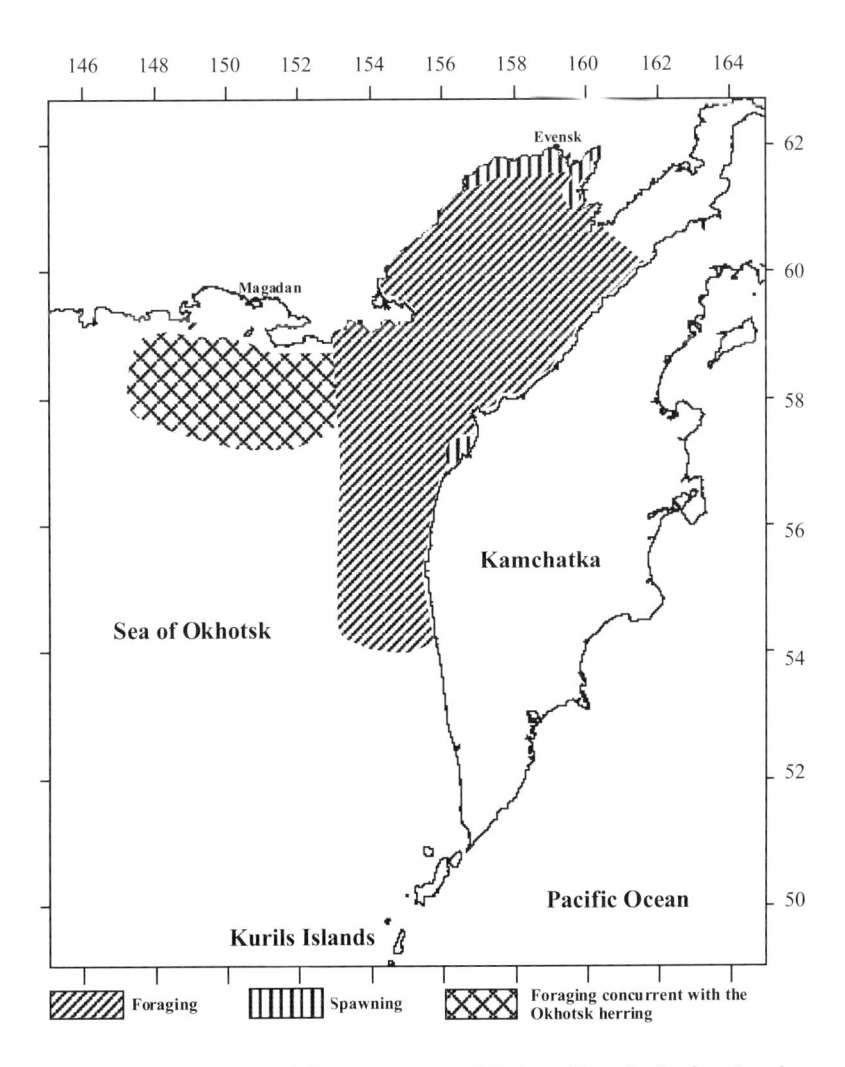

Figure 2. Spawning and foraging areas of Gizhiga-Kamchatka herring during the 1990s.

ing throw nets from boats in the main spawning areas, have also had an important effect on herring abundance.

From 1988 to 1992, the amount of herring catch in the foraging area was up to 1.6% of the spawning stock. In the second half of the 1990s, the population of the Gizhiga-Kamchtka herring became so large that it was recorded near western Kamchatka as an additional catch in a winter catch of pollock. In 1997, such an additional catch (8.1% of the stock) was 6 times greater in comparison with 1996. The higher success rate of the fishing fleet was also favored by a warm winter and reduced ice conditions.

The age composition of the herring stock is one of the main indications of the population. A catch of spawning herring may contain specimens from 3 to 16 years old. As a rule, 3-year-old fish and those older than 10 years are unimportant both for commercial purposes and stock reproduction. The age 6 and 7 herring are usually most abundant, whereas there is a drastic decrease in the numbers of the 8-year-olds, especially from weak year classes. Only strong year classes remain relatively abundant at age 8. (Fig. 3).

In our long-term study results, we classify year classes as "strong," "average," and "poor." The average year class size at age 4 for each category was: 1,343,000,000 for "strong" year classes, 376,000,000 for "average" year classes, and 88,000,000 for "poor" year classes.

Over the period 1931 to 1999, in each 10 years there were typically 4 strong and 6 poor year classes. The appearance of strong year classes is favored by ice melting in early spring in the Shelikof Bay spawning grounds, and also in areas of foraging herring larvae and juveniles. In the last decade of the twentieth century, the year classes of 1993, 1994, and 1997 were strong. It may be possible that the next strong generation will be in 2000. The recent state of the stock of the Gizhiga-Kamchatka herring allows for a 100,000 t catch each year, although the present catch is less than this amount.

We expect a further increase in the numbers of the Gizhiga-Kamchatka herring, related to strong year classes that we expect to recruit in 2003-2004.

The winters of 1998-1999 in the Sea of Okhotsk were cold, with hard ice conditions. The ice closed the sea near northwestern Kamchatka (between 56° and 58°N), where an increased additional catch of herring was recorded in the catch of pollock. The winter catch of herring was 0.9% of the stock in 1998 and only 0.4% in 1999, and the cold winter conditions also negatively influenced the spring catch of herring. The main traditional spawning grounds were filled with ice, and as a result, the spawning was short and on new grounds. The spring catch in 1992-1995 was 1.3-2.5% of the spawning stock; in 1996 and 1997 it was 2.6% and 3.3% respectively; in spring 1998 it was 0.4%; and in spring 1999 it was 0.8%. Compared to 1998 the spawning stock increased 2% in 1999.

The cessation of the catch of foraging herring and the low catch of spawning herring in the last few years have resulted in an older age of the

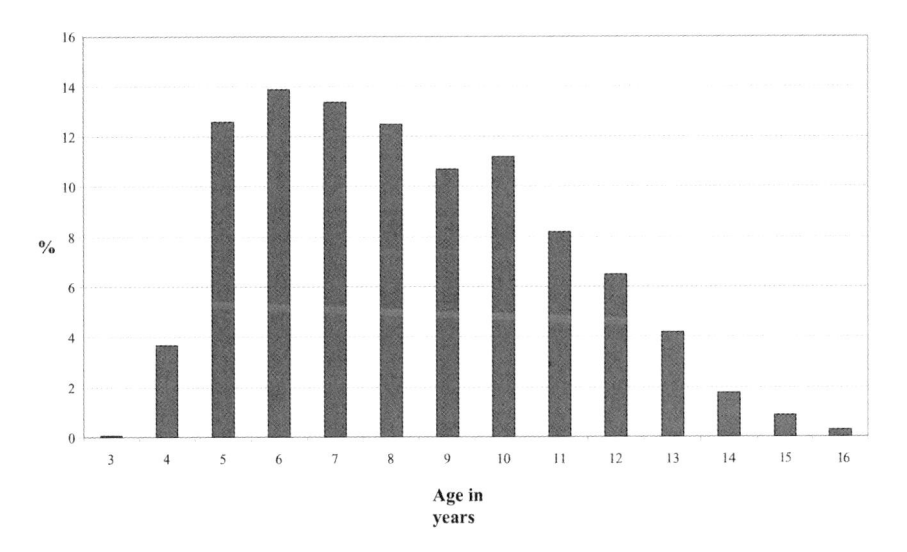

Figure 3. Age composition of Gizhiga-Kamchatka herring: multiyear averages of 1980-1999.

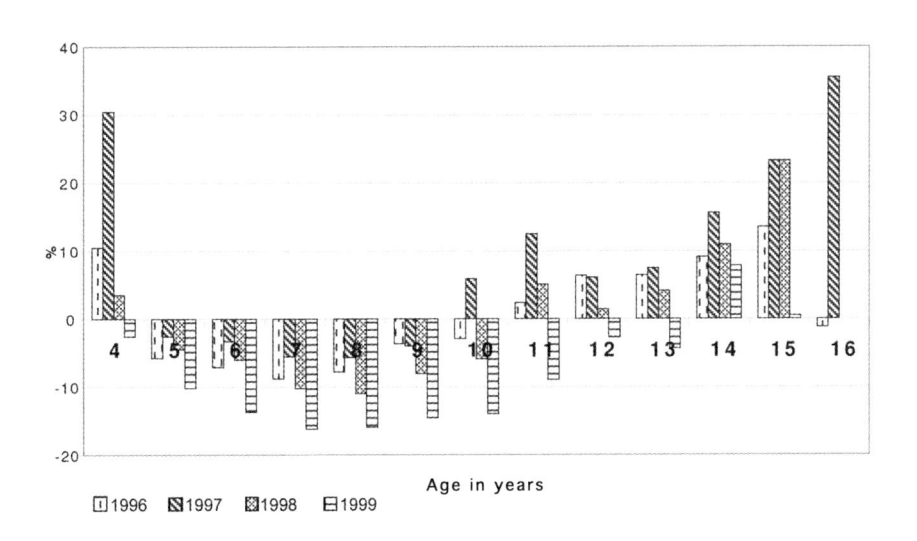

Figure 4. Weight variations of the Gizhiga-Kamchatka herring as multiyear averages (percent).

Table 1. Number of dorsal and pectoral fin rays of Okhotsk herring (60 specimens from the Magadan region) and Gizhiga-Kamchatka herring (90 specimens from the Evensk region), age 4-6 in spring 1999.

Number of fin rays	Gizhiga-Kamchatka herring Range	Gizhiga-Kamchatka herring Average	Okhotsk herring Range	Okhotsk herring Average
Dorsal fin	16.0-18.0	17.2 ± 0.07	18.0-19.0	18.9 ± 0.04
Pectoral fin	15.0-17.0	16.3 ± 0.06	16.0-18.0	17.0 ± 0.04

± standard deviation

stock. The older fish have greater length and weight-at-age characteristics in comparison with multi-year average values. At the same time, the decrease in average age of fish from 5 to 9 years old was noticed (Fig. 4).

In consideration of all these factors, we recommend an increase in overall catch following the concept of reproductive variability of Ye.M. Malkin (1995). Malkin's theory holds that the allowable amount of catch depends on the age of the sexual maturation of females.

In comparison with the Okhotsk herring, the maturation period of the Gizhiga-Kamchatka herring is older. The average age of sexual maturation is 6 years. Proceeding from this, we recommend an allowable commercial catch of 20.7% of the biomass of the Gizhiga-Kamchatka herring, including 6.5% catch during the spawning period.

Because Gizhiga-Kamchatka herring are also reported from the foraging areas of the Okhotsk herring, a method to correctly identify the stocks would be useful for management applications. Okhotsk herring specimens have fewer rays in both the dorsal fin and the pectoral fin than Gizhiga-Kamchatka specimens (Table 1). The two populations have the same number of rays in abdominal fins (9) and anal fins (16). During our autumn studies of the Okhotsk herring foraging areas, approximately 15% of the individuals had characters typical of the Gizhiga-Kamchatka herring.

Conclusion

To provide for an efficient control of the catch, we intend to develop field recommendations for distinguishing between the Okhotsk herring and the Gizhiga-Kamchatka herring. In order to do this, we must continue our studies including marking and morphometric examination; we also plan to use the methods of genetics, otolithometry, and parasite indication.

In order to provide for a higher efficiency of the spring catch, we recommend increasing the share of mobile catching facilities to enter the spawning areas in proper time. In addition, we also recommend to initiate "manufacturing" herring roe on kelp. This product is in a high demand on world markets. There is no demand for herring without roe. Thus, after roe is laid on kelp, which are then gathered and processed, herring can be released and continue spawning during the following years.

The present allowable catch of Gizhiga-Kamchatka herring is lower than necessary. The current state of this stock allows a catch of up to 100,000 t per year. We expect a further increase in numbers of Gizhiga-Kamchatka herring and relate it to suggested strong year classes in 2003-2004.

References

Malkin, Ye.M. 1995. Regulation principle of catch on the basis of the concept of genesial variability of populations. Voprosy Ikhtiol. 35(4):537-540. (In Russian.)

Pravotorova, Ye.P. 1965. Some data on the Gizhiga-Kamchatka herring biology in connection with change of its stock number and foraging area. Izvestiya TINRO 59:102-128. (In Russian.)

Shuntov, V.P., I.V. Volvenko, A.F. Volkov, K.M. Gorbatenko, S.Yu. Shershenkov, and A.N. Starovoytov. 1998. New data about condition of pelagic ecosystems of the Okhotsk and Japan seas. Izvestiya TINRO 124:139–177. (In Russian.)

Vyshegorodtsev, V.A. 1994. The peculiarities of eggs deposit on spawning substrate of Gizhiga-Kamchatka herring. Izvestiya TINRO 115:137–141. (In Russian.)

Management of North Sea Herring and Prospects for the New Millennium

J.H. Nichols
Centre for Environment, Fisheries and Aquaculture Science, Lowestoft Laboratory, Suffolk, United Kingdom

Abstract

North Sea autumn-spawning herring comprise a complex of three separate spawning stocks, which mix and have to be managed as a single unit. Assessment and management is further complicated by the fact that five types of fisheries exploit these stocks during their migrations, two in the North Sea, the others in the Skagerrak and Kattegat. Two fisheries exploit mainly adult fish while the others take juveniles as bycatch in small-mesh fisheries.

Prior to 1981 there was little control, other than market forces, on the catches of North Sea herring. Following the collapse of the stock in the 1970s, and the moratorium on directed herring fishing in the North Sea from 1977 to 1981, a total allowable catch (TAC) constraint was applied to the North Sea directed fishery. The management objective was to maintain the spawning stock biomass (SSB) above 800,000 t.

The target SSB was reached during the 1980s but, with increasing fishing pressure and high juvenile mortality, the stock declined again in the 1990s. By 1994 the SSB had fallen below 800,000 t and in 1995 the ICES Advisory Committee for Fisheries Management recommended a 30% reduction in the TAC. Then, following a new assessment early in 1996, indicating an SSB of around 500,000 t, an immediate, unprecedented, halving of the agreed TAC for 1996 was imposed. This was combined with action to reduce the juvenile mortality, by imposing bycatch limits in the small-mesh fisheries.

This rapid action was in stark contrast with the years that it took to protect the herring in the 1970s. Management objectives are now more firmly defined in a precautionary approach with both biomass and fishing mortality targets which bode well for the future. There is already evidence that the measures taken are beginning to work, although stricter licensing and landing regulations are being introduced to control the quota and curb misreporting.

Introduction

Over many centuries the North Sea herring fishery has been a cause of international conflict sometimes resulting in war, but in more recent times in bitter political argument. Fundamental changes in the nature of the fisheries have been driven both by changes in catching power and in market requirements, particularly the demand for fish-meal and oil. Most of these changes have resulted in greater exploitation pressures that increasingly led to the urgent need to ensure a more sustainable exploitation of North Sea herring. Increased fishing pressures became noticeable for the first time during the 1950s when the spawning stock biomass of North Sea autumn-spawning herring fell from 5 million t in 1947 to 1.4 million t by 1957. That period also witnessed the decline and eventual disappearance of a traditional autumn drift-net fishery in the southern North Sea.

North Sea herring are primarily autumn spawners though some small discrete groups of coastal spring spawners exist in areas such as The Wash and the Thames Estuary. Juveniles of the spring-spawning stocks found in the Baltic, Skagerrak, and Kattegat may also be found in the North Sea as well as Norwegian coastal spring spawners.

The widespread but discrete location of the herring spawning grounds throughout the North Sea has been well documented since the early part of this century. This knowledge has resulted in considerable scientific debate and research on stock definition and identity. The controversy centered on whether or not the separate spawning grounds represented discrete stocks or "races" within the North Sea autumn-spawning herring complex. Resolution of this issue became more urgent as the need for the introduction of management measures increased during the 1950s.

Harden Jones (1968), in his book on fish migration, reviewed all the evidence for the identity of the North Sea spawning stocks and updated the migration patterns previously described by Cushing (1955), Cushing and Bridger (1966), Postuma (1956), and Parrish and Saville (1965). He concluded that from recruitment, a year class appears to maintain its identity and it returns annually to its spawning places until its extinction. These conclusions reiterated the homing hypothesis of Zijlstra (1958, 1963).

These conclusions were the basis for establishing the working hypothesis that the North Sea autumn-spawning herring comprise a complex of three separate stocks each with separate spawning grounds, migration routes, and nursery areas. They are, (1) the Buchan or Scottish group which spawns from July to early September in the Orkney Shetland area and off the Scottish east coast; (2) the Banks or central North Sea group, which derives its name from its former spawning grounds around the western edge of the Dogger Bank; and (3) the Downs group which spawns in very late autumn through to February in the southern Bight of the North Sea and in the eastern English Channel (Fig. 1). Zjilstra (1969) presented meristic and morphometric data to support this division of the North Sea autumn-spawning herring into the three, mainly discrete spawning stocks.

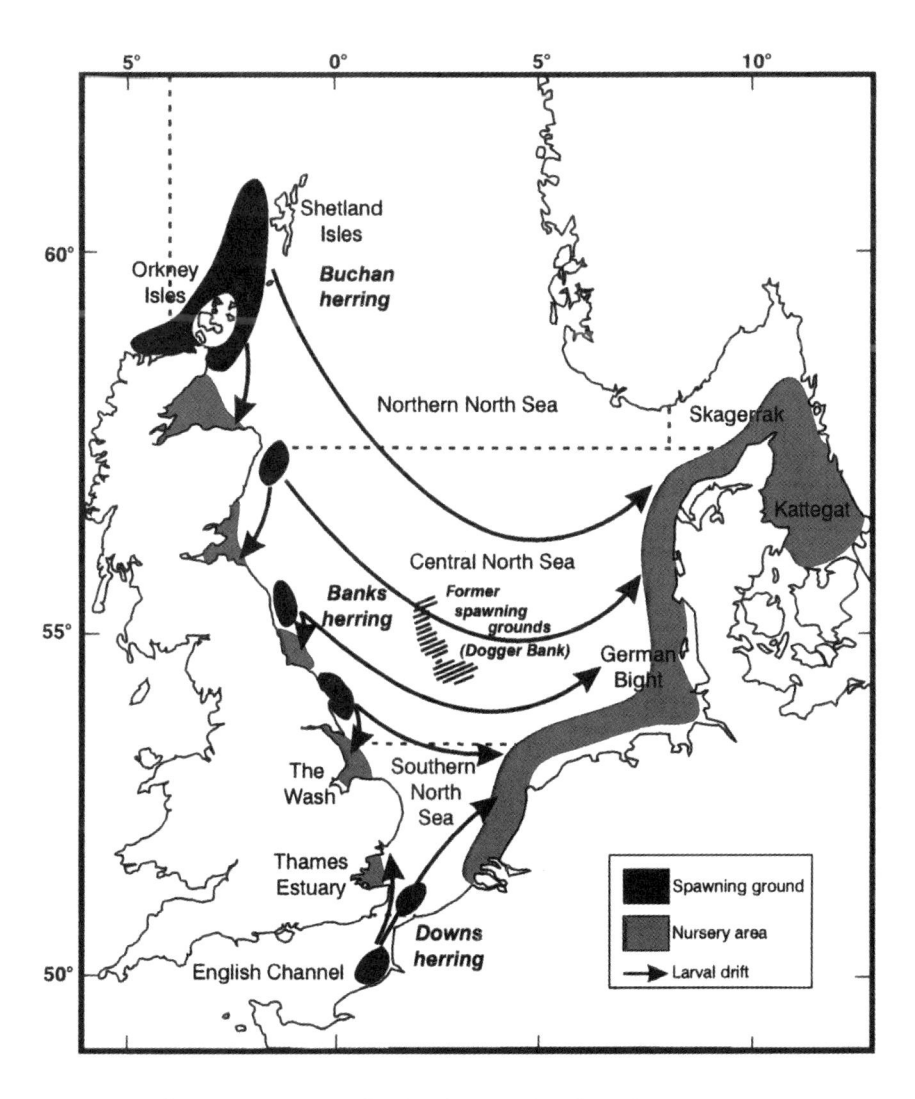

Figure 1. The spawning areas for the three stocks of North Sea autumn-spawning herring showing the direction of larval drift to the nursery areas.

At certain times of the year individuals from the three stocks may mix and are caught together as juveniles and adults but they cannot be readily separated in the commercial catches. As a consequence North Sea autumn-spawning herring have to be addressed as a single management unit.

This paper describes the brief history of the stock complex leading up to the introduction of management measures in the form of a total allowable catch (TAC) constraint for North Sea herring in 1980 and the successes and failures of such measures up to the present day.

Recent History of the North Sea Fishery

The annual landings from 1947 through to the early 1960s were large, but stable, averaging around 650,000 t (Fig. 2). From 1952 to 1962 the high fishing mortality resulted in a rapid decline in the spawning stock biomass from around 5 million t to 1.5 million t (Fig. 3). Recruitment over this period was reasonable, but the number of significant year classes declined in the adult stock. This was a clear indication of overfishing and the impact of the developing industrial fishery in the eastern North Sea. This period witnessed the complete collapse of the historic East Anglian autumn drift-net fishery, which was based entirely on the Downs stock moving south to the Southern Bight of the North Sea and eastern English Channel to spawn. Failure of the Downs stock has been attributed both to high mortality of the juveniles in the North Sea industrial fisheries, and to heavy fishing by bottom trawlers on the spawning concentrations, in the English Channel, during the 1950s.

Fishing mortality on the herring in the central and northern North Sea began to increase rapidly in the late 1960s and had increased to $F_{age\ 2-6} = 1.3$ or a removal of over 70% per year of those age classes, by 1968 (Fig. 2). Landings peaked at over 1 million t in 1965, around 80% of which were juvenile fish. One of the main reasons for this was the development during the 1950s of an industrial fishery for meal and oil, targeted at juvenile herring, off the east coast of Denmark. The fishery developed by a chance encounter of fishing vessels with large shoals of herring in the area. The initial effort was fueled by the increasing demand for fish meal and oil. The development of an industrial fishery for herring removed the one constraint which had protected the stocks from overfishing, namely the limitation on the marketing and disposal of large bulk catches of oily fishes. These small-mesh fisheries, targeted at juvenile herring in the North Sea, expanded rapidly. By 1955 the annual catch of juvenile herring had reached 100,000 t, increasing further to 169,000 t by 1972. By 1976 targeted fishing for herring for industrial purposes in the North Sea was banned but juvenile herring continued to be taken in large numbers in the small-mesh fisheries for sprat and other species as a permitted 10% bycatch.

By 1975 the SSB of North Sea herring had fallen to 83,500 t (Fig. 3) although the total North Sea landings were greater than 300,000 t. At the

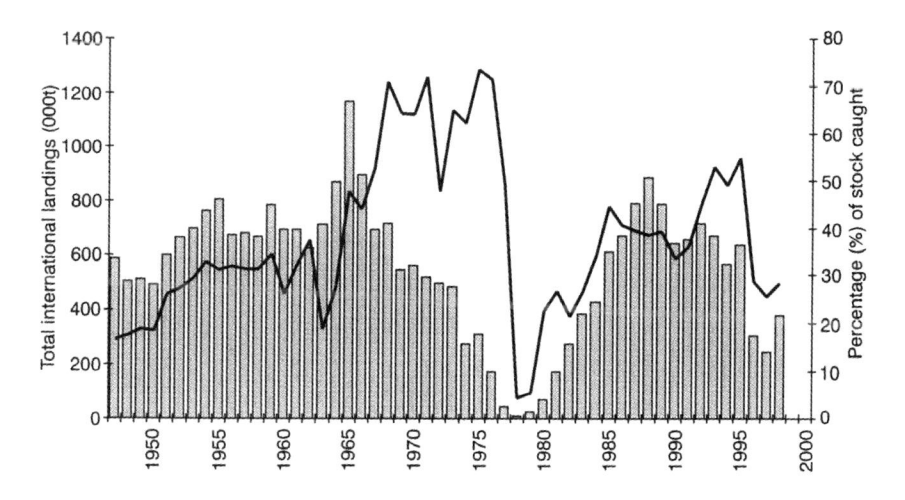

Figure 2. Total annual international landings of North Sea autumn-spawning herring from 1947 to 1998. The trends in fishing mortality over the same period are also shown.

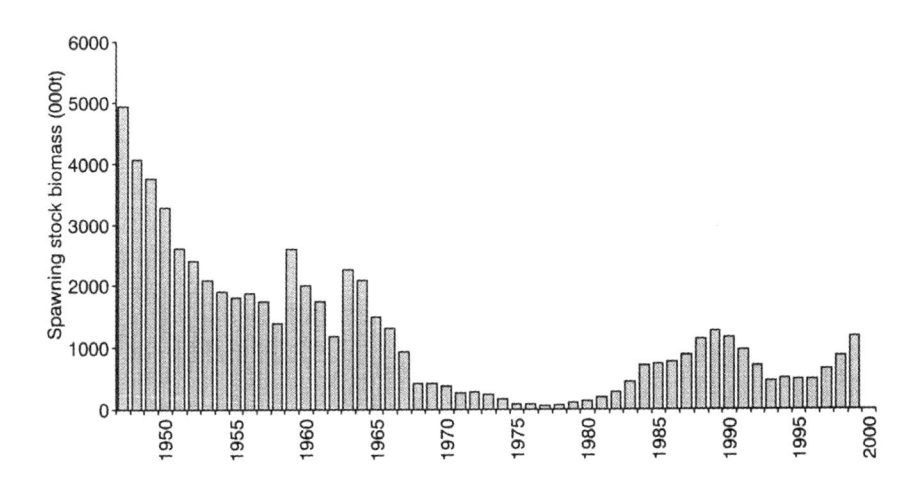

Figure 3. The annual trend in spawning stock biomass for the North Sea autumn-spawning herring over the period 1947-1998 with the predicted value for 1999.

same time, spawning in the central North Sea had contracted to the grounds off the east coast of England while spawning grounds around the edge of the Dogger Bank were abandoned. This heralded the serious decline and near collapse of the North Sea autumn-spawning herring stock which led to a moratorium on directed herring fishing in the North Sea from 1977 to 1981.

By 1980, internationally coordinated larvae and acoustic surveys indicated a modest recovery in the SSB from its 1977 low point of 52,000 t. By 1981 the SSB had increased to over 200,000 t. Prior to the moratorium there was no regulatory control, other than market forces, on catches in the North Sea directed herring fishery. Once the fishery fully reopened in 1983 the North Sea autumn-spawning herring stocks were managed by a TAC. The TAC was only applied to the directed herring fishery in the North Sea, which exploits mainly adult fish for human consumption. Other fisheries also take North Sea autumn spawners. One of these fisheries occurs in the North Sea and exploits the juvenile part of the stock as bycatch in small-mesh fisheries. The others are in nursery areas in the Skagerrak and Kattegat and take juveniles of the North Sea stock as well as Baltic spring spawners. The proportions of North Sea juveniles found in these areas varies with the strength of the year class, with higher proportions in the Skagerrak and Kattegat when the year class is good.

Following the partial reopening of the fishery in 1981 the SSB steadily increased, peaking at 1.3 million t in 1989. Annual recruitment, measured as "0"-group fish, was well above the long-term average over this period (Fig. 4). The 1985 year class was the largest recorded since 1960. Landings also steadily increased over this period, reaching a peak of 876,000 t in 1988. There was a steady increase in fishing mortality to $F_{age\ 2-6} = 0.6$ (ca. 45%) in 1985 and a high bycatch of juveniles in the industrial fishery.

Below-average recruitment of the 1987-1990 year classes resulted in a rapid decrease in SSB to below 500,000 t in 1993 (Fig. 3). Fishing mortality increased rapidly averaging $F_{age\ 2-6} = 0.75$ (ca. 52%) between 1992 and 1995 (Fig. 2) and recorded landings regularly exceeded the TAC. Annual landings from the North Sea industrial fishery for sprat increased from 33,000 t in 1987 to 357,000 t by 1995. There was a resultant high bycatch of immature herring, with juveniles averaging 76% of the total catch in numbers of North Sea herring over this period (Fig. 5).

During the summer of 1991 the presence of the parasitic fungus *Ichthyophonus* spp. was noted in the North Sea herring stock. All the evidence suggested that the parasite was lethal to herring and that its occurrence could have a significant effect on natural mortality in the stock and, ultimately, on spawning stock biomass. High levels of infection were recorded in the northern North Sea north of latitude 60°N while infection rates in the southern North Sea and English Channel were very low. Ultimately it was concluded that, during 1991, the disease had caused high annual mortality in the northern North Sea of up to 16% (ICES 1993). The

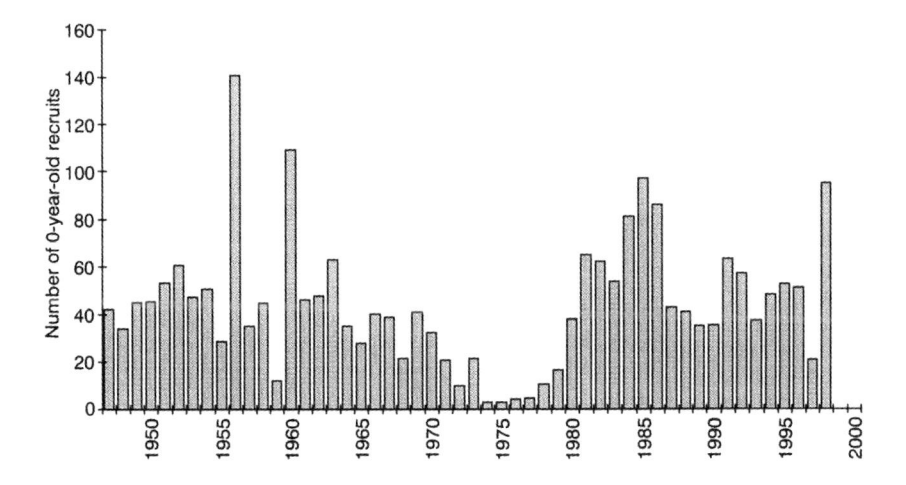

Figure 4. The annual trend in recruitment, as numbers of 0-year-old fish, for the North Sea autumn-spawning herring over the period 1947-1998.

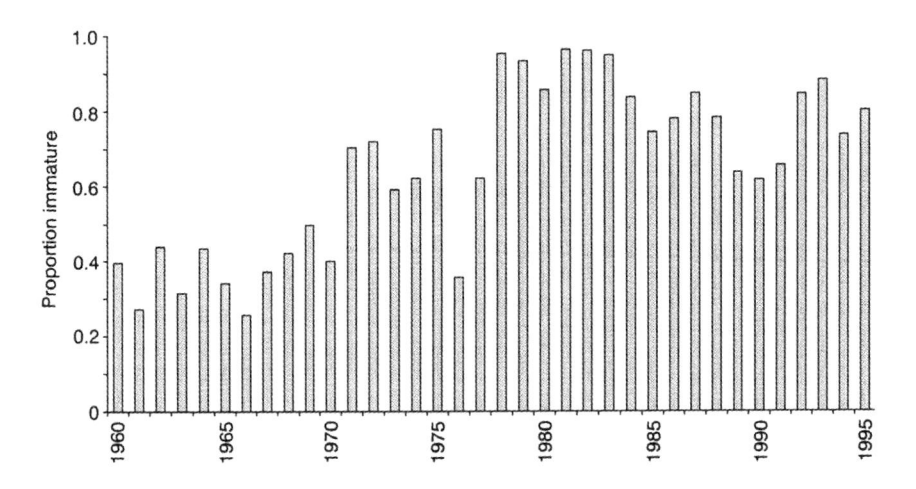

Figure 5. The proportion of immature herring, by number, in the catches of North Sea autumn-spawning herring for the period 1960-1998.

rate of infection subsequently decreased and by 1995 the disease-induced increase in natural mortality was insignificant (ICES 1994).

The increased fishing pressure during the first half of the 1990s and the disease-induced increase in natural mortality led to serious concerns about the possibilities of a stock collapse similar to that in the late 1970s. Reported landings continued at around 650,000 t per year while the spawning stock began to decline again from over 1 million t in 1990. The assessments and predictions at that time overestimated the size of the total North Sea spawning stock. For example, it was not until 1996 that it was realized that the SSB in 1993 had already fallen below 500,000 t (Table 1). The agreed TAC was regularly being exceeded and in 1995 was overshot by 94,000 t (Table 2). This led to very speedy management action during 1996 to address the potential crisis. Strict management measures have since been enforced which are beginning to manifest themselves in greater stability in the catch potential for North Sea herring.

Assessment, Advice, and Management

Assessment

The assessment is carried out annually by a working group of the International Council for the Exploration of the Sea (ICES). The working group comprises a team of fisheries scientists drawn mainly, but not exclusively, from the member states with an interest in the fishery.

From 1972 to 1995 the assessment of the total North Sea stock was done by means of a virtual population analysis (VPA) with ad hoc tuning to data series of larvae production estimates, acoustic surveys, and trawl surveys (ICES 1991). During the early 1990s there was increasing uncertainty about the assessment, which led to the exploration of other assessment models. The uncertainty was generated from differences in the perception of stock size between larvae indices and trawl surveys on the one hand, and acoustic surveys on the other. There were also questions surrounding the level of natural mortality caused by the observed effects of the *Ichthyophonus* fungal disease. An additional problem with the assessment was that the international effort on the larvae surveys covering the whole of the North Sea had halved in 1990 and has continued to decline since then. These surveys had proven a valuable indicator of the recovery of the stock during the moratorium on fishing and had generated a long time series of SSB estimates used for tuning the VPA population trends. By 1993 the temporal and spatial coverage of the larvae surveys was reduced to such a low level that the larvae production estimate could no longer be calculated. In subsequent years the production estimate was replaced by a multiplicative larval index (ICES 1996) which could be calculated on the limited data set.

The assessment working group in 1995 decided to change to an integrated catch analysis (ICA) method (Patterson and Melvin 1996). The method

Table 1 Estimates of the spawning stock biomass (SSB) of North Sea autumn-spawning herring from the assessments over the period 1988-1999, showing the perception of SSB in a particular year against the date of the assessment.

Date of assessment	Spawning stock biomass (× 1,000 t)											
	1988	1989	1990	1991	1992	1993	1994	1995	1996	1997	1998	1999
1989	822											
1990	1,102	1,256	[a]									
1991	1,242	1,549	1,411	1,320	[a]							
1992	1,095	1,340	1,247	1,277	1,346[a]							
1993	1,179	1,456	1,354	1,307	1,320	1,238[a]						
1994	1,146	1,391	1,260	1,149	1,184	1,287[a]	N/A					
1995	1,068	1,286	1,138	993	986	1,055	965[a]	N/A				
1996	1,116	1,240	1,135	939	778	730	974	981[a]	N/A			
1997	1,141	1,265	1,154	950	699	984	790	721[a]	[a]	[a]		
1998	1,138	1,266	1,151	957	692	458	517	496	539			
1999	1,144	1,276	1,169	980	702	465	547	550	519	746	1,145	
2000	*1,106*	*1,249*	*1,149*	*972*	*716*	*471*	*543*	*550*	*488*	*656*	*878*	1,170
					714	*462*	*512*	*500*	*448*	*568*	*701*	*905*
						458	*502*	*474*				

[a] Forecast (Fsq).
Remarks: 1991-1994 assessments include catches of autumn-spawning herring in ICES Division III (Skagerrak and Kattegat). Figures for 2000 (in italics) were added later.

Table 2. The agreed TAC for the North Sea directed fishery, total landings for the North Sea and Eastern English Channel, and the total landings of North Sea autumn-spawning herring for the period 1981-1998.

Year	Agreed TAC (× 1,000 t)	Bycatch limit (× 1,000 t)	Total landings of North Sea autumn spawners — North Sea & Eastern Channel only (× 1,000 t)	All areas (× 1,000 t)
1981	20[a]		52	174
1982	72[a]		116	275
1983	145		148	387
1984	55[a]		320	429
1985	90[a]		536	614
1986	570		547	671
1987	600		625	773
1988	530		698	888
1989	514		700	788
1990	415		553	645
1991	420		566	654
1992	430		549	717
1993	430		524	671
1994	440		468	563
1995	440		534	641
1996	313/156	44	265	307
1997	159	24	209	273
1998	254	22	329	380
1999	265	30	*336*	*372*
2000	265	36		

[a]Southern North Sea and Eastern English Channel only.
(Note: The 1999 landings, in italics, were added later.)

was adopted for the assessment of North Sea herring for 1994 and has been used since then. The method has the advantage of being able to use age-aggregated indices of stock size and also incorporate assumptions about errors, both in survey indices and also in the catch-at-age data set. The method affords an improved estimate of uncertainty in the assessment and in the forward projections of stock size.

Advice

Stock assessment is normally carried out by the working group in March each year. It is subsequently reviewed by the ICES Advisory Committee for Fisheries Management (ACFM) at a meeting in May each year. Following the review, and any recommended changes to the assessment, the ACFM provides advice on the state of the stocks. The advice always includes a series of options for fishery managers to consider in setting the TACs for the following year. Providing catch options for North Sea herring is complicated, not only by the existence of the separate spawning stocks but also by the different fishing fleets exploiting North Sea autumn-spawning herring. As a consequence the management options and advice have to be formulated specific to the separate fleets, each exerting different fishing mortalities on the autumn-spawning stocks. An example of the catch options presented for 1999 is shown in Table 3.

The southern North Sea and English Channel stock has always been considered to be a separate management unit within the North Sea because the population in this area is clearly separated from the other components for most of the year. Historically this component has always been subjected to a higher fishing mortality than the rest of the North Sea and is considered independently of the other spawning stocks. As a consequence, management recommendations are now specific for this stock in order to provide special protection in the form of a separately allocated TAC within the overall North Sea TAC.

The annual advice of the ACFM is subsequently considered at a joint meeting between officials of Norway and the European Union when the TAC for the following year is agreed.

Management

Advice on managing North Sea herring by TAC constraint was first provided in 1981 and was based on maintaining the spawning stock biomass above a minimum biologically acceptable level of 800,000 t, or one-third of the SSB in an unexploited phase. Levels of SSB below 800,000 t had historically been associated with reduced recruitment (Fig. 6) and, in conjunction with high fishing mortality, the increased likelihood of stock collapse.

The TAC was only applied to the directed fishery for herring, mainly for human consumption, in the North Sea. The only constraint on the catch of herring in the small-mesh fisheries, targeted mainly at sprat, was

Table 3. Catch option table for the five fleets that exploit the North Sea autumn-spawning herring, based on the short-term predictions for 1998.

Scenario	F_{juv} (0-1 ring[a])	F_{adult} (2-6 ring)	Fleet Fs $F_{B-E,0-1}$	$F_{A,2-6}$	A	B	C	D	E	Yield	1998
	0.047	0.246	0.046	0.215	254	22	24	6	5	311	1,145
Prediction summary: yields 1999 assuming TAC in 1998
I	0.039	0.2	0.039	0.174	270	19	21	7	3	320	1,518
II	0.047	0.2	0.047	0.172	265	24	21	9	4	322	1,518
III	0.048	0.246	0.048	0.215	325	24	25	9	4	386	1,471
IV	0.06	0.25	0.059	0.215	324	30	25	11	5	395	1,467
V	0.09	0.25	0.089	0.204	309	46	24	17	8	403	1,468
VI	0.12	0.25	0.121	0.193	292	62	23	22	11	410	1,469
VII	0.048	0.25	0.048	0.219	331	23	26	9	4	392	1,467

F-multipliers on all fleets maintained for TAC ratios 1998. Predictions for 1998 based on $F_{98} = F_{97}$ in 1998 (from ICES 1998).
Scenario I Decrease F on all fleets in the same proportion to bring $F_{adult} = 0.2$ and $F_{juv} < 0.1$.
Scenario II Decrease F on fleets A & C to bring $F_{adult} = 0.2$ and $F_{juv} < 0.1$.
Scenario III Status quo F on all fleets.
Scenario IV Increased F_{BDE} until $F_{adult} = 0.25$ Fleet A & C status quo, Fleets B, D & E increased.
Scenario V Increased F_{BDE} until $F_{juv} = 0.09$ Fleet A & C reduced until $F_{adult} = 0.25$, Fleets B, D & E increased.
Scenario VI Increased F_{BDE} until $F_{juv} = 0.12$ Fleet A & C reduced until $F_{adult} = 0.25$, Fleets B, D & E increased.
Scenario VII Increased F_{AC} until $F_{adult} = 0.25$ Fleet A & C increased until $F_{adult} = 0.25$, Fleets B, D & E status quo.
[a]0-1 ring and 2-6 ring refers to the number of "winter" rings on the otolith. Autumn-spawned herring do not lay down a ring in their first winter. As a result the number of winter rings on the otolith does not correspond directly to the age of the fish.

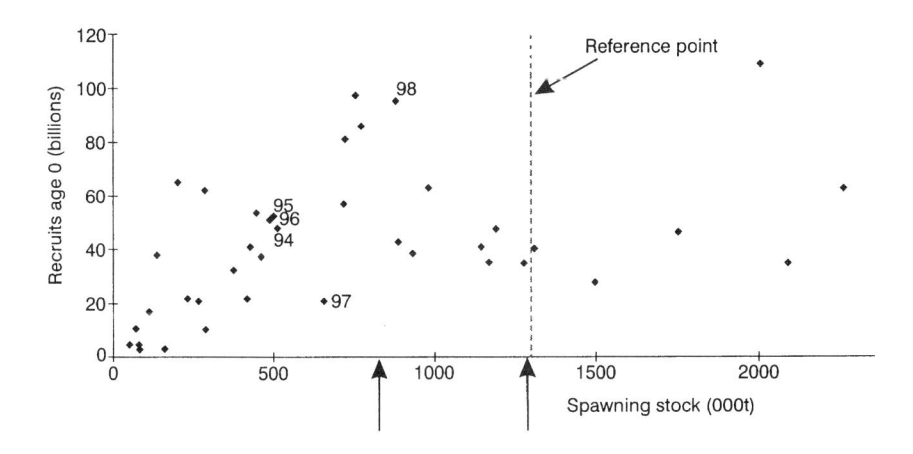

Figure 6. The relationship between spawning stock size and recruitment at age 0 for North Sea autumn-spawning herring with the new reference point for minimum spawning stock biomass

a 10% bycatch limit and some area closures to protect both spawning and juvenile herring. These sprat fisheries are mainly in the central North Sea, and take bulk catches for industrial purposes. The seasonal area closures, which currently apply, restrict fishing off the east coast of England and off the west coast of Denmark.

The landings of autumn-spawned herring from the North Sea regularly exceed the TAC (Table 2) and the total North Sea SSB was seen to be declining from 1990. In spite of this the management advice in 1993 considered the stock to be fairly stable and well within safe biological limits. The very high catches of juveniles was acknowledged and an impact assessment was carried out to examine the effect of preventing all fishing on juveniles. This study indicated the yield would increase by 9% and that the increase in SSB would be approximately 43% (ICES 1993). The TAC eventually agreed for 1994 was forecast to result in an SSB of over 1 million t at spawning time in 1994.

In 1994 the perception of the SSB in 1993 changed and it was believed to be below the minimum biologically acceptable level of 800,000 t (Table.1). The decreasing trend in SSB had been further exacerbated by an unexplained decrease in the mean weight, and proportion mature, of the youngest year classes. The ICES advisory committee also noted that the catches of juveniles had reached the high levels observed in the early 1980s and that this pattern of exploitation would seriously impede the recovery of the spawning stock to safe levels.

The advice for the TAC for 1995 was delayed until October 1994 because of the uncertainties about mean weight and maturity. By then the perception of the stock had changed and it was considered to be above the minimum biologically acceptable level. A reduction in fishing mortality was advised in order to improve the long-term yield. The TAC of 440,000 t set for 1995 was based on the prediction that SSB would continue to be above the minimum biologically acceptable level at spawning time in 1995.

Early in 1995, ICES scientists found that the catches of North Sea herring in 1994 had continued to exceed the TAC and that the SSB had in fact fallen below the minimum acceptable level of 800,000 t. In October 1995, acting on the new information, the ICES ACFM recommended a severe reduction in the TAC for 1996, equivalent to about half the 1995 level. They also recommended that fishing mortality on herring be reduced by at least 50% in the small-mesh fishery for sprat in the North Sea, and in other fisheries that take North Sea herring in the Skagerrak and Kattegat.

The TAC eventually agreed between the EU and Norway for 1996 was 313,000 t, some 80,000 t above the ACFM recommendation. A new assessment early in 1996 showed that the SSB had further declined to less than 500,000 t at spawning time in 1995 (ICES 1996). This was 225,000 t below the forecast from the previous assessment. Concern about this decline, the error in the perception of SSB, and the continuing high fishing mortality rate on juveniles, led to a revision of the ACFM recommendations. In May 1996 the committee advised that urgent, unprecedented action be taken to reduce the TAC for that current year in order to avoid a repeat of the stock collapse of the 1970s. They recommended that the total catch of North Sea autumn-spawning herring should not exceed 298,000 t in 1996 and that the catches by all fleets exploiting this stock should count against this figure. This advice implied an immediate 50% reduction, in the already agreed 1996 TAC for the North Sea directed fishery, to 156,000 t, and a herring bycatch ceiling for the small-mesh fisheries in the North Sea of 44,000 t. They also recommended that as little as possible of the TAC should be taken from the southern North Sea and English Channel areas.

Fishermen faced the very real prospect of the North Sea herring fishery being closed in 1997 unless drastic measures were implemented. The advice was acted on quickly by the Council of Fisheries Ministers of the European Community in conjunction with Norway. As a result immediate action was enforced to halve the TAC for that year and to introduce a bycatch ceiling for herring in the sprat industrial fishery.

In addition to the revised advice for 1996 the ACFM went on to recommend that fishing mortality on all fleets should be reduced by 75% and linked that to an $F_{age\ 2-6} = 0.2$ for the 1997 season. Furthermore, if the catch in 1996 exceeded the agreed reduction, then the North Sea herring fishery should be closed in 1997.

The severe action led to a reduction in exploitation on all age groups. Despite this success the total catch in the North Sea directed fishery in

1996 exceeded the TAC by 65,000 t. A total of 50,000 t was taken from the Downs stock. It was known that some of the over-quota catch of herring had been taken in the North Sea but misreported into other areas. Such misreporting has traditionally been attributable to a number of countries and fishing areas, with the North Sea herring variously reported as West of Scotland herring, Atlanto-Scandian herring, and herring from the Skagerrak and Kattegat. Rigorous licensing arrangements were introduced by some countries to curb this misreporting.

Although the agreed TAC in 1996 was exceeded, the 1997 TAC of 159,000 t in the directed fishery and the bycatch ceiling of 24,000 t were retained for the 1997 fishery. This was because the restrictions in the other four fleets had worked well. The total catch of North Sea autumn spawners had only exceeded the recommended level of 298,000 t by 9,000 t. Furthermore, there was evidence of strong recruit year classes from 1994 to 1996 which would eventually contribute to the SSB.

The advice for 1998 was that the 1997 strategy should be continued. This strategy was considered to be consistent with the objective of keeping the SSB above 800,000 t in the short term and providing an SSB at spawning time in 1998 of 1.2 million t, an increase of around 0.5 million t over the predicted value for 1997. The strict management measures had begun to work, and above-average year classes were entering the fishery. Maintaining this management strategy led to an increase in the TAC for 1998 to 254,000 t in the directed fishery and a bycatch limit of 22,000 t in the other North Sea fisheries.

The ACFM advice in May 1998 for the 1999 season took account of new management objectives which had been defined and agreed at the bilateral meeting between the EU and Norway. The objective reiterated the minimum biological level below which the SSB should not be allowed to fall but also defined a new SSB reference point of 1.3 million t. Above this reference point the stock could be harvested at levels consistent with a fishing mortality of $F = 0.25$ for adult herring and $F = 0.12$ for juveniles. When the SSB is below that reference point then other measures should be agreed consistent with increasing the SSB. This strategy was considered by ICES to be consistent with the new *precautionary approach* to fisheries management then under discussion within ICES.

The recorded landings from the North Sea directed fishery continued to exceed the agreed TAC in 1997 and again in 1998, thus impeding the rapid recovery of the stock to above the newly defined reference point. In 1997 the overshoot on the total recommended catch of North Sea autumn spawners was 26,000 t increasing to 53,000 t in 1998. In both years most of the overshoot was entirely attributable to the North Sea directed fishery.

The TAC advice for 1999 was based on continuing the strategy to increase the SSB to above the new reference point from its predicted level of 1.15 million t in 1997. The recommended fishing mortality of $F = 0.2$ for adults was continued and a recommendation to keep the fishing mortality on age classes 0 and 1 to below $F = 0.1$ was included for the first time. This resulted in an agreed TAC for the directed fishery of 265,000 t and a 30,000 t bycatch in the other North Sea fisheries.

At the present time the total catch of North Sea autumn-spawning herring for 1999 is not known (see Appendix). The advice for the management of the fishery for the year 2000 was to continue the measures agreed for 1999. It was based on a likely overshoot of the 1999 TAC of 55,000 t in the North Sea directed fishery. A series of options were provided for the various fleets, consistent with this strategy. The TAC eventually agreed for 2000 was for 265,000 t in the North Sea directed fishery and a North Sea bycatch limit of 36,000 t.

While the TAC regulations in recent years have been based on the total North Sea autumn-spawning herring there has been separate consideration for the state of the Downs stock component which spawns in the southern North Sea and eastern English Channel. This stock has shown independent trends in exploitation and recruitment but cannot be assessed separately. An area TAC constraint has been ineffective in controlling the fishing mortality on this stock with the agreed TAC being exceeded in every year since 1987 (Table 4). From 1991 to 1996 the area TAC was set at 50,000 t but this was reduced to 25,000 t midway through the 1996 season in line with the emergency action taken for the rest of the North Sea stock complex. At the same time the recommendation from ICES was that no directed fishery for herring should take place in this area. By that time the Downs stock was considered to be outside safe biological limits and at its lowest level since 1980. An annual TAC of 25,000 t has been set since 1997 but the recorded catches have been around 50,000 t per year.

The only indication of the state of this component has been the annual larvae surveys carried out during December and January. Up to 1994 the larvae surveys indicated stability in the SSB in this area but in 1995 they indicated the lowest SSB since 1980. Since 1996 the surveys have indicated a slow increase in the SSB.

Discussion and Future Prospects

It is clear from the recent history of the North Sea herring fisheries that many of the problems can be traced back to overexploitation of the juvenile component of the stocks. Pressure on the juvenile component began in the late 1950s with the advent of the meal and oil fisheries and has continued to the present day. During the 1960s the proportion of juvenile herring in the catches was less than 50% but this increased to over 70% in the 1970s. Since then the contribution of juvenile fish has remained high despite the total ban on directed fishing for herring for industrial purposes in 1976. This is partly attributable of course to the erosion of the older year classes from the population. Recent assessments have shown that the proportion of juveniles in the catch has reduced considerably to less than 50% again, confirming the success of the new management measures.

Table 4. The agreed TAC and the recorded catch of herring from the Southern North Sea and Eastern English Channel (Downs stock component) for the period 1981-1998.

Year	Agreed TAC (t)	Catch (t)
1981	20,000	46,000
1982	72,000	72,000
1983	73,000	64,000
1984	55,000	46,000
1985	90,000	70,000
1986	70,000	51,000
1987	40,000	45,000
1988	30,000	52,000
1989	30,000	79,000
1990	30,000	61,000
1991	50,000	61,000
1992	50,000	74,000
1993	50,000	85,000
1994	50,000	74,000
1995	50,000	63,000
1996	25,000[a]	50,000
1997	25,000	51,000
1998	25,000	48,000
1999	25,000	*54,000*

[a]Reduced in midseason from original 50,000 t.
Note: 1999 landings figure, in italics, added later.

Prior to the moratorium on fishing in 1977 there was no effective control, other than market forces, on the catches of North Sea herring. This was undoubtedly the result of the lack of an international perspective on the need for specific control measures. The shock waves which the subsequent moratorium sent through the industry helped to ensure that some stricter controls would be in place before the fishery was reopened. The specific regulations implemented were limited to a TAC, which was only applied to the North Sea directed fishery, a 10% herring bycatch limit in the North Sea small-mesh fisheries and some area and seasonal closures of herring fishing in the central North Sea.

Over the following years a series of good recruitments and enforcement of the regulations resulted in a steadily increasing SSB. However, the warning signs began appearing with fishing mortality going up and the annual catch increasing to over 800,000 t in 1988, of which 78% by number were juveniles. The management strategy was proved ineffective in controlling the fishing mortality on juveniles. The harvest trend was dictated mainly by the size of the sprat catch in the small-mesh fisheries. Some area closures for this fishery, off the west coast of Denmark and the northeast coast of England, began to introduce some measure of control in the early 1990s but mortality remained high. While the fishing mortality on adults over the period 1991-1995 increased from $F = 0.49$ to $F = 0.8$ the mortality on juveniles was more stable between $F = 0.23$ and $F = 0.3$.

The management problem at this time was hampered by unreliable estimates of the catch and problems with the assessment methodology. These factors combined to produce a poor perception of the actual stock size, resulting in the annual regulatory advice being based on an overestimate of SSB. This was the scenario that ultimately led to the advice for the 1996 fishery being rescinded midway through the season and the severe action being taken to halve the TAC. At this time the fishery faced the very real danger of another complete moratorium on directed fishing. This provided the catalyst for a serious reexamination by the EU and Norway of the methods for the management of the North Sea herring.

The resultant management strategy is based on the new precautionary approach which is applied to stocks within the ICES area. It is targeted at achieving a high probability that the SSB will not fall below 800,000 t but will in the short term increase to above a new reference limit of 1.3 million t. Until that reference point is reached the F on adult herring should not be greater than $F = 0.2$ and should be less than $F = 0.1$ on the juveniles. A proposed strategy for harvesting the stocks at SSB levels above the reference point would permit the fishing mortality on adults to increase to $F = 0.25$ and on juveniles to $F = 0.12$.

It is fair to conclude that the stringent management measures introduced since the 1996 crisis have begun to have a positive effect. The fishing mortality on juvenile herring in the North Sea has been reduced. The proportion of immature fish in the catch in 1997 and 1998 decreased to less than 50% compared with 70-80% observed in previous years. This success is entirely attributable to the strict enforcement of the bycatch limit, in the small-mesh fishery for sprat in the eastern North Sea. Catches of North Sea autumn-spawning herring in the Skagerrak and Kattegat have also remained at or below the predicted levels. However, the TAC in the directed fishery in the North Sea is still being exceeded and it is a cause for some concern that this overshoot increased again in 1998. This contributed to the perception of SSB being lower than that predicted in the previous year by 267,000 t.

New management measures such as designated ports for landing and satellite tracking of vessels are being introduced in an attempt to cut down on misreporting, seen as one of the main causes of the TAC overshoot in recent years.

On a more optimistic note, surveys indicate that following a very poor recruitment from the 1997 year class the early indications are that the 1998 year class may be the second highest for over 30 years.

In addition to the TAC and bycatch limit, other regulations to protect herring will continue. These include the seasonal closures to protect spawning on the east coast of England and the area closures for small-mesh fisheries where juvenile herring are likely to be abundant.

The most recent report of the ICES herring assessment working group for 1999 (ICES 1999) states that "Projections show that this stock (North Sea herring) may have a positive development in the near future up to 1.17 million t in 1999." If the assumptions in the short term (about recruitment and the catch in 1999) are correct and if the strategy agreed by the parties dealing with this stock is followed (EU-Norway agreement) an SSB level of about 1.3 million t may be reached at spawning time in 2000 (see Appendix). The agreed TAC for 2000 does indeed follow that advice and is projected to result in an SSB of around 1.3 million t at spawning time in 2000.

The prospects for the North Sea stock complex in general are improving but there has still been no recovery of the former spawning grounds around the edge of the Dogger Bank. There is also a continuing need for vigilance over the state of the southern component, the Downs stock. This stock component is not following the same pattern of recovery as those in the central and northern North Sea and the larvae surveys in recent years have indicated only a small improvement in the state of this stock.

Management of the North Sea herring is now firmly based on the new precautionary approach with upper and lower reference points for both SSB and fishing mortality. This coupled with stricter enforcement measures should ensure that the stock does not fall to the crisis levels seen in the mid-1970s and again in 1995.

References

Cushing, D.H. 1955. On the autumn-spawned herring races of the North Sea. J. Cons. Perm. Int. Explor. Mer 21:44-60.

Cushing, D.H., and J.P. Bridger. 1966. The stock of herring in the North Sea and changes due to the fishing. Fishery Invest. Lond. (Ser. II) 25(1). 123 pp.

Harder-Jones, F.R. 1968. Fish migration. Edward Arnold, London. 325 pp.

ICES. 1991. Report of the Herring Assessment Working Group for the Area South of 62°N. ICES C.M. 1991/Assess:15.

ICES. 1992. Report of the Herring Assessment Working Group for the Area South of 62°N. ICES C.M. 1992/Assess:9. 114 pp.

ICES. 1993. Report of the Herring Assessment Working Group for the Area South of 62°N. ICES C.M. 1993/Assess:15. 245 pp.

ICES. 1994. Report of the Herring Assessment Working Group for the Area South of 62°N. ICES C.M. 1994/Assess:13. 249 pp.

ICES. 1996. Report of the Herring Assessment Working Group for the Area South of 62°N. ICES C.M. 1996/Assess:10.

ICES. 1998. Report of the Herring Assessment Working Group for the Area South of 62°N. ICES C.M. 1998/ACFM:14.

ICES. 1999. Herring Assessment Working Group Report. ICES C.M. 1999/ACFM:12. 391 pp.

Parrish, B.B., and A. Saville. 1965. The biology of the north-east Atlantic herring populations. In: H. Barnes (ed.), Oceanography and marine biology. An annual review. Vol. 3. George Allen and Unwin Ltd., London, pp. 323-375.

Patterson, K.R., and G.D. Melvin. 1996. Integrated catch at age analysis, Version 1.2. Scottish Fisheries Research Report No. 38. Aberdeen.

Postuma, K.H. 1956. Meeting on herring races. ICES C.M. 1956:16. 8 pp.

Zijlstra, J.J. 1958. On the "races" spawning in the southern North Sea and English Channel. Rapp. P.-V. Reun. Cons. Perm. Int. Explor. Mer 143:134-145.

Zijlstra, J.J. 1963. On the recruitment mechanisms of the North Sea autumn spawning herring. Rapp. P.-V. Reun. Cons. Perm. Int. Explor. Mer 154:198-202.

Zijlstra, J.J. 1969. On the "racial" structure of North Sea autumn-spawning herring. J. Cons. Int. Explor. Mer 33:67-80.

Appendix

(This appendix was produced after presentation of the paper.)

The ICES working group assessment in March 2000 showed that the total catch of North Sea autumn-spawning herring in 1999 exceeded the TAC by 41,000 t (Table 2). The estimate of SSB at spawning time in 1999 was 265,000 t below the predicted value. The new assessment showed that the 1998 SSB had been overestimated by 177,000 t and had actually been below the minimum acceptable level at 701,000 t (Table 1). The main reasons for the downward correction were an overestimate of the strength of the 1996 year class and a reduction in the mean weights at age in the stock.

Because of the reduced estimate of SSB and continuing overshoot of the TAC, the fishing mortality on adult herring continued to be well above the recommended level of $F_{age2-6} = 0.2$ (1998, $F = 0.45$; 1999, $F = 0.38$). The SSB in 2000 is now expected to be around 900,000 t. Recovery of the stock, to above the 1.3 million t reference point, in the short term, is heavily dependent on the 1998 year class being as strong as current predictions and also on reducing the continuing trend of overshooting the TAC.

Addendum (July 2001)

At the most recent ICES working group meeting, in March 2001, it became clear that the fishing mortality on adults in 2000 ($F_{age\ 2-6}$ of 0.42) had been well above the target $F_{age\ 2-6}$ of 0.2. The mortality on juveniles still remains low.

As a consequence of the high adult fishing mortality and a smaller than predicted 1996 year class, the SSB fell to 772,000 t in 2000, which is below the precautionary biomass limit of 800,000 t.

In spite of more rigorous enforcement, area misreporting was still a major problem and the TAC in the North Sea in 2000 had been exceeded by 28,000 t.

The working group admitted that overly optimistic assessments have contributed to TACs that in retrospect have delayed the recovery of the stock. They have given an assurance that the current assessment and recommended advice for 2002 does not suffer from the same problem.

Current recommended advice should ensure an SSB of 1.1 million t in 2001 increasing to 1.7 million t in 2002.

A New Approach to Managing a Herring Fishery: Effort vs. Quota Controls

Denis Tremblay
Fisheries and Oceans Canada, Quebec City, Quebec, Canada

Abstract

Following the collapse of the spring herring fishery in the Gulf of St. Lawrence in the early 1980s, all spawning areas recovered within a few years except for the Magdalen Islands (Fig. 1). TACs are established for both spring and fall spawners using standard analytical assessment techniques such as virtual population analysis. Allocations are then provided to each area based on recent historical landings. In 1997 spring spawning started to occur again in the Magdalen Islands, and has increased steadily since then. Because of a lack of recent history, they could not be given a sufficient allocation. This paper describes how we were able to get around the allocation process to allow a fishery to be prosecuted without compromising our conservation objectives while remaining cautious overall. It also describes how we are working to get better assessments of local abundance within the stock area.

Introduction

The Magdalen Islands spring herring fishery was once a major component of the southern Gulf of St. Lawrence (Northwest Atlantic Fisheries Organization Div. 4T) fishery (Fig. 1). Statistical data for this fishery have been collected since 1917 (Fig. 2). The period is characterized by five peak years: 1918, 1959, and 1967-1969, when landings exceeded 20,000 t. Those periods coincide with the influx of major year classes in the fishery. In 1965 a large purse seine fleet coming from both the Atlantic and Pacific Canadian Coast started the exploitation of herring in the Gulf of St. Lawrence. Overexploitation and poor recruitment lead to the collapse of that fishery in the early 1980s.

Prior to the end of the 1960s the fishery was principally prosecuted with trap nets. The herring were processed as bloaters to be exported, mostly to Caribbean countries and, to a lesser extent, were used as lobster

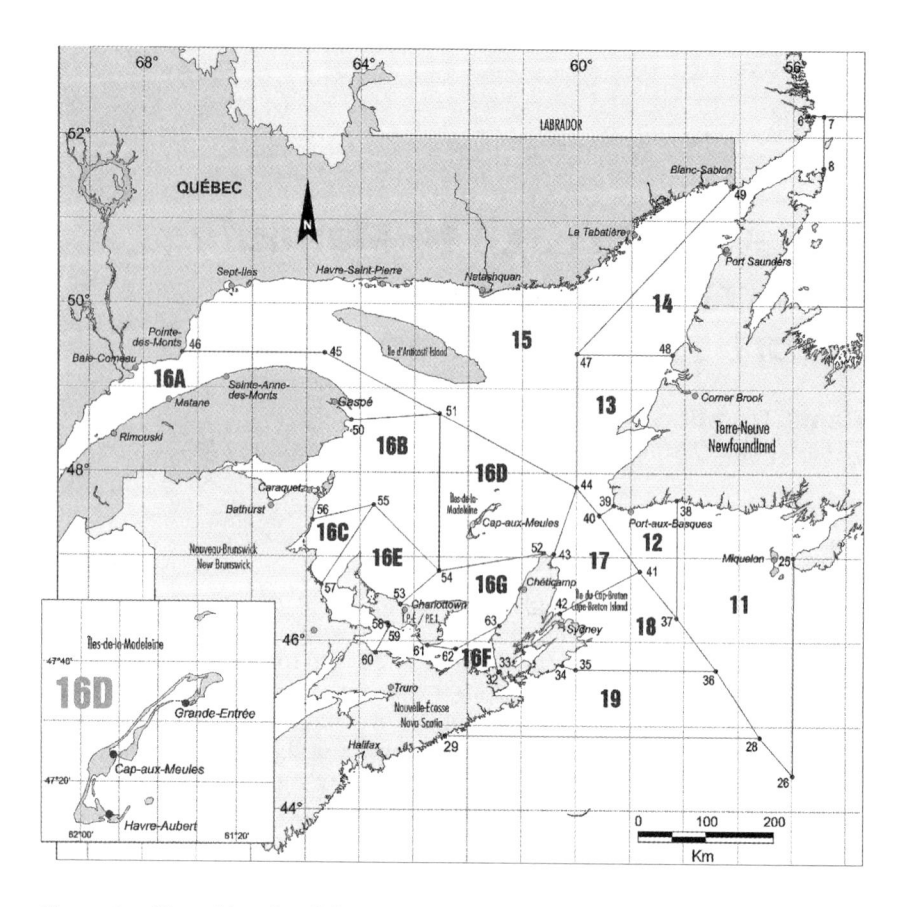

Figure 1. Map of herring fishing areas in the Gulf of St. Lawrence.

bait. Between 1965 and 1980 purse seine catches were mainly processed as fish meal and oil (Integrated Fisheries Management Plan 1998).

In the 1980s, TACs were introduced for both fall and spring components of the fishery. Allocations to local areas, such as the Magdalen Islands, were provided, based mainly on recent catch history (3-5 years) in the mid-1980s. Because the herring resource around the Magdalen Islands had not recovered to previous levels, this area received a very small allocation. In the mid-1990s herring started to show up in greater abundance around the islands, but fishing activity was reduced due to the low allocation provided. This fishery is now totally prosecuted with gillnets.

This paper will explain how both scientists and managers arrived at a method to allow catches to increase in the Magdalen Islands without having to reallocate from other areas. That method should be used until bet-

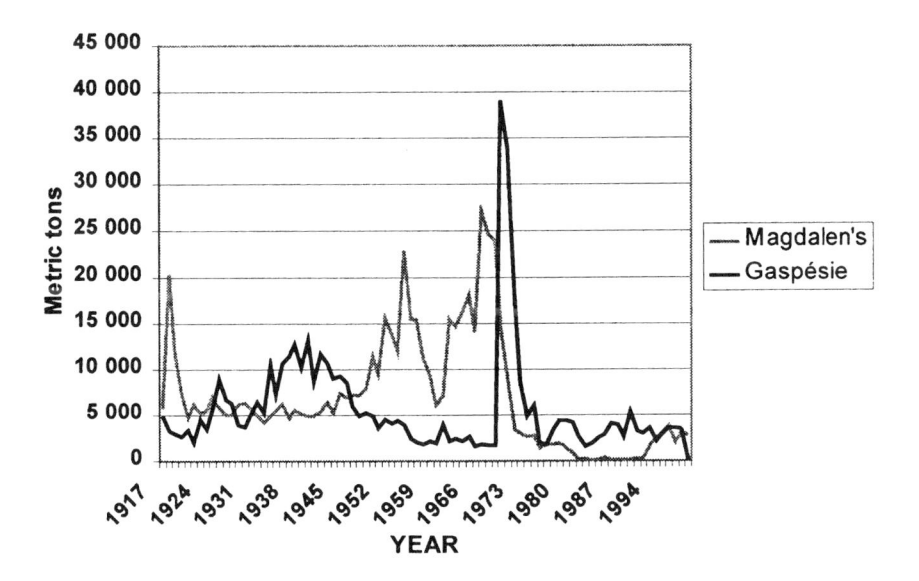

Figure 2. Herring landings in the Magdalen Islands from 1917 to 1999.

ter knowledge of local abundance variability can be obtained using standard techniques.

In absence of any other means, fishing effort is used to regulate fisheries (Hilborn and Walters 1992). Even though the 4T stock is analytically assessed as a whole, the very limited fishing activities that took place around the Islands did not allow for any local abundance estimation. Based on recent and past landings local variations, it was hypothesized that local abundance was somewhat asynchronous (Figs. 3 and 4). Therefore by allowing more to be taken in an area than its initial allocation, there would be a low risk of damaging the population overall.

Instead of going back to the negotiation table, which can be unproductive and difficult in an attempt to reallocate the TAC among local areas, it was decided to use an effort target instead of a straight quota.

Method

The target was established by using the average effort (net-days) of the last three years (1996-1998) coupled with a target catch that should not exceed twice the initial quota allocated. The effort for 1996-1997 is derived from a telephone survey done annually covering approximately 25% of the active fishermen. In 1998, the survey reached 98% of the active

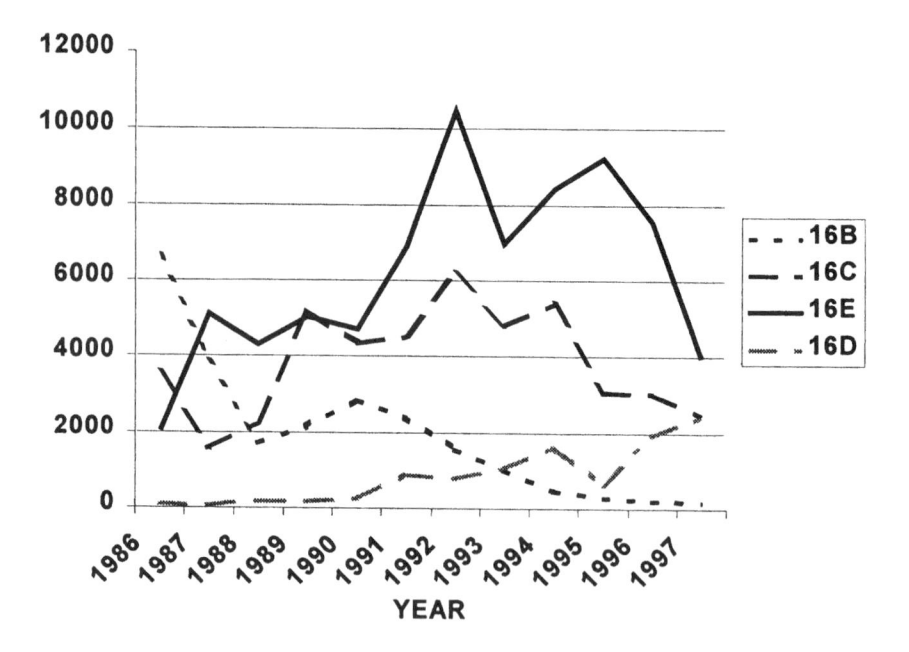

Figure 3. Herring landings in herring areas 16B to 16E (1986-1998).

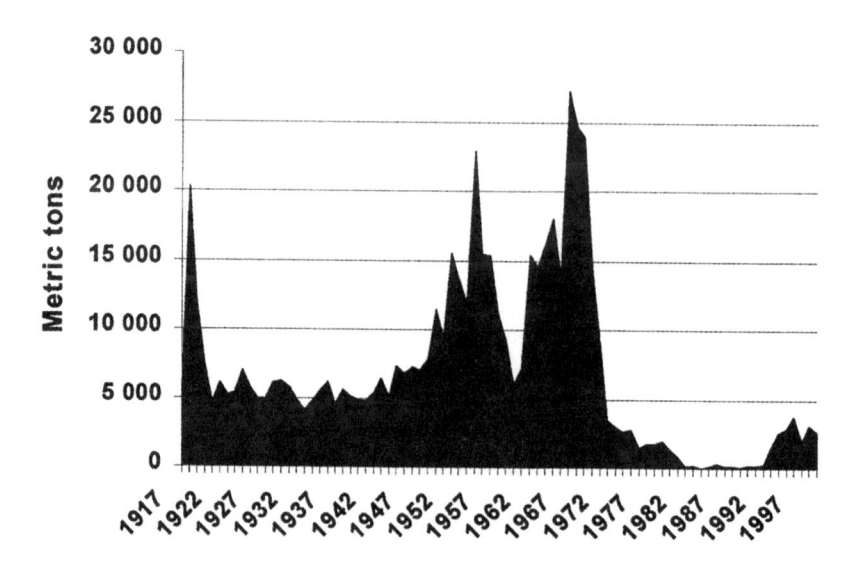

Figure 4. Herring landings in the Magdalen Islands and Gaspesie (Gaspé Peninsula) (1917-1998).

fishermen in an effort to gather more precise information prior to the implementation of the project.

Experimental gillnets of various mesh sizes were also used throughout the fishing season, and a commercial vessel gathered acoustic data recorded on a portable computer to be analyzed later. In order to properly control fishing activities in 1999, fishermen were required to report both the catch and number of nets they used in any trip. Catches were randomly verified, in at least 25% of the cases, by an independent dockside observer. Monitoring at sea was performed by fishery officers to validate the effort data.

Results

The targets were established at 6,400 net-days for the effort and 2,500 t for the quota. The spring fishery began on April 11 and ended on April 18. The number of net-days was estimated at 6,700 and the catch at 2,350 t at the time of closure. Final figures showed a total of 7,131 net-days for a catch of 2,602 t, the difference being explained by the fact that 10 fishermen did not report during the fishing season. One of the concerns expressed by our biologists was that multiple day trips could lead to overexploitation. The detailed data showed that multiple trips were counted in 1999, which was not necessarily the case in previous years. In 1999 a total of 467 trips were made, out of which 42 were multiple day trips.

In comparison 8,089 net days were needed to catch 1,702 t in 1998, while in 1997 only 3,724 net days were used to catch 1,521 t. Table 1 provides a short description of the fishery between 1996 and 1999. Catch per unit of effort (CPUE) increased by 60% from 1998 to 1999, but showed a slight decrease compared to 1997 (16%) in the Magdalen Islands. In comparison with other areas of the southern Gulf of St. Lawrence, such as Baie des Chaleurs (16B), CPUEs have decreased steadily since the early 1990s while in area 16C the decline started in 1997 (Claude Leblanc, DFO, Moncton, pers. comm.).

The use of six different mesh sizes (51-70 mm) provided useful information about the size distribution of the population (Sylvain Hurtubise, DFO, Mont-Joli, pers. comm.). There are three peaks in the size distribution of the herring sampled (Fig. 5), one at 27 cm, one at 32 cm, and one at 33 cm. Commercial gillnets are 57 mm in mesh size according to federal regulations.

Concurrently, an acoustic survey was performed before, during, and after the fishing season for a total of 40 days. Herring were seen every day in variable concentrations. The results have not yet been analyzed, but it is hoped that they will provide a useful relative index of abundance in the near future.

Table 1. Description of the Magdalen Islands herring fishery between 1996 and 1999.

	1996	1997	1998	1999
Number of gillnet licenses	341	341	341	340
Effort (net-days)	7,270	3,784	8,089	7,131
Landings (t)	1,615	1,521	1,702	2,602
Number of fishing-days	25	18	19	11
Average daily effort (nets)	291	210	426	648

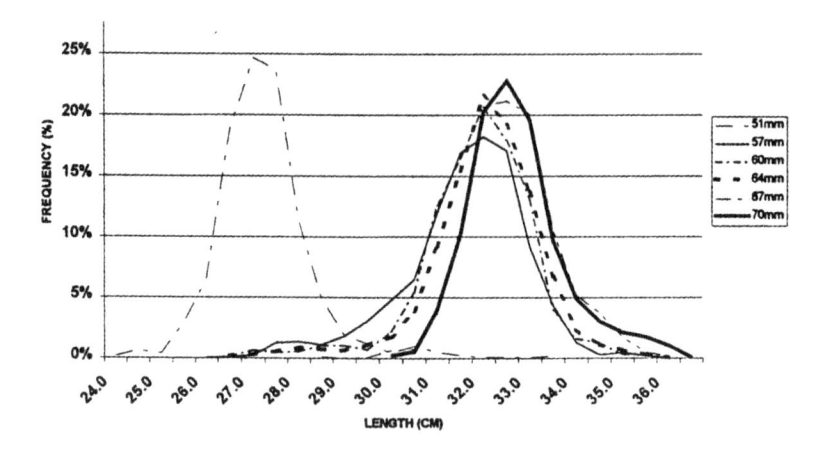

Figure 5. Herring length frequencies in the 1999 experimental fishery.

Discussion

Results from the 1999 fishing season are promising. The control mechanisms that were put in place allow for a very conservative approach by ensuring that all trips are counted and all the effort is taken into consideration. The methodology used seems to fit well for fisheries where the effort can be easily calculated and where changes in gear efficiency over a short period of time are nonexistent. Effort control should only be used on a short time basis, until such time as a conventional approach can be developed.

Tools include recruitment forecasting using CPUE and age compositions from the experimental gillnets, local abundance estimates determined by acoustic surveys, and age structured models to assess overall local abundance independently.

Finally, given the success of the pilot project that was used in only one area of the southern Gulf of St. Lawrence, it is likely that political pressures will be exerted again, this time to apply the approach to all areas of NAFO Division 4T or else to go back to the previous management regime.

Acknowledgment

I wish to thank Mr. Ghislain Chouinard from DFO in Moncton who provided the idea for this project. I also thank Mr. Claude Leblanc, the herring biologist responsible for the assessment of 4T herring, who provided statistical and biological data for the manuscript. He also coordinated the training for the collection of acoustic data from the industry survey, as well as the collection of biological data from the experimental gillnet survey. Many thanks to all personnel from DFO and the Province of Quebec, who actively participated in the implementation of the project. To name a few, I would like to thank Sylvette Leblanc, Carole Turbide, Sylvain Hurtubise, Odile Légaré, and Francis Coulombe. In closing I am indebted to all industry representatives and fishermen from the Magdalen Islands without whom this study would not have been possible.

References

Cleary, L. 1984. The herring *(Clupea harengus harengus)* gillnet fishery of the Magdalen Islands. Results of a fishermen survey (1980, 1981 and 1982). Can. Tech. Rep. Fish. Aquat. Sci. 1244. (In French.)

Cleary, L., and J. Worgan. 1981. Changes in effort, catch-per-unit-effort and biological characteristics in the inshore herring fishery at the Magdalen Islands (1970-1980). CAFSAC (Canadian Atlantic Fisheries Science Advisory Committee) Research Document 81/32.

Hilborn, R., and C.J. Walters. 1992. Quantitative fisheries stock assessment choice: Dynamics and uncertainty. New York, London: Chapman and Hall, 1992, pp. 453-470.

Hodder, V.M., and L.S. Parsons. 1970. A comparative study of herring taken at the Magdalen Islands and along southwestern Newfoundland during the 1969 autumn fishery. ICNAF (International Commission of Northwest Atlantic Fisheries) Research Document 70/77.

Integrated Fisheries Management Plan. 1998. Integrated Fisheries Management Plan: Herring Area 16 1998. Department of Fisheries and Oceans Canada, p.3.

Messieh, S.N. 1989. Changes in the Gulf of St. Lawrence herring populations in the past three decades. NAFO (Northwest Atlantic Fisheries Organization) SCR (Scientific Commission Report) Document 89/74.

Spenard, P. 1979. The herring fishery in the Magdalen Islands, 1900-1978. Rapport technique (Canada, Service des pêches de la mer) 876F. (In French.)

Industry Acoustic Surveys as the Basis for In-Season Decisions in a Comanagement Regime

Gary D. Melvin, Robert L. Stephenson, Mike J. Power, F.J. Fife, and Kirsten J. Clark
Department of Fisheries and Oceans, Biological Station,
St. Andrews, New Brunswick, Canada

Abstract

This paper examines the success of several innovative assessment and management procedures, which have been implemented in the Canadian 4WX herring fishery in recent years. Major progress has been made in the quantification of industry information and in the use of this information as the basis for assessment, through the development of quantitative acoustic surveys for both the purse-seine and gillnet fleets. Industry vessel acoustic systems have been calibrated and are used to record quantitative data from both surveys and fishing operations. Rapid analysis of the results allows in-season decisions on the spatial and temporal suballocation of the total allowable catch (TAC).

Recognizing that the management area contains several spawning components, and that these are subject to erosion through disproportionate fishing effort, management of the fishery now explicitly protects the spatial and temporal diversity of spawning through the use of pre-fishing surveys and the limitation of the fishery to a small fraction of documented biomass. The use of quantitative, industry-based, acoustic equipment is critical to the accurate determination of the size of these spawning components.

A collaborative management approach was implemented initially in 1995 in response to considerable uncertainty and concern about resource status. This approach involves numerous in-season management meetings and conference calls to review biological signals in the fishery, and it allows rapid within season management adjustments when required. This approach has demanded a high level of involvement and cooperation from all participants in the management process, and has led to improved effectiveness of management and care for the resource.

Introduction

Currently, the 4WX herring stock complex is divided into four management units: the New Brunswick coastal component, the SW Nova Scotia/Bay of Fundy component, the 4VWX coastal component, and the outer Scotian Shelf Banks component (Fig. 1). Of these, only the SW Nova Scotia/Bay of Fundy component is subject to an extensive annual assessment and to quota restrictions (Stephenson et al. 2000). In 1995 the outlook for the 4WX herring stock was less than optimistic. There was widespread concern about the near absence of herring from some of the main spawning areas, poor (i.e., low) fat content, and the lack of older fish in the 1994 fishery (Stephenson et al. 1995). These concerns were reiterated at the 1995 assessment meetings and supported by a strong downward trend in the VPA analysis (Stephenson et al. 1996). Representatives of herring industry also expressed uneasiness about the level of effort being exerted on individual spawning components within the stock complex. It was postulated that although the stock is managed on a global TAC, individual major and minor spawning components may be adversely affected, even with a substantially reduced quota, by concentrated fishing effort. Past experiences, such as the collapse of Trinity Ledge spawning grounds, the 1993-1994 decline in availability of herring on German Bank, and the rapid decrease in catches from the Little Hope area, have clearly demonstrated the negative effects of concentrated fishing effort on individual components (Stephenson et al. 1997).

To address these concerns the herring industry and the Department of Fisheries and Oceans (DFO) implemented several initiatives. Between 1994 and 1996 the total allowable catch (TAC) was reduced from 151,000 t to 57,000 t (Fig. 2) and a system of within-season monitoring was implemented. This "in-season" management involved industry-based surveys, increased sampling, changes to data-processing procedures for rapid summary of results, and regular discussions (weekly) with stakeholders (Stephenson et al. 1999). Key to the in-season management was the development of an automated acoustic logging system (AALS) for deployment on commercial fishing vessels and the "Survey, assess, then fish" protocol with target levels based on historical landings from the main spawning components within the stock. The basic principle was that sufficient quantities of fish had to be observed or documented on the spawning grounds to warrant fishing. This provided an additional level of protection for individual spawning components within the global TAC. The purpose of this paper is to describe the development, implementation, and current functioning of these initiatives involving acoustic surveys by the herring industry.

Initial Phase

Prior to the commencement of the 1995 summer/fall herring fishing season, several meetings were held with vessel captains, association repre-

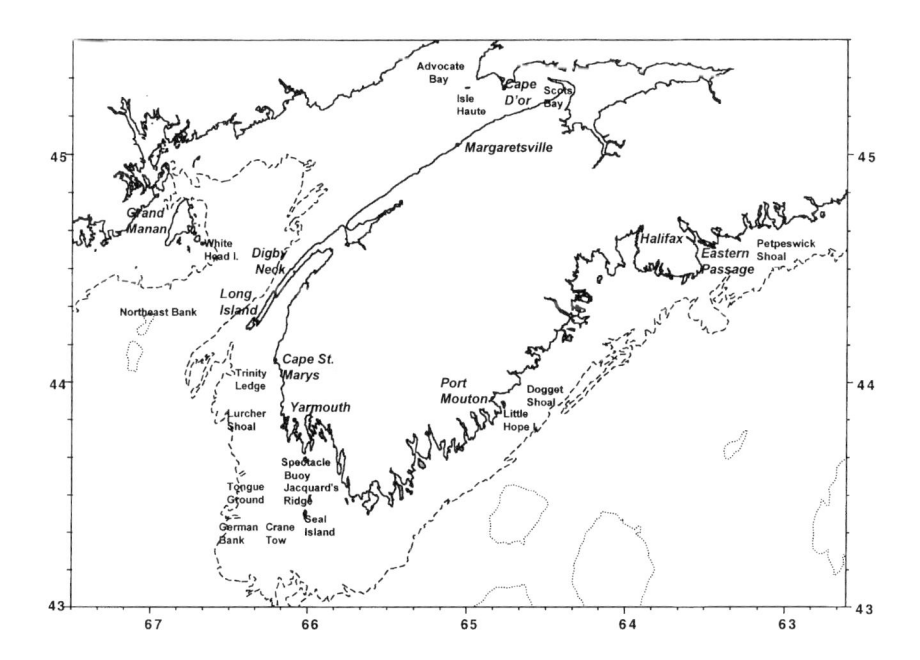

Figure 1. Generalized map of the 4WX herring stock complex.

sentatives, herring processors, DFO Science, and DFO Fisheries Management Sector to address the potential erosion of spawning components through disproportionate fishing effort. As a result of these meetings a set of criteria for fishing on the spawning fish, in addition to the quota, were established for the three main spawning grounds in the SW Nova Scotia/Bay of Fundy area. These criteria required that prior to fishing on a spawning ground during the spawning season (defined as the first appearance of fish suitable for the roe market) surveys must be undertaken to estimate the abundance of herring. Removal by the herring fleet was limited to 20% of the observed biomass. Once this quantity of fish was removed the fleet had to demonstrate, through another survey, sufficient herring to justify further removals or the area was closed to fishing and the fleet required to move to another fishing ground. The area was only open to further fishing during the spawning season when additional fish were observed and a sufficient time period had elapsed to ensure the turnover of fish (2 weeks). The process became known as the "Survey, assess, then fish" protocol.

The "assess" component of the protocol refers to decision-making process in the "in-season" comanagement regime. A working group composed of representatives of the many stakeholders and government was estab-

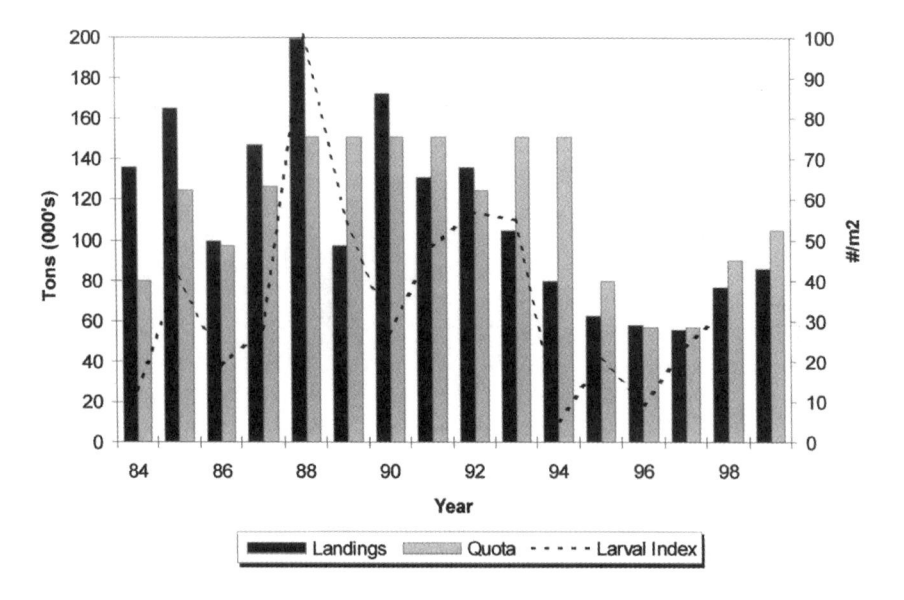

Figure 2. Summary of the 4WX herring landings, quota, and larval abundance index from 1984 to 1999. The larval survey was terminated in 1998.

lished to monitor the surveys and assess the results and empowered to make decisions regarding the distribution of quota within the season. The process involved industry-based subjective in-season surveys of spawning herring aggregations using the acoustic hardware aboard commercial fishing vessels (i.e., sounder and sonar), the captain's experience, and scientific staff to estimate biomass from visual observations of distribution and density. Fish were categorized as light, moderate, or heavy in density and conversion values were applied to estimate biomass (Table 1). Upon completion of a survey, which usually involved 4-6 boats, all participants were consulted, the fish sample data (fish size and roe hardness) were reviewed, and a consensus on the observed area and density, therefore biomass, was reported to the working group. The time interval from the survey to a decision was usually less than 24 hours.

In reality, most surveys either found very few fish, which meant the area remained closed, or they observed large aggregations with biomass estimates exceeding market requirements or the remaining quota. For the few surveys where opening the fishery was questionable, the fishery remained closed until more fish arrived on the spawning grounds or catches from the area were reviewed on a daily basis to ensure the 20% maximum was not exceeded. Survey timing was under the direction of the fishing fleet. If the fishers observed what they considered to be more fish than

Table 1. Summary of weightings for each category used in mapping surveys.

Category	Tons/set	Tons/km²	Acoustic (tons/km²)
No Fish	0	0	0
Light	5	200	230-250
	10	400	
Moderate	25	1,000	600-1,300
	50	2,000	
Heavy	100	4,000	2,000-11,000
	200	8,000	
	250	10,000	
	500	21,000	

The tons/set is based on the fishermen's estimate of their catch if they set on the school of fish, converted to km². The acoustic values are the range of tonnage estimated from acoustic recordings and categorized by the observers.

previously documented, another survey was organized and the results were presented to the working group for review.

While the surveys served their primary role to limit fishing on individual spawning grounds, they did not render themselves to review by non-participants because of their subjective nature. It soon became apparent that if the initiative was going to become part of the formal assessment process, where the sum of estimates from appropriately spaced surveys was an estimate of total spawning stock biomass (SSB), a more objective and quantitative approach was required. This led to the development of an automated acoustic logging system for deployment on commercial fishing vessels (Melvin et al. 1998).

Development of the AALS

The concept of using commercial fishing vessels to document fish distribution and abundance originated from observations made during fishing excursions. Commercial herring seiner captains rely extensively on acoustic hardware to locate fish schools, to estimate quantities, and to set their gear for optimal catches. In doing so, they have gained, through practical experience, the ability to extract information critical for a successful set (i.e., species, school size, and position in the water column). While efforts had been made to summarize captains' observations through logbooks, their observations were undocumented and considered anecdotal until the development of the AALS.

The development of the automated acoustic logging system was a critical step in the integration of industry-based biomass estimates into the in-season decision-making and assessment process. Under funding from the Canadian National Hydroacoustic Program, the logging module of HDPS (Hydroacoustic Data Processing Software, Femto Electronics Inc.) was enhanced to record quantitative acoustic data from off-the-shelf commercial sounders with a calibrated transducer. A module to capture the video output from single, and multi-beam, sonars and the vessel's Global Positioning System/Differential Global Positioning system (GPS/DGPS) telegram was also developed for the system. The main objective of the project was to develop a logging system that would allow fishing captains to accurately record the acoustic information they use on a daily basis to catch fish, without the major expense of purchasing new hardware. An additional requirement was that the system was simple to operate and reliable over an extended period of time because of the unattended nature of data collection.

In 1999 there were six AALS in use in 4WX. Five systems were installed on commercial herring seiners and a portable system was available for deployment from almost any size vessel. The two original systems installed and tested throughout the fall of 1996, under a voluntary program, were still fully functional after four seasons of use. In 1998, the Pelagics Research Council (PRC) purchased four new systems for deployment on commercial seiners. The portable unit is a self-contained system complete with a towed body for deployment, GPS, and power generator if required. Activation of the system is the on/off switch of a power bar located near the sounder.

Data Sources

Over the past 5 years, data to estimate spawning stock biomass have been collected during standard fishing excursions and structured (i.e., organized) surveys using the AALS. Structured surveys are further subdivided into mapping and quantitative acoustic surveys.

The acoustic data collected by the AALS during a standard fishing excursion do not follow a standardized survey design. These data represent the search pattern or preparation to set of the vessel once a school of fish is observed (Fig. 3). However, in many cases the coverage or line through a school of fish is sufficient to estimate the observed biomass. Transects are obtained by selecting segments of the vessel's track, average area backscatter (Sa), by estimating the distance weighted mean of fish per m^2 under the vessel, and multiplying by the area covered. Target strength estimates are based on the fish length and weight of herring sampled from the fishing vessel or vessels fishing in the area (Foote 1987).

Transects for fishing excursions were obtained by dividing the vessel track into a series of nonintersecting lines. In addition, portions of the track were removed to prevent overweighting of areas of heavy fish concentrations when it was obvious the vessel looped back to take a second look at a small area of the fish. The average Sa was computed for each segment (transect) and weighted by the transect distance. Area covered

Herring: Expectations for a New Millennium 681

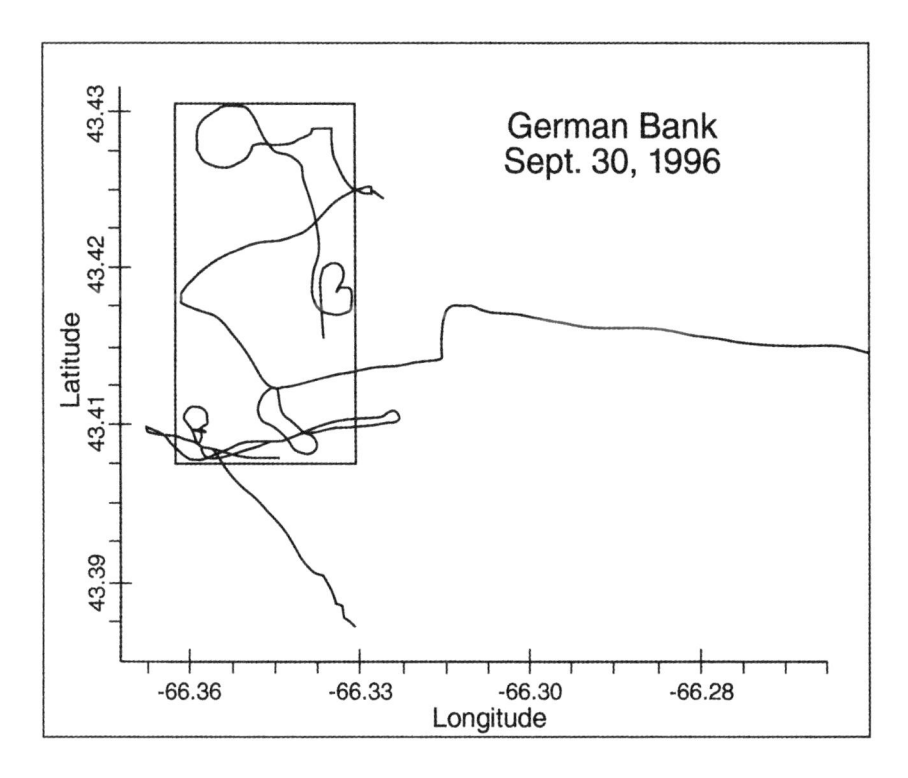

Figure 3. An example of a vessel track from German Bank on the night of September 30, 1999. The rectangle represents the area for which a biomass estimate was made. To avoid double counting due to set patterns and looping only segments of the vessel track from within the rectangle were used in the estimate.

by the vessel was determined by fitting a rectangle or polygon over the vessel track and estimating the area covered. Sonar data were also used to estimate the outer boundaries of the area surveyed. The area estimate was then multiplied by the biomass density (kg per m^2) to determine the tonnage in the area covered by the fishing vessel. Error estimates were derived from transect biomass densities.

During the early years of the in-season decision-making process (i.e., before the AALS) structured surveys were undertaken using a mapping approach. Each vessel participating in a survey would follow predefined transects and record their observations (including position and time) every 10 minutes or less. Observations were subjectively scaled to density categories of no fish, light, moderate, or heavy. In cases where large schools were encountered, the start/stop positions and depth of fish were also recorded. Upon completion of the survey a quick tally of the area surveyed and the observed densities was made and a preliminary estimate was made available to the working group. Later the data were coded and plotted and the area of each scaling category was estimated using a DFO in-house software package called ACON (Black 2000). A tonnage of fish per km^2 based on the fishermen's estimate of the tons of fish per set, expanded to 1 km^2, was assigned to each category (Table 1). An allowance was also made for the depth of fish. The area for each category was then multiplied by the density estimate and summed to determine the observed biomass. Once the AALS were operational, density estimates were made for each category from the acoustic data (Table 1). Figure 4 displays the results of a typical mapping survey involving six vessels.

Structured acoustic surveys using commercial fishing vessels have, in general, followed standard acoustic survey procedures (Dunn and Hanchet 1998, MacLennan and Simmonds 1992, Simmonds et al. 1992). The surveys used randomly selected transects in a two-phase design to cover the area of interest which was defined by the fishing captains. The first phase involved searching a specific area to locate the fish using transects and the second phase concentrated on the documentation of fish aggregations. In all cases the recording vessels worked in conjunction with other vessels to document the distribution and abundance of herring. The number of vessels involved in a survey ranged from 2 to 8, but was generally 4. Upon completion of the survey, the data from the acoustic recording vessels were combined to provide an estimate of the minimum observed biomass for the covered area. Mapping records were also kept for vessels with and without AALS. Data from the nonrecording vessels (i.e., mapping data) were used to produce a mapping estimate for the night, confirm the location of fish aggregations, and determine the boundaries of the fish aggregations (Figs. 4 and 5).

After each survey a conference call with the working group and the survey participants was set up (usually the morning after) to review the preliminary results. In the event that sufficient quantities of fish were observed to permit the fishery to start or continue, data analysis was delayed until a few days later. However, in the event that a number was needed in order for the working group to make a decision, a biomass estimate from the AALS was completed within 24 hours. During the early years of surveys, if insufficient fish were observed a tentative date for the next survey was established. More recently, a survey schedule for the major spawning components was established with flexibility for weather and fish occurrence built in, although the actual survey date was left to the recommendation of the fishing industry. Additional surveys have been undertaken only a few days after a survey because the industry felt more fish were present on the spawning grounds than previously documented. In this case the largest observed biomass estimate from the two surveys was taken for management purposes.

Herring: Expectations for a New Millennium 683

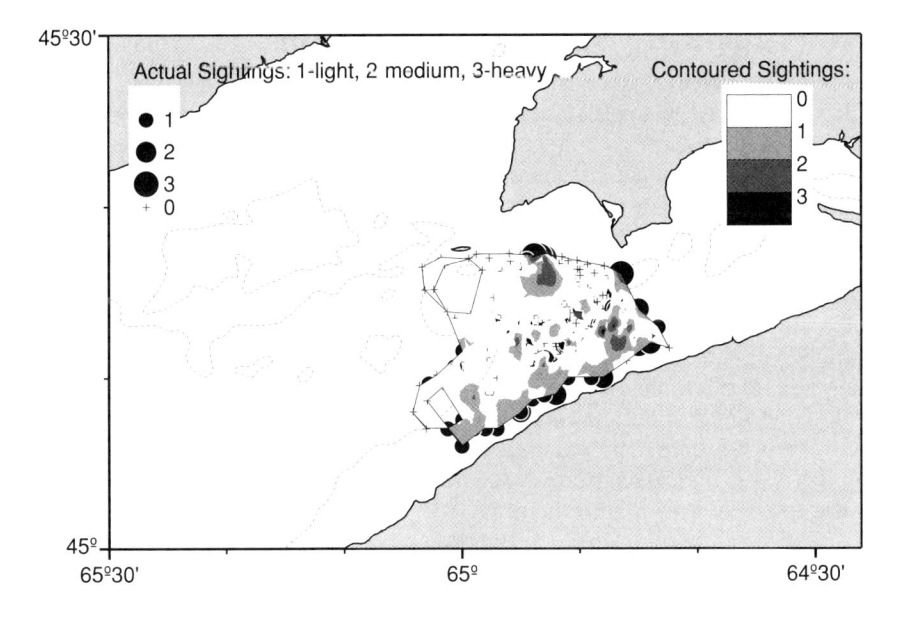

Figure 4. The July 26, 1998, Scots Bay mapping survey vessel track and density contours involving six herring purse-seiners.

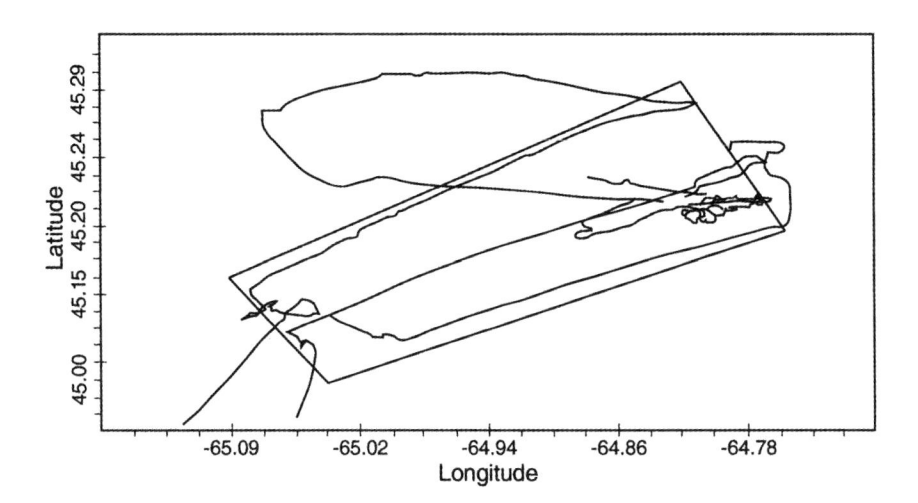

Figure 5. Vessel track of the three vessels equipped with AALS for the Scots Bay, July 26, 1998, survey. The rectangular box defines the area used to estimate SSB.

Nonspawning Survey Application

For the past several years fishermen have reported that the waters around Chebucto Head, just off Halifax, Nova Scotia, support several large overwintering aggregations of herring during December-January. However, the spawning origin of these fish is unknown. It is suspected that the fish are of mixed origin. The area is also inside the 25-mile coastal zone (Fig. 1) which is closed to the commercial seiner fleet. In 1998, at industry's request, a proposal was made to undertake a tagging program to identify the origin of these fish. It was also recommended that an acoustic survey be conducted to estimate the biomass. The program was implemented in January of 1999. During a 3-day period, more than 10,000 herring were tagged and a biomass of greater than 400,000 t was observed in surveys. All surveying was conducted from two herring seiners, one with an AALS and the other using the portable acoustic system. Each participating vessel received compensation of 100 t per night of surveying, but no fish were to be removed from the area.

For January 2000, inshore fishermen were concerned that the mobile fleet may have a serious impact on the small coastal spawning stock, if insufficient quantities of herring showed up in the overwintering area and a fishery was permitted. There was, however, acknowledgment of the need for more information to better understand the distribution and movement of these fish. Considering the precautionary approach and the unknown mixture of the aggregation, the seiner fleet was allowed to survey the Chebucto Head area and to fish to collect herring for tagging at a compensation rate of 100 t per night to a maximum of 1,000 t. No fish could be retained until a minimum of 100,000 t of herring had been observed and a minimum 10,000 fish had been tagged. In addition, if more than 100,000 t of fish were observed the seiner fleet could remove 1% of the observed biomass to a maximum of 5,000 t.

Over a 3-week period in January 2000 six estimates of fish biomass, which ranged from 25,000 to 103,000 t, were made. Figure 6 displays the vessel track and the transects used to estimate biomass on the night of January 23. The time interval from data collection to biomass estimate was 20-36 hours. The working group was activated and local interest groups were involved to monitor the situation and make recommendations as time progressed. Based on the observed biomass only 1,000 t of herring were removed from the area.

Annual Stock Assessment

For assessment purposes it is important to recognize that the approach taken up to 1999 of summing surveys of spawning fish provides only an estimate of the minimum spawning stock biomass for each component. As such it provides an indication of the minimum observed, but says very little about the maximum. It also does not provide an index of abundance

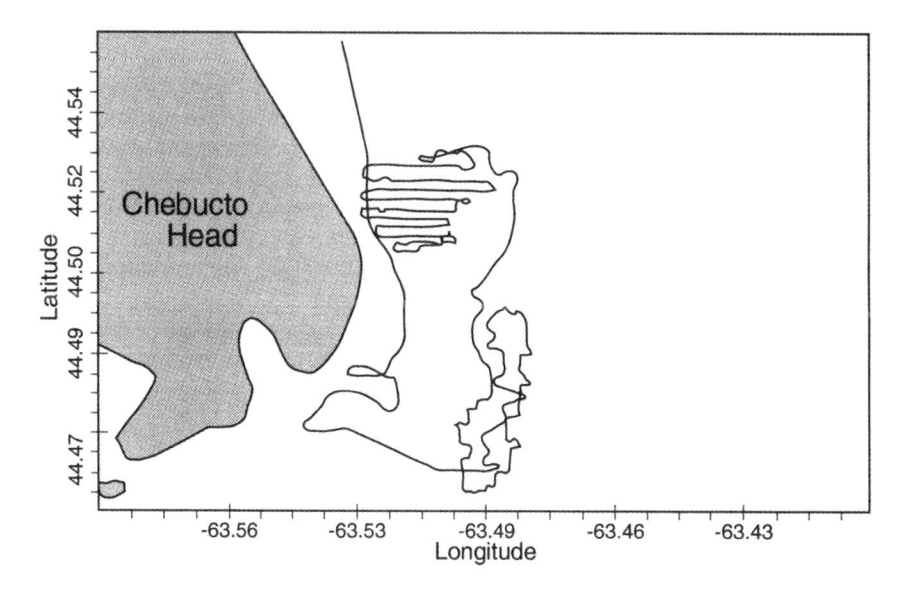

Figure 6. Survey track of the Margaret Elizabeth used to estimate herring biomass off Chebucto Head on January 23, 2000.

in the conventional sense, in that the biomass estimates only cover the survey area, which on any given night may include all or a very small portion of the spawning area. Therefore, the estimates cannot be directly compared between nights or years. The biomass estimates have, however, been used to establish a minimum SSB for the 4WX stock in 1997 and 1998 when a great deal of uncertainty surrounded the VPA. For the 1999 fishing season it was the main quantitative estimate of stock status. In all 3 years the acoustic data were an important factor in determining the recommended quota (Melvin et al. 1999, Stephenson et al. 1999).

A number of issues have arisen regarding the assumptions made to estimate biomass from acoustic surveys and fishing excursions. Of particular concern was the possibility of double counting fish on the same spawning ground when the survey results were summed over the season to estimate SSB, and variability in survey coverage area making direct comparison between years unrealistic. The issue of double counting has been addressed in several reports and is part of a continuing investigation. Currently, a 2-week period is scheduled between surveys. With respect to the limited coverage area, a survey design has been developed, based on the distribution of reported catches during the spawning season, to provide consistent and comparable survey coverage for each of the major spawning grounds (Melvin and Power 1999). The program was partly implemented in 1999 and full implementation is expected for the 2000 spawning season.

Discussion

Information collected by the AALS during standard fishing excursions provides valuable data on the distribution and abundance of herring for the area covered. In addition, those vessels equipped with the acoustic recording systems have the ability to document their observations at any point in time. Typically the captains activate the AALS when they commence their search pattern and turn it off once a set is made. Analysis of these data requires the extraction of representative lines (transects) from the vessel track and the estimation of the coverage area of the search pattern to determine the biomass. Several captains have designed and implemented their own surveys when they encountered a large aggregation of herring they wish to document. The first such event occurred on the night of October 9, 1997 when the *Island Pride* documented 194,000 t of herring on German Bank after completing its night fishing. The captain independently determined the boundaries of the school and proceeded to run a series of transects. This survey played a key role in evaluating the 1997 4WX stock status. Since then several large aggregations of herring, which would have previously gone undetected, have been documented.

Assuming a standardized protocol can be established, industry-based structured acoustic surveys are a key component of a move toward responsible assessment and management strategies for herring. A comparison of the traditional versus the new (in-season) approach is presented in Stephenson et al. (1999). The undertaking of surveys by fishing vessels provides a means for the fishing industry to have direct input into the annual assessment process and stock evaluation, and represents an important contribution to the comanagement process/strategy. The ability to document their observations in a reproducible and quantifiable way gives credibility to captains' observations. Reaction time to investigate unusual events is extremely fast when commercial vessels are used. No longer must the industry wait for science to deploy a vessel, to observe a large aggregation of fish, only to discover that the fish have moved on or dispersed. If something important is observed the fishing industry can document its occurrence with the AALS, then notify DFO scientific staff, who will collect and analyze the data. Using commercial fishing vessels provides the ability to collect, in the case of purse seiners, a representative sample of acoustic targets with the commercial fishing gear for groundtruthing and target strength (TS) distribution (Melvin et al. 2000).

Industry-based surveys, with in-season decision-making prospects, were initially established to provide a second level of protection to individual spawning components regulated by a global stock TAC during a period of uncertainty. Implementation of the process resulted in the near real-time opening and closing of spawning areas during the spawning season, increased cooperation among the stakeholders, and, more important, a turn-around in fish abundance during what appeared to be a period of rapid decline (1995-1996). It is the success of the program which has led to the investigation of additional applications of the approach. During the initial stages of the program, it was never considered that these surveys would become the key approach to estimating SSB for the 4WX stock complex. Today they are the main source of quantitative data used to assess this stock. Another application of the technology includes the use of these data to measure effort and catch rates of gillnet vessels in the Gulf of St. Lawrence (Claytor et al. 2000).

Acknowledgments

The authors would like to thank all members of the Scotia Fundy Region herring in-season management working group for their support and cooperation during the implementation of this program. We would also like to express our sincere thanks to the fishing captains who participated in the survey programs. A special note of thanks goes out to Captains Delma Doucette and Gerry Gorham who permitted the initial installation and testing of the AALS on their vessels. Funding for the technology development was provided by the Canadian National Hydroacoustic Program, a joint initiative of the Department of Fisheries and Oceans and the Canadian Hydrographic Services. Technical support and field personnel were provided by the Pelagics Research Council.

References

Black, G. 2000. ACON Data Visualization Software Version 8.29. Department of Fisheries and Oceans, Halifax, N.S. See: www.mar.dfo-mpo.gc.ca/science/acon.

Claytor, R.R., J. Allard, A. Clay, C. LeBlanc, and G. Chouinard. 2000. Fishery acoustic indices for assessing Atlantic herring populations. ICES C.M. 2000/W:02. 22 pp.

Dunn, A., and S.M. Hanchet. 1998. Two-phase acoustic survey designs for southern blue whiting on the Bounty Platform and the Pukaki Rise. NIWA (National Institute of Water and Atmospheric Research) Technical Report 28, New Zealand. 29 pp.

Foote, K.G. 1987. Fish target strengths for use in echo integrator surveys. J. Acoust. Soc. Am. 82:981-987.

MacLennan, D.N., and E.J. Simmonds. 1992. Fisheries acoustics. Fish and Fisheries Series 5. Chapman and Hall, London. 325 pp.

Melvin, G.D., and M.J. Power. 1999. A proposed acoustic survey design for the 4WX herring spawning components. DFO Atlantic Fisheries Research Document 99/63. 15 pp.

Melvin, G.D., Y. Li, L.A. Mayer, and A. Clay. 1998. The development of an automated sounder/sonar acoustic logging system for deployment on commercial fishing vessels. ICES C.M. 1998/S. 14 pp.

Melvin, G.D., T. Scheidl, F.J. Fife, M.J. Power, S. Boates, K.J. Clark, and R.L. Stephenson. 1999. Evaluation of the 1998 4WX herring acoustic surveys. DFO Atlantic Fisheries Research Document 99/180. 32 pp.

Melvin, G.D., T. Scheidl, F.J. Fife, M.J. Power, K.J. Clark, R.L. Stephenson, C.L. Waters, and S.D. Arsenault. 2000. Summary of the 1999 herring acoustic surveys in NAFO Division 4VWX. DFO Atlantic Fisheries Research Document 2000/66. 40 pp.

Simmonds, E.J., N.J. Williamsin, F. Gerlotto, and A. Aglen. 1992. Acoustic survey design and analysis procedure: A comprehensive review of current practice. ICES Coop. Res. Rep. 187. 131 pp.

Stephenson, R.L., K. Rodman, D.G. Aldous, and D.E. Lane. 1999. An in-season approach to management under uncertainty: The case of the SW Nova Scotia herring fishery. ICES J. Mar. Sci. 56:1005-1013.

Stephenson, R.L., M.J. Power, J.F. Fife, G.D. Melvin, and S.D. Paul. 1997. 1997 evaluation of the stock status of 4WX herring. DFO Atlantic Fisheries Research Document 97/61. 28 pp.

Stephenson, R.L., M.J. Power, J.F. Fife, G.D. Melvin, K.J. Clark, and S. Gavaris. 1996. Evaluation of the stock status of 4WX herring. DFO Atlantic Fisheries Research Document 96/28. 71 pp.

Stephenson, R.L., M.J. Power, K.J. Clark, G.D. Melvin, J.F. Fife, T. Scheidl, C.L. Waters, and S. Arsenault. 2000. 2000 evaluation of 4VWX herring. DFO Atlantic Fisheries Research Document 2000/65. 103 pp.

Stephenson, R.L., M.J. Power, J.B. Sochasky, F.J. Fife, G.D. Melvin, S. Gavaris, T.D. Iles, and F. Page. 1995. Evaluation of the stock status of 4WX herring. DFO Atlantic Fisheries Research Document 95/83. 72 pp.

Present State of the Okhotsk Herring Population after Large-Scale Fishery Resumption

V.I. Radchenko and I.V. Melnikov
Pacific Scientific Research Fisheries Center (TINRO-Center), Vladivostok, Russia

Introduction

Pacific herring have always been one of the main target species for commercial fisheries in Russia. In 1997-1998, it was the second harvest species in the Russian Far East fisheries after walleye pollock. In the far-eastern region, the herring harvest was based on the Okhotsk population. In the mid-1990s, it was undoubtedly the largest herring stock in the world. In 1997, the Okhotsk herring catch amounted to 289,200 t while the total Russian catch of herring was 337,200 t. In 1998 those figures were 343,800 and 407,800 t, respectively. The Okhotsk herring stock is infrequently called the Okhotsk-Ayansky population due to the vast spawning area in the northwestern Okhotsk Sea: from the Tauy Inlet in the north to the Tugur River delta in the south (Fig. 1). Since 1999, the Okhotsk herring share in total Russian catch for this species decreased in 1999 down to 46.8% (203,500 from 434,800 t). The main cause involved another large herring stock (the Korf-Karagin population) into the large-scale fishery. The spawning area of the Korf-Karagin herring covers small bays and inlets of the southwestern Bering Sea.

According to the TINRO-Center forecasts made in the mid-1980s, the Okhotsk herring stock was expected to grow during the 1990s. This forecast was based on integrated studies of climate-oceanographic processes in the North Pacific, marine ecosystems, composition and structure of hydrobiont communities, and in view of stock dynamics in some abundant pelagic fish (Shuntov 1987). Studies conducted in the late 1980s revealed an interesting trend: periods of low stock abundance of walleye pollock (*Theragra chalcogramma* Pallas) coincided with periods of increase in herring stock, and vice versa (Shuntov 1986, 1998a; Naumenko et al. 1990). Hence, the 1980s could be called the "pollock epoch," while the beginning of the 1990s was the" herring epoch."

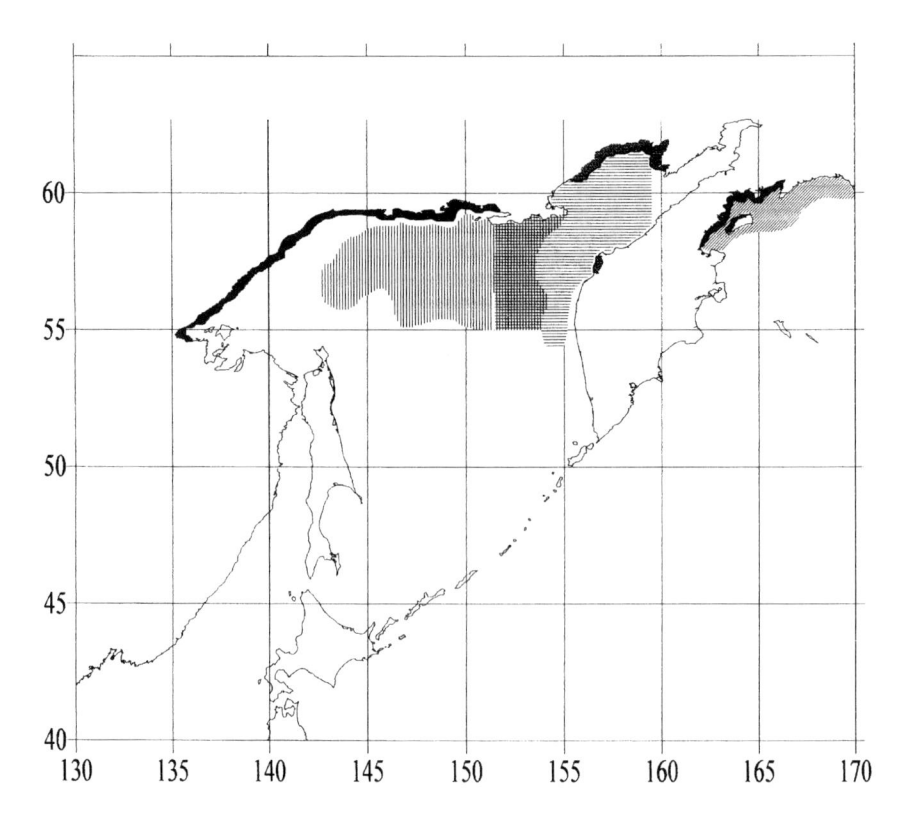

Figure 1. Generalized map of Pacific herring distribution for the largest Asian stocks in the northern Okhotsk and Bering seas. Areas of feeding migration route are indicated by hatching: vertical lines = Ohkotsk population; horizontal lines = Gizhygan-Kamchatsky population; and diagonal lines = Korf Karagin population. Coastal spawning areas are black for all stocks.

Since 1993, TINRO-Center has conducted complex expeditions in the northern Okhotsk Sea twice a year. Pelagic trawl surveys for stock assessment of common fish and squid species are made following a standardized scheme. In 1993-1994, two high-yield year classes of herring born in 1988 and 1989 recruited in the exploitable stock (Radchenko and Glebov 1995). Individuals from these two high-yield year classes were at nearly maximal age (9+ to 10+) in autumn 1998, and soon they will become extinct for the most part. Significant biomass decline was expected due to high natural mortality rates of old herring. Data from the 1998 and 1999 autumn expeditions confirmed our expectations of the Okhotsk herring stock decrease. In this paper, we present data for the purpose of assessing the modern state of the Okhotsk herring population and the perspective for the fishery in the future.

Material and Methods

Herring stock abundance has been estimated from pelagic trawl surveys. Surveys were conducted on stern trawlers with middle gross tonnage. Hauls were performed by pelagic rope trawl, mainly of 108/528 type, during the day. Trawl bags 30-40 m long had an inset with mesh size 6-12 mm, usually 10 mm. The inset was 15 m long. The horizontal trawl opening (estimated from the model) varied from 50 m to 55 m, and the vertical opening from 45m to 50 m. Tow velocity was 4.5-5 knots. The entire epipelagic layer (down to a depth of 200 m or from the bottom to sea surface at lower depth) was gradually trawled for 1 hour.

Fish number and biomass were estimated using a square method (Shuntov et al. 1988):

$$B = \frac{Sq}{sk}, \quad or \quad N = \frac{Sq}{sk} \tag{1}$$

where B = biomass, N = numbers, S = investigated area, q = average catch on the investigated area (in number or weight), s = area covered by trawl during 1-hour haul, and k = trawl catchability coefficient. Standard k value, 0.4 for herring (Shuntov et al. 1988), was accepted after long-term observations, and verification in practice for stock assessment and fishery management.

Herring age was determined by ring counting from scales, using an image-producing visual system for the PC.

Results

In 1998, the trawl survey was executed in the northern Okhotsk Sea epipelagic layer from 6 September to 13 October. During this season, herring usually finish feeding migrations and begin forming of wintering aggregations. The densest aggregations and high trawl catches usually occur in a shelf zone near the Tauy Inlet (Shuntov 1998a). It was also confirmed by our survey (Fig. 2). In addition, the trawl survey showed that there was a general trend of walleye pollock stock decline, and growth of the percentage of herring in the total fish biomass in the northern Okhotsk Sea. Despite some decrease in herring abundance since 1997, the species played a significant part in the pelagic fish community (Table 1). Thus, in 1988 herring contributed 4.9% to the total fish biomass in the northern Okhotsk Sea. In 1998, the percentage of herring was 3.5 times higher, and in 1999 it reached 25.5%.

Analysis of length and age composition indicated that in early October the herring aggregation in the region near the Tauy Inlet consisted mainly of large adult fish aged over 6+ (Fig. 3). Herring aggregations had an evident double-layer vertical structure there. Fatter and less mobile herring occurred in the near-bottom layer in relatively small schools (no

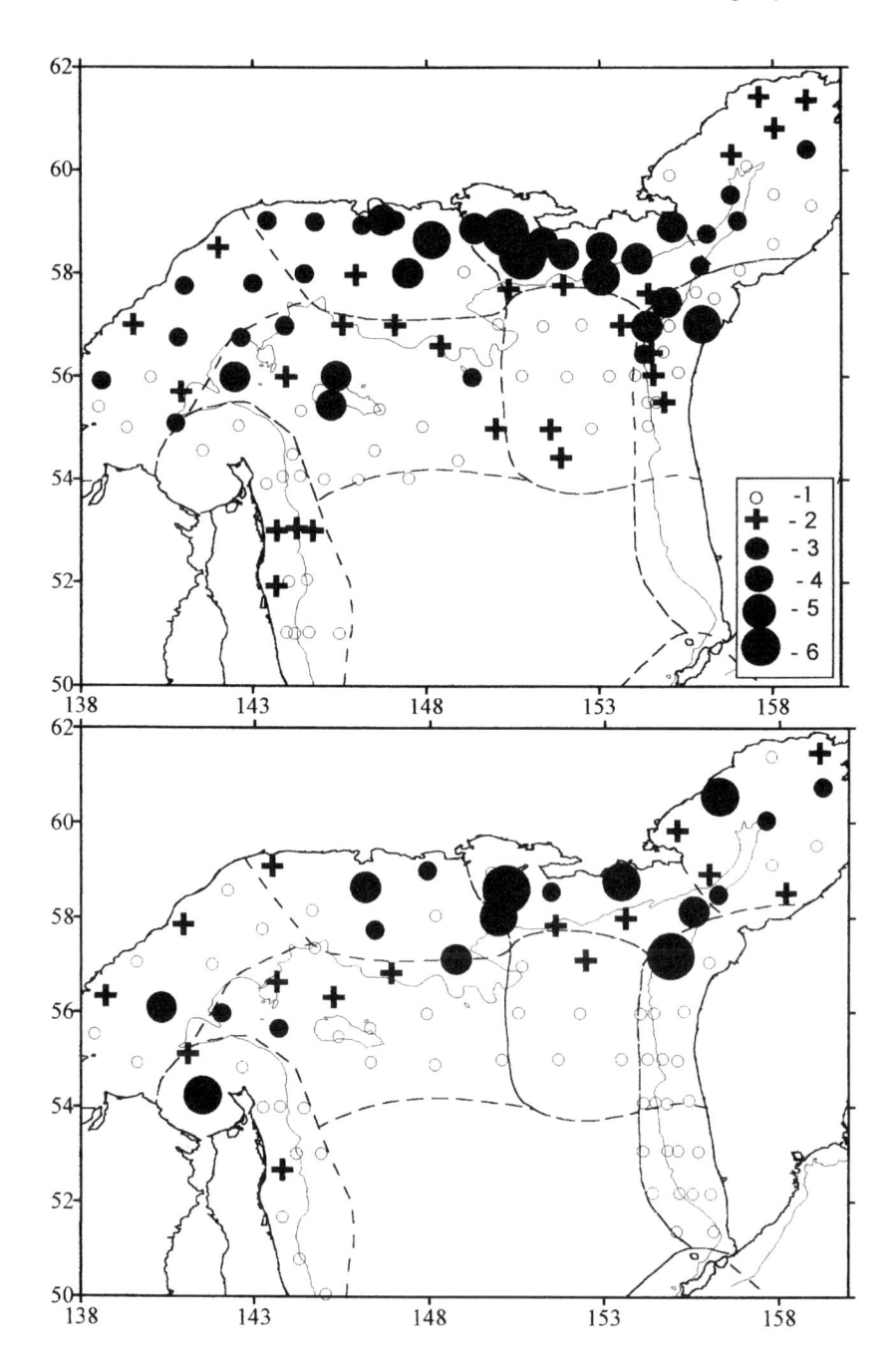

Figure 2. Herring catch distribution (without fish aged 0+) in the northern Okhotsk Sea in August-October 1998 and 1999. 1, no catch; 2, <0.01; 3, 0.01-0.1; 4, 0.11-1.00; 5, 1.01-10.00; 6, >10.00 tons per hour trawl haul.

Table 1. Biomass and ratio of species in the pelagic fish community in the northern Okhotsk Sea in late summer and autumn of 1988 and 1997-1999.

	1988		1997		1998		1999	
Species	10^{3a}	%	10^3	%	10^3	%	10^3	%
Pollock	9,475.2	94.2	4,391.9	57.0	3,636.5	51.5	1,278.0	17.4
Herring	497.2	4.9	2,492.9	32.3	1,209.7	17.1	1,877.6	25.5
Capelin	11.3	0.1	31.0	0.4	1,002.7	14.2	929.3	12.6
Sandlance	7.8	0.1	30.3	0.4	21.2	0.3	1.5	+
Cyclopterids	20.8	0.2	37.1	0.5	127.1	1.8	41.2	0.6
Other	46.9	0.5	727.7	9.4	1,064.0	15.1	3,223.9	43.9
Total	10,059.2	100%	7,710.9	100%	7,061.2	100%	7,351.5	100%

[a]Metric tons.

more than 5-7 t, assessed from acoustic estimations). Those herring presented a low feeding rate. Larger schools (of 25-30 t) occurred in the midwater layer. They consisted of less fat fish which fed intensively. Underyearlings and immature fish, which were feeding in adjacent areas, also occurred in trawl catches. Those fish were more frequent beyond the area of adult herring aggregations, and were feeding intensively judging from high stomach fullness.

In September 1999, trawl surveys were done during a period when autumn redistribution of the adult herring occurs in the northern Okhotsk Sea. Some portion of the adult herring have already aggregated at that time near the Tauy Inlet, where the fishery fleet started the "fat herring" season (Fig. 2). Another portion of the fish stock continued migrations into this region or their feeding route through the adjacent areas. Herring wintering aggregations just began to form. Feeding intensity of herring did not decrease in the region near the Tauy Inlet. Indices of herring stomach fullness averaged 157-169 ‰. Euphausiids contributed 72% of the stomach contents. Herring schools were very mobile. At night, they sank to the near-bottom layers, possibly following euphausiids.

Herring aged 1+, i.e., fish of the high-yield 1998 year class, predominated in numbers in the northern Okhotsk Sea (Fig. 3). Herring of the third year of life (1997 year class) were less in numbers as in biomass. At the same time, the abundance of herring yearlings (aged 1+) could be underestimated since some of those fish usually enter inshore waters at this time of the year (Shuntov 1998b). It is interesting that observed herring of the same age differed by growth rate in inshore and offshore regions. This was observed in all age groups of herring, and all statistical regions. Slow-growing fish occurred in the coastal waters, and fast-growing herring predominated above depths of more than 100 m (Table 2). Herring wintering

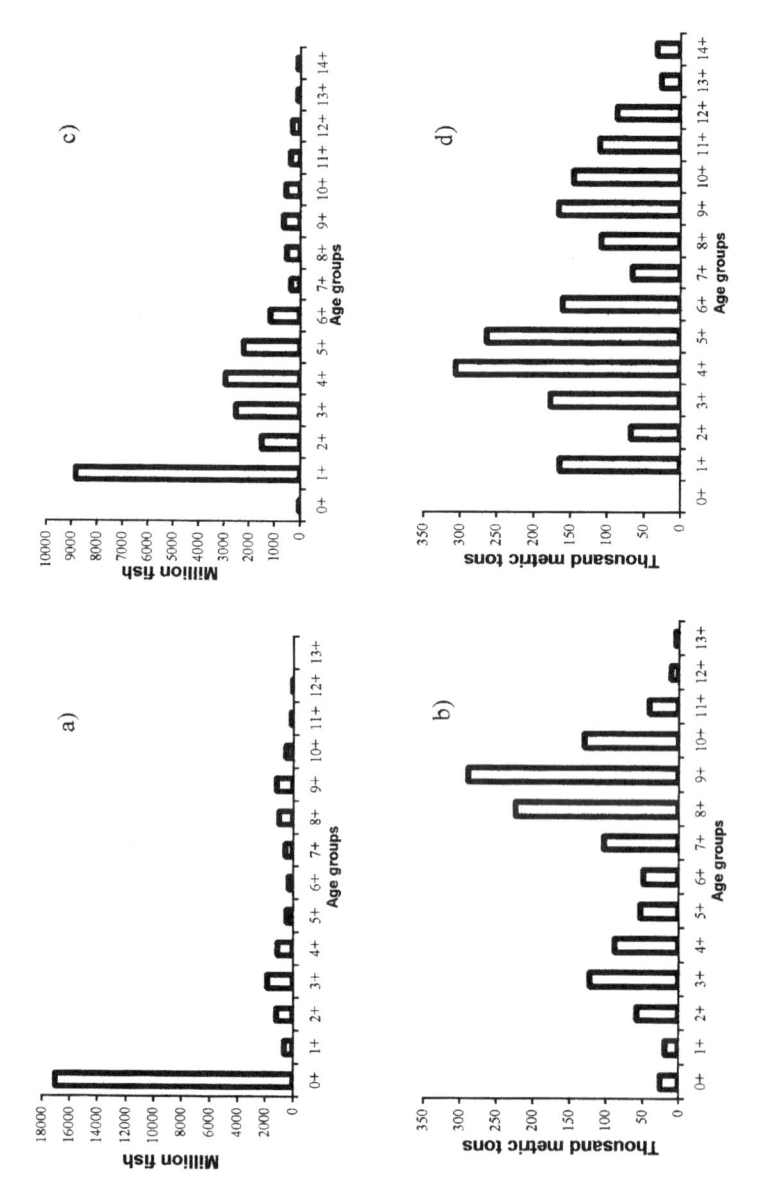

Figure 3. Herring age composition (by numbers and biomass) in the northern Okhotsk Sea in September-October of 1998 (a, b) and in September of 1999 (c, d).

Table 2. Average herring length (cm) in the northern Okhotsk Sea by age groups and depth ranges in September 1999.

Species	0+	1+	2+	3+	4+	5+	6+	7+	8+	9+	10+	11+	12+	13+	14+	Sum
<100 m	7.7	13.5	17.4	19.2	21.4	22.9	23.9	25.7	26.4	28.7	28.2	31.1	31.5	32.1	32.3	15.9
100-200	–	15.2	18.6	19.7	22.2	24.0	25.7	26.5	28.0	29.4	29.6	30.1	30.0	30.4	30.8	27.2
>200 m	–	15.5	19.7	22.3	22.9	23.2	23.8	25.7	26.8	29.4	29.3	31.3	31.5	32.1	32.3	23.8
All	7.7	13.5	17.6	19.8	22.3	23.2	24.6	26.3	27.7	29.4	29.5	30.5	30.5	31.0	31.4	19.4

aggregations consisting of the fattest (and best-kept) fish were found in the depth range from 100 m to 200 m. Positive correlation between the growth rate within an age group and depth was reported for other species, e.g., walleye pollock dwelling in midwater and near-bottom layers (Karp and Traynor 1989).

During September-October 1998, the total herring biomass for the Okhotsk and neighboring Gizhygin-Kamchatsky population was estimated at 1.28 million t, according to trawl surveys. Existence of another large stock in the adjacent area makes it difficult to conduct monitoring on the Okhotsk herring population, particularly due to stock intermixing during feeding migrations. During the last years, the degree of herring stock mixing was estimated using morphometric analysis (Fig. 4). There are some doubts that the Okhotsk herring biomass was likely underestimated in autumn 1998, as will be shown below.

The total Okhotsk herring biomass was estimated at 1.0 million t in autumn 1998. It is approximately two times lower than in August 1997 (Table 1). As expected, biomass decline was almost entirely due to high natural mortality of elder mature herring (Table 3). In 1997, mature herring contributed 70.3% of population numbers (without underyearlings, fish aged below 1 year); and in 1998, only 46.6%. The percentage of the high-yield year classes of 1988-1989 has decreased from 40.6% to 18.9% (Table 3 data on these generations are in bold type).

The Okhotsk herring biomass assessed for summer 1997 and autumn 1998 decreased by 1.06 million t. The decrease was even higher for the mature part of the herring population (fish aged 5+ and older), 1.22 million t. About 0.6 million t should be added to this biomass loss due to somatic and generative production of herring population during one year (1997-1998) calculated with annual P/B-coefficient value at 0.3. Numbers have decreased less than the biomass (from 12.2 to 7.4 billion fish without underyearlings), since losses in the numbers occurred in older age groups. According to trawl survey data, spawning stock numbers also decreased heavily, by 60%, from 8.58 billion to 3.45 billion fish, as well as the average herring body weight: from 172.1 g down to 135.4 g.

Considerably lower estimates of the Okhotsk herring stock abundance had been made for the early and mid-1990s (Yolkin and Farkhutdinov

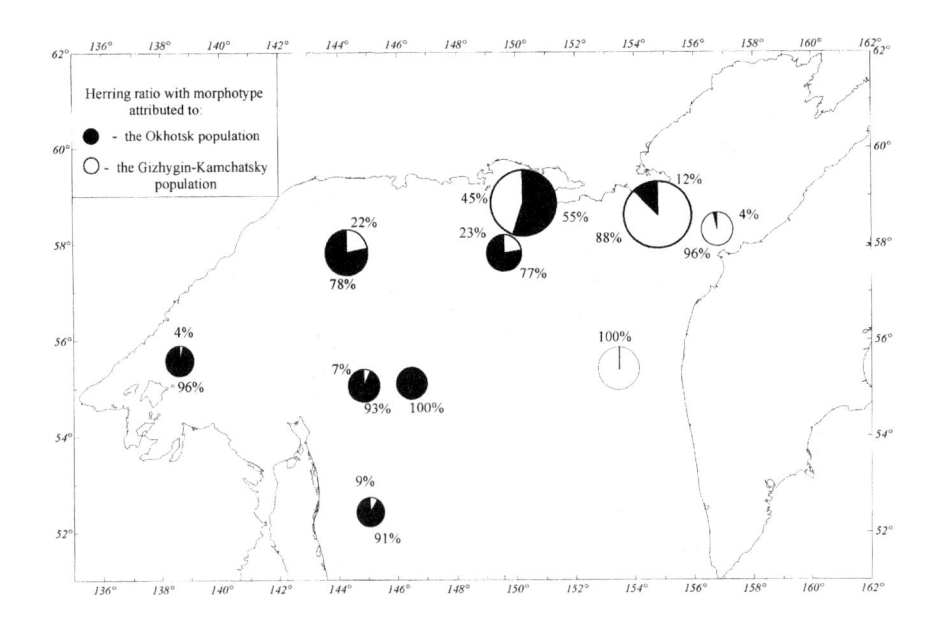

Figure 4. Spatial distribution of herring with morphotype inherent to Gizhygin-Kamchatsky and Okhotsk populations, 10.06-16.08.1997.

1998), and it is important to verify such a sharp herring biomass decline. The decrease in numbers of two high-yield year classes of 1988 and 1989 by 3.55 billion fish corresponded to the total biomass decline by 0.8 million t. According to Tyurnin's (1975) estimates, the average natural mortality rates equal 55.3-67.5% for herring aged 8+ and 9+. In our case this means a biomass loss of about 700,000 t. The resultant discrepancy between the factual and estimated values can be explained by possible underestimation of the herring biomass in autumn 1998. Another possible cause is intensive fishery for the Okhotsk herring in 1997-1998 that was chiefly based on these two year classes. The total herring harvest reached approximately 0.4 million t in September 1997-October 1998.

As for the other age groups of adult Okhotsk herring, elder age groups of 1985-1987 decreased by 1 billion fish (about 0.3 million t), and younger age groups of 1990-1991 by 0.95 billion fish (about 0.2 million t). For these age groups, natural mortality rates were equal to the estimated rate. Average mortality rates were 86.0% for the first, and 33.5% for second group. However, numbers of recruits and immature herring (aged from 2+ to 4+) increased, basing on the 1998 trawl survey. These observations confirm that there was underestimation of immature herring abundance during the summer and autumn pelagic trawl surveys (Shuntov 1998b).

Table 3. Herring age composition (without fish aged 0+) in the northern Okhotsk Sea in the 1990s.

Year and season	Age groups											
	1+	2+	3+	4+	5+	6+	7+	8+	9+	10+	11+	12+
1993 autumn	–	0.5	1.3	**24.0**	**49.9**	11.7	10.5	2.1	+	+	–	–
1994 autumn	–	+	0.2	5.9	**61.5**	**28.4**	3.7	0.3	–	–	–	–
1997 summer	27.7	0.9	0.9	0.2	2.8	6.7	10.8	**24.8**	**15.8**	6.8	2.4	0.2
1998 spring	69.4	4.1	4.8	3.1	2.9	2.6	3.2	6.4	**3.1**	**0.2**	0.2	–
1998 autumn	6.9	13.3	20.8	12.4	5.0	3.3	6.1	11.2	**13.4**	**5.5**	1.6	0.5
1999 autumn	40.2	6.8	11.4	13.3	10.0	5.2	1.6	2.3	2.9	**2.5**	**1.7**	2.1

Data on high-yield year classes of 1988 and 1989 are highlighted by bold type.

We will not point here to the details of herring abundance estimations in spring season (usually from late April to May). It is worth mentioning that length and age composition of herring catches differed significantly in spring and autumn (Fig. 5). In 1999, estimated numbers of herring were practically the same during both seasons, about 16.7 billion fish. However, recruits with body length 19-21 cm predominated in catches in the ice-free area in spring. That age group is most likely to stay near the wintering grounds longer than mature herring. In spring, mature fish migrate to the coastal zone to spawn, whereas most of the underyearlings spend the winter in the shallow waters. In autumn, adult herring are caught as frequently as juveniles that migrate from the coastal zone.

Marine mammal predation on herring, primarily by minke whales, should be considered as a factor effecting herring abundance decline. Minke whales consumed about 440,000 t of the Arctic-Norwegian herring in the North Atlantic annually (Tamura et al. 1998). The whale abundance is estimated at 25,000 in the Okhotsk Sea and adjacent waters (Buckland et al. 1992). With the diet ration in 4% from mean body weight at 5 t (Sobolevsky 1983), an individual minke whale could consume 6 t of food during 1 month. Gregarious pelagic fishes contribute about 93.5% of the average minke whale diet (Tamura et al. 1998). There were 9 months of relatively warm season between August 1997 and October 1998. If about 8,000-10,000 minke whales spent a feeding route in the region of the Okhotsk herring feeding aggregation, and the percentage of herring in the whale diet was about 80%, consumption of herring could be not less than 0.35 million t. Additionally, in 1997 V. Shuntov (1999) observed significant increases of the minke whale and Dall porpoise numbers there. Besides, other marine mammals, including fin whale, beluga whale (during herring spawning migration in coastal zone), dolphins, and seals, can contribute significantly to total herring stock consumption.

In September 1999, the Okhotsk herring biomass was estimated at 1.3 million t, in particular spawning biomass, about 0.8 million t. Significant biomass surplus was determined by the somatic growth of the high-yield year class of 1998. This year class contributed 40.2% of population num-

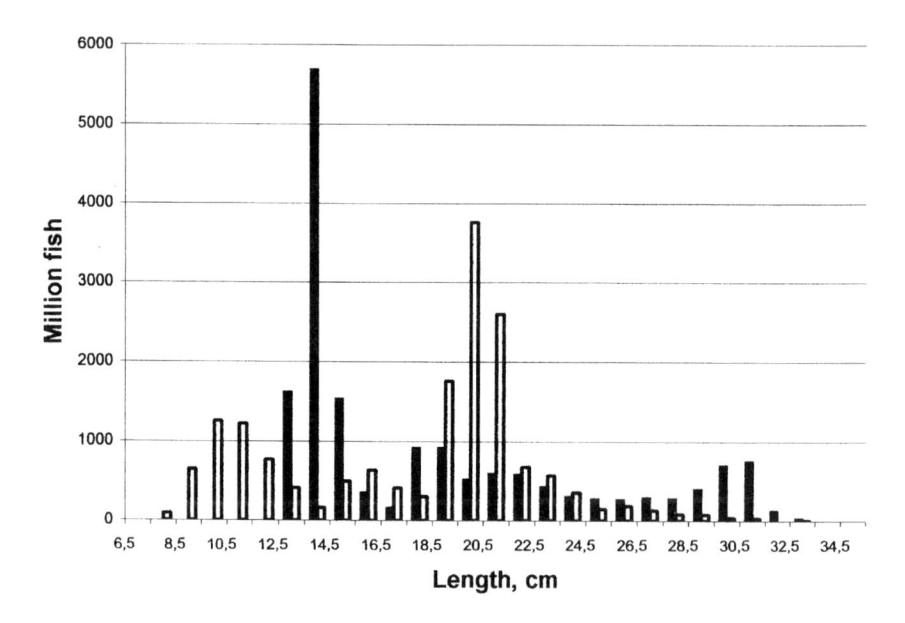

Figure 5. Numbers of herring length groups in the northern Okhotsk Sea in spring (light columns) and autumn (dark) of 1999. Total herring numbers were estimated at 16,733.67 million fish in spring and 16,759.1 million fish in autumn.

bers during the autumn survey. Usually, somatic growth is rather high during the first years of herring life. It determined growth of annual P/B-coefficient for the Okhotsk stock up to 0.45. However, such a significant increase in herring abundance could not be explained by the productivity of the herring population alone. Not only biomass but estimated numbers of younger age classes (from 2+ to 7+, inclusively) have also increased. It can be explained by their underestimation during he autumn survey of 1998. Numbers of older age classes from 8+ to 11+ decreased noticeably. In 1999, they were 28.2% of the total Okhotsk herring biomass, whereas 1 year earlier they were 61.5%. From October 1998 to September 1999, average losses in these age groups were 33% for herring aged 8+ to 9+, and 50.9% for fish aged 10+ to 11+. Such mortality rates are less than rates calculated by Tyurnin (1975). Low estimations of the herring mortality rates also indicated that herring abundance was underestimated by the trawl survey in October 1998. Adjustment by the VPA method with usage of Tyurnin's (1975) natural mortality coefficients suggests that the Okhotsk herring biomass was approximately on the same level in 1998 as in 1999. In this case, herring biomass losses totaled 600,000 t during last year, in particular 150,000 t, due to fishery mortality.

Discussion

There is a hypothesis (Yolkin and Farkhutdinov 1998) on the gradual growth of the Okhotsk herring spawning stock from 0.6 million t in 1995 to 0.9 million t in spring and 1.1 million t in autumn of 1997 (Fig. 6). However, the pattern of biomass changes in separate year classes presented here evidently contradicts this hypothesis. Predomination of two high-yield year classes (1988 and 1989) in the spawning stock suggests an achievement of the maximal population biomass level in the year of their complete maturation (90-97% aged 6 and 7 years), i.e., approximately in 1995.

There were some other effects of the Okhotsk herring biomass and distribution density increase on the population dynamics in the 1990s. The cyclic interchange of high-yield year classes was disturbed. Previously, cycles of 5 (or 6) and 10 (or 12) years were defined for herring population dynamics during the period from the 1930s to he 1980s (Tyurnin 1975, 1980). At this aspect, growth of herring abundance persisted for 2 years and decreased for 3 years. However, the long-term trend of population dynamics was saved inside the 10-year cycle, since interannual changes appeared on relatively lower (or higher) layer than during the previous cycle. There was no high-yield year class in the Okhotsk herring population from 1990 to 1996. Low reproduction rates can be regarded as an effect of density-dependent processes in herring population in those years. In 1992-1993, when the pair of abundant year classes (1988 and 1989) began to recruit in the spawning stock, the biomass of adult Okhotsk herring already reached 2.5 million t, according to the most modest estimations (Radchenko and Glebov 1995, Shuntov 1998a). Namely, the Okhotsk herring spawning biomass exceeded a level for the optimal filling of spawning grounds and successful reproduction at several times. Tyurnin (1975) estimated such optimal level at 1 million t. Productivity of further generations and recruit abundance has decreased in condition of density-dependent factor effects. Only in 1998 has a high-yield year class appeared (at least, on a level of 1+ age) after elimination of most of 1988-1989 herring broods (Figs. 3 and 5). In autumn of 1998, numbers of herring aged 0+ were estimated at 17 billion fish, or 25,300 t.

The high population density of the Okhotsk herring and the low exploitation rates of herring resources by the fishery led to noticeable decreases of the growth rate, maturation, fatness, and fecundity. Even in 1994, the percentage of first-maturing fish in year classes born in 1990 and 1991 (i.e., in the next generations after the high-yield ones) decreased by 18.4-43.0% in comparison with the long-term average value. Mean herring length decreased by 13-15 mm in the chief age groups of the spawning stock (aged 5+ to 7+). Such decrease continued and differences from the average value reached 18-20 mm in later years.

At the present time, the age structure of the Okhotsk herring population has a normal shape inherent for commercial fish stock, despite some biomass decrease in the last 2 years. According to given estimates, there

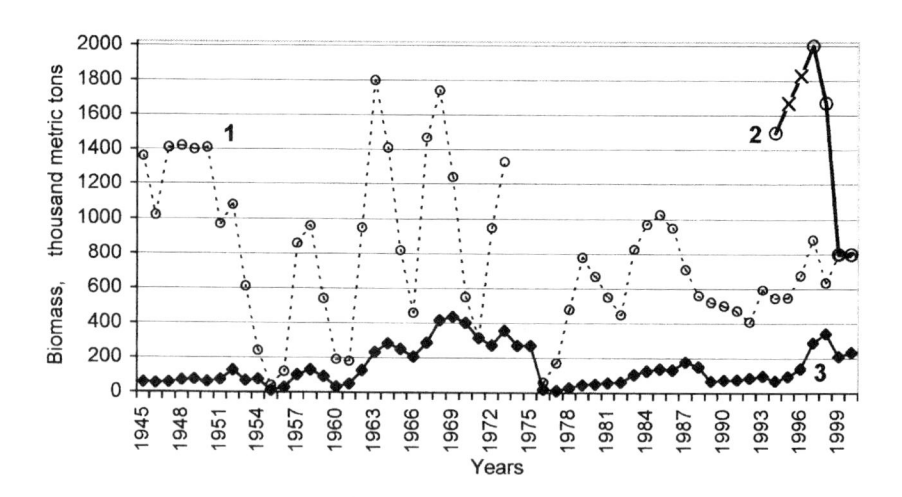

Figure 6. Spawning stock fluctuations and catch dynamics of the Okhotsk herring population, 1945-2000. 1, spawning stock estimations from TINRO-Center Magadan division; 2, corrections made by the TINRO-Center based on pelagic trawl survey data (crosses represent an assumed level for years between autumnal integrated surveys); 3, total annual catch.

are enough numbers of adult fish (0.8 million t), recruits and juveniles (about 0.5 million t), most of which consist of high-yield year classes of 1998 and, likely, 1997. Thus, the fishery stock stabilization in the near future is predicted at a level of 0.9-1.1 million t for the Okhotsk herring and tendency of its growth in the beginning of next millennium.

The Okhotsk herring stock dynamics in the 1990s completely confirmed a conclusion that retention of the optimal number of spawners is one of the most important conditions of herring stock management. Significant exceeding of optimal level by the spawning stock impedes an emergence of high-yield year classes and has the same negative effects as shortage of spawners. Spawner numbers remaining for reproduction cannot be a constant value. It could be decreased in periods of stock abundance growth and vice versa. Therefore, rational exploitation of the Okhotsk herring stock requires incessant monitoring, including large-scale pelagic trawl surveys.

Autumn trawl surveys likely underestimate juvenile herring dwelling in coastal waters. In spring adult herring are underestimated due to prespawning migration inshore, often under ice. High density of herring wintering aggregations inside limited areas is another potential source of interpretation errors of trawl survey data. Biomass of these aggregations could be also underestimated, which likely took place in autumn of 1998. Such situations can appear in periods of stock decrease, especially if the survey grid is not detailed enough. Taking all of it into account, stock biomass estimations given by different methods must be adjusted by VPA. It is very important for the maximal sustainable yield calculation and development of fishery management measures.

References

Buckland, S.T., K.L. Cattanach, and T. Miyashita. 1992. Minke whale abundance in the Northwest Pacific and the Okhotsk Sea, estimated from 1989 and 1990 sighting surveys. Rep. Int. Whal. Comm. 42:387-392.

Karp, W.A., and J.J. Traynor. 1989. Assessments of the abundance of eastern Bering Sea walleye pollock stocks. In: Proceedings of the International Symposium on the Biology and Management of Walleye Pollock. University of Alaska Sea Grant, AK-SG-89-01, Fairbanks, pp. 433-456.

Naumenko, N.I., P.A. Balykin, E.A. Naumenko, and E.P. Shaginyan. 1990. Perennial changes in pelagic fish community in the western Bering Sea. Izv. TINRO 111:49-57. (In Russian.)

Radchenko, V.I., and I.I. Glebov. 1995. Present stock conditions and fishery perspectives of the Okhotsk herring. Rybn. Khoz. 3:23-27. (In Russian.)

Shuntov, V.P. 1986. State of studies on perennial cyclic changes of pelagic fish abundance in the far-eastern seas. Biol. Morya 3:3-14. (In Russian.).

Shuntov, V.P. 1987. On fish productivity of far-eastern seas. Vopr. Ikhtiol. 27(5):747-754. (In Russian.)

Shuntov, V.P. 1998a. Present state of the Okhotsk Sea biological resources. Rybn. Khoz. 4:40-42. (In Russian.)

Shuntov, V.P. 1998b. Reorganizations in the Okhotsk sea pelagic ecosystems: Real fact. Rybn. Khoz. 1:25-27. (In Russian.)

Shuntov, V.P. 1999. Features of current distribution of whales and dolphins in the Sea of Okhotsk. Oceanology 39(2):253-257. (In Russian.)

Sobolevsky, E.I. 1983. Significance of marine mammals in trophic webs of the Bering Sea. Izv. TINRO 107:120-132. (In Russian.)

Tamura, T., Y. Fujise, and K. Shimazaki. 1998. Diet of minke whales *Balaenoptera acutorostrata* in the northwestern part of the North Pacific in summer, 1994 and 1995. Fish. Sci. 64(1):71-76.

Tyurnin, B.V. 1975. Structure of herring spawning stock in the north-western Okhotsk Sea, its dynamics and biological fundamentals of catch forecasting. Master's thesis, TINRO, Vladivostok. Archive No. 14343. 221 pp. (In Russian.)

Tyurnin, B.V. 1980. On reasons of the Okhotsk herring stock decline and measures on its restoration. Biol. Morya 2:69-74. (In Russian.)

Yolkin, E.Ya., and R.K. Farkhutdinov. 1998. Stock state and fishery perspectives of the Okhotsk herring. In: V.I. Goncharov (ed.), Proceedings of Scientific-Practical Conf. of Northeast Russia: Problems of economy and population. Magadan, 1:79. (In Russian.)

Baltic Herring Fisheries Management in Estonia: A Biological, Technical, and Socioeconomic Approach

Tiit Raid and Ahto Järvik
Estonian Marine Institute, Tallinn, Estonia

Abstract

The historical development and current status of herring fishery management in Estonia are presented in the light of biological stock parameters, technical characteristics of fishing technology, and socioeconomic effects. High variability in herring catch structure, CPUE (catch per unit of effort), and profitability of fishery depending on vessel and gear type, as well as seasonal and area effects, introduce significant uncertainties in estimation of efficiency of different management options. The fleet/gear-dependent selectivity, low survival rate of escapees, as well as high quantity of fish enmeshed in the cod-end of trawls and discards additionally hamper the successful implementation of fishery regulatory measures.

After the period of relatively low catches in 1992-1995 the intensity of the herring trawl fishery increased greatly in 1996, resulting in catches fully covering the national quota. Estimations of short- and long-term effects of change in the present ratio of trawl and pound-net fisheries (40,000-45,000 and 12,000-12,500 t, respectively), did not allow recommending a further increase in the share of trawl catches, even in the conditions of increasing TAC (total allowable catch).

The efficiency of regulation via mesh size seems to be low in the herring trawl fishery. Therefore, the management of the herring fishery should preferably be based on regulation of effort in conjunction with introduction of temporary bans on area and/or fishing method. The reduction of effort in the trawl fishery, by decreasing the number of fishing vessels in operation, was recommended to ensure the stability of fishermen income and persistence of herring stocks on a sustainable level.

Herring Fishery in Estonia: Historical Developments and Present Situation

The Baltic herring (*Clupea harengus membras* L.) has been one of the most important commercial fish in all countries around the Baltic Sea for centuries due to its abundance and wide distribution. Fishing techniques and intensity have changed during the long history of the herring fishery. Since the 1920s, in accordance with increasing markets and developments in fishing technology, the pressure on the Baltic herring started to increase rapidly, resulting in peak landings of about 460,000 t in the early 1980s (ICES [International Council for the Exploration of the Sea] 1988, 1989). After that, in the 1990s, herring catches stabilized and started to decrease gradually, reflecting mostly the absence of major market demand but also declining trend in some stock components (ICES 1997).

In Estonia, there are approximately 3,500-4,000 professional fishermen and about 7,500-8,500 people are occupied with fish processing at present, making altogether 0.8% of the total population. In many rural areas, particularly in the western Estonian archipelago, fishing is the main source of income. The Baltic Sea is the main fishing area of Estonia, where the total catch of herring, sprat, cod, and salmon exceeded 77,547 t in 1998 (Ministry of Environment 1999).

The following four major periods in the Estonian herring fishery can be distinguished for the last century:

1. Prior to World War II: mostly coastal fishery with gillnets, trap-nets, and pound-nets (since 1939). Catches were stable at around 10,000 t.

2. 1945 to early 1950s: fast development of the pound-net fishery. Annual catches increased up to 40,000 t.

3. Mid-1950s to early 1960s: development of a bottom trawl fishery and decrease in the coastal fishery. Catches fluctuated around 30,000 t.

4. Since the early 1960s: introduction and development of a pelagic trawl fishery and further decline in the coastal herring fishery. Catches increased up to 45,000-50,000 t in the late 1970s, decreasing to 30,000-35,000 t in the late 1980s. After an abrupt decline to about 25,000 t in the early 1990s as a result of an economic depression, the catches increased again and reached 45,000 t in 1996. The catch in 1998 exceeded 42,000 t (ICES 1997, 1999).

More than 75-80% of the annual herring catches are taken by the trawl fleet, which consists of 233 mostly small or medium-size (16-35 m) vessels equipped with 150-300 hp engines (Table 1). However, due to overcapacity, about one-third of the fleet is not operating presently.

Approximately 30% of the annual herring catch is taken in ICES subdivision 28 (mainly the Gulf of Riga), 35-40% in subdivision 32 (Gulf of Finland), and the rest in subdivision 29 (northeastern Baltic proper and the

Table 1. Estonian Baltic Sea fishing fleet structure (1998).

Vessel category and type	Engine power, hp	Length, m	Quantity
Small vessels (15-20 m)			
Wooden trawl boats (stern trawlers)	54-90	12.6-17	84
MSTB (stern trawler)	90-150	17.6	14
STB (side trawler)	150	18.5	2
MRS (side trawler)	150	16	1
Total			**101**
Cod gillnetters			
SCS	150	25.2	13
Total			**13**
Medium-size vessels (20-30 m)			
MRTK-Baltica (stern trawler)	300	25.5	45
TB, PTS (side trawlers)	150-300	27.3	59
RS (side trawler)	300	35.7	1
MKRTM-Laukava (stern trawler)	800	35.7	1
Big trawlers of various types	500-800	20-32	13
Total			**119**

western Estonian archipelago; Fig. 1). In most areas, herring is taken together with a large bycatch of sprat, which is important to take into account when assessing management measures. The Baltic herring fishery in the Estonian exclusive economic zone (EEZ) currently exploits three big local stocks: the Gulf of Riga stock, the Gulf of Finland stock, and the eastern central Baltic stock (Ojaveer 1981,1991). Due to the low salinity and short growing season, the herring of those stocks is significantly smaller and slow growing, when compared to the stocks from the central or western parts of the Baltic Sea. As a result, the catch in numbers taken in the northeastern part of the Baltic Sea is considerably higher than in the western Baltic, even if the catch in tons is lower (Figs. 2 and 3). Accordingly, the bulk of herring fishing mortality takes place in the northern and northeastern part of the sea; therefore, the effects of management measures in that area are particularly important. The market demand is relatively high for small (total length = 11-14 cm) herring and sprat, since

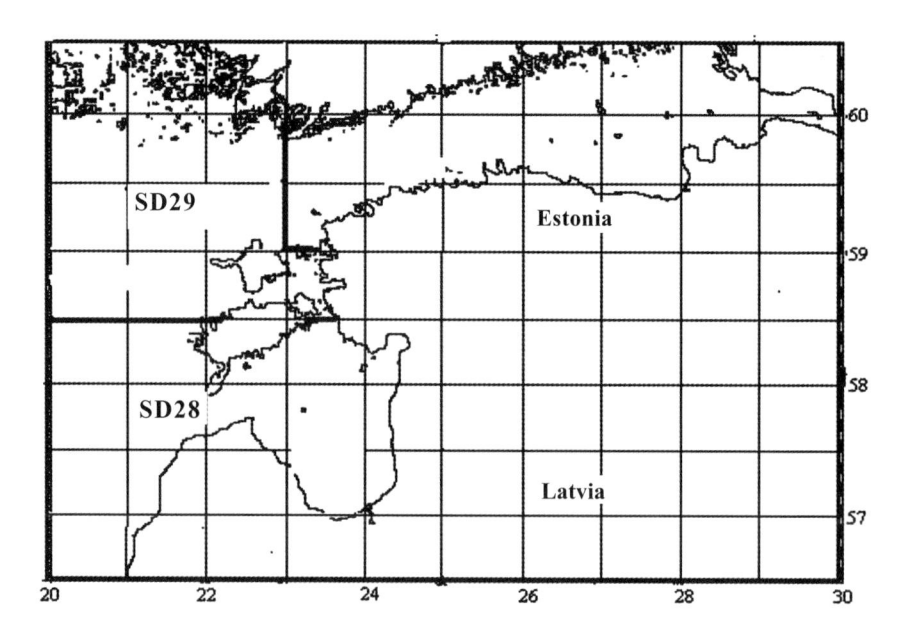

Figure 1. Location of study area in the northeastern Baltic in ICES subdivisions 28, 29, and 32.

more than 80% of the catch is processed by the canning industries for export into eastern European markets (Ministry of Environment of Estonia 1999).

Management of Herring Fishery on Local Level

Need for regulatory measures arose in the Estonian herring fishery in the 1950s, when the extended development of a coastal fishery using big pound-nets of Japanese origin (cacuamis), introduced in Estonia in 1939, caused overfishing on the spawning grounds and, consequently, the landings decreased (Ojaveer 1999). To ease fishing pressure, catch size and gear number limitations were first introduced in the Gulf of Riga in the 1950s (Ojaveer and Järvik 1996).

A new era in the management of the Baltic herring fishery started in the mid-1970s, when the Baltic Sea Fishery Commission introduced a herring catch quota throughout the Baltic Sea at an international level and the first TAC was recommended for the Baltic herring in 1976. Still, the need for local management of the herring fishery within the limits of national quota remained. As it has been shown in numerous cases in the history of human fisheries, the decisions taken in fisheries management do not have an effect on fish stocks only, but inevitably on fishing communities as well (Cushing 1988). Therefore, when considering management measures

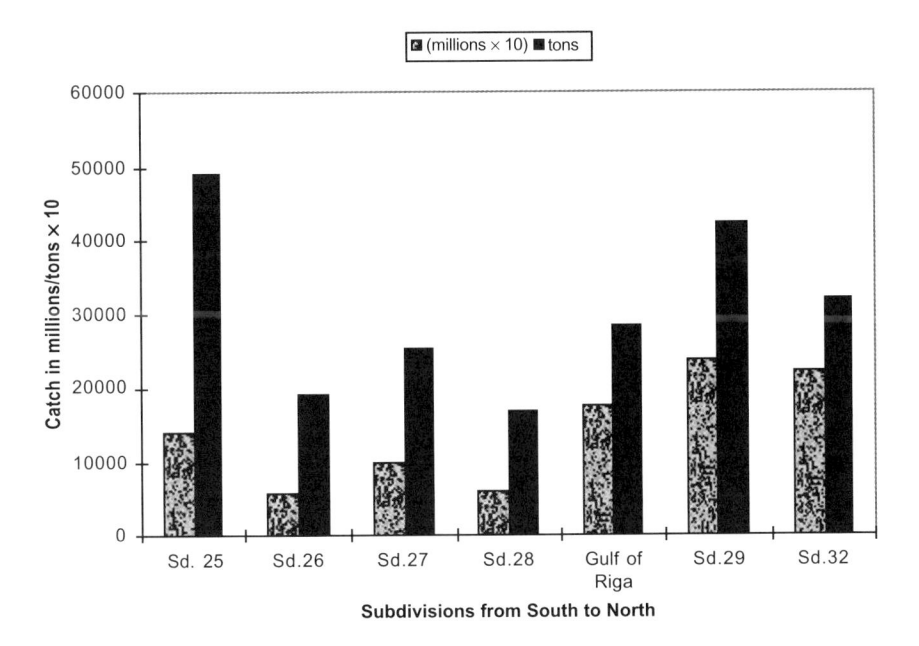

Figure 2. Herring catches in numbers and tons in different subdivisions in 1998 (ICES 1999).

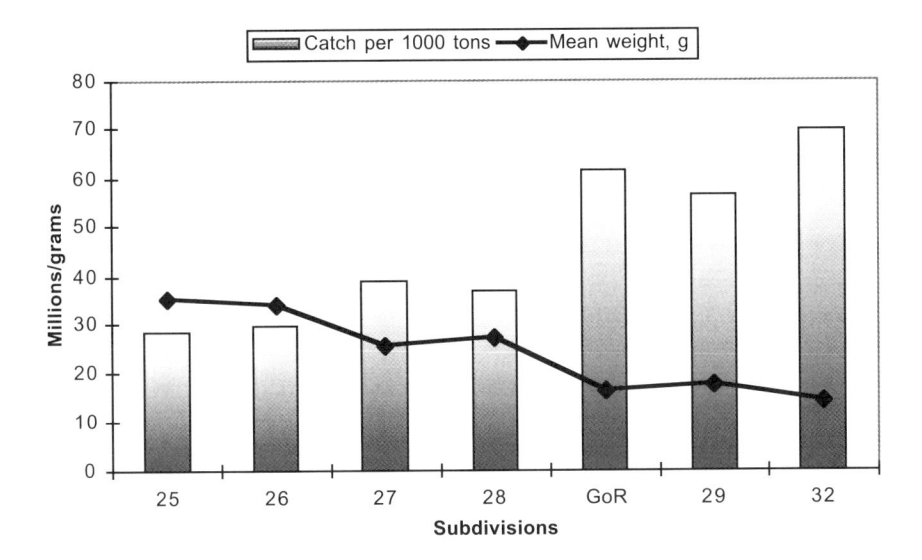

Figure 3. Catch in numbers (10^6) per 1,000 t and mean weight of herring in catches in subdivisions 25-32 and in the Gulf of Riga (GoR) in 1998.

in herring fisheries, the effect on fishing communities depending largely on the state of the stocks should be considered. Consequently, the local fisheries management should ensure, through the implementation of regulatory measures and policies, the optimum exploitation of fish stocks in order to sustain stability in fisheries as well as to allow maximum social stability in fishing communities in the short and long term.

To achieve that objective, the following main regulatory measures have been suggested and implemented in the Estonian herring fishery for different periods, and with different success:

- Minimum landing size and mesh-size regulation
- Introduction of closed areas and periods
- Effort regulation
  Capacity regulation
  Changes in gear composition

Efficiency of Cod-End Mesh Size Regulation in the Herring Trawl Fishery

The mesh size regulation seems to be and is widely recognized as a rather powerful tool for achieving proper catch composition through the use of different mesh sizes (e.g., Gulland 1960, Burd 1991, and others). The mesh size regulation in the Estonian herring trawl fishery (A_{min} = 24 mm in the Gulf of Finland and A_{min} = 28 mm in the Gulf of Riga) was implemented by the authorities in 1976 as a result of selectivity studies (Järvik 1975, Jefanov 1983). However, despite the widespread implementation of the trawl mesh size regulation, the three major shortcomings of that measure can be outlined as follows (Suuronen et al. 1991; Järvik and Raid, in press):

- Effect of spatial and temporal variability in stock structure
- Effect of changes in species composition of pelagic schools
- Effect of unaccounted mortality

Spatial and Temporal Variability in Stock Structure and Composition of Pelagic Schools

Commonly, the minimum allowed mesh size is fixed at the level ensuring 50% (or 75%) escapement of caught fish below the minimum permitted size. For instance, l_{min} = 10 cm (l = body length without caudal fin), for herring in the Estonian EEZ. The selective performance of the trawl fishery (the efficiency of selectivity regulation under the conditions of a fixed mesh size) depends, in addition, on trawling speed and duration, and mesh configuration, amount of fish in the cod-end, and other factors, such as fish body size (girth) (e.g., Treschev 1974, Suuronen et al. 1991).

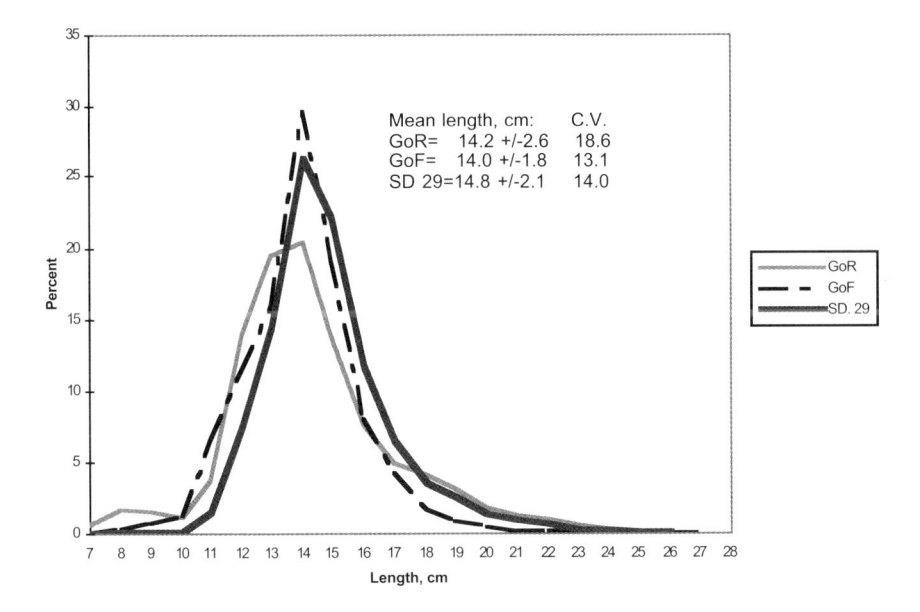

Figure 4. Length distribution of herring in the Gulf of Finland (GoF), Gulf of Riga (GoR), and subdivision 29 (mean of 1995-1996).

As was mentioned above, at least three large herring populations are fished by the Estonian herring fishery. Despite the relatively close location to each other (not more than 100-200 km), the body parameters of herring from the different stocks are rather different (e.g., Ojaveer 1991). Figure 4 presents length distributions of herring in the main fishing areas of the Estonian EEZ in 1995-1996, showing a rather different size composition. That means that using the same fixed mesh size for those different populations will result in a different selection pattern (Järvik and Raid, in press).

The body size of herring has shown remarkable variability in the northeastern Baltic during recent decades. Like elsewhere in the Baltic Sea, both mean weight at age and length at age of herring have changed substantially in the Gulf of Finland during recent decades. Having been rather stable in the 1950s and 1960s (Ojaveer and Rannak 1980), mean weight increased in the 1970s and reached its peak values in the late 1970s and early 1980s. Since then, mean weight at age has decreased rapidly in all age groups, now reaching in the most abundant age groups about 70% of the level in the 1970s and 45-50% of the maximum level in the mid-1990s (Tables 2 and 3, Parmanne et al. 1997). These changes inevitably caused alterations in body girth, playing a key role in cod-end selectivity and the effect of a fixed mesh size. Along with changes in mean weight and length

Table 2. Mean weight at age of herring in Estonian trawl catches in the Gulf of Finland 1970-1995.

	1970-1975		1976-1985		1986-1990		1991-1995	
Age	W, g	S.D.	W, g	S.D.	W, g	S.D.	W, g	S.D.
1	15.4	1.87	15.8	1.46	13.2	1.61	10.1	2.64
2	19.1	0.92	22	3.45	17.2	0.79	15.5	1.48
3	23.7	1.61	29.3	3.05	23.2	1.28	18.4	1.71
4	28.1	2.12	36.5	2.28	29.1	2.90	19.6	2.25
5	30.3	3.33	42.7	4.89	34.3	3.81	22.4	1.51
6	32.5	1.07	48.7	10.69	45.5	5.03	24.3	1.66
7	35.2	3.46	50.8	7.33	53.2	6.45	27	4.61
8	38.9	3.39	56.5	9.53	65.2	14.2	31.2	5.09
9	39.2	3.42	65.6	22.1	65.7	11.8	36	6.88
10+	49.2	13.74	78.3	24.06	86.1	5.58	44.8	8.01

Table 3. Mean length at age of herring in Estonian trawl catches in the Gulf of Finland in 1979-1980, 1984-1985, and 1989-1990.

	1979-1980		1984-1985		1989-1990	
Age	L, cm	S.D.	L, cm	S.D.	L, cm	S.D.
1	13.0	0.07	11.7	0.42	12.8	0.35
2	15.4	0.71	14.2	0.28	14.3	0.21
3	17.4	0.5	16.1	0.71	15.5	0.14
4	18.5	0.42	17.3	0.49	16.2	0.16
5	19.9	0.57	18.7	0.14	16.8	0.15
6	19.5	0.28	19.8	0.71	17.5	0.28
7	19.9	0.07	20.2	1.27	18.2	0
8	19.9	0.42	20.4	20.4	18.7	0.5
9	21.6	1.91	19.9	19.9	20.3	1.27
10+	21.1	0.49	22.2	22.2	24.3	3.46

at age, the size distribution itself has changed toward a narrower size range in the northeastern Baltic, which also influenced the selectivity pattern of herring (Table 4).

Besides long-term alterations in herring body parameters, seasonal changes in size composition due to migration of herring are also an important factor. As shown in Table 5, the length distribution of herring shows remarkable seasonal variability, reflecting the spawning migrations to the Gulf of Riga and the Gulf of Finland in the first half-year, and emigration of bigger herring from those areas after the spawning season.

Since the pelagic trawl fishery takes most of the herring catch, the species composition of pelagic schools also plays an important role in the regulatory process. The main bycatch in the herring pelagic fishery is the Baltic sprat, whose abundance and biomass are extremely variable (Fig. 5). Therefore, the high percentage of sprat in mixed pelagic catches can significantly affect the selective performance of trawls, particularly due to massive enmeshing in cod-ends (Järvik and Raid 1991).

The above allows us to conclude that temporal and spatial variability in herring shoals does not allow the implementation of a mesh size regulation in the herring trawl fishery as a stable measure even in such a limited sea area as the Estonian EEZ (Fig. 1).

Effect of Unaccounted Mortality in Trawl Fishery

Fishing mortality has a significant effect on fish stocks. A schematic description of the mortality process of the herring trawl fishery is presented in Fig. 6. Herring found in the trawling zone will principally end up in one of four main categories: in landing, in discard, in meshing, or in escapees. In general, only landed fish are accounted for in estimates of fishing mortality, leaving discards and enmeshed fish unaccounted for. Minimizing of unaccounted mortality is one of the main tasks of fisheries regulations.

To estimate the effect of mesh size regulation on unaccounted mortality in a pelagic trawl fishery, the theoretical catch in numbers of herring, and number of herring dead in meshing and after escape were calculated for mesh size $A = 28$ mm, and compared to those calculated on the basis of real catch results with mesh size $A = 20$ mm in the Gulf of Finland in the second quarter of 1998. To estimate unaccounted dead fish, the results of meshing experiments and selectivity investigations of Baltic herring (Järvik and Suuronen 1990; Järvik and Raid 1991; Järvik and Raid, in press) and the predicted mortality rates for escaped Baltic herring (Suuronen 1995) were used. The results (Table 6) show that using mesh size $A = 28$ mm instead of $A = 20$ mm, the catch loss in weight would have been 6.7%. The estimates of unaccounted mortality as a result of meshing and escape in number of fish were 2.6 times higher in $A = 28$ mm mesh than they were in $A = 20$ mm mesh. According to the results of Suuronen (1995), the survival rate of escapees is very low (8.8%) independently of mesh size in length groups below 12 cm (total length). That leads to the

Table 4. Length distribution of herring in trawl catches (%) in the Gulf of Finland, 1980-1996.

Length, cm	1980	1981	1982	1983	1984	1985	1986	1987	1988	1989	1990	1991	1992	1993	1994	1995	1996
≤7					0.1				0.8			0.1					
8	0.2	0.4	0.3	0.1	0.1	0.2	0.1	0.1	0.3	0.2	0.1	0.2	0.3	0.1	0.3	0.3	0.3
9	0.9	0.8	0.5	0.4	0.1	0.6	0.2	0.8	0.3	0.7	0.8	0.8	1.6	0.6	0.4	0.8	0.5
10	1.7	2.0	1.4	1.1	0.8	1.2	0.5	2.5	0.8	1.5	1.9	1.0	2.1	1.6	0.7	0.5	1.9
11	1.9	2.9	2.8	1.5	1.2	3.1	1.4	9.6	1.4	1.5	2.3	0.8	3.7	3.4	0.8	2.3	10.3
12	3.8	3.6	2.7	0.8	1.3	16.7	0.7	17.4	2.1	2.9	2.1	1.7	12.2	16.0	3.9	6.0	16.5
13	12.8	15.8	11.6	1.6	3.7	34.5	3.2	10.8	13.3	10.0	13.0	15.2	11.6	23.4	24.4	14.5	17.2
14	18.5	27.9	29.3	10.6	13.0	19.3	19.3	15.7	20.8	28.5	26.1	28.2	19.6	16.5	34.5	36.6	22.8
15	15.8	16.2	18.9	21.3	16.7	8.8	30.6	20.7	24.4	23.8	21.4	25.2	23.9	14.4	16.1	21.1	17.3
16	15.4	10.2	11.3	19.3	16.3	5.0	18.9	10.7	13.1	13.1	12.0	13.8	14.0	12.2	9.0	9.1	7.1
17	11.3	7.7	6.7	15.2	14.1	3.8	9.7	5.9	7.3	6.3	6.7	6.5	6.7	6.6	4.9	4.8	3.8
18	6.3	4.6	5.5	12.0	11.8	2.8	5.9	3.0	4.2	4.5	4.5	3.0	2.5	2.6	2.5	1.9	1.3
19	3.5	2.9	3.6	7.7	7.3	1.7	3.5	1.1	3.7	2.5	2.8	1.6	0.9	1.2	1.4	1.1	0.5
20	3.0	1.4	2.4	3.7	5.1	0.8	2.3	0.4	2.5	1.7	2.5	0.9	0.4	0.7	0.5	0.4	0.3
21	2.2	1.1	1.3	2.2	2.9	0.8	1.4	0.4	2.1	1.2	1.8	0.5	0.4	0.3	0.2	0.2	0.1
22	1.3	1.1	0.6	1.2	2.4	0.2	1.1	0.3	1.1	0.7	0.8	0.4	0.2	0.1	0.2	0.1	0.1
23	0.8	0.7	0.5	0.7	1.4	0.2	0.5	0.3	1.0	0.4	0.8	0.1	0.1	0.1	0.1	0.2	0.1
24	0.5	0.3	0.2	0.3	0.8	0.2	0.3	0.3	0.3	0.2	0.5	0.1	0.1	0.1	0.1		
25	0.1	0.3	0.2	0.2	0.5	0.2	0.1	0.1	0.3	0.1	0.4		0.0	0.1			
26	0.1	0.2	0.0	0.0	0.2	0.1	0.1		0.2	0.1	0.1						
27			0.1	0.1	0.2	0.1	0.1		0.2	0.1	0.1						
≥28			0.2	0.1	0.1					0.1							
n	4,200	4,922	4,988	5,300	4,000	3,849	6,197	3,000	3,800	5,964	4,607	3,800	5,917	8,387	5,080	5,192	5,397
Mean L	15.58	14.85	15.20	15.20	15.65	13.81	14.71	14.00	14.35	15.01	15.07	14.76	14.27	14.09	14.32	14.30	13.60
S.D.	2.66	2.55	2.52	2.53	2.73	2.13	2.14	2.28	2.60	2.26	2.44	1.86	2.05	2.05	1.73	1.74	1.90
CV (%)	17.08	17.15	16.59	16.63	17.46	15.44	14.5	16.30	18.13	15.06	16.21	12.62	14.34	14.58	12.11	12.16	13.95

Table 5. Seasonal length distribution of herring in trawl catches (%) in the Gulf of Finland, 1981-1983.

Length, cm	1981 1 qtr	1981 2 qtr	1981 3 qtr	1981 4 qtr	1982 1 qtr	1982 2 qtr	1982 3 qtr	1982 4 qtr	1983 1 qtr	1983 2 qtr	1983 3 qtr	1983 4 qtr
7	1				0.6	0.1		0.1	0.3	0.1		0.1
8		0.2	0.2	0.3	0.6	0.2	0.3	0.6	0.5	0.7	1.3	0.5
9	1.7	0.5	0.7	0.9	0.6	0.3	0.2	1.7	1.4	0.7	1.5	1.4
10	3.4	1.6	0.6	2.3	1.6	0.3	0.3	1.6	1.6	1.2	1	1.9
11	1.3	2	1.1	5	2.8	2	0.3	0.4	0.2	0.6	1	2
12	1.9	2.3	3.2	2.5	1.9	2.4	5.2	2.5	1.1	1	9	0.9
13	7.4	14	8.7	15.8	8.6	6.3	24.8	16.5	8.4	7.9	32.3	11.2
14	16.6	28.3	26.2	36.4	26.3	24.8	33.8	26.8	20.4	20.5	24	22.8
15	15.1	15.3	26.5	18.1	18.4	22.4	11	18.4	21	18.8	19.8	18
16	14.1	10.7	16.6	9	12.6	12.4	9.9	10.7	19	14	7	15.5
17	13.8	8.8	7.4	4.4	7.3	8	5.8	8.3	12.6	13.7	1.8	11.5
18	6	6.1	4.3	3	6.6	7.6	2.5	5.8	7.6	9.4	0.7	6.9
19	5.8	3.7	2.3	1.3	4.5	5	2.7	2.8	3.1	5	0.2	3
20	2.5	1.9	0.9	0.5	2.7	3.5	1.9	1.7	1	3.2	0.2	2.3
21	2.4	1.4	0.7	0.3	1.7	1.9	0.8	0.8	0.8	1.7	0.2	1
22	2.8	1.4	0.2	0	1	0.7	0.3	0.5	0.3	1.3		0.4
23	1.5	0.8	0.3	0.2	0.9	0.7	0	0.4	0.3	0.3		0.2
24	1.5	0.2	0.1		0.4	0.3	0.2	0.2	0.1	0.2		0.2
25	0.9	0.4			0.2	0		0	0.2	0.1		0.1
26	0.2	0.3			0.3	0.1		0.1	0	0.1		0
27	0.1	0.1			0.2	0.1		0.1	0	0.1		0.1
28					0.8	0.1			0.1	0.1		
Sum	100.0	100.0	100.0	100.0	100.0	100.0	100.0	100.0	100.0	100.0	100.0	100.0
No. of fish	828	1,900	994	1,200	1,100	600	600	1,488	1,400	2,300	400	1,200
Mean	16.1-16.6	15.5-15.8	14.3-14.5	14.5-14.8	15.7-16.1	15.8-16.1	14.8-15.1	15.2-15.4	15.6-15.9	16.1-16.3	13.9-14.3	15.5-15.7
	19.71	16.65	13.25	12.73	20.06	15.6	13.21	15.29	13.95	14.96	11.58	15.23
CV(%)	3.22	2.6	1.91	1.86	3.2	2.49	1.98	2.34	2.2	2.43	1.63	2.37
S.D.												

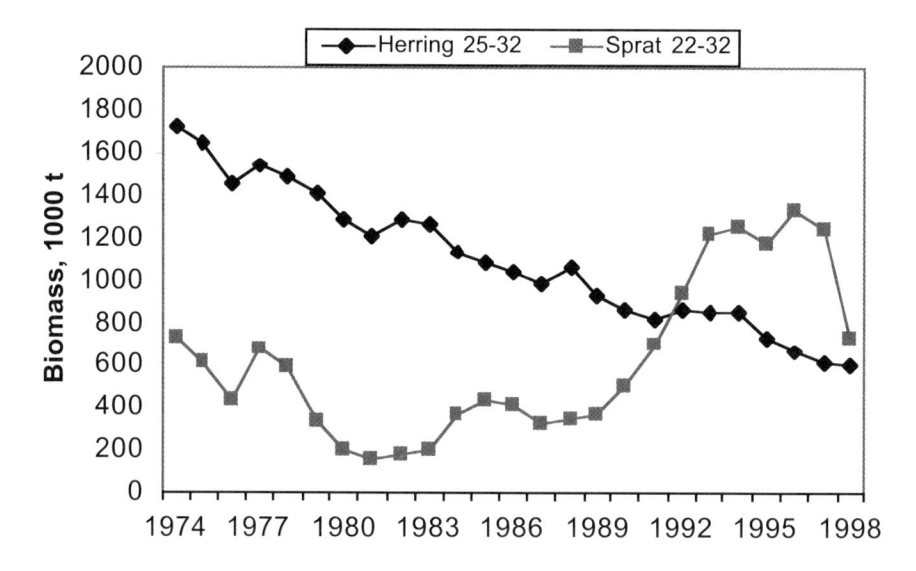

Figure 5. Spawning stock biomass of Baltic herring and sprat in 1974-1998 (ICES 1999).

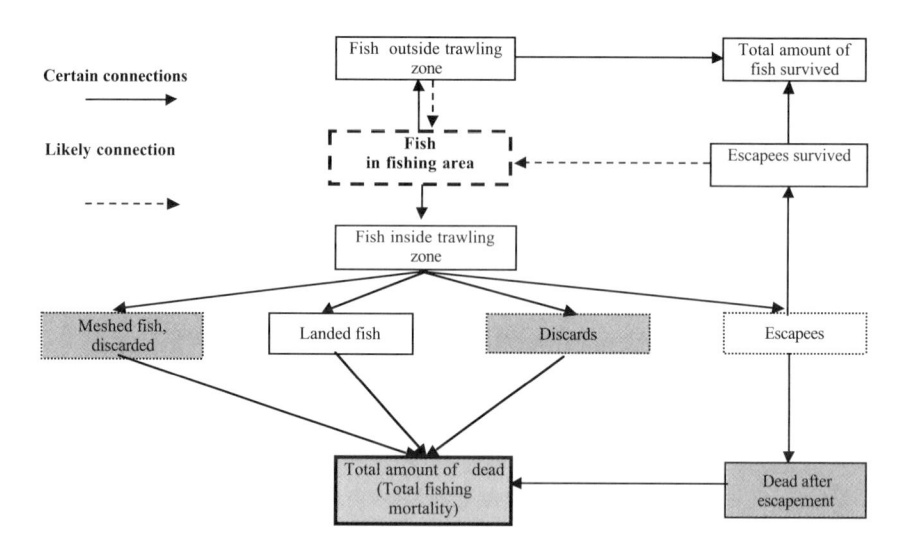

Figure 6. Schematic description of the processes in the herring pelagic trawl fishery. Dark fields refer to sources of unaccounted mortality.

Table 6. Estimated catch in numbers (10^6), losses in meshing and escapees for cod-end $A=28$ mm compared to those calculated per 1,000 fish for actual catches using cod-end $A=20$ mm in the western Gulf of Finland (zone 32.1), 2nd quarter 1998.

Length l, cm	Body-weight, g	Catch in numbers $A=20$ mm	Catch in numbers $A=28$ mm	Unaccounted dead $A=20$ mm	Unaccounted dead $A=28$ mm	Survived $A=20$ mm	Survived $A=28$ mm	Catch weight, t $A=20$ mm	Catch weight, t $A=28$ mm
6	4.03	2.9	0	35	37.7	3.5	3.7	11.6	
7	4.03	18.6	3	28.8	43	2.9	4.3	74.4	12.1
8	4.03	15.7	4.6	4	14.2	0.4	1.3	62.8	18.5
9	4.03	11.4	2.5	0.6	9.3	0	0.1	45.6	10.1
10	4.03	1.4	0.3	0.3	1.5	0		5.6	1.2
11	12	25.7	7.3	4.5	22.9	0		308.4	87.6
12	12	118.6	71.5	0.8	47.9	0		1,423.2	858
13	15.6	304.3	286		18.3			4,747.1	4,461.6
14	18	275.7	275.7					4,962.6	4,962.6
>15	30	225.7	225.7					6,771	6,771
Total		1,000	876.6	74	194.8	6.8	9.4	18,412.3	17,182.7

conclusion that in the present conditions the use of mesh size $A = 20$ mm is justified both biologically and economically, because there is high market demand for canned small herring, and there is no price difference for various herring sizes. On the basis of the described assumptions, a decrease of minimum mesh size to $A = 20$ mm in pelagic trawl cod-ends was suggested in the late 1980s and later implemented by the authorities.

Still, the shortcomings highlighted above allow for the conclusion that implementation of mesh size restrictions as the main regulatory measure in the herring trawl fishery in the northeastern Baltic is rather problematic and an additional set of measures like regulation on fleet and/or area is needed.

At the same time, the minimum mesh size ($A = 24$ mm) enforced in the 1960s has proved to be a successful tool for protecting undersized herring in the pound-net fishery (Fig. 7, Järvik 1985).

Effort and Fleet Regulation

The international fisheries regulation system of the Baltic herring relies on the presumption that TACs, allocated by conventional assessment units, ensure a safe exploitation level for the stock(s) in the assessment units involved. Leaving fleets free to operate within the limits of national quotas is also presumed to have an equal effect on fish stock(s). In fact, as it was shown above, a remarkable variability exists in the catch structure of herring even in a rather limited area such as the northeastern Baltic.

The comparison of the spatial distribution of catches indicates that spatial and temporal distribution of the catch in numbers is not always proportional to that in tons. Since fishing mortality has a major impact on fish stocks, the mean catch in numbers per 1,000 t of landings seems to be a rather good tool for estimation of that effect. The respective mean data for the period 1993-1995 (Table 7) allow for the conclusion that the most rational herring fishery takes place in the second quarter of the year, because of low relative losses in abundance. As shown by Raid (1996), this parameter strongly depends on the share of pound-net catches in the second quarter because of differences in structure of the trawl and pound-net catches (Fig. 7). In the third and fourth quarters, the estimated mean catches in numbers per 1,000 t were at the level of the first quarter in subdivision 32 or even above it (in subdivisions 28 and 29). The fishery in the fourth quarter produced the highest losses per 1,000 t of landings: 67.2 million in subdivision 28, 63.7 million in subdivision 32, and 58.9 million in subdivision 29 in 1993-1995.

Weighted annual mean catches in numbers show that even slight alterations in the allocation of fishing effort can significantly change the exploitation pattern of herring in each particular area of the sea. For example, a catch reallocation of 1,000 t from subdivision 28 to subdivision 29 would result in a gain in stock abundance of approximately 4 million fish. Reallocation of the same quantity from the Gulf of Finland to subdivision 29 would save almost 20 million fish per 1,000 t of landings (Fig. 8). Therefore, local distribution of national quotas between different fishing areas and fleets should deserve more attention as a tool for improving of management quality of the Baltic herring stocks. Moreover, further attempts should be made to optimize the definition of assessment units of Baltic herring.

Economic Considerations

Since 1993 a system of licenses for the herring trawl fishery was implemented in Estonia. Since then, the number of licenses has been quite stable. However, technical improvement of fish search equipment, trawls, and also the introduction of more powerful vessels (Table 1), together with a decreasing national quota for herring by more than 20% during recent years have caused an excess in fishing capacity. Since the herring pound-net fishery provides the main income for coastal fishermen, the increasing share of trawl catches in the utilization of national herring quota has also had a negative socioeconomic impact.

According to the estimates, the average operating cost for catching 1 kg of herring and sprat in the mixed trawl fishery was approximately 2.05 EEK (0.12 U.S. dollars) in 1999, consisting of 45% for fuel costs, 25% for wages, 15% for gear and fittings costs, and 15% for overhead. The average market price of both herring and sprat was 2.5 EEK. According to personal communication with the Estonian Fishermen Association, the average fixed

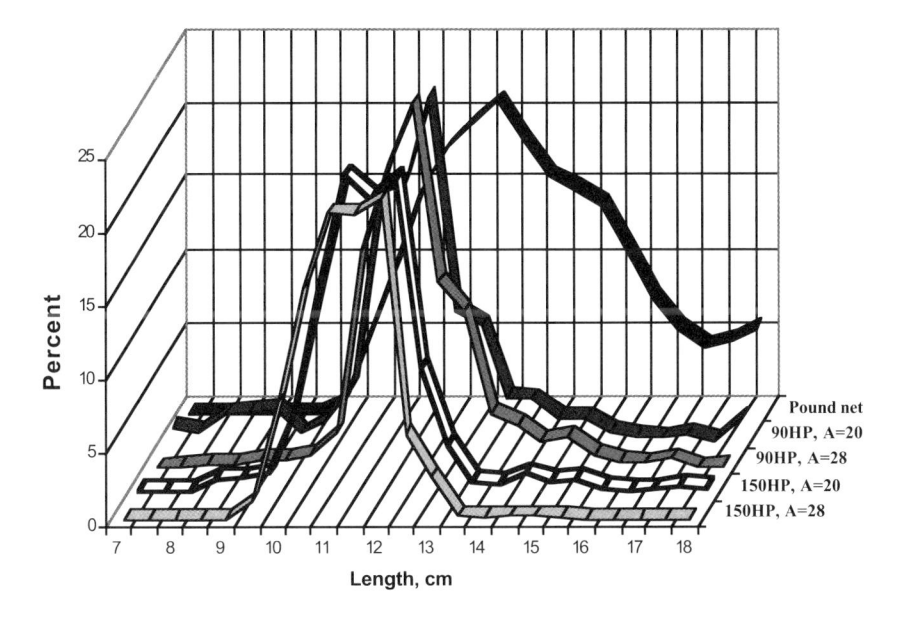

Figure 7. Length distribution of herring catches in pelagic (150 hp) and bottom trawl (90 hp) fisheries with different mesh sizes compared to similar data from pound-nets. Gulf of Finland, second quarter of 1986.

Table 7. Mean catch of herring by subdivision and quarter in the Estonian fishery (1993-1995).

Subdiv.	1st quarter Catch, t	Catch, millions	Numbers per 1,000 t	2nd quarter Catch, t	Catch, millions	Numbers per 1,000 t
28	1,801	111.9	62.1	8,742	316.5	36.2
29	1,442	67.5	46.8	3,472	111.6	32.1
32	5,146	318.9	62.0	5,188	246.8	47.6

Subdiv.	3rd quarter Catch, t	Catch, millions	Numbers per 1,000 t	4th quarter Catch, t	Catch, millions	Numbers per 1,000 t
28	241	15.5	64.3	1,181	79.4	67.2
29	220	12.2	55.5	445	26.2	58.9
32	2,406	146.9	61.1	6,153	391.9	63.7

Mean numbers per 1,000 t of catch Subdivision 28: 43.7 million
Subdivision 29: 39 million
Subdivision 32: 58.5 million

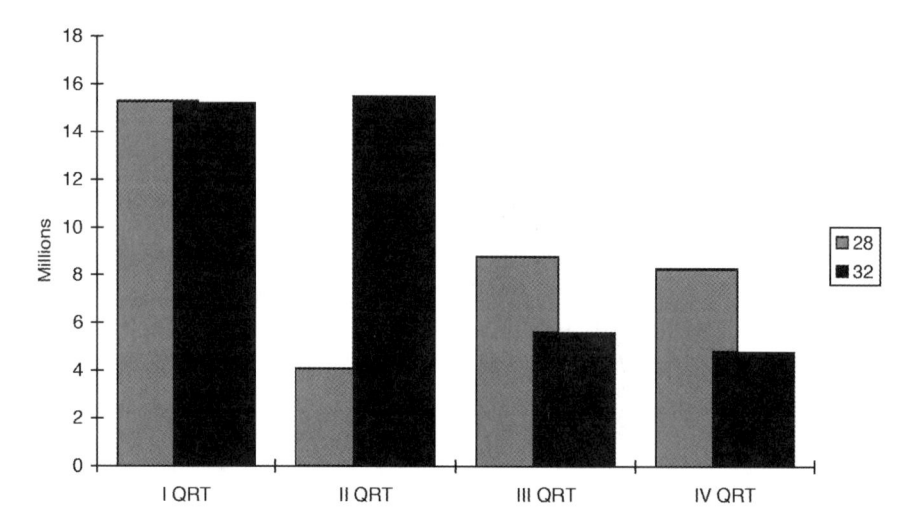

Figure 8. Potential reduction in number of fish per 1,000 t of catch if taken in subdivision 29 instead of subdivisions 28 (Gulf of Riga) and 32 (Gulf of Finland).

costs (depreciation and others basic costs) for medium (300 hp) and small vessels (150 hp) were 500,000 EEK and 150,000 EEK, respectively. Therefore, to ensure a profitability of at least 5%, the estimated amount of required annual catch of herring and sprat equals 300-400 t for a small vessel and over 1,400 t for a medium-size vessel. Taking into account the national quotas for herring and sprat, set at 48,270 and 48,210 t for 1999 and 2000, respectively, not likely increasing in the near future (ICES 1999), the number of both small and medium-size vessels should be decreased to half the 1998 level (Table 1), to ensure a minimum profitability in the trawl fishery. That cut would mean job losses for 500 fishermen.

Conclusions

The management history of the Baltic herring stock(s) has shown that such a widespread fishery regulation measure as minimum mesh size is effective in the inshore fishery with pound-nets and traps only. The mesh size regulation experience in the herring trawl fishery has been mostly unsuccessful. The high variability in selection parameters of trawls, the effect of unaccounted mortality, and the market situation have reduced the biological effectiveness and economical efficiency of that measure.

Optimization of the ratio of trawl and pound-net (including traps) fisheries, temporary closing of some fishing areas to the trawl fishery, and the limitation of the capacity of the trawl fleet are probably the most effective measures for managing the Baltic herring fishery in Estonia.

References

Burd, A.C. 1991. The North Sea herring fishery: An abrogation of management. In: Proceedings of the International Herring Symposium. University of Alaska Sea Grant, AK-SG-91-01, Fairbanks, pp. 1-21.

Cushing, D.H. 1988. The provident sea. Cambridge University Press. 329 pp.

Gulland, J.A. 1964. Variations in selection factors and mesh differentials. J. Cons. Perm. Int. Explor. Mer 29:158-165.

ICES. 1987. Co-operative Research Report No. 161. 417 pp.

ICES. 1988. Co-operative Research Report No. 146. 388 pp.

ICES. 1989. Report of the Working Group on Assessment of Pelagic Stocks in the Baltic. ICES C.M. 1989/Assess:14. 251 pp.

ICES. 1997. Report of the Baltic Fisheries Working Group. ICES C.M. 1999/Assess:15. 504 pp.

ICES. 1999. Report of the Baltic Fisheries Working Group. ICES C.M. 1999/ACFM:15. 555 pp.

Järvik, A. 1975. Selectivity of trawls. Fisherman Handbook 5:40-47. (In Estonian.)

Järvik, A. 1985. Possibilities of rationalization of spring-spawning herring pound-net fishery. Finn. Fish. Res. 6:118-126.

Järvik, A., and T. Raid. 1991. The problem of mesh size in the Baltic herring trawl fishery. In: Proceedings of the International Herring Symposium. University of Alaska Sea Grant, AK-SG-91-01, Fairbanks, pp. 533-541.

Järvik, A., and T. Raid. 2000. Size and species selectivity studies of hauling and fixed gears in Estonia in 1970-1990s: Implementation of survey results. Proceedings of the Polish-Swedish Symposium on Fisheries Selectivity. Publ. Centre of the Sea Fisheries Institue, Gdynia, pp. 94-109.

Järvik, A., and P. Suuronen. 1990. Selectivity of Baltic herring trawls studied by covered codend and twin codend methods. ICES C.M. 1990/B:27. 10 pp.

Jefanov, S. 1983. The effect of different cod-ends on trawl selectivity. Fisherman Handbook 3:10-18. (In Estonian.)

Ministry of Environment of Estonia. 1999. Estonian fishery 2000. Ministry of Environment of Estonia, Tallinn. 50 pp.

Ojaveer, E. 1981. Marine pelagic fishes. In: A. Voipio (ed.), The Baltic Sea. Elsevier Scientific Publishing Company, Amsterdam, pp. 276-292.

Ojaveer, E. 1991. On the condition and management of herring stocks in the Baltic. In: Proceedings of the International Herring Symposium. University of Alaska Sea Grant, AK-SG-91-01, Fairbanks, pp. 521-531.

Ojaveer, H. 1999. Exploitation of biological resources of the Baltic Sea by Estonia in 1928-1995. Limnologica 29:224-226.

Ojaveer, E., and A. Järvik. 1996. Development of management of marine living resources in Estonia since the 1920s. Proceedings of the Polish-Swedish Symposium on Baltic Coastal Fisheries, Gdynia, 2-3 April 1996. pp. 165-174.

Ojaveer, E., and L. Rannak. 1980. Dynamics of some parameters of some herring populations in the northeastern Baltic. ICES C.M. 1980/H:22. 20 pp.

Parmanne, R., A. Popov, and T. Raid. 1997. Fishery and biology of herring (*Clupea harengus* L.) in the Gulf of Finland: A review. Boreal Environ. Res. 2:217-227.

Raid, T. 1996. Trawl or trapnet: Two strategies in herring fishery in the northeastern Baltic. Proceedings of the Polish-Swedish Symposium on Baltic Coastal Fisheries, Gdynia, 2-3 April 1996. Publ. Centre of the Sea Fisheries Institute, pp. 203-214.

Suuronen, P. 1995. Conservation of young fish by management of trawl selectivity. Academic dissertation, University of Helsinki. 158 pp.

Suuronen, P., R.B. Millar, and A. Järvik. 1991. Selectivity of diamond and hexagonal mesh codends in pelagic herring trawls: Evidence of catch size effect. Finn. Fish. Res. 12:143-156.

Treschev, A.I. 1974. Fundamentals of selective fishery. Pischevaya Promyshlennost, Moscow. 446 pp. (In Russian.)

Changing Markets for Alaska Roe Herring

Terry Johnson
University of Alaska Fairbanks, Marine Advisory Program, Homer, Alaska

Gunnar Knapp
University of Alaska Anchorage, Anchorage, Alaska

Abstract

The Pacific herring fishery is one of Alaska's most important commercial fisheries, with average annual landings during the 1990s of 47,100 tons, an average ex-vessel value of $28.9 million, and an average first wholesale value of $80.1 million. Although total landings have been relatively stable, ex-vessel prices and ex-vessel value are highly variable, and declined during the 1990s. This paper examines factors affecting the prices paid for Alaska herring.

The primary product from Alaska's commercial herring fisheries is salted herring roe, a traditional seasonal delicacy in Japan. Roe herring are harvested just prior to spawning, and most are frozen round and exported to Japan for processing. Japan also imports both Pacific and Atlantic herring and roe from several other countries, for production of both salted roe as well as a newer flavored roe product.

Japanese herring roe supply, production, and consumption have declined during the 1990s despite falling prices—reflecting declining demand for traditional herring roe products as lifestyles and tastes change. Changes in wholesale prices from year to year reflect changes in the supply of herring roe to the Japanese market. Changes in prices paid to Alaska processors and fishermen reflect changes in Japanese wholesale prices and the exchange rate between the yen and the dollar.

If Pacific herring harvests remain at levels of the past decade, it is likely that Japanese wholesale prices for salted herring roe—and prices paid to Alaska herring fishermen—will remain low and may fall further. The long-term economic outlook for Alaska herring fishermen is uncertain. Alaska herring may serve as a reminder that viable commercial fisheries depend not only on fishery resources, but also on the markets from which resources derive their value.

Introduction

The commercial fishery for Pacific herring (*Clupea pallasii*) is one of Alaska's most important. As many as 3,000 vessels fish each spring in 20 separate districts. During the 1990s, annual landings of herring in Alaska averaged 47,100 tons, with an average ex-vessel value of $28.9 million and an average first wholesale value of $80.1 million.

The primary product from Alaska's commercial herring fisheries is salted herring roe, or *shio kazunoko*, a seasonal delicacy in Japan. The kazunoko trade has supported valuable roe fisheries from San Francisco to Alaska's Norton Sound, as well as in Russia, Atlantic Canada, and northern Europe. Virtually all the products are consumed in Japan.

Total Alaska herring landings have been relatively stable since the early 1980s. In all but three years since 1983, total landings have been between 40,000 and 55,000 t. However, the ex-vessel value of landings has fluctuated much more widely. For example, total ex-vessel value was $26 million in 1994, $60 million in 1996, and just $11 million in 1998 (Fig. 1).

These extreme fluctuations in ex-vessel value have resulted primarily from fluctuations in ex-vessel prices. For example, the average statewide ex-vessel price paid for roe herring was $470 per ton in 1994, $1,130 per ton in 1996, and $283 per ton in 1998. Ex-vessel prices differ from region to region, but reflect the same upward and downward trends from year to year (Fig. 2).

This paper examines the factors responsible for changes in Alaska herring roe prices and value over time. We begin by describing Alaska herring fisheries and primary processing. Next we discuss Japanese herring roe product forms, processing, distribution. We then discuss trends in Japanese herring roe supply and demand, and the relationship between supply and prices paid for herring roe in Japan and Alaska. We conclude with observations on the importance of fisheries markets, and the implications of changing Japanese markets for the Alaska herring roe industry in the future.

The paper is based on a review of a wide variety of published sources, discussions with fishermen and processors, five seasons as a herring fisherman experienced by author T. Johnson, and interviews conducted during a trip to Tokyo, Sapporo, and Rumoi during November 1998. Data for the paper are from a wide variety of sources, many of which are unpublished. Data sources are described in greater detail in the data appendix.

Alaska Herring Fisheries

Pacific herring (*Clupea pallasii*) is a schooling pelagic fish that occurs in great numbers in the north Pacific Ocean and Bering Sea, and is a close relative of the Atlantic herring (*Clupea harengus*). Herring are believed to return to the same rocky shoreline where they hatched to spawn. They spawn sub-tidally or inter-tidally, depositing their adhesive eggs mostly

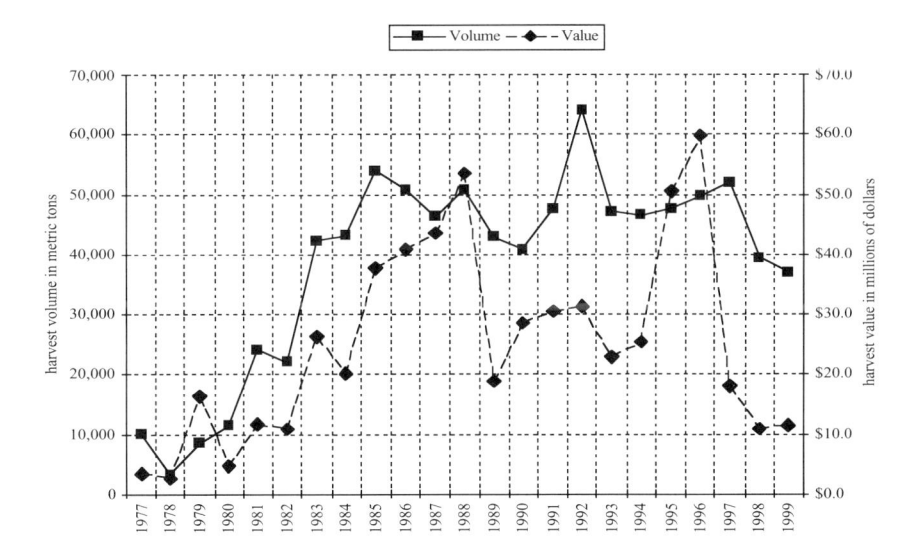

Figure 1. Alaska herring harvest volume and ex-vessel value.

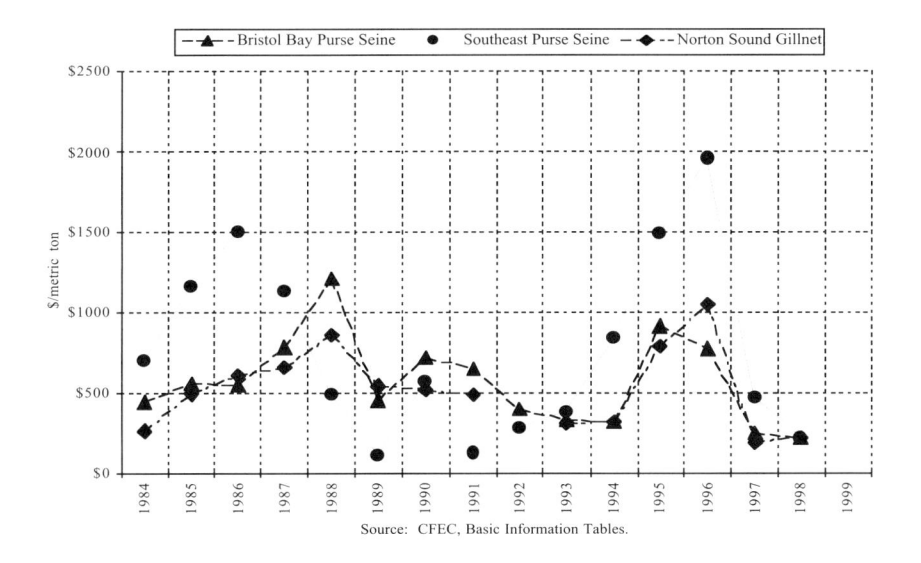

Figure 2. Average ex-vessel prices in selected Alaska herring roe fisheries.

on living substrate such as popweed, ribbon kelp, or bull kelp. After maturity (at age 3-4), spawning occurs every year, with each female fish laying an average of approximately 20,000 eggs.

Great schools of fish migrate shoreward in late winter and spring, first to spawn and then to feed in the nearshore waters until autumn, when they apparently migrate back off shore. Spawning begins as early as December in California, and March in Southeast Alaska and Prince William Sound, and progresses northward to finish as late as mid-June in Norton Sound.

In most areas herring mature and recruit into the fishery at three or four years of age, and may remain in the spawning population for as long as 12 or more years if not taken by the fishery or a predator. Fish in some southern districts reach a harvestable age at approximately 25 cm and 100 grams in weight, whereas catches in the western Bristol Bay and Norton Sound districts typically are 300, 350, or even 400 g at harvest and as much as 40 cm in length.

Alaska's commercial herring fishery dates back to 1878, and by the 1880s was feeding a reduction industry producing herring oil and meal. The reduction fishery peaked in 1929 at more than 78,000 tons (Skud 1960), and continued until 1966, when an oversupplied market no longer supported it.

Alaska sac-roe herring fisheries began in the 1970s and were fully developed as far north as Norton Sound by the early 1980s. Roe fisheries occur in numerous separate districts. Most years the production from the Togiak district in southwestern Alaska is nearly equivalent to the total of all the other districts combined.

Each year, Alaska's sac-roe fishery begins in March in the Southeast Panhandle and progresses to the north and west, ending in Norton Sound usually in early June. Alaska openings are preceded by the San Francisco sac-roe herring fishery which begins in December, and British Columbia openings in February and March.

The objective of the fishery is to produce roe-bearing females from which the egg sacs are removed and processed. Openings in each district are timed to occur just before the bulk of the stock in that area starts to spawn, and may last for as much as several days or as little as 20 minutes, depending on the "ripeness" of the herring and the efficiency of the fleet.

Fishing in some districts is done with gillnets, in some districts with purse seines, and in some districts with both gear types. The relative quality of fish produced by the two gear types varies. Generally gillnets are harder on the fish and lower the quality of carcasses for fillets, but produce higher roe content because they select more effectively for size and maturity due to the fisherman's ability to choose mesh size. In some areas seiners are restricted to fishing deeper waters that are farther from the actual spawning sites, so they tend to catch more immature fish. An advantage of seining, however, is that schools can be sampled without killing many fish, since the whole set can be released if the fish are found to be "green" (immature), or if the ratio of females to males is too low.

Tenders deliver herring from fishing vessels to floating or shore-based processing plants. Payments to fishermen are based on the roe percentage as well as technicians' subjective determinations about the quality of the roe. For example, they decide if the skeins are mature or green, based on color and presence of visible blood vessels. They also can downgrade roe if it is misshapen, puffy, twisted, or otherwise imperfect.

Alaska based processors (referred to in the trade as "packers") freeze most Alaska sac roe herring in the round. In a few Alaska locations, the roe is extracted from the females in local processing plants (as is done in British Columbia and San Francisco). In this case the roes may be brined and partially frozen ("semi-finished") before shipping, or shipped in the raw, unfinished state.

Most frozen roe herring goes either to Japan for thawing and roe extraction, or to a third country (most commonly China, followed by Korea, Thailand, and Malaysia), where the roe is extracted and re-exported to Japan after processing. Most Bering Sea roe herring goes directly to Japan because the larger carcasses are valuable in the dried herring trade.

Sac-roe herring account for the bulk of Alaska herring harvests. However, Alaska's herring also support small roe-on-kelp and food and bait fisheries.

Japanese Herring Roe Products

A wide variety of herring roe products are consumed in Japan. These include several different forms of *kazunoko*, a general term for products made with herring roe.

The traditional Japanese herring roe product is salted herring roe (*shio kazunoko*). Kazunoko means "prolific eggs," and the sacs of densely packed tiny golden eggs symbolize fertility and prosperity in traditional Japanese culture. The skeins or sacs are similar in shape and color to, and slightly larger than, orange or grapefruit sections. The eggs are heavily salted, and when bitten into each cell pops individually, giving the sensation of "crunchiness."

Salted kazunoko has been an important part of the *shogatsu* or new year celebration in Japan for centuries. It is traditionally given as a gift to friends, relatives, and especially to business associates in December prior to the three-day new year holiday as a way of wishing one another prosperity and happiness in the new year. Until fairly recently Japanese people normally stayed at home during the three-day new year holiday; women were not supposed to do any cooking, and salted kazunoko was one of the special holiday foods.

To be edible salted kazunoko must first be soaked overnight in fresh water to remove most of the salt; then with some ceremony it is consumed raw or dipped in soy sauce, in small pieces, using chopsticks.

Kazunoko is never a meal—it is a delicacy normally consumed in amounts of 50 grams or less, perhaps one or two skeins at a time. The flavor is slightly fishy but not particularly rich. Partaking of kazunoko is to partake of the wish for many children, success in business, and good luck in life.

For salted kazunoko, as with other traditional gift products, the cost of the product is deemed a reflection on the esteem in which the giver holds the receiver, so the price has always been high. Salted kazunoko commonly retails for $20-$30 per pound ($9-$13.65 per kg), depending on grade and packaging. Nearly all retail sales of salted kazunoko occur during the last quarter of each year, most of it in the last month, and consumption occurs mainly during the new year holiday.

Top quality salted kazunoko is perfectly formed, bright color, firm consistency, and is attractively packaged in decorative trays usually of 500 g or 300 g net weights. These are known as "branded gift packs" and represent the standard for the trade. Salted kazunoko gift packs are sold mainly through department store seafood or gift food sections, or in other gift item outlets, such as are found in railway stations and big underground shopping malls. "Home packs" are purchased for home consumption rather than gifting. Home pack roes may be more variable in size and quality, and packaging is less decorative, or product may be sold in bulk packaging such as tubs.

Beginning in the early 1980s the Japanese herring roe processing industry developed a new product that could be made with lesser quality herring roe, called *ajitsuke kazunoko*, meaning flavored or seasoned herring roe. Rather than being heavily salted, it is seasoned or marinated in soy sauce, sake lees, peppers, brandy, or other substances, which flavor and to some extent preserve the roe. Because it does not require freshwater leaching before use, flavored herring roe is portable and convenient. The variety of flavors is attractive to younger Japanese people. Because the sauces discolor and obscure the roes in the package, there is no need to use perfect skeins, and broken and misshapen roes are acceptable.

Atlantic herring roe, which is generally smaller and softer than Pacific herring roe, has never been considered suitable for salted kazunoko. But seeking a substitute for high-priced Pacific roe in the 1980s, processors tried Atlantic roe for flavored kazunoko, and found it acceptable. Today most flavored kazunoko is made from Atlantic roe, with lesser amounts coming from smaller and lower grade Pacific herring roes.

Unlike salted kazunoko, most flavored kazunoko is purchased in supermarkets, and it is significantly less expensive. Consumption is year-round rather than confined to the holiday season, although sales peak at the end of the year. Since the product is not heavily salted it appeals to health-conscious consumers who avoid traditional salted foods, as well as to younger people who care more for variety and convenience than tradition. Flavored kazunoko now accounts for about one-third of the kazunoko sold in Japan.

Other herring roe products consumed in Japan include herring roe-on kelp (*kazunoko kombu*), a traditional Japanese delicacy, as well as a wide variety of products in which kazunoko is mixed with other kinds of fish and vegetables. A by-product of larger sized Alaska roe herring frozen in the round is a dried herring meat product (*migaki nishin*). While of considerable absolute value, these other products are a relatively small part of the herring industry or production from Alaska herring.

Japanese Herring Roe Processing and Distribution

Almost all herring roe is processed on the northern island of Hokkaido. Salted kazunoko processing involves soaking in concentrated brine, then in a bleaching tank of hydrogen peroxide, and then in a catalase (an enzyme) that breaks down the hydrogen peroxide to hydrogen and water, which is tasteless and harmless to people. Roes for the gift pack market are sorted and laid out in trays, and then boxed in attractive packages. Processing occurs year-round but intensifies in the fall.

Product may flow from Alaska packers to Japanese retailers in many different ways, but typically through most of the following stages:

Packer,
to trading company,
to wholesaler or distributor (often a processors association),
to processor,
to wholesale market,
to secondary wholesaler,
to retailer.

Trading firms establish purchase arrangements with packers and buy the product, ship it to Japan or another location for stripping, and inspect and ensure quality. Trading firms sell to wholesalers (which are in some cases associations of processors), and sometimes directly to processors. About 80% of kazunoko is sold by processors through wholesale markets (Mitsuhashi), but this share is gradually declining. Anderson (p. 56) estimated that about 42% of salted kazunoko retail sales occur in department stores, 26% in supermarkets, and 16% each in retail fish stores and restaurants, while flavored kazunoko sales are predominantly (75%) in supermarkets and 20% in fish stores.

Trends in Japanese Herring Roe Supply

Herring and herring roe products have been part of the Japanese diet for centuries. The heyday of the commercial fishery, which occurred primarily around the northwest coast of Hokkaido, was the mid-1950s to the mid-1960s when the Japanese public, emerging from wartime destitution, resumed consumption of traditional favorites. In the 1960s the herring resource began a precipitous decline. Various explanations are offered, including overfishing, the destructive effects of a major typhoon, and changes in ocean climate regime and current patterns. Whatever the cause or causes, the herring fishery crashed and has never recovered (H. Ando, Japan Marine Products Importers Association, pers. comm.).

As the stocks began to decline in the early 1960s processors had to look elsewhere for sources of herring to supply the growing demand for kazunoko. Japanese herring fishermen turned first to Soviet waters for herring to replace those which had vanished from their shores. When they were unable to meet the demand, their cooperative association, Dogyoren, began importing herring from Soviet suppliers with the help of a few Japanese trading companies. However, in the late 1960s the resource diminished so much that the Soviet government prohibited herring exports.

Soon after, importers began looking to North America for herring, both because of abundant supplies, and because of the belief that the Western political climate and business ethic would be more compatible with theirs. Early importation was from British Columbia, whose herring industry quickly and successfully shifted from a reduction fishery to take advantage of the lucrative new demand for roe herring. Alaska fishermen and processors got into the roe herring industry in the early 1970s, with Japanese importers establishing business relationships with American packers to supply fish.

During the 1980s and 1990s, Alaska accounted for slightly more than half of North American roe herring landings (Fig. 3). However, annual trends in Alaska harvests did not necessarily correspond with trends in the overall supply of Pacific herring roe to the Japanese market. For example, during the period 1993-1996 Alaska landings increased each year while total North American landings declined each year due to declining Canadian harvests.

In the late 1980s a shortage of Pacific herring caused some importers to look to eastern Canada and western Europe for sources of Atlantic herring. The roes are somewhat different, and most Atlantic roe proved to be unsuitable for high quality salted kazunoko. Skeins are smaller, less brightly colored and, most significantly, the eggs are softer and less crunchy. The combination of a different raw material and shifting consumer preferences led to development of flavored kazunoko.

For a time herring import quotas served to limit the supply of raw material, which helped to stabilize the market. Over the years, however, the total import quota was raised incrementally, at first to satisfy growing demand by the processors for raw material, and later in response to political pressure from the United States (Hastings 1995).

As early as the 1960s Japanese importers bought roe herring from the Russian (then Soviet) Far East. For a time this export was prohibited by the Soviet government to protect depleted stocks, but in recent years the trade has resumed, and is believed to be growing. Much of the roe from the Russian Far East is also used to make flavored kazunoko. Russian Far East

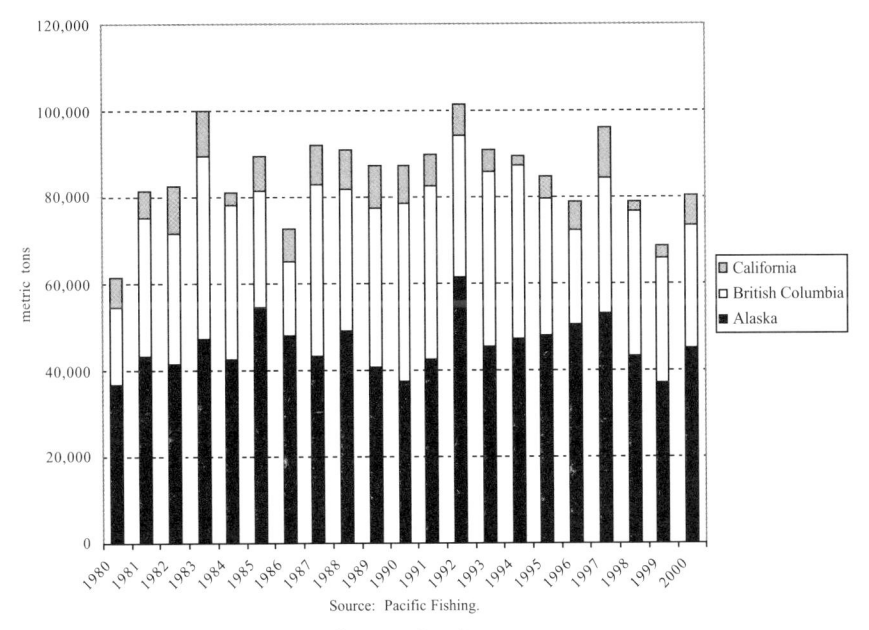

Figure 3. North American roe herring landings.

herring stocks are huge and support landings that have been as high as 250,000 tons in recent years, although most of this catch goes to the domestic "food" (meat) market within Russia rather than to roe export (BANR 1998).

Trends in Japanese Herring Roe Production

Figure 4 shows Japanese trade press estimates of the annual total new supply (excluding carryover inventory) of Pacific and Atlantic herring roe as well as estimates of annual production of salted and flavored herring roe. After increasing rapidly during the 1980s, total herring roe supply and production declined significantly during the 1990s. In general, trends in salted herring roe production are similar to trends in Pacific herring roe supply, since more than 80% of Pacific herring roe is used for salted kazunoko. However, part of the decline in salted roe production during the 1990s was due to a decline in the share of Atlantic herring roe used for salted production from 18% in 1992 to 9% in 1999.

As production declined during the 1990s, so did Japanese consumption of both salted and flavored herring roe. Observers of the Japanese herring roe industry attribute declining herring roe consumption in part to reduced demand resulting from a variety of factors:

- A prolonged recession in Japan.

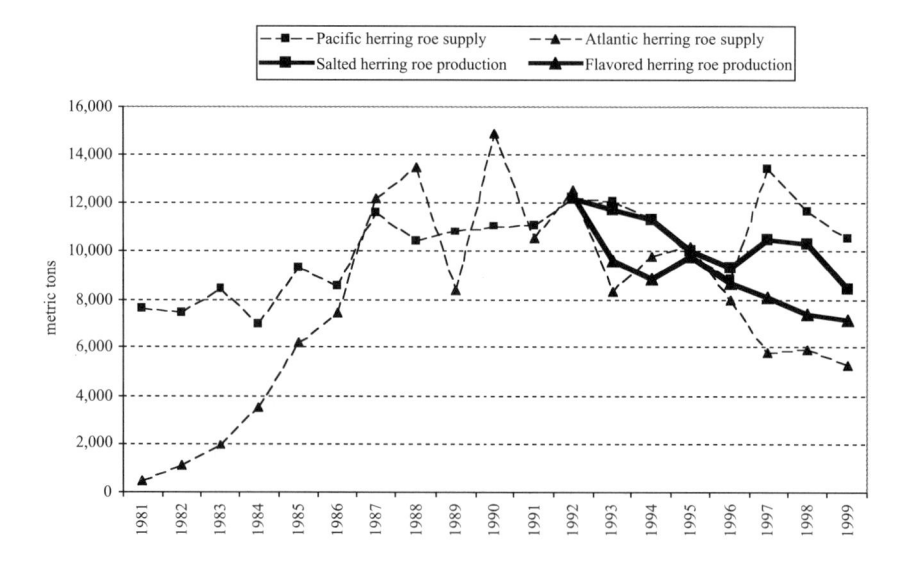

Figure 4. Japanese herring roe supply and estimated production.

- More personal travel during the holidays, resulting in less time at home to participate in traditional activities such as those associated with kazunoko.
- Development of substitute products as holiday gifts, including salted and smoked salmon, smoked ham, and liquor (Anderson et al. 1989).
- Changing corporate culture resulting in a curtailment of company gift-giving, including what one industry participant calls the "moral recomposition of government and industry" (Yasukata Kato, Kato Suisan, pers. comm.) which discourages or prohibits officials from accepting gifts from the corporations they regulate—such as high value branded kazunoko gift packs.
- Changing demographics of Japan's population. As the generation of people who were born before World War II passes out of the population, so does the biggest group of people who enjoy traditional foods. The younger generation enjoys a much wider variety of foods than their predecessors, and traditional Japanese favorites such as salted kazunoko occupy a smaller part of their annual diet. The younger generation tends to be more health conscious, seeking diets lower in salt. The younger generation is also more convenience-oriented and less patient with the rituals of traditional cooking.

Importers and distributors interviewed for this research generally were pessimistic about the future of herring roe products, expressing a widely held view that demand will continue to diminish into the foreseeable future. However, some of them, and more commonly kazunoko processors, believe that demand can be stimulated by lowering the price, developing new products, or employing new ways of marketing familiar products.

Salted Herring Roe Wholesale Prices

Figure 5 shows the average annual wholesale price for salted (*shio kazunoko*) and flavored (*ajitsuke kazunoko*) herring roe at the Tokyo Central Wholesale Market for the years 1984-1999. While actual wholesale prices paid for specific salted and flavored herring roe products vary widely depending upon factors such as grade, size, and origin, average Tokyo wholesale prices serve as indicators of herring roe price trends over time in Japan.

Prices for salted herring roe fluctuate significantly from year to year. Prices exhibit several multi-year price cycles with high points in 1990 and 1996 and low points in 1988, 1993, and 1998. There has been a clear long-term downward trend in prices, which have declined from more than 4,000 yen per kilo in most of the 1980s to less than 3,000 yen per kilo in the late 1990s.

Wholesale prices for flavored herring roe are much lower than for salted herring roe. Flavored herring roe prices also trended downward in the 1980s but were relatively stable in the 1990s. As a result, the price difference between salted herring roe and flavored herring roe declined in the 1990s. Year-to-year changes in flavored herring roe prices appear to be only weakly correlated with changes in salted herring roe prices—suggesting that these products are only weak substitutes.

Figure 6 compares salted herring roe wholesale prices with the total supply of Pacific herring roe and with estimated production and consumption of salted herring roe. (Estimated consumption was calculated by subtracting net inventory accumulation from production.) There has been an inverse relationship between wholesale prices and supply, production, and consumption. The large growth in total supply from 1984 to 1987 was accompanied by a fall in wholesale prices. A period of relatively stable supply was accompanied by relatively stable prices from 1987 to 1991. Higher supply in 1992-1994 was accompanied by another period of low prices. Prices rose as supply fell in 1995 and 1996, and fell sharply as supply, production, and consumption increased in 1997. In 1996 and 1998, the inverse relationship with price is most apparent for consumption—as economic theory would suggest.

In the short-term, salted herring roe wholesale prices are clearly influenced by supply conditions. Lower supply tends to drive prices up while higher supply tends to drive prices down. Prices serve to balance supply

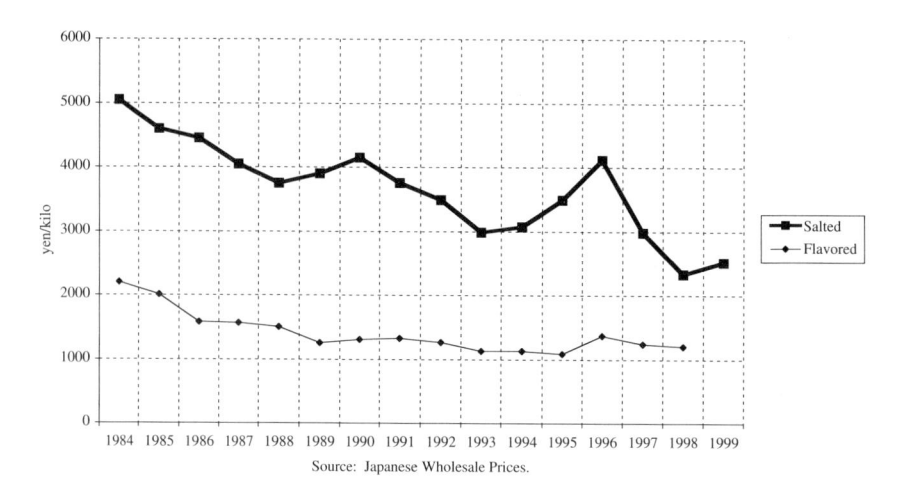

Figure 5. Japanese wholesale prices for herring roe.

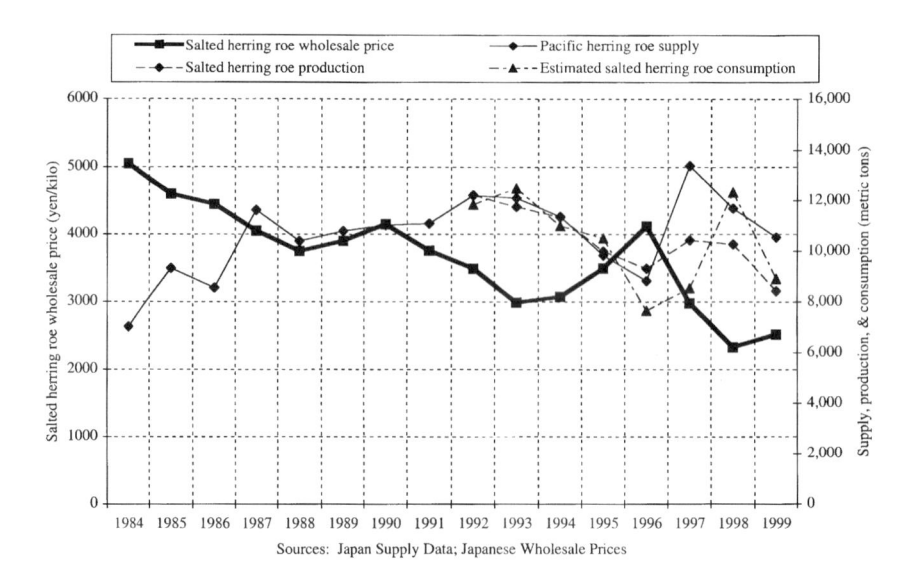

Figure 6. Japanese salted herring roe wholesale price, Pacific herring roe supply, and salted herring roe production and consumption.

and consumption over time (although not necessarily each year, as reflected in changing inventories). Higher prices tend to reduce consumption in years when supply is low; lower prices tend to increase consumption in years when supply is high.

The effects of supply on prices is also evident from frequent articles in the Japanese trade press that closely track Pacific herring roe harvests, imports, and production from American, Canadian, and Russian sources, and attribute changes in market expectations and prices to changes in supply.

The supply-price relationship is more complicated than a simple inverse correlation between total Pacific herring roe supply (or salted roe production or consumption) and average Tokyo wholesale prices. Supply conditions for specific types of Pacific herring roe (such as high-quality Canadian roe) affect prices for specific salted herring roe products. However, detailed supply and price data are not available to investigate these relationships for specific roe products empirically.

Above we discussed industry perceptions that Japanese demand for traditional salted herring roe is declining. The fact that salted herring roe wholesale prices have exhibited a declining trend since the late 1980s, without a corresponding upward trend in average consumption, provides empirical support for this perception.

Because changes in prices from year to year may reflect changes in both supply and demand, it is technically difficult to measure separately the effects of changes in demand and supply on prices. (Here we are using "demand" with the technical economic meaning of the price that buyers are willing to pay for a given total volume, and "supply" with the technical economic meaning of the price that sellers are willing to accept for a given total volume.) However, if prices decline while quantity sold remains constant or decreases, this indicates that demand must have declined, regardless of whether declining supply may also have contributed to the reduction in quantity sold. (In technical economic terms, a decline in both price and quantity sold implies an inward shift in the demand curve, regardless of whether an inward shift in the supply curve has also occurred.)

Table 1 shows average estimated consumption of salted herring roe and average wholesale prices for salted herring roe for four three-year periods. While average prices were 9% lower in 1994-1996 than in 1988-1990, average consumption was 10% lower. Similarly, even though average prices were 23% lower in the period 1997-1999 than in 1991-1993, average consumption was 16% lower. These comparisons suggest that the Japanese demand curve for salted herring roe is indeed shifting inward. Put differently, the prices that Japanese buyers are willing to pay for any given volume of salted herring roe are declining.

Table 1. Average Japanese salted herring roe consumption and wholesale prices of salted herring roe: three-year averages.

	Estimated consumption of salted herring roe (metric tons)	Average wholesale price of salted herring roe (yen per kilo)
Average, 1988-1990	10,737	3,933
Average, 1994-1996	9,713	3,561
Percent change	–10%	–9%
Average, 1991-1993	11,803	3,413
Average, 1997-1999	9,925	2,612
Percent change	–16%	–23%

Note: Estimated average consumption of salted herring roe was assumed to be equal to average supply of Pacific herring roe for years prior to 1992.

Alaska Herring Prices

Alaska herring prices are influenced not only by Japanese wholesale prices but also by the exchange rate between the yen and the dollar. As a result, trends in Japanese wholesale prices measured in dollars differ from trends measured in yen (Fig. 7). The value of the yen relative to the dollar increased sharply between 1984 and 1987 and again from 1992 to 1994. During both of these periods, although wholesale prices in yen declined, wholesale prices measured in dollars increased. In contrast, a fall in the value of the yen between 1995 and 1997 led to a much steeper decline in wholesale prices measured in dollars than in prices measured in yen. More generally, the long-term downward trend in Japanese wholesale prices for salted herring roe was offset for the Alaska herring roe industry in part by an upward trend in the value of the yen relative to the dollar.

Figure 8 compares Japanese wholesale prices for salted herring roe (measured in dollars per ton of herring round weight equivalent, assuming 10% yield) with prices paid for herring to Alaska processors and fishermen. Trends over time in prices paid to Alaska fishermen and processors clearly reflect trends in dollar wholesale prices in Japan. For example, the increase in Japanese wholesale prices between 1994 and 1996 was reflected in higher prices paid to Alaska processors and fishermen, while the decrease in Japanese wholesale prices between 1996 and 1998 was reflected in lower prices paid to Alaska processors and fishermen. The difference or "spread" between Japanese wholesale prices and Alaska ex-vessel prices has remained relatively constant over time.

The correlation between the three price series in Fig. 8 is consistent with a simple model of price determination for Alaska herring. Japanese wholesale prices—as measured in yen—are driven by supply and demand conditions for Pacific herring roe in Japan. Japanese wholesale prices as

Herring: Expectations for a New Millennium 735

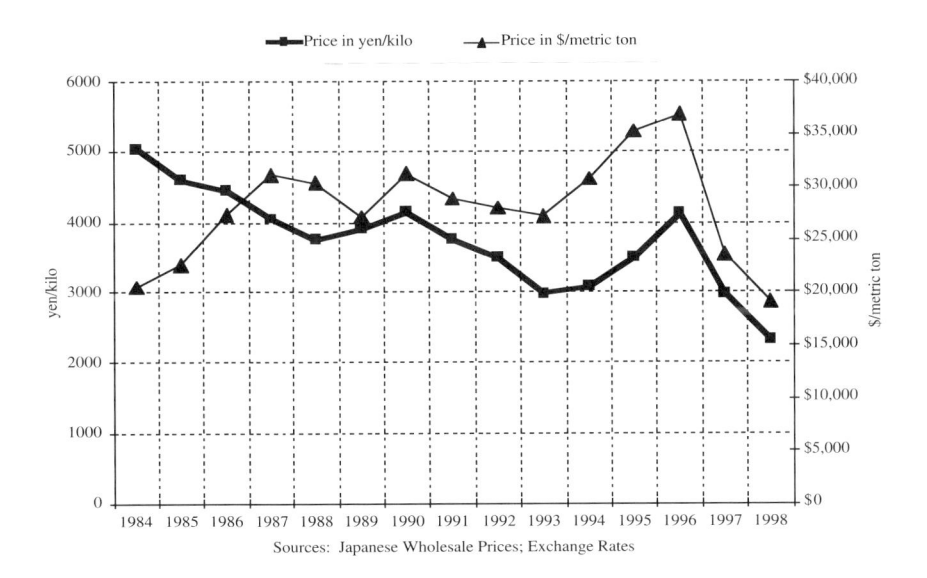

Figure 7. Japanese salted herring roe wholesale prices in yen and dollars.

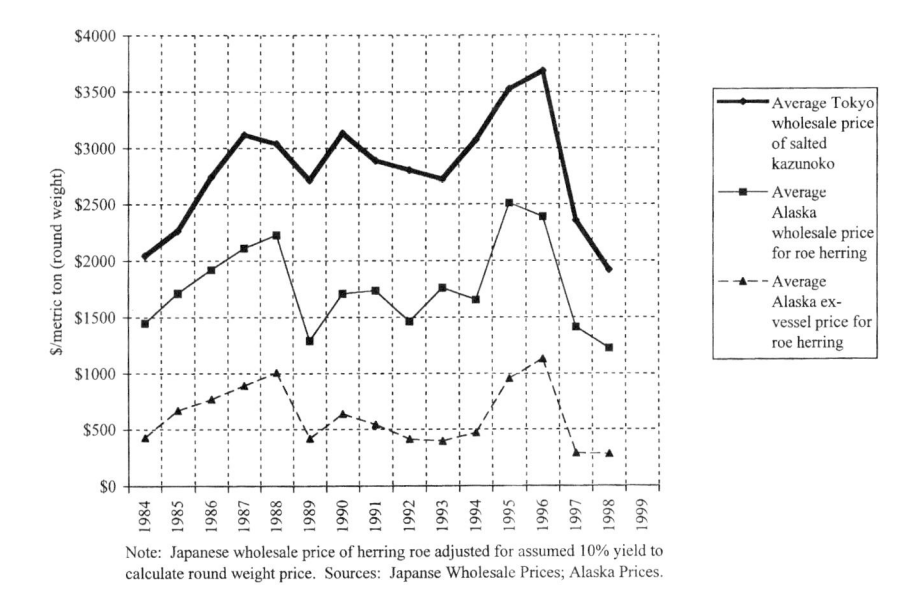

Figure 8. Herring prices: Japanese wholesale, Alaska wholesale, and Alaska ex-vessel.

measured in dollars—reflecting the value of the yen relative to the dollar—drive what Japanese importers are willing to pay Alaska processors for herring. In turn, the prices paid to processors drive what they are willing to pay to fishermen.

In this model, changes in Alaska ex-vessel herring prices reflect changes in Japanese wholesale prices for herring roe. The volatility of Alaska ex-vessel prices reflects the fact that because wholesale prices are much higher than ex-vessel prices (reflecting costs of processing and transportation) a given absolute change in wholesale and ex-vessel prices represents a much greater *relative* change in ex-vessel prices. Thus for example, both the wholesale price and the Alaska ex-vessel price increased by about $650 per ton between 1994 and 1996. However, in relative terms, the wholesale price increased by only 20%, while the ex-vessel price increased by 140%.

This simple model explains most of the year-to-year variation in Alaska herring ex-vessel prices. (A least squares regression of the annual change in the Alaska ex-vessel price against the annual change in the Japanese wholesale price for the period 1985-1998 had an R^2 value of 74%, indicating that 74% of the variation in the change in ex-vessel prices can be explained by variation in the change in wholesale prices.) However, the actual determination of Alaska ex-vessel prices is of course much more complex, reflecting year-to-year changes in not just wholesale prices but many other factors such as the relative mix of different kinds of herring roe in the Japanese market, processing and transportation costs, and changes in market conditions between the times at which processors buy from fishermen, importers buy from processors, and herring roe is sold at wholesale.

In addition, as was shown in Fig. 2, herring prices differ between Alaska roe herring fisheries, reflecting fishery-specific factors such as roe percent and egg size (which affect the roe value per round pound), fishery location, and harvest volume (which affect processing costs), the number of participating processors (which affects competition), and season timing (which affects when the fish are bought).

If Alaska herring harvests were highly responsive to ex-vessel prices, our simple "top-down" derived-demand model in which Alaska ex-vessel prices reflect Japanese market conditions would be theoretically inadequate to characterize Alaska ex-vessel price determination, because changes in derived demand would be partially offset by supply responses. For example, the price-dampening effects of a reduction in Japanese demand would be partially offset by reduced Alaska herring supply. However, in most districts, herring harvests appear to be price-inelastic, with quotas fully harvested even when prices are low.

Ideally, with enough data and resources it would be possible to develop an econometric model that could quantify the different ways in which supply and demand factors interact to determine prices at different levels in the herring roe market, and to explain and predict how each factor affects Alaska ex-vessel herring prices. In practice, however, this goal is probably elusive. The Japanese herring roe market is too complex—with too many sources of supply, too many end products, too many end markets, too many competing products, and too many factors affecting consumer demand—and too little data—to formally model this market with any great degree of statistical accuracy.

Conclusions

The Alaska roe herring fishery is almost completely dependent on the Japanese market for salted herring roe. Changes in prices paid to Alaska fishermen for herring reflect changes in Japanese wholesale prices for herring roe (as measured in dollars), but relative changes in prices to fishermen (in percentage terms) are much greater

The factors driving Alaska herring prices are for the most part beyond the control of Alaska herring fishermen or processors. Changes in the exchange rate between the yen and the dollar can magnify or offset changes in Japanese wholesale prices expressed in yen. The supply of Pacific herring roe to the Japanese market, which may vary widely from year to year, reflects not only Alaska harvests but also Canadian harvests—and Russian harvests may play an important role in the future as well. Thus Japanese market prices do not necessarily rise or fall to help offset the revenue impacts of lower or higher Alaska harvests.

Perhaps of greatest concern to the Alaska herring industry should be declining Japanese demand for traditional salted herring roe. If Pacific herring harvests remain at levels of the past decade, it is likely that Japanese wholesale prices for salted herring roe—and prices paid to Alaska fishermen—will remain low and may fall further.

Alaska herring roe harvests have been relatively stable from year to year. But the relative stability of Alaska's herring resource does not guarantee economic stability of the industry. Ex-vessel value has varied widely from year to year, and declined dramatically in the late 1990s. The long-term economic outlook for Alaska herring fishermen is uncertain. Alaska herring may serve as a reminder that viable commercial fisheries depend not only on fishery resources, but also on the markets from which resources derive their value.

Acknowledgments

In addition to materials listed in the references, this paper is based on discussions with numerous individuals familiar with many different aspects of the Alaska and Japanese herring industries. In the United States these include Bill Atkinson, editor of *Bill Atkinson's News Report*; Jay Hastings, attorney for the Japan Fisheries Association; Nancy Babcock and Ruichi (Rudy) Tsukada, trade specialists for the Alaska Division of Trade and Development; Herman Savikko and Fritz Funk of the Alaska Department of Fish and Game; and Yuko Kusakabe. In Japan they include Mineo Ouchi of Tokyo Seafoods; Junichiro Yoneoka of Maruha Corp.; Ichiro Sawaki and Shin-Ichiro Kaneko of K.K. Ocean Beauty Co.; Noriko Ebuchi, Shin-Ichi Gorai, and Masatoshi Mitsuhashi of Tohto Suisan Co.; Tetsuhiro Goto of Toshin Seafoods Co.; Koji Otsuka and Takoati Okada or Nichiro Corp.; Toshihiko Kajiwaki, Daishiro Nagahara, Osamu Ishikawa, and Toshiaka Takahashi of the Japan Fisheries Agency; H. Tsuda of Ito-Yokado Co.; Tomohiro Asakawa of the U.S. Embassy in Tokyo; Shinichi Tonochi and Makato Ishihara of Kyokuyo Co.; Mitsuyoshi Murakami and Hiroshi Ando of Japan Marine Products Importers Association; Masatsugu Nakagawa of Mitsubishi Corp.; Lee Sung Hwa representing Alaska Seafood Marketing Institute in Tokyo; Kenji Sakota, Makoto Matsumura, and Masatoshi Nishikawa of Daito Gyorui Co.; Katsushi Fukumoto and Masahiro Abe of Nichirei Corp.; Takumi Kuono of The Daiei, Inc.; Takeo Hamamoto of Hamamoto Shoten Co.; Keji Ihara and Makato Sasaki of Ihara Suisan Co.; Yasukata Kato and Hisashi Tsuzuki of KatoSuisan Co.; Hiroshi Yogo of Rumoi fisheries Products Processors Cooperative Association; Jyunya Takeuchi, Ryo Mayama, and Takeshi Tamamura with the Marine Products Division of the Hokkaido government; and Kujiro Abe and Mariko Kuroda, contractors for the Alaska Division of Trade and Development in Tokyo.

References

Anderson, J.L., and Y. Kusakabe. 1993. Analysis of the effect of the *Exxon Valdez* oil spill on the Japanese herring roe market. Report prepared by J.L. Anderson Associates, Inc., Narragansett, Rhode Island.

Anderson, J.L., J.T. Gledhill, and Y. Kusakabe. 1989. The Japanese seafood market: Herring roe. Report prepared for the Canadian Department of Fisheries and Oceans, Economic and Commercial Analysis Directorate.

BANR (Bill Atkinson's News Report). 1992. Bill Atkinson's News Report, Seattle. Oct. 14, 1992.

BANR. 1993. Bill Atkinson's News Report, Seattle. Jan. 27, 1993.

BANR. 1994. Bill Atkinson's News Report, Seattle. Mar. 16, 1994.

BANR. 1996. Bill Atkinson's News Report, Seattle. Oct. 16, 1996.

BANR. 1998. Bill Atkinson's News Report, Seattle. Jan. 21, 1998.

BANR. 1999. Bill Atkinson's News Report, Seattle. Feb. 17, 1999.

BANR. 2000a. Bill Atkinson's News Report, Seattle. Jan. 26, 2000.

BANR. 2000b. Bill Atkinson's News Report, Seattle. Feb. 2, 2000.

Hastings, J. (ed.). 1995. Evolution of the Pacific roe herring fisheries and the Japanese market. Japan Fisheries Association Newsletter. Tokyo. August 1995.

Skud, B., H. Sakuda, and G. Reid. 1960. Statistics of the Alaska herring fishery, 1878-1956. U.S. Fish and Wildlife Service Statistical Digest 48.

Talley, K. 1999. Seafood Trend, January 25, 1999.

Appendix I: Data Sources

This appendix describes sources for all data provided in the text, tables and figures for this paper. Data for Alaska herring harvest volume and ex-vessel value for the years 1977-1998 (Figs. 1 and 8) were calculated for the years 1977-1998 by summing data for individual herring fisheries from the "Basic Information Tables" posted on the Web site of the Alaska Commercial Fisheries Entry Commission (www.cfec.state.ak.us); data for 1999 are preliminary data from the Alaska Department of Fish and Game Web site (www.cf.adfg.state.ak.us). Data for average ex-vessel prices in selected fisheries (Fig. 2) are also from the Commercial Fisheries Entry Commission Basic Information Tables. Data for North American Roe Herring Landings (Fig. 3) are from the annual "Statspack" issue of Pacific Fishing Magazine. Data for Japanese Pacific and Atlantic herring roe supply (Fig. 4) for 1981-1989 are from Anderson and Kusakabe (1993), Table 2.7, p. 2.17; data for 1990 are Japanese trade press estimates reprinted in BANR 1992; data for 1991-1999 are Japanese trade press estimates reprinted in BANR 2000a. Data for salted and flavored herring roe production (Figs. 4 and 6) were calculated from Japanese trade press estimates of herring roe utilization reprinted in BANR 1993, 1994, 1996, 1999, and 2000a.

Data for Japanese wholesale prices for salted herring roe (Figs. 5, 6, and 7 and Table 1) for 1994-1990 are estimated from a graph of monthly Tokyo Central Wholesale Market Prices provided in Anderson and Kusakabe (1993, p. 2.22), weighted 25% by November prices and 75% by December prices; data for 1991-1998 were calculated from Tokyo Central Wholesale Market (TCWM) monthly data by dividing the total annual sales value by total annual sales volume. Data for 1999 were estimated by adding to the 1998 price the increase between 1998 and 1999 in the average salted herring roe wholesale prices for six Japanese wholesale markets for September-December, reported in BANR 2000b. Data for Japanese wholesale prices for flavored herring roe (Fig. 5) for 1994-1990 are from BANR 1996; data for 1991-1998 were calculated from Tokyo Central Wholesale Market (TCWM) monthly data by dividing the total annual sales value by total annual sales volume. Estimated salted herring roe consumption (Fig. 6 and Table 1) was calculated by subtracting net inventory accumulation (the year-to-year increase or decrease in end-of-January salted herring roe inventories) from salted herring roe production.

Average annual Japanese salted herring roe wholesale prices in dollars (Figs. 7 and 8) were calculated in the same way as average annual prices in yen, except that monthly prices were first converted to dollars per metric ton using monthly exchange rate data from the Web site of the Federal Reserve Bank of St. Louis (www.stls.frb.org/fred/data/exchange/exjpus). Data for first wholesale value and average wholesale prices paid to Alaska processors for roe herring (Fig. 8) were provided by the Alaska Department of Fish and Game and are based on annual Commercial Operator Annual Reports filed by Alaska herring roe processors.

Linking Biological and Industrial Aspects of the Finnish Commercial Herring Fishery in the Northern Baltic Sea

Robert Stephenson
Finnish Game and Fisheries Research Institute, Helsinki, Finland and Fisheries and Oceans Canada, St. Andrews Biological Station, St. Andrews, Canada

Heikki Peltonen
Finnish Game and Fisheries Research Institute and Finnish Environment Institute, Helsinki, Finland

Sakari Kuikka
Finnish Game and Fisheries Research Institute, Helsinki, Finland

Jukka Pönni
Finnish Game and Fisheries Research Institute, Kotka, Finland

Mika Rahikainen
Finnish Game and Fisheries Research Institute and University of Helsinki, Department of Limnology and Environmental Protection, Helsinki, Finland

Eero Aro
Finnish Game and Fisheries Research Institute, Helsinki, Finland

Jari Setälä
Finnish Game and Fisheries Research Institute, Turku, Finland

Abstract

Herring is marketed for both fodder and human consumption in Finland. For the human consumption market, herring exceeding 36 g weight is preferred, whereas the fodder herring can be of any size. The main demand of fodder is based on fur production, and the rapid changes of fur markets have had a large impact on the herring fishery. There are variations not only in the markets but also in biological factors. The growth rate of northern Baltic herring has declined substantially in recent years due primarily to changes in prey of herring. The key factors determining the prey resources are the changing hydrography in the Baltic Sea and competition with sprat. It is also possible that changes in the cod stock, modification of fishing gear, and changes in fishing effort may have influenced the mortality rates of herring. Changing biological and market conditions have resulted in significant unforeseen changes in the location, composition, and behavior of the herring fisheries. Fishing in recent years has been concentrated in the Bothnian Sea, and an increasing share has been taken by large trawlers. As the demand of herring for fodder declined the unreported rejection of catches (discarding) may have increased. In addition, because of increases in trawl size and increase in fishing effort in the most important fishing area, there has been an increase in the number of fish escaping through the mesh—but this may have resulted in a higher subsequent mortality rate of herring due to trawl injury. These factors have reduced the production capacity of herring for the human consumption market during the 1990s. In this review, we formulate the driving bioeconomic factors of Finnish herring fishery and demonstrate the magnitude of changes, which are largely unpredictable and uncontrollable by fisheries management.

Introduction

Baltic herring is a dominant species in the Finnish fishery. Approximately 90,000 t of herring have been landed annually in the Finnish fishery in recent years (Fig. 1; Ahvonen et al. 1999). This represents by far the largest landings of a single species in Finland, and makes up over 50% of the total (marine and freshwater) landings. Recent landed value in the herring fishery has been on the order of 80 million Finnish marks (14 million euro); over 40% of the value of the commercial catch and 16% of the total landed value (commercial plus recreational) of the Finnish fishery.

Baltic herring assessments are conducted annually within the International Council for the Exploration of the Sea (ICES) Baltic Fisheries Assessment Working Group (ICES 1999a). The assessment is peer reviewed by the ICES Advisory Committee on Fisheries Management (ACFM) and advice is provided annually to the International Baltic Sea Fisheries Commission (IBSFC) (e.g., ICES 1999b). Quotas and management strategies are discussed and agreed within the IBSFC, and implemented by national governments.

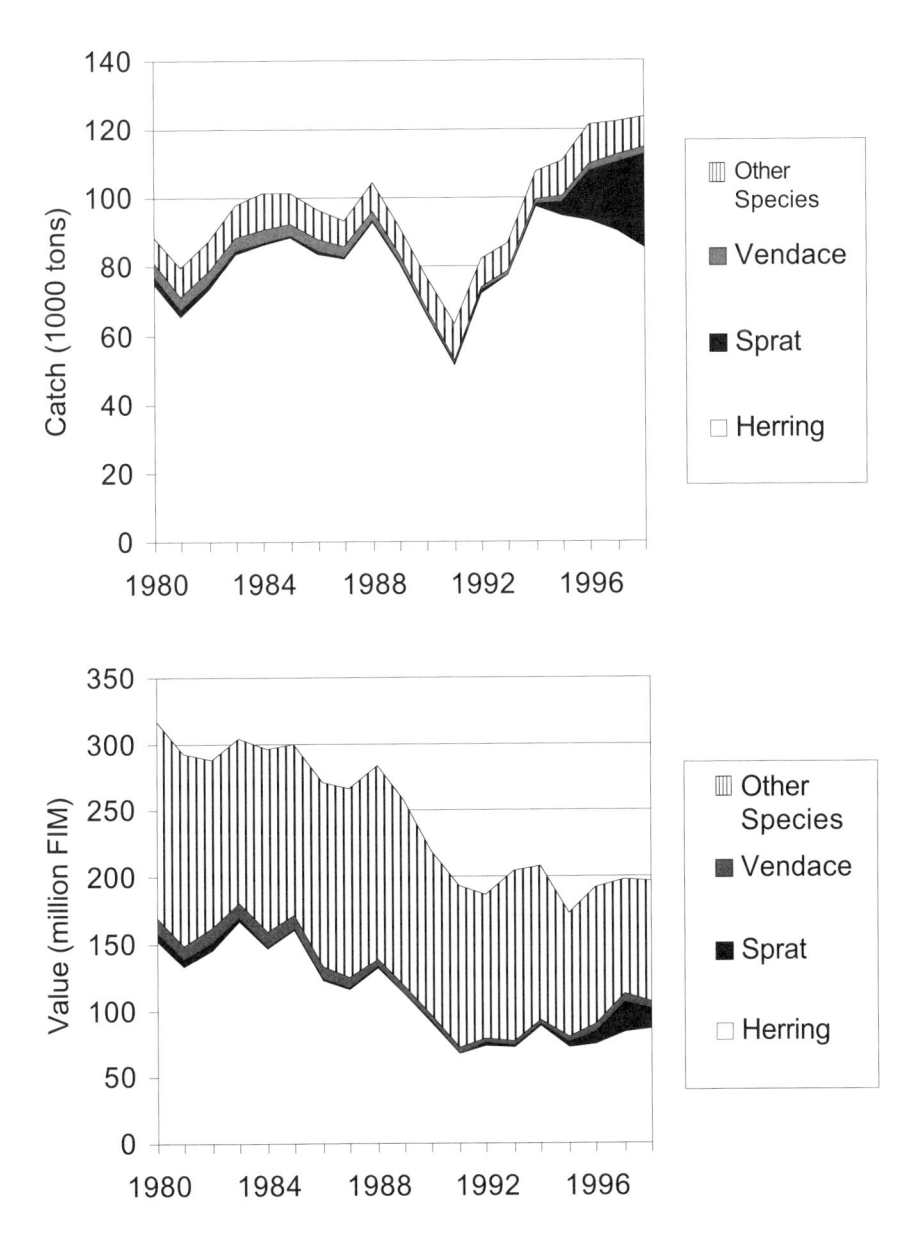

Figure 1. Annual landings and landed value of herring and of other major species in the Finnish fishery, 1980-1998. Vendace = Coregonus alba; sprat = Sprattus sprattus.

Regulation and management are complicated by the multinational jurisdiction over management units of the Baltic Sea, but even more by major perturbations in the Baltic ecosystem which are impacting herring, and by market factors. Past fishery evaluations focusing primarily (almost exclusively) on traditional biological aspects of stock assessment have been imprecise, and have had little predictive capability. We contend that this is due largely to the impact of changing biological and industrial aspects of the fishery that are currently not incorporated into evaluation and management. There is a need for not only improved understanding of the underlying biology, but also a greater understanding and appreciation of the bioeconomic context, issues, and constraints influencing this fishery.

In this paper, we summarize and link the key biological and industrial aspects of the Finnish herring fishery in the northern Baltic Sea.

The Finnish Herring Fishery

The Finnish herring fishery currently involves approximately 225 trawlers (mean length = 16.2 m, mean gross registered tonnage (GRT) = 8 t) specializing in herring and sprat, and a substantial number of smaller coastal vessels (mean length = 6.6 m) using traps. Specialized trawlers fish for a variety of human consumption markets (fillet, whole fish/fish markets, and export) and provide "fodder" for the fur industry. Smaller vessels of the coastal fishery supply herring primarily for fodder.

Almost all of the Finnish commercial herring fishery takes place in the northern Baltic Sea (ICES Subdivisions [SD] 29, 30, 31, 32) (Fig. 2a). Herring fishing takes place all along the Finnish coast, but is currently concentrated in the southern part of the Gulf of Bothnia, (SD 30) and the Archipelago sea (SD 29N) (Fig. 2b). Total Finnish commercial herring landings have fluctuated in the past two decades, between 52,000 t (1991) and 98,000 t (1994) (Fig. 1). There have been substantial changes in the relative contributions of various areas along the coast. Landings from the Bothnian Sea (SD 30) have increased over most of the 20-year period, and there has been substantial reduction in landings from SD 29+32 particularly in the early 1990s (Fig. 3).

Changes in Abundance of Herring and in Relative Abundance of Sprat

The herring of the Gulf of Finland and Archipelago Sea are assessed together with the areas to the south in a combined assessment of SD 25-29 (i.e., Main Basin, including Gulf of Riga; and SD 32, Gulf of Finland). Spawning stock biomass (SSB) of this complex in 1998 was estimated to be on the order of 600,000 t (ICES 1999a,b), and to have decreased steadily from about 1.5 million t in the past two decades for which there is an assessment. Recent annual landings from this combined area (all nations) have been approximately 200,000 t.

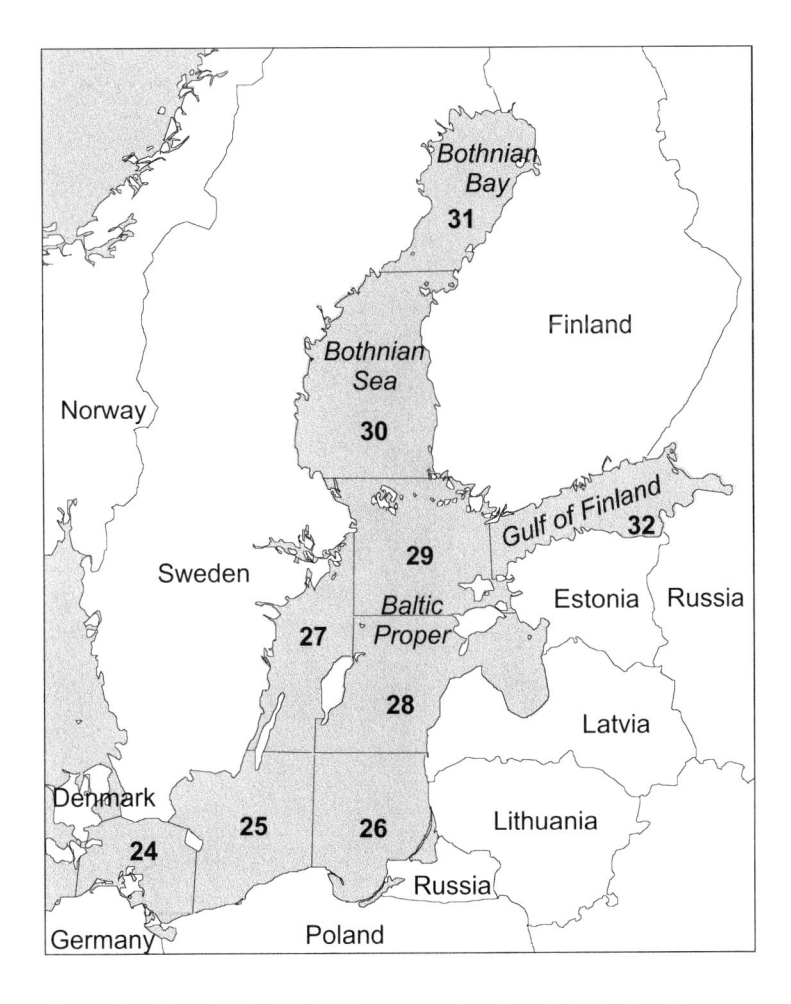

Figure 2a. Map of Finnish fishing areas showing statistical units.

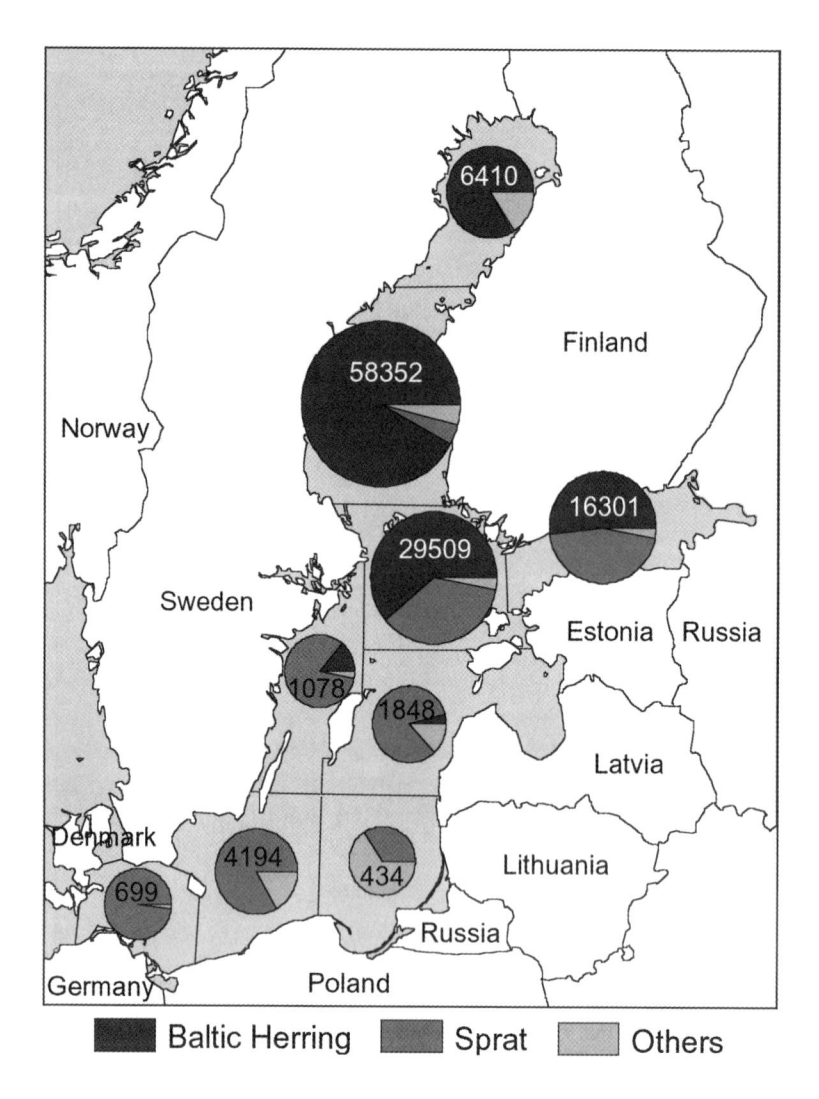

Figure 2b. Relative distribution of Finnish commercial herring fishery landings (metric tons in 1998).

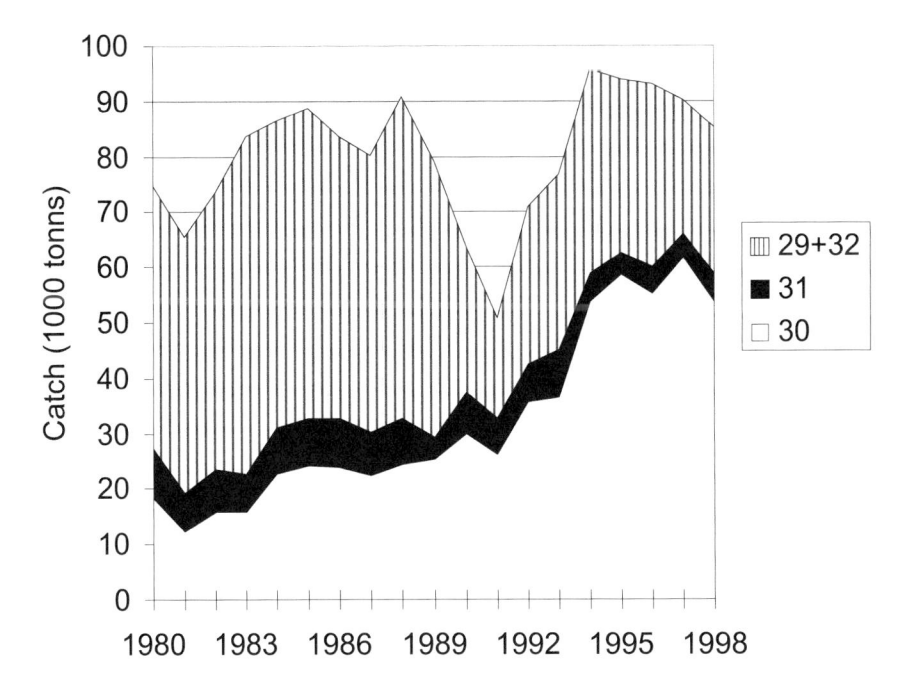

Figure 3. Finnish herring landings from the Bothnian Bay (SD 31), Bothnian Sea (SD 30), Archipelago Sea (SD 29), and Gulf of Finland (SD 32), 1980 to present.

Separate assessments of the Gulf of Finland, Archipelago Sea, and Gulf of Bothnia, undertaken by the Baltic Fisheries Assessment Working Group, are not precise. Assessments in 1998 (ICES 1998) indicated, however, that the three units of particular relevance to the Finnish fishery are of different size (in area and in resource) and that the abundance of herring has fluctuated differently in the three areas over the past two decades. The most recent assessments indicate that SSB in SD 29+32 fluctuated between 250,000 and 300,000 t between 1980 and 1994, but that it has decreased to below 200,000 t in recent years. SSB of herring in SD 31 has been an order of magnitude smaller (20,000-30,000 t) throughout the same period. The herring of SD 30 apparently increased from about 100,000 t in the early 1980s to 300,000 t in 1994, but has subsequently decreased to about 200,000 t (Fig. 4).

Sprat, which is taken together with herring in the trawl fishery, has increased in abundance over the past two decades, and particularly in the 1990s (Fig. 5). Sprat have been particularly prevalent in SD 29 and 32 and have increased in catches in recent years.

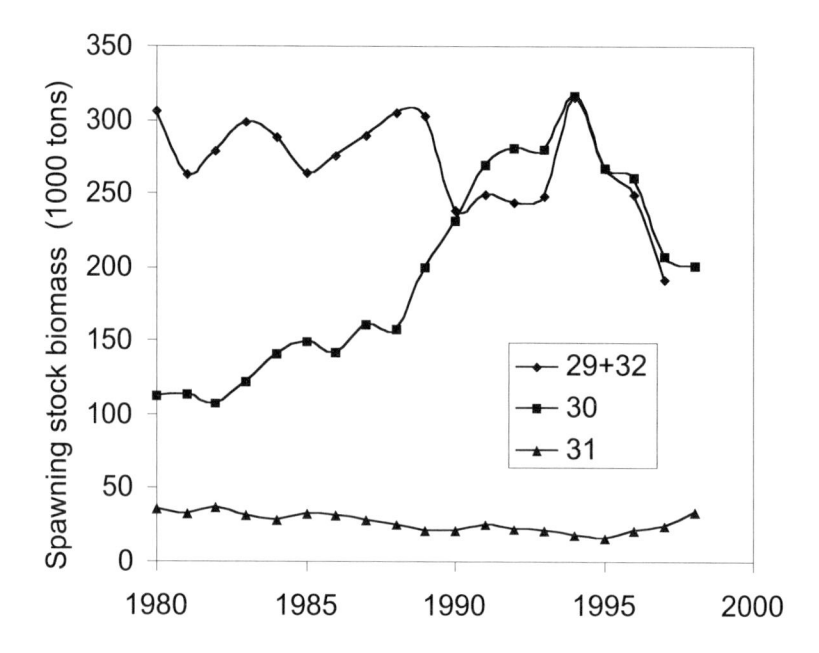

Figure 4. Recent estimates of herring spawning stock biomass (SSB) for Baltic SD 29+32, 30 and 31. SSB estimates from ICES 1998, 1999a.

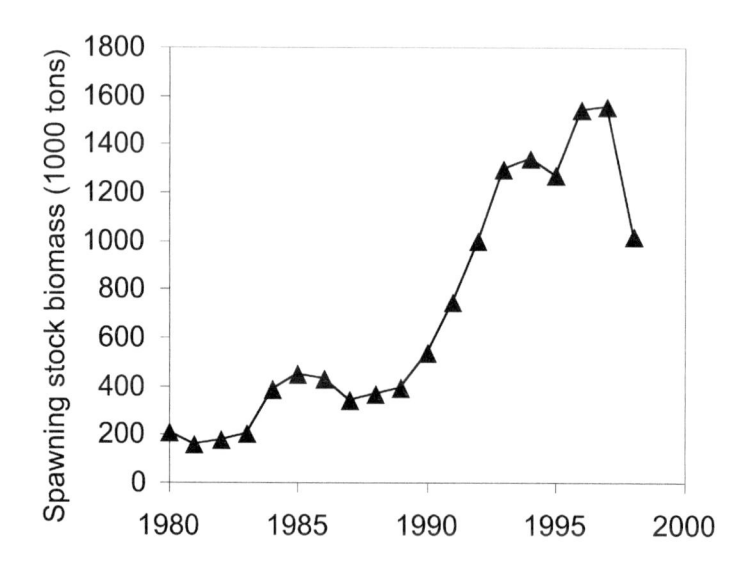

Figure 5. Estimated SSB of sprat in the Baltic Sea; 1980-1999 (ICES 1999a).

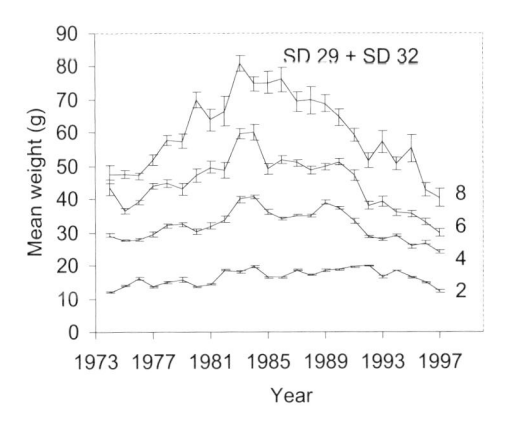

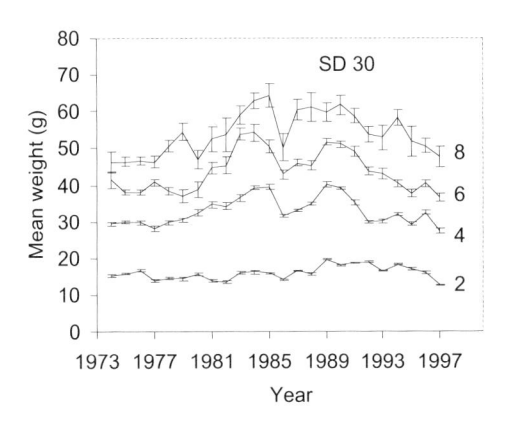

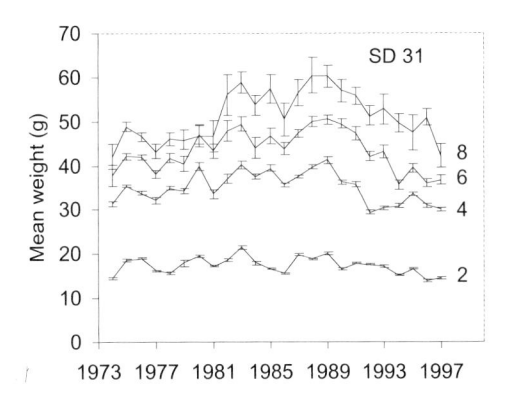

Figure 6. Mean weight at age of herring (quarter 2) in SD 29+32 (top), 30 (middle) and 31 (bottom), 1974-1997.

Changes in Growth of Herring

A dramatic feature of the biology of herring of the Baltic has been a large change in size at age. During the past 25 years in which there has been consistent and intensive sampling of the commercial fishery, mean weight at age has shown a progressive increase from the mid-1970s to the mid-1980s and a subsequent decrease during the late 1980s and 1990s (Fig. 6). The change in growth has been most pronounced in the Gulf of Finland and Archipelago Sea (SD 32 and 29), and in older ages (where the weight at age at the end of the 1990s is approximately half of what it was in the mid-1980s), but the trend is apparent in all areas and ages.

Recent studies have hypothesized links between the pronounced reduction of herring growth in recent years and changes in the zooplankton community related to changing hydrographic conditions, primarily reduced salinity. A prominent feature of the Baltic ecosystem is low salinity. The relatively high volume of fresh water inflow and restricted connection to the Atlantic Ocean result in salinity of about 8 parts per thousand (ppt) in the surface water of the Baltic proper, and below 6 ppt in the Gulf of Bothnia. Annual variations in the intrusions of saline water from the North Sea have caused periods of relatively higher and lower salinity which have had a profound impact on the Baltic ecosystem. In the twentieth century, relatively low salinity prevailed until the 1930s. Salinity began to increase slowly in 1930s and the increase continued for about two decades (Alenius and Haapala 1992). There was some decline in salinity during 1960s but in the 1970s another increase occurred particularly in the Gulf of Bothnia (Samuelsson 1996). In the 1980s and 1990s salinity has decreased almost continuously and reached low levels compared to the earlier decades of the twentieth century (Alenius and Haapala 1992, Samuelsson 1996, Hänninen 2000). Low saline water inflow in recent years, for example, has meant an increase in stagnation and depletion of oxygen resources in the lower layer of the Baltic Main Basin, and this is thought to have reduced suitable spawning habitat for cod. The papers of Flinkman et al. (1998), Vuorinen et al. (1998), Hänninen et al. (2000), and Hänninen (1999) have proposed linkages between reduced herring growth and changes in zooplankton size and species composition. Changes in freshwater runoff and saline water inflow have been linked to changes in the North Atlantic oscillation. The resulting reduced salinity in recent years is hypothesized to have caused a reduction in large neritic copepods, the preferred food of herring, leading to reduced herring growth rate. Flinkman et al. (1998), and Flinkman (1999) have extended the possible linkage of hydrography to reduced reproduction of cod, a reduction in predation pressure on herring, and subsequent density-dependent reduction in herring growth—suggesting, therefore, the possibility of both "bottom up" and "top down" mechanisms influencing herring growth. It is also reasonable to hypothesize some competition between herring and sprat (which has increased

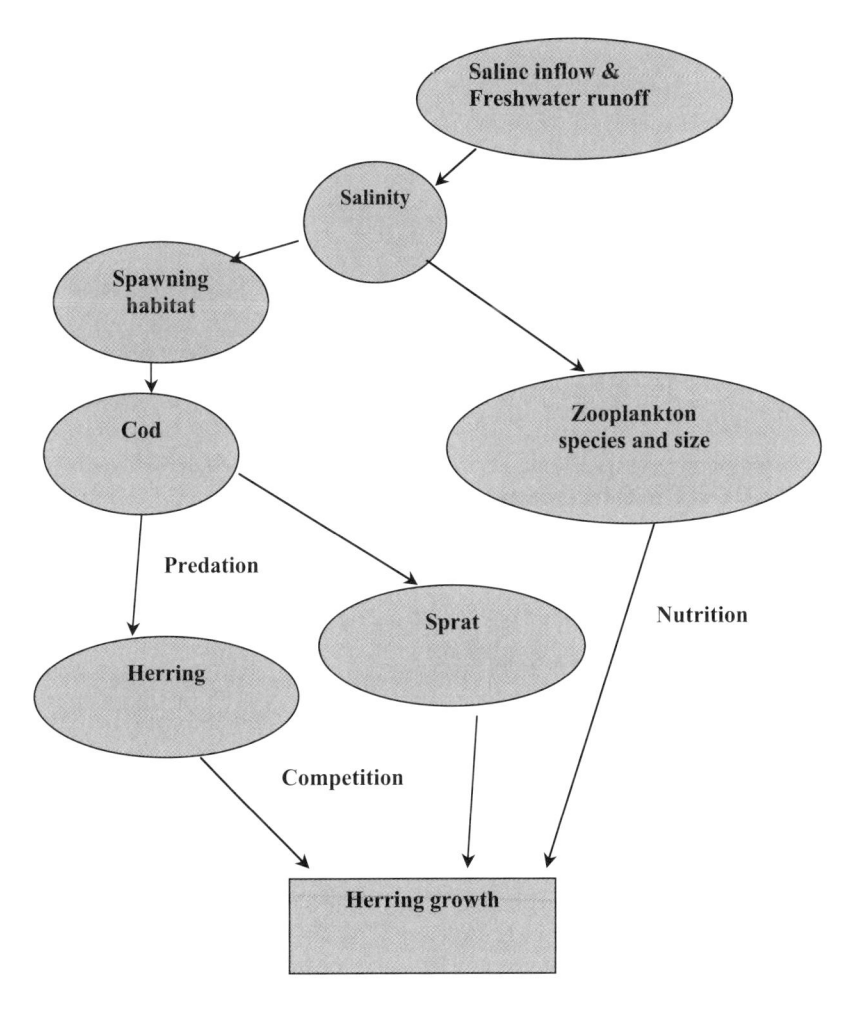

Figure 7. Major hypothesized linkages to herring growth in the Baltic Sea ecosystem.

substantially in recent years), and sticklebacks in the Gulf of Bothnia. These major hypothesized links are shown conceptually in Fig. 7.

Growth changes have had a large and negative effect on the ability to assess herring and to provide advice. The unpredictable growth rate has a major impact on the prediction of future spawning capacity. A usual practice is that the mean weight at age used for projections is based on the last year or an average of the last few years. In a case that there is a decreasing trend in the growth rate, this leads to an overly optimistic biomass prediction for a given TAC, i.e., fishing mortality will be higher than expected. In addition, if one assumes that the maturity at age is independent of weight at age (as in main basin stock assessment, ICES 1999a), it is obvious that the biomass and catch projections are too optimistic which leads to an underestimation of risk of stock collapse.

Market Preferences and Prices

There are three major markets for herring (Fig. 8): human consumption, fodder, and export. The "human consumption" market includes fish landed for domestic fish markets, and for processing (primarily fillets). This market prefers fish greater than 36 g. Herring landings for human consumption in the Finnish commercial fishery have decreased from 29,000 t to 17,000 t. Demand has decreased as market share has been lost to farmed and imported fish; however, the share of processed fish has increased over the past 20 years.

The "fodder" market is fish landed for animal feed (primarily mink and fox fur farms). This market will accept fish of any size. Fodder has been the largest market over the past two decades, but has been variable (between 30,000 to 65,000 t per year).

During the 1990s a large new export market opened for the Finnish herring fishery. There has been an export market of frozen and fresh herring primarily to Russia (but recently also to Estonia) in some years. Russian processors prefer herring smaller than 32 g. This market has become very important in recent years and the quantity exported in 1997 (about 14,300 t) exceeded domestic processing that year.

Figure 9 shows prices (adjusted to the consumer price index) of herring landed to Finnish markets over the past two decades. The removal of price subsidies from the fodder market (1986) and from the "human consumption" market (1995) were followed by substantial reductions in price. Prices have fluctuated within years but have generally decreased (adjusted to consumer price index) over this period.

Impact of Biological and Market Factors on the Finnish Herring Fishery

The management system in place for Baltic herring has been largely unrestrictive. Management strategies, including quotas, are developed by the

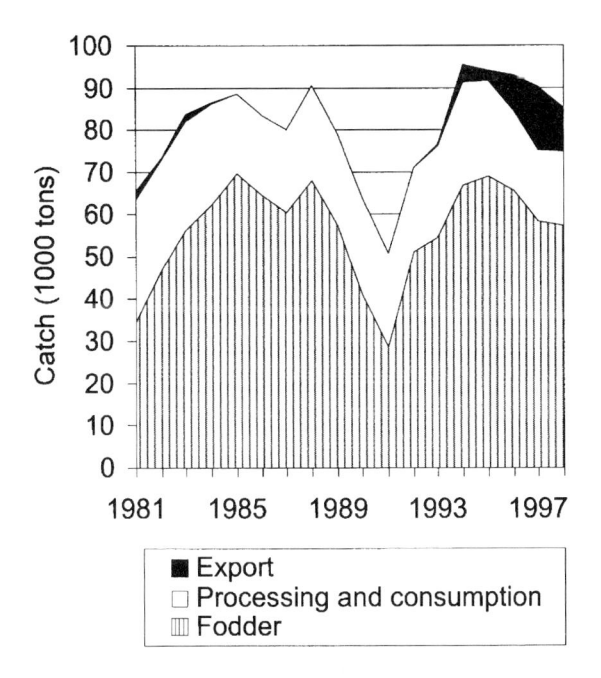

Figure 8. Finnish landings of Baltic herring by market, 1981-1998.

International Baltic Sea Fisheries Commission (IBSFC) and implemented by national governments. To accommodate sharing arrangements in these multinational fisheries, quotas have in many cases been set well above the scientific advice from ICES, and have been so high that they have not restricted the fishery. In addition, monitoring and enforcement mechanisms have been insufficient to control the fisheries. Aside from the national quotas, there have been few management measures in Finland.

In spite of the lack of management regulation, however, other elements have restricted the fishery. Some of these have been related to aspects of herring biology, but the greatest impacts have been related to market demand and prices.

Requirements of the processing industry for herring of particular sizes have had large impact on the location and amount of landings. The fillet market prefers herring greater than 36 g. As the growth rate of herring declined in the Gulf of Finland and Archipelago Sea in the late 1980s and early 1990s herring of suitable size for that market (>36 g) decreased, forcing the fishery to move into the adjacent Bothnian Sea (SD 30). Records of the number of trawl hauls by area (Fig. 10) show a reduction in hauls in SD 29+32 and an increase in the adjacent area 30. Figure 11 shows that

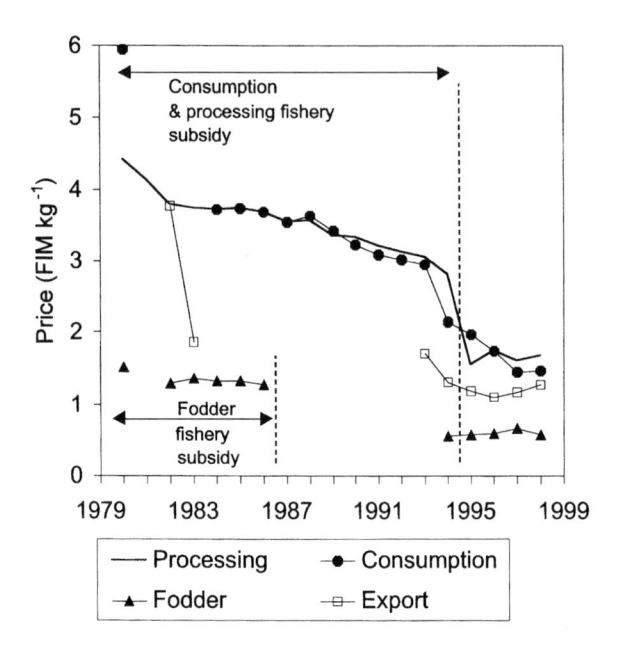

Figure 9. Prices (landed value adjusted to the consumer price index) of herring landed to Finnish markets, 1980-1998.

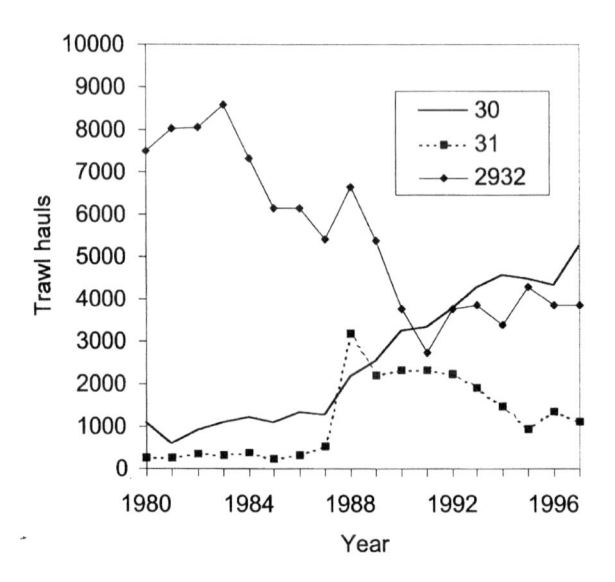

Figure 10. Annual number of pelagic trawl hauls by subdivision of the Finnish herring fishery, 1980-1997

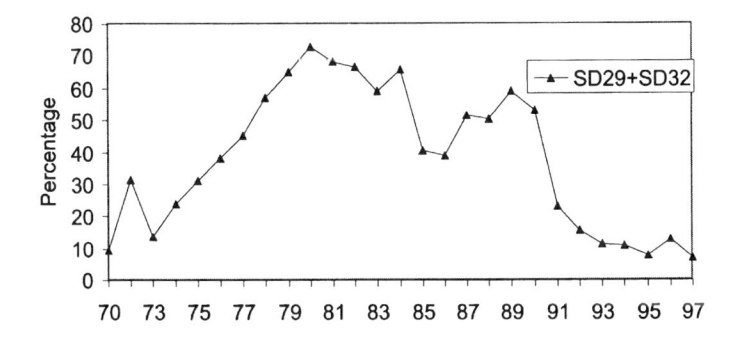

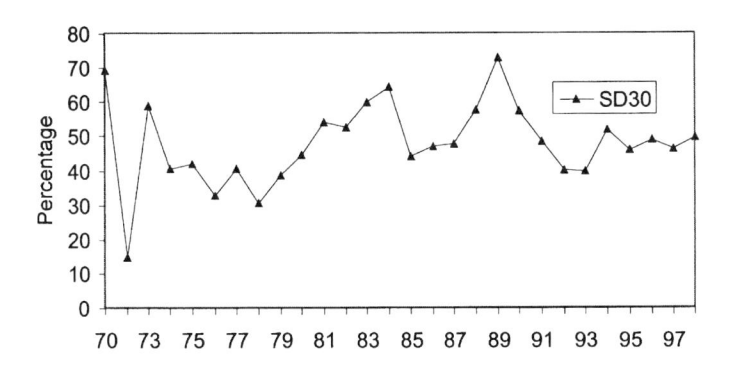

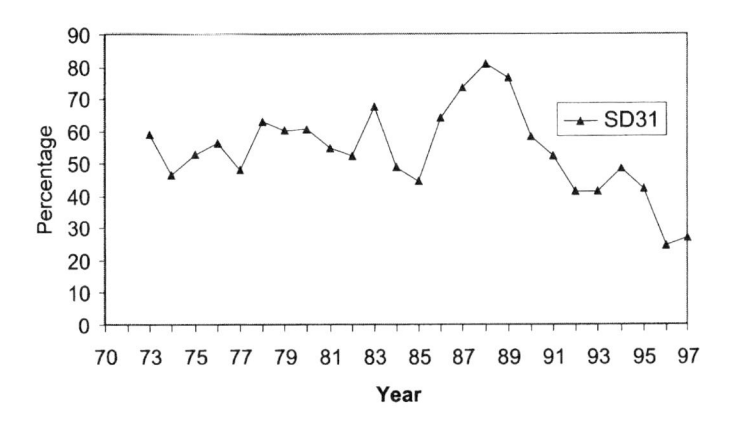

Figure 11. Annual percentage of herring greater than 36 g by weight from pelagic trawl samples in SD 29+32, 30 and 31 in the Finnish herring fishery, 1970-1997.

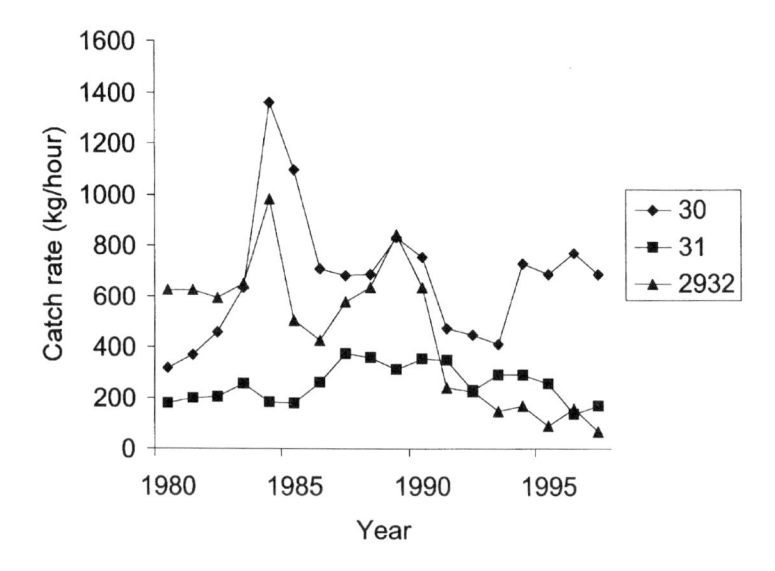

Figure 12. Catch rate of herring suitable for the processing market in three areas of the Finnish herring fishery, 1980-1997.

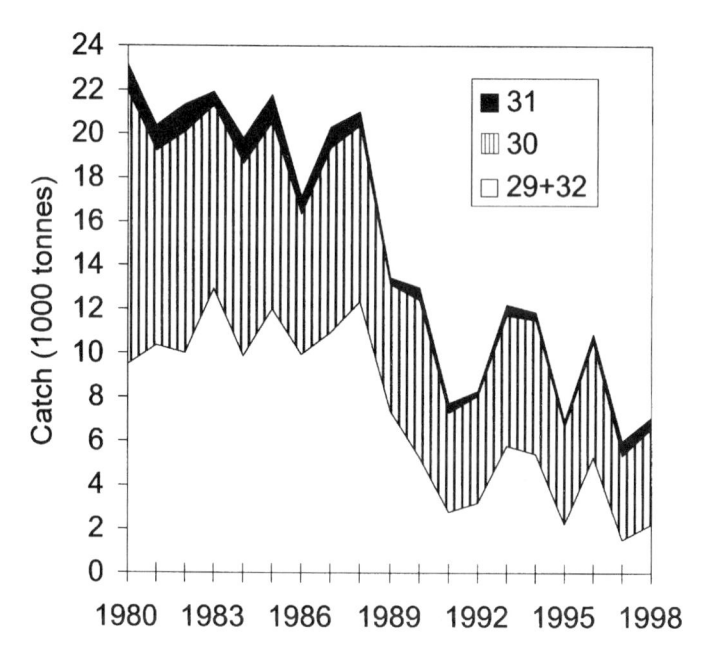

Figure 13. Annual landings from the herring trap fishery in SD 29+32, 30, and 31 of the Finnish herring fishery, 1980-1998.

the fraction of herring greater than 36 g was very low in SD 29+30 after 1990. Figure 12 summarizes the situation, by showing that the catch rate of herring suitable for the market remained high in SD 30 but dropped to a very low level in SD 29+32 between 1989 and 1992.

Perhaps the greatest market impact has been the variations in the demand for fodder caused by the demand in the fur trade. Reduction and expansion of the fur industry were responsible for the large drop in herring landings in 1989 and the increase after 1991. Herring for the fodder market is used almost exclusively for fur production. Finland is a world leader in the production of farmed fox and mink furs, and the state of that industry is sensitive to world demand and prices for fur.

There have been significant changes in the herring fishery, related in part to the factors mentioned above. There has been a move by some trawlers to the Bothnian Sea in recent years. This began as an extension of an annual movement. Prior to 1990, several vessels of this fleet would move in autumn from the eastern Gulf of Finland to the west and from the Bothnian Bay to the Bothnian Sea to avoid early winter ice cover and to fish large herring that moved to those areas in autumn. At the same time fishermen have invested in larger vessels and effective chilling techniques to ensure steady supply to markets.

A number of factors are thought to have contributed to a decline in landings by the trapnet fishery (Fig. 13). The abolishment of the price support system for fodder herring (processors had been subsidized to freeze and store herring for fodder) reduced the profitability of this fishery about 1990. The seasonal trapnet fishery, along with small seasonal trawlers, were less competitive in the market than the large trawlers that could supply fish to processors more steadily year round. At the same time there was a decrease in the catches, especially of large herring in the Archipelago Sea, and many trapnet fishers went out of business or retired without passing on the fishing rights to a family member or selling them to others. The trapnet fishery did not recover when the fodder market improved.

The herring export market has been growing rapidly, but it has exhibited rapid fluctuations in size due to changes in the economies of the principal market countries.

There have been a number of changes in the fishery in recent years in response to the general decline in prices, market requirements, and changing catch per unit effort of large (> 36 g) herring. As the demand for herring fodder decreased, the unrecorded rejection of catches (discarding) is thought to have increased. In an attempt to improve profitability, most trawlers have increased the size of trawls being used, increased trawling time, and changed areas of operation. Due to the mortality of herring escaping from trawls, increased swept areas may be causing additional unseen mortality (e.g., Suuronen et al. 1996), and reduced productivity of the stock.

Table 1. Timeline of major events in the Finnish herring fishery.

1980	Salinity begins decline of two decades
1982	Growth rate peaks, and begins to decline
1986	Removal of fodder subsidy; price declines
1989	Reduction in fur industry reduces fodder market
1989	Trapnet fishery declines, and does not recover
1989	Sprat stock begins to increase rapidly
1990	Absence of large fish (>36 g) in southern area, trawl fishery effort becomes greatest in northern areas
1991	Growth in fur industry expands fodder market
1993	Export market (foreign trade) begins to improve
1995	Removal of food market subsidy; price declines

Discussion

Several factors external to the current fisheries management system have had large impacts on the Finnish herring fishery in recent years (Table 1). Environmental changes influencing herring growth rate, and market demand for amount and size of fish have been the major driving factors. A large fraction of the total landings has been for use as fodder, and total landings therefore have been particularly sensitive to the demand for fodder herring, which is directly linked to changes in the demand of the fur industry. Changing environmental conditions and ecosystem structure (primarily the impact of the change in salinity and resulting changes in predator and prey abundance) have apparently resulted in a change in the growth rate of herring, and the fishery has been forced to adapt to the quick changes of the proportion of herring of suitable size for filleting.

In addition to the obvious impact on total landings, there have been a number of more subtle changes to the fishery including changes in location of fleets, changes in location of processing plants, and changes in the type of fish available to markets. Market demand for large fish of suitable size for filleting was largely responsible for a shift in the location of the fishery, to the Bothnian Sea.

These significant events in the fishery and changes in the industry have been unforeseen. They have been the result of environmental perturbations and market considerations that are largely beyond management control. They are important to the understanding and management of the Finnish fishery, yet to this point they have not been studied, and are not addressed in evaluation or in advice.

This raises an essential question about the role of fisheries management in this type of a situation. The quick decrease of large herring was to large extent unpredictable, and the management system in place has no real tools to respond even if predictions were perfect. One could argue that the total allowable catch (TAC) could be used to prevent the unsustainable use, but it is, at least if used alone, insufficient to control this system on the spatial and temporal scale that is required, and cannot be used to achieve socioeconomic objectives which appear to be relevant. There appear to be three reasons that the current TAC-based system is insufficient. First, the assessment of the present state of the population is imprecise. Second, even if the present state was known precisely, the management measure (TAC) is too aggregated and unenforceable to be effective. Third, the calculation of any socioeconomic valuation must be based on uncertain biological as well as economic variables. These variables change so quickly and unpredictably that it is difficult to target an appropriate sustainable stock level. In the short term, after an extended period of low stock size, it is expected that adjustment costs to the industry would be high relative to future uncertain benefits.

We suggest that the primary role of the current stock assessment and management system, faced with this type of situation, is to define the biological limits. Even this is difficult; (M. Rahikainen and others, unpubl. manuscript) have shown that the ecosystem effects have a major impact on the estimated risk of collapse caused by fishing. At present, we simply lack knowledge on the possible combinations of stock size and environmental factors that would be required to establish combined biological and environmental thresholds.

We have demonstrated that the Finnish herring fishery is driven largely by uncontrollable environmental and economic forces. We suggest a better understanding of these considerations is required at all stages in the evaluation and management of this and other fisheries. This case study emphasizes the need for development of the context and tools for evaluation and management of complete fisheries systems.

Acknowledgments

We are grateful to Aune Vihervuori and Pirkko Söderkultalahti for providing historical information related to the Finnish herring fishery, and to Professor Dan Lane, University of Ottawa, for useful comments on a draft of this paper.

References

Ahvonen, A., E. Nylander, J. Stigzelius, P. Söderkultalahti, A.-L. Tuunainen, and A. Vihervuori (eds.). 1999. Regional fisheries in Finland. Agriculture, Forestry and Fishery Report 1999/10. Finnish Game and Fisheries Research Institute. 58 pp.

Alenius, P., and J. Haapala. 1992. Hydrographic variability in the northern Baltic in the twentieth century. ICES Mar. Sci. Symp. 195:478-485.

Flinkman, J. 1999. Interactions between plankton and planktivores of the northern Baltic Sea: Selective predation and predation avoidance. Ph.D. thesis, University of Helsinki. Walter and Andree de Nottbeck Foundation Scientific Reports 18.

Flinkman, J., E. Aro, I. Vuorinen, and M. Viitasalo. 1998. Changes in northern Baltic zooplankton and herring nutrition from 1980s to 1990s: Top-down and bottom-up processes at work. Mar. Ecol. Progr. Ser. 165:127-136.

Hänninen, J. 1999. Consequences of large-scale perturbations to marine ecosystems and their species composition. Ph.D. thesis, University of Turku, Finland.

Hänninen, J., I. Vuorinen, and P. Hjelt. 2000. Climatic factors in the Atlantic control the oceanographic and ecological changes in the Baltic Sea. Limnol. Oceanogr. 45:703-710.

ICES. 1998. Report of the ICES Baltic Fisheries Assessment Working Group. ICES C.M. 1998/ACFM:16.

ICES. 1999a. Report of the ICES Baltic Fisheries Assessment Working Group. ICES C.M. 1999/ACFM:15.

ICES. 1999b. Report of the ICES Advisory Committee on Fisheries Management, 1998. ICES Cooperative Research Report 229 (Part 2).

Samuelsson, M. 1996. Interannual salinity variations in the Baltic Sea during the period 1954-1990. Continental Shelf Research 16:1463-1477.

Suuronen, P., D.L. Erickson, and A. Arrensalo. 1996. Mortality of herring escaping from pelagic trawl codends. Fish. Res. 25:305-321.

Vuorinen, I., J. Hänninen, M. Viitasalo, U. Helminen, and H. Kuosa. 1998. Proportion of copepod biomass declines with decreasing salinity in the Baltic Sea. ICES J. Mar. Sci. 55:767-774.

Participants

Armstrong, Michael
Massachusetts Div. Marine Fisheries
30 Emerson Avenue
Gloucester, MA 01930
michael.armstrong@state.ma.us

Arrhenius, Fredrik
Institute of Marine Research
National Board of Fisheries
P.O. Box 4
Lysekil, S-453 21
SWEDEN
f.arrhenius@imr.se

Bargmann, Greg
Washington Dept. Fisheries and Wildlife
600 Capitol Way North
Olympia, WA 98501-1091
bargmggb@dfw.wa.gov

Beamish, Richard
Dept. Fisheries and Oceans
Pacific Biological Station
Nanaimo, BC V9R 5K6
CANADA
beamish@dfo-mpo.gc.ca

Bonk, Alexander
KamchatNIRO
18 Naberezhnaya
Petropavlovsk, Kamchatsky 683600
RUSSIA
mail@kamniro.kamchatka.su

Brannian, Linda
Alaska Dept. Fish and Game
333 Raspberry Road
Anchorage, AK 99518-1599
linda_brannian@fishgame.state.ak.us

Brennan, Kevin
Alaska Dept. Fish and Game
211 Mission Road
Kodiak, AK 99615
kevin_brennan@fishgame.state.ak.us

Brown, Evelyn
University of Alaska Fairbanks
Institute of Marine Science
P.O. Box 757220
Fairbanks, AK 99775-7220
ebrown@ims.uaf.edu

Brown, Heather
University of California
Bodega Marine Lab
P.O. Box 1045
Bodega Bay, CA 94923
hmbrown@ucdavis.edu

Brown, Liz
P.O. Box 266
Unalaska, AK 99685
liz_brown@envircon.state.ak.us

Browning, James
Alaska Dept. Fish and Game
333 Raspberry Road
Anchorage, AK 99518-1599
james_browning@fishgame.state.ak.us

Burkey Jr., Charles
Alaska Dept. Fish and Game
P.O. Box 1467
Bethel, AK 99559-1467
charlie_burkey@fishgame.state.ak.us

Campbell, Rod
Alaska Dept. Fish and Game
211 Mission Road
Kodiak, AK 99615
Rod_Campbell@fishgame.state.ak.us

Cardinale, Max
Institute of Marine Research
National Board of Fisheries
P.O. Box 4
Lysekil, S-453-21
SWEDEN
m.cardinale@imr.se

Clayton, Linda
University of Alaska Fairbanks
Seward Marine Center
P.O. Box 730
Seward, AK 99664

Connolly, Daniel
Alaska Dept. Fish and Game
211 Mission Road
Kodiak, AK 99615
daniel_connolly@fishgame.state.ak.us

Craig, Robi
Sitka Tribe of Alaska
456 Katlian Street
Sitka, AK 99835

Dewey, Michelle
Washington Dept. Natural Resources
919 N. Township Street
Sedro-Wooley, WA 98284
michelle.dewey@wadnr.gov

DuBois, Larry
Alaska Dept. Fish and Game
333 Raspberry Road
Anchorage, AK 99518
larry.dubois@fishgame.state.ak.us

Eklund, Jan
Vaasa Administrative Court
P.O. Box 204
Vaasa, FIN-65101
FINLAND
jan.eklund@om.fi

Fey, Dariusz
Sea Fisheries Institute
Dept. Oceanography
ul. Kollataja 1
81-332 Gdynia,
POLAND
dfey@mir.gdynia.pl

Foy, Bob
University of Alaska Fairbanks
FITC
900 Trident Way
Kodiak, AK 99615
foy@ims.uaf.edu

Funk, Fritz
Alaska Dept. Fish and Game
P.O. Box 25526
Juneau, AK 99802
fritz_funk@fishgame.state.ak.us

Gates, John
University of Rhode Island
Dept. Environ. and Natural Res. Ec.
302 Lippitt Hall, 5 Lippitt Road
Kingston, RI 02881-0814
jga8714@postoffice.uri.edu

Gay, Shelton
Prince William Sound Science Center
P.O. Box 705
Cordova, AK 99574
shelton@pwssc.gen.ak.us

Gerke, Brandee
University of Alaska Fairbanks
School of Fisheries and Ocean Sci.
11120 Glacier Highway
Juneau, AK 99801
ftblg@uaf.edu

Gray, Daniel
Alaska Dept. Fish and Game
333 Raspberry Road
Anchorage, AK 99518-1599
daniel_gray@fishgame.state.ak.us

Gretsch, Dennis
Alaska Dept. Fish and Game
211 Mission Road
Kodiak, AK 99615
dennis_gretsch@fishgame.state.ak.us

Hammer, Nils
Institute for Sea Fisheries
Palmaille 9
Fed. Research Centre for Fisheries
Hamburg, 22767
GERMANY
hammer.ish@bfa-fisch.de

Hamner, Helen
Alaska Dept. Fish and Game
333 Raspberry Road
Anchorage, AK 99518
Helen_Hamner@fishgame.state.ak.us

Hansen, Chris
NorQuest Seafood Co.
4225 - 23rd Avenue West
Seattle, WA 98199
chansen@norquest.com

Hauser, Bill
Alaska Dept. Fish and Game
Habitat and Restoration
333 Raspberry Road
Anchorage, AK 99518
bill_hauser@fishgame.state.ak.us

Hay, Doug
Pacific Biological Station
Dept. Fisheries and Oceans
Nanaimo, BC V9R 5K6
CANADA
HayD@pac.dfo-mpo.gc.ca

Hoshikawa, Hiroshi
Hokkaido Central Fisheries
Experimental Station
Yoichi, Hokkaido 046-8555
JAPAN
hosikawah@fishexp.pref.hokkaido.jp

Ishida, Ryotaro
Hokkaido Central Fisheries
Experimental Station
238 Hamanaka
Yoichi, Hokkaido 046-8555
JAPAN
ishidar@fishexp.pref.hokkaido.jp

Ivshina, Elsa
SakhNIRO
Komsomolskaya St. 196
Juzhno, Sakhalinsk 693016
RUSSIA
elsa@tinro.sakhalin.ru

Jacques, Martin
NorQuest Seafood Co.
4225 - 23rd Avenue West
Seattle, WA 98199
mjacques@norquest.com

Jarvik, Ahto
Estonian Marine Institute
Viljandi Rd 18B
Tallinn, 11216
ESTONIA
ahto@ness.sea.ee

Jørstad, Knut
Inst. of Marine Research
Postboks 1870
5024 Bergen,
NORWAY
knut.jorstad@imr.no

Kitto, Beth
U.S. Forest Service
P.O. Box 129
Girdwood, AK 99587
bethkitto@yahoo.com

Kline, Tom
Prince William Sound Science Center
P.O. Box 705
Cordova, AK 99574
tkline@pwssc.gen.ak.us

Knapp, Gunnar
University of Alaska Anchorage
Inst. for Social and Econ. Research
3211 Providence Drive
Anchorage, AK 99508
afgpk@uaa.alaska.edu

Kobayashi, Tokimasa
Hokkaido National Fisheries
Research Institute
116 Katsurakoi
Kushiro, Hokkaido 085-0802
JAPAN
tokikoba@hnf.affrc.go.jp

Kruse, Gordon
Alaska Dept. Fish and Game
Division of Commercial Fisheries
P.O. Box 25526
Juneau, AK 99802
gordon_kruse@fishgame.state.ak.us

Kyte, Michael
Golder Associates
18300 NE Union Hill Road
Redmond, WA 98052
mkyte@golder.com

Larson, Robert
Alaska Dept. Fish and Game
P.O. Box 667
Petersburg, AK 99833

Lindstrom, Ulf
Norwegian Inst. Fisheries and Aquaculture
Tromso, N-9005
NORWAY
ulf.lindstrom@fiskforsk.norut.no

Lloyd, Denby
Alaska Dept. Fish and Game
211 Mission Road
Kodiak, AK 99615
denby_lloyd@fishgame.state.ak.us

Marty, Gary
University of California
1 Shields Avenue
Davis, CA 95616-8732
gdmarty@ucdavis.edu

McFarlane, Sandy
Dept. Fisheries and Oceans
Pacific Biological Station
Nanaimo, BC V9R 5K6
CANADA
mcfarlanes@pac.dfo-mpo.gc.ca

McPherson, Arran
Dalhousie University
Dept. Oceanography
Halifax, NS B3H 4J1
CANADA
aaam@phys.ocean.dal.ca

Melvin, Gary
Dept. Fisheries and Oceans
St. Andrews Biological Station
531 Brandy Cove Road
St. Andrews, NB E5B 2L9
CANADA
melving@mar.dfo-mpo.gc.ca

Menard, Jim
Alaska Dept. Fish and Game
P.O. Box 1467
Bethel, AK 99559-1467
jim_menard@fishgame.state.ak.us

Moreland, Stephanie
University of Alaska Fairbanks
Seward Marine Center
P.O. Box 730
Seward, AK 99664
fnsmm@uaf.edu

Mundy, Phil
Exxon Valdez Trustee Council
645 G Street, Suite 401
Anchorage, AK 99501-3451
phil_mundy@oilspill.state.ak.us

Murphy, Bob
Alaska Dept. Fish and Game
211 Mission Road
Kodiak, AK 99615
bob_murphy@fishgame.state.ak.us

Nakayama, Yuu
University of Tokyo
Ocean Research Institute
1-15-1 Minamidai, Nakano-ku
Tokyo, 164-8639
JAPAN
yu@ori.u-tokyo.ac.jp

Naumenko, N.I.
KamchatNIRO
18 Naberezhnaya
Petropavlovsk, Kamchatsky 683600
RUSSIA
mail@kamniro.kamchatka.su

Nelson, Patricia
Alaska Dept. Fish and Game
211 Mission Road
Kodiak, AK 99615
patti_nelson@fishgame.state.ak.us

Nichols, John
17 Barkis Meadow
Blundeston
Lowestoft, Suffolk NR23 5AL
UK
j.h.nichols@barkis.junglelink.co.uk

Nielsen, Rasmus
Danish Inst. for Fisheries Research
North Sea Centre
P.O. Box 101
Hirtshals, DK-9850
DENMARK
rn@dfu.min.dk

Norcross, Brenda
University of Alaska Fairbanks
Institute of Marine Science
P.O. Box 757220
Fairbanks, AK 99775-7220
norcross@ims.uaf.edu

O'Connell, Jacqueline
University of British Columbia
Dept. Earth and Ocean Sciences
1461-6270 University Blvd.
Vancouver, BC V6T 1Z4
CANADA
oconnell@zoology.ubc.ca

Ona, Egil
Institute for Marine Research
P.O. Box 1879
Bergen, 5024
NORWAY
Egil.Ona@imr.no

Oskarsson, Gudmundur
Dalhousie University
Dept. Oceanography
Halifax, NS B3H 4J1
CANADA
gjos@phys.ocean.dal.ca

Otis, Ted
Alaska Dept. Fish and Game
3298 Douglas Street
Homer, AK 99603-8027
ted_otis@fishgame.state.ak.us

Paul, A.J.
University of Alaska Fairbanks
Seward Marine Center
P.O. Box 730
Seward, AK 99664
ffajp@uaf.edu

Paul, Judy
University of Alaska Fairbanks
Seward Marine Center
P.O. Box 730
Seward, AK 99664
fnjm1@uaf.edu

Peltonen, Heikki
Finnish Game and Fisheries Research Institute
P.O. Box 6
00721 Helsinki,
FINLAND
heikki.peltonen@rktl.fi

Peterson, Sara
California Dept. Fish and Game
284 Harbor Blvd.
Belmont, CA 94002
sarap@sfsu.edu

Power, Michael
Dept. Fisheries and Oceans
Biological Station
531 Brandy Cove Road
St. Andrews, NB E5B 2L9
CANADA
powermj@mar.dfo-mpo.gc.ca

Pritchett, Marc
Alaska Dept. Fish and Game
P.O. Box 667
Petersburg, AK 99833
marc_pritchett@fishgame.state.ak.us

Quinn, Terry
University of Alaska Fairbanks
JCFOS
11120 Glacier Highway
Juneau, AK 99801
fftjq@uaf.edu

Radchenko, Vladimir I.
SakhNIRO
196 Komsomolskaya Street
Yuzhno-Sakhalinsk
Russia 693023
vlrad@tinro.sakhalin.ru

Rahikainen, Mika
Finnish Game and Fisheries Research Inst.
P.O. Box 6
00721 Helsinki,
FINLAND
mika.rahikainen@rktl.fi

Raid, Tiit
Estonian Marine Institute
Viljandi Rd 18 b
Tallinn, EE-11216
ESTONIA
raid@sea.ee

Rajasilta, Marjut
University of Turku
Archipelago Research Institute
Turku, FIN-20014
FINLAND

Rowell, Kathy
Alaska Dept. Fish and Game
333 Raspberry Road
Anchorage, AK 99518-1599
kathy_rowell@fishgame.state.ak.us

Ryan, Connie
California Dept. Fish and Game
284 Harbor Blvd.
Belmont, CA 94002
cryan@dfg2.ca.gov

Salomone, Paul
Alaska Dept. Fish and Game
P.O. Box 1467
Bethel, AK 99559-1467

Sasaki, Masayoshi
Hokkaido Central Fisheries
Experimental Station
Yoichi, Hokkaido 046-8555
JAPAN
sasakimy@fishexp.pref.hokkaido.jp

Schweigert, Jake
Dept. Fisheries and Oceans
Pacific Biological Station
3190 Hammond Bay Road
Nanaimo, BC V9R 5K6
CANADA
schweigertj@dfo-mpo.gc.ca

Seeb, Jim
Alaska Dept. Fish and Game
Div. of Commercial Fisheries
333 Raspberry Road
Anchorage, AK 99518
jim_seeb@fishgame.state.ak.us

Seiser, Pam
University of Alaska Fairbanks
P.O. Box 85348
Fairbanks, AK 99708
ftpes@uaf.edu

Shotwell, Kalei
University of Alaska Fairbanks
School of Fisheries and Ocean Sci.
11120 Glacier Highway
Juneau, AK 99801
ftkss@uaf.edu

Slotte, Aril
Marine Research Institute
P.O. Box 1870-Nordnes
N-5817 Bergen,
NORWAY
aril.slotte@imr.no

Smirnov, Andrey
TINRO
Nagaevskaya St., 51
Magadan, 685024
RUSSIA
tinro@online.magadan.su

Spies, Bob
Applied Marine Sciences
4749 Bennet Drive #L
Livermore, CA 94550-0100
spies@amarine.com

Stephenson, Robert
Dept. Fisheries and Oceans
St. Andrews Biological Station
531 Brandy Cove Road
St. Andrews, NB E5B 2L9
CANADA
stephensonr@mar.dfo-mpo.gc.ca

Stevenson, David
NOAA/NMFS
Northeast Regional Office
1 Blackburn Drive
Gloucester, MA

Sykas, Dave
NorQuest Seafood Co.
4225 23rd Avenue West
Seattle, WA 98199
dykas@norquest.com

Taylor, Capt Michael
Maritime International Inc.
P.O. Box 7745
New Bedford, MA 02742
mhmtaylor@aol.com

Toresen, Reidar
Marine Research Institute
P.O. Box 1870-Nordnes
N-5817 Bergen,
NORWAY
reidar.toresen@imr.no

Tremblay, Denis
Dept. Fisheries and Oceans
104, Rue Dalhousie
3rd floor
Quebec, Quebec G1K 7Y7
CANADA
tremblden@dfo-mpo.gc.ca

Trofimov, Igor
KamchatNIRO
18 Naberezhnaya
Petropavlovsk, Kamchatsky 683600
RUSSIA
mail@kamniro.kamchatka.su

Vines, Carol
University of California Davis
Bodega Marine Laboratory
P.O. Box 247
Bodega Bay, CA 94923
cavines@ucdavis.edu

Watanabe, Yoshiro
University of Tokyo
Ocean Research Institute
1-15-1 Minamidai, Nakano
Tokyo, 164-8639
JAPAN
ywatanab@ori.u-tokyo.ac.jp

Witherell, Dave
North Pacific Fishery Management Council
605 W 4th Ave., Suite 306
Anchorage, AK 99501
David.Witherell@noaa.gov

Yang, Mei-Sun
NMFS
Alaska Fisheries Science Center
7600 Sand Point Way NE
Seattle, WA 98115-0070
Mei-Sun.Yang@racesmtp.afsc.noaa.gov

Yund, Phil
University of Maine
Darling Marine Center
193 Clarks Cove Road
Walpole, ME 04573
philyund@maine.edu

Zimmermann, Christopher
Institute for Sea Fisheries
Palmaille 9
Fed. Research Centre for Fisheries
Hamburg, 22767
GERMANY

Index

Page numbers in italic indicate figures and tables.

A

AALS (automated acoustic logging system), 676, 679-682, *683,* 686
abundance. *See also* biomass
 Barents Sea, and cod predation, 91-99
 Bering Sea, *82*
 British Columbia, 425-426
 California, 429
 and coho salmon in prey competition, 37-50
 estimation. *See* hydroacoustics; spawning stock biomass
 Georges Bank, 416-417
 Gulf of Alaska, 1977-1987, *182*
 Gulf of Maine, 326
 Gulf of St. Lawrence, 412
 historical background of research, 14-15
 larval, Sound (ICES SD23), 354, *355*
 modeling, 367
 Norwegian spring-spawning herring
  in southern areas, *303,* 308, 309
  and water temperature, 279, 281, *282,* 283, 299
  by year-class, *268*
 peak aerial survey index, *338,* 339
 Sakhalin coast, Russia, 247, *248-250,* 250
Acartia. See under prey
ACON software, 682
acoustics. *See* hydroacoustics
ADAPT-VPA, 411
Advisory Committee for Fishery Management (ACFM)
 Baltic Sea stock assessment, 742
 northern Atlantic stock assessment, 318, 397, 398
 North Sea stock assessment, 645, 655, 658, 659
aerial surveys
 Gizhiga-Kamchatka stock, 637
 mile-days of milt, 366, *370, 374*
 peak aerial survey index, *338, 339*
 Prince William Sound, 539
age
 determination, historical background, 12
 lack of data standardization, 384
 relationship to body length, *187-188,* 193, *388-389*
 relationship to body weight. *See* growth rate
 relationship to scale length, 189, *190-191,* 192

age-frequency analysis, 173
age-structured assessment model, 366, 425
ajitsuke kazunoko, 726
Alaska. *See also* Prince William Sound
 abundance, 1977-1987, *182*
 age-frequency data, *178*
 anatomy of a strong year-class, 171-198
 catch statistics, 421, *423, 723*
 roe herring market, 721-739
 site maps, *174, 419*
 stock inventory 2000, 419-422, *423*
Alaska Department of Fish and Game (ADFG)
 aerial survey by, 539, *540*
 age-frequency data, 173
 Prince William Sound population management, 336, *337,* 366, 369, 539, *540*
ALC. *See* alizarin complexone
Aleutian Low Pressure Index (ALPI), 54, 61, *64*
algae. *See also* spawn-on-kelp
 of British Columbia, 495
 diatoms, *108*
 of northern Japan, *209-212*. See also *Cystoseira; Laminaria*
 phytoplankton, 75, 93, *108*
 primary productivity and recruitment dynamics, 529, 531
 red, herring egg attachment, 218, 431, 495
 roe-on-kelp fishery, 725
 of the Sakhalin coast, 250
 and spawning bed selection, northern Japan, 199, 202, 204-205, *207,* 208, *209-212,* 216, 218-221
alizarin complexone marking (ALC)
 increments and age determination, 231, *232-234,* 235, 239, 242
 methods, 228, 230
 with northern Japan herring, 104
ALPI (Aleutian Low Pressure Index), 54, 61, *64*
Amphipoda. *See under* prey
ANCOVA, size analysis, 54, 55
Anderson, Johann, 4, 9
André, Carl, 611-613
Andreeva, A., 591-597, 629-633
Aneer, G., 599-609
anesthesia, 465

768   Index

Anker-Nilssen, T., 357-361
Archipelago Sea, 600, *601*, 604, 606, 744, *747*, 750
Arctic Ocean
    site map, *438*
    stock inventory 2000, 437-442
ARIMA (Auto Regressive Integrated Moving Average), 54
Armstrong, Michael P., 575-590
Aro, Eero, 741-760
Arrhenius, F., 153-154, 611-613
Artemia, 228
artificially produced herring, *vs.* wild herring, 104
assimilation rates, 32
Atlantic herring. See *Clupea harengus*
Atlantic Ocean
    currents, *263, 317*
    eastern, stock inventory 2000, 392-405
    North
        catch statistics, 617
        correlation of water temperature with biomass and recruitment, 315-334
    northwestern North, stock inventory 2000, 406-412
    southwestern, stock inventory 2000, 412-418
    western, stock inventory 2000, 414-418. See also Bay of Fundy; Georges Bank; Gulf of Maine; Nova Scotia
    western North, site map, *407*
Atlantic puffin. See *Fratercula arctica*
automated acoustic logging system (AALS), 676, 679-682, *683*, 686
Auto Regressive Integrated Moving Average (ARIMA), 54

B

Baffin Bay, 441, 442
Balls, Ronald, 13
Baltic Fisheries Assessment Working Group, 394, 742, 747
Baltic herring, 385. See also *Clupea harengus membras*
Baltic Sea
    Estonian fishery, 703-720
    Finnish fishery, *741-760*
    growth rate of herring
        and density-dependence, 153-154
        and spawning time, 155-169
    length-at-age, *388*
    microsatellite polymorphism and genetic structure, 611-613

Baltic Sea *(continued)*
    northern, biological and industrial aspects, 741-760
    site description, 156
    site maps, *158, 706, 745*
    spawning beds, 200
    spawning period, 156
    stock inventory 2000, 392-396
    weight-at-age, *390*
Barents Sea, 439
    herring abundance, food supply, and distribution, 91-99
    intermingling of stocks, 629-633
    as nursery area, 262, 264, 302
    oocyte resorption, 285-295
    site map, *438*
Barkley Sound, British Columbia, 498-508
Bathyraja, 82
    aleutica, 84-86
    minispinosa, 84-86, 87
    parmifera, 82, *84-86*, 87
    violacea, 84-86
Bay of Fundy. See also Atlantic Ocean, western
    catch statistics, *408*
    effects of water temperature on recruitment, 315, *326*, 328-329
    length-at-age, *388*
    long-term sampling program, 135-152
    management areas, *137*
    migration, 10
    range of herring, *317*
    spawning stock biomass, *319, 320, 323, 408*
    stock decline, 1966-1980, 326
    weight-at-age, *391*
Beamish, R.J., 37-50, 51-68
Beaufort Sea
    length-at-age, *388*
    stock inventory 2000, *438*, 440-441
    weight-at-age, *391*
Bering Sea
    eastern
        age-frequency data, *178*
        role of herring in predatory fish diet, 87
        site map, *430*
    length-at-age, *389*
    life history strategy, 420
    stock inventory 2000, 420-421, 422
    weight-at-age, *391*
    western. See also Karagin Bay
        role of herring in predatory fish diet, 81-89
        site map, *82*

biodiversity, 16, 17, 309, 568
biological characteristics
    Atlantic herring, 135-152
    Lake Furen herring, 573
    low- vs. high-vertebrae herring, *592,* 592-593, *594,* 630
biomass. *See also* spawning stock biomass
    estimates. *See* hydroacoustics, surveys; virtual population analysis
    loss due to somatic and generative production, 695, 698
    northern Atlantic stocks, 318, *319, 320,* 323
    Nova Scotia stocks, 678
    Okhotsk Sea stocks, 695, 699, *700*
    Prince William Sound stocks, 536
    Sound (ICES SD23), *348-349,* 351, *351,* 354, *354*
    square method for estimation, 691
    Strait of Georgia stocks, 40-41, *41, 60,* 64
birds, predatory
    British Columbia spawning beds, 490, 494, 505, 506
    of Norwegian spring-spawning herring, 357, *358,* 359, 360
    Prince William Sound, 543
    WBEC value of herring for, 132
blackcod, 172, 193
body length of herring
    Baltic Sea, *159*
    Bering Sea, western, 81-82, *83, 84*
    and estimation of trophic level, *75*
    Gulf of Finland, *709, 710, 712, 713, 717*
    Gulf of St. Lawrence, 672
    Japan, northern, 101, 104, *105,* 110-113, 114
    larval, and smelt predation, 117
    Okhotsk Sea, *695,* 697, *698*
    relationship to age, *187-188,* 193, *388-389*
    relationship to body weight, 121, 124, *127,* 146, *147,* 188
    relationship to echo sounding target strength, 462, 680
    relationship to hatching date, *238*
    relationship to maturity, *144*
    relationship to otolith radius, 157, 160, *160*
    relationship to scale length, 189, *189*
    Scotia-Fundy area, 140, *142, 143, 144, 145,* 146, 147
body length of prey, 109

body weight of herring
    at age. *See* growth rate
    Baltic Sea, *749, 755*
    Bering Sea, western, 82, *84*
    dry, and whole body energy content, 121, 124, *127, 128, 131,* 132
    Gizhiga-Kamchatka stock, *641*
    Gulf of Finland, *710*
    relationship to body length, 121, 124, *127,* 146, *147,* 188
    Scotia-Fundy area, 140, *142, 143, 145,* 146, *147*
bootstrap procedure, 369, 374, *375*
Bothnian Sea, 742, 744, *745-747,* 757. *See also* Baltic Sea
Box-Jenkins ARIMA, 54
British Columbia. *See also* Fraser River; Juan de Fuca Strait; Strait of Georgia
    abundance, 1977-1987, *182*
    age-frequency data, *179*
    anatomy of a strong year-class strength, 171-198
    egg loss estimates and stock assessment, 489-508
    herring in predatory diet, 87
    length-at-age, *389*
    roe market, *729*
    site maps, *174-175, 419, 491*
    spawning bed selection, 262
    stock inventory 2000, 422, *423,* 424-427
    weight-at-age, *391*
British Columbia Department of Fisheries, 425
Brown, Evelyn D., 335-345, 535-558
buoyancy, 481-482, 510
bycatch
    Baltic Sea, 711, *714*
    North Sea, 645, 650, 657, 658, 661

C

Cadrin, Steven X., 575-590
California, *419*
    catch statistics, *423,* 428
    roe market, *729*
    San Francisco Bay, egg-on-kelp survival, 455-460
    stock inventory 2000, 427-429
    California Department of Fish and Game, 427
Canada. *See also* specific provinces
    Arctic Ocean herring, *438,* 440, 441-442
    Department of Fisheries and Oceans ACON software, 682

Canada *(continued)*
  coho salmon database, 39
  Nova Scotia stock management, 676, 677
  Environment Canada, Water Resources Branch, 40
Canadian Hydroacoustic Program, 680
capelin, 91, 94-96, *693*
Cape Navarin. *See* Bering Sea, western
carbon
  carbon/nitrogen ratio, 121, 124, *129-131*
  stable isotope abundance, 69-80, 123
Cardinale, M., 153-154
catch per unit effort (CPUE)
  Bering Sea, western, *82*
  Gulf of St. Lawrence, 671
  Japan, 521, 525-526, 529, *530*
catch restrictions. *See* quotas; total allowable catch
catch statistics
  Alaska, 421, *423, 723*
  Atlantic Ocean, western, 136
  Baltic Sea–Gulf of Finland, *396, 715, 717, 743, 753*
  Bering Sea, eastern, *423*
  British Columbia, *423*, 425
  California, *423*, 428
  Celtic Sea, 402
  Estonia, 703, 704, *707*
  Finland, *607, 743, 747, 753, 756*
  Georges Bank, 415-416
  Gizhiga-Kamchatka herring, 635, 636, 640, 642-643
  Gulf of Maine, 418
  Gulf of St. Lawrence, 411, 667, *669, 670*, 671, *672*
  Iceland, 405
  Japan, 200, 228, *523*, 525-526, *527*
  jurisdictional issues, 384
  Magdalen Islands, Quebec, 667, *669, 670*, 671, *672*
  Newfoundland, 406, 410
  North Atlantic, *408*, 617
  North Sea, 400, 648, *649*
  Nova Scotia, *408*, 678
  Prince William Sound, 536, *538*
  Russia, 432, *433,* 635, 636, 640, 642-643, 689, 696, *700*
  Sakhalin coast, 245, *250*
  Sakhalin-Hokkaido stock, *433,* 434, *523,* 524, 526, *527*
  Scotia-Fundy herring, 413
  Scotland, west of, 403, 404
  Strait of Georgia, 40-41, *41,* 55, *59*
  Yellow Sea, *433*

Celtic Sea
  length-at-age, *388*
  stock inventory 2000, 401-402
  weight-at-age, *390*
Chernobyl, *7,* 8
Chesson's selectivity (a) index, 25, 31
Chimura, Masayuki, 521-533
Chukchi Sea, *178, 438,* 440
chum salmon. See *Oncorhynchus keta*
Cladocera. *See under* prey
Clark, K.J., 559-571, 675-688
Clausocalanus pergens. See under prey
Claytor, R., 381-454
climate, effects on herring. *See also* El Niño
  British Columbia, 51-67
  historical background of research, *7,* 15
  Japan, northern, 200
  North Atlantic Oscillation winter index, 360
  Norway, 279-284
  Norwegian spring-spawning stock, 357-361
  Sea of Okhotsk, 640
  Strait of Georgia, 45-47, 48
  and worldwide stock abundance, 443
Clupea harengus
  age estimation by otolith increments, 239
  biological characteristics, 135-152
  effects of hydrography, the Sound (ICES SD23), 347-356
  effects of water temperature on larval growth, 241
  effects of water temperature on recruitment, 279-284, 315-334
  factors influencing spawning, Norway, 255-278
  feeding dynamics, Prince William Sound, 22-31
  genetic analysis, 617-628
  growth rate, Baltic Sea
    decrease in, 153-154, 156
    and spawning time, 155-169
  historical background of fisheries, 3-4
  maturation rate, 271-274
  microsatellite polymorphism, 611-613
  morphometric variation among spawning groups, 575-590
  nomenclature, 591, *592*
  Norwegian spring-spawning. *See* Norwegian spring-spawning herring
  oocyte resorption, 285-295
  predators. *See* predators and predation

Clupea harengus (continued)
    regional analysis. *See* stock
        assessment, inventory 2000
    roe market, 726, 728, *730*
    survival factors, first-year, 549
    uses for. *See* herring
    vertebrae, 591, *592*
    Clupea harengus harengus, 591, *592*.
        See also *Clupea harengus*
Clupea harengus maris-albi, 591, *592*
Clupea harengus membras
    management in Estonia, 703-720
    nomenclature, 385, 591, *592*
    tagging experiment, Finland, 599-609
Clupea harengus pallasii, 591. See also
    Clupea pallasii
Clupea harengus suworowi, 591
Clupea pallasii
    anatomy of a strong year-class, 171-198
    disease and population assessment,
        Prince William Sound, 363-379
    egg loss estimates, British Columbia,
        489-508
    egg survival on kelp, San Francisco
        Bay, 455-460
    feeding dynamics
        competition with coho salmon for
            prey, 37-50
        northern Japan, 101-115
        Prince William Sound population,
            21-35
    first-year survival and stage-specific
        estimates, 535-558
    genetic analysis, 617-628
    hatching date, relationship to larval
        growth, 227-243
    maturation rate, 272
    migration between two ecosystems,
        British Columbia, 51-67
    nomenclature, 591, *592*
    predators. *See* predators and predation
    recruitment dynamics, latitudinal
        differences in, 521-533
    regional analysis. *See* stock
        assessment, inventory 2000
    Sakhalin-Hokkaido herring. See
        Sakhalin-Hokkaido stock
    spawning bed selection, northern
        Japan, 199-226
    stable isotope analysis, Prince William
        Sound, 69-80
    trophic position
        in Prince William Sound, 69-80
        in western Bering Sea, 81-89
    uses of. *See under* herring

Clupea pallasii (continued)
    vertebrae, 591, *592*
    whole body energy content, 121-133
Clupea pallasii marisalbi, 591, *592*, 630
Clupea pallasii pallasii, 591, *592*
Clupea pallasii suworowi, 591, *592*
Clupeidae, overview, 385
cod. See also *Gadus*
    in the Baltic Sea, 153-154, 395, 704,
        750, *751*
    depth distribution and herring
        presence, 97, 99
    echo sounding target strength, 480, 481
    effects on herring year-class, 329-330
    good year-class and recruitment,
        simultaneous with other
        species, 328
    herring predation by, Barents Sea,
        91-92, 96-99
    historical background of fisheries, 3
    in the northeast Arctic Ocean, 329
    Pacific. See *Gadus macrocephalus*
coho salmon. See *Oncorhynchus kisutch*
collapse. *See* stock collapse
comanagement, 15, 150, 675-688
commercial fisheries. *See also* technology
    development
    automated acoustic logging system in,
        680-681
    designated ports for landing, 662
    economic aspects, 716, 718
    Estonia, 704, 705, 716, 718
    Finland, 744, *746*
    involvement in data collection
        Gulf of St. Lawrence, 412, 669, 671
        Nova Scotia, 138, *141*, 146, 150, 577,
            676, 678, 679, 680, 686
    satellite tracking of vessels, 662
competition of herring
    with cod in the Barents Sea, 91-99
    intraspecific, 545
    with pollock in Prince William Sound,
        545
    with salmon in Georgia Strait, 37-50
components. *See* population components
condition
    in disease modeling, 377
    and echo sounding target strength,
        465, *470, 471*
    and oocyte resorption, 290
    relationship to whole body energy
        content, 121, 124, *125, 126*,
        132
    Scotia-Fundy area, 146, *148-149*
    and spawning ground selection,
        Norway, 267, *268*, 269, 308

conductivity and temperature at depth
(CTD)
Prince William Sound, 23, *26*
Sound (ICES SD23), 347, *352, 353*
conservation. *See also* quotas
bycatch limits, North Sea, 645, 650,
657, 658, 661
future research, 16
Gizhiga-Kamchatka stock, 637
mesh-size regulations, 708-711, 715
seasonal area closures
Baltic Sea, 708
North Sea, 657, 662, 663
spawning ground protection, Nova
Scotia, 562-564
"convention on biological diversity," 17
Convention on Fishing and Conservation
of the Living Resources in the
Baltic Sea and the Belts, 394
Cook, Doug, 615-628
Copepoda. *See under* prey
"critical period" hypothesis, 392, 522
cross-correlation, 336
Crustacea. *See under* prey
currents
Alaska coastal, 70
Atlantic, *263, 317*
Japan, *524*
Kamchatka, Russia, 637
North Cape Current, 93
Norway, *263,* 358
Prince William Sound, *340*
Taylor columns, 359
Cystoseira
crassipes, 250
as egg substrate, Russia, 431
hakodatensis, and spawning bed
selection, 199, 204, *207,* 208,
209, 210, 211, 216, 218, *220*

D

data collection
by commercial fisheries
Gulf of St. Lawrence, 412, 669, 671
Nova Scotia area, 138, *141,* 146,
150, 577, 676, 678, 679,
680, 686
dissemination of information, 150
lack of standardization, 384
Davis Strait, 441, 442
Decapoda. *See under* prey
deformities, 543-544
Dekastri stock, 430-432, *433,* 444
Deltagraph 3.1, 72
Denmark, the Sound (ICES SD23), 347-356

density-dependent processes
decrease in growth rate, Baltic Sea,
153-154
and eggs. *See* eggs, density in
spawning bed
mortality, 260, 301, 308
DFO. *See* Canada, Department of Fisheries
and Oceans
disease, 194. See also *Ichthyophonus
hoferi;* viral hemorrhagic
septicemia virus
in Alaska, 421
effects on recruitment, 369, 374, *376,*
376-377
in the North Sea, 365
in Prince William Sound, 22, 363-379,
536, 545, 550-551
ulcers, 377
distribution
Barents Sea, and availability for cod,
91, 92-93, 96-99
Bering Sea, western, *82*
of eggs in spawning beds. *See* eggs,
distribution in spawning bed
historical background of research,
11-12, 14
Japan, 101-115, *524*
link with zooplankton distribution,
93-96
overview, in stock inventory, 385-392
polar migration theory, 4, 9-10, 392
Dogyoren, 728
Dolgov, Andrey V., 91-99
double counting, in acoustic surveys, 685
dry weight, 121, 124, *127, 128, 131,* 132

E

Eaglek Bay, Alaska, *23,* 548
echo sounding, 324
backscattering area, 509, 680
calibration, 479-480
effects of swimming angle, 473, *475,*
479
effects of swimming depth, 483, 510
effects of swimming speed, *478*
and gonad development. *See*
gonadosomatic index
herring *vs.* other species, 480-481
Marconi 424 echo sounder, 13
Nova Scotia survey, 678
SIMRAD EK 500 echo sounder, 466, *467*
and the swim bladder, 462-463
target strength, seasonal variation in,
461-487
tilt angle *vs.* swimming angle, 509-519

ecological aspects
  in fisheries management, 65, 309
  future research, 16
  trophic enrichment of ^{15}N, 70, 72
  trophic position of herring, 69-80
economic aspects
  Estonian fishery, 716, 718
  ex-vessel prices and value, 721, 722, *723*, 736
  Finnish fishery, *743*, 752, *754*, 757
  future research, 17
  roe market, 721
    Alaska herring prices, 734-737
    retail prices, 726
    wholesale prices, 731-733, *734, 735*
    yen-dollar exchange rate, 734, 736, 737
Ecopath model, 70, 78
eelgrass. See *Phyllospadix iwatensis*; *Zostera marina*
eggs
  density in spawning bed
    British Columbia, *496-497*, 498, *499, 500-501*
  Gizhiga-Kamchatka stock, 637
  Ishikari Bay, Japan, 199, 216, *217*, 222
  Sakhalin coast, Russia, *252*, 253
  deposition on macroalgae. See spawn-on-kelp
  development, *7, 205*
  distribution in spawning bed
  depth and predation, 543
  effects on year-class strength, 335-345
  relationship to topography, 216-218, *217*, 221
  vertical, 204-205, *206*, 218, *219-220*
  oocyte resorption, 285-295
  overview, 386
  Prince William Sound, 366, *370*, 371, *374*, 536, 539, 543, *548*
  for roe market, 725, 728
  survival
    in British Columbia, 489-508
    estimation, 535
    on kelp, 455-460, 495
    in Prince William Sound, 536, *538*, 543, *548*
Eklund, J., 155-169, 599-609
electrophoresis, starch gel, 587, 630, 632
El Niño, *7*, 8, 426, 428
energy content
  cost of migration, 258-259, 261, 270, 308
  in disease modeling, 377
  and timing of metamorphosis, 546
  whole body, 121-133

enforcement, 662
Envall, Mats, 611-613
environmental factors. *See also* climate; water temperature
  effects on year-class strength, 335-345
  exposure, of eggs in spawning beds, 490, 543
  in overwintering survival, 549-550
  wave action, 490, 505, 535, 543
  wind, 359-360
enzyme staining, 630
escapement model, 425
Estonia, fisheries management, 703-720
Estonian Fishermen Association, 716
Etrumeus teres
  distribution, *524*
  recruitment dynamics, 521, 525-526, 528-529
Euparal, 157
Euphausiacea. *See under* prey
European Union, target strength project AIR3-CT94-2142, 463
Eurytemora. See under prey
evolution
  divergence, 616, 620
  member/vagrant hypothesis, 14, 258
  selection, and "informative" loci, 625
  size-dependent maturation rate, 271
EVOS. See *Exxon Valdez* oil spill
EVOS Trustee Council, 364, 377
Exact test, 618, 620, 624
Exclusive Economic Zone (EEZ), 81, 705, 708, 711
exposure, of eggs in spawning beds, 490, 543
ex-vessel prices and value, 721, 722, *723*, 736
Exxon Valdez oil spill (EVOS), 8, 444
  catch data, *538*
  effects, 69-70, 364, 536-537, *538*, 551
  historical background, *7*
  larval deformity from, 543-544

F

failures. *See* stock collapse
FAO (Food and Agriculture Organization), 384, 617
Faroe Islands, 269, 398
fat content
  and echo sounding target strength, 461, *470, 471*, 472, 483
  herring in Barents Sea, 92, 94, 95-96
fecundity, 436

feeding dynamics. *See also* food supply
  body length of prey, 109
  Japan, northern, 101-115
  and oocyte resorption, 290
  and temperature effects on zooplankton, 21-35
Fife, F.J., 559-571, 675-688
Finland. *See also* Baltic Sea
  Baltic herring market, 3
  biological and industrial aspects of the fishery, 741-760
  catch statistics, *607*
  spawning bed topography, 221
  tagging, spring-spawning Baltic herring, 599-609
Finland. Game and Fisheries Research Institute, 602
fin rays, 636, 642, *642*
fisheries management
  catch limitations. *See* moratoriums; restrictions; total allowable catch
  comanagement, 15, 150, 675-688
  developments in, 14-15, 150
  ecosystem organization in, 65
  effort *vs.* quota controls, 667-673
  enforcement of restraints, 662
  first total allowable catch limit, 136
  historical background of research, *6-7*, 12, 15, 387, 392
  in-season approach, 564-565, 623, 675-688
  interdisciplinary approach, 16
  international approach, 16
  jurisdiction issues, 16, 384, 744
  by region. *See* stock assessment, inventory 2000; *specific geographic names*
  relevance of former spawning area to, 301-313
fisheries science, role of herring investigations in, 1-20
fishing effort
  Baltic Sea, 742
  Estonia, 708, 715-716
  Nova Scotia, 676
  vs. quota controls, 669-673
fishing industry. *See* commercial fisheries
fishing methods. *See also* technology development
  cacuamis (pound-nets), 706, 718
  dominant mesh sizes, 411, 671
  introduction of echo sounding and sonar, 324
  nylon nets, 324

fishing methods *(continued)*
  power-block, 324
  torches, 15
fitness, physiological. *See* condition
Food and Agriculture Organization (FAO), 384, 617
food supply. *See also* prey; zooplankton
  captive larvae, 228
  effects on recruitment level, 329
  effects on spawning time, 222, 228
  overview of diet, 386
  starvation, 194, 535, 546-547, *548*
  stomach contents
    of juveniles, northern Japan, 104, *108,* 109
  Prince William Sound, *30*
Fosseidengen, J.E., 461-487
Fossum, P., 357-361
Foy, Robert J., 21-35
Fraser River, British Columbia, 40, 45, *46,* 47, 65
Fratercula arctica, 357, *358,* 359, 360
freshwater runoff
  Baltic Sea, 750, *751*
  effects on fertilization and hatching success, 223
  Prince William Sound, 31, 343
F-STAT software, 618
fungal disease. *See Ichthyophonus hoferi*
Funk, F., 381-454

G

Gadus, 386
  macrocephalus, 87, 545
  morhua, 1, 481, 522
gam (generalized additive models), 336, 341, *342*
Gdansk convention, 394
generalized additive model (gam), 336, 341, *342*
generalized linear model (GLM), *352,* 354, 469, 489, 493-494
genetic drift, 616-617
genetics
  Gulf of Maine-Georges Bank stocks, 588
  lactate dehydrogenase loci
    Barents Sea herring, *631,* 632
    White Sea herring, 593, 595, *595*
  microsatellite markers, 7
    mutation at, 623-624
    polymorphism of Atlantic herring, 611-613
    at three spatial scales, 615-628
  mitochondrial DNA, 264-265, 624

Georges Bank
    catch statistics, *408*
    genetic aspects of stocks, 588
    length-at-age, *388*
    morphometric variation among
        spawning groups, 575-590
    population decline, 1978-1987, 326
    range of herring, *317*
    site map, *407*
    spawning stock biomass, *319, 320,*
        323, 408
    stock inventory 2000, 414-418
    stock recovery, 417, 444
    weight-at-age, *391*
Georgia Strait. *See* Strait of Georgia
gills, regulation of spawning, 167
Gizhiga-Kamchatka stock, 430-432, *433,*
        444, 635-643, *690, 696*
GLM. *See* generalized linear model
Global Positioning System (GPS), 680
Golubev, A.N., 629-633
gonadosomatic index (GSI), 465, 469, *470,*
    475, *478,* 479, *479,* 482-483
gonads
    deformed, 289
    development, and migration distance,
        269, 270
    and echo sounding target strength,
        461, 473-474, 483-484
    ovaries, *291-293*
    weight
        British Columbia herring, 426
        Scotia-Fundy herring, 146, *149*
        and spawning time, 167
GPS (Global Positioning System), 680
Greenland Sea, *438,* 439
growth rate, *390-391*
    Baltic Sea, 750-752, *758*
    decrease in, and density-dependence,
        153-154
    feeding dynamics and, 22
    larval, relationship to hatching date,
        northern Japan, 227-243
    long-term sampling program, Scotia-
        Fundy area, 140, *143*
    relation to spawning time, Baltic Sea,
        155-169
    strong year-class, Gulf of Alaska, 180,
        183, *186-189*
    in a strong year-class, British Columbia
        and Alaska, 171-198
GSI. *See* gonadosomatic index

Gulf of Alaska
    anatomy of a strong year-class, 171-198
    change in oceanographic conditions,
        1977, 172
    herring in predatory diet, 87
    herring recruitment for Prince William
        Sound, 70, 74
    influence on Prince William Sound
        hydrography, 339
    length-at-age, *389*
    life history strategy, 420
    site map, *174-175*
    stable carbon isotope signature, 70
    weight-at-age, *391*
Gulf of Finland, 703-720, 750
Gulf of Maine
    catch and SSB statistics, *408*
    length-at-age, *388*
    long-term sampling program, 135-152
    management areas, *137*
    morphometric variation among
        spawning groups, 575-590
    site map, *407*
    stock decline, 1960s to 1970s, 326
    stock inventory 2000, 414-415, 418
    water temperature, *317*
    weight-at-age, *391*
Gulf of St. Lawrence
    catch and SSB statistics, *408*
    site maps, *407, 668*
    stock collapse, 667
    stock inventory 2000, 411-412
    weight-at-age, 390

H

Haegele, Carl, 489-508
hake, Pacific. *See Merluccius productus*
Halley, Edmond, 4
Hanseatic League, 384
Hardy-Weinburg Equilibrium, 620, *621, 622*
Harpacticoida. *See under* prey
hatching date/period
    determination of, methodology, 203
    Ishikari Bay, Japan, 234, *236-237*
    northern Japan, 203, *208,* 212-216,
        227-243
    relationship to larval growth, 227-243,
        238
hatching success and low salinity, 223
Hay, D.E., 171-198, 381-454
Heinke, Fr., theory development by, *5,* 10-11
herring
    Atlantic. *See Clupea harengus; Clupea*
        harengus harengus
    as bait, 406, 422, 428, *538,* 668

herring *(continued)*
   Baltic, 385. See also *Clupea harengus membras*
   canned, as substitute for sardines, 428
   for fish-meal and oil, 646, 648, 668, 724
   for fodder in fur production, 742, 744, 752, *753, 754,* 757, *758*
   for human consumption, 428, 742, 752, 753, *753, 754, 758.* See also roe market
   king. See *Clupea harengus*
   nomenclature, 386, 591-592
   Norwegian spring-spawning. See Norwegian spring-spawning herring
   Pacific. See *Clupea pallasii*
   as pet food, 428
   role in fisheries science, 1-20
   round. See *Etrumeus teres*
   Sakhalin-Hokkaido. See Sakhalin-Hokkaido stock
   Scotia-Fundy. See Bay of Fundy; Nova Scotia; Scotia-Fundy herring
   Tzosha-Pechora. See *Clupea pallasii suworowi*
   White Sea. See *Clupea pallasii marisalbi*
   Yellow Sea, 436, *437*
Herring Assessment Working Group, 400, 404, 663
Herring Committee, 6
"herring hypothesis" (larval retention), 14-15, 136, 150, 262, 392
Herring 2000 Symposium, 15, 444
historical background
   Estonian fishery, 704
   fishing technology development, 13, 324
   global use of herring, 384
   herring fishery by region. See stock assessment, inventory 2000; *specific geographic names*
   herring research, 1-15
   international agreements, 136
   North Sea fishery, 648-652
   Norwegian spring-spawning herring fishery, 266, 297-300, 304
   research, 4-8, 387-392
   roe market to Japan, 727-729
   Sakhalin coast herring and spawning beds, 245-246, 247, *249-250, 251-252,* 253
   subsistence fisheries, 421
Hokkaido Central Fisheries Experimental Station, 228
Hokkaido Island. See Japan, Ishikari Bay; Sakhalin-Hokkaido stock

homing. See natal homing
Hoshikawa, Hiroshi, 199-226
Hudson Bay, 441, 442
Hydroacoustic Data Processing Software, 680
hydroacoustics. *See also* echo sounding
   double counting, 685
   historical background, 7
   surveys
      for in-season management decisions, 675-688
      nonspawning application, 684
      Norwegian spring-spawning herring, 397
      Nova Scotia, 675-688
      Sound (ICES SD23), 347, *348-349*
      transect methodology, 680-681
   technology development, 13, 324
   variation in target strength, 461-487
hydrography
   Baltic Sea, 750
   historical background of research, 15
   Prince William Sound, 339, *340*
   Sound between Denmark and Sweden, (ICES SD23), 347-356
   White Sea, 593
Hypomesus japonicus, 117-120

I

IBSFC (International Baltic Sea Fishery Commission), 394, 706, 753
ICA. See integrated catch at age/analysis model
Iceland
   effects of water temperature on recruitment, 315, 324, *326, 327,* 328-329
   length-at-age, *388*
   migration distance and condition, 269
   spawning stock biomass, *319, 320,* 323
   stocks in, 397
   summer-spawning herring. See also *Clupea harengus*
   stock inventory 2000, 404-405
   weight-at-age, *390*
ICES. See International Council for the Exploration of the Seas
Ichthyophonus, 650, 652
   hoferi, 363-379, 551
ICNAF (International Commission of Northwest Atlantic Fisheries), 136, 562
Iles, T. Derrick, 135-152

Infinite Allele Model, 618
Institute of Marine Research, 630
Institute of Ocean Sciences, 54
integrated catch at age/analysis model (ICA), 401, 402, 406, 652, 655
Integrated Fisheries Management Plan, 668
international agreements
    for Baltic herring, 394
    first total allowable catch limit, 136
    Georges Bank, 418
    North Sea, 658, 663
    for Norwegian spring-spawning herring, 398
International Baltic Sea Fishery Commission (IBSFC), 394, 706, 753
International Commission of Northwest Atlantic Fisheries (ICNAF), 136, 562
international conflict, North Sea fishery, 646
International Council for the Exploration of the Seas (ICES), 384
    ACFM. See Advisory Committee for Fishery Management
    assessment and monitoring
        Baltic Sea stocks, 394, 742, 747
        northern Atlantic stocks, 318, 397, 398
        North Sea stocks, 399, 400, 645, 652, 655, 658, 659, 663
        Sound (ICES SD23) stock, *354*
    historical background, 6, 12
    monitoring protocol, 137-138
    target strength regression analysis, 462
    Working Group of Atlanto-Scandian Herring and Capelin, 630
International Herring Symposium, 1990, 560
international standards, 16
International Symposium on the Biological Characteristics of Herring and Their Implications for Management, 1983, 3
invertebrates
    benthic, northern Japan, *209*
    eggs, as food source, *30*
    as predators, 490, 543
Ireland. See Celtic Sea
Ishida, Ryotaro, 101-115, 227-243
Ishikari Bay. See Japan, Ishikari Bay
Ishikari Fisheries Extension Office, 228
isoenzyme analysis, in stock identification, 318
isotopic analysis, 69-80, 123
Ivshina, E., 245-254, 381-454

J

Jakobsson, J., 381-454
Japan
    and the Alaskan roe herring market, 721-739
    catch statistics, *523,* 525
    Ishikari Bay
        distribution and feeding habits, 101-115
        relationship of hatching date to larval growth, 227-243
        site description, 102
        site maps, *103, 201, 213, 229*
        spawning bed selection, 199-226
    Lake Furen, Hokkaido, *389,* 573
    Sakhalin-Hokkaido herring. See Sakhalin-Hokkaido stock
    stock enhancement, 101-102, 573
Japanese sardine. See *Sardinops melanostictus*
Japanese surf smelt. See *Hypomesus japonicus*
Järvik, Ahto, 703-720
Jeffrey's Ledge, 575-590
Jensen, Torben F., 347-356
Johnson, Terry, 721-739
Jönsson, N., 599-609
Jørstad, K.E., 629-633
Juan de Fuca Strait, 51-67
jurisdiction issues, 16, 384, 744
juveniles
    habitat, *vs.* adults, 193
    northern Japan, distribution and feeding habits, 101-115
    Prince William Sound, 545-551
    whole body energy content, estimation, 121-133

K

Kääriä, J., 599-609
Kamchatka. See Bering Sea, western; Gizhiga-Kamchatka stock
Kanin Bank. See Barents Sea
Karagin Bay, 117-120. See also Bering Sea, western
Karpov, A.K., 591-597, 629-633
Kawai, Tadashi, 199-226
kazunoko, 725-727, *730. See also* roe market
kelp. *See* algae; spawn-on-kelp
king herring. See *Clupea harengus*
Kline, T.C., 69-80, 121-133
Knapp, Gunnar, 721-739
Kobayashi, T., 381-454, 573
Kodiak area, *178*
Korean Peninsula, 113

Korf-Karagin stock, 430-432, *433,* 444, 689, *690. See also* Bering Sea, western
kriging techniques, 153
Kuikka, Sakari, 741-760

L

lactate dehydrogenase loci, 593, 595, *595, 631,* 632
"lagoon" form, 430, 431
Laine, Päivi, 155-169
Lake Furen, Hakkaido, *389,* 573
Laminaria
   cichorioides, 205
   ochotensis, 205
   religiosa, 205
   and spawning bed selection, 199, 204-205, *207,* 208, *209,* 216, 218
   as substrate. *See* spawn-on-kelp
Laptev Sea, *438,* 439-440
larvae
   deformities, 543-544
   modeling of advection, 15
   multiplicative larval index, 652
   survival, Prince William Sound, 535, 543-545, *548*
larval drift
   effects of wind, Norway, 359-360
   modeling, 544-545
   in the North Sea, *647*
   Norwegian spring-spawning herring, 262, *263,* 264, 301-302
   in Prince William Sound, 544-545, *548*
larval retention hypothesis. *See* "herring hypothesis"
least-squares regression, 336
length. *See* body length of herring
life history
   adaptation to latitudinal differences, 531
   Bering Sea *vs.* Gulf of Alaska, 420
   "lagoon" form, 430, 431
   mixing at stages, *561*
   relevance of former spawning area, 297-313
lingcod, strong year-class, 172, 193
literature review, themes in research, 4-8
LOBE V 5 software, 466
locally weighted smoothing procedure (LOWESS), *188*
Lundgren, Bo, 347-356

M

macrobenthos. *See also* algae; spawn-on-kelp
   British Columbia, 495
   Japan, northern, 208, *209-212*

macrobenthos (continued)
   Sakhalin coast, Russia, 250
Macrocystis, egg survival on, 455-460
Magdalen Islands, Quebec, 667-673
Maine. See Gulf of Maine
malformations, 543-544
management. *See* Fisheries management
Marconi 424 echo sounder, 13
marine mammals, 9, 12, 410, 697
marine vascular plants. See *Phyllospadix iwatensis; Zostera marina*
Marty, Gary D., 363-379
"match/mismatch" hypothesis, 14, 392, 522
maturation of herring
   in British Columbia, 172
   Gizhiga-Kamchatka stock, 642
   in Ishikari Bay, Japan, 228
   in Norway, *273*
   Scotia-Fundy stock, 140, *144*
   size-dependent, 271, 272-274
   at "two winter rings," 403
maximal sustainable yield (MSY), 701
Mazhirina, G.P., 285-295
McCarter, P.B., 171-198
McFarlane, G.A., 37-50, 51-68
McPherson, Arran A., 615-628
McQuinn, I., 381-454

Melnikov, I.V., 689-701
Melvin, G., 381-454, 559-571, 675-688
member/vagrant hypothesis, 14-15, 258, 262
Merluccius productus
   abundance, Strait of Georgia, 52, 54-55, *56-58,* 62, 64-65
   competition for prey, 48
   metabolic demands, basal, 32
metamorphosis period, 122
metapopulations
   Nova Scotia, 561, 568
   Prince William Sound, lack of evidence, 335, 343
methodologies
   acoustic surveys, 680-681
   age determination from scales, 176-177, *177*
   age-frequency analysis, 173
   benthic surveys, 200, 202-205
   determination of spawning and hatching period, 203
   DNA analysis, 618-619
   fishing effort estimation, 669, 671
   historical background of research, *5-7*
   otolith analysis, 157, 163, 228, 230, 577

methodologies *(continued)*
  pelagic trawl surveys, 691
  sampling and treatment
    in the Barents Sea, 81
    eggs, 492-493
    gonads, 286
    in the Gulf of Maine, 137-138,
      139-140
    in northern Japan, 101, 102, 104,
      106-107
    in Prince William Sound, 122-124
  tagging, 602, *603, 606*
  target strength measurement, 463,
    464, 465, 466, 680
  video camera, with caged herring, 511,
    512, 513
  year-class strength construction, 39
Metomidate, 465
microsatellite markers. *See* genetics,
    microsatellite markers
migration. *See also* natal homing
  Baltic Sea, 711
  energy loss from, 258-259, 261, 270, 308
  Finland, spring-spawning Baltic
    herring, 599-609
  larval drift. *See* larval drift
  "learned," 259, 260, 266-267, 308
  of low-vertebrae herring, and the
    White Sea, 592, *594*
  northern Japan, 241
  North Sea herring, 645, 646, *647*
  Norway, 256, *257,* 270-274, 301-308,
    397-398
  Norwegian spring-spawning herring,
    262, *263, 264,* 301-302, *305,*
    305-306, *306, 307*
  overview, 1, 386
  polar migration theory, 4, 9-10, 392
  by region. *See* stock assessment,
    inventory 2000; *specific
    geographic names*
  Rügen spring-spawning herring, 354,
    355
  seasonal, within the Barents Sea, 92
  state-dependent model, 258-259, 260,
    261, 267, 270, *271*
  vertical, 48, 92, 359
mile-days of milt, 366, *370, 374*
mitochondrial DNA analysis, 264-265, 624
modeling
  abundance, 425
  ADAPT-VPA, 411
  advection of fish larvae, first three-
    dimensional hydrodynamic, 15
  age-structured assessment model, 366

modeling *(continued)*
  Alaska roe market, supply and
    demand, 736-737
  echo sounding target strength, 469,
    475, 479
  Ecopath, 70, 78
  egg loss in spawning beds, 498, *502,*
    504
  escapement model, 425
  Infinite Allele Model, 618
  integrated catch at age/analysis model,
    401, 402, 406, 652, 655
  larval drift, 535, 544-545
  life history mixing at stages, *561*
  mortality
    natural, 365-367, *368,* 369, 371, *372*
    overwinter, 546-547, *548*
  in Prince William Sound stock
    assessment, 366-376
  Ricker model, 339
  spawning bed selection, 258-270, *271*
  spawning waves, 271-274
  Stepwise Mutation Model, 618
  VPA. *See* virtual population analysis
  year-class survival, 547
Moiseev, Sergei I., 81-89
Molloy, J., 381-454
monitoring
  dock-side, 413
  hydroacoustic, Sound (ICES SD23), 347,
    348-349
  ICES protocol, 137-138
  recommendations, 550
  by vessels. *See* commercial fisheries,
    involvement in data
    collection
moratoriums
  Iceland, 405
  North Sea, 645, 650, 662
morphometrics, 575-590, *579,* 616
mortality
  in the Baltic Sea, 711, *714*
  density-dependent, 260, 301, 308
  egg
    on *Macrocystis* kelp, 456-457
    Prince William Sound, 543, *548*
  juvenile, Prince William Sound, 545-547,
    548
  larval, Prince William Sound, 535,
    543-545, *548*
  natural, modeling, 365-367, *368,* 369,
    371, *372*
  in the Okhotsk Sea, 697-698
  overwinter, 546-547, *548*
  in pelagic trawl fishery, 711, *714*

moratoriums *(continued)*
  in Prince William Sound, 22, 32, 365-371, *372*, 535-558
  relationship to strong-year class, Gulf of Alaska, 194
  size-selective, 241
Moscow State University, 630
Mount St. Helens, *7, 8*
multiple regression, herring recruitment, 336
multivariate analysis of variance, herring morphometrics, 578, 580
Murmansk Bank. *See* Barents Sea
mythology, 10

N

Naarminen, M., 599-609
NAO (North Atlantic Oscillation winter index), 360
natal homing
  Finland, 606, 607
  North Sea, 646
  Norwegian spring-spawning herring, 255, 258, 259, 262, 264, *271*
  and stock "cells," 559
National Marine Fisheries Service, 415, 416, 577, 578
Naumenko, N., 381-454
needles, used in tagging, *603*
Neocalanus
  cristatus, 71
  flemingeri, 71
  plumchrus, 40
Nesterova, Valentina N., 91-99
net-days. *See* fishing effort
Newfoundland
  east and southeast, stock inventory 2000, 406-407, *408*
  length-at-age, *388*
  weight-at-age, *390*
  west, stock inventory 2000, *408*, 409-411
Nichols, J.H., 645-665
Nielsen, J. Rasmus, 347-356
nitrogen
  carbon/nitrogen ratio, 121, 124, *129-131*
  stable isotope abundance, 69-80, 123
  trophic enrichment factor, 70, 72
NMFS (National Marine Fisheries Service), 415, 416, 577, 578
nomenclature, 591-592
Norcross, Brenda L., 21-35, 335-345, 535-558

Nordkyn Bank. *See* Barents Sea
North Atlantic Oscillation winter index (NAO), 360
North Cape Current, 93
North Sea
  catch option table, *656*
  catch statistics, 400, 648, *649*
  discrete spring-spawning stocks, 646
  effects of water temperature on recruitment, 315, 324, *326, 327,* 328-329
  fisheries management, 645-665
  length-at-age, *388*
  migratory mixes with Baltic Sea herring, 395
  spawning stock biomass, *319, 320,* 323
  stock inventory 2000, 399-401
  weight-at-age, *390*
Norway. *See also* Barents Sea; Norwegian spring-spawning herring
  currents, *263,* 358
  effects of climate, 279-284
  factors influencing spawning, 255-278
  fjord populations, 629-633
  historical background of fisheries science, 3
  length-at-age, *388*
  microsatellite polymorphism, 611-613
  relevance of former spawning area in life history and management, 297-313
  seasonal climate changes, 360
  site maps, *298, 358*
  weight-at-age, *390*
Norwegian spring-spawning herring (NSSH). *See also Clupea harengus*
  biology of stock, 397-398
  effects of puffins, 357-361
  effects of seasonal climate changes, 357-361
  effects of water temperature, 279-284
  factors influencing spawning, Norway, 255-278
  greatest stock size, 442
  oocyte degeneration in female recruits, 285-295
  range, *317*
  relevance of former spawning area, 297-313
  spawning stock biomass, *319, 320,* 323
  stock identification, 303-304
  stock inventory 2000, 397-399
  use of target strength analysis, 462

Nova Scotia. *See also* Atlantic, western; Bay of Fundy; Gulf of Maine; Scotia-Fundy herring
  acoustic surveys of in-season management, 675-688
  catch statistics, 408, 678
  length-at-age, *388*
  long-term sampling program, 135-152
  management areas, *137, 563,* 677
  spawning areas, 562-563, 568-569
  weight-at-age, *391*
Novikov, G.G., 591-597, 629-633
NSSH. *See* Norwegian spring-spawning herring
nursery areas
  estuaries, 113
  Irish Sea, 401
  Japan, northern, 101, 113
  in prediction of spawning ground selection, 259
  Prince William Sound, 21, 544-545
  survival in, 535

O

Oda, K.T., 381-454
Ohtsuki, Tomohiro, 199-226
Oithona. See under prey
Okhotsk Sea, 430-432, *433,* 444, 635-643
Okhotsk stock, 430-432, *433,* 444, 689-701
Ona, E., 461-487, 509-519
Oncorhynchus
  herring in diet, *84*
  keta, 48, *84*
  kisutch
    competition for prey with herring, Strait of Georgia, 37-50
    diet, Strait of Georgia, *44*
    seasonal mortality patterns, 37-38
Onset Computer Corporation, 23
oocytes. *See* eggs
O'Reilly, Patrick T., 615-628
Orlov, Alexei M., 81-89
Orlova, Emma L., 91-99
Østvedt, Ole Johan, 279-284
otoliths. *See also* growth rate
  analysis, relationship between hatching date and larval growth, 227-243
  formation of first increment, 227
  general patterns from, 156
  geographic variation in growth, 160-163
  increment formation and age determination, 231, *232-234,* 235, 239, 242

otoliths *(continued)*
  marking. *See* alizarin complexone marking
  methods of analysis, 157, 163, 228, 230
  micrograph of, *233*
  relationship of radius to body length, 157, 160, *160*
  temporal variation in growth, 163, *164-165,* 166-167
overfishing. *See also* stock collapse
  Estonia, 706
  historical background, 2, *6*
  North Sea, 648
  Norway, 283, 297
  Sakhalin-Hokkaido stock, 435
oxygen capacity, regulation of spawning time by, 167

P

Pacific cod. *See Gadus macrocephalus*
Pacific Decadal Oscillation (PDO or PMDO), 339, 343, *344,* 420
Pacific hake. *See Merluccius productus*
Pacific herring. *See Clupea pallasii*
Pacific Ocean
  eastern
    site map, *419*
    stock inventory 2000, 418-429
  northwestern North, stock inventory 2000, 429-432, *433*
  southwestern North, stock inventory 2000, 432, *433,* 434-435
  western North
    site map, *430*
    stock inventory 2000, 429-437
Pacific sleeper shark. *See Somniosus pacificus*
Paracalanus. See under prey
parasites, 616. *See also Ichthyophonus hoferi*
Parmanne, R., 381-454
participants, 761-766
pas (peak aerial survey index), *338,* 339
Paul, A.J., 121-133
Paul, J.M., 121-133
Paulsen, O.I., 629-633
P/B-coefficient, 695, 698
PDO (Pacific Decadal Oscillation), 339, 343, *344,* 420
peak aerial survey index (pas), *338,* 339
Pelagics Research Council (PRC), 680
pelagic trawl surveys, 577, 578, 690
Peltonen, Heikki, 741-760
Peterson, Sara, 455-460
Peter the Great Bay stock, 430-432, *433,* 444

Phyllospadix iwatensis
   and spawning bed selection, 199, *202,* 204-205, *207,* 208, *209, 210,* 212, 216-218, *219*
   spawning beds of Sakhalin coast, Russia, 250, *251*
phytoplankton, 75, 93, *108*
PINRO, 630, 632
plants. See algae; *Phyllospadix iwatensis;* spawn-on-kelp; *Zostera marina*
polar migration theory, 4, 9-10, 392
Pollachius pollachius, 481
pollock, 545, 637. See also *Pollachius pollachius; Theragra chalcogramma*
   echo sounding target strength, 481
   Okhotsk Sea, *693,* 695
pollution, *7,* 76
Pönni, Jukka, 741-760
population components, 10, 14. See also stock discreteness
   "complexes," of Scotia-Fundy herring, 414, 559, 566, *567,* 568
   Gulf of Maine–Georges Bank herring, 575-590
   metapopulations, 335, 343, 561, 568
   in southwestern Atlantic, 412-413
population dynamics
   density. See density-dependent processes
   gene flow, 611, 616
   historical background of research, *5-7,* 10-12, 14-15
   inter- and intra-stock structure, 616
   stock discreteness, 559-571
   subunits. See population components
"population thinking," 10-12
Power, M., 135-152, 381-454, 559-571, 675-688
PRC (Pelagics Research Council), 680
precautionary approach
   historical background and trends in use, 7, 17, 560, 568
   for North Sea herring, 659, 662, 663
   use by ICES, 659
predators and predation
   in the Bering Sea, 81-89
   birds. See birds, predatory
   cod. See cod; *Gadus*
   effects of alternate food source to herring, 194, 195
   escape from some predators. See polar migration theory
   invertebrates, 490, 543
   in Japan, 221

predators and predation *(continued)*
   marine mammals, 9, 12, 410, 490, 697
   of Norwegian spring-spawning herring, 357, *358,* 359, 360
   in the Okhotsk Sea, 697
   shallow spawning beds to reduce, *221*
   smelt in Karagin Bay, 117-120
   and strong year-class, Gulf of Alaska, 194, 195
prey. See also feeding dynamics; food supply
   Acartia, 109, *110*
      omorii, 108
      steueri, 108
   Amphipoda (amphipods), 30
      northern Japan, *108, 110, 112,* 114
      stable isotope analysis, 76
      Strait of Georgia, 43, *44,* 45, 48
   Annelida, *30*
   Arthropoda, *30*
      in the Baltic Sea, 742
      in the Barents Sea, 91-99
   Bivalvia, *30*
   Branchiopoda, *30*
   Bryozoa, *30*
   Calanoida, *30*
   Calanus, 94-96, 97, 98
   Centropages abdominalis, 108
   Chaetognatha, *30,* 44
   Cirripedia, 28, *30, 108*
   Cladocera (cladocerans), 28, *30, 108,* 109, *111, 112*
   Clausocalanus pergens, 101, *108,* 109, *110, 111*
   Copepoda (copepods), 28, *30,* 329, 386
      Barents Sea, 91
      British Columbia, 426
      Gulf of Alaska, and strong year-class, 195
      northern Japan, 101, *108,* 109, 114
      Norway, 359
      stable isotope analysis, Prince William Sound, 71-72, *73,* 73-74, 76, *77*
      Strait of Georgia, 43, *44*
      west Newfoundland, 409
   Crustacea (crustaceans), 94, *108,* 114, *209*
   Cyclopoida (cyclopoids), *30, 108,* 114
   Cyphonautes, *30*
   Decapoda, *30,* 43, *44,* 45, 48, *108*
   diatoms, *108*
   eggs, *108.* See also prey, invertebrate eggs

prey *(continued)*
  Euphausiacea (euphausiids;krill), 28, *30*, 386
    Barents Sea, 91, 93, 94, 95, 98
    British Columbia, 426
    Okhotsk Sea, 693
    stable isotope analysis, 76
    Strait of Georgia, 43, *44*, 45, 48
    west Newfoundland, 409
  Euphausia pacifica, 40
  Eurytemora, 101, *111*, *112*
    affinis, *108*, 109, *110*
    herdmanii, *108*
    pacifica, *108*, 109, *110*, *111*
  Evadne, *108*, 109, *110*
  fish, *108*, *112*, 114
    eggs, *30*, *110*
    larvae, *30*, *110*
    teleost, *44*, 45, 72, 74, *77*
  Gammaridae, *108*
  Gammarus, *110*
  Gastropoda (gastropods), *30*, *108*
  Harpacticoida (harpacticoids), *30*, *108*, 109, 110, *110*, *111*, *112*, 114
  insects, *44*
  invertebrate eggs, *30*, 43, *44*
  Larvacea, 28, *30*
  Malacostraca, *30*
  Mesocalanus tenuicornis, *108*
  Microsetella, *108*, 109, *110*, *111*
  Mollusca, *30*, *209*
  Mysidacea, *30*, *108*
  Mysidae (mysids), *110*, *112*, 114
  Neocalanus
    cristatus, 71, 72, *73*, 73-74
    flemingeri, 71
    plumchrus, 40, 43, 71
  Oikopleura, *108*, 109, *111*, *112*
  Oithona
    atlantica, 101, *108*, 109, *110*, *111*
    similis, 101, *108*, 109, 110, *110*, *111*, *112*, 114
  Paracalanus, *108*
    parvus, 31, 101, *108*, 109, 110, *110*, *111*, *112*, 114
  Pelecypoda, *108*
  Podon, *108*, 109, *110*
  Polychaeta, *30*, *108*
  Pseudocalanus, *108*
    minutus, 40, 43
  Tortanus
    derjugini, *108*, 109, 110, *110*, *112*, 114
    discaudatus, *108*, 109, 110, *110*, *112*, 114

prey *(continued)*
  Urochordata, *30*
  zooplankton. *See* zooplankton
Prince William Sound, Alaska
  age-frequency data, *178*
  catch statistics, 536, *538*
  currents, *340*
  disease and population assessment, 363-379
  factors that affect year-class strength, 335-345
  first-year survival and stage-specific estimates, 535-558
  five spawning regions, 336, *337*, 345
  site description, 22
  site maps, *23*, *123*, *337*, *537*
  stable carbon isotope signature, 70
  temperature effects on zooplankton, and herring feeding, 21-35
  trophic position of herring, 69-80
  water temperature, *340*
  whole body energy content of herring, 121-133
principal component analysis, morphometrics, 577, 578, 580, 581, *581*, *582-585*, 586
probability density function (PDF), target strength, 473, *476*, 482
PROCOMM software, 466
productivity, 32, 69, 529, 531
puffin, Atlantic. *See Fratercula arctica*
PWS. *See* Prince William Sound

Q

Quebec, Magdalen Islands, 667-673
Quinn II, Terrance J., 363-379
quotas
  British Columbia, 425
  California, 428
  catch. *See* total allowable catch
  for Gizhiga-Kamchatka stock, 636
  Gulf of Finland, 716
  Gulf of St. Lawrence, 411
  individual (ITQ), 405
  national, 394
  Newfoundland, 406
  Nova Scotia, 564-565, 677, *678*
  over-quota catches and misreporting, North Sea, 659, 663
  vessel, 15, 136, 716
  vs. fishing effort, 667-673

R

races, in fish stocks, 5
Radchenko, V., 381-454, 689-701
Rahikainen, Mika, 741-760
Raid, Tiit, 703-720
Rajasilta, M., 155-169, 599-609
recovery. *See* stock recovery
recruitment
    in Alaska, 420-421
    correlation with spawning, 341, *657*
    correlation with water temperature, 281-283, 315-330, 324, *326, 327,* 328-329
    "critical period" hypothesis, 392, 522
    definition, 319, 321
    dynamics, latitudinal differences in, 521-533
    effects of disease on, 369, 374, *376,* 376-377
    "herring hypothesis" (larval retention), 14-15, 136, 150, 262, 392
    "match/mismatch" hypothesis, 14, 392, 522
    member/vagrant hypothesis, 14-15, 258, 262
    in the North Sea, *651,* 657
    in Norway, 285, 289, 301-303
    to Prince William Sound, 69-70
    variability in, 14-15, 522
recruitment success, northern Atlantic herring, 318, *320,* 323
regime shifts, 51-52, 65
    British Columbia, 1989, 64-65
    effects on year-class strength, 335-345
    northern Japan, 200
repeat homing. *See* spawning, return/repeat
reproduction. *See also* eggs; spawning; spawning period
    deformities, 543-544
    fecundity, 436
    fertilization and low salinity, 223
    historical background of research, *5-7*
research
    chronology, *5-7*
    environmental impact on stock fluctuation, review, 316
    future role of herring investigations, 16-17
    historical background, 4-8, 387, 392
    historical role of herring investigations, 1-15
    residual sum of squares (RSS), 369, *370*

restrictions. *See also* quotas; total allowable catch
    Alaska, 422
    Gulf of St. Lawrence, 411
    Iceland, 405
"retention area," 359
return spawning. *See* spawning, return/repeat
Ricker model, 339
rockfish, shortraker. *See Sebastes borealis*
roe market
    Alaska, to Japan, 721-739
    Beaufort Sea, feasibility study, 440
    British Columbia, 172, 425, *729*
    California, 428, 429, *729*
    Celtic Sea, to Japan, 401, 402
    distribution, 727
    gonad weight to body weight, 146, *149*
    overview of the process, 724-725
    pricing, 731-737
    Prince William Sound, 364, 536, *538*
    processing, 725, 727
    products, 725-727, 729-731
    Russia, 728
    Sea of Okhotsk, Russia, 642
    trends in supply, 727-729
roe-on-kelp fishery, Alaska, 725
Røttingen, I., 297-313, 629-633
Rügen spring-spawning herring
    biomass data, *348-349,* 351, *351,* 354, *354*
    effects of hydrography, 347-356
runoff. *See* freshwater runoff
Russia. *See also* Barents Sea; Bering Sea; Gizhiga-Kamchatka stock; Sakhalin-Hokkaido stock; Sea of Okhotsk
    catch statistics, 432, *433*
    roe market, 728-729
    stock inventory 2000, 430-432, *433*

S

Sætre, R., 357-361
St. Andrews Biological Station, 136
saithe, echo sounding target strength, 481
Sakhalin-Hokkaido stock
    catch statistics, *433,* 434, *523,* 524, 526, *527*
    length-at-age, *389*
    site map, *248*
    stock collapse, 245-254, 382
    stock inventory 2000, 432, *433,* 434-435
    weight-at-age, *391*

salinity
   decreasing, Baltic Sea, 750, *751*, 758
   low, effect on fertilization and hatching success, 223
   regional effects in Prince William Sound, 343
   Sound between Denmark and Sweden (ICES SD23), *353*, 354, 355
salmon
   Baltic Sea, 704
   chum. See *Oncorhynchus keta*
   coho. See *Oncorhynchus kisutch*
   competition for prey with herring, 37-50
   detritus, after spawning, 76
   spawning, 12
salt preservation, 384
sardine, Japanese. See *Sardinops melanostictus*
sardine industry, 428
Sardinops melanostictus, recruitment dynamics, 521-533
Sargassum, 251
   confusum, and spawning bed selection, 207, *209*, 221
   miyabei, *209*, 250
   pallidum, 250
   thunbergii, *209*
Sasaki, Masayoshi, 101-115, 227-243
scales
   age determination from, 176-177, *177*
   interpretation of growth from, 156
   length
      relationship to age, 189, *190-191*, 192
      relationship to body length, 189, *189*
   and nomenclature, *592*, *594*
schooling behavior, 260, 386, 510
Schweigert, J., 37-50, 51-68, 381-454, 489-508
SCI (stomach content index), 118, *119*
Scotia-Fundy herring. See *also* Bay of Fundy; Nova Scotia
   catch statistics, *408*, 413
   length-at-age, *388*
   long-term sampling program, 135-152
   management, 135-152, *563*, 676
   spawning stock biomass, *408*, 413
   stock inventory 2000, 413-414
   stock structure, discreteness and biodiversity, 559-571
   weight-at-age, *391*

Scotland
   microsatellite polymorphism and genetic structure, 611-613
   west of
      length-at-age, *388*
      stock inventory 2000, 402-404
      weight-at-age, *390*
SEA (Sound Ecosystem Assessment program), 122, 537, 543, 545
seabirds. See birds
seagrass. See *Phyllospadix iwatensis*; *Zostera marina*
Sea of Japan. See Japan, Ishikari Bay; Sakhalin-Hokkaido stock
Sea of Okhotsk, 430-432, *433*, 444, 635-643
seasonal aspects. See *also* spawning period
   in Alaska field data, "summer bias," 78
   distribution within the Barents Sea, 91, 92-93, 96-99
   polar migration theory, 4, 9-10, 392
   variation in echo sounding target strength, 461-487
   within-season management regimes, 15
sea surface temperature index (SST), 54
Sebastes borealis, 82, *84-86*, 87
selectivity index, 25, 31
Seliverstova, E.I., 91-99, 285-295
Semenova, A.V., 591-597
Setälä, Jari, 741-760
SevPINRO, 630, 632
sex ratio, Scotia-Fundy area, 146, *147-148*
shark, Pacific sleeper. See *Somniosus pacificus*
Shaw, Paul W., 615-628
Shetland Islands, 399
Shirahuji, Norio, 521-533
shortraker rockfish. See *Sebastes borealis*
Siberian Sea, *438*, 439-440
signal-to-noise ratio, 617
Simmonds, J., 381-454
Simpson Bay, Alaska, *23*, 121-133, *548*
size. See body length; body weight
Sjöstrand, B., 381-454
Slotte, Aril, 255-278, 297-313
smelt, Japanese surf, 117-120
Smirnov, Andrey A., 635-643
socioeconomic aspects, 17, 716, 718
software
   ACON, 682
   F-STAT, 618
   Hydroacoustic Data Processing, 680
   image analysis, 176, 577
   LOBE V 5, 466
   Minitab statistical, 173

software *(continued)*
  PROCOMM, 466
  SAS, 577
  SYSTAT, 469
Somniosus pacificus, 84-86
sonar. *See* hydroacoustics
Sound (ICES SD23)
  biomass data, *348-349*, 351, *351*, 354, *354*
  hydrography and herring occurrence, 347-356
  site map, *350*
Sound Ecosystem Assessment program (SEA), 122, 537, 543, 545
spawn indices, Gulf of Alaska data, 176, 180, *184-185*
spawning. *See also* stock assessment, inventory 2000
  Atlantic vs. Pacific herring, 385-386
  bed selection. *See also* spawn-on-kelp
  Baltic Sea, 200
  factors influencing, 255-278, 600
  Finland, 221, 600
  influence of the first, 259, 265, *271*, 305
  northern Japan, 199-226
  Norway, historical changes, 297-299, *300*
  correlation with recruitment, Prince William Sound, 341
  effects of the *Exxon Valdez* oil spill, 538-543, 549
  historical background of research, 5-7
  on kelp. *See* spawn-on-kelp
  low- vs. high-vertebrae herring, 593, *594*
  Nova Scotia areas, 562-563, 568-569
  and population in adjacent seas, 11-12
  Prince William Sound areas, *540-542*
  regulation by oxygen availability and gill surface, 167
  return/repeat
    coho salmon, Strait of Georgia, 39
    Finland, 600
    Norway spring-spawning herring, 265, 304, 305-308, 309
    relevance in life history and management, 297-313
  vs. natal homing, 258
  Sakhalin coast beds, 245-254
spawning period
  Alaska, 724
  Archipelago Sea, 600
  determination of, methodology, 203
  Gizhiga-Kamchatka stock, 637

spawning period *(continued)*
  Gulf of Alaska, in strong year-class, 180, *184-185*, 192-193
  Japan, northern, 200, 203, *208*, 212-216, 221
  models of spawning waves, 271-274
  North Sea, 646
  Norway, 255, 270-274, 285
  Nova Scotia, 565
  Prince William Sound, 122, 538, *539*
  regional analysis. *See* stock assessment, inventory 2000; *specific geographic names*
  relation to growth rate, Baltic Sea, 155-169
  Sakhalin coast, Russia, 247
spawning stock biomass (SSB)
  Alaska, *423*
  Atlantic
    northern, 318, *319, 320*, 323, 324, *325*, 328-329
    western North, *408*
  Baltic Sea, *396, 714*, 747, *748*
  British Columbia, *423*
  effects of egg loss on, 489-508
  Georges Bank, 417
  Iceland, 405
  Newfoundland, 410
  North Sea, 400, 401, 645-660, 663
  Norwegian spring-spawning herring, *319, 320*, 323
  Scotia-Fundy herring, *408*, 413
  Scotland, west of, 404
spawn-on-kelp
  Alaska, 422
  British Columbia, 425, 495
  California, 428, 455-460
  egg density, 431
  egg survival, 455-460, 495
  Prince William Sound, 536, *538*
  Russia, 431
species richness, 14, 309
sprat. *See also Sprattus sprattus*
  Baltic Sea, 153-154, 704, 711, 742, 744, *746*, 747, *748*, 750-751, 758
  echo sounding target strength, 481
  North Sea, 648, 658, 662
  spawning, 12
Spratelloides gracilis, 521, *524*, 526, 529
Sprattus sprattus, 462, *743*
SSB. *See* spawning stock biomass
SST (sea surface temperature index), 54
Stæhr, Karl-Johan, 347-356
starvation, 194, 535, 546-547, *548*
Stasenkov, V.A., 629-633

Stasenkova, N.J., 629-633
Statview 4.5, 72
Stephenson, Robert L., 1-20, 381-454, 559-571, 675-688, 741-760
Stepwise Mutation Model, 618
Stevenson, D.K., 381-454
stock assessment
 and egg loss estimates, British Columbia, 489-508
 historical background of research, 5-7, 12
 inventory 2000
  Arctic Ocean, 437-442
  eastern Atlantic and adjacent seas, 392-405
  eastern Pacific, 418-429
  overview of distribution and biology, 385-392
   synopsis, summary, and conclusions, 442-445
   western North Atlantic, 405-418
   western North Pacific, 429-437
 inventory by Federov, 385
 modeling, Prince William Sound, 366-376
 Nova Scotia, acoustic survey for, 684-685
 size and structure, influence of spawning, 259-260, 266-267, 271
 virtual population analysis. *See* virtual population analysis
 western Atlantic herring, 136
Stock Assessment Review Committee, 415
stock collapse
 Atlantic herring, 1950s-1960s, 136
 Baltic Sea herring, current, 395
 Barents Sea, 1970s, 91
 Bering Sea, western, 81
 British Columbia, 1960s, 424
 Georges Bank, late 1970s, 416
 Gizhiga-Kamchatka herring, 635, 636
 Gulf of Maine, 1960s to 1970s, 326
 Gulf of St. Lawrence, 667
 historical background of research, 15
 Iceland, late 1960s, 404, 405
 Japan, 727-728
 Newfoundland
  1970s, 406
  western, 1999, 410
 North Sea, 400-401, 645, 646
 Norway, 1965 to mid-1980s, 283, 297, 397, 398
 Prince William Sound, 32, 69-70, 364, 365, *538*, 551

stock collapse *(continued)*
 Sakhalin-Hokkaido population, 245-254, 382, 434, 435, 444
 Strait of Georgia, 1967-1971, 47, 62
 timeframe to recovery, *5,* 382
 use of term, 382
stock discreteness, 559-571
 identification methods. *See also* genetics
  fin rays, 636, 642, *642*
  isoenzyme analysis, 318
  morphometric analysis, 575-590
  with Norwegian spring-spawning herring, 303-304
stock enhancement, Japan, 101-102, 573
stock recovery
 Georges Bank, 417, 444
 Gizhiga-Kamchatka herring, 636
 Gulf of St. Lawrence, 667
 Iceland, 405
 North Sea, 401
 Norwegian spring-spawning herring, 397, 399
 timeframe from collapse, *5,* 382
stocks. *See* Dekastri stock; Gizhiga-Kamchatka stock; Korf-Karagin stock; Norwegian spring-spawning herring; Okhotsk stock; Rügen spring-spawning herring; Sakhalin-Hokkaido stock; Scotia-Fundy herring
stock structure
 four questions to address, 560-561
 influence of spawning on, 259-260, 266-267, *271*
 microsatellite markers at three spatial scales, 615-628
 Nova Scotia herring, 559-571
 variability in, and mesh-size regulation, 708-711, 715
stomach content index (SCI), 118, *119*
stomach contents. *See* food supply, stomach contents
Strait of Georgia, British Columbia
 coho salmon, and prey competition, 37-50
 egg loss on spawning beds, 493, 499-500, *502-503*
 predator interactions, and migration, 51-67
 site description, 52
 site maps, *53,* 491
Strait of Juan de Fuca, 51-67
Stroganov, A.N., 629-633
subsistence fisheries, Alaska, 421, 536

Sundberg, Per, 611-613
"survey, assess, then fish" protocol, 677
surveys. *See* aerial surveys; hydroacoustics, surveys; pelagic trawl surveys
survival
　egg-on-kelp, 455-460
　eggs, on spawning beds, 489-508
　first-year, 535-558
　modeling by year-class, 547
　in nursery areas, 535
sustainability, future research, 16
Svellingen, I., 461-487
Sweden, the Sound (ICES SD23), 347-356
swim bladder index (SBI), 465, *470, 471,* 472, 475, *478,* 480-482
swim bladder physiology, and echo sounding target strength, 462-463, 472, *473,* 481, 509
swimming angle, effects on echo sounding, 473, *475, 479,* 482, 509-519
swimming speed, effects on echo sounding, *478*
SYSTAT software, 469

T

TAC. *See* total allowable catch
Taggart, Christopher T., 615-628
tagging
　Finland, spring-spawning Baltic herring, 599-609
　historical background of research, *5-7*
　methods, 602, *603,* 606
　Norwegian spring-spawning herring, *305,* 305-306, *306, 307*
　Nova Scotia, Chebucto Head herring, 684
Tajima, Ken-ichiro, 199-226
Takayanagi, Shiro, 101-115, 227-243
Tanasichuk, R., 381-454
Tang, Q., 381-454
target strength, seasonal variation in, 461-487
Taylor columns, 359
technology development, 13, 324, 435, 716
teeth on vomer, *592*
Telitsina, L. A., 629-633
temperature. *See* water temperature
temperature loggers, 23
temperature time series, 328
Theragra, 82
　chalcogramma, 74
　　climate-related changes, 193

Theragra (continued)
　effects on herring abundance, Russia, 432, 689
　herring in diet, *87, 85*
　trophic level, 74
Thompson, M., 171-198, 381-454
tilt angle, *vs.* swimming angle, 509-519. *See also* swimming angle
TINRO-Center, Russia, 689, 690, 700
Togiak, Alaska, *178, 182*
Tokyo Central Wholesale Market, 731
topography of spawning beds. *See* eggs, distribution in spawning bed
Toreson, R., 279-284, 315-334, 357-361, 381-454
Tortanus. See under prey
total allowable catch (TAC)
　of Baltic herring, 394, 759
　Celtic Sea, 402
　Estonia, 703
　Finland, 759
　first international agreement, 136
　Gulf of St. Lawrence, 411
　North Sea, 645, 650-663
　Norway, 302
　of Norwegian spring-spawning herring, 398, 399
　Nova Scotia, 559, 564, 676
tracking. *See* tagging
Tremblay, Denis, 667-673
trophic level
　calculation by nitrogen stable isotope analysis, 72
　herring in Prince William Sound, 69-80
truss measurements, 575
tuna. See *Spratelloides gracilis*

U

United States. *See also* Gulf of Alaska; Gulf of Maine; specific states
　east, catch and SSB statistics, *408*
　Juan de Fuca Strait, 51-67
upwelling, 339, 343
U.S. National Marine Fisheries Service, 415, 416, 577, 578

V

Vancouver Island, British Columbia
　east coast. *See* Strait of Georgia
　site map, *491*
　west coast, 51-67, 424, 489-508
Vasilets, Petr M., 117-120
vertebrae, 591, *592,* 616, 630, 632
vertical distribution of eggs. *See under* distribution

vertical migration
  in the Barents Sea, 92
  as a mechanism in larval retention. See
    "herring hypothesis"
  in Norway, and wind-induced
    turbulence, 359
vessels. See commercial fisheries
viral hemorrhagic septicemia virus (VHSV),
  363-379, 536-537, 545, 550-551
virtual population analysis (VPA)
  Baltic Sea, 395
  Georges Bank, 416
  Gulf of Maine, 418
  Gulf of St. Lawrence, 411, 412
  North Sea, 652
  of Norwegian spring-spawning herring,
    279-284, 397
  Okhotsk Sea, 698
  use of target strength analysis, 462

W

Watanabe, Yoshiro, 521-533
water temperature
  Atlantic Ocean, northern, *317*
  Barents Sea, 91, *322*
  Beaufort Sea, 441
  effects on biomass and recruitment,
    279-284, 281-283, 315-334
  effects on larval growth, northern
    Japan, 227, 235, *239-240*, 241
  effects on Norwegian spring-spawning
    herring, 279-284
  effects on Sakhalin-Hokkaido
    population, 246, 435
  effects on spawning ground selection,
    Norway, 261, *268, 271,* 302
  effects on spawning time, northern
    Japan, 222-223
  effects on zooplankton, and herring
    feeding, 21-35
  Gulf of Maine, *317*
  Ishikari Bay, Japan, *239*
  Prince William Sound, *340*
  Sound (ICES SD23), *353,* 354, 355
  Strait of Georgia, 45, *45,* 48, 54, 61,
    63, 65
Watters, D.L., 381-454
wave action, 490, 505, 535, 543
WBEC (whole body energy content), 121-133
weight. See body weight of herring
Wennevik, V., 629-633
Whale Bay, Alaska, *23,* 548
whales, 9, 12, 697
Wheeler, J., 381-454
White Sea, 392, *438,* 439, 591-597

White Sea herring. See *Clupea pallasii marisalbi*
whole body energy content (WBEC), 121-133
Wilcock, John, 363-379
Willette, Mark, 363-379
wind, 359-360
Working Group of Atlanto-Scandian
  Herring and Capelin, 630

Y

year-class
  abundance, Norway, *268, 269*
  anatomy of a strong, British Columbia
    and Alaska, 171-198
  Barents Sea, 92-93
  effects of egg distribution and
    environmental factors on,
    335-345
  effects of first six months of life, 329,
    522
  feeding dynamics and, 22, 522
  Gizhiga-Kamchatka stock, 640
  historical background of research, 1, 12
  methodology for constructing, 39
  modeling survival, 547
  Okhotsk Sea, *694, 697,* 699
  Russia, 431-432
  strong, simultaneous for several
    species, 328
  survival, Prince William Sound, 547, *548*
Yellow Sea
  biology of herring, 436
  catch statistics, *433*
  length-at-age, *389*
  site map, *430*
  stock inventory 2000, 436, *437,* 444
  weight-at-age, *391*
Yoshida, Hideo, 227-243

Z

Zaikof Bay, Alaska, *23,* 548
Zhao, X., 461-487
zooplankton. See also food supply; prey
  Alaska, relationship to herring density,
    420
  Baltic Sea, relationship to herring
    biomass, 153-154, 750
  Barents Sea, 91
  Prince William Sound, 21-35, 343
  sampling methodology, 23-24
  temperature effects on, and herring
    feeding, 21-35, 48
  vertical migration, 48
Zostera marina, 431, 505